n-Dimensional Quasiconformal (QCf) Mappings

n-Dimensional Quasiconformal (QCf) Mappings

by
Petru Caraman

EDITURA ACADEMIEI
BUCUREȘTI
ROMÂNIA

ABACUS PRESS
TUNBRIDGE WELLS, KENT
ENGLAND

1974

First published in Romanian in 1968
under the title HOMEOMORFISME CVASICONFORME *n*-DIMENSIONALE by
EDITURA ACADEMIEI ROMÂNE,
str. Gutenberg 3 bis, București, România.

Revised, enlarged and translated from the Romanian by the author.
This English edition first published in 1974
under the joint imprints of

EDITURA ACADEMIEI ROMÂNE
and
ABACUS PRESS, TUNBRIDGE WELLS, KENT

ISBN 0 85626 005 3

Printed in Romania

CONTENTS

Part 2

Definitions of QCfH and their equivalence

6.

PREFACE

This monograph is the English translation of my Romanian book "Homeomorfisme cvasiconforme n*-dimensionale", published in 1968, which had its origin in the lectures I delivered at the Stoilow Seminar in Bucharest (January 1—April 1, 1964), and several of its sections constitute a course given by me at the university "Al. I. Cuza" in Jassy (1966—1971). It is intended for post graduates and research scientists who are interested in quasiconformal* (QCf) *mappings, but it is accessible even to students of advanced courses of mathematics.*

There are already several books for the case $n = 2$ *on this topic: L. Bers [6], L. I. Volkovyskiĭ [7], I. N. Vekua [9], H. Kunzi [3], Cabiria Andreian-Cazacu [20] and [2], O. Lehto and K. I. Virtanen [2], G. N. Položii [22], G. D. Suvorov [2] and L. Ahlfors [20], but my monograph is the first to be devoted to* n*-dimensional* QCf *mappings. This is understandable as their systematic study only began in 1960.*

The bidimensional QCf *mappings represent the natural extension of the bidimensional conformal mappings, which form a very rich class (as we can see from Riemann's theorem), but (and this makes them more important) the* n*-dimensional quasiconformal homeomorphisms* (QCfH) *represent the natural extension of a very poor class, because the* n*-dimensional conformal mappings* ($n > 2$)*, by Liouville's theorem, come to a finite product of inversions in spheres.*

Different authors who dealt with QCfH *in* n*-space* ($n > 2$) *generally confined themselves to the case where* $n = 3$*. However, I prefer to present the results in this monograph in the general form corresponding to an arbitrary* n*, because in this way the results are more comprehensive. For instance, most of the numerical coefficients and exponents which occur in the various formulae in the theory of two- and three-dimensional* QCfH *become, in the case of an arbitrary* n*, functions of* n*, but this general form cannot be deduced from their value for* $n = 3$*; and moreover, for an arbitrary* n*, the proofs are generally not more complicated than those for* $n = 3$*.*

*The monograph is in three parts. After a short historical account, where the various kinds of bidimensional extensions of the class of analytic functions of a complex variable (*QCf *mappings, α-monogenic functions, polygenic functions, etc.) are reviewed, the preliminary chapters (Part 1), define terms and give results (part of them without proof), which are not directly related to the* QCfH*, but which arise in the proofs of the other two parts.*

In Part 2, which is the principal part of the book, all the definitions of the K-QCfH *in* n*-space are reviewed and their equivalence with or without the same* K *(in the latter we find the relation between various* K*) is established. By proving* *first the equivalence of the various definitions of the* QCfH, *the book becomes more unitary; then, it has the advantage that it is sufficient to prove a certain property of the* QCfH *according to one of the definitions and to conclude that it still holds for the* QCfH *according to all the other definitions. All these definitions of the* K-QCfH *(referred to as of Grötzsch's type) are of the following form: "A certain pair of functions (which depend on the point, the dimension, the definition and the homeomorphisms) are bounded by* K *in the domain of definition* D"*. Another class of* QCfH *(I refer to them as of Lavrantev's type) is characterized by the continuity of a certain pair of functions. Each* QCfH *according to a definition of Lavrantev's type in* D *is* K_Δ-QCfH, *according to the corresponding definition of the Grötzsch's type in* Δ, *for every compact subdomain* $\Delta \subset\subset D$, *i.e. with* $\bar{\Delta} \subset D$. *If the pair of functions which arise in the definition of Grötzsch's type are bounded by a pair of functions continuous in* D, *we obtain a new class of* QCfH *(referred to as of Teichmüller's type) wich is more general and which contains the two preceding ones. Finally, a new still more general class, of* QCfH *(I refer to them as of Andreian's type) which contains the three others is characterized by the property that the pair of functions is bounded by a constant* K *in every compact subdomain* $\Delta, \subset\subset D$. *Next, the various equivalent definitions of the* QCfH *between two* $(n-1)$*-dimensional surfaces (given by F. Gehring and J. Väisälá [1] for* $n - 1 = 2$*) are also presented. Another chapter of the second part is devoted to the connection between the* QCfH *and the functions of several complex variables. Also dealt with are the quasi-pseudoconformal mappings (Bergman [1], [2]) and the discwise* QCf *diffeomorphisms (Hitotumatu [1]). An important part of the book is the chapter on conformal mappings in* n*-space* $(n > 2)$.

Part 3 contains the study of several important properties of the QCfH *(other important properties have been established in Part 2 and used in the proof of the equivalence of the various definitions).*

All the results concerning QCfH *in* n*-space are presented with a detailed proof.*

The bibliography is divided into three parts: (1) references about the n*-dimensional* QCfH $(n > 2)$, *(2) references about the bidimensional* QCfH, *and (3) other works quoted in the monograph. (In this way the* QCfH *in n-space will not disappear in the many references on the bidimensional* QCfH*). In each part of the bibliography the author's names are arranged alphabetically and their works chronologically. When a work is written by several authors, the names are put in alphabetical order and the work is inserted in the bibliography according to the first author.*

Reference is made to a theorem or a formula in the same chapter by quoting the corresponding number; if the quoted formula or theorem is in another chapter, the number of the chapter precedes the number of the theorem or formula, e.g. formula (3.4), theorem 1.6; finally, if we refer to a formula or theorem from another part, the number of the part appears first, followed by the number of the chapter and finally the number of the formula or theorem, e.g. formula (1. 3. 4).

The most important general notations are given in Chapter 1 of Part 1, however, the book is also provided with an index of symbols and with author and subject indices.

I am grateful to Professor F. W. Gehring, who sent me the manuscripts of several of his papers, e.g. of References [11], [21] and of his joint paper with Agard [1], but especially for the manuscript of his important joint paper with J. Väisälä [1]. (All these papers have since been published.)

PREFACE TO THE ENGLISH LANGUAGE EDITION

In this revised version of the Romanian original, I have taken the opportunity to characterize K-*quasiconformal homeomorphisms (and prove the equivalence to the other definitions) in Chapter 2.5, by means of the modulus of certain surface families, and in Chapter 2.10 by means of angle distortion, according to our latest results published in 1970 and 1971.*

Another important modification has been made to the bibliography, which now contains 1564 titles (an increase of about 550 on the original Romanian version); of these, 443 are on n-*dimensional quasiconformal mappings, 1034 on bidimensional mappings, and 87 on works on other topics which are nevertheless quoted in the book. Included in the part of the bibliography corresponding to the* n-*dimensional quasiconformal mappings are not only the works devoted to these topics, but also those dealing with bidimensional quasiconformal mappings or other topics which at least mention the possibility of extension of some results to* n-*dimensional quasiconformal mappings.*

In the Romanian version, Russian, Bulgarian, Serbo-Croatian and Ukrainian titles were written in the Slavonic alphabet, but in this English version of the book, Russian (and even Bulgarian, Serbo-Croatian and Ukrainian) titles are written in Latin transcription and the English translation is given in brackets. Also, all other works which are not written in English, French, German, Italian or Spanish have been included in the bibliography with the English translation of the title. The abbreviations of the Journals are given according to Mathematical Reviews.

J. Väisälä's book "Lectures on n-*dimensional quasiconformal mappings", which provides a very clear and concise introduction to the subject, was published in 1971.*

At the time of its publication, the Romanian original of the present book was not only exhaustive, but an important part of it (several chapters) represented extensions of results contained in manuscripts. Of course this is not the case today. For instance, the Finnish mathematicians and Ju. G. Rešetnjak have recently studied quasiconformal mappings from a more general point of view, i.e. renouncing the restrictive condition of univalence.

Minneapolis, 23rd February 1972 *Petru Caraman*

HISTORICAL NOTES

Let

$$u = u(x, y), \quad v = v(x, y) \tag{1}$$

be a pair of functions of two real variables. Its simple association with the imaginary symbol i ($i^2 = -1$) in an expression of the form

$$f(z) = f(x + iy) = u(x, y) + iv(x, y) \tag{2}$$

is only a formal modification concerning the study of the pair of functions (2), without realizing however a unitary structure. Then the problem arose to find some conditions, which imposed to the functions (1), give expression (2) a unitary structure in the sense that its study presents an analogy with the function theory of a real variable. Such a unitary structure was obtained asking for instance on the pair of functions (1), or the complex function (2), to satisfy one of the following equivalent conditions, characterizing the *conformal mappings* or the *analytic functions:*

1° The invariance of the infinitesimal circles.
2° The invariance of the modulus of quadrilaterals.
3° The invariance of the modulus of rings.
4° The invariance of the infinitesimal squares.
5° The invariance of the modulus of arc families.
6° The invariance of the angles.
7° The proportionality of the line elements.
8° The differentiability and the Cauchy—Riemann condition.
9° Morera theorem.
10° The existence of the derivative.
11° Taylor expansion.

12° The condition $v = [\vec{k}, \operatorname{grad} u]$, where the brackets denote the vector product and $\vec{k}$ is a vector perpendicular to the plane of the function.

13° The class A_D of the analytic functions in the domain D can be characterized by the four properties:

(i) A_D is a real vector space;

(ii) if $f \in A_D$ and $f(z_0) = 0$ for some $z_0 \in D$, then $\dfrac{f(z)}{z - z_0}$ is continuous at z_0;

(iii) $1, i \in A_D$;

(iv) A_D is maximal with respect to properties (i), (ii) and (iii).

However, the conditions were too restrictive especially because the important functions which were not analytic became more and more numerous. Then the problem of finding some conditions for the pair of functions (1) which allow for $f(z)$ the development of a unitary theory similar to the theory of analytic functions, but more comprehensive, arose again. The solution of this problem was tried in two principal ways:

1) By considering the *binary complex variable* $z=x+\alpha y(\alpha^2=\mu+\alpha\nu)$ instead of the complex variable $z=x+iy\,(i^2=-1)$, which is obtained from the preceding one in the particular case $\nu=0$, $\mu=-1$ (P. Capelli [1], [2]).

2) By weakening some restrictions which occur in the definitions of the analytic functions, or by getting rid of them.

Concerning the first way of generalization of the analytic functions we mention the following two more important particular cases: the *dual complex variable* obtained for $\mu=\nu=0$ (Study [1]) and the *hyperbolic complex variable* obtained for $\nu=0$, $\mu=1$. The class of *dual complex functions* of a dual complex variable is called the *class of para-analytic* (D) *functions* and the class of *hyperbolic complex functions* of a hyperbolic complex variable is called the *class of para-analytic* (P) *functions* (Frechet [1]—[8]).

For the second way of generalization — as it is natural owing to the large number of equivalent definitions of the analytic functions as well as to the different degrees in which the restrictions of these definitions can be weakened — a lot of intermediate classes between the simple association (2) of the pair of functions (1) and the class of analytic functions appeared. Another consequence of this fact is that the theory of these classes of functions appears as a meeting point of various domains of mathematics as, for instance, the theory of functions of a complex variable, the theory of functions of two real variables, the theory of partial differential equations (linear elliptic systems of two equations), topology (the homeomorphisms and the inner mapping in the sence of Stoilow), differential geometry, the theory of functions of a hypercomplex variable and even the theory of categories (the Teichmüller spaces defined by the help of the QCfH between Riemann surfaces, were obtained recently by Grothendieck [1] by means of the theory of the categories).

This process of formation of the classes of functions from above is also reflected in the terminology. Thus, for instance, there are both different terms for the same concept and the same terms for quite different concepts.

And now, we shall begin strictly speaking the historical account. The oldest way of generalizating the analytic functions is by partial differential equations. Thus in 1867 Beltrami [1] considered a surface with the line element

$$ds^2 = E\mathrm{d}u^2 + 2F\mathrm{d}u\mathrm{d}v + G\mathrm{d}v^2 \tag{3}$$

and a function $f(u, v)$ of the complex variable w, where $\mathrm{d}w=\chi\,(U\mathrm{d}u+V\mathrm{d}v)$, χ is an integrating factor and $U\mathrm{d}u+V\mathrm{d}v$ is one of the conjugate complex factors in which (3) is decomposed. The condition that the derivative

$$\frac{\mathrm{d}f}{\mathrm{d}w}=\frac{\dfrac{\partial f}{\partial u}\,\mathrm{d}u+\dfrac{\partial f}{\partial v}\,\mathrm{d}v}{\chi(U\mathrm{d}u+V\mathrm{d}v)}$$

is independent of the direction $\frac{dv}{du}$ implies

$$U\frac{\partial f}{\partial v} - V\frac{\partial f}{\partial u} = 0,$$

hence, setting $f = \varphi + i\psi$ and separating the real and imaginary part, he obtained the system

$$\frac{E\varphi_v - F\varphi_u}{H} = -\psi_u, \quad \frac{G\varphi_u - F\varphi_v}{H} = \psi_v, \quad (H = \sqrt{EG - F^2}),$$

known as *Beltrami's equations.* Later on E. Hedrick and L. Ingold [4] (1925) considered an analytic function $W = U + iV = F(w) = F(u + iv)$, which is no longer defined in a plane domain, but in a surface with the line element (3) and then, transforming Cauchy—Riemann equations to curvilinear coordinates, they obtained equations of the form:

$$V_x = \frac{1}{\sqrt{EG - F^2}}\begin{vmatrix} U_x & E \\ U_y & F \end{vmatrix}, \quad V_y = \frac{1}{\sqrt{EG - F^2}}\begin{vmatrix} U_x & F \\ U_y & G \end{vmatrix},$$

where (1) represents the curvilinear equations of the surface. Beltrami's system is usually known under the canonical form

$$v_y = pu_x + qu_y, \quad -v_x = -qu_x + pu_y, \tag{4}$$

which can be also written in the complex form

$$w_{\bar z} - \sigma(z)\overline{w}_z = 0,$$

where

$$w_z = \frac{1}{2}(w_x - iw_y), \quad w_{\bar z} = \frac{1}{2}(w_x + iw_y), \quad \sigma(z) = \frac{1 - p + iq}{1 + p - iq}.$$

In connection with this, we wish to settle a question of priority. L. Bers [13], I. Cristea [1] and G. N. Položiĭ [9], [11], [22] assert that "E. Picard is the first who observed a possibility of a relation between the theory of analytic functions and the study of the solutions of certain elliptic systems of partial differential equations in his Note [1] of 1891, but his idea passed unobserved until the last decade, the only exception being Beltrami's Note [4] published in 1911". But Beltrami died in 1900, and 1911 represents the year when volume 3 of Beltrami's "Opere matematiche" was printed, the pages quoted by I. Cristea, L. Bers and G. N. Položii corresponding to Beltrami's Notes [2], [3] of 1878. Moreover, these two Notes are preceded, as we have seen above, by his Note [1] of 1867, i.e. 25 years before Picard's Note [1].

The study of the solutions of Beltrami's equations, as well as those of some more general elliptic systems, from a point of view similar to the analytic function theory was done by many other authors, whose works are to be found among the references for $n = 2$ (at the end of our book). We shall confine ourselves to mention L. Bers [6] (pseudo-analytic functions), L. I. Volkovyskiĭ [7] (quasiconformal mappings), I. N. Vekua [9] (generalized analytic functions) and G. N. Položiĭ [22] [p-analytic and (p, q)-analytic functions], who in their monographs deal especially with this aspect of the generalization of analytic functions. The theory of pseudo-analytic functions gives us not only the possibility of a thorough investigation of the solutions of the linear elliptic systems, but also represents a strong mathematical tool, fundamental in the bidimensional mathematical theory of the irrotational compressible fluids.

Some authors did not content themselves with considering only linear elliptic systems; thus for instance Z. Ja. Šapiro [1] considered quasilinear systems and M. A. Lavrent'ev [16], [19] elliptic systems of the general form

$$\Phi_k(x, y, u, v, u_x, u_y, v_x, v_y) = 0 \quad (k = 1, 2).$$

Others renounced even at the ellipticity of the system. Thus G. N. Položiĭ [7], besides the (p, q)-analytic functions, which are the solutions of Beltrami's equations (4) and the p-analytic functions, which are the solutions of the same system with $q = 0$, introduced also some other (p, q)-analytic functions as solutions of the following hyperbolic system:

$$pu_x + qu_y = v_y, \quad qu_x - pu_y = -v_x;$$

B. V. Šabat [14] called the solutions of the hyperbolic system considered by him "conformal" mappings (the inverted commas belong to Šabat and have the rôle to distinguish these mappings from the conformal ones); M. A. Lavrent'ev [20] dealt with QCf mappings corresponding to the equations of the mixed type.

As the solutions u, v of the Cauchy-Riemann equations are the conjugate solutions of Laplace differential equation, so the solutions of Beltrami's equations (4) or of a more general elliptic system are conjugate solutions of a second-order elliptic equation generalizing Laplace equation (S. Bergman [9]). These solutions are called *pseudo-harmonic* and are the real and imaginary part of a pseudo-analytic function (see M. Morse's monograph [2]).

The problem of the extension of the conformal mappings to three dimensions arose even in the last century. Thus, in 1850 Liouville [1] proved that a conformal mapping of class C^3 (characterized by the proportionality of the line elements) is a Möbius transformation (i.e. a finite product of inversions in spheres). (Recently this result was improved by Ju. G. Rešetnjak [2], who removed the restriction $f \in C^3$; he showed that the conformal mappings, i.e. the homeomorphisms which map infinitesimal spheres into infinitesimal spheres, are finite products of inversions in spheres). E. Hedrick and L. Ingold [3] observed that the three-dimensional conformal mappings are the solutions of the second-order elliptic system

$$\begin{cases} u_x^2 + v_x^2 + w_x^2 = u_y^2 + v_y^2 + w_y^2 = u_z^2 + v_z^2 + w_z^2, \\ u_x u_y + v_x v_y + w_x w_y = u_x u_z + v_x v_z + w_x w_z = u_y u_z + v_y v_z + w_y w_z = 0, \end{cases}$$

and in n-space

$$\delta_{pq} x_i^{*p} x_k^{*q} = \delta_{ik} \sqrt[n]{J^2} \qquad (i, k = 1, \ldots, n),$$

where $x^* = f(x) = x^{*i} e_i$, $x_i^{*p} = \dfrac{\partial x^{*p}}{\partial x^i}$, and which, if $n = 2$, reduces to

$$u_x^2 + v_x^2 = u_y^2 + v_y^2, \quad u_x u_y + v_x v_y = 0,$$

which is equivalent with the two linear systems

$$\begin{cases} u_x = v_y, \\ u_y = -v_x, \end{cases} \quad \begin{cases} u_x = -v_y, \\ u_y = v_x, \end{cases}$$

satisfied by the direct or indirect conformal mappings. (These systems were obtained by E. Hedrick and L. Ingold [3] if $n = 3$, and by G. E. Raynor [1] for an arbitrary n.) Since generally the conformal mappings and *a fortiori* the QCfH are not solutions of linear systems, their study from the point of view of the partial differential equations, or the study of the solutions of certain partial differential equations by means of the n-dimensional QCfH is almost inexistent for $n > 2$.

The main way of generalization of the conformal mappings (especially if $n > 2$) is the geometrical one. The first to give a geometric extension of the class of the conformal mappings was H. Grötzsch [1] in 1928. He defined the class A_Q of the non-conformal mappings $f \in C^2$ in a domain D of the complex plane as locally homeomorphic and mapping infinitesimal circles into infinitesimal ellipses with the ratio $\frac{a}{b}$ of the semiaxes verifying the inequality

$$\frac{1}{Q} \leqq \frac{a}{b} \leqq Q$$

everywhere in D, except possibility a countable set with eventual cluster points only in D, and in these points of exception (branched-points) the mapping is of the form $w = z^m$.

Although E. Hedrick, L. Ingold and W. D. A. Westfall [1] defined the *non-analytic functions* only as mappings of the class C^3 (we remind that the mappings of the class C^1 are M. A. Lavrent'ev's [14] QCfH with a pair of characteristics p, ϑ and B. V. Šabat's [10] QCfH with two pairs of characteristics $p, \vartheta, p_1, \vartheta_1$), they also established some geometrical properties of these mappings, as for instance that they map certain infinitesimal ellipses (called Tissot's indicatrices because Tissot [1] used them for the first time in a chartographic study) in infinitesimal circles. In 1923 too, E. Hedrick and L. Ingold [1], dealing with non-analytic functions in 3-space, proved that certain infinitesimal ellipsoids are taken into infinitesimal spheres, and for the conformal mappings the infinitesimal ellipsoids reduce to infinitesimal spheres (and if $n = 2$ the infinitesimal ellipses to infinitesimal circles).

In 1935, M. A. Lavrent'ev [14] introduced the *almost analytic functions* (called by Volkovyskiĭ [7] QCf mappings with a pair of characteristics) characterized by the property that they map certain infinitesimal ellipses into infinitesimal circles. For the determination of the infinitesimal ellipses he uses not only the dilatation $\frac{a}{b}$ (as in Grötzsch's definition of non-conformal mappings), but also the angle ϑ the major axis of the ellipse makes with the positive half of the real axis of the coordinates, and instead of asking p to be bounded in D, he asks the continuity of the characteristics p, ϑ in D. An extension of this definition given by B. V. Šabat [11] is the class of QCfH with two pairs of characteristics (which map infinitesimal ellipses into infinitesimal ellipses). A still more general class of QCfH was defined for $n = 2$ by I. N. Pesin [3] (the corresponding mappings transform regular curve families into regular curve families), (this definition was "in embryo" at A. I. Markuševič [1]). The last three classes of QCfH have also been studied by ourselves in our Notes [1], [2], [4], [6]—[15], [17], [19]—[26], [28].

The term "*quasiconformal*" was introduced for the first time by L. Ahlfors [4], who characterized the QCfH by the property that a simply connected open surface W is mapped onto the plane z and the corresponding metric (3) is positive definite and $E + G \leqq 2k\sqrt{EG - F^2}$, where $k \leqq 1$.

Starting from invariance of squares as a characteristic of conformal mappings, M. A. Lavrent'ev [15] introduced a class of QCfH characterized by the property that certain infinitesimal parallelograms (determined by the side V, the corresponding height W, the angle α between V and the real axis and the acute angle ν between its sides) are mapped into infinitesimal squares. This class of QCfH is used in the theory of elasticity.

From the other geometric definitions of the conformal mappings the invariance of the modulus of the quadrilaterals Q and that of the modulus of the rings A, give rise to QCfH characterized by the inequality (A. Pfluger [3], J. Hersch and A. Pfluger [1] and L. Ahlfors [4])

$$M(Q^*) \leqq KM(Q),$$

respectively by

$$M(A^*) \leqq KM(A)$$

(F. Gehring and J. Väisälä [2] and E. Reich [1]); we also mention the characterization of the QCfH by the angle distortion (for $n=2$, S. Agard and F. Gehring [1] and O. Taari [2] and for an arbitrary n, S. Agard [1], O. Taari [1] and our paper [31]), by the distortion of the harmonic and of the hyperbolic measure (A. Kelingos [4]), of the extremal length of the curve families (J. Väisälä [1]) and by the invariance of the curve families of the extremal length zero (H. Renggli [3] and our paper [33]). For an interesting survey of the plane (especially geometric definitions) see F. Gehring's Note [22].

The QCfH are topological mappings subject to an additional analytic, geometric or metric condition. If one removes the restriction of the univalence, then the QCf mappings become inner transformations (in Stoilow's sense) verifying one of the conditions from above. S. Stoilow [1] gave an extension of the inner transformations also for $n = 3$.

Let us mention, just in passing, the definition of the QCfH on a Riemann surface (R. Cacciopoli [4], Cabiria Andreian-Cazacu [7]), A. Bilimovici's [1]—[13] non-analytic functions characterized by the reflexion measure $\vec{B} = \text{grad } v - [\vec{k}, \text{grad } u] \neq 0$, where $B^2 = (u_x - v_y)^2 + (u_y + v_x)^2$, and where if $\vec{B}_1$ is the component of $\vec{B}$ in the direction of the vector grad u and $\vec{B}_2 \perp \vec{B}_1$, we get for $\vec{B}_1 = 0$ the class of para-analytic (P) functions and for $\vec{B}_2 = 0$ the class of para-analytic (D) functions, the analytic definitions (R. Cacciopoli [1], A. Mori [2] and L. Bers [15]) and finally L. Bers' [13] axiomatic definition of the pseudo-analytic functions of the first and second kind characterized with the help of the generators F, G, which generalize the quantities 1, i occurring in the condition (iii) (from above) of his axiomatic definition of the analytic functions.

Let us say a few words about the extensions of the analytic functions characterized by the existence of the derivative. A function with the limit of the increment ratio depending on the direction is said to be *polygenic* (in distinction from the monogenic functions with the limit of this ratio independent of the direction). This term was introduced by E. Kasner in a communication held at the 257th meeting of October 29, 1927 of the American Mathematical Society.

In this very general case, the polygenic functions were studied by V. S. Fedorov, [5], A. D. Taĭmanov [1] and Ju. Ju. Trohimčuk [1]. N. N. Luzin [1] used for their study the set of the derivative numbers $\mathfrak{M}_z = \bigcap_{n=1}^{\infty} \overline{M}_{\varepsilon_n}$, $\varepsilon_n \to 0$ as $n \to \infty$, where $\overline{M}_{\varepsilon_n}$ is the closure of the set of values of the increment ratio $\dfrac{\Delta f}{\Delta z}$ for all Δz $(0 < |\Delta z| < \varepsilon_n)$. If $f(z)$ is monogenic in z_0 then $\mathfrak{M}_{z_0} = f'(z_0)$ and if $f(z)$ is differentiable in z_0, then $\mathfrak{M}_{z_0}$ is a circumference called Kasner circle. Usually however, the polygenic functions are supposed to be of class C^1 and sometimes even of class C^3 (E. Hedrick, L. Ingold and W. D. A. Westfall [1]).

On the other hand, some mathematicians did not confine themselves simply and solely to renounce at the existence of the derivative and to put in exchange some additional differentiability conditions, but they replaced the classical derivative by a more general one. Thus D. Pompeiu [1] introduced in 1912 the areolar derivative (called also phase derivative) defined by the expression

$$\lim \frac{\int_C f(z)\mathrm{d}z}{\frac{1}{2}\int_C x\mathrm{d}y - y\mathrm{d}x},$$

where the limit is taken for the continuous contraction of the curve C in an interior point of the domain bounded by C. The class of the polygenic functions admitting an areolar derivative is said to be the class of (α)-monogenic or (α)-holomorphic functions. This class of functions was investigated in detail especially by N. Teodorescu [1]—[24] and his pupil D. Pascali [7]—[16], and was extended to higher dimensions by M. Nedelcu [1]—[13] and D. Pascali [1]—[6]. If $f(z)$ is assumed to be differentiable, the areolar derivative is of the form $\frac{\partial f}{\partial \bar{z}} = \frac{1}{2}[u_x - v_y + i(v_x + u_y)]$ and the monogenic functions are clearly characterized by $\frac{\partial f}{\partial \bar{z}} = 0$.

The theory of the QCf mappings in n-space, except for some isolated Notes (E. Hedrick and L. Ingold [1]—[6], M. A. Lavrent'ev [1]—[6], A. I. Markuševič [1], M. Kreines [1] and E. Zimmermann [1], [2]) began to develop in a systematic way only from 1960 onwards, but this happened so impetuously that we were obliged to renounce covering in our monograph the whole of the new territory recently opened up. The theory of QCfH in n-space, which is a double extension of the plane conformal mappings (both concerning the conformality and the dimension) was preceded by the theory of the pseudo-conformal mappings (i.e. sets of n analytic functions of n complex variables). This theory, the origin of which is to be found in P. Cousin's [1] and H. Poincaré's [1] papers, has developed by its own methods completely different from those of the QCfH in n-space. Nevertheless, the n-dimensional pseudo-conformal mappings may be considered as a subclass (of Lavrent'ev's type, see part 2, chapter 11) of $2n$-dimensional QCfH (as it will be proved in part 2, chapter 16). As classes of mappings connected with both, the pseudo-conformal and the QCfH, S. Bergman's [1], [2] quasi-pseudo-conformal mappings and Hitotumatu's [1] discoidal QCf diffeomorphisms deserve to be mentioned too.

Part 1

Preliminary results

CHAPTER 1

Terminology and general notations

In this chapter we shall try to justify some general terms adopted by us and we shall introduce some general notations.

Terminology. As we already said in the historical account, the various ways in which the conformal mappings and, the analytic functions respectively, have been generalized resulted in the fact that, for instance, we have the same term for different classes of mappings or functions, and different terms for the same class. The most wide-spread term for the class of the mappings we deal with, and which was adopted by us too (as it results also from the title of our book), is that of QCF *mappings* (introduced by L. Ahlfors [4]). If $n = 2$ (especially when it comes to extensions from the point of view of the partial differential equations), frequent use is made of the term *pseudo-analytic function* chiefly in Western literature (L. Bers [6]) and the term of *generalized analytic functions* especially in Soviet literature (I. N. Vekua [9]). In the Japanese literature we also find the term *pseudo-regular functions* (Z. Yûjôbô [1]), which is derived from the corresponding term *regular functions* used by the Japanese mathematicians for analytic functions; as "regular functions" is a term less wide-spread than "analytic functions", the corresponding term "pseudo-regular functions" is employed much less than "pseudo-analytic functions". The oldest terms are *non-analytic functions* (E. Hedrick, L. Ingold and W. D. A. Westfall [1]) and *non-conformal mappings* (H. Grötzsch [1]) and were used almost only by their authors; this is due to the following disadvantages: on the one hand they are too vague, implying any mapping which is not conformal and any function which is not analytic, respectively, that is why it would be necessary to use a second term which specifies the class (thus for instance Grötzsch calls his mappings non-conformal of the class A_Q), because otherwise the paradoxical situation would occur of finding mappings which are not conformal, or functions which are not analytic, respectively, without being non-conformal, or non-analytic respectively; on the other hand, it is more natural that the extension of the class of conformal mappings or of analytic functions, respectively, be a class containing them, and not excluding them. Thus, in spite of the fact that the term of non-conformal excludes the conformal mappings from the class considered by Grötzsch, its definition contains the conformal mappings as a particular case, for $Q = 1$, which gives rise to the following incongruity of speech that the conformal mappings are non-conformal. The term of *polygenic functions* introduced by E. Kasner (1927) referred to the functions of the class C^1; later on the definition of the QCfH involved differentiability almost everywhere in the domain of definition and together with the old class the old term disappeared too. The terms of *pseudo-conformal mappings* (R. Cacciopoli [1]) and of *quasi-analytic functions* (I. I. Daniliuk [7]) used for QCfH did not catch on because these terms had other wide-spread meanings. Thus "pseudo-conformal mappings" is an accepted expression for a set of n analytic functions of n complex variables (see S. Bergman [10]) and the term of "quasi-analytic functions" corresponds to the following definition: "A function $f(x)$ of a real variable is said to be quasi-analytic if it is uniquely determined by its derivatives $f^{(n)}(a)$, a fixed and $n = 0, 1, \ldots$" (T. Bang [1]) or "A set of functions defined on a domain D is said to be quasi-analytic if the equality of two functions of the set on no matter how small a subdomain of D implies their equality in D" (S. Kodama [1]). We find also the terms *almost analytic functions* introduced by M. A. Lavrent'ev [14] and *almost conformal mappings* used by S. Kahramaner [1] which are not wide-spread. Finally, let us mention the terms *(p, q)-analytic* and *p-analytic functions* introduced by G. N. Položiĭ in [9] and [1], respectively, for the solutions of Beltrami's equations

$$v_y = pu_x + qu_y, \quad -v_x = -qu_x + pu_y,$$

and of this system with $q = 0$.

Now let us discuss the terms *positive* and *increasing*. For some authors, a positive ε or an increasing $f(x)$ mean $\varepsilon > 0$ and an increasing function, respectively, with the exclusion of the possibility that such a function be constant in some subinterval. (This is also the meaning of the preceding terms in everyday speech, as for instance in the words of "the temperature is increased" or "a positive character" and also in the mathematical phrase of: "For every positive number ε, there exists a positive number $\delta(\varepsilon)$, such that..." which is used for instance to define the continuity) and if it is allowed to the quantity also to be equal to zero (for instance $\varepsilon \geqq 0$), respectively to the function to remain constant in some subintervals, they employ the terms nonnegative and monotonic increasing, respectively. We should prefer to adopt this terminology too, but for the Bourbakists (and this is an argument which counts) a positive quantity or an increasing function requires the possibility of the equality with zero ($a \geqq 0$), respectively the existence of some subintervals where the function is constant, and when they want to exclude this possibility, the terms strictly positive and strictly increasing are used. However, if we meet in a paper the term positive or increasing, we shall not be sure which of the two different meanings is intended by the author. To avoid any confusion, we have decided to use throughout this monograph the pair of terms: *strictly positive — nonnegative, strictly negative — nonpositive, strictly increasing — monotonic increasing and strictly decreasing — monotonic decreasing.*

For simplicity, we shall not use the prefix "hyper" in the words q-dimensional hypersurface or hypersphere and also the words n-dimensional and (n — 1)-dimensional, respectively, in the expressions n-dimensional ball and $(n-1)$-dimensional surface or sphere, respectively.

Abbreviations. a.e. = almost everywhere, AC = absolutely continuous, ACA = absolutely continuous on arcs, ACL = absolutely continuous on lines, cap = conformal capacity, inf = infimum, sup = supremum, lim, lim and lim mean limit, upper and lower limit, respectively.

General notations. The points of the Euclidean n-space R^n and the corresponding vectors are denoted by $x = (x^1, \ldots, x^n)$ or by $x = x^i e_i$, where x^i is the coordinate of x corresponding to the axis Ox^i, e_i is the unit vector in the direction of the Ox^i axis and i is a dummy index which takes on the values $1, \ldots, n$. If E is a set, then ∂E, $\bar{E}$ and CE mean the boundary, the closure and the complement of E, respectively. $x \in E$ means x is a point of E, while $x \bar{\in} E$ means that x does not belong to E and $E' \subset E$ means that E' is contained in E. If $S(x)$ designates a certain statement relating to the point x, then we write $E = \{x;\ S(x)\}$ to state that E is the set of those points x for which the statement $S(x)$ holds. We shall use it to introduce the following notations: $A \cup B = \{x;\ x \in A \text{ or } x \in B\}$, $A \cap B = \{x;\ x \in A,\ x \in B\}$, $A - B = \{x;\ x \in A,\ x \bar{\in} B\}$. $|x|^2 = \delta_{ik} x^i x^k$ (where δ_{ik} is Kronecker delta), then $d(x, y) = |x - y|$ and $d(E_1, E_2) = \inf\limits_{\substack{x \in E_1 \\ x \in E_2}} d(x, y)$. Let us denote by D the domain of the definition of the QCfH $x^* = f(x) = x^{*i} e_i$, by γ, σ and σ_q the curves, the surfaces and the q-dimensional surfaces, respectively, by Γ, Σ and Σ_q the curve, surface and q-dimensional surface families, respectively, by A, C, I, Z the rings, cones, intervals and cylinders, respectively and by the same quantities with asterisk their images under f, i.e. for instance $E^* = f(E)$, $D^* = f(D)$, $A^* = f(A), \ldots$ Let us denote by ξ a boundary point of D. $x \to a$ means x tends to a, $f\colon D \to R^n$ means f maps D into R^n and $f\colon D \rightleftarrows D^*$ means f is a homeomorphism of D onto D^*. If $u(x)$ is a real function, then $\nabla u = \left(\dfrac{\partial u}{\partial x^1}, \ldots, \dfrac{\partial u}{\partial x^n}\right)$, $|\nabla u|^2 = \delta^{ik} \dfrac{\partial u}{\partial x^i} \dfrac{\partial u}{\partial x^k}$ in a point $x \in R^n$ and if $x^* = f(x)$ is a mapping of D in R^n, i.e. $f\colon D \to R^n$, then $x^{*i}_k = \dfrac{\partial x^{*i}}{\partial x^k}$ $(i, k = 1, \ldots, n)$. Let Δx, $\Delta f = f(x + \Delta x) - f(x)$ be the increment of x and $f(x)$, respectively. $f \in C^m$ means that the components of $f(x)$ have continuous partial derivatives of order p $(p = 1, \ldots, m)$. $B^q(x, r)$ and $S^q(x, r)$ are the *q-ball* (q-dimensional ball), respectively the *q-sphere* (q-dimensional sphere) centered at x and of radius r, then $B^n(x, r) = B(x, r)$, $B^q(O, r) = B^q(r)$, $B^q(1) = B^q$, $S^{n-1}(x, r) = S(x, r)$, $S^q(O, r) = S^q(r)$, $S^q(1) = S^q$. Let us denote by m, m_q the n-dimensional and q-dimensional Lebesgue measure, respectively, by $d\sigma$, $d\sigma_q$, $d\tau$, ds the element of surface, of q-dimensional surface, of volume and of arc length (line element), respectively, and by $\omega_q = mB^q$, where $\omega_q = \dfrac{\pi^{\frac{q}{2}}}{\Gamma\left(1 + \dfrac{q}{2}\right)}$, $\Gamma(m+1) = m!$, $\Gamma\left(m + \dfrac{1}{2}\right) = \dfrac{\sqrt{\pi}\,(2m)!}{4^m m!}$ $(m = 0, 1, \ldots)$, hence, since the line element of R^q in polar coordinates is $ds^2 = d\rho^2 + \rho^2 d\sigma_{q-1}$, where ρ is the length of the radius vector and $d\sigma_{q-1}$ is the element of the unit $(q-1)$-sphere, and by Fubini theorem (Saks [2], theorem 8.1, p. 77 or Natanson [1], p. 445) the volume of $B^q(r)$ is

$$\omega_q r^q = \int_{B^q(r)} d\tau = \int_0^r \rho^{q-1} d\rho \int_{S^{q-1}} d\sigma_{q-1} = \frac{ar^q}{q},$$

then the $(q-1)$-dimensional area of S^{q-1} is $a = q\omega_q$. Finally, let us denote by

$$\mathscr{J}(x) = \|x_k^{*i}\| = \begin{pmatrix} x_1^{*1} & \cdots & x_n^{*1} \\ \cdots & \cdots & \cdots \\ x_1^{*n} & \cdots & x_n^{*n} \end{pmatrix}$$

the functional matrix, by

$$J(x) = \det |x_k^{*i}| = \begin{vmatrix} x_1^{*1} & \cdots & x_n^{*1} \\ \cdots & \cdots & \cdots \\ x_1^{*n} & \cdots & x_n^{*n} \end{vmatrix}$$

the Jacobian of $x^* = f(x)$ and by

$$J_G(x) = \overline{\lim_{r \to 0}} \frac{T(x, r)}{mB(x, r)}, \tag{1}$$

where

$$T(x, r) = m\{f[B(x, r)]\}, \tag{2}$$

the extension of the Jacobian given by F. Gehring [5]. In the points of differentiability $J_G(x) = |J(x)|$.

If a statement $S(x)$ holds for every $x \in D - E$ and $mE = 0$, $S(x)$ is said to hold a. e. in D.

CHAPTER 2

Spaces L^p and l^p

Real linear *n*-space. We introduce the concept of linear space following R. Miron [1], because his definition is the only one in which the corresponding system of axioms is independent.

A set L is said to be a *linear space* if for its elements called vectors and denoted by $x, y, \ldots$, an operation $x + y \in L$ called addition is defined and a set R is given, whose elements called scalars and denoted by $\alpha, \beta, \ldots$ are such that for any $\alpha \in R$ and $x \in L$ the product $\alpha x \in L$ is defined, the addition and the product verifying the following axioms:

A_1. $x + (y + z) = (x + y) + z$ for any $x, y, z \in L$.

A_2. $x + O = x$, $O \in L$ for all $x \in L$.

A_3. For each $x \in L$, there exists an $x_1 \in L$ such that $x + x_1 = O$.

(The axioms $A_1 - A_3$ imply that the elements O and x_1 are unique. Let us denote $x_1 = -x$.)

A_4. $\alpha(x + y) = \alpha x + \alpha y$ for any $\alpha \in R$ and $x, y \in L$.

A_5. There exists an $\alpha \in R$ such that $x \neq O$ imply $\alpha x \neq O$.

A_6. $\alpha x = \beta x$, $x \neq O$ implies $\alpha = \beta$ for any $\alpha, \beta \in R$.

A_7. $(\alpha, \beta) \in R \times R$ implies the existence of an element $\gamma \in R$ such that $\alpha x + \beta x = \gamma x$ for any $x \in L$.

A_8. For each $\alpha \in R$ there exists an $\alpha_1 \in R$ such that $\alpha(-x) = \alpha_1 x$ for any $x \in L$.

A_9. If α verifies A_5, then there exists an $\alpha_1 \in R$ such that $\alpha_1(\alpha x) = x$ for any $x \in L$.

A_{10}. For each pair $(\alpha, \beta) \in R \times R$ there exists an element $\gamma \in R$ such that $\alpha(\beta x) = \gamma x$ for any $x \in L$.

The set L is a linear space in the classical sense (i.e. characterized by the systems of axioms in use up to now), since the first 10 axioms above imply the following propositions (some of them are axioms in other systems of axioms) as simple consequences:

a) $\alpha(x - y) = \alpha x - \alpha y$ for any $\alpha \in R$, $x, y \in L$.

b) $O\alpha = O$ for any $\alpha \in R$.

c) For each ordered pair (α, β), $\alpha, \beta \in R$, there exists a unique $\gamma \in R$, such that $\alpha x + \beta x = \gamma x$ for any $x \in L$. This scalar γ is called the sum $\alpha + \beta$ of the elements α, β.

d) There exists at least a $0 \in R$ such that $0x = O$ for any $x \in L$ and it is unique.

e) $\alpha(-x) = -(\alpha x) = \alpha' x$ for any $\alpha \in R$, $x \in L$ and α' is unique.

f) There exists at least an $e \in R$ such that $ex = x$ for any $x \in L$.

g) For any pair $\alpha, \beta \in R$ and $x, y \in L$, $\alpha x + \beta y = \beta y + \alpha x$.

h) For each ordered pair (α, β), $\alpha, \beta \in R$, there exists a unique $\gamma \in R$ such that $\alpha(\beta x) = \gamma x$ for any $x \in L$. This scalar γ is called the product $\alpha\beta$ of the scalars α, β.

i) There exists a unique $e \in R$ such that $ex = x$ for any $x \in L$.

j) R is a skew field with respect to the addition and the multiplication of the scalars.

(R. Miron [1] in his Note proves the independence and the consistency of the axioms A_1—A_{10}.)

The linear space L is a *real linear n-space* if it satisfies in addition the following axioms:

A_{11}. The field R obtained from above is isomorphic to the real number field.

A_{12}. The linear space L over R has the finite dimension $n \geqq 1$.

Normed linear space. A linear space is said to be *normed* if there is a real number $\|x\|$ (called the norm of x) associated with each vector x and

1) $\|x\| \geqq 0$, where $\|x\| = 0$ if and only if $x = O$;

2) $\|\alpha x\| = |\alpha| \, \|x\|$, where $\alpha \in R$, and in particular $\|-x\| = \|x\|$;

3) $\|x_1 + x_2\| \leqq \|x_1\| + \|x_2\|$.

Banach space. A sequence $\{x_m\}$ of points of a normed linear space is said to be a *Cauchy sequence* if $\|x_p - x_q\| \to 0$ as $p, q \to \infty$.

A normed linear space is said to be *complete* if any Cauchy sequence converges to a point of the space. A complete normed linear space is a *Banach space*.

Euclidean n-space R^n. A normed linear n-space on the real number field is an *Euclidean n-space* if the norm of x is $|x| = (\delta_{ik}x^i x^k)^{\frac{1}{2}}$. As an Euclidean n-space is complete, it is a Banach space.

The *scalar product of two vectors* $x, y \in R^n$ is

$$xy = \delta_{ik} x^i y^k. \tag{1}$$

where $x = x^i e_i = |x| \cos(x, x^i) e_i$, $y = y^i e_i = |y| \cos(y, x^i) e_i$. Hence $xy = |x| \, |y| \delta_{ik} \cos(x, x^i) \cos(y, y^k)$. On the other hand x, y are two concurrent vectors (all the vectors x are supposed to have the origin in O); if ϑ is the acute angle between them, then the scalar product in the plane xOy (determined by them) is of the form

$$xy = |x| \, |y| \cos \vartheta. \tag{2}$$

From (1) and (2) we get

$$\cos \vartheta = \delta_{ik} \cos(x, x^i) \cos(y, y^k). \tag{3}$$

Space L^p. Let X be a linear normed n-space and $p > 0$. A mapping f: $R^n \to X$ is said to be *L^p-integrable* if

1° $f(x)$ is Lebesgue measurable,

2° $\int |f(x)|^p \, d\tau < \infty$.

Let us denote by $L^p(R^n, X)$ the set of the classes of p-integrable functions (a class contains all the p-integrable functions equal a.e.). With each function $f \in L^p$ one associates a norm:

$$\|f\|_p = (\int |f|^p \mathrm{d}\tau)^{\frac{1}{p}}.$$

The set $L^p(R^n, X)$ with the above norm may be organized as a normed linear n-space called the *space* $L^p(R^n, X)$. When X is an arbitrary linear normed space or whenever it can be done without ambiguity, for the sake of simplicity, let us set $L^p(R^n, X) = L^p$ and $L^p(E, X) = L^p(E)$.

Let $f(x)$ be a measurable mapping in a domain $D \subset R^n$ and let $h > 0$ be a sufficiently small constant. The mapping

$$f_h(x) = (2h)^{-n} \int_{x-h}^{x+h} f(\xi)\mathrm{d}\tau = (2h)^{-n} \int_{x^1-h}^{x^1+h} \cdots \int_{x^n-h}^{x^n+h} f(\xi^1, \ldots, \xi^n)\mathrm{d}\xi^1 \cdots \mathrm{d}\xi^n$$

is termed the h-*average mapping* of $f(x)$.

PROPOSITION 1. *For each* $f \in L^p$, $f_h \to f$ *in* L^p, i.e.

$$\lim_{h \to 0} \|f_h - f\|_p = 0.$$

(For the proof see M. Nicolescu [5], volume 3, p. 259).

PROPOSITION 2. *If* X *is a Banach space, then* $L^p(R^n, X)$ *is a Banach space too.* (See M. Nicolescu [5], Volume 3, p. 239).

Space l^p. The *space* l^p, $p > 0$, is said to be the set of all real number sequences $x = \{x_m\}$ with the norm

$$\|x\|_p = \left(\sum_{m=1}^{\infty} |x_m|^p\right)^{\frac{1}{p}} < \infty.$$

Characteristic function. The function

$$\chi_E(x) = \begin{cases} 1 \text{ whenever } x \in E, \\ 0 \text{ whenever } x \in CE \end{cases}$$

is called the *characteristic function* of E (in R^n).

Step function. Let P be a family of subsets of a set E and X a linear normed space. Any mapping $f: E \to X$ of the form

$$f(x) = a^i \chi_{E_i}(x),$$

where $a^i \in X$, $E_i \in P$ and χ_{E_i} is the characteristic function of $E_i (i = 1, \ldots, m)$ is termed *a step function* (on the family P).

Clarkson theorem [1]. *For space L^p or l^p, with $p \geqq 2$, the following inequalities between the norms of two arbitrary elements x, y of the space are valid*:

$$2(\|x\|^p + \|y\|^p) \leqq \|x + y\|^p + \|x - y\|^p \leqq 2^{p-1}(\|x\|^p + \|y\|^p), \tag{4}$$

$$2(\|x\|^p + \|y\|^p)^{q-1} \leqq \|x + y\|^q + \|x - y\|^q, \tag{5}$$

$$\|x + y\|^p + \|x - y\|^p \leqq 2(\|x\|^q + \|y\|^q)^{p-1}, \tag{6}$$

where $q = \dfrac{p}{p-1}$. *For* $1 < p \leqq 2$, *these inequalities hold in the reverse sense.*

CHAPTER 3

Measure theory

In this chapter we introduce the concept of Lebesgue, Hausdorff, Carathéodory, Gross and Federer measures, of length and area of a set in R^n and some other results which we shall use in parts II and III, as for instance the theorems of Gross, of Federer—Young, of Radon—Nikodim, of Fubini and of Tonelli and a theorem of change of measure.

Functions of a set. Let $P(E_0)$ be a family of subsets of a set E_0. A function $\varphi: P(E) \to R^1$ will be called *additive function of a set on a set* E' if (*i*) $E' \in P(E_0)$, (*ii*) $\varphi(E)$ is defined for each set $E \subset E'$ and if (iii) $\varphi(\bigcup_m E_m) = \sum_m \varphi(E_m)$ for every sequence $\{E_m\}$ of sets $E_m \subset E'$ $(m = 1, 2, \dots)$ and such that $E_i \cap E_k = \emptyset$ whenever $i \neq k$.

Remark. In his more restrictive definition of a function of a set on a set E', S. Saks ([2], p. 8) puts also the additional condition $\varphi(E) < \infty$ for each set $E \subset E'$. But this would imply for the definition of a measure (which is a particular function of a set) the additional condition $\mu E < \infty$.

Lebesgue measure. It is known that Carathéodory ([3], p. 239) calls *outer measure* a function of a set m^* which verifies the following four axioms:

I. m^*E is unique for any set E of a linear n-space and $0 \leqq m^*E \leqq \infty$. There exist sets E with $m^*E \neq 0, \infty$, and $m^*\emptyset = 0$.

II. $E' \subset E$ implies $m^*E' \leqq m^*E$.

III. $E = \bigcup_k E_k$ implies $m^*E \leqq \Sigma_k m^*E_k$.

IV. $d(E, E') \neq 0$ implies $m^*E + m^*E' = m^*(E \cup E')$.

When the considered n-space is R^n(Euclidean), we shall add the following axiom:

V. m^* is invariant to a motion in R^n.

A set $E \subset R^n$ is said to be *measurable* if for any set $W \subset R^n$, the relation

$$m^*W = m^*E \cap W + m^*(W - E \cap W).$$

holds. In this case m^*E is termed the *measure* of E and is denoted by mE.

As it was proved by C. Carathéodory ([3], theorem 2, p. 246, theorem 1, p. 248 and theorems 2 and 3, p. 251) for any measure which verifies the axioms I—IV, the open and the closed sets, as well as the union and the intersection of a countable sequence of measurable sets is a measurable set.

A measure is called *regular* (Carathéodory [3], p. 258), if it satisfies the following axiom:

VI. $m^*E = \inf_{M \supset E} mM$, where the infimum is taken over all the measurable sets M containing E.

This axiom may be replaced by the following more strict one:

VI′. $m^*E = \inf_{B_n \supset E} mB_n$, where the infimum is taken over all the sets B_n containing E and obtained as intersection of a sequence of sets each of them being a countable union of closed sets.

Lebesgue measure is a measure verifying the axioms I—V, VI′ and

VII. The n-dimensional unit cube has the measure 1.

We shall introduce now some other measures verifying the axioms I—V.

Hausdorff, Carathéodory, Gross and Federer measures. *A Borel set* is said to be a set which can be obtained from the closed and open sets by repeated applications of operations of union and intersection to denumerable numbers of sets.

Let us denote by $\mathfrak{M}_n$ the set of all the subsets of R^n, by B_n and C_n the set of all Borel and convex subsets, respectively and by τ_n and β_n the set of all the domains and balls, respectively.

If $r > 0$, $\mathscr{F} \subset \mathfrak{M}_n$, $g(G)$ is a function of a set for which $0 \leqq g(G) \leqq \infty$ whenever $G \in \mathscr{F}$ and $g(\emptyset) = 0$, $E \in M_n$, then

$$[\Xi_n^r(\mathscr{F}, g)](E) = \inf_{\mathscr{G} \in N_r} \sum_{G \in \mathscr{G}} g(G)\,,$$

where N_r is the set of countable subfamilies $\mathscr{G} \subset \mathscr{F}$ such that $E \subset \bigcup_{G \in \mathscr{G}} G$ and $d(G) < r$. In this connexion we remind that the infimum of the empty set is ∞. Hence if $N_r = \emptyset$, then $[\Xi_n^r(\mathscr{F}, g)](E) = \infty$. Further, if $E = \emptyset$, we can take as $\mathscr{G}$ for instance the family of sets consisting of only one element, namely the empty set, and then $[\Xi_n^r(\mathscr{F}, g)](\emptyset) = 0$.

The limit

$$[\Xi_n(\mathscr{F}, g)](E) = \lim_{r \to 0} [\Xi_n^r(\mathscr{F}, g)(E)] \leqq \infty,$$

which exists, since

$$[\Xi_n^r(\mathscr{F}, g)](E) \leqq [\Xi_n^s(\mathscr{F}, g)](E)$$

whenever $r > s > 0$, is a measure verifying I—V.

And now, if M_n^m is the set of the matrices with m columns and n rows $a \in M_n^m$, k, p integers, $0 < k \leqq m$, $0 < p \leqq n$, then $(a|_p^k)$ is the upper left-hand minor of a with p rows and k columns and

$$P_a^k \qquad (a \in M_n^n.\ k = 1, \cdots, n)$$

is the linear transformation with the matrix $(a|_k^n)$ (if a is an orthogonal matrix, then P_a^k is an orthogonal projection). Let

$$G_n = E\{a;\ a \in M_n^n,\ |a(x)| = |x|,\ x \in R^n\}$$

be the set of all orthogonal matrices of n-space.

If $q \leqq n$ are positive integers, then

$$\gamma_n^q(E) = \sup_{a \in G_n} m_q[P_a^q(E)] \qquad \text{whenever } E \in \mathfrak{M}_n,$$

$$\chi_n^q(E) = \omega_q \left[\frac{d(E)}{2}\right]^q \qquad \text{whenever } E \in \mathfrak{M}_n.$$

$S_n^q = \Xi_n(\beta_n, \chi_n^q)$ is *spherical measure*,

$H_n^q = \Xi_n(\mathfrak{M}_n, \bar{\chi}_n^q)$ is *Hausdorff measure* (both were introduced by Hausdorff [1]).

$C_n^q = \Xi_n(C_n, \gamma_n^q)$ is *Carathéodory* [2] *measure*,

$\Gamma_n^q = \Xi_n(\mathfrak{M}_n, \gamma_n^q)$ is *Gross* [1], [2] *measure*,

$\Phi_n^q = \Xi_n(\tau_n, \gamma_n^q)$ is *Federer* [1]–[3] *measure.*

From now on, for the sake of simplicity, we shall drop the index n in the preceding notations.

These five measures verify axioms I—V (Federer [3]).

Clearly $\mathfrak{M}_n \supset B_n \supset \tau_n \supset C_n \supset \beta_n$ and $\gamma_n^q(G) \leqq \chi_n^q(G)$, hence

$$\Gamma^q(E) \leqq \Phi^q(E) \leqq C^q(E) \leqq S^q(E),\ \Gamma^q(E) \leqslant H^q(E) \leqq S^q(E). \tag{1}$$

Hausdorff measure $H^{n-1}(E) = a(E)$ will be termed *area* of E and Carathéodory and Hausdorff linear measures which are equal

$$C^1(E) = H^1(E) = \Lambda^*(E);$$

will be called *the length* of E.

Sets of Σ-finite length and area. A set E is said to *be of Σ-finite length (area)* if it is the union of a sequence of sets $\{E_m\}$ with $\Lambda(E_m) < \infty$ $[a(E_m) < \infty]$.

Lipschitz condition. A mapping $f(x)$ is said to *satisfy a Lipschitz condition* (or *to be Lipschitzian*) on a set E if there exists a constant $0 \leqq M < \infty$ (known as *the Lipschitz constant*) such that

$$|f(x) - f(x')| \leqq M|x - x'|,$$

or every pair of points $x,\ x' \in E$.

Federer–Young theorem. *If* $y = u(x)$, $u : R^n \to R^1$, *is a Lipschitzian map, then*

$$\int_{-\infty}^{\infty} H^{n-1}[u^{-1}(y) \cap F]\mathrm{d}y = \int_F |\nabla u| \mathrm{d}\tau,$$

whenever F is a compact set of R^n *and consequently*

$$\int_F g(x)|\nabla u(x)|\mathrm{d}\tau = \int_{R^1}[\int_{u^{-1}(y)} g(x)\mathrm{d}\sigma]\mathrm{d}y,$$

whenever g is an H^n*-integrable function on* R^n.

We say that a set is Hausdorff integrable (H^n-integrable) if it is integrable with respect to an integral defined by Hausdorff measure.

And now let us generalize to n-space

Gross theorem. *Let* Π *be an arbitrary but fixed plane. If* $\Gamma^{n-1} E < \infty$ *and* $E' \subset \Pi$ *is the set the points which are the projection of at least N points of E, then*

$$m^*_{n-1}E' \leqq \frac{1}{N}\Gamma^{n-1}E. \tag{2}$$

Let us suppose for the moment E is contained in the unit cube with two faces parallel to Π. Let be $x \in E'$ and let us denote by E_x the subset of E projected in x and by λ the biggest number such that there exist N points $x_1, \ldots, x_N \in E_x$ with $d(x_p, x_q) \geqq \lambda$ ($p \neq q$; $p, q = 1, \ldots, N$). In this way, to each $x \in E'$, there corresponds a unique $\lambda > 0$. Let $E'_k \subset E'$ be the set of those points in which $\lambda > \frac{1}{k}$. Since $E' = \bigcup_k E'_k$ and $E'_k \subset E'_{k+1}$, then $\lim_{k\to\infty} m^*_{n-1}E'_k = m^*_{n-1}E'$, where $m^*_{n-1}E'$ is the outer $(n-1)$-dimensional Lebesgue measure of E'. Hence, we can choose a sufficiently big k such that

$$m^*_{n-1}E'_k > m^*_{n-1}E' - \varepsilon. \tag{3}$$

And now, let us decompose the unit cube by planes parallel to Π in k congruent strata $W_1, \ldots, W_k$, where, to the interior of W_i (i.e. the set of points that have a neighbourhood contained in W_i) there are to be added the points belonging to the upper boundary $(n-1)$-dimensional cube parallel to Π. The points of E corresponding to a point of E'_k are distributed in at least N different strata W_p. If we denote by E'_{kp} those points of E'_k with the property that at least one of the corresponding points of E is contained in W_p $(1 \leqq p \leqq k)$, then

$$Nm^*_{n-1}E'_k \leqq \sum_p m^*_{n-1}E'_{kp}. \tag{4}$$

In order to prove it, let us first show that

$$m^*_{n-1}(E_1 \cup E_2) + m^*_{n-1}(E_1 \cap E_2) \leqq m^*_{n-1}E_1 + m^*_{n-1}E_2. \tag{5}$$

Indeed, let G_i be some measurable sets such that $G_i \supset E_i$ and $m^*_{n-1}E_i = m_{n-1}G_i$ $(i = 1, 2)$. As $E_1 \cup E_2 \subset G_1 \cup G_2$, it follows that $m^*_{n-1}(E_1 \cup E_2) \leqq m_{n-1}(G_1 \cup G_2)$, where it is possible to choose G_1, G_2 in order to have equality; for this it is enough to consider a measurable set G such that $m^*_{n-1}(E_1 \cup E_2) = m_{n-1}G$ and to substitute G_1, G_2 by $G \cap G_1$, $G \cap G_2$, respectively. But, for the sake of simplicity, let us suppose G_1, G_2 were choosen from the beginning to involve equality in the preceding inequality. Then, it clearly follows

$$m_{n-1}(G_1 \cup G_2) + m_{n-1}(G_1 \cap G_2) = m_{n-1}G_1 + m_{n-1}G_2$$

(Carathéodory [3], theorem 6, p. 252), hence and since

$$m_{n-1}(G_1 \cap G_2) \geqq m^*_{n-1}(E_1 \cap E_2),$$

we obtain (5), which implies

$$m^*_{n-1}E_1 + m^*_{n-1}E_2 + m^*_{n-1}E_3 \geqq m^*_{n-1}(E_1 \cup E_2) + m^*_{n-1}E_3 + m^*_{n-1}(E_1 \cap E_2) \geqq$$

$$\geqq m^*_{n-1}(E_1 \cup E_2 \cup E_3) + m^*_{n-1}[(E_1 \cup E_2) \cap E_3] + m^*_{n-1}(E_1 \cap E_2) \geqq$$

$$\geqq m^*_{n-1}(E_1 \cup E_2 \cup E_3) + m^*_{n-1}(E_1 \cap E_2 \cup E_1 \cap E_3 \cup E_2 \cap E_3) +$$

$$+ m^*_{n-1}(E_1 \cap E_2 \cap E_3)$$

and in general

$$\sum_{k=1}^{m} m^*_{n-1}E_k \geqq \sum_{i=1}^{m} m^*_{n-1}K_i,$$

where K_i is the set of those points which belong to at least i sets E_k. Clearly, $K_p \subset K_q$ whenever $q \leqq p$, so that the preceding inequality implies

$$\sum_{k=1}^{m} m^*_{n-1}E_k \geqq \sum_{i=1}^{m} m^*_{n-1}K_i \geqq i m^*_{n-1}K_i.$$

If we choose as E_k even the sets E'_{kp} and remark that each point of E'_k is contained in almost N sets E'_{kp}, then we obtain the inequality (3).

The measurability of sets W_p $(p = 1, \ldots, k)$ yields

$$\Gamma^{n-1}E = \sum_p \Gamma^{n-1}(E \cap W_p) \tag{6}$$

(Carathéodory [3], theorem 18, p. 273). But Γ^{n-1} $(E \cap W_p)$ is at least equal to the $(n-1)$-dimensional Lebesgue measure of the projection of $E \cap W_p$ on Π, as it results from Gross measure definition.

Then, since $m^*_{n-1}(E'_{kp}) \leqq m^*_{n-1}(E \cap W_p)_\Pi$, we obtain

$$\Gamma^{n-1}(E \cap W_p) \geqq m^*_{n-1}E'_{kp},$$

hence, by (4) and (6), $\Gamma^{n-1}E \geqq Nm^*_{n-1}E'_k$, and (3) then implies

$$m^*_{n-1}E' - \varepsilon < m^*_{n-1}E'_k \leqq \frac{1}{N} \Gamma^{n-1}E,$$

fand letting $\varepsilon \to 0$ yields (2) in the hypothesis E is bounded. Let us get rid of it and let Q_p be the cube with the centre at the origin and the edges parallel to the coordinate axes. If E^*_p is the set of Π in which there are projected at least N points of $E \cap W^*_m$, then on the one hand $m^*_{n-1}E' = \lim\limits_{p\to\infty} m^*_{n-1}E^*_p$, and on the other hand $\Gamma^{n-1}E = \lim\limits_{p\to\infty} \Gamma^{n-1}(E \cap W^*_p)$ (Carathéodory [3], theorem 15, p. 270), and since

$$m^*_{n-1}E^*_m \leqq \frac{1}{N}\Gamma^{n-1}(E \cap W^*_m)$$

or any m, applying the first part of the theorem (for bounded E), we obtain Gross theorem in general.

Corollary 1. *If $\Gamma^{n-1}E < \infty$ and E' is the set of those points of a plane in which a finite number of points of E is projected, then $m_{n-1}E' = 0$.*

Corollary 2. *If E is a set of Σ-finite area and E' is the set of those points of a plane in which a countable set of points of E is projected, then $m_{n-1}E' = 0$.*

Indeed, since $E = \bigcup\limits_p E_p$, $H^{n-1}E_p < \infty$, (1) implies $\Gamma^{n-1}E_p \leqq H^{n-1}E_p < \infty$ for any p, hence, by the preceding corollary, for each p, the set of those points in which is projected a countable set of points of E_p is a set of $(n-1)$-dimensional Lebesgue measure zero. But the union of a denumerable set of measure zero is again a set of measure zero, thus each point of a plane — except possibly a set of $(n-1)$-dimensional measure zero — is the projection of at most a finite number of points of E_p, and then a countable set of points of E.

Functions of a set *AC* with respect to a measure. An additive function of a set $\varphi: P(E_0) \to R^1$ is said to be *AC* (*absolutely continuous*) *with respect to a measure μ on a set E'* if for every $\varepsilon > 0$ there exists a $\delta = \delta(\varepsilon) > 0$ such that if $E \subset E'$, $E \in P(E_0)$ and $\mu E < \delta$, then $\varphi(E) < \varepsilon$. In particular, $\mu E = 0$ implies $\varphi(E) = 0$.

Theorem of Radon–Nikodim. *If E is a set of finite measure or, more generally the sum of a sequence of sets of finite measure, then, in order that an additive* *function of a set* $\varphi: P(E_0) \to R^1$ *on a set* E' *be AC on* E', *it is necessary and sufficient that this function of a set be the indefinite integral of some integrable function of a point on E.*

Theorem of change of measure. *Whenever on a set* $E \in P(E_0)$, *we have*

$$\nu(E) = \int_E g(x) d\mu(E) + \vartheta(E), \tag{7}$$

where $\vartheta(E)$ *is a non-negative function, additive and singular with respect to* $P(E_0)$ *and the measure* μ *and where* $g(x) \geqq 0$ *is integrable with respect to the measure* μ *over E, then also*

$$\int_E f(x) d\nu(E) = \int_E f(x) g(x) d\mu(E) + \int_E f(x) d\vartheta(E), \tag{8}$$

for every set $E \in P(E_0)$ *and for every function* $f(x)$ *that possess a definite integral with respect to the measure* ν. (Saks [2], p. 37).

If $f(x) \geqq 0$ on the set E, then the three terms in (8) are non-negative and

$$\int_E f(x) d\nu(E) \geqq \int_E f(x) g(x) d\mu(E). \tag{9}$$

If $\nu(E)$ is AC, then (7) reduces to $\nu(E) = \int_E g(x) d\mu(E)$ and (8) to

$$\int_E f(x) d\nu(E) = \int_E f(x) g(x) d\mu(E). \tag{10}$$

Theorem of Fubini. *If* $u(x)$ *is an integrable function on the interval* $I = \{x;\ a^i \leqq x^i \leqq b^i (i = 1, \ldots, n)\}$, *if* $I_q = \{x;\ a^k \leqq x^k \leqq b^k\ (k = 1, \ldots, q)\}$ *and* $I_{n-q} = \{x;\ a^p \leqq x^p \leqq b^p\ (p = q+1, \ldots, n)\}$, *then*

1. *For almost all points of* I_q, $u(x)$ *is integrable on* I_{n-q}.
2. *If* Δ_q *is the set of these points (clearly* $m\Delta_q = mI_q$), *then*

$$\int_{I_{n-q}} u(x) dx^1 \cdots dx^{n-q}$$

is integrable on Δ_q.

3. *The following formula*:

$$\int_I u(x) dx^1 \cdots dx^n = \int_{I_q} dx^1 \cdots dx^q \int_{I_{n-q}} u(x) dx^{q+1} \cdots dx^n,$$

holds. (Natanson [1], p. 445 or Saks [2], theorem 8.1, p. 77).

Theorem of Tonelli. *If $u(x)$ is a measurable function in $I \subset R^n$ and one of the successive integrals*

$$\int_{a^1}^{b^1} \cdots \int_{a^n}^{b^n} dx^1 \cdots dx^{i-1} dx^{i+1} \cdots dx^n \int_{a^i}^{b^i} u(x) dx^i < \infty \quad (i = 1, \ldots, n),$$

then $u(x)$ is integrable on I, the other $n - 1$ successive integrals do also exist and

$$\int_{a^1}^{b^1} \cdots \int_{a^n}^{b^n} dx^1 \cdots dx^{i-1} dx^{i+1} \cdots dx^n \int_{a^i}^{b^i} u(x) dx^i = \int_I u(x) dx < \infty.$$

(The proof is the same as for $n = 2$, Saks [1], theorem 17, p. 75).

Corollary. *Let be $E \subset I$, $I_n = \{x;\, a^k < x^k < b^k,\, x^n = a^n (k = 1, \ldots, n-1)\}$ and $J_y = \{x;\, x = y + \eta e_n,\, y \in I_n,\, 0 < \eta < b^n - a^n\}$.* If $m_1(E \cap J_y) = 0$ *for almost all $y \in I_n$, then E is either non-measurable or* $mE = 0$ (Carathéodory [3], theorem 3, p. 628).

Spherical n-space is the unit n-dimensional sphere in R^{n+1}.

Spherical distance between two points is the length $\leqq \pi$ of the arc of the great circle joining the two points.

The spherical ball $B^S(x, r)$ is the locus of points in S^n at a spherical distance less than r from the point x.

The set $F_r = \bigcup_{x \in F} \overline{B^S(x, r)}$ is termed *the set parallel to F at the spherical distance* $r (0 < r < \pi)$.

If $E^S \subset S^n$ and E is the stereographic projection of E^S, then *the spherical measure* of E^S is

$$m^S E_S = \int_E |J(\zeta)| d\tau,$$

where J is the Jacobian of the stereographical projection. A set of S^n will be called *measurable with respect to the spherical measure* if its stereographical projection is measurable.

Inequality of Brunn–Minkowski. *If $F \subset S^n$, $F \neq S^n$ is a closed set, F_r the set parallel to F at the distance r and ρ, ρ_r the radii of the spherical balls with the volumes equal to the spherical measures of F and F_r, respectively, then $\rho_r \geqq \rho + r$. There is equality if and only if $F = B^S(x, \rho)$ and then $F_r = \overline{B^S(x, \rho + r)}$. In particular, if* $\rho = 0$, *then F reduces to the centre of F_r.* (This theorem is used by E. Schmidt [1] for the proof of the isoperimetric inequality.)

Let $F \subset S^n$, $F \neq \emptyset$, be a closed and bounded set. Then the area (in the sense of Minkowski) of the boundary of F is given by

$$a(F) = \underline{\lim}_{r \to 0} \frac{m^S F_r - m^S F}{r},$$

In the definition one preferred $\underline{\lim}$ because by establishing the isoperimetric inequality with this definition for the area, it results *a fortiori* for the definition with $\overline{\lim}$.

The isoperimetric inequality for spherical n-space. *Among all the sets of S^n with the same n-dimensional spherical measure, the spherical ball has the smallest area (in Minkowski sense) of its boundary* (Schmidt [1]).

Remark. The isoperimetric inequality will be applied by us for the $(n-1)$-dimensional domains of the sphere, obtained by the intersection of the sphere with a polyhedron. But the $(n-2)$-dimensional area (in Minkowski sense) of the boundary of this domain coincides with its elementary area.

An inequality of Hardy–Littlewood–Smith. *If $u(x) \geqq 0$ belongs to L^p, $p > 1$, then $\bar{u}$ also belongs to L^p and*

$$\int_{R^n} \bar{u}(x)^p d\tau \leqq \left(\frac{4p}{p-1}\right)^p \int_{R^n} u(x)^p d\tau,$$

where $\bar{u}(x) = \sup \dfrac{1}{\omega_n r^n} \displaystyle\int_{B(x,r)} u(\xi) d\xi$, *the supremum being taken over all closed balls centred at x* (K. T. Smith [1]).

CHAPTER 4

Theorem of Rademacher – Stepanov

In this chapter, we shall generalize to n-dimensions (see our Note [23]) the theorem of Rademacher—Stepanov (Rademacher [1], Stepanov [1], [2]) which will be required in Chapter 1 of the second part of our book to prove the differentiability a.e. in D of the QCfH. The proof is based on a theorem of F. Roger [1].

Theorem of Roger. We recall first (G. Bouligand [1], p. 66) that a half-line Ox starting from a cluster point O of a set E is said to be *an intermediate half-tangent at the point O to the set E* if any right circular cone with the vertex O and the axis Ox, which has the angle and the altitude arbitrarily small contains at least a point of E different from O. The set of all intermediate half-tangents of a set E in a fixed cluster point O is called *the contingent of E in O* (by a contingent of an isolated point of E we shall understand the empty set). The set of all lines which can be decomposed in two half-lines both belonging to the contingent of E in O is termed *the bilateral contingent of E in O.*

Theorem of Roger. *In any set $E \subset R^q$, except possibly a set of H^p-measure zero, the subset in which the bilateral contingent does not contain a $(q - p)$-space coincides with the subset of E in which the bilateral contingent reduces to a p-space and the whole contingent to a pencil of $(p + 1)$-spaces with the preceding p-space as basis (if $p = 1$ as axis) of the pencil.*

The proof is similar to that of theorem 3.6 of S. Saks' monograph ([2], p. 266), a theorem which may be obtained from the preceding one for $p = q - 1$, $q = 3$.

Theorem of Rademacher – Stepanov. *Let $u(x)$ be a real-valued continuous function on an open set $D \subset R^n$. Then $u(x)$ is differentiable a.e. in a measurable subset E of D if and only if*

$$\Lambda_u(x) = \overline{\lim_{\Delta x \to 0}} \frac{|\Delta u(x)|}{|\Delta x|} < \infty \tag{1}$$

a.e. in E.

Assume first $u(x)$ is differentiable at x_0. Then

$$\frac{|u(x_0 + \Delta x) - u(x_0)|}{|\Delta x|} = \left| \frac{|u_i(x_0)\Delta x^i|}{|\Delta x|} + \varepsilon(|\Delta x|, x_0 \right|,$$

where $\varepsilon(|\Delta x|, x_0) \to 0$, as $|\Delta x| \to$, hence

$$\Lambda_u(x_0) = \overline{\lim_{\Delta x \to 0}} \frac{|u_i(x_0)\Delta x^i|}{|\Delta x|} = \sup_s |u_i(x_0) \cos(s, x^i)| = \sum_{i=1}^{n} |u_i(x_0)| < \infty.$$

For the sufficiency, let us denote by $\sigma_n \subset R^{n+1}$ the n-dimensional surface $u = u(x)$, which has as an orthogonal projection in the n-dimensional plane $u(x) = 0$ just D, and by σ_n° the set of σ_n which has as orthogonal projection in $u(x) = 0$ the subset E^0 of E in which (1) holds. But this condition implies that at each point $x \in \sigma_n^\circ$ the contingent of σ_n does not contain the semitangent Ou; hence, in such a point, the contingent of σ_n can be neither the $(n+1)$-space nor an $(n+1)$-half-space. Then applying Roger's theorem for $p = n$, $q = n + 1$, it follows that everywhere in σ_n°, except possibly in a set of H^n-measure zero, the contingent reduces to an n-dimensional plane, namely the plane tangent to the n-dimensional surface σ_n at this point. But since a set of H^n-measure zero has as an orthogonal projection on each coordinate plane in a set of n-dimensional Lebesgue measure zero and since the existence at a point of σ_n° of an n-dimensional plane tangent to σ_n and non-parallel to Ou implies the differentiability of $u(x)$ at this point (M. Nicolescu [5], theorem 2, p. 446), it follows that $u(x)$ is differentiable a.e. in E^0 and hence (by the hypothesis of the theorem) a.e. in E, as desired.

CHAPTER 5

The modulus of a curve family

In this chapter we introduce the concept of the modulus of a curve family, of a family of q-dimensional surface, of a ring, of a cylinder and of an interval, we establish some properties of the modulus of some arc families and of some rings. All these preliminary results are to be used in Chapter 1 of part 2 for the equivalence of Väisälä's definitions and in the last chapter of the third part to obtain lower bounds for the coefficients of QCf of the cylinder and of the cone.

Curves. In the usual terminology, we shall say that a set $\gamma \subset R^n$ is *an arc* if it is homeomorphic to the unit interval (0,1), which may be open, half-open or closed. If γ is homeomorphic to a circle, it is called *a closed curve.* In all these cases γ is termed *a curve.* A curve γ is said to be *locally rectifiable* if every compact subcurve of γ is rectifiable. Obviously, the concepts rectifiable and locally rectifiable are different only if γ is an arc corresponding to an open or half-open interval.

A real function $\rho(x)$ is *Borel measurable* if the set $\{x; \rho(x) < \alpha\}$ is a Borel set for any real α.

If $\rho(x) \geqq 0$ is a Borel measurable function defined in a set containing γ, the line integral of ρ over γ is given by

$$\int_\gamma \rho ds = \int_0^l \rho[x(s)] ds,$$

where the integral on the right is the usual Lebesgue integral. If γ is locally rectifiable, we define the line integral as an improper integral. If γ is not rectifiable, then the integral can be defined by means of the linear measure $\Lambda(\gamma)$:

$$\int_\gamma \rho ds = \int_\gamma \rho d\Lambda.$$

In the case of locally rectifiable curves, these definitions are equivalent.

The families Γ_m are called *separate* if there exist disjoint Borel sets B_m, such that $\gamma_m \in \Gamma_m$ implies $\Lambda(\gamma_m - B_m) = 0$ $(m = 1, 2, \ldots)$.

A family Γ_1 is termed *minorized* by Γ_2 if for each $\gamma_1 \in \Gamma_1$ there exist a $\gamma_2 \in \Gamma_2$ such that $\gamma_2 \subset \gamma_1$. We shall write $\Gamma_2 < \Gamma_1$.

The modulus of a family of q-dimensional surfaces. *A q-dimensional surface* $\sigma_q \subset R^n$ is a connected and locally Euclidean set (i.e. each point of σ_q has a neighbourhood $V \subset \sigma_q$ which is homeomorphic to an open set in R^q). Thus, a q-dimensional surface is the continuous image of a domain of R^n. Let us call it *squarable* if $\int_{\sigma_q} d\sigma_q < \infty$.

Let Σ_q be a family of q-dimensional squarable surfaces $\sigma_q \subset R^n$ $(1 \leqq q \leqq n-1)$ and let $F(\Sigma_q)$ be the family of functions $\rho(x)$, defined for all $x \in R^n$, such that

1° $\rho(x) \geqq 0$ for all $x \in R^n$,

2° $\rho(x)$ is Borel measurable,

3° $\int_{\sigma_q} \rho^q d\sigma_q \geqq 1$ for each $\sigma_q \in \Sigma_q$.

Clearly, $F(\Sigma_q) \neq \emptyset$, because it always contains the function $\rho(x) = \infty$.

The quantity

$$M_p(\Sigma_q) = \inf_{\rho \in F(\Sigma_q)} \int_{R^n} \rho^p(x) d\tau$$

is said to be *the p-modulus of* Σ_q. Its inverse

$$\lambda_p(\Sigma_q) = \frac{1}{M_p(\Sigma_q)}$$

is called *the p-extremal length of* Σ_q. We shall drop the index p if $p = n$.

Remarks. 1. If all the surfaces $\sigma_q \in \Sigma_q$ are contained in a Borel set $E \subset R^n$, then it suffices to consider the family $F_E(\Sigma_q)$ of functions $\rho(x)$ which are defined only in E and satisfy the conditions 1°—3° in E. Then the p-modulus can be written in the form

$$M_p(\Sigma_q) = \inf_{\rho \in F_E(\Sigma_q)} \int_E \rho^p(x) d\tau. \tag{1}$$

2. If in (1) E is an $(n-1)$-dimensional surface σ and instead of the volume element $d\tau$ we take the surface element $d\sigma$ and the integration is with respect to Hausdorff measure, we obtain *the surface p-modulus*

$$M_p^\sigma(\Sigma_q) = \inf_{\rho \in F_\sigma(\Sigma_q)} \int_\sigma \rho^p(x) d\sigma.$$

We shall drop the index if $p = n-1$. If $\sigma = S(x, r)$, we have *the spherical p-modulus* $M_p^S(\Sigma_q)$ and the *spherical modulus* $M^S(\Sigma_q) = M_{n-1}^S(\Sigma_q)$.

We shall restrict ourselves specially to two particular kinds of families of q-dimensional surfaces, namely the arc families and surface families.

p-exceptional curve families. A family γ of curves in R^n is termed *p-exceptional* if $M_p(\Gamma) = 0$. A property is said *to hold for p-almost every curve* in a family Γ if the subfamily of Γ in which it does not hold is p-exceptional. The index p is again omitted if $p = n$.

Properties of the p-modulus of an arc family.

PROPOSITION 1. *If* $\Gamma = \bigcup_m \Gamma_m$, *then*

$$M_p(\Gamma) \leqq \Sigma_m M_p(\Gamma_m). \tag{2}$$

Indeed, let us set $\rho(x) = \sup_m \rho_m(x)$, where $\rho_m \in F(\Gamma_m)$ $(m = 1, 2, \ldots)$. Clearly, $\rho \in F(\Gamma)$. In order to show that

$$\int_{R^n} \rho^p d\tau \leqq \Sigma_m \int_{R^n} \rho_m^p d\tau, \tag{3}$$

let us define for an arbitrary index k

$$g_k(x) = \max\{\rho_1(x), \cdots, \rho_k(x)\}, \quad E_m = \{x; x \in R^n, \rho_m(x) = g_k(x)\}.$$

Then $g_k(x)$ is also Borel measurable, E_m are Borel sets and $R^n = \bigcup_m E_m$. Hence

$$\int_{R^n} g_k^p \, d\tau \leqq \Sigma_m \int_{E_m} g_k^p \, d\tau = \Sigma_n \int_{E_m} \rho_m^p \, d\tau \leqq \Sigma_m \int_{R^n} \rho_m^p \, d\tau,$$

and letting $k \to \infty$ yields (3), because $g_k(x) \to \rho(x)$ monotonically, which implies $\int_{R^n} g_k^p d\tau \to \int_{R^n} \rho^p d\tau$.

Now let us choose $\rho_m \in F(\Gamma_m)$ such that

$$\int_{R^n} \rho_m^p d\tau \leq M_p(\Gamma_m) + \varepsilon 2^{-m}.$$

Then $\rho \in F(\Gamma)$ and

$$M_p(\Gamma) \leqq \int_{R^n} \rho^p \, d\tau \leqq \Sigma_m \int_{R^n} \rho_m^p \, d\tau \leqq \Sigma_m M_p(\Gamma_m) + \varepsilon,$$

hence, letting $\varepsilon \to 0$, we obtain (3), as desired.

PROPOSITION 2. *If* $\Gamma = \bigcup_m \Gamma_m$ *and* Γ_m *are separate, then*

$$M_p(\Gamma) = \Sigma_m M_p(\Gamma_m). \tag{4}$$

(2) holds by the preceding proposition and we have only to prove that

$$M_p(\Gamma) \geqq \Sigma_m M_p(\Gamma_m). \tag{5}$$

First, since Γ_m are separate, it follows the existence of a system $\{B_m\}$ of disjoint Borel sets such that $\gamma \in \Gamma_m$ implies $\Lambda(\gamma - B_m) = 0$. Then, let be $\rho \in F(\Gamma)$; evidently, $\rho \in F(\Gamma_m)$ $(m = 1, 2, \ldots)$. Let us define

$$\rho_m(x) = \begin{cases} \rho(x) & if \quad x \in B_m, \\ 0 & if \quad x \bar{\in} B_m. \end{cases}$$

Obviously, $\rho_m \in F(\Gamma_m)$, which implies

$$\int_{B_m} \rho^p d\tau = \int_{R^n} \rho_m^p d\tau \geqq M_p(\Gamma_m),$$

hence

$$\int_{R^n} \rho^p d\tau \geqq \Sigma_m \int_{B_m} \rho^p d\tau \geqq \Sigma_m M_p(\Gamma_m).$$

As ρ was an arbitrary admissible function, we obtain (5), which combined with (2) yields (4), as desired.

PROPOSITION 3. *If $\Gamma_1, \Gamma_2, \ldots$ are separate and $\Gamma < \Gamma_m$, then* (5) *holds.*

The proof is the same as in the preceding proposition.

COROLLARY 1. *If $\Gamma_1 < \Gamma_2$, then $M_p(\Gamma_1) \geqq M_p(\Gamma_2)$.*

COROLLARY 2. *$\Gamma_1 \supset \Gamma_2$ implies $M_p(\Gamma_1) \geqq M_p(\Gamma_2)$.*

PROPOSITION 4. *A family Γ is p-exceptional if and only if there exists a Borel measurable function $\rho \geqq 0$, $\rho \in L^p$, such that*

$$\int_\gamma \rho ds = \infty$$

for all $\gamma \in \Gamma$.

For the sufficiency, let us suppose there exists a ρ satisfying the hypothesis of the proposition. Then

$$\int_\gamma \frac{\rho}{m} ds \geqq 1 \quad (m = 1, 2, \cdots)$$

for all $\gamma \in \Gamma$, hence $\dfrac{\rho}{m} \in F(\Gamma)$, and by virtue of the condition $\rho \in L^p$,

$$M_p(\Gamma) \leqq \lim_{m \to \infty} \int_{R^n} \left(\frac{\rho}{m}\right)^p d\tau = \lim_{m \to \infty} \frac{1}{m^p} \int_{R^n} \rho^p d\tau = 0.$$

For the necessity, let us suppose conversely $M_p(\Gamma) = 0$ and let $\{\rho_m\}$ be a sequence of functions $\rho_m \in F(\Gamma)$ such that

$$\int_{R^n} \rho_m^p d\tau < \frac{1}{4^m}.$$

If we choose ρ of the form

$$\rho(x) = [\Sigma_m 2^m \rho_m(x)^p]^{\frac{1}{p}},$$

we obtain

$$\int_{R^n} \rho^p d\tau = \Sigma_m 2^m \int_{R^n} \rho_m^p d\tau < \infty,$$

and on the other hand

$$\int_\gamma \rho \mathrm{d}s \geqq \int_\gamma 2^{\frac{m}{p}} \rho_m \mathrm{d}s \geqq 2^{\frac{m}{p}} \qquad (m = 1, 2, \ldots)$$

for all $\gamma \in \Gamma$, hence

$$\int_\gamma \rho \mathrm{d}s = \infty.$$

PROPOSITION 5. *If $mE = 0$, then for each $p > 0$, $H^q(E \cap \sigma_q) = 0$ for p-almost every $\sigma_q \subset R^n$.*

Indeed, let us define

$$\rho(x) = \begin{cases} \infty & \text{if } x \in E, \\ 0 & \text{if } x \bar{\in} E. \end{cases}$$

Then

$$\int_{\sigma_q} \rho \mathrm{d}\sigma_q = (\infty) H^q(E \cap \sigma_q) = \infty$$

for each σ_q such that $H^q(E \cap \sigma_q) > 0$. If we denote by Σ_q' the family of these surfaces, then

$$M_p(\Sigma_q') \leqq \int_{R^n} \rho^p \mathrm{d}\tau = \int_E \rho^p \mathrm{d}\tau = 0,$$

i.e. Σ_q' is p-exceptional. (For $q = 1$ we obtain the corresponding result for arc families.)

PROPOSITION 6. *If $\{f_m\}$ is a sequence of functions and $E \subset R^n$ is a measurable set such that*

$$\lim_{m \to \infty} \int_E |f_m - f|^p \mathrm{d}\tau = 0,$$

then there exists a subsequence $\{f_{m_\nu}\}$ such that

$$\lim_{\nu \to \infty} \int_\gamma |f_{m_\nu} - f| \mathrm{d}s = 0 \tag{6}$$

for almost every $\gamma \subset E$.

Let $\{m_\nu\}$ be a sequence of integers such that

$$\int_E |f_{m_\nu}(x) - f(x)|^p \mathrm{d}\tau < 2^{-\nu(p+1)}$$

and let us denote

$$g_\nu(x) = |f_{m_\nu}(x) - f(x)|.$$

Setting

$$\Gamma_\nu = \{\gamma; \gamma \subset E, \int_\gamma g_\nu \mathrm{d}s > 2^{-\nu}\}, \ \Gamma_k' = \bigcup_{\nu > k} \Gamma_\nu, \ \Gamma = \bigcap_k \Gamma_k',$$

we have $2^\nu g_\nu \in F(\Gamma_\nu)$, hence

$$M_p(\Gamma_\nu) \leqq \int_E (2^\nu g_\nu)^p \, d\tau = 2^{p\nu} \int_E g_\nu^p d\tau < 2^{-\nu}.$$

We combine this inequality with the corollary 1 and 2 of the proposition 3 to obtain

$$M_p(\Gamma) \leqq M_p(\Gamma'_k) \leqq \sum_{\nu > k} M_p(\Gamma_\nu) \leqslant \sum_{\nu > k} 2^{-\nu} = 2^{-k},$$

and letting $k \to \infty$ yields $M_p(\Gamma) = 0$.

Let us suppose $\gamma \bar{\in} \Gamma$. Then there exists a k such that $\gamma \bar{\in} \Gamma'_k$, hence

$$\int_\gamma \left| f_{m_\nu} - f \right| ds = \int_\gamma g_\nu ds \leqq 2^{-\nu}$$

for all $\nu > k$, which implies (6), as desired.

Modulus of a ring. *An arc* γ is said to *join two sets* $E_1, E_2 \subset D$ *in* $D - E_1 - E_2$, if the end points a_1, a_2 of γ belong to E_1, E_2, respectively and $\gamma - a_1 - a_2 \subset D - - E_1 - E_2$. By a *ring A* we mean a domain the complement of which consists of two components C_0, C_1 of which C_0 is bounded and C_1 is unbounded. The boundary of A has the components $F_0 = C_0 \cap \bar{A}$ and $F_1 = C_1 \cap \bar{A}$. F_0 is called the *inner boundary* and F_1 the *outer boundary*. Let Γ_A be the family of all arcs which join the boundary components of A in A. Then the quantity

$$M(A) = \frac{n\omega_n}{M(\Gamma_A)} \tag{7}$$

is called the modulus of A.

The spherical ring.

PROPOSITION 7. *Let A be the spherical ring* $r_1 < |x| < r_2$. *Then*

$$M(A) = \left(\log \frac{r_2}{r_1}\right)^{n-1}. \tag{8}$$

Let $\rho \in F(\Gamma_A)$ and let $\gamma_e = \{x; x = re, r_1 < r < r_2\}$ be the radial segment which joins the boundary components of A and is parallel to the unit vector e. Using Hölder's inequality (see Hardy, Littlewood and Polya [1], theorem 189, p. 140) we obtain

$$1 \leqq \left(\int_{\gamma_e} \rho ds\right)^n \leqq \left(\log \frac{r_2}{r_1}\right)^{n-1} \int_{r_1}^{r_2} \rho^n r^{n-1} dr.$$

Integrating over all e we obtain by Fubini's theorem in polar coordinates

$$n\omega_n \leqq \left(\log \frac{r_2}{r_1}\right)^{n-1} \int_A \rho^n \mathrm{d}\tau.$$

The equality holds for

$$\rho = \frac{1}{|x| \log \dfrac{r_2}{r_1}}.$$

Thus

$$M(\Gamma_A) = n\omega_n \left(\log \frac{r_2}{r_1}\right)^{1-n}, \tag{9}$$

which by (7) implies (8).

***p*-modulus of non-rectifiable curves.**

PROPOSITION 8. *For each $p > 0$, p-almost every bounded curve is rectifiable.*

Let Γ be the family of all bounded non-rectifiable curves. For an arbitrary $\varepsilon > 0$, we can find a continuous function $\rho(x)$ such that $\rho(x) > 0$ for all $x \in R^n$ and

$$\int_{R^n} \rho^p \mathrm{d}\tau < \varepsilon. \tag{10}$$

Then

$$\int_\gamma \rho \mathrm{d}s = \infty \tag{11}$$

for every $\gamma \in \Gamma$. Indeed, as γ is bounded we can suppose it closed, otherwise its end points may be added to it. The continuous function $\rho(x)$ attains its minimum $m_\gamma > 0$ on the closed set γ. But if we suppose

$$\int_\gamma \rho \mathrm{d}s = l < \infty,$$

it would follow that

$$l = \int_\gamma \rho \mathrm{d}s \geqq m_\gamma \int_\gamma \mathrm{d}s,$$

whence

$$\int_\gamma \mathrm{d}s \leqq \frac{l}{m_\gamma} < \infty,$$

which would contradict the non-rectifiability of γ. But (11) implies that $\rho \in F(\Gamma)$. Consequently, $M_p(\Gamma) < \varepsilon$. Since ε was arbitrary, this proves the theorem.

PROPOSITION 9. *Almost every curve is rectifiable.* (For the proof see Väisälä [1]).

Inequality of Hardy – Littlewood and Polya on the rearrangement of functions. Suppose that $T = u(t)$ is a non-negative integrable real function of a real variable in $(-a, a)$ $(0 < a \leqq \infty)$, so that it is measurable and finite a.e. in $(-a, a)$. If $M(T)$ is the measure of the set in which $u(t) \geqq T$, then $M(T)$ is a monotonic decreasing function of T. We assume also $M(T) < \infty$ for all $T > 0$. Let us define an even function $\tilde{u}(t)$ by agreeing that $\tilde{u}\left[\frac{1}{2}M(T)\right] = T$ and that $\tilde{u}(-t) = \tilde{u}(t)$. Then $\tilde{u}(t)$ decreases symmetrically on each side of the origin (where

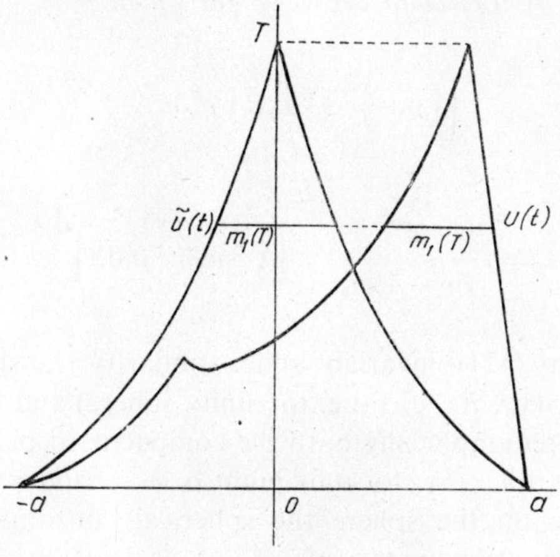

Fig. 1

it has generally a peak) and is defined uniquely in $(-a, a)$ except for at most an enumerable set of values of t, namely those corresponding to intervals of constancy of $M(T)$. It is possible to complete the definition of $\tilde{u}(t)$ by agreeing, for example, that $\tilde{u}(t) = \frac{1}{2}[\tilde{u}(t-0) + \tilde{u}(t+0)]$ at the points of discontinuity. $\tilde{u}(t)$ is called *the rearrangement of $u(t)$ in symmetrically monotonic decreasing order.*

The two (in general quite different) sets in which $T_1 \leqq u(t) \leqq T_2$ and $T_1 \leqq \tilde{u}(t) \leqq T_2$ have the same measure, and the same is true of sets in which $u(t) > T$ and $\tilde{u}(t) > T$. The functions $u(t)$ and $\tilde{u}(t)$ are said to be *equi-measurable.*

Theorem of Hardy – Littlewood – Polya. *Whether a is finite or infinite*

$$\int_{-a}^{a} u(t)v(t)\mathrm{d}t \leqq \int_{-a}^{a} \tilde{u}(t)\tilde{v}(t)\mathrm{d}t,$$

where $\tilde{u}(t)$, $\tilde{v}(t)$ are the rearrangement of $u(t)$, $v(t)$ in symmetrically monotonic decreasing order.

Hardy, Littlewood and Polya ([1], theorem 378, p. 278) established the preceding theorem for $\tilde{u}$, $\tilde{v}$ representing the rearrangment of u, v in monotonic decreasing order (not symmetrically), but the proof is the same.

A property of the modulus of some rings. In order to obtain an upper bound for the modulus of a ring (considered by Teichmüller), we shall previously establish a lower bound depending on the radius for the n-modulus of a curve family on a sphere, and then, by integrating with respect to the radius, we shall obtain a lower bound for the modulus of a curve family in a spherical ring, whence we shall deduce an upper bound for the modulus of the above ring.

PROPOSITION 10. *Suppose that D is an open half space, $\sigma = S(x_0, R) \cap D$ and that $\rho(x) \geqq 0$ is a Borel measurable function in $S(x_0, R)$. Then each pair of points $x_1, x_2 \in \bar{\sigma}$ can be joined by a circular arc $\gamma \subset \sigma$ for which*

$$\left(\int_\gamma \rho \mathrm{d}s\right)^n \leqq A_0 R \int_\sigma \rho^n \mathrm{d}\sigma, \tag{12}$$

where

$$A_0 = \frac{2^{2n-1}}{(n-1)\omega_{n-1}} \left(\int_0^{\frac{\pi}{2}} \sin^{\frac{2-n}{n-1}} \vartheta \, \mathrm{d}\vartheta\right)^{n-1}. \tag{13}$$

Since the inequality (12) is invariant under similarity transformations of R^n onto itself, we may assume $S(x_0, R) = S$ (i.e. the unity sphere) and that $x_1 = (1, 0, \ldots, 0)$. Let $x^* = f(x)$ map S stereographically onto the compactified plane Π passing through O. Then x_1 corresponds to ∞, x_2 to some point $a \neq \infty$ and γ to a half-line starting from a. Let us choose on the sphere the spherical coordinates $\vartheta_1, \ldots, \vartheta_{n-1}$ and in the plane Π the polar coordinates $|x^*|, \vartheta_2, \ldots, \vartheta_{n-1}$. If $S - \sigma$ is non-empty, this set corresponds to a closed $(n-1)$-ball or a closed half plane E (according to the fact that x_1 is an inner point of σ or $x_1 \in \partial\sigma$). Since x_1, x_2 are not inner points of $S - \sigma$ it follows that E does not contain a or ∞ as interior points, hence, the convexity of E implies the existence of an angle α such that for $\alpha < \vartheta_2 < \alpha + \pi$, the ray $x^* = a + re(\vartheta_2)$ $(0 < r < \infty)$ — where $e(\vartheta_2)$ is a unit vector which makes an angle ϑ_2 with Ox_2^* — does not meet E. This ray corresponds to γ, $\mathrm{d}|x^*| = \mathrm{d}r$ and $|x^*| = \tan \frac{\vartheta_1}{2}$, where ϑ_1 is the angle between the radius vector of a point $x \in S$ and the negative part of the x^1-axis. For each such arc we see that

$$\int_\gamma \rho \mathrm{d}s = \int_{\vartheta_1^2}^{\vartheta_1^1} \rho \mathrm{d}\vartheta_1 = \int_{|a|}^{\infty} \rho \frac{\mathrm{d}\vartheta_1}{\mathrm{d}|x^*|} \mathrm{d}|x^*| = 2\int_{|a|}^{\infty} \rho \left(\cos\frac{\vartheta_1}{2}\right)^2 \mathrm{d}|x^*| =$$

$$= 2\int_{|a|}^{\infty} \frac{\mathrm{d}|x^*|}{1 + |x^*|^2} = 2\int_0^{\infty} \frac{\rho \mathrm{d}r}{1 + |x^*|^2},$$

and thus we may choose a particular arc γ joining x_1 and x_2, so that, integrating with respect to $\vartheta_2, \ldots \vartheta_{n-1}$ for $\alpha < \vartheta_2 < \alpha + \pi$, by Tonelli's theorem in polar coordinates and Hölder's inequality, we obtain

$$\int_\gamma \rho ds \leqq \frac{4}{(n-1)\omega_{n-1}} \int_\Omega \frac{\rho d\sigma}{r^{n-2}(1+|x^*|)^2} \leqq$$

$$\leqq \frac{2^{\frac{n+1}{n}}}{(n-1)\omega_{n-1}} \left[\int_\Omega \frac{2^{n-1}\rho^n d\sigma}{(1+|x^*|^2)^{n-1}}\right]^{\frac{1}{n}} \left[\int_\Omega \frac{d\sigma}{\frac{n(n-2)}{r^{n-1}}(1+|x^*|^2)^{\frac{1}{n-1}}}\right]^{\frac{n-1}{n}}, \tag{14}$$

where Ω is the half plane $\alpha < \vartheta_2 < \alpha + \pi$ of Π. But

$$\int_\Omega \frac{2^{n-1}\rho^n d\sigma}{(1+|x^*|^2)^{n-1}} \leqq \int_{f(\sigma)} \frac{2^{n-1}\rho^n d\sigma}{(1+|x^*|^2)^{n-1}} = \int_{f(\sigma)} \frac{2^{n-1}\rho^n |x^*|^{n-2} d|x^*| d\sigma_{n-2}}{(1+|x^*|^2)^{n-1}} =$$

$$= \int_\sigma \frac{\rho^n 2^{n-1}\left(\cos\frac{\vartheta_1}{2}\right)^{2(n-1)}\left(\sin\frac{\vartheta_1}{2}\right)^{n-2} d\vartheta_1 d\sigma_{n-2}}{2\left(\cos\frac{\vartheta_1}{2}\right)^2\left(\cos\frac{\vartheta_1}{2}\right)^{n-2}} =$$

$$= \int \rho^n \sin^{n-2}\vartheta_1 d\vartheta_1 d\sigma_{n-2} = \int_\sigma \rho^n d\sigma,$$

where $d\sigma = \sin^{n-2}\vartheta_1 \ldots \sin\vartheta_{n-2} d\vartheta_1 \ldots d\vartheta_{n-1}$ is *the surface element of the unit sphere*.

Now let us verify that

$$A_0 = \sup_{|a|<\infty} \frac{2^{n+1}}{(n-1)^n \omega_{n-1}^n} \left(\int_\Pi \frac{d\sigma}{|x^*-a|^{\frac{n(n-2)}{n-1}}(1+|x^*|^2)^{\frac{1}{n-1}}}\right)^{n-1} < \infty.$$

But by the preceding theorem and since a may be chosen of the form $a = a^2 e_2$ (which is possible to obtain by a rotation about the origin, because the distance and then also the preceding integral are invariant with respect to such a transformation), it follows

$$\int_{-\infty}^{\infty} \frac{dx^{*_2}}{|(x^{*_2}-a^2)e_2 + x^{*_k}e_k|^{\frac{n(n-2)}{n-1}}(1+|x^*|^2)^{\frac{1}{n-1}}} \leqq$$

$$\leqq \int_{-\infty}^{\infty} \frac{dx^{*_2}}{|x^*|^{\frac{n(n-2)}{n-1}}(1+|x^*|^2)^{\frac{1}{n-1}}},$$

where the dummy index takes on the values $3, \ldots, n$, $(1+|x^*|^2)^{\frac{-1}{n-1}}$ is its own rearrangement in symmetrical non-increasing order and $|x^*|^{\frac{n(2-n)}{n-1}}$ is the rearrangement in symmetrical non-increasing order of $|(x^{*_2} - a^2)\, e_2 + x^{*_k} e_k|^{\frac{n(2-n)}{n-1}}$. Integrating with respect to $x^{*_3}, \ldots, x^{*_n}$ then yields

$$\int_{\Pi} \frac{d\sigma}{r^{\frac{n(n-2)}{n-1}} (1+|x^*|^2)^{\frac{1}{n-1}}} \leqq \int_{\Pi} \frac{d\sigma}{|x^*|^{\frac{n(n-2)}{n-1}} (1+|x^*|^2)^{\frac{1}{n-1}}}$$

and applying Fubini's theorem in polar coordinates (with the radius $|x^*|$), we obtain

$$\int_{\Pi} \frac{d\sigma}{|x^*|^{\frac{n(n-2)}{n-1}} (1+|x^*|^2)^{\frac{1}{n-1}}} = (n-1)\omega_{n-1} \int_0^{\infty} \frac{d|x^*|}{|x^*|^{\frac{n-2}{n-1}} (1+|x^*|^2)^{\frac{1}{n-1}}} =$$

$$= (n-1)\omega_{n-1} \int_0^{\infty} (t^{n-2} + t^n)^{\frac{-1}{n-1}} dt \leqq$$

$$\leqq (n-1)\omega_{n-1} \left(\int_0^1 t^{\frac{2-n}{n-1}} dt + \int_1^{\infty} t^{\frac{-n}{n-1}} dt \right) = 2(n-1)^2 \omega_{n-1} < \infty,$$

where we set $|x^*| = t$. And now, by the substitution $t = \tan \frac{\vartheta}{2}$ in the integral $\int_0^{\infty} (t^{n-2} + t^n)^{\frac{-1}{n-1}} dt$, it follows

$$A_0 = \frac{2^{n+1}}{(n-1)\omega_{n-1}} \left[\int_0^{\infty} (t^{n-2} + t^n)^{\frac{-1}{n-1}} dt \right]^{n-1} =$$

$$= \frac{2^{n+1}}{(n-1)\omega_{n-1}} \left[\int_0^{\pi} \left(\tan^{n-2} \frac{\vartheta}{2} + \tan^n \frac{\vartheta}{2} \right)^{\frac{-1}{n-1}} \frac{d\vartheta}{2 \cos^2 \frac{\vartheta}{2}} \right]^{n-1} =$$

$$= \frac{2^n}{(n-1)\omega_{n-1}} \left(\int_0^{\pi} \sin^{\frac{2-n}{n-1}} \vartheta \, d\vartheta \right)^{n-1} = \frac{2^{2n-1}}{(n-1)\omega_{n-1}} \left(\int_0^{\frac{\pi}{2}} \sin^{\frac{2-n}{n-1}} \vartheta \, d\vartheta \right)^{n-1},$$

i.e. (13) as desired.

In order to obtain (12) for an arbitrary R, it is enough to multiply the inequality by R^n.

COROLLARY. *Let $\rho \geqq 0$ be a Borel measurable function in $S(x, R)$. Then each pair of points $x_1, x_2 \in S(x, R)$ can be joined by a circular arc $\gamma \subset S(x, R)$ for which*

$$\left(\int_\gamma \rho \mathrm{d}s\right)^n \leqq \frac{A_0 R}{2^n} \int_{S(x, R)} \rho^n \mathrm{d}\sigma, \tag{15}$$

and this inequality is the best possible evaluation.

If we appeal to the argument in the proof of the preceding proposition, we observe that the preceding inequality follows immediately by taking into account that when σ is the whole unit sphere, then instead of the first inequality of (14) we have the formula

$$\int_\gamma \rho \mathrm{d}s = \frac{2}{(n-1)\omega_{n-1}} \int_\Pi \frac{\mathrm{d}\sigma}{r^{n-2}(1+|x^*|^2)}.$$

In order to prove that the inequality (15) is the best possible, it is sufficient to show that setting $\rho = |\nabla u|$, where

$$u(x) = \int_0^\vartheta \sin^{\frac{2-n}{n-1}} \vartheta \, \mathrm{d}\vartheta,$$

(15) reduces to an equality.

Indeed, if we denote

$$u_i = \frac{\partial u(x)}{\partial x^i} \quad (i = 1, \ldots, n) \text{ and } u_{\vartheta_1} = \frac{\partial u(R, \vartheta_1, \ldots, \vartheta_{n-1})}{\partial \vartheta_1},$$

and since the polar coordinates are changed into cartesian ones by

$$x^i = R \sin \vartheta_1 \ldots \sin \vartheta_{i-1} \cos \vartheta_i \quad (i = 1, \ldots, n),$$

it follows that

$$u_1 = -\frac{u_{\vartheta_1}}{R} \sin \vartheta_1,$$

$$u_j = \frac{u_{\vartheta_1}}{R} \cos \vartheta_1 \sin \vartheta_2 \ldots \sin \vartheta_{j-1} \cos \vartheta_j \quad (j = 2, \ldots, n)$$

whence

$$|\nabla u|^2 = u_1^2 + \ldots u_n^2 = \frac{u_{\vartheta_1}^2}{R^2}.$$

On the other hand, choosing γ to be the circular arc $0 \leqq \vartheta_1 \leqq \pi$, $\vartheta_k = 0$ $(k = 2, \ldots, n-1)$, and taking into account (13), the inequality (15) reduces to

$$\left(\int_0^\pi \sin^{\frac{2-n}{n-1}} \vartheta_1 \mathrm{d}\vartheta_1\right)^n \leqq \frac{R}{(n-1)\omega_{n-1}} \left(\int_0^\pi \sin^{\frac{2-n}{n-1}} \vartheta_1 \mathrm{d}\vartheta_1\right)^{n-1} \times$$

$$\times \int_0^\pi \cdots \int_0^\pi \int_0^{2\pi} \frac{R^{n-1}(\sin \vartheta_1)^{n-2} \cdots \sin \vartheta_{n-2} \mathrm{d}\vartheta_1 \cdots \mathrm{d}\vartheta_{n-1}}{R^n \sin^{\frac{n(n-2)}{n-1}} \vartheta_1} =$$

$$= \left(\int_0^\pi \sin^{\frac{2-n}{n-1}} \vartheta_1 \mathrm{d}\vartheta_1\right)^n$$

i.e. to an equality, as desired.

PROPOSITION 11. *Let A be the ring $r_1 < |x| < r_2$ and let E_1, E_2 be two disjoint subsets of A such that each sphere $|x| = r$, $r_1 < r < r_2$, contains at least one point of each $E_k (k = 1, 2)$. If Γ is the family of all arcs joining E_1 and E_2 in $A - E_1 - E_2$, then*

$$M(\Gamma) \geqq \frac{2^n}{A_0} \log \frac{r_2}{r_1},$$

where A_0 is the constant in (13).

PROPOSITION 12. *Let x_1, x_2 be two points on the sphere $S(x, r)$. Let A be a ring with complementary components C_0, C_1 such that C_0 contains the points x and x_1 while C_1 contains x_2. Then*

$$M(A) \leqq \chi(n), \tag{16}$$

where

$$\chi(n) = \frac{n\omega_n A_0}{2^{n-1} \log 3}, \tag{17}$$

and A_0 is the constant in (13).

(For the proofs of the propositions 11 and 12 see J. Väisälä [1]).

Other properties of the arc families.

PROPOSITION 13. *Let E_1, E_2 be two sets contained in the plane Π, $d(E_1, E_2) = d$ and let Γ be the family of all arcs which join E_1 and E_2 in $\Pi - E_1 - E_2$. Then*

$$M_n(\Gamma) > \frac{4}{A_0 d} \cdot \tag{18}$$

Let be $x_1 \in E_1$, $x_2 \in E_2$ so that $d(x_1, x_2) = d$ and let us take the origin in $\dfrac{x_1 + x_2}{2}$.

Let $S\left(\dfrac{e_n d}{4}, \dfrac{d}{4}\right)$ be the sphere tangent in the origin to the plane Π, let Γ_1 be the family of all arcs joining x_1 and x_2 in $B^{n-1}\left(\frac{d}{2}\right)$ and let Γ' be its image by the inverse of the stereographic projection. Since the line element grows by stereographic projection, $\rho \in F(\Gamma')$ implies $\rho \in F(\Gamma_1)$, and by proposition 10 applied to the hemisphere C, which is projected in $B^{n-1}\left(\frac{d}{2}\right)$, for any $\rho \in F(\Gamma')$ yields

$$\int\limits_{B^{n-1}\left(\frac{d}{2}\right)} \rho^n d\sigma_\Pi = \int_C \rho^n \frac{d\sigma_\Pi}{d\sigma_S} d\sigma_s \geqq \int_C \rho^n d\sigma_s \geqslant \frac{4}{A_0 d},$$

hence, by virtue of corollary 2 of proposition 3, $M_n(\Gamma) \geqq M_n(\Gamma_1) \geqq \frac{4}{A_0 d}$.

PROPOSITION 14. *Let Γ be the family of all arcs which join the disjoints sets $E_1, E_2 \subset B(x, R)$ in $B(x, R) - E_1 - E_2$,* let Π_h $(0 \leqq h \leqq H)$ *be a one parameter family of cross-sections of the ball $B(x, R)$ by parallel planes which meet the two sets E_1, E_2 and fill in a spherical segment with the altitude H. Then*

$$M(\Gamma) \geqq \frac{4H}{A_0 d}, \tag{19}$$

where $d = \sup\limits_{0 \leqq h \leqq H} d(E_1 \cap \Pi_h, E_2 \cap \Pi_h)$.

Indeed, by (18) for an arbitrary $\rho \in F(\Gamma)$

$$\int\limits_{B(x,R)} \rho^n d\tau \geqq \int_V \rho^n d\tau = \int_0^H \left(\int_{\Pi_h} \rho^n d\sigma \right) dh \geqq \frac{4H}{A_0 d} \tag{20}$$

holds.

PROPOSITION 15. *Let $E_1, E_2 \subset B(x, R)$ be two connected sets with a common limit point x_0. If Γ is the arc family joining E_1 and E_2 in $B(x, R) - E_1 - E_2$, then $M(\Gamma) = \infty$.*

Let us suppose first $x_0 \in B(x, R)$. Then it is possible to find an r_0, such that each sphere $S(x_0, R)$ $(0 < r < r_0)$ intersects the two sets E_1, E_2 and $B(x_0, r_0) \subset B(x, R)$. Let Γ_1 be the family of arcs joining E_1 and E_2 in $B(x_0, r_0) - E_1 - E_2$. Clearly, $\Gamma \supset \Gamma_1$, so that, combining the proposition 11 with corollary 2 of proposition 3, we obtain $M(\Gamma) \geqq M(\Gamma_1) = \infty$.

In the hypothesis $x_0 \in S(x, R)$, let r_0 be chosen in such a way that $S(x_0, r)$ $(0 < r < r_0)$ meet the two sets E_1, E_2 and let Γ' be the family of arcs which join E_1 and E_2 in $B(x_0, r_0) \cap B(x, R) - E_1 - E_2$. Then, by the proposition 10, for each $\rho \in F(\Gamma')$, we have

$$\int\limits_{S(x_0, r) \cap B(x, R)} \rho^n d\sigma \geqq \frac{1}{A_0 r}.$$

Integrating with respect to r on the interval (r', r_0), where $(0 < r' < r_0)$, then, by corollary 2 of proposition 3 and taking the infimum over all such ρ yields

$$M(\Gamma) \geqq M(\Gamma') \geqq \frac{1}{A_0} \log \frac{r_0}{r'}$$

for every $r' \in (0, r_0)$, which implies $M(\Gamma) = \infty$, as desired.

Suppose that D is an open half space and that E_0 and E_1 are disjoint continua in $\bar{D}$. (By *a continuum* we mean a compact connected set in R^n which contains more than one point.) Next for small $t > 0$ let Γ and $\Gamma(t)$ be the families of arcs which join E_1 to E_2 and $E_1(t)$ to $E_2(t)$ in D, respectively, where $E_k(t)$ denotes the closed set of points which lie within distance t of E_k for $k = 0,1$.

The following result yields an important relation between the families of functions $F(\Gamma)$ and $F_{\llcorner}\Gamma(t)]$.

PROPOSITION 16. *If* $\rho \in F(\Gamma)$ *and if* $\rho \in L^n$, *then for each* $a > 1$ *there exists a* $t > 0$ *such that* $a\rho \in F[\Gamma(t)]$.

Choose $b > 0$ so that $a(1 - 2b) = 1$, let $c = \min[d(E_0), d(E_1)] > 0$, $d = d(E_0, E_1) > 0$ and $\delta = \min[d(E_0, \partial D), d(E_1, \partial D)]$. Since $\rho \in L^n$, we can choose t, $0 < t < \min\left(\frac{c}{4}, \frac{d}{6}, \frac{\delta}{2}\right)$, such that

$$\int\limits_{B(x,\,2t)} \rho^n d\tau \leqq \frac{\log 2}{A_0} b^n \tag{21}$$

for all $x \in \bar{D}$, where A_0 is the constant in (13); because if for every $t > 0$ it existed at least an $x \in \bar{D}$ such that the inequality (21) is not verified, then for each sequence $\{t_m\}$, where $t_m \to 0$ as $m \to \infty$, it would be possible to find a sequence $\{x_m\}$, such that

$$\int\limits_{B(x_m,\,t_m)} \rho^n d\tau > \frac{\log 2}{A_0} b^n,$$

and from $\{x_m\}$ to extract a subsequence $\{x_{m_k}\}$, $x_{m_k} \to x_0 \in D$; but then, clearly, for each $t > 0$,

$$\int\limits_{B(x_0,\,t)} \rho^n d\tau > \frac{\log 2}{A_0} b^n$$

which would contradict that, on account of the complete additivity of the Lebesgue integral

$$\sum_{m=1}^{\infty} \int_{A_m} \rho^n d\tau \leqq \int_{R^n} \rho^n d\tau < \infty,$$

where A_m is the spherical ring $\frac{1}{m+1} < |x - x_0| < \frac{1}{m}$, the rest

$$\int_{B\left(x_0, \frac{1}{m}\right)} \rho^n d\tau \to 0.$$

(21) implies that for each $x \in \bar{D}$, we can find a sphere $S(x, u)$ such that $t < u = u(x) < 2t$ and

$$A_0 u \int_{S(x, u)} \rho^n d\sigma \leqq b^n, \tag{22}$$

because otherwise

$$A_0 u \int_{S(x, u)} \rho^n d\sigma > b^n$$

for all $u \in (t, 2t)$, hence, integrating with respect to u, we should obtain

$$\int_{B(x, 2t)} \rho^n d\tau \geqq \int_t^{2t} \int_{S(x, u)} \rho^n d\sigma du > \frac{b^n \log 2}{A_0},$$

contradicting so (21).

To complete the proof we must show that

$$\int_\gamma \rho ds \geqq 1 - 2b$$

for all $\gamma \in \Gamma(t)$, which, taking into account the definition of b, is equivalent to

$$\int_\gamma a\rho ds \geqq 1,$$

i.e. $a\rho \in F[\Gamma(t)]$. Choose $\gamma \in \Gamma(t)$, there are two cases to consider.

Suppose first that there exist finite points $x_k \in \bar{\gamma} \cap E_k(t)$ $(k = 0, 1)$. Then since $x_k \in \bar{\gamma}$, since γ is a connected set in D and since $d(\gamma) \geqq \frac{2d}{3} > 4t > 2u = d[S(x_k, u)]$ $(k = 0,1)$ it follows $\gamma \cap S(x_k, u) \neq \emptyset$. Next because $E_k \subset \bar{D}$ are connected and $d(E_k) \geqq c > 4t > 2u = d[S(x_k, u)]$ a similar argument shows that $E_k \cap S(x_k, u) \neq \emptyset$ $(k = 0,1)$. We conclude from proposition 10 and (22) that for $k = 0,1$, there exists a circular arc γ_k which joins γ and E_k in $S(x_k, u)$ and for which

$$\int_{\gamma_k} \rho ds \leqq b \tag{23}$$

It is then easy to show that $\bar{\gamma}_0 \cup \bar{\gamma}_1 \cup \gamma$ contains an arc β which joins E_0 to E_1 in D and hence, taking into account that $\rho \in F(\Gamma)$, it follows that

$$\int_\gamma \rho ds \geqq \int_\beta \rho ds - \int_{\gamma_0} \rho ds - \int_{\gamma_1} ds \geqq 1 - 2b.$$

Suppose next that one of the sets, say $\bar{\gamma} \cap E_1(t)$, contains only the point ∞. Then $\bar{\gamma} \cap E_0(t)$ contains a point $x_0 \neq \infty$, and arguing as above, we can find a circular arc γ_0, which joins γ to E_0 in $S(x_0, u)$ and for which (23) holds with $k = 0,1$. Since $\infty \in E_1$, $\bar{\gamma}_0 \cup \gamma$ contains an arc β which joins E_0 to E_1 in $D - E_0 - E_1$ and we obtain

$$\int_\gamma \rho ds \geqq \int_\beta \rho ds - \int_{\gamma_0} \rho ds \geqq 1 - b > 1 - 2b.$$

Thus the proof of proposition 16 is complete.

PROPOSITION 17. *Suppose that D is an open set, that E_0 and E_1 are disjoint bounded continua in D, and that for small $t > 0$, Γ and $\Gamma(t)$ are the families of arcs which join E_0 to E_1 and $E_0(t)$ to $E_1(t)$ in D, respectively. Then*

$$M(\Gamma) = \lim_{t \to 0} M[\Gamma(t)]. \tag{24}$$

If $0 < t_1 < t_2$, then $\Gamma \subset \Gamma(t_1) \subset \Gamma(t_2)$ and hence

$$M(\Gamma) \leqq M[\Gamma(t_1)] \leqq M[\Gamma(t_2)]. \tag{25}$$

Thus the limit in (24) exists and

$$M(\Gamma) \leqq \lim_{t \to 0} M[\Gamma(t)]. \tag{26}$$

Suppose now that $0 < M(\Gamma) < \infty$. Fix $a > 1$ and choose $\rho \in F(\Gamma)$ such that

$$\int_{R^n} \rho^n d\tau \leqq aM(\Gamma).$$

By proposition 16, we choose $t > 0$ so that $a\rho \in F[\Gamma(t)]$. Thus

$$M[\Gamma(t)] \leqslant a^n \int_{R^n} \rho^n d\tau \leqq a^{n+1} M(\Gamma),$$

and we obtain

$$\lim_{t \to 0} M[\Gamma(t)] \leqq a^{n+1} M(\Gamma),$$

If we let $a \to 1$, then

$$\lim_{t \to 0} M[\Gamma(t)] \leqq M(\Gamma) \tag{27}$$

and (26) imply (24) in this case.

If $M(\Gamma) = \infty$, then (27) is clearly verified. It remains only to consider the case $M(\Gamma) = 0$. But then for each $\varepsilon > 0$, there exists a $\rho \in F(\Gamma)$ such that

$$\int_{R^n} \rho d\tau < \varepsilon,$$

and, by the preceding proposition, there exists a $t > 0$ such that $a\rho \in F[\Gamma(t)]$, hence

$$0 \leqq M[\Gamma(t)] \leqq a^n \int_{R^n} \rho^n d\tau < a^n \varepsilon,$$

and then, on account of (25),

$$0 \leqq \lim_{t \to 0} M[\Gamma(t)] \leqq a^n \varepsilon.$$

and, since ε is arbitrary,

$$\lim_{t \to 0} M[\Gamma(t)] = M(\Gamma) = 0.$$

Thus the proof is complete.

PROPOSITION 18. *If E_0 and E_1 are disjoint continua, then there exist a domain D and disjoint continua C_0 and C_1 such that C_0 and C_1 are components of CD and $\partial C_k \subset E_k \subset C_k$ for $k = 0, 1$.*

Choose $x_0 \in E_0$, $x_1 \in E_1$ so that $d(x_0, x_1) = d(E_0, E_1)$, let $x = \dfrac{x_0 + x_1}{2}$ and let D be the component of $C(E_0 \cup E_1)$ which contains x. Then D is a domain and each component Δ of $C(\bar{D})$ is a domain with a connected boundary (see for instance Newman [1], theorem 14.2, p. 123 and the remark of p. 137). Next since $E_0 \cap E_1 = \emptyset$ and since

$$\partial \Delta \subset \partial C\bar{D} \subset \partial D \subset E_0 \cup E_1,$$

either $\partial \Delta \subset E_0$ or $\partial \Delta \subset E_1$. Now let

$$C_k = (\bigcup \Delta) \bigcup E_k \ (k = 0, 1),$$

where for each k the union is taken over all components Δ of $C\bar{D}$ for which $\partial \Delta \subset E_k$. Then it is not difficult to see that C_0 and C_1 are continua and that $CD = C_0 \cup C_1$. Since $\partial D \subset E_0 \cup E_1$ and $x_k \in \partial D \cap E_k$ for $k = 0,1$, ∂D is not connected and hence C_0 and C_1 are the components of CD. Finally we see that

$$\partial C_k \subset \partial D \cap C_k \subset E_k \subset C_k \ (k = 0, 1),$$

and the proof is complete.

PROPOSITION 19. *Suppose that D is a half space, that E_0 and E_1 are disjoint continua in $\bar{D}$, and that $\tilde{E}_0$ and $\tilde{E}_1$ are the symmetric images of E_0 and E_1 in the plane ∂D. If Γ is the family of arcs which join E_0 and E_1 in D and Γ_1 the family of arcs which join $E_0 \cup \tilde{E}_0$ and $E_1 \cup \tilde{E}_1$ in R^n, then*

$$M(\Gamma) = \frac{1}{2} M(\Gamma_1). \tag{28}$$

We may assume, for convenience of notation, that D is the half space $x^n > 0$. If we let $\tilde{\Gamma}$ denote the family of arcs which join $\tilde{E}_0$ and $\tilde{E}_1$ in $\tilde{D}$ (the half space $x^n < 0$), then Γ and $\tilde{\Gamma}$ are separate families and $\Gamma \cup \tilde{\Gamma} \subset \Gamma_1$. Obviously $M(\Gamma) = M(\tilde{\Gamma})$ and hence, by proposition 2 and corollary 2 of proposition 3,

$$2M(\Gamma) = M(\Gamma) + M(\tilde{\Gamma}) = M(\Gamma \cup \tilde{\Gamma}) \leqq M(\Gamma_1). \tag{29}$$

Next let $\bar{\Gamma}$ denote the family of arcs which join E_0 and E_1 in $\bar{D}$, let $f\colon R^n \to \bar{D}$ be the continuous mapping given by

$$f(x) = (x^1, \cdots, x^{n-1}, |x^n|)$$

and let $\rho \in F(\bar{\Gamma})$. Set $\rho_1 = \rho \circ f$ ($\circ$ means the functional product, i.e. the successive effectuation of the two mappings) and choose $\gamma_1 \in \Gamma_1$. Then $f(\gamma_1)$ contains an arc $\gamma \in \bar{\Gamma}$ and hence

$$\int_{\gamma_1} \rho_1(x) ds = \int_{\gamma_1} \rho[f(x)] ds \geqq \int_{f(\gamma_1)} \rho(x) d\Lambda \geqq \int_{\gamma} \rho(x) ds \geqq 1$$

(where $d\Lambda$ is the linear Hausdorff measure element). Thus $\rho_1 \in F(\Gamma_1)$,

$$M(\Gamma_1) \leqq \int_{R^n} \rho_1^n d\tau = 2 \int_{\bar{D}} \rho^n d\tau \leqq 2 \int_{R^n} \rho^n d\tau,$$

and, since ρ is arbitrary, we conclude that

$$M(\Gamma_1) \leqq 2M(\bar{\Gamma}).$$

To complete the proof of (28) we must show that

$$M(\bar{\Gamma}) \leqq M(\Gamma). \tag{30}$$

If $M(\Gamma) = \infty$, then the inequality is obvious. Let us suppose then $M(\Gamma) < \infty$. Fix $a > 1$ and choose $\rho \in F(\Gamma)$ so that $\rho \in L^n$. By proposition 16 we can choose $t > 0$

so that $a\rho \in F[\Gamma(t)]$. Set $\rho_1(x) = a\rho(x + te_n)$, let $\gamma_1 \in \overline{\Gamma}$, and let γ be the arc γ_1 translated through the vector te_n. Then $\gamma \in \Gamma(t)$ and we have

$$\int_{\gamma_1} \rho_1(x) ds = \int_{\gamma} a\rho(x) ds \geqq 1.$$

Hence $\rho_1 \in F(\overline{\Gamma})$,

$$M(\overline{\Gamma}) \leqq \int_{R^n} \rho_1^n d\tau = a^n \int_{R^n} \rho^n d\tau,$$

and taking the infimum over all such ρ yields

$$M(\overline{\Gamma}) \leqq a^n M(\Gamma).$$

Finally if we let $a \to 1$, we obtain (30), which together with (29) implies (28) as desired.

PROPOSITION 20. *Let D be a bounded domain in R^{n-1} and let E be a closed subset of D. If Γ is the family of arcs which join E and ∂D in $D - E$, then*

$$M_n(\Gamma) \geqq \frac{\omega_{n-1}^{\frac{n}{n-1}}}{(n-1)^{n-2}(m_{n-1}D)^{\frac{1}{n-1}}}. \tag{31}$$

The equality holds if $D = B^{n-1}(x, R)$ and E is its centre. In this case

$$M_n(\Gamma) = \frac{\omega_{n-1}}{(n-1)^{n-2}R}.$$

We may assume that E contains the origin O. Let γ_e be the segment joining O to ∂D in the direction corresponding to e. Obviously, $\Gamma < \{\gamma_e\}$. For each $\rho \in F(\{\gamma_e\})$ and $e \in S^{n-1}$, we have

$$1 \leqq \int_{\gamma_e} \rho ds.$$

By Hölder's inequality, this implies

$$1 \leqq \left(\int_{\gamma_e} \rho ds\right)^n \leqq \int_{\gamma_e} \rho^n r^{n-2} dr \left(\int_{\gamma_e} r^{\frac{2-n}{n-1}} dr\right)^{n-1} = (n-1)^{n-1} R(e) \int_{\gamma_e} \rho^n r^{n-2} dr,$$

where $R(e)$ is the length of γ_e. Integrating over S^{n-1}, on account of Fubini's theorem, we obtain

$$\int_D \rho^n d\sigma = \int_{S^{n-2}} d\sigma_{n-2} \int_{\gamma_e} \rho^n r^{n-2} dr \geqq \frac{1}{(n-1)^{n-1}} \int_{S^{n-2}} \frac{d\sigma_{n-2}}{R(e)}. \tag{32}$$

On the other hand, Hölder's inequality implies

$$(n-1)^n\omega_{n-1}^n = \left(\int\limits_{S^{n-2}} d\sigma_{n-2}\right)^n \leqq \int\limits_{S^{n-2}} R^{n-1}(e)d\sigma_{n-2}\left[\int\limits_{S^{n-2}} \frac{d\sigma_{n-2}}{R(e)}\right]^{n-1} =$$

$$= (n-1)m_{n-1}D\left[\int\limits_{S^{n-2}} \frac{d\sigma_{n-2}}{R(e)}\right]^{n-1}.$$

Together with (32) this gives

$$\int_D \rho^n d\sigma \geqq \frac{\omega_{n-1}^{\frac{n}{n-1}}}{(n-1)^{n-2}(m_{n-1}D)^{\frac{1}{n-1}}}.$$

Since $\rho \in F(\{\gamma_e\})$ was arbitrary and $\Gamma < \{\gamma_e\}$, this proves (31).

Now let $D = B^{n-1}(R)$ and E the origin. The function

$$\rho \equiv \frac{1}{(n-1)^n R^{\frac{1}{n-1}} r^{\frac{n-2}{n-1}}} \in F(\Gamma).$$

Hence (by Fubini's theorem),

$$M_n(\Gamma) \leqq \int_D \rho^n d\sigma = \int\limits_{S^{n-2}} d\sigma_{n-2}\int_0^R \frac{dr}{(n-1)^n R^{\frac{n}{n-1}} r^{\frac{n-2}{n-1}}} =$$

$$= \frac{\omega_{n-1}}{(n-1)^{n-2}R} = \frac{\omega_{n-1}^{\frac{n}{n-1}}}{(n-1)^{n-2}(m_{n-1}D)^{\frac{1}{n-1}}}.$$

which proves the last part of the theorem.

Remark. With the aid of this proposition we shall obtain a lower bound of the modulus of an arc family corresponding to a cylinder, which in its turn, will be used in the last chapter of the third part to get bounds for the outer coefficient of QCf of an infinite cylinder.

PROPOSITION 21. *Suppose that $a < b$, that Z is the finite part of the cylinder $(x^1)^2 + \ldots + (x^{n-1})^2 < 1$ which is bounded by the planes $x^n = a, b$, and that E is a connected set in Z which joins the bases of Z. If Γ is the family of arcs in Z which join E to the lateral surface of Z, then*

$$M(\Gamma) \geqq \frac{\omega_{n-1}(b-a)}{(n-1)^{n-2}}. \tag{33}$$

There is equality in (33) if E is the segment $x^1 = \ldots = x^{n-1} = 0$, $a < x^n < b$

Choose $\rho \in F(\Gamma)$. For $a < u < b$, the plane $x^n = u$ meets both E and the lateral surface of Z, and we can apply the preceding proposition to obtain

$$\int_{R^n} \rho^n d\tau \geqq \int_a^b du \int_{x^n = u} \rho^n d\sigma \geqq \frac{\omega_{n-1}(b-a)}{(n-1)^{n-2}},$$

This yields (33) as desired.

Now suppose that E is the segment $x^1 = \ldots = x^{n-1} = 0$, $a < x^n < b$, and set

$$\rho(x) = \begin{cases} \dfrac{1}{(n-1)r^{\frac{n-2}{n-1}}} & \text{for } x \in Z, \\ 0 & \text{for } x \in CZ, \end{cases}$$

where $r = d(x, Ox^n)$. Then $\rho \in F(\Gamma)$ and

$$\int_{R^{n-1}} \rho^n d\tau = \int_{B^{n-1}} \frac{d\sigma}{(n-1)^n r^{\frac{n(n-2)}{n-1}}} \int_a^b dx^n = \frac{\omega_{n-1}(b-a)}{(n-1)^{n-1}} \int_0^1 r^{\frac{2-n}{n-1}} dr = \frac{\omega_{n-1}(b-a)}{(n-1)^{n-2}}.$$

Hence in this case there is equality in (33).

The next proposition will serve us to calculate the outer coefficient of QCf of a cone.

PROPOSITION 22. *Suppose that* $0 < \alpha \leqq \dfrac{\pi}{2}$ *and* $0 < a < b$, *that* C *is the part of the cone*

$$x^n > \cot \alpha [(x^1)^2 + \cdots + (x^{n-1})^2]^{\frac{1}{2}}$$

which is bounded by the spheres $S(a)$ *and* $S(b)$, *and that* E *is a connected set in* C *which joins the spherical bases of* C. *If* Γ *is the family of arcs in* C *which join* E *to the lateral surface of* C, *then*

$$M(\Gamma) \geqq \frac{(n-1)\omega_{n-1}}{q(\alpha)^{n-1}} \log \frac{b}{a}, \tag{34}$$

where

$$q(\alpha) = \int_0^\alpha \sin^{\frac{2-n}{n-1}} \vartheta \, d\vartheta. \tag{35}$$

There is equality in (34) *if* E *is the segment* $x^1 = \ldots = x^{n-1}$, $a < x^n < b$.

Choose $\rho \in F(\Gamma)$ and for each $t > 0$ let $\sigma(t) = S(t) \cap C$. We first show that

$$\int_{\sigma(t)} \rho^n \, d\sigma \geqq \frac{(n-1)\omega_{n-1}}{q(\alpha)^{n-1} t} \tag{36}$$

for $a < t < b$.

For this fix $t, a < t < b$. Since E joins $S(a)$ and $S(b)$ in C, we can find a point $x_t \in E \cap \sigma(t)$. Next let Π be any fixed plane tangent at x_t to $\sigma(t)$, and let e be a unit vector with the origin at x_t and contained in Π; e and Ox_t determine a bidimensional plane cutting $\sigma(t)$ along a circular arc γ_e which contains a pair of arcs joining E to the lateral surface of C. Thus

$$\int_{\gamma_e} \rho \, ds \geqq 2.$$

We can fix a point $x \in \gamma_e$ and set $\varphi = \sphericalangle (Ox, Ox_t)$. The preceding inequality, on account of Hölder's inequality, yields

$$2^n \leqq \int_{\gamma_e} \rho^n \sin^{n-2}\varphi \, ds \left(\int_{\gamma_e} \sin^{\frac{2-n}{n-1}}\varphi \, ds \right)^{n-1}. \tag{37}$$

Since $0 \leqq \varphi < 2\pi$, it follows that the length of γ_e is at most $2\alpha t \leqq \pi t$, so that

$$\int_{\gamma_e} \sin^{\frac{2-n}{n-1}}\varphi \, ds = t \int_{\gamma_e} \sin^{\frac{2-n}{n-1}}\varphi |d\varphi| \leqq 2tq(\alpha).$$

Then (37) implies

$$\int_{\gamma_e} \rho^n t^{n-1} \sin^{n-2}\varphi |d\varphi| \geqq \frac{2}{t} q(\alpha)^{1-n},$$

and we obtain (36) by integrating both sides of this inequality over all the unit vectors e situated in one of the half planes of Π. Next if we integrate both sides of the inequality (36) with respect to t over (a, b), we get

$$\int_{R^n} \rho^n \, d\tau \geqq \int_a^b \left[\int_{\sigma(t)} \rho^n \, d\sigma \right] dt \geqq \frac{(n-1)\omega_{n-1}}{q(\alpha)^{n-1}} \log \frac{b}{a},$$

and since ρ is arbitrary, this implies (34).

Finally suppose that E is the segment $x^1 = \ldots = x^{n-1} = 0$, $a < x^n < b$, and let

$$\rho(x) = \begin{cases} \dfrac{1}{tq(\alpha) \sin^{\frac{n-2}{n-1}}\varphi} & \text{for } x \in C, \\ 0 & \text{for } x \in CC, \end{cases}$$

where $t = |x|$, and φ is the acute angle between the segment Ox and the positive x^n-axis. Then $\rho \in F(\Gamma)$ and

$$\int_{R^n} \rho^n \, d\tau = \int_{R^n} \frac{d\tau}{t^n (\sin\varphi)^{\frac{n(n-2)}{n-1}} q^n(\alpha)} = \frac{(n-1)\omega_{n-1}}{q^n(\alpha)} \int_a^b \frac{dt}{t} \int_0^\alpha \frac{\sin^{n-2}\varphi \, d\varphi}{(\sin\varphi)^{\frac{n(n-2)}{n-1}}} =$$

$$= \frac{(n-1)\omega_{n-1}}{q^n(\alpha)} \log\frac{b}{a} \int_0^\alpha \sin^{\frac{2-n}{n-1}}\varphi \, d\varphi = \frac{(n-1)\omega_{n-1}}{q^{n-1}(\alpha)} \log\frac{b}{a}\,.$$

Hence in this case there is equality in (34).

Modulus of a cylinder. The configuration in R^n which corresponds to the quadrilateral in R^2 is the cylinder. A domain Z in R^n, together with two sets $B_1, B_2 \subset \partial Z$ is called *a cylinder* if $\overline{Z}$ can be mapped topologically onto the closure of the unit cylinder $\{x; (x^1)^2 + \ldots + (x^{n-1})^2 < 1, 0 < x^n < 1\}$, so that B_1, B_2 are mapped onto the bases $(x^1)^2 + \ldots + (x^{n-1})^2 < 1$, $x^n = 0,1$. B_1 and B_2 are called *the bases* of Z. If Γ_Z is the arc family joining B_1 and B_2 in Z, then

$$M(\Gamma_Z) = M(Z).$$

A cylinder is called *a right cylinder*, if the bases $B, B + he$ of Z lie in two parallel planes: $\Pi, \Pi + he$, where e is the unit vector perpendicular to Π, and if $Z = \{x; x = y + \alpha e, y \in B, 0 < \alpha < h\}$. h is called *the height* of Z.

PROPOSITION 23. *Let Z be a right cylinder with base B and height h. Then*

$$M(Z) = \frac{m(Z)}{h^n} = \frac{m_{n-1}B}{h^{n-1}}. \tag{38}$$

The function $\rho \equiv \dfrac{1}{h} \in F_Z(\Gamma_Z)$ and thus

$$M(Z) \leqslant \int_Z \rho^n \, d\tau = \frac{mZ}{h^n}. \tag{39}$$

Let then $\rho \in F(\Gamma_Z)$. For each $y \in B$, let $\gamma_y = \{x; x = y + \alpha e, 0 < \alpha < h\}$. Then $\gamma_y \in \Gamma_Z$ and by Hölder's inequality we obtain

$$1 \leqq \Big(\int_{\gamma_y} \rho \, ds\Big)^n \leqq h^{n-1} \int_{\gamma_y} \rho^n \, ds.$$

Integrating over $y \in B$ yields

$$m_{n-1}B \leqq h^{n-1} \int_Z \rho^n \, d\tau,$$

whence

$$\int_Z \rho^n \, d\tau \geqq \frac{m_{n-1}B}{h^{n-1}} = \frac{mZ}{h^n}.$$

This together with (39) imply (38) for the modulus of a right cylinder.

Rengel's inequality (in n-space). Let Z be a cylinder with bases B_1, B_2 and $d = d(B_1, B_2)$. Then $\rho \equiv \dfrac{1}{d} \in F_Z(\Gamma_Z)$. Consequently

$$M(Z) \leqq \frac{mZ}{d^n}. \tag{40}$$

This is a generalization of Rengel's inequality for moduli of quadrilaterals.

Modulus of an interval. *An interval* $I = \{x;\ 0 < x^i < \alpha^i\ (i = 1, \ldots, n)\}$ is a special case of a right cylinder. If α^a is chosen as the height of I, then *the modulus of the interval I with the height α^a is*

$$M(I) = \frac{\alpha^1 \cdots \alpha^{a-1} \alpha^{a+1} \cdots \alpha^n}{(\alpha^a)^{n-1}}. \tag{41}$$

The propositions 14 and 15 have been obtained by B. Zorič [1], the propositions 1—6 by Fuglede [1] (in more general hypotheses) and the rest of the results by J. Väisälä [1] and F. Gehring and J. Väisälä [1].

CHAPTER 6

ACL mappings

Mappings with bounded variation. The variation of a mapping $f(x) < \infty$ on a set $E \subset R^1$ is defined by the formula

$$V(f, E) = \sup \Sigma_m |f(b_m) - f(a_m)|,$$

where the supremum is taken over all sequences of non-overlapping intervals $\{[a_m, b_m]\}$ whose end-points $a_m, b_m \in E$ $(m = 1, 2, \ldots)$. If $V(f, E) < \infty$, the mapping f is said to be VB (*of bounded variation*) *on* E.

The derivative of a mapping of a real variable $f(x)$ relative to a set E in a point $x_0 \in \overline{E}$ is the limit

$$\lim_{x \to x_0} \frac{f(x) - f(x_0)}{x - x_0},$$

which (when it exists) is taken for $x \to x_0$ by values belonging to the set E.

PROPOSITION 1. *A mapping $f(x)$ which is VB on a set $E \subset R^1$, is derivable with respect to the set E at almost all points of E* (S. Saks [2], theorem 4.4, p. 223).

AC **mappings.** A mapping $f(x) < \infty$ is termed AC (*absolutely continuous*) on a set $E \subset R_1$, if for any given $\varepsilon > 0$ there exists an $\eta > 0$ such that for every sequence of non-overlapping intervals $\{[a_m, b_m]\}$ whose end-points $a_m, b_m \in E$, the inequality $\Sigma_m(b_m - a_m) < \eta$ implies that

$$\Sigma_m |f(b_m) - f(a_m)| < \varepsilon.$$

PROPOSITION 2. *Every mapping which is AC on a set $E \subset R^1$ is VB on E* S. Saks [2], p. [23).

PROPOSITION 3. *If $u(x)$ is a bounded integrable function such that*

$$U(x) - U(x_0) = \int_{x_0}^{x} u(x)\, dx \qquad (1)$$

for all $x \in [a, b]$ *and if* $x = v(t)$ *is an AC function such that*

$$a \leqq v(t) \leqq b. \tag{2}$$

for all $t \in [t_0, T]$, *then*

$$U[v(t_2)] - U[v(t_1)] = \int_{t_1}^{t_2} u[v(t)]v'(t)\,\mathrm{d}t \tag{3}$$

holds whenever $t_1, t_2 \in [t_0, T]$ (Carathéodory [3], theorem 1, p. 559).

PROPOSITION 4. *Let* $u(x)$ *be the AC function given by* (1) *and let* $v(t)$ *be a function AC in* $[t_0, T]$ *with the range* (2), *then* $u[v(t)]$ *is AC in* $[t_0, T]$ *if and only if* $u[v(t)]v'(t)$ *is measurable in this interval and then* (3) *holds* (Carathéodory [3], theorem 4, p. 543).

PROPOSITION 5. *An AC function is the definite integral of its derivative* (S. Saks [2], theorem 3, p. 332).

Let $\gamma \in R^n$ be a rectifiable curve of length 1 and let $x = x(s)$ $(0 \leqq s \leqq 1)$ be its representation by means of the arc-length s. We say that a mapping $f: R^n \to R^m$ is *AC along* γ if $f[x(s)]$ is *AC* on the interval (0, 1).

PROPOSITION 6. *If* γ *is a rectifiable curve given by the equation* $x = x(t)$ *and the function* $s = s(t)$ *is its length, then*

(i) *in order that* $s(t)$ *be AC on* (0, 1) *it is necessary and sufficient that* $x(t)$ *should be so*;

(ii) *we have*

$$\frac{\mathrm{d}s}{\mathrm{d}t} = |x'(t)| \tag{4}$$

a.e. in (0,1) (S. Saks [2], p. 123).

ACL **mappings.** A continuous mapping $f(x)$ is said to be *ACL* (*absolutely continuous on lines*) in a domain D, if for each interval $I = \{x;\ \alpha^i < x^i < \beta^i\ (i = 1, \ldots, n)\}$, $\bar{I} \subset D$, $f(x)$ is *AC* on almost every line segment in I, parallel to the co-ordinate axes, i.e. if $I_i = \{x;\ x \in \bar{I},\ x^i = \alpha^i\}$ is a face of I and if E is the set of points $y \in I_i$ such that $f(x)$ is not *AC* on the segment $J_y = \{x; x = y + \xi e_i, 0 < \xi < \beta^i - \alpha^i\}$, then $m_{n-1}E = 0$. (The equivalence of the two preceding definitions is a consequence of the proof of the proposition 5.23, since $E \times J_y$, is measurable.) We see at once that another equivalent condition is that the family $\{J_y;\ y \in E\}$ is exceptional. Consequently, the concept "almost every curve", defined by means of moduli, is in this case the same as in the usual terminology.

ACL_p **mappings.** A mapping $f(x)$ is said to be ACL_p in D if it is ACL and if its partial derivatives are L^p-integrable over each compact subset of D. Thus ACL_1 means AC in the sense of Tonelli.

Properties of ACL mappings.

PROPOSITION 7. *A mapping $u(x)$ continuous and ACL in D has partial derivatives Borel measurable a.e. in D.*

Let $I \subset\subset D$, be an interval and $J_y = \{x;\ x = y + \xi e_n,\ y \in I_n,\ 0 < \xi < \beta^n - \alpha^n\}$ one of the line segments in I parallel to Ox^n and on which $u(x)$ is AC. The propositions 1 and 2 imply the existence of $u_n = \dfrac{\partial u}{\partial x^n}$ a.e. in J_y for almost every $y \in I_n$.

But the set E of those points of I where u_n does not exist verifies the hypotheses of the corollary of Tonelli's theorem (Chapter 3). On the other hand, since the functions

$$\overline{\lim_{h\to 0}} \frac{u(x+he_n)-u(x)}{h},\quad \underline{\lim_{h\to 0}} \frac{u(x+he_n)-u(x)}{h}$$

are Borel measurable, the set of the points where they are equal is measurable too (S. Saks [2], theorem 8.1, p. 14); but this set is just the set where u_n exists, i.e. $I - E$, and the measurability of $I - E$ implies the measurability of E, hence, by the corollary of Tonelli's theorem (Chapter. 3), $mE = 0$.

The same argument allows us to establish the existence a.e. in I of each first order partial derivatives of $u(x)$, hence we conclude their simultaneous existence a.e. in I. But, since by Lindelöf theorem (Carathéodory [3], p. 44), it is possible to cover D by a countable set of intervals $I \subset\subset D$, we obtain the existence a.e. in D of all partial derivatives u_i $(i = 1, \ldots, n)$, which implies the formula

$$u_i(x) = \overline{\lim_{h\to 0}} \frac{u(x+he_i)-u(x)}{h} = \underline{\lim_{h\to 0}} \frac{u(x+he_i)-u(x)}{h} \quad (i = 1, \ldots, n)$$

a.e. in D. Thus u_i $(i = 1, \ldots, n)$ are equivalent to Borel measurable functions (two functions are called equivalent if they are different on a set of measure zero) and then u_i are Borel measurable a.e. in D.

The points of Lebesgue. A point x, where

$$\lim_{h\to 0} 2(h)^{-n} \int_{x-h}^{x+h} |f(\xi) - f(x)|\, d\tau = 0$$

is said to be *a point of Lebesgue* of the mapping $f(x)$.

PROPOSITION 8. *Every point where an integrable mapping is continuous is a point of Lebesgue of this mapping.*

Indeed, given any $\varepsilon > 0$, there exists a $\delta > 0$ such that $h < \delta$, should imply $|f(\xi) - f(x)| < \varepsilon$ whenever $\xi \in [x - h,\ x + h]$, hence

$$(2h)^{-n} \int_{x-h}^{x+k} |f(\xi) - f(x)|\, d\tau < \varepsilon.$$

PROPOSITION 9. *If an integrable mapping $f(x)$ is continuous in a point x, then the formula*

$$\lim_{r \to 0} \frac{1}{mB(r)} \int_{B(r)} f(x + \xi)\, d\tau = f(x)$$

holds.

The same proof as for the preceding proposition.

Theorem of Fuglede [1]. *Let $f(x)$ be ACL_p in a domain $D \subset R^n$. Then for p-almost every curve $\gamma \subset D$, $f(x)$ is AC along each compact subcurve of γ.* (For the proof see J. Väisälä [1]).

ACA **mappings on a surface.** We say that a mapping $f(x)$ is *ACA* (*absolutely continuous on arcs*) in σ if $M(\Gamma) = 0$, where Γ is the family of all locally rectifiable arcs in σ which contain a compact subarc on which $f(x)$ is not *AC*.

COROLLARY. *If σ is a plane domain and if $f(x)$ is ACA in σ, then $f(x)$ is ACL in σ. Conversely, if $f(x)$ is ACL in σ and if the partial derivatives of $f(x)$ are locally L^{n-1}-integrable in σ, then $f(x)$ is ACA in σ.*

The first part is obvious, and the second part is a consequence of the preceding theorem.

CHAPTER 7

Subharmonic and superharmonic functions

In order to establish in the next part Reade's theorem [1], we introduce the concept of subharmonic and superharmonic functions and recall also some of their properties. Taking into account that the proofs of these properties in n-space are similar to the corresponding proofs for $n = 2$, we shall content ourselves to send for the proofs to the monograph of T. Rado [1].

An upper (lower) semi-continuous function in D is said to be a function $u(x)$ $[v(x)]$ defined in D and with the property that for every $x_0 \in D$ and every $\lambda > u(x_0)$ $[\lambda < v(x_0)]$ there exists a $\delta = \delta(x_0, \lambda) > 0$ such that $u(x) < \lambda\,[v(x) > \lambda]$ whenever $d(x, x_0) < \delta$. If $u(x_0) = -\infty$ this means $u(x) \to -\infty$ as $x \to x_0$ [if $v(x_0) = \infty$ this means $v(x) \to \infty$ as $x \to x_0$].

Harmonic functions. *A harmonic function in D* is a function $H(x)$ which verifies in D the equation

$$\Delta H = \frac{\partial^2 H}{(\partial x^1)} + \cdots + \frac{\partial^2 H}{(\partial x^n)^2} = 0.$$

Subharmonic functions. A function $u(x)$ $(-\infty \leqq u < \infty)$ is *subharmonic in D* if it satisfies the following conditions:

a) $u(x) \not\equiv -\infty$ in D;

b) $u(x)$ is upper semi-continuous in D;

c) let $\Delta \subset D$ be a domain and $H(x)$ be harmonic in Δ, continuous in $\bar{\Delta}$ and $H \geqq u$ on $\partial\Delta$. Whenever these assumptions are satisfied, we also have $H \geqq u$ in Δ.

Superharmonic functions. A function $v(x)$ $(-\infty < v \leqq \infty)$ is *superharmonic* in D if it satisfies the following conditions:

a′) $v(x) \not\equiv \infty$ in D;

b′) $v(x)$ is lower semi-continuous on D;

c′) let $\Delta \subset D$ be a domain and $H(x)$ be harmonic in Δ, continuous in $\bar{\Delta}$ and $H \leqq v$ on $\partial\Delta$. Whenever these assumptions are satisfied, we also have $H \leqq v$ in Δ.

Clearly, a function v is superharmonic in a domain D if the function $u = -v$ is subharmonic.

Dirichlet domain is a closed domain $\bar{\Delta}$ such that for every continuous function u on $\partial\Delta$ there exists a function H which is harmonic in Δ, continuous in $\bar{\Delta}$ and equal to u on $\partial\Delta$.

Properties of subharmonic and superharmonic functions.

PROPOSITION 1. *A harmonic function is both subharmonic and superharmonic, and conversely* (T. Radó [1], p. 1).

PROPOSITION 2. *In order that a function $u(x)$ be subharmonic in D it is necessary and sufficient to satisfy the conditions a), b) and that for every point $x_0 \in D$ with $u(x_0) > -\infty$ we have a $\rho(x_0) > 0$ such that for $r < \rho(x_0)$,*

$$u(x_0) \leqq \frac{1}{\omega_n r^n} \int_{B(x_0, r)} u(x)^n \, d\tau \equiv V(u, x_0, r) \tag{1}$$

holds (T. Radó [1], theorem 2.3, p. 7).

COROLLARY 1. *In order that a function $v(x)$ be superharmonic in D it is necessary and sufficient to satisfy the conditions a′), b′) and that for every point $x_0 \in D$ with $v(x_0) < \infty$ we have a $\rho(x_0) > 0$ such that $r < \rho(x_0)$ imply $v(x_0) \geqq$ $\geqq V(v, x_0, r)$.*

COROLLARY 2. *In order that a function $H(x)$ be harmonic in D it is necessary and sufficient to be continuous in D with $|H(x_0)| \neq \infty$ and for every point $x_0 \in D$ with $|H(x_0)| \neq \infty$ we have a $\rho(x_0) > 0$ such that $r < \rho(x_0)$ implies $H(x_0) = V(H, x_0, r)$.*

PROPOSITION 3. *In order that a function $u(x)$ be subharmonic in D it is necessary and sufficient to satisfy the conditions (a) and (b) and that for every point $x_0 \in D$ with $u(x_0) > -\infty$ we have a $\rho(x_0) > 0$ such that $r < \rho(x_0)$ implies*

$$u(x_0) \leqq \frac{1}{n\omega_n r^{n-1}} \int_{S(x_0, r)} u(x)^{n-1} \, d\sigma \equiv S(u, x_0, r). \tag{2}$$

(For the proof see T. Radó [1], theorem 23, p. 7).

COROLLARY 1. *In order that a function $v(x)$ be superharmonic in D it is necessary and sufficient to satisfy the conditions (a′) and (b′) and that for every point $x_0 \in D$ with $v(x_0) < \infty$ we have a $\rho(x_0) > 0$ such that $r < p(x_0)$ imply $v(x_0) \geqq$ $\geqq S(v, x_0, r)$.*

COROLLARY 2. *In order that a function $H(x)$ be harmonic in D it is necessary and sufficient to be continuous in D with $|H(x)| \neq \infty$ and for every point $x_0 \in D$ with $|H(x_0)| \neq \infty$ we have a $\rho(x_0) > 0$ such that $r < \rho(x_0)$ implies $H(x_0) = S(H, x_0, r)$.*

PROPOSITION 4. *In order that a function $u(x)$ be subharmonic in D it is necessary and sufficient that $V(u, x_0, r) \leqq S(u, x_0, r)$ for every ball $B(x_0, r) \subset\subset D$* (T. Radó [1], theorem 3.25, p. 12).

PROPOSITION 5. *If $u(x)$ is subharmonic in D, then* $S(u, x_0, r)$ *and* $V(u, x_0, r)$, *where* $B(x_0, r) \subset D$, *are non-decreasing functions of r* (T. Radó [1], theorems 2.8, 2.9, p. 8).

PROPOSITION 6. *Let $u(x)$ be subharmonic in D and let $\Delta \subset\subset D$, be a domain. If $H(x)$ is harmonic in Δ and continuous in $\bar{\Delta}$, then $u(x) = H(x)$ either everywhere or nowhere in Δ.* (T. Radó [1], theorem 1.15, p. 4).

COROLLARY. *If $u(x)$ is subharmonic in $B(x_0, r)$, continuous on $S(x_0, r)$ and $u(x_0) = V(u, x_0, r)$, then $u(x)$ is harmonic in D.*

PROPOSITION 7. *If* $\lambda(x)^n$ *and* $\lambda(x)^{n-1}$ *are both harmonic in* D, *then* $\lambda(x) =$ $=$ *const. in* D.

Indeed,

$$\Delta[\lambda(x)^n] = n\lambda^{n-2}[(n-1)\delta^{ik}\lambda_i\lambda_k + \lambda\Delta\lambda] = 0,$$

$$\Delta[\lambda(x)^{n-1}] = (n-1)\lambda^{n-3}[(n-2)\delta^{ik}\lambda_i\lambda_k + \lambda\Delta\lambda] = 0,$$

where $\lambda_i = \dfrac{\partial\lambda}{\partial x^i}$ $(i = 1,\ldots, n)$. The two preceding equations imply

$$(n-1)\delta^{ik}\lambda_i\lambda_k = (n-2)\delta^{ik}\lambda_i\lambda_k,$$

hence $\delta^{ik}\lambda_i\lambda_k = 0$ and then $\lambda_i = 0$ $(i = 1,\ldots, n)$, i.e. $\lambda(x) =$ constant in D.

Part 2

Definitions of QCfH and their equivalence

CHAPTER 1

Väisälä's definitions of K-QCfH and their equivalence

In this chapter we shall present first three definitions the equivalence of which was established in the case $n = 3$ by J. Väisälä [1] and for an arbitrary n by Chén Hàng-lén [1] and by ourselves in a series of lectures delivered at the "Stoilow Seminar" in Bucharest.

We shall use the notations:

$$\delta(x) = \max\left[\underline{\delta}(x), \bar{\delta}(x)\right], \tag{1}$$

$$\underline{\delta}(x) = \overline{\lim_{r\to 0}} \frac{T(x, r)}{\omega_n l(x, r)^n}, \tag{2}$$

$$\bar{\delta}(x) = \overline{\lim_{r\to 0}} \frac{\omega_n L(x, r)^n}{T(x, r)}, \tag{3}$$

$$\delta_L(x) = \overline{\lim_{r\to 0}} \frac{L(x, r)}{l(x, r)}, \tag{4}$$

$$l(x, r) = \min_{|x'-x|=r} |f(x') - f(x)|, \tag{5}$$

$$L(x, r) = \max_{|x'-x|=r} |f(x') - f(x)|, \tag{6}$$

$$T(x, r) = m\{f[B(x, r)]\}, \tag{7}$$

where $f(x)$ is a homeomorphism in D and x, $x' \in D$. δ, $\underline{\delta}$, $\bar{\delta}$ and δ_L are called the *maximal*, *inner*, *outer* and *linear local dilatations*. They are all continuous and ≥ 1 in D. Their least upper bounds in D are called the corresponding *dilatations of the mapping* $f(x)$. If $n = 2$, all are equal between them.

Suppose x_0 is a point of differentiability of $f(x)$ with $J(x_0) \neq 0$, i.e. an *A-point* (*point of affinity*). Then

$$f(x_0 + \Delta x) = f(x_0) + f'(x_0)\Delta x + o(|\Delta x|),$$

where $\dfrac{O(|\Delta x|)}{|\Delta x|} \to 0$ as $|\Delta x| \to 0$. The linear transformation $x^* = f'(x_0)\,\Delta x$ with $|\Delta x| = 1$, maps the unit ball B onto an ellipsoid E with semi-axes $a_1(x_0) \geqq \ldots \geqq \geqq a_n(x_0) > 0$, where

$$a_1(x_0) = \max_{|\Delta x|=1} |f'(x_0)\Delta x|, \qquad a_n(x_0) = \min_{|\Delta x|=1} |f'(x)\Delta x|. \tag{8}$$

For the sake of simplicity, let us denote

$$\max |f'(x_0)| = \max_{|\Delta x|=1} |f'(x_0)\Delta x|, \qquad \min |f'(x_0)| = \min_{|\Delta x|=1} |f'(x_0)\Delta x|. \tag{9}$$

Clearly,

$$\max |f'(x_0)| = \sup_s \left| \frac{\partial f(x_0)}{\partial s} \right|, \qquad \min |f'(x_0)| = \inf_s \left| \frac{\partial f(x_0)}{\partial s} \right|, \tag{10}$$

where

$$\left| \frac{\partial f(x_0)}{\partial s} \right| = \lim_{|\Delta x|_s \to 0} \frac{|f(x) - f(x_0)|}{|(x - x_0)|},$$

and $|\Delta x|_s \to 0$ means that x tends to x_0 in the direction s. In an A-point x_0 we have also the relations

$$|J(x_0)| = a_1(x_0) \cdots a_n(x_0), \tag{11}$$

$$\underline{\delta}(x_0) = \frac{|J(x_0)|}{\min |f'(x_0)|^n} = \frac{a_1(x_0) \cdots a_{n-1}(x_0)}{a_n(x_0)^{n-1}}, \tag{12}$$

$$\bar{\delta}(x_0) = \frac{\max |f'(x_0)|^n}{|J(x_0)|} = \frac{a_1(x_0)^{n-1}}{a_2(x_0) \cdots a_n(x_0)}, \tag{13}$$

$$\delta_L(x_0) = \frac{\max |f'(x_0)|}{\min |f'(x_0)|} = \frac{a_1(x_0)}{a_n(x_0)}, \tag{14}$$

$$\delta(x_0) \leqq \delta_L(x_0)^{n-1}, \tag{15}$$

$$\delta_L(x_0) \leqq \delta(x_0)^{\frac{2}{n}}. \tag{16}$$

Metric definition. A homeomorphism $f(x)$ is called K-QCfH ($1 \leqq K < \infty$) in a domain D if δ (x) is bounded in D and

$$\delta(x) \leqq K \tag{17}$$

a.e. in D.

First geometric definition (by modulus of arc families). A homeomorphism $x^* = f(x)$ is called K-QCfH ($1 \leqq K < \infty$) in D if

$$\frac{M(\Gamma)}{K} \leqq M(\Gamma^*) \leqq KM(\Gamma) \tag{18}$$

for each curve family Γ in D, where $\Gamma^* = f(\Gamma)$.

First analytic definition. A homeomorphism $f(x)$ is called K-QCfH ($1 \leqq \leqq K < \infty$) in D if it is ACL, a.e. differentiable and

$$\frac{1}{K} \max |f'(x)|^n \leqq |J(x)| \leqq K \min |f'(x)|^n. \tag{19}$$

a.e. in D.

$f(x)$ is called QCfH (according to each of the preceding definitions) if it is K-QCfH (according to the corresponding definition) for some K.

The class of QCfH is a group. From the geometric definition it follows immediately that the inverse of a K-QCfH is a K-QCfH and the composite mapping of a K'-QCfH and a K''-QCfH is a $K'K''$-QCfH. Clearly, these properties still hold for all the definitions equivalent with the preceding geometric definition.

The equivalence of the three definitions. Let us establish first several preliminary lemmas.

LEMMA 1. *Let $x^* = f(x)$ be a homeomorphism of a domain D such that*

$$M(Q^*) \geqq \frac{1}{K} M(Q) = \frac{1}{K} \tag{20}$$

for each cube $Q, \subset \subset D$. If $f'(x)$ exists at x_0, then

$$\max |f'(x_0)|^n \leqq K|J(x_0)|. \tag{21}$$

Since translations, rotations and reflections have no influence on (20) and (21), we may assume that $x_0 = f(x_0) = O$ and that $f'(O)$ is given by

$$f'(O)e_i = a_i e_i \quad (i = 1, \ldots, n). \tag{22}$$

The differentiability of $f(x)$ at O implies

$$f(x) = f'(O)x + \varepsilon(O, x)|x|,$$

where $\varepsilon(x, O) \to 0$ as $x \to O$ and

$$\max |f'(O)| = a_1, \; J(O) = a_1 \cdots a_n \geqq 0. \tag{23}$$

If $a_1 = 0$, (21) is trivial. If $a_1 > 0$, we choose $0 < \varepsilon < \frac{a_1}{2}$ and then $h > 0$ such that the cube $\overline{Q} = \{x; \; 0 \leqq x^i \leqq h, \; i = 1, \ldots, n\}$ lies in D and

$$|f(x) - f'(O)x| < \varepsilon h$$

for $x \in \overline{Q}$. Let the faces $x^1 = 0$, h be the bases of Q. Then the distance between the bases of Q^* is $\geqq h(a_1 - 2\varepsilon)$, since, the preceding inequality implies

$$f(x) = f'(O)x + \theta\varepsilon h, \quad |\theta| < 1, \tag{24}$$

whence, by (22),

$$|f(h, x^2, \ldots, x^n) - f(0, x^2, \ldots, x^n)| > a_1 h - 2h\varepsilon = h(a_1 - 2\varepsilon).$$

On the other hand, (22) and (24) yield

$$mQ^* < h^n(a_1 + 2\varepsilon) \cdots (a_n + 2\varepsilon).$$

Hence, by the generalized Rengel's inequality (1.5.40),

$$M(Q^*) < \frac{(a_1 + 2\varepsilon) \cdots (a_1 + 2\varepsilon)}{(a_1 - 2\varepsilon)^n}.$$

Letting $\varepsilon \to 0$ and using (20) with $M(Q) = 1$, as it follows from (1.5.41), we obtain

$$a_1^n \leqq K a_1 \ldots a_n.$$

By (23), this gives (21).

LEMMA 2. *Let $x^* = f(x)$ be a homeomorphism of a domain D such that*

$$M(A^*) \leqq KM(A) \tag{25}$$

for all spherical rings $A^* \subset\subset D^*$. *Then*

$$\frac{L(x, r)}{l(x, r)} < e^{(\chi K)^{\frac{1}{n-1}}}$$

provided that the ball

$$\{|f(x') - f(x)| \leqq L(x, r)\} \subset f(D).$$

$\chi = \chi(n)$ *is the constant of proposition* 1.5.12.

If $l(x_0, r) = L(x_0, r)$, there is nothing to prove. Otherwise let A^* be the spherical ring centred at $f(x)$ and with radii $l(x, r)$, $L(x, r)$, where $l(x, r)$, $L(x, r)$ are given by (5) and (6), respectively. By (25) and propositions 1.5.7 and 1.5.12, we have

$$\left[\log \frac{L(x, r)}{l(x, r)}\right]^{n-1} = M(A^*) \leqq KM(A) < \chi K,$$

since the inverse images of the spheres $|f(x') - f(x)| = L(x, r)$ and $|f(x') - f(x)| = l(x, r)$ are tangent to the sphere $|x' - x| = r$ at two different points, being so in the hypotheses of proposition 1.5.12.

If $\varphi(E)$ is an additive function of a set in R^n then *the upper* and *the lower symmetrical derivatives of* $\varphi(E)$ *at a point* x are defined by

$$\overline{D}_{sym}\varphi(x) = \overline{\lim_{r\to 0}} \frac{\varphi[B(x, r)]}{\omega_n r^n}, \quad \underline{D}_{sym}\varphi(x) = \underline{\lim_{r\to 0}} \frac{\varphi[B(x, r)]}{\omega_n r^n},$$

respectively. When these two derivatives are equal, their common value $D_{sym}\varphi(x) = \overline{D}_{sym}\varphi(x) = \underline{D}_{sym}\varphi(x)$ is called *the symmetrical derivative of* $\varphi(E)$ *at* x. In particular, if $\varphi(E) = mf(E)$, then

$$J_G(x) = \overline{\lim_{r\to 0}} \frac{T(x, r)}{mB(x, r)} = \overline{D}_{sym}\varphi(x).$$

An additive function of a set will be called *monotone* on a set E if its values for the subsets of E are of a constant sign. A non-negative additive function of a set will also be termed as monotonic increasing, on account of the fact that, for each pair of measurable sets E_1 and E_2, the inclusion $E_1 \subset E_2$ implies $\varphi(E_2) = \varphi(E_1) + \varphi(E_2 - E_1) \geqq \varphi(E_1)$. For the same reason non-positive functions of a set will be termed as *monotonic decreasing*.

If $\varphi(E)$ is an additive function of a set on a set E, then *the upper* and *lower variation* of the function $\varphi(E)$ on the set E_0 are defined by

$$\overline{W}(\varphi, E_0) = \sup_{E\subset E_0} \varphi(E), \quad \underline{W}(\varphi, E_0) = \inf_{E\subset E_0} \varphi(E),$$

respectively. Since $\varphi(\emptyset) = 0$, it follows that $\overline{W}(\varphi, E_0) \geqq 0 \geqq \underline{W}(\varphi, E_0)$. The number

$$(\varphi, E_0) = \overline{W}(\varphi, E_0) - \underline{W}(\varphi, E_0)$$

will be called *absolute variation* of the function $\varphi(E)$ on E_0. If $W(\varphi, E) < \infty$, the function $\varphi(E)$ is said to be with *bounded variation.*

Theorem of Jordan. *In order that an additive function of a set be with bounded variation, it is necessary and sufficient that it should be the difference of two monotone functions* (S. Saks [1], theorem 1, p. 10).

COROLLARY. *A non-negative function (monotonic creasing) is with bounded variation.*

Theorem of Lebesgue. *An additive function of a set with bounded variation on a set E has a finite symmetrical derivative a.e. in E* (S. Saks [1], theorem 2, p. 49).

The decomposition theorem of de la Vallèe-Poussin. *If $\varphi(E)$ is an additive function of a set in R^n and if E_∞ denotes the set of those points x at which one at least of the derivatives $\overline{D}_{\mathrm{sym}}\varphi(x)$ and $\underline{D}_{\mathrm{sym}}\varphi(x)$ is infinite, then for any bounded set E, Borel measurable, we have*

$$\varphi(E) = \varphi(E \cap E_\infty) + \int_E D_{\mathrm{sym}}\varphi(x)\, d\tau.$$

(For the proof see S. Saks [2], theorem 14.5, p. 151).

If in particular $\varphi(E) = mf(E)$, then $\varphi(E \cup E_\infty) > 0$, hence

$$\varphi(E) \geqq \int_E D_{\mathrm{sym}}\varphi(x)\, d\tau. \tag{26}$$

LEMMA 3. *Let $x^* = f(x)$ be a homeomorphism of a domain D such that $\overline{\delta}(x) < \infty$ a.e. in D. Then f(x) is differentiable a.e. in D.*

For each Borel set $E \subset D$, $\varphi(E) = mf(E)$ is a non-negative additive function of a set and possesses, by the Lebesgue theorem, a.e. in E a symmetrical derivative $D_{\mathrm{sym}}\varphi(x) < \infty$. Let x_0 be a point where $D_{\mathrm{sym}}\varphi(x_0)$, $\overline{\delta}(x_0) < \infty$. Then, clearly, $J_G(x_0) = D_{\mathrm{sym}}\varphi(x_0) < \infty$, and

$$\Lambda_f(x_0)^n = \overline{\lim_{\Delta x \to 0}} \left[\frac{|f(x_0 + \Delta x) - f(x_0)|}{|\Delta x|} \right]^n = \overline{\lim_{r \to 0}} \left[\frac{L(x_0, r)}{r} \right]^n \leqq$$

$$\overline{\lim_{r \to 0}} \frac{\omega_n L(x_0, r)^n}{T(x_0, r)} \overline{\lim_{r \to 0}} \frac{T(x_0, r)}{\omega_n r^n} = \overline{\delta}(x_0) J_G(x_0) < \infty, \tag{27}$$

where $r = |\Delta x|$. Hence

$$\Lambda_{x^{*i}}(x_0) \leqq \Lambda_f(x_0) < \infty \quad (i = 1, \cdots, n).$$

But the preceding inequality holds a.e. in D so that, by Rademacher–Stepanov's theorem (Chapter 4 of part I), it follows that $x^{*i}(x)$ $(i = 1, \ldots, n)$ are differentiable a.e. in D separately and then also simultaneously, which clearly implies that $f(x)$ is differentiable a.e. in D, as desired.

LEMMA 4. *Suppose that F is a compact set in the x^n-axis. For each $\varepsilon > 0$ we can find a $\delta > 0$ such that for each r, $0 < r < \delta$, there exist points $x_1, \ldots, x_N \in F$ such that*

$$F \cap \bigcup_{n=1}^{N} B(x_m, r), \tag{28}$$

$$|x_i - x_k| \geqq |i - k| r \quad (i, k = 1, \ldots, N), \tag{29}$$

$$Nr < m_1 F + \varepsilon. \tag{30}$$

For each $r > 0$ let $F(r) = \{x;\ x \in Ox^n,\ d(F, x) < r\}$. Then, because F is compact,

$$F = \bigcap_{r>0} F(r), \quad m_1 F = \lim_{r \to 0} m_1 F(r)$$

(see C. Carathéodory [3], theorem 9, p. 254), hence we can pick a $\rho > 0$ so that

$$m_1 F(r) < m_1 F + \varepsilon \tag{31}$$

for $0<r<\rho$. Now fix such an r and let $x_1 = \inf_{x \in F} x$; $x_1 \in F$ because F is compact. Clearly x_1 is the most left point of F. Next let x_2 be the most left point of the compact set

$$F_1 = F - B(x_1, r).$$

Since $x_1 < x_2$, and since $B(x_1, r)$ contains all points whose distance from x_1 is less than r, it follows that $x_1 + r \leqq x_2$. Continue in this way letting x_3 be the most left point of the compact set

$$F_2 = F_1 - B(x_2, r) = F - [B(x_1, r) \cup B(x_2, r)],$$

and terminate the process when

$$F_N = F - \bigcup_{m=1}^{N} B(x_m, r) = \emptyset.$$

This must happen in a finite number of steps because F is bounded, and we obtain (28) and (29). For (30) let I_m denote the linear interval $(x_m,\ x_m + r)$ for

$m = 1, \ldots, N$. By (29) these intervals are disjointed and we conclude, by (31), that $Nr = m_1 \bigcup_{m=1}^{N} I_m < m_1 F(r) < m_1 F + \varepsilon$, as desired.

We recall that a set which is the union of a countable set of closed sets is said to be an *F_σ-Borel set*, and a set which is the intersection of a countable set of open sets is said to be a *G_δ-Borel set*. Every open set, and generally the complement of a G_δ-Borel set, is an F_σ-Borel set; similarly, every closed set, and in general the complement of an F_σ-Borel set, is a G_δ-Borel set.

LEMMA 5. *Let $x^* = f(x)$ be a homeomorphism of a domain D such that $\bar{\delta}(x)$ is bounded. Then $f(x)$ is ACL in D.*

Let be $\bar{I} \subset D$, and let I_n be its base. For each Borel set $E_0 \subset I_n$, let be $Z_{E_0} = \{x;\ x = y + \xi e_n,\ y \in E_0,\ \alpha^n < \xi < \beta^n\}$. The function

$$\varphi(E_0) = mf(Z_{E_0}) \tag{32}$$

is a non-negative additive function of a set and possesses (by the Lebesgue theorem quoted above) a symmetrical derivative $D_{\text{sym}}\varphi(y) < \infty$ for almost every $y \in I_n$. Let y_0 be a point of I_n where $D_{\text{sym}}\varphi(y_0) < \infty$. It suffices to show that $f(x)$ is AC on the segment Z_{y_0} in order to obtain the same result for almost all the segments parallel to each of the coordinate axes.

Suppose $\bar{\delta}(x) \leqq K_1$. Let us show that for every F_σ-Borel set $E \subset Z_{y_0}$, the following inequality holds

$$\Lambda(E^*)^n \leqq \frac{2^{n+1} K_1 \omega_{n-1}}{\omega_n} D_{sym}\varphi(y_0)(m_1 E)^{n-1}, \tag{33}$$

where $\Lambda(E^*)$ is the length of the set E^* (Chapter 1.3). Since E is the union of a countable sequence of compact sets $F_m \subset Z_{y_0}$, $F_1 \subset F_2 \subset \ldots$, then

$$\lim_{m\to\infty} F_m = E, \quad \lim_{m\to\infty} F_m^* = E^*,$$

implies (see C. Carathéodory [3], theorem 8, p. 254):

$$\lim_{m\to\infty} m_1 F_m = m_1 E, \quad \lim_{m\to\infty} \Lambda(F_m^*) = \Lambda(E^*).$$

In other words, without restricting the generality, we can assume even E is compact. For the sake of simplicity, let us set $y_0 = 0$.

Since $f(x)$ is continuous on the compact set E, we can choose a $\rho_1 > 0$ such that $B(x, r) \subset I$, $L(x, r) < \frac{1}{2}\varepsilon$ whenever $x \in E$, $0 < r < \rho_1$. Then, by the preceding lemma applied to the compact set E, we can find a $\rho_2 (0 < \rho_2 \leqq \rho_1)$ such that for every $r \in (0, \rho_2)$ there exist points $x_1, \ldots, x_N \in E$ which verify the conditions

(28), (29), (30) of the preceding lemma. Fix such an r and $B_m = B(x_m, r)$, $L_m = L(x_m, r)$, $l_m = l(x_m, r)$. Every $B_m^* = f(B_m)$ is contained in a ball with a diameter $2L_m < \varepsilon$ and since these balls cover E^*, the definition of the length implies

$$\Lambda^\varepsilon(E) \leqq \sum_{m=1}^{N} 2L_m.$$

and hence by Hölder's inequality

$$\Lambda^\varepsilon(E^*)^n \leqq 2^n\left(\sum_{m=1}^{N} L_m\right)^n = 2^n\left(\sum_{m=1}^{N} \frac{L_m}{T_m^{\frac{1}{n}}} T_m^{\frac{1}{n}}\right)^n \leqq \frac{2^n}{\omega_n}\sum_{m=1}^{N} \frac{\omega_n L_m^n}{T_m}\left(\sum_{m=1}^{N} T_m^{\frac{1}{n-1}}\right)^{n-1} \leqslant$$

$$\leqslant \frac{2^n}{\omega_n}\sum_{m=1}^{N} \frac{\omega_n L_m^n}{T_m}\sum_{m=1}^{N} T_m N^{n-2}. \tag{34}$$

The balls B_m lie in the cylinder $Z_r = \{x;\ x = y + \xi e_n,\ |y| < r,\ \alpha^n < \xi < \beta^n\}$, and by (29) they cover points of this cylinder at most twice. $\{B_m^*\}$ therefore cover points of the image of this cylinder at most twice and hence, on account of (32),

$$\sum_{m=1}^{N} T_m = \sum_{m=1}^{N} mB_m^* \leqq 2\varphi(|y| < r).$$

Combining this with (34) we obtain

$$\Lambda^\varepsilon(E^*)^n \leqq \frac{2^{n+1}}{\omega_n}\left[\max_m\left\{\frac{\omega_n L_m^n}{T_m}\right\}\right]\varphi(|y| < r)N^{n-1} =$$

$$= 2^{n+1}\frac{\omega_{n-1}}{\omega_n}\left[\max_m\left\{\frac{\omega_n L_m^n}{T_m}\right\}\right]\frac{\varphi(|y| < r)}{\omega_{n-1}r^{n-1}}(Nr)^{n-1},$$

hence, by (30),

$$\Lambda^\varepsilon(E^*)^n \leqslant 2^{n+1}\frac{\omega_{n-1}}{\omega_n}\left[\max_m\left\{\frac{\omega_n L_m^n}{T_m}\right\}\right]\frac{\varphi(|y| < r)}{\omega_{n-1}r^{n-1}}(m_1 E + \varepsilon)^{n-1}$$

for $0 < r < \rho_2$, and letting $r \to 0$ we get

$$\Lambda^\varepsilon(E^*)^n \leqq 2^{n+1}\frac{\omega_{n-1}}{\omega_n}\sup_{x\in D}\bar{\delta}(x)D_{sym}\varphi(O)(m_1 E + \varepsilon)^{n-1}.$$

Since this inequality holds for all $\varepsilon > 0$ we conclude that

$$\Lambda(E^*)^n \leqq 2^{n+1}\frac{\omega_{n-1}}{\omega_n}\sup_{x\in D}\bar{\delta}(x)D_{\text{sim}}\,\varphi(O)(m_1E)^{n-1},$$

which implies (33) with $y_0 = 0$. Then choosing E to be a sequence of disjoint intervals $[y_0 + a_m e_n, y_0 + b_m e_n] \subset Z_{y_0}$, it follows

$$[\sum_n |f(y_0 + b_m e_n) - f(y_0 + a_m e_n)|]^n \leqq \frac{2^{n+1}K_1\omega_{n-1}}{\omega_n} D_{\text{sim}}\,\varphi(y_0)\,[\Sigma(b_m - a_m)]^{n-1}, \tag{35}$$

hence $f(x)$ is AC on Z_{y_0} and then ACL in D.

LEMMA 6. *Let $x^* = f(x)$ be a homeomorphism in D. Then the following conditions are equivalent*:

(i) *For each curve family Γ in D*

$$M(\Gamma^*) \geqq \frac{1}{K}M(\Gamma). \tag{36}$$

(ii) *$f(x)$ is ACL in D, differentiable a.e. in D and*

$$\max|f'(x)|^n \leqq K|J(x)|. \tag{37}$$

a.e. in D.

Suppose first that (i) holds. From (3), (4) and taking into account that $\dfrac{\omega_n l(x, r)^n}{T(x, r)} \leqq 1$, it then follows that

$$\bar{\delta}(x) = \overline{\lim_{r\to 0}}\frac{\omega_n L(x, r)^n}{T(x, r)} = \overline{\lim_{r\to 0}}\left\{\left[\frac{L(x, r)}{l(x, r)}\right]^n \frac{\omega_n l(x, r)^n}{T(x, r)}\right\} \leqq \delta_L(x)^n \tag{38}$$

for all $x \in D$. Hence by lemma 2 and on account of (1.5.7),

$$\bar{\delta}(x) \leqq \delta_L(x)^n \leqq e^{n(\chi K)^{\frac{1}{n-1}}} \tag{39}$$

for all $x \in D$, so that, by lemmas 3 and 5, $f(x)$ is ACL in D and differentiable a.e. in D, and from lemma 1, since $M(Q) = M(\Gamma_Q)$, where Γ_Q is the curve family joining two opposite faces of the cube, we get (37) at every point of differentiability.

To prove that (ii) implies (i), we first show that the partial derivatives of $f(x)$ are L^n-integrable over each compact subset $F \subset D$, i.e. $f(x)$ is ACL_n in D. From the first formula (10) we obtain

$$\left|\frac{\partial f(x)}{\partial x^i}\right| \leqq \max |f'(x)|$$

hence, by (37),

$$\int_F \left|\frac{\partial f(x)}{\partial x^i}\right|^n d\tau \leqq K \int_F |J(x)|\, d\tau.$$

At every point x where $f(x)$ is differentiable,

$$|J(x)| = J_G(x) = D_{\mathrm{sym}}\varphi(x), \tag{40}$$

where $\varphi(E) = mf(E)$ is defined over the set of Borel sets $E \subset D$ (see T. Rado and P. V. Reichelderfer [1], p. 334 or O. Lehto and K. I. Virtanen [2], lemma 3.2, p. 136). By (26), the preceding inequality implies

$$\int_F \left|\frac{\partial f}{\partial x^i}\right|^n d\tau \leqq K \int_F D_{\mathrm{sim}}\varphi(x)\, d\tau \leqslant K\varphi(F) = Kmf(F) < \infty. \tag{41}$$

Let then Γ be an arbitrary family of curves in D and let Γ_0 be the subfamily of Γ consisting of all rectifiable curves $\gamma \in \Gamma$ such that $f(x)$ is AC along each compact subcurve of γ. Since $f(x)$ is ACL_n, Fuglede's theorem (Chapter 1.6) implies $M(\Gamma_0) = M(\Gamma)$. On the other hand, $\Gamma_0^* \subset \Gamma^*$ implies $M(\Gamma_0^*) \leqq M(\Gamma^*)$. Hence it suffices to prove that (36) holds for Γ_0.

Let E be the subset of D where $f(x)$ is not differentiable. By assumption $m(E) = 0$. Let E_0 be a Borel set such that $E \subset E_0 \subset D$ and $mE_0 = 0$ (such a set exists, see Natanson [1], p. 118).

Let $\rho^* \in F(\Gamma_0^*)$. Define the corresponding function $\rho(x)$ in D

$$\rho(x) = \begin{cases} \rho^*[f(x)] \max |f'(x)| & \text{if } x \in D - E_0, \\ \infty & \text{if } \quad x \in E_0. \end{cases} \tag{42}$$

We show that $\rho \in F_D(\Gamma_0)$. First, $\rho(x) \geqq 0$. Next, since the partial derivatives of an ACL function are Borel measurable (by the proposition 1.6.7), we see that $\rho(x)$ is Borel measurable. Now let $\gamma \in \Gamma_0$ and let γ_0 be a compact subcurve of γ. Since $f(x)$ is AC along γ_0, γ_0^* is rectifiable by proposition 1.6.2 (S. Saks [1], theorem 8, p. 58). We represent γ_0 and γ_0^* by means of their arc-lengths s, t:

$$\gamma_0 : x = x(s)\,(0 \leqq s \leqq l), \quad \gamma_0^* : x^* = x^*(t)\,(0 \leqq t \leqq l^*)$$

and choose the orientation so that t is an increasing function of s. Since $\varphi(s) = f[x(s)]$ is AC on $(0, l)$, $t(s)$ is also AC in (0, 1) and $|\varphi'(s)| = \frac{dt}{ds}$ (proposition 1.6.6).

If $\varphi(s)$ is differentiable at s_0 and $f(x)$ is differentiable at $x(s_0)$, then

$$|\varphi'(s_0)| = \lim_{s\to s_0} \frac{|f[x(s)] - f[x(s_0)]|}{|s - s_0|} \leqq \lim_{s\to s_0} \frac{|f[x(s)] - f[x(s_0)]|}{|x(s) - x(s_0)|} \leqq \max |f'[x(s_0)]|.$$

Thus, by proposition 1.6.1 and 1.6.2,

$$\rho[x(s)] \geqq \rho^*\{f[x(s)]\} \frac{dt}{ds}$$

or almost every s in (0, 1). Since $t(s)$ is AC, this implies,

$$\int_{\gamma} \rho \, ds \geqq \int_{\gamma_0} \rho \, ds \geqq \int_0^l \rho^* \frac{dt}{ds} ds = \int_0^{l^*} \rho^* \, dt = \int_{\gamma_0^*} \rho^* \, dt,$$

by (1.3.9) and (1.6.4), hence

$$\int_{\gamma_0} \rho \, ds \geqq \sup_{\gamma_0^* \subset \gamma^*} \int_{\gamma_0^*} \rho^* \, dt = \int_{\gamma^*} \rho^* \, dt \geqq 1.$$

Consequently, $\rho \in F_D(\Gamma_0)$.

We estimate $M(\Gamma_0)$ by means of ρ. By virtue of (37) and (42) we obtain

$$M(\Gamma_0) \leqq \int_D \rho^n \, d\tau \leqq K \int_D \rho^*[f(x)]^n |J(x)| \, d\tau.$$

Since (40) holds a.e. in D, we obtain

$$M(\Gamma_0) \leqq K \int_D \rho^*[f(x)]^n |J(x)| \, d\tau \leqq K \int_{D^*} \rho^{*n} \, d\tau, \tag{43}$$

by (1.3.8) and the decomposition theorem of de la Vallée-Poussin (quoted above). Since (43) holds for each $\rho^* \in F(\Gamma_0^*)$, it follows

$$M(\Gamma_0) \leqq KM(\Gamma_0^*),$$

which implies (36), as desired.

Corollary. *A homeomorphism $x^* = f(x)$ of a domain D is K-QCfH in D according to the geometric definition if and only if both $x^* = f(x)$ and its inverse $x = f^{-1}(x^*)$ verify one of the conditions* (i), (ii) *of the preceding lemma.*

LEMMA 7. *If $f(x)$ is a homeomorphism of a domain D, then $\bar{\delta}(x)$ is bounded in D if and only if one of the conditions* (i), (ii) *of the preceding lemma holds for some K.*

The sufficiency of the condition (i) is an immediate consequence of lemma 2 and of inequalities (38), while the sufficiency of the condition (ii) follows from its equivalence with (i).

For the necessity of (i), (ii), let be $\bar{\delta}(x) \leqq K$ in D. The lemmas 3 and 5 then imply that $f(x)$ is ACL in D and differentiable a.e. in D and in a point of differentiability the preceding inequality becomes of the form $\max |f'(x)|^n \leqq K|J(x)|$ by formula (13). These allow us to conclude that the condition (ii), and also (i) (equivalent with it), are verified.

COROLLARY. *In order that a homeomorphism $x^* = f(x)$ of a domain D be* QCfH *in D according to the geometric definition it is necessary and sufficient that is outer dilatation $\delta(\bar{x})$ and that of its inverse $x = f^{-1}(x^*)$ be bounded in D.*

THEOREM 1. *A QCfH according to the geometric definition maps sets of measure zero into sets of measure zero.*

Suppose $f(x)$ is QCfH in D and let F be a closed subset of D such that $mF = 0$. Let Γ be the family of all the curves $\gamma \subset D$ such that either $\Lambda(\gamma \cap F) > 0$ or $f(x)$ is not AC along every compact subcurve of γ. By proposition 1.5.5 and the theorem of Fuglede -- which is possible to apply because a QCfH is ACL_n in view of the formula (41) — it follows that Γ is exceptional. Then Γ^* is exceptional too by the quasiconformality. Since $f(x)$ is AC along compact subcurves of all curves not in Γ, $\Lambda(\gamma^* \cap F^*) = 0$ for all $\gamma^* \in \Gamma^*$. Thus $\Lambda(\gamma^* \cap F^*) = 0$ for almost every curve $\gamma^* \subset D^*$. In particular, $m_1(J^* \cap F^*) = 0$ for almost every line segment J^* parallel to one of the coordinate axes. By Tonelli's theorem $mF^*=0$. Thus a QCfH maps every closed set of measure zero onto a set of measure zero.

Let us show now, more generally, that the same is true for every set G of the class of G_δ-Borel sets (i.e. of the form $G = \bigcap_k D_k$, where the sets D_k are open) and such that $mG = 0$. Indeed, a set G of the class G_δ is mapped by a homeomorphism again into a set of the class G_δ. Suppose, to prove it is false, that $mG^* = \alpha > 0$. G^* as a set of the class G_δ is Borel measurable, hence, by a criterion of de la Vallée-Poussin (Natanson [1], p. 117), given any $\varepsilon > 0$ there exists a closed set $F^* \subset G^*$ such that $m(G^* - F^*) < \varepsilon$. If we choose $\varepsilon < \frac{\alpha}{2}$, the corresponding inequality yields $mF^* < \frac{\alpha}{2}$. On the other hand, the inverse image of F^* is a closed set $F \subset G$, hence $mF = 0$, and from the above $mF^* = 0$, which contradicts the inequality $mF^* > \frac{\alpha}{2}$, and then the hypothesis $mG^* > 0$. Thus $mG^*=0$.

To complete the proof of the theorem, we show that a QCfH maps every set of measure zero again onto a set of measure zero. But there exists a set G of the class G_δ, such that $E \subset G$, $mG = 0$ (see Natanson [1], p. 118), hence, since the theorem is true for the sets of the class G_δ, we obtain $mG^* = 0$, hence, since $E^* \subset G^*$, it follows $mE^* = 0$, as desired.

Remark. A homeomorphism which maps sets of measure zero into sets of measure zero, maps also measurable sets onto measurable sets and is called *measurable.* The corresponding function of a set $\varphi(E) = mf(E)$ is clearly *AC* (Chapter 1.3).

COROLLARY. *Let* $x^* = f(x)$ *be a* QCfH (*according to the geometric definition*) *of a domain D. If E is a measurable subset of D then*

$$mE^* = \int_E |J(x)| \, d\tau. \tag{44}$$

Moreover $J(x) \neq 0$ *a.e. in D.*

The first part is an immediate consequence of the decomposition theorem of de la Vallée-Poussin, the preceding theorem and formula (40).

For the second part, let us observe that, by proposition 1.6.7, $x_k^{*i}(x)$ are Borel measurable, hence also $J(x)$ is Borel measurable, and clearly the set $E_1 = \{x;\ x \in D,\ J(x) = 0\}$ is Borel measurable too. Then, we can apply the first part of the theorem, to obtain

$$mE_1^* = m\left[f(E)\right] = \int_{E_1} |J(x)| \, d\tau = 0.$$

But applying the preceding theorem for the inverse (which is also QCfH), we obtain $mE_1 = 0$, as desired.

THEOREM 2. *Let* $f(x)$ *be a homeomorphism which satisfies one of the conditions* (i), (ii) *of lemma* 6 *or has the dilatation* $\bar{\delta}(x)$ *bounded in D. Then* $f(x)$ *is QCfH in D according to the geometric definition.*

Suppose $\bar{\delta}(x)$ is bounded in D. We prove that $f(x)$ is QCfH in D by showing that the outer dilatation $\delta^*(x^*)$ of the inverse mapping is bounded in D^*. Let $x_0^* = f(x_0) \in D^*$ and choose $r > 0$ such that $\overline{B[x_0, L^*(x_0^*, r)]} \subset D$, where we denote by $L^*(x^*, r)$ and $l^*(x^*, r)$ the quantities (5) and (6) for $f^{-1}(x^*)$. Assuming $l^*(x_0^*, r) < L^*(x_0^*, r)$, we denote by A the spherical ring centered at x_0 and with radii $l^*(x_0^*, r)$, $L^*(x_0^*, r)$. Let x_1^* be the point on the sphere $|x^* - x_0^*| = r$, where $|x - x_0|$ attains its minimum value $l^*(x_0^*, r)$.

Furthermore, let C_1^* and C_2^* be the half-lines

$$C_1^* = \{x^*;\ x^* = x_0^* + \alpha(x_1^* - x_0^*),\ \alpha \geqq 1\},$$
$$C_2^* = \{x^*;\ x^* = x_0^* + \alpha(x_0^* - x_1^*),\ \alpha \geqq 0\}$$

and put $F_1^* = A^* \cap C_1^*$, $F_2^* = A^* \cap C_2^*$. Consider the family Γ^* of all arcs which join F_1^* and F_2^* in A^*. Since $d(F_1^*, F_2^*) > r = l[x_0, L^*(x_0^*, r)]$, we have

$$M(\Gamma^*) \leqq \frac{mA^*}{r^n} \leqq \frac{T[x_0, L^*(x_0^*, r)]}{l[x_0, L^*(x_0^*, r)]^n}.$$

Indeed, $\rho^*(x^*) = \dfrac{1}{r} \in F(\Gamma^*)$, hence

$$M(\Gamma^*) \leqq \int_A \frac{d\tau}{r^n} = \frac{mA^*}{r^n}.$$

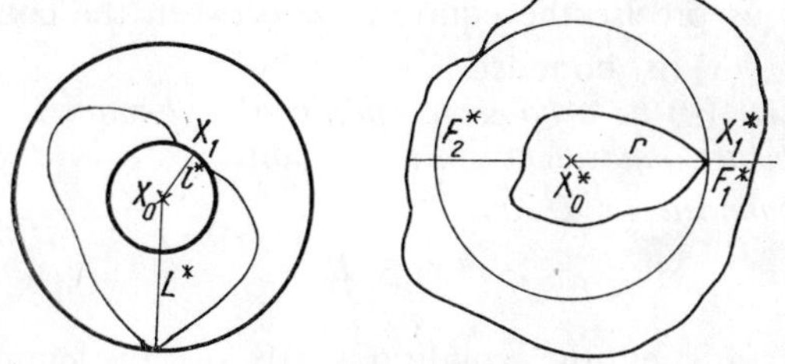

Fig. 2

On the other hand, by theorem 1.5.11 and taking into account that, by the preceding lemma, since $\bar{\delta}(x)$ is bounded, it follows the inequality (36) for a certain K and that Lebesgue measure of the image of a spherical ring is at most equal to the measure of the image of the ball with the radius equal to that of the outer sphere which bounds the ring, we obtain

$$\log \frac{L^*(x_0^*, r)}{l^*(x_0^*, r)} \leqq \frac{A_0}{2^n} M(\Gamma) \leqq \frac{A_0 K}{2^n} M(\Gamma^*) \leqq \frac{A_0 K T[x_0, L^*(x_0^*, r)]}{2^n l[x_0, L^*(x_0^*, r)]^n}.$$

Obviously, this holds also if $l^*(x_0^*, r) = L^*(x_0^*, r)$. As $r \to 0$, it implies

$$\delta_L^*(x_0^*) \leqq e^{\omega_n \frac{A_0 K}{2^n} \underline{\delta}(x_0)}.$$

Since $f(x)$ satisfies one of the two equivalent conditions (i), (ii), it follows that $\underline{\delta}(x)$ is bounded in D, because it suffices to choose the family Γ^* in (36) such that $\Gamma^* = \Gamma_A^*{}^*$, where A^* is an arbitrary spherical ring in D^*, and then, by lemma 2 and on account of (2), (4) and $\dfrac{T(x, r)}{\omega_n\, L(x, r)^n} \leqq 1$, we obtain

$$\underline{\delta}(x) = \overline{\lim_{x \to 0}} \frac{T(x, r)}{\omega_n l(x, r)^n} = \overline{\lim_{r \to 0}} \left\{ \left[\frac{L(x, r)}{l(x, r)} \right]^n \frac{T(x, r)}{\omega_n L(x, r)^n} \right\} \leqq \delta_L(x)^n \leqq e^{n(K\chi)^{\frac{1}{n-1}}}, \tag{45}$$

hence it follows that also $\delta_L^*(x^*)$ is bounded in D^*. But the inequalities (38) hold also for $f^{-1}(x^*)$, i.e.

$$\bar{\delta}^*(x^*) \leqq \delta_L^*(x^*)^n$$

and then $\bar{\delta}^*(x^*)$ is also bounded in D^*, which implies, by virtue of the corollary of lemma 7, that $f(x)$ is QCfH (according to the geometric definition) in D, as desired.

Since the conditions (i), (ii) are equivalent to the condition that $\bar{\delta}(x)$ be bounded in D, it follows that the conclusion of the theorem still holds if we suppose that one of the conditions (i), (ii) is satisfied.

And now, let us precise the equivalence between the conditions (i), (ii) and the condition that $\bar{\delta}(x)$ is bounded.

LEMMA 8. *Let $f(x)$ be a homeomorphism of a domain D. Then the conditions* (i), (ii) *of lemma 6 are equivalent to the condition*

(iii) $\bar{\delta}(x)$ *is bounded in D and*

$$\bar{\delta}(x) \leqq K \tag{46}$$

a.e. in D.

Suppose $f(x)$ satisfies the conditions (i), (ii) of lemma 6. By lemma 7, $\delta(x)$ is bounded in D, and by lemma 3, $f(x)$ is differentiable a.e. in D. By the preceding theorem $f(x)$ is QCfH, and hence, by the corollary of theorem 1, $J(x) = 0$ a.e. in D. But at a point of differentiability with $J(x) \neq 0$ hold (37) and (13) which yield (46) a.e. in D.

Conversely, if $\bar{\delta}(x)$ is bounded in D, then by lemmas 3,5 and corollary of theorem 1, $f(x)$ is differentiable with $J(x) \neq 0$ a.e. in D and ACL in D. Let x_0 be a point of differentiability with $J(x_0) \neq 0$, where $\bar{\delta}(x) \leqq K$. Then (13) implies (37).

THEOREM 3. *Geometric and analytic definitions of K-QCfH are equivalent.*

First, let $f(x)$ be K-QCfH according to the geometric definition in D. By lemma 6, it is ACL, a.e. differentiable and the left-hand inequality of (19) holds a.e. Moreover, by the corollary of theorem 1, $J(x) \neq 0$ a.e. in D. At each point x_0 where $f(x)$ is differentiable and $J(x_0) \neq 0$, by (12) and (13) we obtain

$$\underline{\delta}(x_0) = \frac{|J(x_0)|}{\min|f'(x_0)|^n} = \frac{\max|f^{-1'}(x_0^*)|^n}{|J(x_0^*)|} = \bar{\delta}^*(x_0^*). \tag{47}$$

Indeed, in such a point the maximal dilatation is proportional to the semi-axis a_1 and the minimal dilatation to the semi-axis a_n of the corresponding infinitesimal ellipsoid, as clearly follows from (8). Obviously, at $x_0^* = f(x_0)$ the maximal dilatation of the inverse homeomorphism $f^{-1}(x)$ is proportional to $\frac{1}{a_n}$ and the minimal to $\frac{1}{a_1}$. (11) implies that $|J(x_0)|$ is proportional to the product $a_1 \dots a_n$ and since $J(x_0)\, J^*(x_0^*) = 1$ it follows that $|J^*(x_0^*)|$ is proportional to $\frac{1}{a_1 \dots a_n}$. Thus, (11), (12), (8) and (13) yield

$$\bar{\delta}^*(x_0^*) = \frac{\max|f^{-1}(x_0^*)|^n}{|J^*(x_0^*)|} = \frac{(a_1^*)^n}{a_1^* \cdots a_n^*} = \frac{a_1 \cdots a_n}{(a_n)^n} = \underline{\delta}(x_0).$$

But, by (47) and lemma 6 for $f^{-1}(x)$, the inequality

$$M(\Gamma^*) \leqq KM(\Gamma), \tag{48}$$

which can be written also

$$M(\Gamma) \geqq \frac{1}{K} M(\Gamma^*), \tag{49}$$

implies that

$$|J(x)| \leqq K \min |f'(x)|^n, \tag{50}$$

which holds a.e. in D on account of theorem 1 and of the fact that the inverse of a QCfH in D according to the geometric definition is again a QCfH in D according to the geometric definiton.

Conversely, let $f(x)$ be K-QCfH according to the analytic definition. Then it satisfies the condition (ii) of lemma 6, and hence also the condition (i) and on account of theorem 2, $f^{-1}(x^*)$ verifies (49) (in general with another K), so that by lemma 8, $\bar{\delta}^*(x^*)$ is bounded in D^*. On the other hand, inequality (50), which holds a.e. in D, implies, on account of (47) and theorem 1, the inequality

$$\bar{\delta}^*(x^*) \leqq K \tag{51}$$

a.e. in D^*. Hence, by the preceding lemma, it follows that $f^{-1}(x^*)$ satisfies the condition (i), which may be written also under the form (48), which together with (36) characterize the K-QCfH according to the geometric definition.

THEOREM 4. *Geometric and metric definitions of K-QCfH are equivalent.*

First, let $f(x)$ be K-QCfH according to the geometric definiton. By the preceding lemma applied to $f(x)$ and $f^{-1}(x)$, condition (i) implies condition (iii), hence by (45), (47) and theorem 1, it follows the boundedness of $\bar{\delta}(x)$, and $\underline{\delta}(x)$ in D and $\bar{\delta}(x)$, $\underline{\delta}(x) \leqq K$ a.e. in D, i.e. $f(x)$ is K-QCfH in D according to the metric definition.

Conversely, let $f(x)$ be K-QCfH according to the metric definition. Then, by lemma 8, it satisfies the condition (ii) of lemma 6. Moreover, (12) implies (50) a.e. in D, which together with (ii) characterizes the K-QCfH according to the analytic definition. Hence, the equivalence between the geometric and analytic definition, allows us to conclude that $f(x)$ is K-QCfH in D according to the geometric definition, as desired.

Second analytic definition. The analytic definition of QCfH can be formulated in a more general way by renouncing the hypothesis of the differentiability a.e. in D, and at the same time changing the condition ACL by the condition ACL_n (Väisälä [3]). We begin with some notations.

Suppose $f(x)$ possesses the first order partial derivatives at a point x. For every such x we define a linear mapping $Df(x)$: $R^n \to R^m$ by

$$Df(x)e_k = \frac{\partial f(x)}{\partial x^k} \quad (k = 1, \ldots, n),$$

where $Df(x)\, e_k$ is the product between the tensor $Df(x)$ [with a contravariant index i $(i = 1, \ldots, n)$ and a covariant index $k (k = 1, \ldots, m)$] and the unit vector e_k. Furthermore, we set

$$L(x,f) = \max_{|\Delta x|=1} |Df(x)\Delta x|, \quad l(x,f) = \min_{|\Delta x|=1} |Df(x)\Delta x|.$$

Obviously, $Df(x)$, $L(x,f)$, $l(x,f)$ and $J(x)$ exist a.e. in a domain where $f(x)$ is ACL. At a point of differentiability $Df(x)$ is the derivative $f'(x)$ of f at x.

A homeomorphism $f(x)$ is said to be K-QCfH $(1 \leqq K < \infty)$ in D if it is ACL, a.e. differentiable and satisfies the double inequality

$$\frac{L(x,f)^n}{K} \leqq |J(x)| \leqq Kl(x,f)^n \tag{52}$$

a.e. in D.

At a point of differentiability of $f(x)$ the preceding inequality reduces to (19).

Equivalence of analytic definitions. In order to establish that this analytic definition is equivalent to the preceding one, let us first introduce some concepts and to prove some lemmas.

Suppose that $f(x)$ is a mapping of an open set $D \subset R^n$ into R^m. We say that $f(x)$ is *partially differentiable at $x \in D$ with respect to* $R_i^{n-1} = \{x; x^i = 0\}$ if there exists a linear mapping $D_i f(x)$: $R_i^{n-1} \to R^m$ such that for $\Delta x \in R_i^{n-1}$, we have

$$f(x + \Delta x) = f(x) + D_i f(x)\Delta x + |\Delta x|\varepsilon(x, \Delta x),$$

where $\varepsilon(x, \Delta x) \to 0$ as $\Delta x \to 0$. The mapping $D_i f(x)$ is called *the partial derivative of $f(x)$ with respect to* R_i^{n-1}. If $D_i f(x)$ exists, then the ordinary partial derivatives $\frac{\partial f(x)}{\partial x^k}$ exist for $k \neq i$ and satisfy the equality

$$\frac{\partial f(x)}{\partial x^k} = D_i f(x) e_k. \tag{53}$$

If, in addition, $\frac{\partial f(x)}{\partial x^i}$ exists, then

$$Df(x)(k + be_i) = D_i f(x)k + \frac{\partial f(x)}{\partial x^i}\, b \tag{54}$$

for all $k \in R_i^{n-1}$ (see F. and R. Nevanlinna [1], p. 95, or J. Dieudonné [1], p. 167).

Given a measurable set $E \subset R$ and a point $x \in E$, if

$$\lim_{r \to 0} \frac{m[B(x, r) \cap E]}{\omega_n r^n} = 1,$$

then x is said to be *a point of density of E*. Let us denote by $(E; x_0^1, \ldots, x_0^{i-1}, x_0^{i+1}, \ldots, x_0^n)$ the intersection of E with the line $x^k = x_0^k$ $(k \neq i; k = 1, \ldots, n)$; if

$$\lim_{r \to 0} \frac{m\{[B(x_0, r) \cap E]; x_0^1, \ldots, x_0^{i-1}, x_0^{i+1}, \ldots, x_0^n\}}{2r} = 1 \qquad (i = 1, \cdots, n),$$

then x is termed *a point of linear density in the direction of the coordinate axes.*

LEMMA 9. *If $f(x)$ is an open continuous mapping of a domain $D \subset R^n$ into R^n such that the $(n-1)$-dimensional partial derivatives $D_i f(x)$ exist a.e. in D, then $f(x)$ is differentiable a.e. in D.*

We may assume that D is bounded. And now, by Egorov's theorem (see for instance S. Saks [1], theorem 24, p. 43), given a sequence of functions finite a.e. in D and measurable

$$g_k(x) = \sum_{i=1}^{n} \sup_{\substack{0 < |\Delta x| < \frac{1}{k} \\ \Delta x \in R_i^{n-1}}} \frac{|f(x + \Delta x) - f(x) - D_i f(x) \Delta x|}{|\Delta x|}$$

and which converges a.e. in D to zero, then for each $\eta > 0$ we can find a closed set $F \subset D$, such that $m(D - F) < \eta$ and such that the sequence $\{g_k(x)\}$ converges to zero uniformly in E. From this it follows that the partial derivatives $D_i f(x)$ are continuous in F. Let x_0 be a point of linear density in the direction of each coordinate axis. Since almost all points of a measurable set possess this property (S. Saks [2], p. 298), it suffices to show that $f(x)$ is differentiable at x_0. For convenience of notation, we put $x_0 = O$.

For each $0 < \varepsilon < 1$ there exists a $\delta > 0$ such that the following conditions are satisfied:

(i) The cube $Q = \{x; |x^i| \leqq \delta \ (i = 1, \ldots, n)\} \subset D$.

(ii) $|D_i f(x) - D_i f(O)| < \varepsilon$ whenever $x \in Q \cap F$ and $1 \leqq i \leqq n$.

(iii) $|g_k(x)| < \varepsilon$ for $x \in E$ and $\dfrac{1}{k} < \delta$.

(iv) If J_0 is a linear segment containing O and lying in Q and in some coordinate axis, then $[m_1(J_0 \cap F)](1 + \varepsilon) > m_1 J_0$.

The first three conditions are satisfied by the definition of F, and the fourth condition is an immediate consequence of the property of the origin to be a point of linear density in the direction of each coordinate axis.

Now let $h \in R^n$ be such that $0 < |h| < \frac{\delta}{2}$. Then we can find a closed interval $I = \{x; a^i \leqq x^i \leqq b^i \ (i = 1, \ldots, n)\}$ such that

$$h^i - \varepsilon|h| \leqq a^i < h^i < b^i \leqq h^i + \varepsilon|h|, \tag{55}$$

and such that the points $a^i e_i$, $b^i e_i \in F$ (i is not a summatory index) for $1 \leqq i \leqq n$.

Indeed, condition (iv) implies that each pair of linear intervals $[(1 - \varepsilon)h^i e_i, h^i e_i)$ and $[h^i e_i, (1 + \varepsilon) h^i e_i]$ $(i = 1, \ldots, n)$ contains at least one point of F, so that we can find for each i $(i = 1, \ldots, n)$ a pair of points $a^i e_i$, $b^i e_i \in F$ such that $a^i < h^i < b^i$ and $b^i - a^i < 2\varepsilon\,|h^i|$.

Since $f(x)$ is open, the function $|f(x) - f(0) - Df(0)|$ attains its maximum in I at some boundary point, say at $y = b^i e_i + k$ (i is not a summatory index), where $k \in R_i^{n-1}$. Thus, by (53), (54), (ii), (iii) and since (55) implies $|k| < n|h|$ and $|y - h| = |b^i e_i + k - h| \leqq |b^i - h^i| + \varepsilon\,|h|\,(n - 1) < \varepsilon\,|h| + (n - 1)\,\varepsilon\,|h| = n\varepsilon\,|h|$, we have

$$\begin{aligned} &|f(h) - f(O) - Df(O)h| \leqq |f(y) - f(O) - Df(O)h| \leqq |f(y) - f(b^i e_i) - \\ &- D_i f(b^i e_i)k| + \left|f(b^i e_i) - f(O) - \frac{\partial f(O)}{\partial x} b^i\right| + |D_i f(b^i e_i)k - D_i f(O)k| + \\ &+ |Df(O)y - Df(O)h| \leqq 2n\varepsilon|h| + 2\varepsilon|h| + n|Df(O)|\varepsilon|h| < \\ &< \frac{\delta}{2}[2(n + 1) + n|Df(O)|]\varepsilon. \end{aligned}$$

Hence $f(x)$ is differentiable at O, and the lemma is proved.

Theorem of Calderon—Cesari. *If $u(x)$ is an ACL_p-function $(p > n)$ in an open set $D \subset R^n$, then $u(x)$ is differentiable a.e. in D.* (L. Cesari [1] for $n = 2$, A. P. Calderon [1] for an arbitrary n.)

LEMMA 10. *Let $f(x)$ be an open mapping of an open set $D \subset R^n$ into R^n. If $f(x)$ is* ACL_p *for some $p > n - 1$, then $f(x)$ is differentiable a.e. in D.*

By the preceding lemma, it suffices to show that the $(n-1)$-partial derivatives $D_i f(x)$ exist a.e. in D.

Since f is ACL_p, it follows from Fubini's theorem that for almost every real number t, the restriction of f to the plane $\Pi_t = \{x; x^i = t\}$ is ACL_p. By the aforementioned result of Cesari—Calderon, the partial derivative $D_i f(x)$ exists in such a $D \cap \Pi_t$ except for a set of $(n-1)$-dimensional measure zero. The lemma will thus follows from Fubini's theorem if we show that the set $E \subset D$ in which $D_i f(x)$ exists in D is measurable. In fact we shall prove that E is a Borel set.

Let Δ be the set of all points x in D at which the ordinary partial derivatives $\frac{\partial f(x)}{\partial x^m}$ exist for $m \neq i$. Then Δ is a Borel set and the functions $\frac{\partial f(x)}{\partial x^i}$ are Borel measurable in Δ (proposition 1.6.7). For each $x \in \Delta$ we define a linear mapping $T(x)\colon R_i^{n-1} \to R^n$ by $T(x)\, e_m = \dfrac{\partial f(x)}{\partial x^m}$, where $T(x)e_m$, is the product of the tensor $T(x)$ [contravariant in $m\, (m = 1, \ldots, n-1)$ and covariant in $k\, (k = 1, \ldots, n)$] and the unit vector e_m. Next for each $x \in \Delta$ and each positive integer $k > 0$, we set

$$g_k(x) = \sup \frac{|f(x+h) - f(x) - T(x)h|}{|h|},$$

where the supremum is taken over all $h \in R_i^{n-1}$ such that $x + h \in D$, the coordinates h^i of h are rational and $0 < |h| < \dfrac{1}{k}$. Then the functions g_k are Borel measurable in Δ, and the same is true for the function $g = \inf g_k = \lim\limits_{k \to \infty} g_k$. Since $E = \{x;\, g(x) = 0\}$, E is a Borel set. Hence, since the subset of D in which $D_i f(x)$ exists, and then also the set $E' \subset D$ in which $D_i f(x)$ does not exist, are measurable, it follows, by Fubini's theorem applied to the characteristic function of the measurable set E', that

$$mE' = \int m_{n-1}(E';\, x^i)\, \mathrm{d}x^i = 0,$$

where we denoted by $(E';\, x^i)$ the intersection between the set E' and the plane $x^i = \text{const}$. Thus $D_i f(x)$ exists a.e. in D, and this allows us to conclude, on account of the preceding lemma, the differentiability of f a.e. in D.

THEOREM 5. *The second analytic definition of the K-QCfH is equivalent to the other three definitions.*

A K-QCfH $f(x)$ according to each of the first three definitions (which are equivalent) is ACL_n, as it follows from (41), and satisfies a.e. in D the double inequality (52), because this inequality reduces, at the points of differentiability, to the double inequality (19). We conclude that $f(x)$ is K-QCfH according to the second analytic definition.

Conversely, let $f(x)$ be a K-QCfH according to the second analytic definition. Since $f(x)$ is ACL_n, it follows, by the preceding lemma, that it is differentiable a.e. in D, which implies that it is K-QCfH according to the first analytic definition (and then also according to the geometric and metric definition).

Remarks. 1. If $n = 2$, it is possible to renounce the condition of the differentiability a.e. in D (without replacing it by the condition ACL_n) by virtue of the following result of F. Gehring and O. Lehto [1]: "Let $w(z)$ be continuous and open and possess finite partial derivatives a.e. in D. Then $w(z)$ is differentiable a.e. in D". But this theorem is no more true if $n > 2$ as it follows from the following Serrin's [1] example. Set $a_k = \dfrac{k}{2^k}$, $\omega(t) = \dfrac{\log \dfrac{t+\varepsilon}{1+\varepsilon}}{\log \dfrac{\varepsilon}{1+\varepsilon}}$, where $\varepsilon = \varepsilon_k$ is such that $\log \varepsilon_k = 2^k$. Let us decompose a cube Q into N_k^n cubes with the edges equal to 2^{-k} and let $\{x_{k_m}\}$ be the centers of these cubes. Next, let us define the function

$$u_k(k) = \begin{cases} \omega\left(\dfrac{|x-x_k|}{\rho_k}\right) & \text{whenever } |x - x_k| < \rho_k < 2^{-2k}, \\ 0 & \text{otherwise.} \end{cases}$$

Geometrically, a surface $u_k(x)$ consists in a family of circular cones of height 1, the radius of the $(n-1)$-dimensional balls lying in the base plane equal to ρ_k, and the centre x_k. Let $u = \sum_{k=1}^{\infty} a_k u_k(x)$ and $g(x, t) = [x, t + u(x)]$. The function $g(x, t)$ is a homeomorphism of $Q \times R \subset R^{n+1}$ onto itself, which is ACL_n but is nowhere differentiable. This example shows also that the estimation $p > n - 1$ of the preceding lemma cannot be improved.

2. It follows from the preceding lemma that we can replace the condition ACL_n in the preceding theorem by the condition ACL_p for some $p > n - 1$.

3. From the preceding lemma we infer that the second analytic definition can be formulated also in the following other form:

A homeomorphism $f(x)$ is said to be K-QCfH ($1 \leqslant K < \infty$) in D if it is ACL_n in D and satisfies the double inequality (19).

Second geometric definition (by modulus of rings). A homeomorphism $f(x)$ is said to be K-QCfH ($1 \leqq K < \infty$) in D if

$$\frac{1}{K} M(A) \leqq M(A^*) \leqq KM(A) \tag{56}$$

for all rings A with $\bar{A} \subset D$.

The equivalence with the other definitions.

THEOREM 6. *The definition of the* QCfH *by modulus of rings is equivalent to the other four definitions.*

Clearly, the first four definitions imply the preceding one, because it can appear as a particular case of the first geometric definition where we consider only the arc families Γ_A.

Conversely, assume that $f(x)$ satisfies (56). Then from (25), on account of lemma 2 and inequalities (38) we infer that $\bar{\delta}(x)$ is bounded in D, and, by theorem 2, $f(x)$ is QCfH in D according to the geometric definition. Thus, by lemma 2 and corollary of theorem 1, it follows a.e. in D the differentiability of $f(x)$ with $J(x) \neq 0$. Let x_0 be such a point. Since $f^{-1}(x^*)$ also satisfies (56), it suffices to prove that

$$\max |f(x_0)|^n \leqq K|J)x_0)|. \tag{57}$$

As in the proof of lemma 1, we may assume that $x_0 = f(x_0) = O$ and $f'(O)$ is given by (22).

Let $h, \eta > 0$ be such that the closure of the ring A, bounded by the $(n-1)$-dimensional cube $\bar{Q} = \{x; x^1 = 0, -h \leqq x^k \leqq h, k = 2, \ldots, n\}$ and the planes

$$x^1 = \pm \frac{\eta h}{a_1}, \quad x^k = \pm \left(1 + \frac{\eta}{a_k}\right) h \quad (k = 2, \ldots, n) \tag{58}$$

lie in D. Consider the subfamilies $\Gamma_1, \Gamma_2 \subset \Gamma_A$ whose curves γ lie in the two right cylinders with bases $Q, Q + \frac{\eta h e_1}{a_1}$ and $Q, Q - \frac{\eta h e_1}{a_1}$, respectively and joining these bases. Then Γ_1, Γ_2 are separate and $\Gamma_A < \Gamma_1, \Gamma_2$. By the propositions 1.5.3 and 1.5.23,

$$M(\Gamma_A) \geqq M(\Gamma_1) + M(\Gamma_2) = 2^n \left(\frac{a_1}{\eta}\right)^{n-1}. \tag{59}$$

Consider then the image ring A^*. First fix $\eta < 1$ and let $0 < \varepsilon < \frac{\eta}{2}$. Next choose h so small that (24) holds for $x \in \bar{A}$. Then the distance between the boundaries of A^* is $\geqq (\eta - 2\varepsilon) h$, since

$$\left| f\left(-\frac{\eta h}{a_1}, x^2, \ldots, x^n\right) - f(0, x^2, \ldots, x_n) \right| \geqq a_1 \frac{\eta h}{a_1} - 2\varepsilon h = (\eta - 2\varepsilon) h,$$

$$\left| f\left(x^1, \ldots, x^{i-1}, h\left(1 + \frac{\eta}{a_i}\right), x^{i+1}, \ldots, x^n\right) - f(x^1, \ldots, x^{i-1}, h, x^{i+1}, \ldots, x^n) \right| \geqq$$

$$\geqq h(\eta - 2\varepsilon),$$

$$\left| f\left(x^1, \ldots, x^{i-1}, -h\left(1 + \frac{\eta}{a_i}\right), x^{i+1}, \ldots, x^n\right) - f(x^1, \ldots, x^{i-1}, -h, x^{i+1}, \ldots, x^n) \right| \geqq$$

$$\geqq h(\eta - 2\varepsilon).$$

Now, if $\rho^* \equiv \frac{1}{(\eta - 2\varepsilon) h}$, then $\rho^* \in F(\Gamma_{A^*})$, and

$$M(\Gamma_{A^*}) \leqq \int_{A^*} \rho^{*n} \, d\tau = \frac{1}{(\eta - 2\varepsilon)^n h^n} \int_{A^*} d\tau = \frac{mA^*}{(\eta - 2\varepsilon)^n h^n}. \tag{60}$$

On the other hand, since, by theorem 1, $mf(Q) = 0$, it follows $mA^* = m[A^* \cup f(Q)]$, i.e. mA^* is equal to the measure of the image of the interval definite by the planes (58). Thus, by (22) and (24),

$$mA^* \leqq (2h)^n (\eta + \varepsilon)(a_2 + \eta + \varepsilon) \cdots (a_n + \eta + \varepsilon).$$

Combining (56), (59) and (60) we obtain

$$2^n\left(\frac{a_1}{\eta}\right)^{n-1} \leqq M(\Gamma_A) \leqq KM(\Gamma_{A^*}) \leqq K\frac{(2h)^n(\eta+\varepsilon)(a_2+\eta+\varepsilon)\cdots(a_n+\eta+\varepsilon)}{(\eta-2\varepsilon)^n h^n},$$

and letting $\varepsilon \to 0$,

$$a_1^{n-1} \leqq K(a_2+\eta)\cdots(a_n+\eta).$$

Since this holds for all $0<\eta<1$, then letting $\eta \to 0$ and taking into account (13), yields (57), as desired.

Other characterization of the K-QCfH is the following:

THEOREM 7. *A homeomorphism $x^* = f(x)$ of a domain D is K-QCfH ($1 \leqq K < \infty$) in D according to the geometric definition if and only if the ratio $\dfrac{M(A^*)}{M(A)}$ is bounded for all spherical rings $A^* \subset D^*$* and

$$M(Q_1^*) \geqq \frac{1}{K}, \quad M(Q_2) \geqq \frac{1}{K}, \tag{61}$$

for all cubes Q_1, Q_2^* such that $\bar{Q}_1 \subset D$, $Q_2^* \subset D^*$.

If $f(x)$ is K-QCfH according to the first geometric definition, then taking $\Gamma = \Gamma_A$, $\Gamma = \Gamma_{Q_1}$, $\Gamma = \Gamma_{Q_2}$, respectively, we obtain $\dfrac{M(A^*)}{M(A)} \leqq K$ for all spherical rings $A^* \subset D^*$ and (61), respectively.

Conversely, let $f(x)$ be a homeomorphism which satisfies the condition of the theorem. Then, by lemma 2, $f(x)$ is QCfH according to the metric definition (and on account of the equivalence it is QCfH also according to the other definitions). Next, by lemma 1, applied to the homeomorphisms $f(x)$ and $f^{-1}(x^*)$ and taking into account the inequalities (61), we conclude that $f(x)$ is K-QCfH in D according to geometric definition (and then also according to the other).

Third geometric definition (by modulus of cylinders). (Väisälä [3]). A homeomorphism $x^* = f(x)$ is said to be K-QCf ($1 \leqq K < \infty$) in D if

$$\frac{M(Z)}{K} \leqq M(Z^*) \leqq KM(Z) \tag{62}$$

for all cylinders $Z \subset D$.

Let us first introduce the notation

$$E + E' = \{x + x';\ x \in E,\ x' \in E'\},$$

Thus, for instance $E + B(r)$ is the set of all points $x \in R^n$ such that $d(x, E) < r$.

The equivalence with the other definitions. We begin with some preliminary lemmas (Väisälä).

LEMMA 11. *If* α *and* β *are rectifiable arcs in* R^n $(n \geqq 3)$, *with lengths* a *and* b, *respectively, then*

$$m[\alpha + \beta + B(r)] \leqq \omega_{n-2}r^{n-2}ab + \omega_{n-1}r^{n-1}(a+b) + \omega_n r^n. \tag{63}$$

We first establish (63) for the case where the arcs α and β are polygonal (self-intersections allowed). In this case we can verify (63) by elementary methods. Let $x_0, \ldots, x_p$ and $y_0, \ldots, y_q$ be the successive vertices of α and β, respectively. By performing a preliminary translation, we may assume that $x_0 = y_0 = O$. Let σ_m and τ_m be the line segments joining x_{m-1} to x_m and y_{m-1} to y_m, respectively, and set $s_m = |x_m - x_{m-1}|$, $t_m = |y_m - y_{m-1}|$. We shall construct the set $\alpha + \beta + B(r)$ stepwise as follows.

Set $E_1 = B(r)$. Thus $mE_1 = \omega_n r^n$. Next let $E_2 = \sigma_1 + B(r)$. Then $m(E_2 - E_1) = \omega_{n-1}r^{n-1}s_1$, because the volume of $E_2 - E_1$ is equal to the volume of the cylinder with the axis σ_1 and the base an $(n-1)$-dimensional ball of radius r. In general, $E_{m+1} = (\sigma_1 \cup \ldots \cup \sigma_m) + B(r)$. We easily see that $m(E_{m+1} - E_m) \leqq \omega_{n-1}r^{n-1}s_m$ (we have strict inequality for instance when two non-consecutive segments are situated at a distance less than r). Setting $F_1 = \alpha + B(r)$, we thus obtain

$$m[\alpha + \beta + B(r)] \leqq \omega_{n-2}r^{n-2}ab + \omega_{n-1}r^{n-1}(a+b) + \omega_n r^n. \tag{64}$$

Now let $E_1' = \tau_1 + B(r)$. Then $m(E_1' - F_1) \leqq \omega_{n-1}r^{n-1}t_1 - mU_1$, where $U_1 = (E_2 \cap E_1') - E_1$. Next let $E_2' = \sigma_1 + \tau_1 + B(r)$, where $\sigma_1 + \tau_1$ is the parallelogram with the sides σ_1 and τ_1. Then $m(E_2' - E_1' - F_1) \leqq \omega_{n-2}r^{n-2}t_1 s_1 + mV_1 - mU_2$, where $E_2' - E_1' - F_1 = E_2' - (E_1' \cup F_1)$, $V_1 = x_1 + U_1$ and $U_2 = [\sigma_2 + B(r)] \cap [x_1 + \tau_1 + B(r)] - [x_1 + B(r)]$. Since $mV_1 = mU_1$ we have: $m(E_2' - F_1) = m(E_2' - E_1' - F_1) + m(E_1' - F_1) \leqq \omega_{n-1}r^{n-1}t_1 + \omega_{n-2}r^{n-2}t_1 s_1 - mU_2$. In general, setting $E_{m+1}' = (\sigma_1 \cup \ldots \cup \sigma_m) + \tau_1 + B(r)$, we get

$$m(E_{m+1}' - F_1) \leqq \omega_{n-1}r^{n-1}t_1 + \omega_{n-2}r^{n-2}t_1(s_1 + \cdots + s_m) - mU_{m+1},$$

where $U_m = [\sigma_m + B(r)] \cap [x_{m-1} + \tau_1 + B(r)] - [x_{m-1} + B(r)]$, and $U_{p+1} = \emptyset$. Setting $F_2 = \alpha + \tau_1 + B(r)$ we thus obtain $m(F_2 - F_1) \leqq \omega_{n-1}r^{n-1}t_1 + \omega_{n-2}r^{n-2}t_1 a$. Similarly, setting $F_{m+1} = \alpha + (\tau_1 \cup \ldots \cup \tau_m) + B(r)$ we get

$$m(F_{m+1} - F_m) \leqq \omega_{m-1}r^{n-1}t_m + \omega_{n-2}r^{n-2}t_m a. \tag{65}$$

Since $\alpha + \beta + B(r) = F_{q+1}$, (64) and (65) imply (63) in the case where α, β are polygonal arcs.

We consider now the general case where α, β are arbitrary rectifiable arcs. For an $\varepsilon > 0$ choose inscribed polygons α_0 and β_0 for α and β such that their lengths a_0, b_0 satisfy the inequalities $a_0 > a - \varepsilon$, $b_0 > b - \varepsilon$. It is easy to see that α is a subset of $\alpha_0 + B(q)$ and β of $\beta_0 + B(q')$, where $q^2 = (2a + \varepsilon)\frac{\varepsilon}{4}$, $q'^2 = (2b + \varepsilon)\frac{\varepsilon}{2}$. Hence

$$\alpha + \beta + B(r) \subset \alpha_0 + \beta_0 + B(r + q + q'),$$

where q, q' tend to zero as $\varepsilon \to 0$. Applying the lemma to α_0, β_0 and letting $\varepsilon \to 0$ yields (63) in the general case.

LEMMA 12. *If $f: D \rightleftarrows D^*$ such that*

$$M(Z) \leqq KM(Z^*)$$

for every cylinder Z in D, then, for any spherical ring $A: a < |x - x_0| < b \subset\subset D$, $f(x)$ is AC on almost all radial segments $J_0 = \{x;\ x = x_0 + te,\ a < t < b,\ |e| = 1\}$.

Let $\xi = \psi^{-1}(y)$ be the stereographical projection of the sphere $S(a)$ on the plane $\Pi = \{x^n = 0\}$. For each Borel set $E \subset \Pi$, let $Q_E = \{x\,; x = y + te,\ y \in \psi(E),\ a < t < b,\ |e| = 1\}$. The function $\varphi(E) = mf(Q_E)$ is a non-negative additive function of a set and possesses a symmetrical derivative $D_{\text{sym}}\varphi(\xi) < \infty$ a.e. in Π. Fix such a point ξ_0, and let J_0 be the segment Q_{ξ_0}. It suffices to show that f is AC on J_0. We may assume (without loss of generality) that $\xi_0 = O$ and $y_0 = \psi(O) = (0, \ldots, 0, -a)$.

Let $\varepsilon > 0$ and let $\delta_1, \ldots, \delta_p$ be closed disjoint intervals on J_0 such that their total length be less than ε. Let t_m, t'_m be the endpoints and d_m the length of $\delta_m (m = 1, \ldots, k)$. We denote by $B^{n-1}(r)$ the ball $\{x;\ |x| < r,\ x^n = 0\}$. Choose $r > 0$ such that $\bigcup_m \delta_m + B^{n-1}(r) \subset D$ and consider the cylinder

$$Z_m = [\delta_m + B^{n-1}(r),\ B^{n-1}(t_m, r),\ B^{n-1}(t'_m, r)] \qquad (m = 1, \ldots, p).$$

By (1.5.38), $M(Z_m) = \dfrac{\omega_{n-1} r^{n-1}}{d_m^{n-1}}$. We next estimate $M(Z^*)$ by means of the generalized Rengel's inequality (1.5.40). If l_m is the Euclidean distance between the bases of Z_m^* and if $V_m = mZ_m^*$, we have $M(Z_m^*) \leqq \dfrac{V_m}{l_m^n}$ by (1.5.40). Since $M(Z_m) \leqq KM(Z_m^*)$, we obtain

$$\frac{KV_m}{\omega_{n-1} r^{n-1}} \geqq \frac{l_m^n}{d_m^{n-1}}.$$

Summing over all m we further obtain

$$\frac{K\Sigma_m V_m}{\omega_{n-1} r^{n-1}} \geqq \sum_m \frac{l_m^n}{d_m^{n-1}}.$$

hence, by Hölder's inequality,

$$(\Sigma_m l_m)^n = \left(\Sigma_m d^{\frac{n-1}{n}} \frac{l_m}{d^{\frac{n-1}{n}}}\right)^n \leqq \Sigma_m \frac{l_m^n}{d_m^{n-1}} (\Sigma_m d_m)^{n-1},$$

and then

$$\frac{K\Sigma_m V_m}{\omega_{m-1} r^{n-1}} \geqq \Sigma_m \frac{l_m^n}{d_m^{n-1}} \geqq \frac{(\Sigma_m l_m)^n}{(\Sigma_m d_m)^{n-1}},$$

whence

$$\varepsilon^{n-1} > (\Sigma_m d_m)^{n-1} \geqq \frac{\omega_{n-1} r^{n-1} (\Sigma_m l_m)^n}{K\Sigma_m V_m}.$$

Since $\Sigma_m V_m \leqq \varphi[B^{n-1}(r)]$, this implies

$$(\Sigma_m l_m)^n \leqq \frac{K\varepsilon^{n-1}\varphi[B^{n-1}(r)]}{\omega_{n-1} r^{n-1}}.$$

Letting $r \to 0$ yields

$$[\Sigma_m |f(t_m') - f(t_m)|]^n \leqq K\varepsilon^{n-1} D_{sym}\varphi(O).$$

This inequality proves the lemma.

THEOREM 8. *The third geometric definition of the K-QCfH is equivalent to the other Väisälä's definitions.*

For $n = 2$ the cylinders reduce to quadrilaterals. (We recall that a *quadrilateral* is a Jordan domain with four distinguished points called *the vertices of the quadrilateral* on its boundary; these points decompose the boundary in four parts termed *the sides of the quadrilateral.*) For quadrilaterals the theorem is well known (see for instance H. Künzi [3], p. 80), and we therefore assume that $n \geqq 3$.

Clearly, a K-QCfH according to the first geometric definition is K-QCfH also according to the third geometric definition (by the modulus of cylinders).

Conversely, suppose $f: D \rightleftharpoons D^*$ is K-QCfH according to the third geometric definition and let be $A = \{x; a < |x - x_0| < b\} \subset\subset D$ and Γ the family of all arcs which join the boundary spheres of A in A. According to the preceding theorem it suffices to show that $\frac{M(\Gamma^*)}{M(\Gamma)}$ is bounded for all such rings A.

From lemma 12 it easily follows that there exists a radial segment $J_0 = \{x; x = x_0 + te, a \leqq t \leqq b, |e| = 1\}$, such that f is AC on J_0. Thus $J_0^* = f(J_0)$ is rectifiable (S. Saks [1], theorem 8, p. 58), and has a length $l < \infty$.

Fix r such that $0 < r < \frac{b-a}{2}$ and such that the modulus of the family Γ_r of all arcs joining the boundary spheres of the ring $A_r = \{x; a + r < |x - x_0| < b - r\}$ in A_r be less than $2M(\Gamma)$, i.e.

$$M(\Gamma_r) < 2M(\Gamma). \tag{66}$$

Next choose a number $h > 0$ such that $|f(x_1) - f(x_2)| > h$ whenever $|x_1 - x_2| \geqq r$ and $x_1, x_2 \in \bar{A}$. For any $0 < s < a$, let $L_s = [J_0 + \overline{B(s)}] \cap \bar{A}$ and $\delta(s) = \sup d(x^*, J_0^*)$ over all $x^* \in L_s^*$. Thus

$$L_s^* \subset J_0^* + \overline{B[\delta(s)]}. \tag{67}$$

Since f is continuous, $\delta(s) \to 0$ as $s \to 0$. Putting

$$q = \left[\frac{mA^*}{M(\Gamma)}\right]^{\frac{1}{n}}, \tag{68}$$

we can choose s so small that

$$\omega_{n-1}\delta(s)^{n-2}lq + \omega_{n-1}\delta(s)^{n-1}(l + q) + \omega_n\delta(s)^n < \frac{\omega_n h^n}{2}. \tag{69}$$

Let Z_{rs} be the cylinder (Q, E_1, E_2) where $\bar{Q} = A_r - L_s$, $E_1 = \{x; x \in \bar{Q}, |x - x_0| = a + r\}$, and $E_2 = \{x; x \in \bar{Q}, |x - x_0| = b - r\}$. Setting $\Gamma_{rs} = \Gamma_{rs}$, then by (66) and corollary 2 of proposition 1.5.3, we have

$$M(\Gamma_{rs}) \leqq M(\Gamma_r) < 2M(\Gamma).$$

On the other hand, since $M(Z_{rs}) = M(\Gamma_{rs})$, then the double inequality (62), which characterizes the K-QCfH according to the third geometric definition, implies $M(\Gamma_{rs}^*) \leqq KM(\Gamma_{rs})$, hence

$$M(\Gamma_{rs}^*) \leqq 2KM(\Gamma). \tag{70}$$

We next derive a lower bound for $M(\Gamma_{rs}^*)$. Let $\rho^* \in F(\Gamma_{rs}^*)$. We may assume $\rho^*(x^*) = 0$ on CZ_{rs}^*.

We construct the h-average function ρ_h^*

$$\rho_h^*(x^*) = \frac{1}{\omega_n h^n}\int_{B(h)} \rho^*(x^* + y^*)\, d\tau.$$

Let Γ_q^* be the family of all arcs in Γ^* whose length is less than q. We next show that $2\rho_h^* \in F(\Gamma_q^*)$.

Using Fubini's theorem we first write the above integral as follows:

$$\int_{\gamma^*} \rho_h^* \, ds = \frac{1}{\omega_n h^n} \int_{B(h)} d\tau \left(\int_{\gamma^* + y^*} \rho^* \, ds \right) \tag{71}$$

for all $\gamma^* \in \Gamma_q^*$. If $(\gamma^* + y^*) \cap L_s^* = \emptyset$, then $\gamma^* + y^*$ contains a subarc which belongs to Γ_{rs}^*. Hence, in this case

$$\int_{\gamma^* + \gamma^*} \rho^* \, ds \geqq 1. \tag{72}$$

Let E^* be the set of all $y^* \in B(h)$ such that $(\gamma^* + y^*) \cap L_s^* \neq \emptyset$. Hence we infer that there exist at least a point $x_{\gamma^*}^* \in \gamma^*$ and a point $x_L^* \in L_s^*$ such that $x_{\gamma^*}^* + y^* = x_L^*$, hence we get $y^* = -x_{\gamma^*}^* + x_L^* \in \tilde{\gamma}^* + L_s^*$ (where by $\tilde{\gamma}^*$ we denoted the arc symmetric of γ^* with respect to the origin). Then (67) implies $E^* \subset \tilde{\gamma}^* + J_0^* + \overline{B[\delta(s)]}$. By lemma 11 and the inequality (69) it follows $mE^* < \dfrac{\omega_n h^n}{2}$, and this together with (71) and (72), yields

$$2 \int_{\gamma^*} \rho_h^* \, ds \geqq 1,$$

hence $2\rho_h^* \in F(\Gamma_q^*)$. We thus obtained the estimate

$$M(\Gamma_q^*) \leqq 2^n \int_{A^*} \rho_h^{*n} \, d\tau.$$

By *Minkowski's inequality* (see G. H. Hardy, J. E. Littlewood and G. Polya [1], theorem 202, p. 148):

$$\left\{ \int_{E_1} \left[\int_{E_2} f(x, y) \, dy \right]^p dx \right\}^{\frac{1}{p}} \leqq \int_{E_2} \left[\int_{E_1} f(x, y)^p dx \right]^{\frac{1}{p}} dy, \quad p > 1, \tag{73}$$

this implies

$$M(\Gamma_q^*) \leqq 2^n \int_{A^*} \rho_h^{*n} \, d\tau = \frac{2^n}{(\omega_n h^n)^n} \int_{A^*} \left[\int_{B(h)} \rho^*(x^* + y^*) \, d\tau \right]^n d\tau \leqq$$

$$\leqq \frac{2^n}{(\omega_n h^n)^n} \left\{ \int_{B(h)} \left[\int_{A^*} \rho^*(x^* + y^*)^n \, d\tau \right]^{\frac{1}{n}} d\tau \right\}^n =$$

$$= \frac{2^n}{(\omega_n h^n)^n} \left\{ \int_{B(h)} \left[\int_{A^* + y^*} \rho^*(x^*)^n \, d\tau \right]^{\frac{1}{n}} d\tau \right\}^n \leqq$$

$$\leqq \frac{2^n}{(\omega_n h^n)^n} \left\{ \int_{B(h)} \left[\int_{Z_{rs}^*} \rho^*(x^*)^n \, d\tau \right]^{\frac{1}{n}} d\tau \right\}^n = 2^n \int_{Z_{rs}^*} \rho^{*n} \, d\tau.$$

Since this holds for every $\rho^* \in F(\Gamma_{rs}^*)$, we obtain $M(\Gamma_q^*) \leqq 2^n M(\Gamma_{rs}^*)$. By (70) this implies

$$M(\Gamma_q^*) \leqq 2^{n+1} K M(\Gamma). \tag{74}$$

Next define a function ρ_q^*

$$\rho_q^*(x^*) = \begin{cases} \dfrac{1}{q} & \text{for } x^* \in A^*, \\ 0 & \text{for } x^* \bar{\in} A^*. \end{cases}$$

Then, clearly, $\rho_q^* \in F(\Gamma^* - \Gamma_q^*)$. Thus, on account of (68),

$$M(\Gamma^* - \Gamma_q^*) \leqq \int_{A^*} \rho_q^{*n} \, d\tau = \int_{A^*} \frac{d\tau}{q^n} = \frac{mA^*}{q^n} = M(\Gamma).$$

Hence, by (74) and taking into account proposition 1.5.1, it follows

$$M(\Gamma^*) \leqq M(\Gamma_q^*) + M(\Gamma^* - \Gamma_q^*) \leqq (2^{n+1} K + 1) M(\Gamma),$$

and since $M(A) = \dfrac{n\omega_n}{M(\Gamma)}$, we obtain $\dfrac{M(A)}{M(A^*)} \leqq 2^{n+1} K + 1$. This inequality holds for any spherical ring $A \subset D$. Since the cubes are particular cylinders, then, if we apply the inequality (62) to the cubes $Q_1 \subset\subset D$, $Q_2^* \subset\subset D^*$, we obtain even the inequalities (61). But Q_1, Q_2^* are arbitrary cubes with the aforenamed property, so that, by the preceding theorem, $f(x)$ is K-QCfH also according to the other definitions of Väisälä.

CHAPTER 2

Gehring's geometric definition and Loewner's definition of K-QCfH. Their equivalence

Conformal capacity of a ring A (cap A). Gehring's [5] geometric definition of the K-QCfH is stated in terms of rings and characterized by the same double inequality (1.56) as in Väisälä's corresponding definition [1]. The difference between them is that Väisälä [1] introduces the modulus of a ring by means of the relation (1.5.7), while Gehring [5] defines it as

$$\operatorname{mod} A = \left[\frac{n\omega_n}{\operatorname{cap} A}\right]^{\frac{1}{n-1}}, \tag{1}$$

where cap A is the conformal capacity of a ring A and mod A denotes (following Gehring [5]) the modulus of a ring according to his definition, because the two definitions are not equivalent except for $n = 2$ (as it will be seen in this chapter).

Then following Loewner [1], we define *the conformal capacity of A* as

$$\operatorname{cap} A = \inf_u \int_A |\nabla u|^n \, d\tau, \tag{2}$$

where the infimum is taken over all functions u which are continuous and ACL in A with boundary values 0 on F_0 and 1 on F_1 and where the gradient $\nabla u = \left(\frac{\partial u}{\partial x^1}, \ldots, \frac{\partial u}{\partial x^n}\right)$ exists whenever all the partial derivatives exist, i.e. a.e. n A (proposition 1.6.7). We call such a function $u(x)$ *an admissible function for A.*

Cap $A = M(\Gamma_A)$. This relation will allow us to conclude that the class of K-QCfH characterized by the double inequality

$$\frac{\operatorname{mod} A}{K} \leqq \operatorname{mod} A^* \leqq K \operatorname{mod} A$$

coincides with the class of K^{n-1}-QCfH according to Väisälä's definitions.

Now let us point out the reason for which Gehring preferred to define the modulus of a ring by means of relation (1). In the particular case of a bidimensional circular ring bounded by the two spheres $S(r_1)$, $S(r_2)$ $(0 < r_1 < r_2)$, mod $A = \log \frac{r^2}{r_1}$. In the n-space $M(A) = \left(\log \frac{r_2}{r_1}\right)^{n-1}$ (proposition 1.5.7), which for $n = 2$ reduces to $\log \frac{r_2}{r_1}$. However Gehring wanted that the modulus conserve its form from $n = 2$ also for an arbitrary n, i.e.

$$\operatorname{mod} A = \log \frac{r_2}{r_1} \tag{3}$$

and that is why he preferred the expression (1).

In order to prove that $M(\Gamma_A) = \operatorname{cap} A$, let us establish some preliminary lemmas (obtained for $n = 3$ by F. Gehring [7]).

LEMMA 1. *Let A be a ring with non-degenerate boundary components and let $\rho \geqq 0$ be Borel measurable and $\rho \in L^n$. For each $a > 0$ there exists a $b > 0$ with the following property: if x_0 and x_1 are the endpoints of a rectifiable curve β, if $d(x_0, F_0) < b$, and if either $d(x_1, F_1) < b$ or $|x_1| > \frac{4}{b}$, then*

$$\int_\beta \rho \, ds \geqq \inf_\gamma \int_\gamma \rho \, ds - a, \tag{4}$$

where the infimum is taken over all locally rectifiable curves $\gamma \in \Gamma_A$.

Fix $a > 0$ and choose $c > 0$ so that

$$\left(\frac{a}{2}\right)^n = \frac{A_0 c}{\log 2}. \tag{5}$$

where A_0 is the constant of proposition 1.5.10 given by (1.5.13). Then we can find a number b $\left(0 < b < \frac{1}{\sqrt{2}}\right)$ such that the following is true:

$$C_0 \subset B\left(\frac{1}{b}\right), \tag{6}$$

$$d(C_0) > 4b, \tag{7}$$

$$C_1 \cap S\left(\frac{2}{b}\right) \neq \emptyset, \tag{8}$$

$$3b < d(F_0, F_1), \tag{9}$$

$$\int_{|x-y|<2b} \rho^n \, d\tau \leqq c \tag{10}$$

for every point y by Radon—Nikodim's theorem (Chapter 1.3) and

$$\int_{|x|>\frac{2}{b}} \rho^n \, d\tau \leqq c \tag{11}$$

by virtue of the complete additivity of Lebesgue integral, since if A_m denotes the ring bounded by the spheres centred at the origin and with the radii m and $m+1$, then

$$\infty > \int_{R^n} \rho^n \, d\tau \geqq \sum_{m=1}^{\infty} \int_{A_m} \rho^n \, d\tau,$$

where the rest of the series tends to zero as $m \to \infty$.

Now let β be a rectifiable curve joining x_0 and x_1, where $d(x_0, F_0) < b$, and where either $d(x_1, F_1) < b$ or $|x_1| > \frac{b}{4}$. In order to establish (4), it suffices to show for $k = 1, 2$ that either β meets F_k or there exists a circular arc α_k joining β to F_k such that

$$\int_{\alpha_k} \rho \, ds \leqq \frac{a}{2} \quad (k = 0, 1).$$

Then, for example, if $\beta \cap \partial A = \emptyset$, $\beta \cup \alpha_0 \cup \alpha_1$ would contain a rectifiable curve $\gamma \in \Gamma_A$ and

$$\int_\beta \rho \, ds \geqq \int_\gamma \rho \, ds - \int_{\alpha_0} \rho \, ds - \int_{\alpha_1} \rho \, ds \geqq \inf_\gamma \int_\gamma \rho \, ds - a,$$

as desired.

Suppose that $\beta \cap F_0 = \emptyset$ and choose $y_0 \in F_0$ so that

$$|x_0 - y_0| \leqq b. \tag{12}$$

Then condition (10) implies that there exists a sphere $S_0 = S(y_0, r_0)$ $(b < r_0 < 2b)$ such that

$$r_0 \int_{S_0} \rho^n \, d\sigma \leqq \frac{c}{\log 2}, \tag{13}$$

since if

$$\int_{S_0} \rho^n \, d\sigma > \frac{c}{r \log 2}$$

for all $r(b \leqq r \leqq 2b)$, then, by integrating with respect to r, yields

$$\int\limits_{|y_0-x|<2b} \rho^n \,\mathrm{d}\tau \geqq \int\limits_{b\leqq|y_0-x|\leqq 2b} \rho^n \,\mathrm{d}\tau > \frac{c}{\log 2}\int_b^{2b}\frac{\mathrm{d}r}{r} = c,$$

which contradicts inequality (10). Now, by (12), x_0 lies inside S_0, and by (9), in the hypothesis $d(x_1, F_1) < b$ or by (6) in the hypothesis $|x_1| > \frac{4}{b}$, it follows that x_1 lies outside S_0. Furthermore, by (7) $d(S_0) < 4b < d(F_0)$, hence $S_0 \cap \beta$, $S_0 \cap F_0 \neq \emptyset$ and by the proposition 1.5.10 and relation (13) we can find a circular arc $\alpha_0 \subset S_0$ that joins β to F_0 and for which

$$\int_{\alpha_0} \rho \,\mathrm{d}s \leqq \left(A_0 r_0 \int_{S_0} \rho^n \,\mathrm{d}\sigma\right)^{\frac{1}{n}} \leqq \left(\frac{A_0}{\log 2}\right)^{\frac{1}{n}} = \frac{a}{2}.$$

Now suppose

$$\beta \cap F_1 = \emptyset. \tag{14}$$

Then we can find a sphere $S_1 = S(y_1, r_1)$ such that

$$r_1 \int_{S_1} \rho^n \,\mathrm{d}\sigma \leqq \frac{c}{\log 2}. \tag{15}$$

When $d(x_1, F_1) < b$, we choose $x_1 \in F_1$ so that $|x_1 - y_1| \leqq b$, and r_1 $(b < r_1 < 2b)$ so that (15) holds by virtue of (10). When $|x_1| > \frac{4}{b}$, we take y_1 as the origin and on the basis of (11), choose $r_1 \left(\frac{2}{b} < r_1 < \frac{4}{b}\right)$ so that (15) is valid. This is possible, because otherwise, if

$$\int_{S_1} \rho^n \,\mathrm{d}\sigma > \frac{c}{r\log 2},$$

for all $r\left(\frac{2}{b} < r < \frac{4}{b}\right)$, then by integrating with respect to r, we should obtain

$$\int\limits_{|x|>\frac{2}{b}} \rho_n \,\mathrm{d}\tau \geqq \int\limits_{\frac{2}{b}<|x|<\frac{4}{b}} \rho^n \,\mathrm{d}\tau > c,$$

which contradicts (11).

In each case it is easy to see that x_0 and x_1 are separated by S_1 and hence $S_1 \cap \beta \neq \emptyset$. Hypothesis (14) further implies $\beta \subset CC_1$, and by (8) $S_1 \cap C_1 \neq \emptyset$. Thus $S_1 \cap F_1 \neq \emptyset$. Proposition I.5.10 and relation (5) now yield a circular $\alpha_1 \subset S_1$, joining β to F_1 such that

$$\int_{\alpha_1} \rho \, ds \leqq \frac{a}{2}$$

and the proof of lemma 1 is complete.

LEMMA 2. Cap A $= \inf \int_A |\nabla w|^n \, d\tau$, *where the infimum is taken over all the functions* $w \in C^1$ *which are zero on* F_0 *and* 1 *on* F_1.

Let us denote

$$\text{cap}' A = \inf_w \int_A |\nabla w|^n \, d\tau.$$

Clearly, cap $A' \geqq$ cap A, where cap A is given by (2). It suffices to show that

$$\text{cap}' A \leqq \int_A |\nabla u|^n \, d\tau, \tag{16}$$

for all u admissible for the ring A. We may assume $|\nabla u| \in L^n$ over A, for otherwise there is nothing to prove. Next fix $0 < a < \frac{1}{2}$, let

$$v = \begin{cases} 0 & \text{if } u < a, \\ \dfrac{u-a}{1-2a} & \text{if } a \leqq u \leqq 1-a, \\ 1 & \text{if } 1-a < u \end{cases} \tag{17}$$

and extend v to be 0 on C_0 and 1 on C_1. The set $\{x; a \leqq u(x) \leqq 1 - \mathrm{a}\}$ is a compact subset of A and lies at a distance b from ∂A. Let $c < b$ and let

$$w(x) = \frac{1}{\omega_n c^n} \int_{B(c)} v(x + y) \, d\tau$$

or all $x \in A$. This function has boundary values 0 on F_0 and 1 on F_1.

Now let us prove the existence of $\dfrac{\partial w}{\partial x^i}$. First, let us denote

$$\tilde{w}(x) = \int_{Q(2c)} v(x + y) \, d\tau,$$

where $Q(2c)$ is the cube with centre O and the length of the edges $2c$. If we take a new variable of integration $\eta = x + y$ and $Q(x, 2c)$ is the cube with centre x and the length of the edges $2c$, then, by Fubini's theorem, we get

$$\tilde{w}(x) = \int_{Q(x,\,2c)} v(\eta)\, d\tau =$$

$$= \int_{x^i-c}^{x^i+c} d\eta^i \int_{x^1-c}^{x^1+c} d\eta^1 \cdots \int_{x^{i-1}-c}^{x^{i-1}+c} d\eta^{i-1} \int_{x^{i+1}-c}^{x^{i+1}+c} d\eta^{i+1} \cdots \int_{x^n-c}^{x^n+c} v(\eta)\, d\eta^n < \infty \quad (i = 1, \ldots, n),$$

which implies the existence and the continuity of $\dfrac{\partial \tilde{w}(x)}{\partial x^i}$ $(i = 1, \ldots, n)$ in A. Hence and since from (17) we see that v is ACL everywhere and then $\dfrac{\partial v(\eta)}{\partial x^i}$ exists a.e., we deduce

$$\frac{\partial \tilde{w}}{\partial x^i} = \int_{x^1-c}^{x^1+c} d\eta^1 \cdots \int_{x^{i-1}-c}^{x^{i-1}+c} d\eta^{i-1} \int_{x^{i+1}-c}^{x^{i+1}+c} d\eta^{i+1} \cdots \int_{x^n-c}^{x^n+c} [v(\eta^1, \ldots, \eta^{i-1}, x^i +$$

$$+ c, \eta^{i+1}, \cdots, \eta^n) - v(\eta^1, \cdots, \eta^{i-1}, x^i - c, \eta^{i+1}, \ldots, \eta^n)]\, d\eta^n =$$

$$= \int_{x^1-c}^{x^1+c} d\eta^1 \cdots \int_{x^{i-1}-c}^{x^{i-1}+c} d\eta^{i-1} \int_{x^{i+1}-c}^{x^{i+1}+c} d\eta^{i+1} \cdots \int_{x^n-c}^{x^n+c} d\eta^n \left[\int_{x^i-c}^{x^i+c} \frac{\partial v(\eta)}{\partial \eta^i}\, d\eta^i \right] d\eta^n.$$

Whence, since from (17), $|\nabla V|^n \in L^n$ and Hölder's inequality implies that $|\nabla V|$, and then also $\dfrac{\partial v(\eta)}{\partial x^i}$ is integrable in $Q(x, 2c)$ and by virtue of Tonelli's theorem (Chapter 1.3), we obtain

$$\frac{\partial \tilde{w}(x)}{\partial x^i} = \int_{Q(x,\,2c)} \frac{\partial v(\eta)}{\partial \eta^i}\, d\tau = \int_{Q(2c)} \frac{\partial v(x+y)}{\partial x^i}\, d\tau \quad (i = 1, \cdots, n)$$

for each $x \in A$. But then, clearly, there exists

$$\frac{\partial \tilde{w}(x)}{\partial x^i} = \lim_{\Delta x^i \to 0} \frac{\tilde{w}(x^1, \ldots, x^{i-1}, x^i + \Delta x^i, x^{i+1}, \ldots, x^n) - \tilde{w}(x)}{\Delta x^i} =$$

$$\int_{Q(2c)} \lim_{\Delta x^i \to 0} \frac{v(x^1+y^1, \ldots, x^{i-1}+y^{i-1}, x^i+\Delta x^i+y^i, x^{i+1}+y^{i+1}, \ldots, x^n+y^n) - v(x+y)}{\Delta x^i}\, d\tau =$$

$$\int_{Q(2c)} \frac{\partial v(x+y)}{\partial x^i}\, d\tau,$$

hence

$$\lim_{\Delta x^i \to 0} \left| \frac{w(x^1, \ldots, x^{i-1}, x^i + \Delta x^i, x^{i+1}, \ldots, x^n) - w(x)}{\Delta x^i} \right| =$$

$$\lim_{\Delta x^i \to 0} \frac{1}{\omega_n c^n} \left| \int_{B(c)} \frac{v(x^1 + y^1, \ldots, x^{i-1} + y^{i-1}, x^i + \Delta x^i + y^i, x^{i+1} + y^{i+1}, \ldots, x^n + y^n)}{\Delta x^i} - \right.$$

$$\left. - \frac{v(x+y)}{\Delta x^i} \, d\tau \right| \leqq$$

$$\leqq \lim_{\Delta x^i \to 0} \frac{1}{\omega_n c^n} \int_{B(c)} \left| \frac{v(x^1 + y^1, \ldots, x^{i-1} + y^{i-1}, x^i + \Delta x^i + y^i, x^{i+1} + y^{i+1}, \ldots, x^n + y^n)}{\Delta x^i} - \right.$$

$$\left. - \frac{v(x+y)}{\Delta x^i} \right| d\tau \leqq$$

$$\leqq \lim_{\Delta x^i \to 0} \frac{1}{\omega_n c^n} \int_{Q(2c)} \left| \frac{v(x^1 + y^1, \ldots, x^{i-1} + y^{i-1}, x^i + \Delta x^i + y^i, x^{i+1} + y^{i+1}, \ldots, x^n + y^n)}{\Delta x^i} - \right.$$

$$\left. - \frac{v(x+y)}{\Delta x^i} \right| d\tau$$

and letting $\Delta x^i \to 0$, since from above we are allowed to interchange the limit and the integral sign in the last side of the preceding inequalities, it follows that we are allowed to do the same with respect to the first integral, obtaining:

$$\frac{\partial w(x)}{\partial x^i} = \frac{1}{\omega_n c^n} \int_{B(c)} \frac{\partial v(x+y)}{\partial x^i} \, d\tau \quad (i = 1, \ldots, n). \tag{18}$$

$\int_F \frac{\partial v(x')}{\partial x^i} \, d\tau$ is bounded and, by Radon—Nikodim theorem (Chapter 1.3), it follows that this integral is AC, hence $\frac{\partial w(x)}{\partial x^i}$ is continuous, since, given $\varepsilon > 0$, there is a $\delta > 0$ sufficiently small such that $|x' - x| < \delta$ implies

$$\left| \frac{\partial w(x')}{\partial x^i} - \frac{\partial w(x)}{\partial x^i} \right| = \frac{1}{\omega_n c^n} \left| \int_{B(c)} \frac{\partial v(x'+y)}{\partial x^i} \, d\tau - \int_{B(c)} \frac{\partial v(x+y)}{\partial x^i} \, d\tau \right| =$$

$$\frac{1}{\omega_n c^n} \left| \int_{B(c)} \frac{\partial v[x + (x' - x + y)]}{\partial x^i} \, d\tau - \int_{B(c)} \frac{\partial v(x+y)}{\partial x^i} \, d\tau \right| =$$

$$= \frac{1}{\omega_n c^n} \left| \int_{B(x'-x,\, c)} \frac{\partial v(x+y)}{\partial x^i} \, d\tau - \int_{B(c)} \frac{\partial v(x+y)}{\partial x^i} \, d\tau \right| =$$

$$= \frac{1}{\omega_n c^n} \left| \int_{B(x'-x,\, c) - B(c)} \frac{\partial v(x+y)}{\partial x^i} \, d\tau - \int_{B(c) - B(x'-x,\, c)} \frac{\partial v(x+y)}{\partial x^i} \, d\tau \right| < \varepsilon.$$

Next, from (18), we obtain

$$\nabla w(x) = \frac{1}{\omega_n c^n} \int_{B(c)} \nabla v(x + y) \, d\tau$$

for each $x \in A$. Then applying Minkowski's inequality (see G. H. Hardy, J. E. Littlewood and G. Polya [1], (6.13.9), p. 148):

$$\left\{ \int_{E_1} \left[\int_{E_2} f(x, y) \, dy \right]^p dx \right\}^{\frac{1}{p}} \leqq \int_{E_2} \left[\int_{E_1} f(x, y)^p \, dx \right]^{\frac{1}{p}} dy,$$

we get

$$\left(\int_A |\nabla w(x)|^n \, d\tau \right)^{\frac{1}{n}} \leqq \frac{1}{\omega_n c^n} \int_{B(c)} \left[\int_A |\nabla v(x + y)|^n \, d\tau \right]^{\frac{1}{n}} d\tau.$$

But

$$\int_A |\nabla v(x + y)|^n \, d\tau = \int_A |\nabla v(x)|^n \, d\tau \leqq \frac{1}{(1 - 2a)^n} \int_A |\nabla u(x)|^n d\tau$$

for each $y \in B(c)$. Hence

$$\int_A |\nabla w|^n \, d\tau \leqq (1 - 2a)^{-n} \int_A |\nabla u|^n \, d\tau.$$

Thus

$$\text{cap}' A \leqq \int_A |\nabla w|^n \, d\tau \leqq (1 - 2a)^n \int_A |\nabla u|^n \, d\tau.$$

Inequality (16) is now obtained by letting $a \to 0$.

LEMMA 3. *Let γ be a locally rectifiable arc joining two points x_0, $x_1 \in D$, $u \in C^1$ in D and $u_0 = u(x_0)$, $u_1 = u(x_1)$. Then*

$$\int_\gamma |\nabla u| \, ds \geqq |u_1 - u_0|.$$

Indeed,

$$\int_{u_0}^{u_1} du = \int_\gamma \frac{du}{ds} \, ds = \int_\gamma \frac{\partial u}{\partial x^i} \frac{dx^i}{ds} \, ds = \int_\gamma \frac{\partial u}{\partial x^i} \cos(s, x^i) \, ds.$$

It suffices to show that

$$\frac{\partial u}{\partial x^i} \cos(s, x^i) \leqq |\nabla u| = \left[\sum_{i=1}^n \left(\frac{\partial u}{\partial x^i} \right)^2 \right]^{\frac{1}{2}}, \tag{19}$$

which holds by the Schwartz inequality

$$\left[\frac{\partial u}{\partial x^i}\cos(s, x^i)\right]^2 \leqq \sum_{i=1}^{n}\left(\frac{\partial u}{\partial x^i}\right)^2.$$

We have equality in (19) if $\cos(s, x^i) = \dfrac{\dfrac{\partial u}{\partial x^i}}{|\nabla u|}$ $(i = 1, \ldots, n)$.

COROLLARY. *If u is ACL_p in D, then the conclusion of the preceding lemma holds for p-almost all arcs $\gamma \subset D$ joining x_0 and x_1.*

The relations (19) in this case are a consequence of the proof given by J. Väisälä [1] of the theorem of B. Fuglede (Chapter 1.6) and the rest of the proof is the same.

LEMMA 4. *Let $A: a < |x| < b$. Then*

$$\operatorname{cap} A = \frac{n\omega_n}{\left(\log\dfrac{b}{a}\right)^{n-1}}. \tag{20}$$

Let $u \in C^1$, $u = 0$ on $|x| = a$ and $u = 1$ on $|x| = b$. Then integrating along a radius and applying Hölder's inequality yields

$$1 \leqq \left(\int_a^b |\nabla u| \, dr\right)^n \leqq \left(\int_a^b |\nabla u|^n r^{n-1} \, dr\right)\left(\log\frac{b}{a}\right)^{n-1}.$$

Hence, integrating over S and applying Tonelli's theorem (Chapter 1.3).

$$\frac{n\omega_n}{\left(\log\dfrac{b}{a}\right)^{n-1}} \leqq \int_A |\nabla u|^n \, d\tau.$$

On the other hand choosing

$$u(x) = \frac{\log\left(\dfrac{|x|}{a}\right)}{\log\dfrac{a}{b}}, \tag{21}$$

we have equality and the proof is complete.

LEMMA 5. *If at least one of the components of the boundary ∂A of a ring A degenerates in a point, then* $\operatorname{cap} A = 0$.

Suppose first C_0 reduces to a point, which we may assume to be the origin. Let be $S = S(r)$ where $r = d(O, C_1)$. The spherical ring A', with O and S as components of the boundary $\partial A'$, has, by the preceding lemma, $\operatorname{cap} A' = 0$.

Fix $\varepsilon > 0$ and let $u(x)$ be an admissible function for the ring A' such that

$$\int_{A'} |\nabla u|^n \, d\tau < \varepsilon.$$

Extend this function outside A' by $u = 1$ outside S. Then we obtain a function continuous and ACL everywhere, with $|\nabla u| = 0$ outside S and $u = 1$ on F_1 (the outer component of ∂A) because F_1 is outside S. Thus $u(x)$ is an admissible function for A and we have

$$\operatorname{cap} A \leqq \int_A |\nabla u|^n \, d\tau = \int_{A'} |\nabla u|^n \, d\tau < \varepsilon.$$

Letting $\varepsilon \to 0$ yields $\operatorname{cap} A = 0$.

Suppose now C_1 reduces to a point, which we may assume to be ∞. Let S be a sphere centred at the origin and containing inside C_0. Let A'' be the spherical ring bounded by S and ∞. By the preceding lemma $\operatorname{cap} A'' = 0$. Fix $\varepsilon > 0$ and let $u(x)$ be an admissible function for A'' such that

$$\int_{A'} |\nabla u|^n \, d\tau < \varepsilon.$$

Extending $u(x)$ outside A'' by $u = 0$ inside S, we obtain a function continuous and ACL in R^n with $|\nabla u| = 0$ inside S and $u = 0$ on the boundary F_0 of A, because F_0 lies inside S. Thus $u(x)$ is an admissible function for A and we get

$$\operatorname{cap} A \leq \int_A |\nabla u|^n \, d\tau = \int_{A''} |\nabla u|^n \, d\tau < \varepsilon.$$

Letting $\varepsilon \to 0$ yields $\operatorname{cap} A = 0$ in this case too.

Set

$$L(\rho) = \inf_{\gamma} \int_{\gamma} \rho \, ds, \tag{22}$$

where the infimum is taken over all locally rectifiable curves $\gamma \in \Gamma_A$, and

$$V(\rho) = \int_A \rho^n \, d\tau.$$

In both cases $\rho \geqq 0$ is a Borel measurable function defined in A.

LEMMA 6. Cap $A = \inf \dfrac{V(\rho)}{L(\rho)^n}$, *where the infimum is taken over all Borel measurable functions* $\rho \geqq 0$ *for which* $V(\rho)$ *and* $L(\rho)$ *are not simultaneously* 0 *or* ∞.

Fix $a > 0$ and let be $u \in C^1$ on A, $u = 0$ on F_0, $u = 1$ on F_1 and such that

$$\int_A |\nabla u|^n \, d\tau < \operatorname{cap} A + a. \tag{23}$$

Set $\rho = |\nabla u|$. Then, by lemma 3 where $u_0 = 0$ and $u_1 = 1$, it follows

$$\int_\gamma \rho \, ds = \int_\gamma |\nabla u| \, ds \geqq 1$$

for each locally rectifiable arc $\gamma \in \Gamma_A$. Hence

$$L(\rho) \geqq 1, \tag{24}$$

and (23) may be written under the form

$$V(\rho) < \operatorname{cap} A + a. \tag{25}$$

Letting $a \to 0$, the preceding inequality and (24) yield

$$\inf_\rho \frac{V(\rho)}{L(\rho)^n} \leqq \operatorname{cap} A.$$

To complete the proof, we must show that

$$\operatorname{cap} A \leqq \frac{V(\rho)}{L(\rho)^n} \tag{26}$$

for all Borel measurable functions $\rho \geqq 0$ with $V(\rho)$ and $L(\rho)$ not simultaneously 0 or ∞. Now inequality (26) is trivial in the case $\operatorname{cap} A = 0$, $L(\rho) = 0$ or $V(\rho) = \infty$. Hence we may assume without loss of generality that $\operatorname{cap} A > 0$, $L(\rho) > 0$ and $V(\rho) < \infty$. But we may assume even $L(\rho) \geqq 1$. Indeed, fix ρ_0 such that $L(\rho_0) \geqq \alpha$. $0 < \alpha < 1$. Next choose $\rho_1 = \dfrac{\rho_0}{\alpha}$. For such a ρ_1 we have

$$L(\rho_1) = L\left(\frac{\rho_0}{\alpha}\right) = \frac{1}{\alpha} \inf_\gamma \int_\gamma \rho_0 \, ds = \frac{L(\rho_0)}{\alpha} \geqq 1,$$

and

$$\frac{V(\rho_1)}{L(\rho_1)^n} = \frac{V\left(\dfrac{\rho_0}{\alpha}\right)}{L\left(\dfrac{\rho_0}{\alpha}\right)^n} = \frac{\dfrac{1}{\alpha^n} \quad (\rho_0)}{\dfrac{1}{\alpha^n} L(\rho_0)^n} = \frac{V(\rho_0)}{L(\rho_0)^n}.$$

Thus $\inf\limits_{\rho} \dfrac{V(\rho_0)}{L(\rho_0)^n}$ is the same if instead of taking the infimum over all Borel measurable functions $\rho \geqq 0$ such that cap $A > 0$, $L(\rho)>0$ and $V(\rho) < \infty$, we restrict ourselves to the subclass of those ρ for which $L(\rho) \geqq 1$ (the other conditions remaining unchanged). Finally, from the preceding lemma we infer that cap $A>0$ implies that A is non-degenerate.

Let $0 < a < 1$ and extend ρ to be equal to zero in CA. Then

$$\int\limits_{R^n} \rho^n \, d\tau = \int\limits_{A} \rho^n \, d\tau = V(\rho) < \infty, \tag{27}$$

hence $\rho \in L^n$, and ρ satisfies the hypotheses of lemma 1, and we can find a number b $(0 < b < 1)$ for which the conclusions of lemma 1 hold.

Next, let

$$g(x) = \frac{1}{mU} \int_U \rho(x + y) d\tau,$$

where $U = B(b)$. By Hölder's inequality and taking into account that $\rho \in L^n$, it follows that $g(x)$ is bounded in R^n, because

$$g(x) \leqq \frac{1}{mU} \left[\int_U \rho^n(x + y) \, d\tau \right]^{\frac{1}{n}} (mU)^{\frac{n-1}{n}} \leqq \left[\frac{1}{mU} \int_{R^n} \rho^n(x) \, d\tau \right]^{\frac{1}{n}} = M < \infty.$$

Hence and by Radon—Nikodim's theorem (Chapter 1.3) it follows that the integral $\int_U \rho(x + y) d\tau$ is AC, hence $g(x)$ is continuous, since given $\varepsilon > 0$ there exists a $\delta > 0$ sufficiently small such that $|x' - x| < \delta$ imply

$$|g(x') - g(x)| = \frac{1}{mU} \left| \int_U \rho(x' + y) \, d\tau - \int_U \rho(x + y) d\tau \right| =$$

$$= \frac{1}{mU} \left| \int_U \rho[x + (x' - x + y)] \, d\tau - \int_U \rho(x + y) \, d\tau \right| =$$

$$= \frac{1}{mU} \left| \int_{U'} \rho(x + y) d\tau - \int_U \rho(x + y) \, d\tau \right| =$$

$$= \frac{1}{mU} \left| \int\limits_{U'-U} \rho(x + y) \, d\tau - \int\limits_{U-U'} \rho(x + y) \, d\tau \right| < \varepsilon,$$

where $U' = B(x' - x, b)$.

Moreover

$$\int\limits_{\beta} g \, ds \geqq L(\rho) - a \tag{28}$$

for each polygonal arc β joining two points x_0, x_1, where $x_0 \in F_0$ and where either $x_1 \in F_1$ or $|x_1| > \frac{5}{b}$. For by Tonelli's theorem,.

$$\int_\beta g(x)\,\mathrm{d}s = \int_\beta \left[\frac{1}{mU}\int_U \rho(x+y)\,\mathrm{d}\tau(y)\right]\mathrm{d}s(x) =$$

$$= \frac{1}{mU}\int_U \left[\int_{\beta_y} \rho(x)\,\mathrm{d}s(x)\right]\mathrm{d}\tau(y),$$

where β_y denotes the translation of β through the vector y. Lemma 1 then implies that

$$\int_{\beta_y} \rho\,\mathrm{d}s \geqq L(\rho) - a, \tag{29}$$

for all $|y| < b$, because the endpoints x_0^y, x_1^y of the arc β_y satisfy the conditions $d(x_0^y, F_0) < b$ and either $d(x_1^y, F_1) < b$ or $|x_1^y| \geqq |x_1| - |x_1^y - x_1| > \frac{5}{b} - b > \frac{4}{b}$ (on account of $0 < b < 1$) and F_0, F_1 are non-degenerate (i.e. the hypotheses of lemma 1 are fulfilled). Clearly (29) implies (28), hence, since $g(x)$ is bounded and F_0, F_1 are non-degenerate, it follows $L(\rho) < \infty$.

Now for each x let

$$u(x) = \inf_{\gamma(x)} \int_{\gamma(x)} g\mathrm{d}s, \tag{30}$$

where $\gamma(x)$ is a polygonal arc joining x to F_0, and set

$$v(x) = \min\left[1, \frac{u(x)}{L(\rho) - a}\right]. \tag{31}$$

It is easy to see that $u(x)$ satisfies a uniform Lipschitz condition. Indeed

$$|u(x') - u(x)| = \left|\inf_{\gamma(x')} \int_{\gamma(x')} g\mathrm{d}s - \inf_{\gamma(x)} \int_{\gamma(x)} g\mathrm{d}s\right| \leqq \left|\inf_{\gamma'(x')} \int_{\gamma'(x')} g\mathrm{d}s -\right.$$

$$\left. - \inf_{\gamma(x)} \int_{\gamma(x)} g\mathrm{d}s\right| \leqq \left|\inf_{\gamma(x,\,x')} \int_{\gamma(x,\,x')} g\mathrm{d}s\right| \leqq M|x' - x|,$$

where we assumed $\inf_{\gamma(x')} \int_{\gamma(x')} g\mathrm{d}s \geqq \inf_{\gamma(x)} \int_{\gamma(x)} g\mathrm{d}s$, where $\gamma'(x')$ is a polygonal arc joining x, to F_0 and passing through x, $\gamma(x, x')$ is a polygonal arc joining x to x' and M is the constant bounding $g(x)$.

Then $v(x)$ satisfies a uniform Lipschitz condition too. Moreover, $v = 0$ on F_0 and $v = 1$ on F_1 and outside the sphere $S\left(\frac{5}{b}\right)$. Since a function which satisfies a uniform Lipschitz condition is ACL, it follows that $v(x)$ is admissible for the ring A. Next, by (30), $\frac{\partial u}{\partial s} \leqq g$, hence, by (19), $|\nabla u| \leqq g$, and on account of (31), $|\nabla v| \leqq \frac{g}{L(\rho) - a}$ a.e. in A (i.e. where the partial derivatives exist), whence

$$\operatorname{cap} A \leqq \frac{1}{[L(\rho) - a]^n} \int_A g^n \, d\tau.$$

The Minkowski inequality (see G. H. Hardy, J. E. Littlewood and G. Polya [1] (6.13.9), p. 148), taking into account (27), implies that

$$\int_A g(x)^n \, d\tau \leqq \left\{\frac{1}{mU} \int_U \left[\int_A \rho(x + y)^n \, d\tau(x)\right]^{\frac{1}{n}} d\tau(y)\right\}^n \leqq$$

$$\leqq \left\{\frac{1}{mU} \int_U \left[\int_{R^n} \rho(x)^n \, d\tau(x)\right]^{\frac{1}{n}} d\tau(y)\right\}^n \leqq V(\rho).$$

Thus

$$\operatorname{cap} A \leqq \frac{v(\rho)}{[L(\rho) - a]^n}$$

and letting $a \to 0$,we obtain (26). This completes the proof of lemma 6.

Corollary. Cap $A = M(\Gamma_A)$.

Indeed, inequality (23) holds for all $a > 0$, hence letting $a \to 0$ we get

$$M(\Gamma_A) = \inf_\rho \int_A \rho^n \, d\tau \leqq \operatorname{cap} A,$$

where the infimum is taken over all $\gamma \in F(\Gamma_A)$. On the other hand, inequality (26), where we may assume (as above) $L(\rho) \geqq 1$, implies

$$\operatorname{cap} A \leqq \inf_\rho \frac{V(\rho)}{L(\rho)^n} \leqq \inf_\rho \int_A \rho^n \, d\tau = M(\Gamma_A),$$

where $\rho \in F(\Gamma_A)$.

The equivalence of Gehring's geometric definition of K-QCfH and Väisälä's definitions of K^{n-1}-QCfH.

THEOREM 1. *Gehring's geometric definition of* K-QCfH *is equivalent with Väisälä's definitions of* K'-QCfH *with* $K' = K^{n-1}$.

This theorem is an immediate consequence of the preceding corollary, combined with formulas (1.5.7) and (1) (which give the definition of the modulus of a ring according to Väisälä and to Gehring) and with the inequalities (1.56) and

$$\frac{\operatorname{mod} A}{K} \leqq \operatorname{mod} A^* \leqq K \operatorname{mod} A, \tag{32}$$

(which characterize the K-QCfH in the two corresponding cases).

Indeed, if $f(x)$ is K-QCfH according to Gehring's definition, then, in view of (1), (32) yields

$$\frac{1}{K}\left[\frac{n\omega_n}{\operatorname{cap} A}\right]^{\frac{1}{n-1}} \leqslant \left[\frac{n\omega_n}{\operatorname{cap} A^*}\right]^{\frac{1}{n-1}} \leqslant K\left[\frac{n\omega_n}{\operatorname{cap} A}\right]^{\frac{1}{n-1}}$$

and rising it to $n - 1$, we get

$$\frac{1}{K^{n-1}} \cdot \frac{n\omega_n}{\operatorname{cap} A} \leqslant \frac{n\omega_n}{\operatorname{cap} A^*} \leqslant K^{n-1} \frac{n\omega_n}{\operatorname{cap} A},$$

hence, on account of corollary of the preceding lemma,

$$\frac{M(A)}{K^{n-1}} \leqslant M(A^*) \leqslant K^{n-1} M(A),$$

i.e. just (1.56), where, however, instead of K, we have K^{n-1}.

Loewner's definition of K-QCfH [1]. A homeomorphism $f(x)$ is said to be K-QCfH $(1 \leqq K < \infty)$ in D if

$$\frac{\operatorname{cap} A}{K^{n-1}} \leqslant \operatorname{cap} A^* \leqslant K^{n-1} \operatorname{cap} A \tag{33}$$

for all rings $A \subset\subset D$.

Its equivalence with Gehring's definition.

THEOREM 2. *Loewner's definition is equivalent to Gehring's geometric definition.*

This is an immediate consequence of the inequalities (32) and (33) on account of (1).

Gehring's metric definition, Markušević – Pesin's definition and the definition of the K-QCfH with one and with two systems of characteristics. Their equivalence.

Gehring's metric definition [5]. A homeomorphism $f(x)$ of a domain D is said to be K-QCfH $(1 \leqq K \leqq \infty)$ in D if $\delta_L(x)$ is bounded in D and

$$\delta_L(x) \leqq K$$

a.e. in D.

Remark. In our monograph, when we define a concept or obtain a result in n-space, in fact in general, we do it for $n > 1$, especially as most of the definitions would appear meaningless if we try to interpret them for $n = 1$. Almost all the results in our book represent a generalization of the corresponding results of the plane (of the theory of functions of a complex variable). There are however definitions and theorems, even some of the definitions of the QCfH, which conserve the meaning also for $n = 1$ (see for instance L. Ahlfors and A. Beurling [1]). This is true also for the preceding definition, because, for $n = 1$, the linear local dilatation $\delta_L(x)$ of a strictly monotone function $f(x)$ (i.e. strictly increasing or strictly decreasing) may be expressed by

$$\delta_L(x) = \overline{\lim_{h\to 0}} \frac{f(x+h)-f(x)}{f(x)-f(x-h)},$$

where h is a real number sufficiently small so that $x - h$ or $x + h$ may not be out of the interval of the definition of $f(x)$. When $f(x)$ is defined on the interval $[-\infty, \infty]$ L. Ahlfors and A. Beurling give the following more restrictive condition:

$$\frac{1}{\rho} \leqq \frac{f(x+h)-f(x)}{f(x)-f(x-h)} \leqq \rho, \qquad (1)$$

for all x, $h \in (-\infty, \infty)$ and they call it the *ρ-condition.*

Connexion between Gehring's metric definition and the preceding definitions.

THEOREM 1. *The class of K-QCfH according to Gehring's metric definition is contained in the class of K-QCfH according to his geometric definition and in the class of K^{n-1}-QCfH according to Väisälä's definitions. The bounds K, K^{n-1} are respectively the best possible.*

Indeed, the inequality (1.15) implies that

$$\delta(x) \leqq \delta_L(x)^{n-1} \leqq K^{n-1},$$

hence the class of K-QCfH according to Gehring's metric definition is contained in Väisälä's class of K^{n-1}-QCfH. The inclusion of the class of K-QCfH according to Gehring's metric definition in the class of K-QCfH according to his geometric definition follows immediately from the preceding inclusion and theorem 2.1.

For the second part of the theorem it is enough to consider the affine mapping $f(x) = \left(\frac{x^1}{K}, x^2, \ldots, x^n\right)$. By (1.1), (1.2), (1.3), (1.4), (1.12), (1.13), (1.14), the corresponding dilatations are $\delta_L = K$, $\bar{\delta} = K^{n-1}$, $\underline{\delta} = K$, $\delta = K^{n-1}$, hence $\delta = \delta_L^{n-1} = K^{n-1}$. Thus $f(x)$ is K-QCfH according to Gehring's metric definition and K^{n-1}-QCfH according to Väisälä's definitions, and then K-QCfH according to Gehring's geometric definition.

THEOREM 2. *Väisälä's class of K-QCfH is contained in the class of $K^{\frac{2}{n}}$-QCfH according to Gehring's metric definition. The bound $K^{\frac{2}{n}}$ is the best possible.*

The first part of the theorem is an immediate consequence of the inequality (1.16) and for the second part it suffices to consider the affine mapping

$$x^{*i} = K^{\frac{2(i-1)}{n(n-1)}} x^i \qquad (i = 1, \ldots, n),$$

which is K-QCfH according to Väisälä's definitions by virtue of (1.12), (1.13) and taking into account that $a_i(x) = K^{\frac{2(i-1)}{n(n-1)}}$. But (1.14) implies

$$\delta_L(x) = \frac{a_1(x)}{a_n(x)} = K^{\frac{2}{n}},$$

and hence $f(x)$ is $K^{\frac{2}{n}}$-QCfH according to Gehring's metric definition.

THEOREM 3. *The class of K-QCfH according to Gehring's geometric definition is contained in the class of $K^{\frac{2(n-1)}{n}}$-QCfH according to his metric definition. The bound $K^{\frac{2(n-1)}{n}}$ is the best possible.*

This theorem is an immediate consequence of the preceding theorem and of theorem 2.1.

The first part of theorems 1 and 3 was established for $n = 3$ by F. Gehring [5].

Definition of Markuševič–Pesin. We shall now give a characterization of K-QCfH which is equivalent to Gehring's metric definition.

Let us first introduce some preliminary concepts.

A family of surfaces $\{\sigma_\alpha\}$ $(0 < \alpha < 1)$ is called *regular of parameter* k $(1 \leqq k < \infty)$ relatively to a point x_0, if the surfaces σ_α are the images of the spheres $|t| = \alpha$ by the homeomorphism $x = \varphi(t)$ of the ball $|t| < 1$ onto a neighbourhood of the point $x_0 = \varphi(0)$ and

$$\overline{\lim_{\alpha\to 0}} \frac{\max\limits_{x\in\sigma_\alpha} |x - x_0|}{\min\limits_{x\in\sigma_\alpha} |x - x_0|} = k.$$

A homeomorphism $f(x)$ is called *regular at* x_0, if it maps a regular family of parameter k relative to x_0 in a regular family of parameter k' $(1 \leqq k' < \infty)$ relative to $f(x_0)$. The quantity

$$q(x_0) = \inf k(x_0)k'(x_0),$$

where the infimum is taken over all regular families relative to x_0, is called *the characteristic of the mapping $f(x)$ in* x_0.

A homeomorphism $f(x)$ is called *regular in a domain D* if it is regular at each of its points.

A homeomorphism $f(x)$ of a domain D is called K-QCfH $(1 \leqq K < \infty)$ in D if the characteristic $q(x)$ of $f(x)$ is bounded in D and

$$q(x) \leqq K$$

a.e. in D.

Remark. This definition was given for $n = 2$ by Pesin [3]. Markušević definition [1] is more general and is obtained from the preceding one if $f(x)$ is no longer a homeomorphism but only a continuous mapping, and $q(x) < \infty$ in D instead of $q(x) < K'$ in D.

***K*-QCfH with two sets of characteristics.** We recall first that the characteristics of an ellipsoid E are the quantities

$$\text{(C)} \qquad \gamma_k^i,\ p_m = \frac{a_m}{a_n} \quad (i, k = 1, \ldots, n;\ m = 1, \ldots, n-1),$$

where γ_k^i are the directing cosines of the axes of E and a_m, a_n $(a_1 \geqq \ldots a_n > 0)$ are its semi-axes. The quantity p_1 is called the principal characteristic.

We say that a mapping $f(x)$ *maps an infinitesimal ellipsoid* $E[(C), x]$ *into an infinitesimal ellipsoid* $E[(C^*), f(x)]$, if it is one-to-one and continuous in a neighbourhood of x and maps every ellipsoid $E_h[(C), x]$ with the centre x, the characteristics (C) and the semiminor axis $a_n = h$ sufficiently small onto a Jordan surface $f(E_h)$ comprised between two homothetical ellipsoids $E_{h_1'}[(C^*), f(x)]$ and $E_{h_2'}[(C^*), f(x)]$ with the centre $f(x)$, the characteristics (C^*) and the semiminor axes h_1', h_2', so that

$$\lim_{h \to 0} \frac{h'_1}{h'_2} = 1$$

as E_h shrinks itself homothetically to the point x.

A homeomorphism $f(x)$ of a domain D is called K-QCfH with two sets of characteristics (C), (C^*), if it maps every infinitesimal ellipsoid $E[(C), x]$ into an infinitesimal ellipsoid $E[(C^*), f(x)]$, where the principal characteristics $p_1(x)$, $p'_1(x)$ are bounded in D and

$$p_1(x) p'_1(x) \leqq K.$$

a.e. in D.

Remark. As it is easy to see, the class of K-QCfH with two sets of characteristics is a subclass of the class of K-QCfH of Markuševič–Pesin.

K-QCfH with a set of characteristics. A homeomorphism $f(x)$ is called K-QCfH *with a set of characteristics* (C) if it maps every infinitesimal ellipsoid $E[(C), x]$ into an infinitesimal sphere, where the principal characteristic $p_1(x)$ is bounded in D and $p_1(x) \leqq K$ a.e. in D.

The condition that infinitesimal ellipsoids are mapped into infinitesimal spheres is analytically characterized by the condition

$$\lim_{h \to 0} \frac{\max\limits_{x' \in E_h} |f(x') - f(x)|}{\min\limits_{x' \in E_h} |f(x') - f(x)|} = 1.$$

Remarks. 1. This class of K-QCfH is obviously obtained from the preceding one if the infinitesimal ellipsoids $E[(C^*), f(x)]$ reduce to infinitesimal spheres, i.e. $p'_1(x) \equiv 1$. in D.

2. If the infinitesimal ellipsoids $E[(C), x]$ also reduce to infinitesimal spheres [i.e. $p_1(x) = p'_1(x) \equiv 1$ in D], then the class of K-QCfH reduces to the class of 1-QCfH, i.e. the class of conformal mappings.

3. The class of K-QCfH with one or two sets of characteristics was introduced for $n = 2$ by M. A. Lavrent'ev [14] and by M. A. Lavrent'ev and B. V. Šabat [1], respectively; however in their definitions, they assumed the continuity of all the characteristics (C) and (C), (C^*), respectively instead of the condition $p_1(x)$, respectively $p_1(x)$, $p'_1(x)$, be bounded in D.

$K_1, \ldots, K_{n-1}$-QCfH. Another particular case of the K-QCfH with two sets of characteristics is the following Krivov's definition:

A homeomorphism $f(x)$ is said to be $K_1, \ldots, K_{n-1}$-QCfH ($\infty > K_1 \geqq \ldots \ldots K_{n-1} \geqq 1$) in D if it maps infinitesimal spheres into infinitesimal ellipsoids with p_m ($m = 1, \ldots, n-1$) bounded in D and $p_m \leqq K_m$ ($m = 1, \ldots, n-1$) a. e. in D.

Remark. V. V. Krivov's [1] definition is more restrictive ($n = 3$, $f \in C^1$ and $J \neq 0$).

Some properties of the diffeomorphisms. By theorem 1, a QCfH according to Gehring's metric definition is QCfH according to Väisälä's definitions and then is differentiable a.e. in D. That is why, in order to establish the equivalence between Gehring's metric definition and Markuševič—Pesin's we need several properties of the differentiable homeomorphisms (diffeomorphisms). All the results of this chapter have been established by ourselves in [6], [20] and [23].

LEMMA 1. *Let $f(x)$ be an affine transformation with the principal characteristic p_1. Then $p_1 = q$ (where q is the characteristic in the definition of Markuševič—Pesin).*

An affine transformation is clearly a K-QCfH with a set of characteristics. Let $E_h = E_h[(C), x]$ be the ellipsoid mapped by $f(x)$ into a sphere. E_h is comprised between two concentric spheres S_1, S_2 of radii r_1, r_2, so that $\frac{r_1}{r_2} = p_1$, i.e. E_h lies in S_1 and is tangent to it and lies outside S_2 and is tangent to it. For the family of ellipsoids $\{E_h\}$ we have $k(x)k'(x) = p_1$, where k, k' are the parameters of the regular families involved in the definition of the characteristic q and $k'(x) = 1$ because $f(E_h)$ is a sphere. Hence clearly $p_1 \geqq q$.

We shall prove that $p_1 \leqq q$. If we consider a regular family of parameter $k \geqq p_1$ in a point x, then obviously $kk' \geqq p_1$. Let us consider now a regular family Σ of parameter $k < p_1$ and let be $\tilde{\sigma} \in \Sigma$ and comprised between the two spheres $\tilde{S}_1$, $\tilde{S}_2$ of radii $\tilde{r}_1$, $\tilde{r}_2$ so that $\frac{\tilde{r}_1}{\tilde{r}_2} = \tilde{k} < p_1$. Let $E_{\tilde{h}}$ be an ellipsoid comprised between $\tilde{S}_1$, $\tilde{S}_2$ with the same directing cosines γ_k^i $(i, k = 1, \ldots, n)$ and the same distribution of the semi-axes as E_h. It is easy to see that $\tilde{r}_1$, $\tilde{r}_2 = \tilde{k}\tilde{r}_1$ are respectively the semiminor and the semimajor axes of $E_{\tilde{h}}$. But $\tilde{f}(x)$ maps E_h into a sphere by stretching the semiminor axis $p_1\lambda$ times and the semimajor axis only λ times. Hence, since the direction of the semi-axes of $E_{\tilde{h}}$ is the same as those of E_h, we conclude that $f(x)$ maps $E_{\tilde{h}}$ into another ellipsoid $f(E_{\tilde{h}})$ by stretching the semiminor axis $p_1\lambda$ times and the semimajor axis λ times. Thus, the semimajor axis of $f(E_{\tilde{h}})$ is $p_1\lambda\tilde{r}_2$ and the semiminor $\tilde{k}\lambda\tilde{r}_2$. Hence, the principal characteristic of $f(E_{\tilde{h}})$ is $\tilde{p}_1 = \frac{p_1}{\tilde{k}}$ and $f(E_{\tilde{h}})$ is comprised between the spheres of radii $\tilde{R}_2 = \tilde{k}\lambda\tilde{r}_2$ and $\tilde{R}_1 = p_1\lambda\tilde{r}_2$. Thus, $\frac{\tilde{r}_1}{\tilde{r}_2} \cdot \frac{\tilde{R}_1}{\tilde{R}_2} = p_1$. If instead $E_{\tilde{h}}$ we consider $\tilde{\sigma}$, then obviously $\frac{\tilde{r}_1}{\tilde{r}_2} = \tilde{k}$ is unchanged, but $\frac{\tilde{R}_1}{\tilde{R}_2}$ does not decrease, so that, in this case, $\frac{\tilde{r}_1}{\tilde{r}_2} \cdot \frac{\tilde{R}_1}{\tilde{R}_2} \geqq p_1$. Then, since this inequality holds for all $\tilde{\sigma} \in \Sigma$ with the corresponding $\tilde{k} < p_1$ and since $\tilde{k} < p_1$ for all $\tilde{\sigma} \in \Sigma$ with a diameter sufficiently small, we conclude that $kk' \geqq p_1$, where k' is the parameter of the family $f(\Sigma)$. Thus in the two cases ($k \geqq p_1$ and $k < p_1$) we have

$kk' \geqq p_1$ and since $kk' = p_1$ for the family of ellipsoids $\{E_h\}$, we conclude that $q(x) = \inf kk' = p_1$, where x is an arbitrary point.

We recall that in a point of affinity (A-point) of a mapping $f(x)$ the corresponding linear transformation

$$f_1(x, x_0) = f(x_0) + \frac{\partial f(x_0)}{\partial x^i}(x^i - x_0^i) \tag{2}$$

maps certain ellipsoids into spheres.

Let us call *dilatation modulus* the quantity

$$\left|\frac{\partial f}{\partial s}\right| = \lim_{|\Delta x|_s \to 0} \left|\frac{\Delta f}{\Delta x}\right|,$$

where $|\Delta x|_s \to 0$ means $x \to x_0$ in the direction s.

LEMMA 2. *In an A-point of a mapping f(x) the double inequality*

$$\frac{p_1 \cdots p_{n-1}|J|}{p_1^n} \leqq \left|\frac{\partial f}{\partial s}\right|^n \leqq p_1 \cdots p_{n-1}|J|$$

holds. There is equality in the first part if s is the direction of the major axis of the ellipsoid E_h corresponding to (2) *and there is equality in the second part if s is the direction of the minor axis of E_h.*

The equation of the infinitesimal ellipsoid $E[(C), x]$ corresponding to the linear transformation (2) can be written under the form

$$\alpha_{ik}\, dx^i\, dx^k = \sqrt[n]{p_1^2 \cdots p_{n-1}^2}\, a_n^2,$$

where $\det|\alpha_{ik}| = 1$ and

$$\sum_{\substack{i_1, \ldots, i_m \\ i_1 < \ldots < i_m}}^{n} \begin{vmatrix} \alpha_{i_1 i_1} & \cdots & \alpha_{i_1 i_m} \\ \cdots & \cdots & \cdots \\ \alpha_{i_m i_1} & \cdots & \alpha_{i_m i_m} \end{vmatrix} > 0 \qquad (m = 1, \ldots, n-1)(\alpha_{i_a i_b} = \alpha_{i_b i_a}).$$

But, since, by hypothesis, $f(x)$ maps $E[(C), x]$ into infinitesimal spheres, it follows that

$$\rho^2 = \delta_{mp}\, dx^{*m}\, dx^{*p} = \delta_{mp} x_k^{*m} x_q^{*p} dx^k dx^q,$$

where δ_{mp} is the Kronecker delta and

$$\delta_{mp} x_k^{*m} x_q^{*p}\, dx^k\, dx^q = \rho^2$$

is the equation of $E[(C), x]$ written under another form. The proportionality of the coefficients, taking into account that det $|\alpha_{ik}| = 1$, yields

$$\frac{\alpha_{qk}}{\delta_{mp} x_q^{*m} x_k^{*p}} = \frac{1}{\sqrt[n]{J^2}} = \frac{\sqrt[n]{p_1^2 \cdots p_{n-1}^2 a_n^2}}{\rho^2}.$$

The last equation of the preceding system allows us to conclude that the dilatation in the direction of the minor axis is

$$\frac{\rho}{a_n} = \sqrt[n]{p_1 \cdots p_{n-1} |J|},$$

and in the direction of the major axis is

$$\frac{\rho}{a_n p_1} = \sqrt[n]{\frac{p_2 \cdots p_{n-1} |J|}{p_1^{n-1}}}.$$

To complete the proof it is enough to observe that the maximal dilatation of a mapping which transforms infinitesimal ellipsoids into infinitesimal spheres is in the direction of the minor axis, and the minimal dilatation is in the direction of the major axis.

LEMMA 3. *If $f(x)$ is a mapping, then*

$$\Lambda_f(x_0) = \overline{\lim_{x \to x_0}} \frac{|f(x) - f(x_0)|}{|x - x_0|} = \sup_s \left| \frac{\partial f(x_0)}{\partial s} \right|, \tag{3}$$

$$\lambda_f(x_0) = \underline{\lim_{x \to x_0}} \frac{|f(x) - f(x_0)|}{|x - x_0|} = \inf_s \left| \frac{\partial f(x_0)}{\partial s} \right| \tag{4}$$

hold whenever $f(x)$ is differentiable.

Let us establish (3). Clearly

$$\Lambda_f(x_0) \geqq \sup_s \left| \frac{\partial f(x_0)}{\partial s} \right|.$$

The inequality of opposite sense will be proved by *reductio ad absurdum.* Suppose then, to prove it is false, that

$$K = \sup_s \left| \frac{\partial f(x_0)}{\partial s} \right| < \Lambda_f(x_0) = K' = K + \alpha, \quad \alpha > 0. \tag{5}$$

at a point of differentiability x_0. Hence, there exists a sequence $\{x_m\}$, $x_m \to x_0$, such that

$$\lim_{n\to\infty} \frac{|f(x_m)-f(x_0)|}{|x_m - x_0|} = K + \alpha,$$

and then, for a sufficiently large index m,

$$\frac{|f(x_m)-f(x_0)|}{|x_m - x_0|} > K + \frac{\alpha}{2}. \tag{6}$$

On the other hand, the differentiability of $f(x)$ implies that

$$\frac{|f(x_m)-f(x_0)|}{|x_m - x_0|} \leqq \frac{|f_1(x_m, x_0)-f(x_0)|}{|x_m - x_0|} + |\varepsilon(x_m, x_0)|. \tag{7}$$

But the point x_m corresponding to this index m lies on a ray starting from x_0, so that, since

$$\frac{|f_1(x, x_0)-f(x_0)|}{|x - x_0|_s} = \left|\frac{\partial f(x_0)}{\partial x^i}\frac{x^i - x_0^i}{|x - x_0|_s}\right| = \left|\frac{\partial f(x_0)}{\partial x^i}\cos(s, x^i)\right|$$

and since it depends only on s, it follows that

$$\frac{|f_1(x_m, x_0)-f(x_0)|}{|x_m - x_0|} = \frac{|f_1(x, x_0)-f(x_0)|}{|x - x_0|_s} =$$

$$= \left|\frac{\partial f_1(x, x_0)}{\partial s}\right| = \left|\frac{\partial f(x_0)}{\partial s}\right| \leqq \sup_s \left|\frac{\partial f(x_0)}{\partial s}\right| = K.$$

Hence, by (7) and taking into account that, for a sufficiently large index m, $|\varepsilon(x_m, x_0)| < \frac{\alpha}{2}$, we obtain

$$\frac{|f(x_m)-f(x_0)|}{|x_m - x_0|} < K + \frac{\alpha}{2},$$

which contradicts (6), and therefore (5), and allows us to conclude that (3) holds. A slight modification of the above argument establishes formula (4).

As an immediate consequence of lemmas 2 and 3 we have

LEMMA 4. *If $f(x)$ is a mapping, then*

$$\Lambda_f = \sup_s \left| \frac{\partial f}{\partial s} \right| = \sqrt[n]{p_1 \cdots p_{n-1}|J|}, \tag{8}$$

$$\lambda_f = \inf_s \left| \frac{\partial f}{\partial s} \right| = \sqrt[n]{\frac{p_2 \cdots p_{n-1}|J|}{p_1^{n-1}}}. \tag{9}$$

hold at every A-point of $f(x)$.

COROLLARY. *At an A-point of a mapping $f(x)$*

$$\frac{\Lambda_f}{\lambda_f} = p_1, \tag{10}$$

holds, where p_1 is the principal characteristic of the ellipsoid corresponding to the linear transformation (2).

A singular linear transformation is defined by

$$\begin{cases} x^{*k} = a_i^k x^i + b^k & (k = 1, \cdots, r), \\ x^{*m} = \lambda_k^m a_i^k x^i + b^m & (m = r+1, \cdots, n), \end{cases}$$

where the matrix $\|a_i^k\|$ has the rank r $(1 \leqq r \leqq n-1)$, $\delta^{ip} a_i^k a_p^k \neq 0$, $\delta^{kq} \lambda_k^m \lambda_q^m \neq 0$ $(m, k = 1, \ldots, r)$ and the dummy indices k, q take on the values $1, \ldots, r$ and i, p the values $1, \ldots, n$.

LEMMA 5. *The dilatation modulus of a singular linear transformation $f(x)$ with the matrix of rank r verifies the formula*

$$\left| \frac{\partial f}{\partial s} \right| = \left| (e_k + \lambda_k^m e_m) \sqrt{\delta^{ip} a_i^k a_p^k} \cos(s, s_k) \right|,$$

where m takes on the values $r+1, \ldots, n$ and i, p the values $1, \ldots, n$, and where s_k $(k = 1, \ldots, r)$ is a direction normal to the plane $a_i^k x^i =$ const.

Indeed,

$$\left| \frac{\partial f}{\partial s} \right| = \left| (e_k + \lambda_k^m e_m) \frac{\partial (a_i^k x^i)}{\partial s} \right| =$$

$$\left| (e_k + \lambda_k^m e_m) \sqrt{\delta^{ip} a_i^k a_p^k} \sum_{i=1}^{n} \cos(x, s^i) \cos(s_k, x^i) \right| =$$

$$= \left| (e_k + \lambda_k^m e_m) \sqrt{\delta^{ip} a_i^k a_p^k} \cos(s, s_k) \right|.$$

Hence, it follows that the dilatation modulus vanishes in any direction normal to the r-dimensional plane determined by the directions s_k $(k = 1,\ldots, r)$, whereas it is different of zero in all the other directions.

LEMMA 6. *At a point of differentiability* x_0 *of a mapping* $f(x)$, *only one of the following three possibilities may exist:*

1. $J(x_0) = 0$ *and the rank* $r = 0$; *then* $\left|\dfrac{\partial f(x_0)}{\partial s}\right| = 0$ *for all s.*

2. $J(x_0) = 0$ *and the rank* $r > 0$; *then the corresponding linear transformation* (2) *is singular and*

$$\left|\frac{\partial f}{\partial s}\right| = \left|(e_k + \lambda_k^m e_m)\sqrt{\delta^{ip} a_i^k a_p^k}\cos(s, s_k)\right|,$$

where s_k $(k = 1,\ldots, r)$ *are directions normal to the plane* $a_i^k x^i =$ constant. *Hence there exist* ∞^{n-r-1} *directions* s' *normal to the r-dimensional plane determined by the directions* s_k $(k = 1,\ldots, r)$ *such that* $\left|\dfrac{\partial f(x_0)}{\partial s'}\right| = 0.$

3. $J(x_0) \neq 0$. *Then the linear transformation* $f_1(x, x_0)$ *is non-singular and in any direction s*

$$\left|\frac{\partial f}{\partial s}\right| = \left|\frac{\partial f_1}{\partial s}\right| \neq 0.$$

Case 1 is obvious, case 2 is an immediate consequence of the preceding lemma and case 3 follows from lemma 2.

Some properties of the K-QCfH in Markuševič—Pesin sense. These properties will be used to prove the equivalence between Gehring's metric definition and Markuševič—Pesin's.

THEOREM 4. *Let* x_0 *be a point of differentiability of a* K-QCfH (*in Markuševič—Pesin's sense*) $f(x)$ *and let* $\left|\dfrac{\partial f(x_0)}{\partial s_0}\right|$ *be the dilatation modulus in the direction* s_0. *Then in every direction s, the dilatation modulus* $\left|\dfrac{\partial f(x_0)}{\partial s}\right|$ *exists and satisfies the double inequality*

$$\frac{1}{q(x_0)}\left|\frac{\partial f(x_0)}{\partial s_0}\right| \leqq \left|\frac{\partial f(x_0)}{\partial s}\right| \leqq q(x_0)\left|\frac{\partial f(x_0)}{\partial s_0}\right|, \tag{11}$$

where $q(x)$ *is the characteristic in Markuševič—Pesin's definition.*

Let $\{\sigma_\alpha\}$ be a regular surface family of parameter k relatively to x_0. According to the hypotheses of the theorem, $f(x)$ maps $\{\sigma_\alpha\}$ into a regular surface family of parameter k'. Hence

$$\overline{\lim_{\alpha\to 0}} \frac{|f(x_\alpha)-f(x_0)|_s}{|f(x_\alpha)-f(x_0)|_{s_0}} \leqq k', \quad \overline{\lim_{\alpha\to 0}} \frac{|x_\alpha - x_0|_{s_0}}{|x_\alpha - x_0|_s} \leqq k,$$

where $x_\alpha \in \sigma_\alpha$ and $|x_\alpha - x_0|_s$, $|f(x_\alpha) - f(x_0)|_s$ are the norm of the vector $x_\alpha - x_0$ of a direction s and of its image by $f(x)$, respectively. Then

$$\left|\frac{\partial f(x_0)}{\partial s}\right| = \overline{\lim_{\alpha\to 0}} \left[\frac{|f(x_\alpha)-f(x_0)|_s}{|f(x_\alpha)-f(x_0)|_{s_0}} \frac{|f(x_\alpha)-f(x_0)|_{s_0}}{|x_\alpha - x_0|_{s_0}} \frac{|x_\alpha - x_0|_{s_0}}{|x_\alpha - x_0|_s}\right]$$

$$\leqq k(x_0)k'(x_0)\left|\frac{\partial f(x_0)}{\partial s_0}\right|.$$

But this inequality holds for every regular family relatively to x_0, so that

$$\left|\frac{\partial f(x_0)}{\partial s}\right| \leqq q(x_0)\left|\frac{\partial f(x_0)}{\partial s_0}\right|.$$

Now, interchanging s and s_0, we also obtain the first part of inequality (11).

THEOREM 5. *The Jacobian of a K-QCfH (in Markuševič—Pesin's sense) $f(x)$ is zero at a point of differentiability x_0 if and only if all its first order partial derivatives vanish at x_0.*

Let be $J(x_0) = 0$ and suppose, to prove it is false, that at least one first order partial derivative, say $x_q^{*p}(x_0) \neq 0$. Then $\left|\frac{\partial f(x_0)}{\partial x^q}\right| \neq 0$ and the preceding theorem yields $\left|\frac{\partial f(x_0)}{\partial s}\right| \neq 0$ in all the directions s. Hence, by the preceding lemma, $J(x_0) \neq 0$. The contradiction obtained establishes the theorem.

THEOREM 6. *Suppose the K-QCfH (in Markuševič—Pesin's sense) $f(x)$ is differentiable at a point x_0. Then formulae (8) and (9) hold.*

If $J(x_0) \neq 0$, the theorem is an immediate consequence of lemma 4 and if $J(x_0) = 0$ it is a consequence of the preceding theorem and lemma.

COROLLARY. *In a point of differentiability x_0 of a K-QCfH (in Markuševič—Pesin's sense) $f(x)$, the inequality*

$$|J(x_0)| \geqq \frac{\Lambda_f(x_0)^n}{K^{n-1}} \tag{12}$$

holds.

The corollary is an immediate consequence of the preceding theorem and lemma 4.

THEOREM 7. *The inverse of a K*-QCfH *in Markuševič — Pesin sense is again a K*-QCfH *in Markuševič—Pesin's sense, and the composite mapping of a* K'-QCfH *and a* K''-QCfH *(both in Markuševič—Pesin's sense) is a* $K'K''$-QCfH *in Markuševič—Pesin's sense.*

It follows immediately from the definition.

Covering in the sense of Vitali. Let $E \subset R^n$ be a set with $m^*E < \infty$. For every point $x \in E$, let us associate a sequence of closed sets $\{E_k(x)\}$. We shall say that the set of all these sequences *covers E in the sense of Vitali* if for each $k = 1, 2, \ldots$ it satisfies the following conditions:

1° $E_k(x) \subset Q_k(x)$, where $Q_k(x)$ is a cube centered at x and with the edge $a_k(x)$ such that $a_k(x) \to 0$ as $k \to \infty$.

2° There exists an $\alpha(x) > 0$ such that

$$\frac{mE_k(x)}{mQ_k(x)} > \alpha(x).$$

Vitali's theorem. If $E \subset R^n$, $m^*E < \infty$, and $\{E_m(x)\}$, where x runs over E, is a covering in the sense of Vitali of the set E then there exists a finite number of disjoint sets $E_p = E_{mp}(x_p)$ $(p = 1, \ldots, q)$ so that

$$m\left(\sum_{p=1}^{q} E_p\right) > \frac{1}{2} m^*E.$$

In the same hypothesis, there exists a countable set of sets $E_p = E_{mp}(x_p)$ $(p = 1, 2, \ldots)$ so that

$$m(E \cap \Sigma_p E_p) = m^*E.$$

(For the proof see C. Carathéodory [3], theorem 2, p. 304 and theorem 3, p. 305.)

THEOREM 8. *A* QCfH *in the sense of Markuševič—Pesin and its inverse have the maximal dilatations* $\Lambda_f(x) < \infty$ *and* $\Lambda_{f^{-1}}(x^*) < \infty$, *a.e. in D and a.e. in* D^*, *respectively.*

The first part of the theorem will be proved by *reductio ad absurdum.* Suppose, to prove it is false, that

$$\Lambda_f(x) = \infty \tag{13}$$

holds a.e. in a set $E \subset D$ with $m^*E > 0$. Hence, there exists a domain $D_1, \subset\subset D$, such that if $E_1 = E \cap D_1$, then $m^*E_1 > 0$. Indeed, fix $x \in E$ and denote by $V_x \subset D$ a neighbourhood of x. Since the space R^n is regular, we can choose a neighbourhood $U_x, \subset\subset V_x$. The set $\{U_x\}$ where x runs over E is a covering of E. By Lindelöf's theorem we may pick out a countable covering $\{U_{x_k}\}$. Then, there exists at least a point $x_k \in E$ such that $m^*(U_{x_k} \cap E) > 0$, as it clearly follows from the inequality

$$0 < m^*E \leqq \sum_{k=1}^{\infty} m^*(U_{x_k} \cap E).$$

Such a neighbourhood U_{x_k} satisfies all the conditions imposed to D_1. The contradiction will follow from the fact that on the one hand we obtain $mf(D_1) < \infty$, since $f(x)$ is continuous on $\bar{D}_1$, and on the other hand we prove (using an appropriate covering of E_1 in the sense of Vitali and Vitali's theorem) that $mf(D_1)$ can be made to become as large as one wishes.

Fix a sufficiently large number N. Condition (13) implies that for each $x_0 \in E$, there exists a sequence of points $x_m = x_m(x_0) \in D_1$ $(m = 1, 2, \ldots)$ so that

$$\frac{|f(x_m) - f(x_0)|}{|x_m - x_0|} > N, \tag{14}$$

holds and $\rho_m = |x_m - x_0|$ steadily decreases to the limit zero.

Let $\{\sigma_\alpha\}$ $(0 < \alpha < 1)$ be a regular surface family of parameter k relatively to the point x_0, let $\{\bar{G}_\alpha\}$ be the family of closed domains bounded by the corresponding surfaces $\{\sigma_\alpha\}$, let be $E^* = f(\bar{G}_\alpha)$ and let $\{f(\sigma_\alpha)\}$ be the corresponding surface family regular of a parameter k'. Suppose the family $\{\sigma_\alpha\}$ is chosen so that $kk' < K_1 + 1$, where the constant K_1 is a bound for $q(x)$ in D. Each x_m lies on a surface σ_m comprised between two spheres centred at x_0 and of the radii R_m and r_m $(R_m > r_m)$. Similarly the surfaces $\sigma_m^* = f(\alpha_m)$ are comprised between two spheres centred at $f(x_0)$ and of radii R_m^* and r_m^* $(R_m^* > r_m^*)$. For a sufficiently large index m, $\sigma_m \subset D_1$ and

$$\frac{R_m}{r_m} < k + 1, \qquad \frac{R_m^*}{r_m^*} < k' + 1. \tag{15}$$

hold. The set of all sequences $\{\bar{G}_m\}$ of closed domains covers E_1 in the sense of Vitali, because

$$\frac{mG_m}{\omega_n R_m^n} \geqslant \frac{\omega_n r_m^n}{\omega_n R_m^n} > \frac{1}{(k+1)^n}.$$

holds for each point which belongs to E_1. Next appealing to Vitali's theorem we can pick out from $\{\bar{G}_m\}$ a finite set of disjoint closed domains $\Delta_p = \bar{G}_{m_p}(x_p)$ $(p = 1, \ldots, q)$, so that

$$\sum_{p=1}^{q} m\Delta_p > \frac{1}{2} m^*E.$$

Since $f(x)$ is one-to-one, it follows that the domains $f(\Delta_p)$ are disjoint, hence

$$mf(D_1) > \sum_{p=1}^{q} mf(\Delta_p).$$

and by (15) and (14)

$$mf(\Delta_p) \geqslant \omega_n r_p^{*n} > \frac{\omega_n R_p^{*n}}{(k'+1)^n} > \frac{\omega_n \rho_p^n N^n}{(k'+1)^n} > \frac{N^n \omega_n r_p^n}{(k'+1)^n} >$$

$$> \frac{N^n \omega_n R_p^n}{(k+1)^n (k'+1)^n} > \frac{N^n m\Delta_p}{(K_1+2)^{2n}}$$

and then

$$mf(D_1) > \frac{N^n}{(K_1+2)^{2n}} \sum_{p=1}^{q} m\Delta_p > \frac{N^n}{2(K_1+2)^{2n}} m^* E_1,$$

which holds for N as large as one pleases, so contradicting the inequality $mf(D_1) < \infty$.

The second part of the theorem is an immediate consequence of the preceding theorem.

THEOREM 9. *A* QCfH *(in Markuševič—Pesin's sense) $f(x)$ is differentiable a.e. in D.*

The preceding theorem implies that the n functions $x^{*i}(x)$ $(i = 1, \ldots, n)$ of the mapping $x^* = f(x)$ satisfy the hypothesis of the theorem of Rademacher–Stepanov. Hence every $x^{*i}(x)$ is differentiable a.e. in D and then all the n functions are differentiable simultaneously a.e. in D, which is the same with the differentiability of $f(x)$ a.e. in D.

THEOREM 10. *Let $f(x)$ be a K-*QCfH *(in Markuševič — Pesin's sense) of the unit ball B into itself so that $f(O) = O$ and let $\sigma \subset B$ be a Jordan surface containing the origin,* $B_0 = \left[\frac{2^{(n-3)}3\omega_n(\pi K)^{n-1}}{(n-1)\omega_{n-1}}\right]^{\frac{1}{n}}$, $r_0^* = \min_{x \in \sigma} |f(x)|$ *and* $r_0 =$ $d(O, \sigma)$. *Then*

$$r_0^* e^{-\left(\frac{B_0}{|x|}\right)^n} < |f(x)| < \frac{B_0}{\left(\log \frac{r_0}{|x|}\right)^{\frac{1}{n}}} \tag{16}$$

for every $x \in \{f^{-1}[B(r_0^*)] \cap B(r_0)\}$. *In particular, if* $f: B \rightleftarrows B$, *then*

$$e^{-\left(\frac{B_0}{|x|}\right)^n} < |f(x)| < \frac{B_0}{\left(\log \frac{1}{|x|}\right)^{\frac{1}{n}}}$$

whenever $x \in B$.

Fix $x_1 \in B(r_0)$, $x_1 \neq 0$ and set $|x_1| = r_1$. Let us estimate the volume of $f(r_1 \leqq |x| \leqq r_0)$. The preceding theorem implies that $f(x)$ is differentiable a.e. in B. Let E be the set of the points of differentiability of $f(x)$. From the inequality (1.40), which holds in every point of differentiability, and appealing to de la Vallée-Poussin's decomposition theorem (Chapter 1), we obtain

$$\omega_n > mf(r_1 < |x| < r_0) \geqq mf[(r_1 < |x| < r_0) \cap E] \geqq \int\limits_{(r_1<|x|<r_0)\cap E} |J|\, d\tau. \quad (17)$$

Hence, by (12) and since $mE = \omega_n$, we get

$$\omega_n > \frac{1}{K^{n-1}} \int\limits_{(r_1<|x|<r_0)\cap E} \Lambda_f^n\, d\tau = \frac{1}{K^{n-1}} \int\limits_{r_1<|x|<r_0} \Lambda_f^n\, d\tau = \frac{1}{K^{n-1}} \int_{r_1}^{r_0} r^{n-1}\, dr \int_S \Lambda_f^n\, d\sigma =$$

$$= \frac{1}{K^{n-1}} \int_{r_1}^{r_0} r^{n-1}\, dr \int_0^{\pi} \cdots \int_0^{\pi}\int_0^{2\pi} \Lambda_f^n \sin^{n-2}\vartheta_1 \cdots \sin\vartheta_{n-2}\, d\vartheta_1 \cdots d\vartheta_{n-1} = \quad (18)$$

$$= \frac{1}{K^{n-1}} \int_{r_1}^{r_0} r^{n-2}\, dr \int_0^{\pi} \cdots \int_0^{\pi} \sin^{n-2}\vartheta_1 \cdots \sin\vartheta_{n-2}\, d\vartheta_1 \cdots d\vartheta_{n-2} \int_0^{2\pi} \Lambda_f^n r\, d\vartheta_{n-1}.$$

Hölder's inequality yields

$$\left(\int_0^{2\pi} \Lambda_f r d\vartheta_{n-1}\right)^n \leqq \int_0^{2\pi} \Lambda_f^n r d\vartheta_{n-1} \left(\int_0^{2\pi} r d\vartheta_{n-1}\right)^{n-1} = (2\pi r)^{n-1} \int_0^{2\pi} \Lambda_f^n r d\vartheta_{n-1}, \quad (19)$$

and the theorems 6 and 9 imply that

$$\int_0^{2\pi} \Lambda_f r d\vartheta_{n-1} \geqq \int_\gamma \Lambda_f\, ds \geqq \int_\gamma \frac{ds^*}{ds}\, ds = \int_{\gamma^*} ds^* = l(r) > 2|f(x_1)| \quad (20)$$

in E, i.e. for almost all $r(r_1 < r < 1)$, or for at least one $k(k = 1, \ldots, n-2)$ for almost all $\vartheta_k(0 < \vartheta_k < \pi)$ and where $l(r)$ is the length of the image of

$$|x| = r,\ \vartheta_k = \text{const.} \qquad (k = 1, \ldots, n-2).$$

Inequalities (18), (19) and (20) imply that

$$\omega_n > \frac{2|f(x_1)|^n}{(\pi K)^{n-1}} \int_{r_1}^{r_0} \frac{dr}{r} \int_0^{\pi} \cdots \int_0^{\pi} \sin^{n-2}\vartheta_1 \cdots \sin\vartheta_{n-2}\, d\vartheta_1 \cdots d\vartheta_{n-2} >$$

$$> \frac{2|f(x_1)|^n}{(\pi K)^{n-1}} \log\frac{r_0}{r_1} \int_{\frac{\pi}{6}}^{\frac{5\pi}{6}} \int_0^{\pi} \cdots \int_0^{\pi} \sin^{n-2}\vartheta_1 \cdots \sin\vartheta_{n-2}\, d\vartheta_1 \cdots d\vartheta_{n-2} >$$

$$> \frac{|f(x_1)|^n}{2^{n-3}(\pi K)^{n-1}} \log \frac{r_0}{r_1} \int_{\frac{\pi}{6}}^{\frac{5\pi}{6}} \int_0^{\pi} \int_0^{\pi} \sin^{n-3} \vartheta_2 \cdots \sin \vartheta_{n-2} \, d\vartheta_1 \cdots d\vartheta_{n-2} =$$

$$= \frac{(n-1)\,\omega_{n-1} |f(x_1)|^n}{2^{n-3}\, 3(\pi K)^{n-1}} \log \frac{r_0}{r_1},$$

hence

$$|f(x_1)| < \left[\frac{2^{n-3}\, 3\omega_n (\pi K)^{n-1}}{(n-1)\,\omega_{n-1} \log \dfrac{r_0}{|x_1|}} \right]^{\frac{1}{n}},$$

which proves the second part of the inequality (16) for every $x_1 \in B(r_0)$. But we have seen that the inverse mapping $x = f^{-1}(x^*)$ is also K-QCfH in Markuševič–Pesin's sense (with the same K) and applying the preceding inequality to $x = f^{-1}(x^*)$ we obtain

$$|x_1| < \left[\frac{2^{(n-3)}\, 3\omega_n (\pi K)^{n-1}}{(n-1)\,\omega_{n-1} \log \dfrac{r_0^*}{|f(x_1)|}} \right]^{\frac{1}{n}}$$

for every $x \in f^{-1}[B(r_0^*)]$, hence

$$|f(x_1)| > r_0^* e^{\frac{-2^{n-3}\, 3\omega_n (\pi K)^{n-1}}{(n-1)\,\omega_{n-1} |x_1|^n}},$$

which establishes also the second part of the inequality (16).

Remark. This theorem was established by us in [23]. In the case $n=3$, J. Väisälä [1] obtained for his K-QCfH with the additional conditions $f(B) = B$ and $f(0) = 0$, the inequality

$$|f(x)| \leqq \frac{8K}{\left(\log \dfrac{1}{|x|}\right)^2}.$$

I. S. Ovčinikov and G. D. Suvorov [1] proved that for the homeomorphisms $f \in C^1$ with the additional conditions $f(B) = B, f(0) = 0$ and $I(f, B) \leqq k$, $I(f^{-1}, B) \leqq k$, where

$$I(f, B) = \int_B |\nabla f|^3 \, d\tau_3,$$

the double inequality

$$\frac{1}{2}\exp\left[-\frac{N_1(k)}{d(x',x'')^3}\right] \leqq d[f(x'),f(x'')] \leqq \frac{N_1(k)^{\frac{1}{3}}}{\left[\log\frac{1}{d(x',x'')}\right]^{\frac{1}{3}}},$$

holds whenever $d(x', x'') \leqq M(k)$, where M, N_1 are constants depending only on k.

THEOREM 11. *Let $f(x)$ be a K-QCfH in Markuševič—Pesin's sense in D. Then*

$$\delta_L(x) = \overline{\lim_{r\to 0}} \frac{\max\limits_{|x'-x|=r} |f(x')-f(x)|}{\min\limits_{|x'-x|=r} |f(x')-f(x)|} < B_0(K_1+2)^3\, e^{[B_0(K_1+2)^3]^n}$$

holds in D, where B_0 is the constant in the preceding theorem and K_1 is a bound for $q(x)$ in D.

Fix $x_0 \in D$ and let $\{\sigma_\alpha\}$ $(0 < \alpha < 1)$ be a surface family regular of parameter k relative to x_0. The family $\{f(\sigma_\alpha)\}(0 < \alpha < 1)$ is regular of a parameter k'. Suppose now we choose $\{\sigma_\alpha\}$, so that $kk' < K+1$. For every $\alpha \leqq \alpha_0$ (α_0 sufficiently small), $\sigma_\alpha \subset D$ and

$$\frac{\max\limits_{x\in\sigma_\alpha} |x-x_0|}{\min\limits_{x\in\sigma_\alpha} |x-x_0|} < k+1, \quad \frac{\max\limits_{x\in\sigma_\alpha} |f(x)-f(x_0)|}{\min\limits_{x\in\sigma_\alpha} |f(x)-f(x_0)|} < k'+1. \tag{21}$$

Set

$$\rho_0 = \max_{x\in\sigma_\alpha} |x-x_0|,\ \rho_0^* = \max_{x\in\sigma_\alpha} |f(x)-f(x_0)|,$$

and

$$y^* = \psi(y) = \frac{1}{\rho_0^*}\varphi(\rho_0 y) = \frac{1}{\rho_0^*}\left\{f\left[\frac{\rho_0(x-x_0)}{\rho_0}+x_0\right]-f(x_0)\right\} = \frac{1}{\rho_0^*}[f(x)-f(x_0)], \tag{22}$$

where $y = \dfrac{x-x_0}{\rho_0}$. Obviously, $\psi(0) = 0$, $|x - x_0| = \rho_0$ corresponds to the sphere $|y| = 1$, $|x^* - x_0^*| = \rho_0^*$ to the sphere $|y^*| = 1$ and σ_{α_0} to a surface σ'_{α_0}. Finally set

$$r_0 = \min_{y\in\sigma'_{\alpha_0}} |y| > \frac{1}{k+1} > \frac{1}{K_1+2},\ r_0^* = \min_{y\in\sigma'_{\alpha_0}} |\psi(y)| > \frac{1}{k'+1} > \frac{1}{K_1+2}.$$

Evidently, since $x^* = f(x)$ is K-QCfH (in Markuševič—Pesin's sense) in D it follows that $\psi(y)$ is K-QCfH (in Markuševič—Pesin's sense) in the domain bounded by σ'_{α_0}, because we obtain y from x and y^* from x^* by composition of translations and magnifications. Since for an α sufficiently small $B(x_0, \rho_0) \subset D$, $\psi(y)$ satisfies the hypotheses of the preceding theorem and then

$$\frac{e^{-\left(\frac{B_0}{|y|}\right)^n}}{K_1+2} < r_0^* e^{-\left(\frac{B_0}{|y|}\right)^n} < |\psi(y)| < \frac{B_0}{\left(\log \frac{r_0}{|y|}\right)^{\frac{1}{n}}} < \frac{B_0}{\left[\log \frac{1}{(K_1+2)|y|}\right]^{\frac{1}{n}}} \tag{23}$$

holds for every

$$y \in \left\{\psi^{-1}[B(r_0^*)] \cap B\left(\frac{1}{K_1+2}\right)\right\}.$$

Set $r_0' = \min\limits_{|y^*|=r_0^*} |\psi^{-1}(y^*)|$. We distinguish two cases: 1° $r_0' \geqq \dfrac{1}{(K_1+2)^2}$ and 2° $r_0' < \dfrac{1}{(K_1+2)^2}$.

1° $r_0' \geqq \dfrac{1}{(K_1+2)^2}$. In this case, (23) and $e < K_1 + 2$ imply that

$$\frac{e^{-[B_0(K_1+2)^3]^n}}{K_1+2} < \frac{e^{-\left(\frac{B_0}{|y|}\right)^n}}{K_1+2} < |\psi(y)| < \frac{B_0}{\left[\log \frac{1}{(K_1+2)|y|}\right]^{\frac{1}{n}}} < \frac{B_0}{[\log (K_1+2)]^{\frac{1}{n}}} < B_0$$

for every y in

$$\frac{1}{(K_1+2)^3} \leqq |y| \leqq \frac{1}{(K_1+2)^2}, \tag{24}$$

hence

$$\frac{|\psi(y_1)|}{|\psi(y_2)|} < B_0(K_1+2)\, e^{[B_0(K_1+2)^3]^n}, \tag{25}$$

for all y_1, y_2 in the ring (24).

2° $r_0' < \dfrac{1}{(K_1+2)^2}$. In this case we distinguish two possibilities which we examine successively: $A \cdot r_0' \leqq \dfrac{1}{(K_1+2)^4}$ and $B \cdot \dfrac{1}{(K_1+2)^4} < r_0' < \dfrac{1}{(K_1+2)^2}$.

$A \cdot r_0' \leqq \dfrac{1}{(K_1+2)^4}$. Let σ'_{α_1} be the surface of the family $\{\sigma'_\alpha\}$ which lies inside the surface $\psi^{-1}(|x^*| = r_0^*)$ and is tangent to it. In this case, (21) yields by one hand

$$\min_{y \in \sigma'_{\alpha_1}} |y| \leqq \frac{1}{(K_1+2)^4}, \quad \max_{y \in \sigma'_{\alpha_1}} |y| < \frac{1}{(K_1+2)^3},$$

and by the other hand, taking into account that $f(\sigma'_{\alpha_1})$ is tangent to the sphere $|y^*| = r_0^* > \dfrac{1}{K_1+2}$,

$$\min_{y \in \sigma'_{\alpha_1}} |\psi(y)| > \frac{1}{(K_1+2)^2}.$$

Thus, in this case,

$$\frac{|\psi(y_1)|}{|\psi(y_2)|} < (K_1+2)^2 \tag{26}$$

for all y_1, y_2 in the ring with the boundary components σ'_{α_0} and σ'_{α_0}. But this ring contains the ring (24), so that (26) holds for all y_1, y_2 in the ring (24).

$B \cdot \dfrac{1}{(K_1+2)^4} < r_0' < \dfrac{1}{(K_1+2)^2}$. Arguing as in A of 2°, we conclude that in this case (26) holds for all y_1, y_2 in the intersection E of the ring (24) with the ring having the boundary components σ'_{α_0}, σ'_{α_1}. Besides, (23) implies that (25) holds for all y_1, y_2 in the intersection of the ring (24) with $\overline{CE}$.

Let now y_1, y_2 be in the ring (24), but $y_1 \in E$, $y_2 \in CE$. Let also y_3 be in

$$\sigma'_{\alpha_1} \cap \left\{ \frac{1}{(K_1+2)^3} \leqq |y| \leqq \frac{1}{(K_1+2)^2} \right\}.$$

Then (25) and (26) yield

$$\frac{|\psi(y_1)|}{|\psi(y_2)|} = \frac{|\psi(y_1)|}{|\psi(y_3)|} \cdot \frac{|\psi(y_3)|}{|\psi(y_2)|} \leqq (K_1+2)^3 B_0\, e^{[B_0(K_1+2)^3]^n}. \tag{27}$$

But

$$(K_1+2)^2 \leqq B_0(K_1+2)^3 e^{[B_0(K_1+2)^3]^n},$$

and (27) holds for all y_1, y_2 in ring (24) for the two cases 1° and 2°.

But (27) also holds for all y_1, y_2 in any ring of the following sequence:

$$\frac{1}{(K_1+2)^{m+3}} \leqq |y| \leqq \frac{1}{(K_1+2)^{m+2}} \cdot \quad (m = 0, 1, 2, \ldots). \tag{28}$$

Indeed, let be

$$S_m = \left\{ |y| = \frac{1}{(K_1+2)^m} \right\}, \; r_m = \min_{y \in \sigma'_{\alpha m}} |y|,$$

$$r_m^* = \min_{y \in \sigma'_{\alpha m}} |\psi(y)|, \; r'_m = \min_{|y^*| = r_m^*} |\psi^{-1}(y^*)|,$$

where σ'_{α_m} is the surface of the family $\{\sigma'_\alpha\}$ which lies inside S_m and is tangent to it. Let us consider, as above, the cases $r'_m \geqq \dfrac{1}{(K_1+2)^{m+2}}$ and $r'_m < \dfrac{1}{(K_1+2)^{m+2}}$. Let $x'' = (K_1+2)^m y$ be a homothety with the ratio of similitude $(K_1+2)^m$ and set

$$\psi(y) = \psi\left[\frac{x''}{(K_1+2)^m}\right] = \chi(x'').$$

Then

$$\frac{|\psi(y_1)|}{|\psi(y_2)|} = \frac{|\chi(x'_1)|}{|\chi(x''_2)|} < B_0(K_1+2)^3 \, e^{[B_0(K_1+2)^3]^n},$$

holds for all y_1, y_2 in the corresponding ring of sequence (28), because x''_1, x''_2 lie in ring (24) and $\chi(x'')$ satisfies the conditions of the case 1° for $r'_m \geqq \dfrac{1}{(K_1+2)^{m+2}}$ and of the case 2° for $r'_m < \dfrac{1}{(K_1+2)^{m+2}}$.

But for every $r'' < \dfrac{1}{(K_1+2)^2}$, the sphere $S(r'')$ is contained in one of the rings (28), thus

$$\frac{\max\limits_{|y|=r''} |\psi(y)|}{\min\limits_{|y|=r''} |\psi(y)|} < B_0(K_1+2)^3 \, e^{[B_0(K_1+2)^3]^n}$$

and on account of the notations (22)

$$\frac{\max\limits_{|x-x_0|=r} |f(x)-f(x_0)|}{\min\limits_{|x-x_0|=r} |f(x)-f(x_0)|} = \frac{\max\limits_{|y|=r''} |\psi(y)|}{\min\limits_{|y|=r''} |\psi(y)|} < B_0(K_1+2)^3 \, e^{[B_0(K_1+2)^3]^n},$$

where $r = \rho_0 r''$. As this inequality holds for every $r < \dfrac{1}{(K_1+2)^2}$, it holds also for $r \to 0$ and this completes our proof.

The equivalence between Gehring's metric definition of K-QCfH and Markuševič—Pesin's. The results of this paragraph were established by us in [23].

THEOREM 12. *The definition of QCfH in Markuševič—Pesin's sense is equivalent to those of Gehring and Väisälä.*

The preceding theorem implies that a QCfH in Markuševič—Pesin's sense has $\delta_L(x)$ bounded in D and thus is QCfH according to Gehring's metric definition (and then also to the other Gehring's and Väisälä's definitions).

The converse inclusion is obvious, because $\delta_L(x) \leqq K$ means that the family of the spheres (which is a regular family of parameter 1) is mapped in a regular family of parameter $k'(x) \leqq K$. Thus the definition of the QCfH in Markuševič—Pesin's sense is equivalent to Gehring's definitions and hence also to Väisälä's definitions.

COROLLARY. *A* QCfH *in Markuševič—Pesin's sense in D is measurable, has* $J(x) \neq 0$ *a.e. in D and satisfies* (1.44) *whenever E is a measurable subset of D.*

This is an immediate consequence of the preceding theorem combined with the corollary of theorem 1.1

LEMMA 7. *Let* $x^* = f(x)$ *be a regular homeomorphism. Then any regular surface family* $\{\sigma_\alpha\}$ $(0 < \alpha < 1)$ *of parameter* $k (1 \leqq k < \infty)$ *relative to an A-point* x_0 *is mapped into a regular surface family of parameter* k' $(1 \leqq k' < \infty)$ *if and only if the affine mapping* $f_1(x, x_0)$ *given by* (2) *maps* $\{\sigma_\alpha\}$ $(0<\alpha<1)$ *into a regular family of parameter* k'.

For the sufficiency, let be

$$\frac{\max\limits_{x \in \sigma_\alpha} |x - x_0|}{\min\limits_{x \in \sigma_\alpha} |x - x_0|} = k + \varepsilon_\alpha,$$

where $\lim\limits_{\alpha \to 0} \varepsilon_\alpha = 0$. In the A-point x_0, (8) and (9) imply that

$$\sqrt[n]{\frac{p_2(x_0) \cdots p_{n-1}(x_0)|J(x_0)|}{p_1(x_0)^{n-1}}} = \lambda_f(x_0) \leqq$$

$$\leqq \frac{|f_1(x, x_0) - f(x_0)|}{|x - x_0|} = \left|\frac{\partial f(x_0)}{\partial s}\right| \leqq \Lambda_f(x_0) =$$

$$\sqrt[n]{p_1(x_0) \cdots p_{n-1}(x_0)|J(x_0)|},$$

where s is the direction of the vector $x - x_0$. Hence

$$\overline{\lim_{\alpha\to 0}} \frac{\max\limits_{x\in\sigma_\alpha} |f(x)-f(x_0)|}{\min\limits_{x\in\sigma_\alpha} |f(x)-f(x_0)|} \leqq$$

$$\leqq \overline{\lim_{\alpha\to 0}} \frac{\max\limits_{x\in\sigma_\alpha} |f_1(x,x_0)-f(x_0)| + \max\limits_{x\in\sigma_\alpha}\left[|\varepsilon(x,x_0)||x-x_0|\right]}{\min\limits_{x\in\sigma_\alpha} |f_1(x,x_0)-f(x_0)| - \max\limits_{x\in\sigma_\alpha}\left[|\varepsilon(x,x_0)||x-x_0|\right]} \leqq$$

$$\leqq \overline{\lim_{\alpha\to 0}} \frac{\max\limits_{x\in\sigma_\alpha} |f_1(x,x_0)-f(x_0)|}{\min\limits_{x\in\sigma_\alpha} |f_1(x,x_0)- \ (x_0)|} \lim_{\alpha\to 0} \frac{1+\dfrac{\dfrac{\max\limits_{x\in\sigma_\alpha}|\varepsilon(x,x_0)|}{\max\limits_{x\in\sigma_\alpha}|f_1(x,x_0)-f(x_0)|}}{\max\limits_{x\in\sigma_\alpha}|x-x_0|}}{1-\dfrac{\dfrac{\max\limits_{x\in\sigma_\alpha}|\varepsilon(x,x_0)|}{\min\limits_{x\in\sigma_\alpha}|f_1(x,x_0)-f(x_0)|}}{\max\limits_{x\in\sigma_\alpha}|x-x_0|}} =$$

$$= k'\overline{\lim_{\alpha\to 0}} \frac{1+\dfrac{\dfrac{\max\limits_{x\in\sigma_\alpha}|\varepsilon(x,x_0)|}{\max\limits_{x\in\sigma_\alpha}|f_1(x,x_0)-f(x_0)|}}{\min\limits_{x\in\sigma_\alpha}|x-x_0|(k+\varepsilon_\alpha)}}{1-\dfrac{\dfrac{\max\limits_{x\in\sigma_\alpha}|\varepsilon(x,x_0)|}{\min\limits_{x\in\sigma_\alpha}|f_1(x,x_0)-f(x_0)|}}{\min\limits_{x\in\sigma_\alpha}|x-x_0|(k+\varepsilon_\alpha)}} \leqq$$

$$\leqq k'\overline{\lim_{\alpha\to 0}} \frac{1+\dfrac{\dfrac{(k+\varepsilon_\alpha)\max\limits_{x\in\sigma_\alpha}|\varepsilon(x,x_0)|}{|f_1(x',x_0)-f(x_0)|}}{|x'-x_0|}}{1-\dfrac{\dfrac{(k+\varepsilon_\alpha)\max\limits_{x\in\sigma_\alpha}|\varepsilon(x,x_0)|}{|f_1(x'',x_0)-f(x_0)}}{|x''-x_0|}} \leqq$$

$$\leqq k' \lim_{\alpha\to 0} \frac{1 + \dfrac{(k+\varepsilon_\alpha)\max\limits_{x\in\sigma_\alpha}|\varepsilon(x,x_0)|}{\sqrt[n]{\dfrac{p_2(x_0)\cdots p_{n-1}(x_0)\,|J(x_0)|}{p_1(x_0)^{n-1}}}}}{1 - \dfrac{k+\varepsilon_\alpha)\max\limits_{x\in\sigma_\alpha}|\varepsilon(x,x_0)|}{\sqrt[n]{\dfrac{p_2(x_0)\cdots p_{n-1}(x_0)\,|J(x_0)|}{p_1(x_0)^{n-1}}}}} = k',$$

where

$$|f_1(x',x_0) - f(x_0)| = \max_{x\in\sigma_\alpha}|f_1(x,x_0) - f(x_0)|,$$

$$|f_1(x'',x_0) - f(x_0)| = \min_{x\in\sigma_\alpha}|f_1(x,x_0) - f(x_0)|.$$

With the same notations and arguing as above, we have

$$\overline{\lim_{\alpha\to 0}} \frac{\max\limits_{x\in\sigma_\alpha}|f(x) - f(x_0)|}{\min\limits_{x\in\sigma_\alpha}|f(x) - f(x_0)|} \geqq$$

$$\geqq \overline{\lim_{\alpha\to 0}} \frac{\max\limits_{x\in\sigma_\alpha}|f_1(x,x_0) - f(x_0)|}{\min\limits_{x\in\sigma_\alpha}|f_1(x,x_0) - f(x_0)|} \overline{\lim_{\alpha\to 0}} \frac{1 - \dfrac{\dfrac{\max\limits_{x\in\sigma_\alpha}|\varepsilon(x,x_0)|}{\max\limits_{x\in\sigma_\alpha}|f_1(x,x_0) - f(x_0)|}}{\max\limits_{x\in\sigma_\alpha}|x - x_0|}}{1 + \dfrac{\dfrac{\max\limits_{x\in\sigma_\alpha}|\varepsilon(x,x_0)|}{\min\limits_{x\in\sigma_\alpha}|f_1(x,x_0) - f(x_0)|}}{\max\limits_{x\in\sigma_\alpha}|x - x_0|}} =$$

$$= k' \lim_{\alpha\to 0} \frac{1 - \dfrac{\dfrac{(k+\varepsilon_\alpha)\max\limits_{x\in\sigma_\alpha}|\varepsilon(x,x_0)|}{\max\limits_{x\in\sigma_\alpha}|f_1(x,x_0) - f(x_0)|}}{\min\limits_{x\in\sigma_\alpha}|x - x_0|}}{1 + \dfrac{\dfrac{(k+\varepsilon_\alpha)\max\limits_{x\in\sigma_\alpha}|\varepsilon(x,x_0)|}{\min\limits_{x\in\sigma_\alpha}|f_1(x,x_0) - f(x_0)|}}{\min\limits_{x\in\alpha}|x - x_0|}} \geqq$$

$$\geqq k' \overline{\lim_{\alpha\to 0}} \frac{1-\dfrac{(k+\varepsilon_\alpha)\max\limits_{x\in\sigma_\alpha}|\varepsilon(x,x_0)|}{\dfrac{|f_1(x',x_0)-f(x_0)|}{|x'-x_0|}}}{1+\dfrac{(k+\varepsilon)\max\limits_{x\in\sigma_\alpha}|\varepsilon(x,x_0)|}{\dfrac{|f_1(x'',x_0)-f(x_0)|}{|x''-x_0|}}} \geqq$$

$$\geqq k' \overline{\lim_{\alpha\to 0}} \frac{1-\dfrac{(k+\varepsilon_\alpha)\max\limits_{x\in\sigma_\alpha}|\varepsilon(x,x_0)|}{\sqrt[n]{\dfrac{p_2(x_0)\cdots p_{n-1}(x_0)\,|J(x_0)|}{p_1(x_0)^{n-1}}}}}{1+\dfrac{(k+\varepsilon_\alpha)\max\limits_{x\in\sigma_\alpha}|\varepsilon(x,x_0)|}{\sqrt[n]{\dfrac{p_2(x_0)\cdots p_{n-1}(x_0)\,|J(x_0)|}{p_1(x_0)^{n-1}}}}} = k',$$

Thus

$$\overline{\lim_{\alpha\to 0}} \frac{\max\limits_{x\in\sigma_\alpha}|f_1(x,x_0)-f(x_0)|}{\min\limits_{x\in\sigma_\alpha}|f_1(x,x_0)-f(x_0)|} = k' \tag{29}$$

implies that

$$\overline{\lim_{\alpha\to 0}} \frac{\max\limits_{x\in\sigma_\alpha}|f(x)-f(x_0)|}{\min\limits_{x\in\sigma_\alpha}|f(x)-f(x_0)|} = k'. \tag{30}$$

The necessity is a consequence of the sufficiency. Indeed, suppose that a homeomorphism $f(x)$ maps the family $\{\sigma_\alpha\}$ in a regular family of parameter k', i.e. satisfies (30), and $f_1(x, x_0)$ would map the same family in a regular family of a parameter $k'' \neq k'$, i.e.

$$\overline{\lim_{\alpha\to 0}} \frac{\max\limits_{x\in\sigma_\alpha}|f_1(x,x_0)-f(x_0)|}{\min\limits_{x\in\sigma_\alpha}|f_1(x,x_0)-f(x_0)|} = k'' \neq k'.$$

But then, by the sufficiency, we should have also

$$\overline{\lim_{\alpha\to 0}} \frac{\max\limits_{\alpha\in\sigma_\alpha}|f(x)-f(x_0)|}{\min\limits_{x\in\sigma_\alpha}|f(x)-f(x_0)|} = k'' \neq k',$$

which would contradict (30). This completes our proof.

COROLLARY. *A QCfH in Markuševič—Pesin's sense maps infinitesimal ellipsoids centred at an A-point* x_0 *and with the characteristics (C) into infinitesimal spheres* *if and only if the linear transformation $f_1(x, x_0)$ given by* (2) *maps the ellipsoids $E_h[(C), x_0]$ in spheres.*

THEOREM 13. *The definition of the K-QCfH in Markuševič—Pesin's sense is equivalent to Gehring's metric definition (with the same K).*

Suppose $f(x)$ is a K-QCfH in Markuševič—Pesin's sense in D. By theorem 11, $\delta_L(x)$ is bounded in D.

Let $q(x_0) = \inf k(x_0)k'(x_0)$ be the characteristic of $f(x)$ at an A-point x_0, where the infimum is taken over all surface families regular relative to x_0. Let $q_1(x_0) = \inf k_1(x_0)k_1'(x_0)$ be the characteristic of the affine mapping $f_1(x, x_0)$ given by (2). But the preceding lemma and lemma 1 imply that $q(x_0) = q_1(x_0) = p_1(x_0)$, where $p_1(x_0)$ is the principal characteristic of $f_1(x, x_0)$. Hence, $q(x) \leqq K$ a.e. in D implies $p_1(x) \leqq K$ a.e. in D. Thus, the formula

$$\delta_L(x_0) = \lim_{r\to 0}\frac{\dfrac{|f(x')-f(x_0)|}{|x'-x_0|}}{\dfrac{|f(x'')-f(x_0)|}{|x''-x_0|}} \leqq \frac{\overline{\lim\limits_{x\to x_0}}\dfrac{|f(x)-f(x_0)|}{|x-x_0|}}{\underline{\lim\limits_{x\to x_0}}\dfrac{|f(x)-f(x_0)|}{|x-x_0|}} = \frac{\Lambda_f(x_0)}{\lambda_f(x_0)}, \tag{31}$$

where we set

$$\max_{|x-x_0|=r} |f(x)-f(x_0)| = |f(x')-f(x_0)|,$$

$$\min_{|x-x_0|=r} |f(x)-f(x_0)| = |f(x'')-f(x_0)|,$$

yields, on account of (10),

$$\delta_L(x_0) \leqq \frac{\Lambda_f(x_0)}{\lambda_f(x_0)} = p_1(x_0) = q(x_0) \leqq K$$

a.e. in D. Thus $f(x)$ is K-QCfH in D according to Gehring's metric definition.

Conversely, suppose $f(x)$ is K-QCfH according to Gehring's metric definition. Then

$$q(x) < \delta_L(x)$$

holds, because, by definition, $q(x) = \inf k(x)k'(x)$, while $\delta_L(x)$ is the only product $k(x)\,k'(x)$ where $\{\sigma_\alpha\}$ is the family of spheres $\{S(x, r)\}$. Thus $q(x)$ is bounded in D and $q(x) \leqq K$ a.e. in D. Hence we have established the equivalence of the two definitions.

The connection between Gehring's metric definition, Markuševič—Pesin's definition and Gehring's and Väisälä's other definitions.

THEOREM 14. *A* K-QCfH *according to Gehring's geometric definition is* K_1-QCfH *according to his metric definition (and then also according to Markuševič—Pesin's), where K, K_1 satisfy the inequalities*

$$K \leqq K_1 \leqq K^{\frac{2(n-1)}{n}}, \quad K_1^{\frac{n}{2(n-1)}} \leqq K \leqq K_1,$$

and the estimations are the best possible.

This is an immediate consequence of the theorems 1, 3 and of the preceding one.

Remark. The theorem 14 implies that for $n = 2$ all the preceding definitions of K-QCfH are equivalent (with the same K).

COROLLARY. *A K-QCfH according to Väisälä's definitions is K_2-QCfH according to Gehring's metric definition (and then also to Markuševič—Pesin's), where K, K_2 satisfy the inequalities*

$$K^{\frac{1}{n-1}} \leqq K_2 \leqq K^{\frac{2}{n}}, \quad K_2^{\frac{n}{2}} \leqq K \leqq K_2^{n-1},$$

and the estimations are the best possible.

This is an immediate consequence of theorem 2.1 and of the preceding one.

Kopylov's definition. It is easy to see that the class of K-QCfH with two sets of characteristics is contained in the class of K-QCfH in Markuševič—Pesin's sense. This inclusion is strict, as it follows from the following example.

The homeomorphism

$$f(z) = \begin{cases} \left(K + \dfrac{|z|}{\arg z}\right) z & \text{whenever } 0 < |z| < \arg z \leqq 2\pi, \\ \left[K + \dfrac{\arg z}{|z|} + \left(1 - \dfrac{\arg z}{|z|}\right)\dfrac{|z|}{2\pi}\right] z & \text{whenever } 0 < \arg z \leqq |z| \leqq 2\pi, \\ O & \text{at } z = 0, \end{cases} \tag{32}$$

where $z = x + iy$, $k > 0$, has $\delta_L(O) = 1 + \dfrac{1}{K}$ and then is $\left(1 + \dfrac{1}{K}\right)$-QCfH at O according to Gehring's metric definition and, by theorem 13, also according to Markuševič—Pesin's; however $f(z)$ is not QCfH with one or two sets of characteristics in O. Moreover, $f(z)$ is differentiable a.e. in $B(2\pi)$ and has all first order partial derivatives bounded and $K^2 \leqq J \leqq (K+1)(K+2)$ in $B(2\pi)$, hence, by formula (7) of L. I. Volkovyskii's monograph ([7], p. 19);

$$p_1 + \frac{1}{p_1} = \frac{u_x + u_y + v_x + v_y}{|J|},$$

which holds at an A-point, it follows that $p_1(x)$ is bounded a.e. in $B(2\pi)$, and then (1.12) and (1.13) yield (1.19) (with some other K) a.e. in $B(2\pi)$. Since $f \in C^1$ in $B(2\pi)$ except on the curve $|z| = \arg z$, it is easy to see that $f(z)$ is ACL in $B(2\pi)$. Taking into account that $f(z)$ is differentiable a.e. in $B(2\pi)$ and satisfies (1.19) a.e. in $B(2\pi)$, we conclude that $f(z)$ is QCfH in $B(2\pi)$ according to Väisälä's definition, and then also according to Markuševič—Pesin's, without being QCfH with one or two sets of characteristics.

However, by a slight modification, the definition of K-QCfH with one or two sets of characteristics can be made to become equivalent to the K-QCfH in Markuševič—Pesin's sense. We give first the following Kopylov's [1] definition:

A K-QCfH in D is a QCfH in D, which for almost all the points of differentiability $x \in D$, maps infinitesimal spheres into infinitesimal ellipsoids with the principal characteristic $p_1'(x) \leqq K$.

On the model of Kopylov, we can define the following class of K-QCfH corresponding to the class of K-QCfH with two sets of characteristics:

A K-QCfH *with two principal characteristics* in D is a QCfH in D which maps infinitesimal ellipsoids in infinitesimal ellipsoids with the principal characteristics satisfying the inequality $p_1(x)p_1'(x) \leqq K$ a.e. in D.

If $p_1'(x) = 1$ a.e. in D (i.e. the images of the infinitesimal ellipsoids by f reduce to spheres a.e. in D), we obtain the class of K-QCfH *with a principal characteristic* (corresponding to the class of K-QCfH with a set of characteristics).

Remark. Both in Kopylov's definition and in the two others given on its model, it is not necessary to specify with respect to which definition the quasiconformality in D is intended, since we proved that all the definitions of the K-QCfH (except those with one or two sets of characteristics) are equivalent if we do not specify the constant K.

Equivalence between Kopylov's definition, the definitions of *K*-QfCH with one or two principal characteristics and Markuševič—Pesin's definition.

THEOREM 15. *The definition of the* K-QCfH *with a principal characteristic is equivalent to Markuševič—Pesin's.*

It is trivial that every K-QCfH with a principal characteristic is a K-QCfH in Markuševič—Pesin's sense.

Conversely, suppose $f(x)$ is K-QCfH in Markuševič—Pesin's sense. Then, combining theorem 9 and corollary of theorem 12, we infer that $f(x)$ is differentiable with $J(x) \neq 0$ a.e. in D, and lemma 1 and corollary of lemma 7 imply $q(x_0) = p_1(x_0)$ whenever x_0 is an A-point of D. Hence, $f(x)$ which is QCfH in D, satisfies also the inequality $p_1(x) \leqq K$ a.e. in D, i.e. $f(x)$ is K-QCfH with a principal characteristic in D.

COROLLARY 1. *The definition of* K-QCfH *with two principal characteristics is equivalent to Markuševič—Pesin's.*

Since the class of K-QCfH with a principal characteristic is contained in the class of K-QCfH with two principal characteristics, and this one in its turn is contained in the class of K-QCfH in Markuševič-Pesin's sense, it follows that the equivalence between the definitions of the K—QCfH with a principal characteristic and that of Markuševič—Pesin implies the equivalence of all the three definitions.

COROLLARY 2. *Kopylov's and Markuševič—Pesin's definitions are equivalent.*

This is a consequence of the preceding theorem, taking into account that if a linear transformation maps the ellipsoids with the principal characteristic p_1 into spheres, then it will map the spheres into ellipsoids with the principal characteristic p_1.

CHAPTER 4

K-QCfH in Kreines' sense and Callender's K-QCfH

Sobolev's derivatives. Let $u(x)$ be a real function integrable in every bounded domain Ω contained in D. If there exists an integrable function $\omega_{\alpha_1 \dots \alpha_n}$ such that

$$\int_\Omega \left[u(x) \frac{\partial^p v(x)}{(\partial x^1)^{\alpha_1} \cdots (\partial x^n)^{\alpha_n}} + (-1)^{p+1} v(x) \omega_{\alpha_1, \dots, \alpha_n}(x) \right] d\tau = 0,$$

for all functions $v \in C^p$, with compact support in D, then the quantity $\omega_{\alpha_1, \dots, \alpha_n}$ is said to be Sobolev derivative of order p of the function $u(x)$ in D (Sobolev [1], p. 39).

A function $u(x)$ which possesses Sobolev derivatives of the first order is called to be *strongly differentiable* (Serrin [1], p. 39).

Analytic definition of K-QCfH in Kreines' sense (see M. Kreines [1], Ju. G. Rešetnjak and B. V. Šabat [1], p. 673).

A homeomorphism $x^* = f(x)$ is said to be K-QCfH ($1 \leqq K < \infty$) in D, if its components $x^{*i}(x)$ ($i = 1, \dots, n$) have first order Sobolev derivatives L^n-integrable on every compact subset of D and the inequality

$$|\nabla f|^n \leqq n^{\frac{n}{2}} K |J(x)|, \tag{1}$$

where

$$|\nabla f|^2 = \sum_{i,k=1}^{n} (x_k^{*i})^2. \tag{2}$$

holds a.e. in D.

Equivalence between Kreines' definition and the other definitions (This equivalence was proved by ourselves in [25]).

THEOREM 1. *A K-QCfH according to Väisälä's definitions is K-QCfH in Kreines' sense.*

Suppose $f(x)$ is a K-QCfH according to Väisälä's definitions. Then $f(x)$ is ACL_n in D. We show that this implies the existence of Sobolev L^n-integrable first order derivatives in D.

Indeed, since $x^{*i}(x)$ $(i = 1,\ldots, n)$ are *ACL*, they are *AC* on almost all line segments in *D* parallel to each of the coordinate axes. In order to simplify the proof, let $u(t)$ denote any component of $f(x)$ considered as a function of the variable t on a line segment $(a, b) \subset D$ parallel to the x^i-axis. Since $u(t)$ is *AC* on (a, b), we conclude that $u(t)$ is the indefinite integral of an integrable function (see Natanson [1], theorem 3, p. 332), i.e. $u(t) = \int_a^t \omega(\xi)d\xi + C$. Set $\Omega(t) = \int_a^t \omega(\xi)d\xi$ and suppose $v \in C^1$ in $(a + \varepsilon, b - \varepsilon)$, where $\left(0 < \varepsilon < \dfrac{b-a}{2}\right)$ and $v(t) \equiv 0$ outside this interval, i.e. $v(t)$ is an arbitrary function in the class considered in the above definition of Sobolev derivative for $p = 1$. Taking into account the expression of u, we get

$$\int_a^b u \frac{dv}{dx} dx = \int_a^b \Omega \frac{dv}{dx} dx + \int_a^b C \frac{dv}{dx} dx = \Omega v\Big|_a^b - \int_a^b \omega v \, dx + Cv\Big|_a^b = -\int_a^b \omega v \, dx.$$

Since the preceding relation holds on almost all line segments in *D* parallel to each of the coordinate axes and for every v in the considered class and since $u(t)$ may be every component of $f(x)$ and t every $x^i (i = 1,\ldots, n)$, we conclude the existence in *D* of first order Sobolev derivatives of each $x^{*i}(x)$ with respect to each variable $x^k (i, k = 1,\ldots, n)$.

Next, the uniqueness of Sobolev ([1], p. 40) derivatives and the fact that the ACL condition implies the existence of first order partial derivatives a.e. in *D* infer that Sobolev first order derivatives of $x^{*i}(x)$ $(i = 1,\ldots, n)$ coincide a.e. in *D* with the corresponding $x_k^{*i}(x)$ $(i, k = 1,\ldots, n)$ (in the classical sense) and since the latter are L^n-integrable, then so are also the former.

A *K*-QCfH according to Väisälä's definitions is differentiable a.e. in *D*. Next, it is clear that such a homeomorphism maps certain infinitesimal ellipsoids with the centre in an *A*-point of $f(x)$ into infinitesimal spheres. The squares of the semi-axes of such an ellipsoid are proportional to the roots of the following equation in *S*:

$$\begin{vmatrix} \sum_{i=1}^n (x_1^{*i})^2 - S \cdots \sum_{i=1}^n x_1^{*i} x_n^{*i} \\ \cdots\cdots\cdots\cdots\cdots\cdots \\ \sum_{i=1}^n x_1^{*i} x_n^{*i} \cdots \sum_{i=1}^n (x_n^{*i})^2 - S \end{vmatrix} = 0,$$

which may be written also as

$$S^n - \sum_{i,k=1}^n (x_k^{*i})^2 S^{n-1} + \cdots + J^2 = 0,$$

hence, recalling the notation (2),

$$a_1^2 + \cdots + a_n^2 = \sum_{i,\,k=1}^{n} (x_k^{*i})^2 \equiv |\nabla f|^2, \quad a_1 \cdots a_n = |J|, \tag{3}$$

where $a_i(a_1 \geqq \ldots \geqq a_n)$ are the semi-axes of the ellipsoid mapped into the unit sphere S by the linear transformation $f_1(x, x_0)$ given by (3.2) and corresponding to $f(x)$ and an A-point x_0. Since $f(x)$ is K-QCfH according to Väisälä's definitions, the formulae (1.13), (3) imply

$$\frac{|\nabla f|^n}{|J|} = \frac{(a_1^2 + \cdots + a_n^2)^{\frac{n}{2}}}{a_1 \cdots a_n} \leqq \frac{(na_1^2)^{\frac{n}{2}}}{a_1 \cdots a_n} = n^{\frac{n}{2}} \frac{a_1^{n-1}}{a_2 \cdots a_n} \leqq n^{\frac{n}{2}} K$$

a.e. in D. Hence $f(x)$ is K-QCfH in Kreines' sense and the proof is complete.

THEOREM 2. *The definition of QCfH in Kreines' sense is equivalent to all the other definitions.*

The preceding theorem shows that a K-QCfH according to Väisälä's definitions is K-QCfH in Kreines'sense. Now let $f(x)$ be K-QCfH in Kreines' sense. We want to show that it is also QCfH according to Väisälä's definitions.

We begin by establishing that a homeomorphism $f(x)$ with Sobolev first order derivatives L^n-integrable is ACL_n. In order to simplify the proof, let $u(t)$ denote (as in the proof of the preceding theorem) a component of $f(x)$ considered asa function of one of the n variables $x^i(i = 1, \ldots, n)$. Since $u(t)$ has a Sobolev first order derivative, its definition ensures the existence of an integrable function ω such that

$$\int_a^b u \frac{\mathrm{d}v}{\mathrm{d}x} \,\mathrm{d}x = -\int_a^b \omega v \,\mathrm{d}x, \tag{4}$$

where v is an arbitrary function of the class considered in the definition of Sobolev derivatives, hold on almost all line segments in D parallel to the x^i-axis. Let $\Omega = \int_a^t \omega \mathrm{d}x$. Then

$$-\int_a^b \omega v \,\mathrm{d}x = \int_a^b \Omega \frac{\mathrm{d}v}{\mathrm{d}x} \,\mathrm{d}x,$$

and (4) yield

$$\int_a^b (u - \Omega) \frac{\mathrm{d}v}{\mathrm{d}x} \,\mathrm{d}x = 0,$$

and $\int_a^b \frac{dv}{dx}\, dx = 0$ gives $u - \Omega = C$ (see for example Gelfand and Fomin [1], p. 15).

Hence, $u = \Omega + C = \int_a^t \omega dx + C$ and we conclude that $u(t)$ is AC on (a, b). Since this is true for all the components of $f(x)$ and for almost all line segments in D parallel to the coordinate axes, we conclude that $f(x)$ is ACL in D. Hence, $f(x)$ has a.e. in D first order partial derivatives respectively equal to the corresponding Sobolev first order derivatives. Thus the L^n-integrability of the latter implies the same property for the former. Then $f(x)$ is ACL_n and by lemma 1.10 $f(x)$ is differentiable a.e. in D. In order to prove that $f(x)$ is QCfH according to Väisälä's analytic definition, it suffices to show (by virtue of theorem 1.2) that $f(x)$ satisfies the inequality

$$\max |f'(x)|^n \leqq K_1 |J(x)| \tag{5}$$

a.e. in D. First suppose x_0 is a point of differentiability of $f(x)$ with $J(x_0) = 0$. Then (1) and (2) imply that $x_k^{*i}(x_0) = 0$ $(i, k = 1, \ldots, n)$, hence $\max |f'(x_0)| = 0$ and (5) holds for any K. Finally suppose x_0 is an A-point of $f(x)$. From (1) and (3) we obtain

$$\frac{a_1^n}{a_1 \cdots a_n} \leqq \frac{|\nabla f|^n}{|J|} \leqq n^{\frac{n}{2}} K$$

and then, on account of (1.13), the inequality (5) is satisfied again, this time with $K_1 = n^{\frac{n}{2}} K$.

Callender's definition [2]. A homeomorphism $f(x)$ is K-QCfH $(1 \leqq K < \infty)$ in D, if it is ACL and

$$|\nabla f|^2 \leqq n K J^{\frac{2}{n}}$$

a.e. in D.

Equivalence between the definitions of the K-QCfH in Kreines' sense and Callender's definition of the $K^{\frac{n}{2}}$-QCfH.

THEOREM 3. *The definition of the* K-QCfH *in Kreines' sense is equivalent to Callender's definition of* $K^{\frac{n}{2}}$-QCfH.

The equivalence of the condition ACL_n to that of the existence of Sobolev first order derivatives L^n-integrable was established in the course of the proof of the preceding two theorems, whereas the equivalence of the corresponding inequalities is trivial.

Generalizing his preceding definition, D. Callender [2] introduces also the class of *almost* QCfH.

A homeomorphism $f(x)$ is said to be *almost* QCfH in D, if it is ACL_n in D and

$$|\nabla f|^n \leqq (nK)^{\frac{n}{2}}|J| + K_1$$

a.e. in D, where $1 \leqq K < \infty$ and $0 \leqq K_1 < \infty$.

Remark. This definition is a generalization of the *elliptic mappings* given for $n = 2$ by L. Nirenberg [2]. His definition [1] corresponding for a general n is the following:

A homeomorphism $f(x)$ is said to be an *elliptic mapping* in D if it is ACL_n in D and

$$|\nabla f|^n \leqq K \sum_{i,k=1}^{n} (x_k^{*i} x_i^{*k} - x_i^{*i} x_k^{*k}) + K_1,$$

where $0 < K < \dfrac{n-1}{n-2}$ and $0 < K_1 < \infty$.

In fact, Nirenberg's definition is more restrictive, since he assumes that $f \in C^1$.

CHAPTER 5

Šabat's definition of the K-QCfH and Kühnau's definitions of K_I, K_O-QCfH

Šabat's definition. In this paragraph we shall introduce the K-QCfH characterized by the distortion of the modulus of the q-dimensional surface families (Ju. G. Rešetnjak and B. V. Šabat [1])

An interval $I = \{\xi; a^i < \xi^i < b^i \ (i = 1, \ldots, n)\}$ may be written as the Cartesian product $I = I_q \times I_{n-q}$, where $I_q = \{\eta; a^k < \eta^k < b^k \ (k = 1, \ldots, q)\}$, $I_{n-q} = \{\zeta; a^{m+q} < \zeta^m < b^{m+q} \ (m = 1, \ldots, n-q)\}$. Let us denote by η and ζ also the corresponding points of R^n, i.e. $\eta = (\eta^1, \ldots, \eta^q, 0, \ldots, 0)$, $\zeta = (0, \ldots, 0, \zeta^{q+1}, \ldots, \zeta^n)$. Let $\{\sigma_q^\zeta\}$ be the family of the q-dimensional surfaces $\sigma_q^\zeta = \varphi(I_q \times \zeta)$, $\zeta \in I_{n-q}$, where $\varphi\colon I \to R^n$ is a QCfH.

For Šabat's definition we call *admissible* a subfamily of $\{\sigma_q^\zeta\}$, or, if $q = n-1$, a family of concentric spheres or a quasiconformal image of such a family, or even a contably union of such families.

And now, we shall give the definition of K-QCfH by means of the distortion of the modulus of q-dimensional surface families as follows:

A homeomorphism $f\colon D \rightleftharpoons D^*$ is said to be K-QCfH $(1 \leqq K < \infty)$ in D if

$$\frac{M(\Sigma_q)}{K^{n-q}} \leqq M(\Sigma_q^*) \leqq K^{n-q} M(\Sigma_q), \tag{1}$$

holds for all admissible surface families $\Sigma_q \subset D$.

Remark. For $q = 1$, we have first Väisälä's geometric definition (by means of arc families) for K^{n-1}-QCfH. For $q = n-1$, we obtain Šabat's definition given in [3]. (We must specify that the arc families involved in Väisälä's definition are arbitrary, and the surface families in Šabat's definition are piecewise smooth, f being assumed to belong to class C^1)

Connexion between Šabat's and Markuševič—Pesin's definitions

In the first part of this chapter, we establish the K-quasiconformality in Šabat's sense of a K-QCfH according to Markuševič—Pesin's definition. We begin with some preliminary results.

LEMMA 1. *Suppose that E is a G_δ-Borel set of Σ-finite q-dimensional Hausdorff measure. Then*

$$H^q(E) = \sup\{H^q(F);\ F \text{ compact},\ F \subset E\}.$$

A set E is said to be of Σ-finite q-dimensional Hausdorff measure if $E = \bigcup_p E_p$ and $H^q(E_p) < \infty$ $(p = 1, 2, \ldots)$.

Suppose that $H^q(E) < \infty$. Then it is sufficient to exhibit, for each $\varepsilon > 0$, a compact set $F \subset E$ such that $H^q(F) > H^q(E) - \varepsilon$. Because E is a G_δ-Borel set, we can write E as the intersection of a contracting sequence of open sets $\{G_m\}$ $(G_{m+1} \subset G_m)$. But each G_m can be expressed as the union of an expanding sequence of compact sets $\{F_{m_k}\}$ $(F_{m_{k+1}} \supset F_{m_k})$ and, since (see C. Carathéodory [3], theorem 8, p. 254)

$$\lim_{k \to \infty} H^q(E \cap F_{m_k}) = H^q(E \cap G_m) = H^q(E),$$

we can find, for each m, a compact $F_m \subset G_m$ such that

$$H^q(E - F_m) = H^q(E) - H^q(E \cap F_m) < \frac{\varepsilon}{2^m}.$$

If we let $F = \bigcap_{m=1}^{\infty} F_m$, then F is compact, $F \subset E$, and

$$H^q(E) - H^q(F) = H^q(E - F) \leqq \sum_{m=1}^{\infty} H^q(E - F_m) < \varepsilon,$$

as desired, for $H^q(E) < \infty$.

When $H^q(E) = \infty$, we can express E as the union of a sequence of sets $\{E_m\}$ of finite q-dimensional Hausdorff measure. For each m, we can find a G_δ-Borel set $E'_m \supset E_m$ such that $H^q(E'_m) = H^q(E_m)$ and hence, if we let

$$E''_m = \bigcup_{k=1}^{m} (E'_k \cap E),$$

then $\{E''_m\}$ is an expanding sequence of G_δ-Borel sets whose union is E. Since $H^q(E''_m) < \infty$ and since $H^q(E''_m) \to \infty$, the argument of the first part exhibits a sequence of compact sets $\{F_m\}$, such that $F_m \subset E''_m \subset E$ and $H^q(F_m) \to \infty$.

A set of $S(r)$ will be called *measurable with respect to the spherical measure* (see Chapter 1.3) if its stereographical projection is measurable with respect to Lebesgue's $(n - 1)$-dimensional measure. A mapping $f_{|S(r)}$ [i. e. the *restriction of* f *to* $S(r)$] is said to be H^{n-1}-AC *with respect to the spherical measure* (see Chapter 1.3) if $H^{n-1}[f_{|S(r)}(E^S)] = 0$ whenever $E^S \subset S(r)$ and $m^S E^S = 0$.

LEMMA 2. *Let f be a* QCfH *in a domain $D \subset R^n$ and $A = \{x; r_0 < |x - y| < r_1\} \subset D$, then, for almost all $r \in (r_0, r_1)$,* f *is H^{n-1}-AC in $S(r)$ with respect to spherical measure.*

The proof is based on a method used by F. Gehring [20] and by S. Agard [2]. For simplicity let $y = O$.

Taking into account the equivalence of the different definitions of the QCfH, we may suppose, without loss of generality, that f is QCfH according to Gehring's metric definition. Hence $\delta_L(x) < \infty$ in A.

For positive integers μ, ν, let $E_{\mu,\nu}$ be the set of points $x_0 \in A$ such that

$$|f(x_1) - f(x_0)| \leqq \mu |f(x_2) - f(x_0)|,$$

whenever

$$|x_1 - x_0| = |x_2 - x_0| \leqq \frac{1}{\nu}.$$

Because $\delta_L(x) < \infty$ at every point of A, it follows that

$$A = \bigcup_{\mu, \nu} E_{\mu, \nu},$$

and since f is continuous, it is clear that $E_{\mu,\nu}$ is closed. For sufficiently small $t > 0$, let the modulus of continuity $b(t)$ be defined by

$$b(t) = \max \{|f(x_1) - f(x_2)|;\ |x_1 - x_2| \leqq t\},$$

where we require $d(x_k, A) \leqq t$ and $x_k \in D$ $(k = 1, 2)$. By uniform continuity on compact sets, it is clear that $b(t) \to 0$ as $t \to 0$.

The measure function Φ, defined for a measurable $E \subset (r_0, r_1)$ by

$$\Phi(E) = mf[\bigcup_{r \in E} S(r)]$$

has Lebesgue derivative $\Phi'(r) < \infty$ for almost every r, that is to say, that letting $I_\rho(r)$ denote the segment $[r - \rho, r + \rho]$ in R^1, the limit

$$\Phi'(r) = \lim_{\rho \to 0} \frac{\Phi[I_\rho(r)]}{2\rho}$$

is finite a.e. in (r_0, r_1).

Next, let E^S be a closed subset of $S(r) \cap E_{\mu,\nu}$. We shall prove that there exists a constant $Q(r)$, such that

$$H^{n-1}[f(E^S)] \leqq Q(r)\mu^{n(n-1)} m^S E^S. \tag{2}$$

Fix $0 < t < \frac{1}{\nu}$, and a covering of E^S with spherical balls $B^S(x, t)$, so that the centre of each ball belongs to E^S but does not belong to another ball. Such a covering is clearly finite. Let $B(x_k, t)$ $k = 1, \ldots, q(t)$, be the spherical balls. There exists an integer N_{n-1} such that no point of $S(r)$ can belong to more than N_{n-1} of these balls. (For instance $N_1 = 2$, $N_2 = 4$, etc.). Let $B(x_k, t')$ $(k = 1, \ldots, q)$ be the full open balls of radius $t' < t$ centred at x_k and such that $B^S(x_k, t') = B(x_k, t') \cap S(r)$. Clearly,

(i) $E^S \subset \bigcup_{k=1}^{q} B^S(x_k, t) \subset \bigcup_{k=1}^{q} B(x_k, t')$,

(ii) $\bigcup_{k=1}^{q} B^S(x_k, t) \subset E^S(t) = \{x;\ x \in S(r),\ d^S(x, E^S) \leqq t\}$, where d^S is the spherical distance,

(iii) No point of $S(r)$ lies in more than N_{n-1} of $B^S(x_k, t)$ and no point of R^n lies in more than N_{n-1} of the $B(x_k, t')$.

By (i), the images $f[B(x_k, t)]$ cover $f(E)$ and are open sets of diameter not exceeding $b(2t)$. Let

$$L_k = L(x_k, t), \quad l_k = l(x_k, t).$$

Since $x_k \in E^S \subset E_{\mu,\nu}$ and $0 < t < \frac{1}{\nu}$, it follows that $L_k \leqq \mu l_k$. The images $f[B(x_k, t')]$ lie in (or contain) the balls of centre $f(x_k)$ and radius $L_k(l_k)$. Set

$$\Xi_{n-1}^{b(2t)}(\mathfrak{M}_n, \chi_n^{n-1})(E^S) = \inf \Sigma_p\, \omega_{n-1}\left[\frac{d(E_p)}{2}\right]^{n-1},$$

where the infimum is taken over all countable coverings $\{E_m\}$ of E such that $d(E_p) < b(2t)$. Then

$$\Xi_{n-1}^{b(2t)}(\mathfrak{M}_n, \chi_n^{n-1})[f(E^S)] \leqq \sum_{k=1}^{q} \omega_{n-1} L_k^{n-1} \leqq \omega_{n-1}\mu^{n-1} \sum_{k=1}^{q} l_k^{n-1} \tag{3}$$

and, by Hölder's inequality,

$$\left(\sum_{k=1}^{q} l_k^{n-1}\right)^n \leqq q\left(\sum_{k=1}^{q} l_k^n\right)^{n-1}. \tag{4}$$

Because f is a homeomorphism, the images $f[B(x_k, t')]$, like the balls $B(x_k, t')$, can cover any point at most N_{n-1} times. Therefore

$$\omega_n \sum_{k=1}^{q} l_k^n \leqq \sum_{k=1}^{q} mf[B(x_k, t')] \leqq N_{n-1} m\{\bigcup_{k=1}^{q} f[B(x_k, t')]\} \leqq$$

$$\leqq N_{n-1} mf[I_t(r) \times S] = N_{n-1}\Phi[I_t(r)].$$

Combining the previous inequality with (3), (4), we find

$$\Xi_{n-1}^{b(2t)}(\mathfrak{M}_n, \chi_n^{n-1})[f(E^S)]^n \leq \frac{\omega_{n-1}^n}{\omega_n^{n-1}} \mu^{n(n-1)} q N_{n-1} \Phi[I_t(r)] =$$

$$= \left(\frac{2\omega_{n-1}}{\omega_n}\right)^{n-1} \mu^{n(n-1)} N_{n-1} \left\{\frac{\Phi[I_t(r)]}{2t}\right\}^{n-1} (\omega_{n-1} t^{n-1} q).$$

But by (ii) and (iii),

$$\omega_{n-1} t^{n-1} q \leq \sum_{k=1}^{q} m^S[B^S(x_k, t)] \leq N_{n-1} m^S E^S(t),$$

and hence it follows that

$$\Xi_{n-1}^{b(2t)}(\mathfrak{M}_n, \chi_n^{n-1})[f(E^S)]^n \leq \left(\frac{2\omega_{n-1}}{\omega_n}\right)^{n-1} \mu^{n(n-1)} N_{n-1} \left\{\frac{\Phi[I_t(r)]}{2t}\right\} m^S E^S(t).$$

Letting $t \to 0$, we know that,

(a) $\dfrac{\Phi[I_t(r)]}{2t} \leq \Phi'(t)$ a.e. in (r_0, r_1),

(b) $\lim_{t \to 0} E(t) = \bigcap_{t>0} E^S(t) = E^S$ (because E^S is compact) and therefore $m^S[E^S(t)] \to m^S E^S$ [because $m^S E^S(t) < \infty$],

(c) $b(2t) \to 0$, and therefore $\Xi_{n-1}^{b(2t)}(\mathfrak{M}_n, \chi_n^{n-1})\,[f(E^S)] \to H^{n-1}[f(E^S)]$.

Therefore from (2) it follows that

$$Q(r) = \left(\frac{2\omega_{n-1}}{\omega_n}\right)^{n-1} N_{n-1} \Phi'(r).$$

To complete the proof, let E^S be any compact subset of $S(r)$, with $m^S E^S = 0$. Then the previous case applies to $E_{\mu,\nu}^S = E^S \cap E_{\mu,\nu}$ and since $m^S E_{\mu,\nu}^S = 0$, it is quite clear that $H^{n-1} f(E_{\mu,\nu}^S)] = 0$, hence

$$H^{n-1}[f(E^S)] \leq \sum_{\mu,\nu} H^{n-1}[f(E_{\mu,\nu}^S)] = 0.$$

On the other hand, taking $E^S = S(r)$, the same argument shows that the image of any measurable subset $E^{\tilde{S}}$ of $S(r)$ has Σ-finite H^{n-1}-measure, i.e. $E^{\tilde{S}} = \bigcup_k E_k^S$ so that $H^{n-1}(E_k^S) < \infty$, and then, by (2), $H^{n-1}[f(E_k^S)] < \infty$ $(k = 1, 2, \ldots)$. Therefore, if E^S is a G_δ-set in $S(r)$ with $m^S E^S = 0$, then

$$H^{n-1}[f(E^S)] = \sup \{H^{n-1}[f(F^S;\ F^S \text{ compact},\ F^S \subset E^S]\} = 0,$$

by lemma 1, since $f(F^S)$ is compact and $f(E^S)$ a G_δ-set.

Finally if E^S is any measurable set in $S(r)$ with $m^S E^S = 0$, there exists a G_δ-set $E_0^S \supset E^S$, with $m^S E_0^S = m^S E^S = 0$. (This is true for sets in R^{n-1} compactified by adjoining the point designated by the symbol ∞, and since the stereographic projection is a homeomorphism, and then transforms open sets onto open sets and G_δ-sets onto G_δ-sets, and also maps null sets into null sets, we conclude that the above relation for spherical sets is also true.) Therefore $H^{n-1}[f(E^S)] \leqq \leqq H^{n-1}[f(E_0^S)] = 0$, which completes the proof of the lemma.

A mapping $\varphi_{|I_q}$ is said to be *H^q - AC with respect to Lebesgue q-dimensional measure* if $H^q[\varphi_{|I_q}(E)] = 0$ whenever $E \subset I_q$ and $m_q E = 0$.

LEMMA 3. *Let φ be a QCfH in a domain $D \subset R^n$ and $I \subset D$, then, for almost all $\zeta \in I_{n-q}$, φ is H^q - AC in I_q with respect to q-dimensional Lebesgue measure.*

The argument is similar to that of the preceding lemma.

THEOREM 1. *Every K-QCfH according to Markuševič—Pesin's definition is K-QCfH according to Šabat's definition for $q = 1, \ldots, n-1$.*

Suppose $\varphi\colon I \to D$ be QCfH and $f: D \to R^n$ be K-QCfH $(1 \leqq K < \infty)$ in Markuševič—Pesin's sense. Let us show first that for almost all $\zeta \in I_{n-q}$, $f_{|\sigma_q^\zeta}$ (i.e. the restriction of f to σ_q^ζ) is differentiable a.e. in σ_q^ζ and *bimeasurable* (i.e. measurable together with its inverse.)

The differentiability of f with $J \neq 0$ a.e. in σ_q^ζ for almost all ζ in I_{n-q} and of f^{-1} with $J^* \neq 0$, a.e. in $f(\sigma_q^\zeta)$ for almost all ζ in I_{n-q} is a consequence of theorem 3.9, corollary of theorem 3.12 and of proposition 1.5.5. The same is true for φ.

Then, at an A-point $\xi_0 \in I_q \times \zeta_0$ of φ its principal part maps the ellipsoid E with centre ξ_0 and semi-axes $a_1 \geqq \ldots \geqq a_n > 0$ onto the unit ball centered at x_0. The q-plan $I_q \times \zeta_0$ cuts from E a q-dimensional ellipsoid E_q, with the semi-axes $b_1 \geqq \ldots \geqq b_q > 0$, which is mapped by the principal part of φ onto the unit q-ball B_q. Clearly,

$$J_{\xi x} = \frac{1}{a_1 \cdots a_n}, \quad J_{\eta x} = \frac{1}{b_1 \cdots b_q},$$

and since the area of E_q attains its maximum (minimum) when the directions of the semi-axes $b_{\cdot}, \ldots, b_q$ coincide with those of $a_1, \ldots, a_q$ $(a_{n-q+1}, \ldots, a_n)$, we have

$$a_{n-q+1} \cdots a_n \leqq b_1 \cdots b_q \leqq a_1 \cdots a_q,$$

hence, and on account of theorem 3.5, we conclude that $J_{\eta x} \neq 0$ if and only if $\zeta \in J_{\xi x} \neq 0$ and then, since $J_{\xi x} \neq 0$ a.e. in σ_q^ζ for almost every $\zeta \in I_{n-q}$, the same conclusion still holds for $J_{\eta x}$. The same argument may be used also for $J_{\eta x^*}$.

Since φ is QCfH, then, by lemma 3, φ is H^q - AC on $I_q \times \zeta$ for almost all $\zeta \in I_{n-q}$, i.e. for almost all $\zeta \in I_{n-q}$, $m_q E = 0$ (where $E \subset I_q$) implies $H^q[\varphi(E)] = 0$.

Now let $I_q \times \zeta$ be a q-plane on which φ is measurable and $J_q(\eta) \neq 0$. a.e., where $J_q(\eta) = \sqrt{\sum_{i_1 < \ldots < i_q} \frac{\partial(x^{i_1}, \ldots, x^{i_q})}{\partial(\eta^1, \ldots, \eta^q)}}$, and let $E \subset I_q \times \zeta$. Then

$$H^q[\varphi(E)] = \int_E J_q(\eta) \, d\sigma_q.$$

If $H^q[\varphi(E)] = 0$, since $J_q(\eta) \neq 0$ a.e. in E, it follows that $m_q E = 0$, i.e. $\varphi(\xi)$ is bimeasurable on $I_q \times \zeta$ for almost all $\zeta \in I_{n-q}$.

This conclusion still holds for $f \circ \varphi$, which is also QCfH.

But then we conclude that f is measurable on σ_q^ζ for almost all $\zeta \in I_{n-q}$.

Now, let us prove that the family Σ_q^0 of surfaces σ_q^ζ where ζ runs over the set E_{n-q}^0 with $m_{n-q}E_{n-q}^0 = 0$, is an exceptional family. Clearly, if $E^0 = \bigcup_{\zeta \in E_{n-q}^0} I_q \times \zeta$, then $mE^0 = 0$ and since φ is measurable (corollary of theorem 3.12), it follows $m\varphi(E^0) = 0$. Next, since φ is a homeomorphism, $H^q(\sigma_q^\zeta) > 0$, and then

$$\rho(x) = \begin{cases} \infty & \text{if } x \in \varphi(E^0), \\ 0 & \text{if } x \in \varphi(E^0), \end{cases}$$

is clearly admissible for $\{\sigma_q^\zeta\}$, with $\zeta \in E_{n-q}^0$ and $\int_{\sigma_q^\zeta} \rho(x)^q d\sigma_q = \infty$. But $m\varphi(E^0) = 0$ yields $\int_{\varphi(E^0)} \rho^n \, d\tau = 0$, and then $\rho \in L^n$, allowing us to conclude that $\{\sigma_q^\zeta\}(\zeta \in E_{n-q}^0)$ is an exceptional family.

Thus, the measurability of f on σ_q^ζ for almost all $\zeta \in I_{n-q}$ implies the measurability of f on almost all the surfaces of the family $\{\sigma_q^\zeta\}$ $(\zeta \in I_{n-q})$. But the same argument as above, proves that the family $\{f(\sigma_q^\zeta)\}$ is exceptional too, and then, we have $M(\Sigma_q^0) = M[f(\Sigma_q^0)] = 0$.

Now, let us consider a family $\{S(x_0, r)\}$ $(0 < r_0 < r < r_1 < \infty)$ of concentric spheres centred at a point x_0 and with the radius r. For simplicity, we suppose $x_0 = O$. We may assume, that the family $\{S(r)\}$ and the closure of the corresponding spherical ring $A = \{x;\ r_0 < r < r_1\}$ are contained in D. We observe that for a.e.r in (r_0, r_1), $f_{|S(r)}$ is differentiable with $J(\xi) \neq 0$ a.e. in $S(r)$, since the QCf of f in A implies its differentiability with $J(\xi) \neq 0$ a.e. in A, i.e. except in a set E with $mE = 0$, and then, for almost all the surfaces $S(r)$ we have $m^S[S(r) \cap E] = 0$, as it follows from proposition 1.5.5. By lemma 2, for almost all r in (r_0, r_1), f is H^{n-1} - AC in $S(r)$ with respect to the spherical measure.

Next, let $S(r)$ be a sphere on which f is H^{n-1} - AC and is differentiable with $J(x) \neq 0$ a.e. in $S(r)$, and let φ be the inverse of the stereographic projection. Since the Jacobian J_φ of φ is > 0 everywhere in the corresponding plane Π, φ is differentiable everywhere in Π and φ, φ^{-1} transform null sets into null sets (see for instance E. Schmidt [1]), we conclude that if f is differentiable with $J(x) \neq 0$ a.e. in $S(r)$, then $f \circ \varphi$ is differentiable with $J_{f \circ \varphi}(\xi) \neq 0$ a.e. in Π. If E is a measurable set of Π then

$$H^{n-1}[f \circ \varphi(E)] = \int_E J_{f \circ \varphi}(\xi) \, d\sigma,$$

where $d\sigma$ is the surface element of Π. If $H^{n-1}[f \circ \varphi(E)] = 0$, since $J_{f \circ \varphi}(\xi) \neq 0$ a.e. in E, it follows that $m_{n-1}E = 0$, and by the AC property of φ also $m^S E^S = 0$, where $E^S = \varphi(E)$, which allows us to conclude that for almost all r in (r_0, r_1), f^{-1} is H^{n-1} - AC on $f[S(r)]$. And now, as we did above, we shall show that the family $\{f[S(r)]\}$, where $r \in E$ and $m_1 E = 0$, is an exceptional family.

Clearly, $E^0 = \bigcup_{r \in E} S(r)$ is a null set. Indeed, since E is a null set, it follows that $m_1 E = \inf_{\delta \supset E} m_1 \delta = 0$, where δ are open sets. Hence, there is a sequence $\{\delta_k\}$ of open, sets, so that

$$\lim_{k \to \infty} m_1 \delta_k = 0. \tag{5}$$

For each open set $\delta \subset R^1$, let us consider the n-dimensional open set $\Delta = \bigcup_{r \in \delta} S(r)$ If $\delta \subset (r_0, r_1)$, then

$$n\omega_n r_0^{n-1} m_1 \delta < m\Delta < n\omega_n r_1^{n-1} m_1 \delta,$$

hence, on account of (5), and since we can suppose, without loss of generality, $\delta_{k+1} \subset \delta_k \subset (r_0, r_1)$ $(k = 1, 2, \ldots)$, it follows that

$$\lim_{k \to \infty} \Delta_k = \lim_{k \to \infty} \bigcup_{r \in \delta_k} S(r) = mE^0 = 0,$$

i.e. E^0 is a null set.

Since f is QCfH and then measurable (theorem 1.1), it follows that $mf(E^0) = 0$. Next, since since f is a homeomorphism, we have $H^{n-1}\{f[S(r)]\} > 0$ and then

$$\rho(x) = \begin{cases} \infty & \text{if } x \in E^0, \\ 0 & \text{if } x \bar{\in} E^0 \end{cases}$$

is clearly admissible for $\{f[S(r)]\}$, where $r \in E$ and $m_1 E = 0$, and evident $\int_{f[S(r)]} \rho^n \, d\sigma = \infty$. But $mf(E^0) = 0$ yields $\int_{f(E^0)} \rho^n \, d\tau = 0$, and then $\rho \in L^n$, hence $\{f[S(r)]\}$, where $r \in E$ and $m_1 E = 0$, is an exceptional family (proposition 1.5.4), whence we conclude that f^{-1} is AC on almost every surface $f[S(r)]$ i.e. the family of surfaces $f[S(r)]$ on which f^{-1} is not AC is exceptional. Since $mE^0 = 0$ and $\int_{S(r)} \infty \, d\sigma = \infty$, it follows that also the family $\{S(r)\}$, where $r \in E$ and $m_1 E = 0$, is exceptional (proposition 1.5.5). Thus, we established that f is H^{n-1}-AC on almost all surfaces $S(r)$, and the exceptional family (on which f is not H^{n-1} - AC) is transformed again into an exceptional family. Hence, and since f is differentiable with $J \neq 0$ a.e. in almost all surface $S(r)$ (as we established above), we conclude that, for almost all surfaces $S(r)$, $f(x)$ is H^{n-1} - AC and is differentiable with $J \neq 0$ a.e. in $S(r)$.

Now, let Σ_q be a family of surfaces σ_q^ζ and set $\Sigma_q' = \Sigma_q - \Sigma_q^0$ (where Σ_q^0 is defined as above). If $q = n - 1$, we may suppose that Σ_q is a family $\{S(r)\}$ and Σ_q^0 is the corresponding exceptional subfamily. Fix a surface $\sigma_q \in \Sigma_q'$ and a point $x \in \sigma_q$, which is an A-point of f. Let $E[(C), x]$ be the infinitesimal ellipsoid centred at x and with the characteristics (C), mapped by f into an infinitesimal sphere. In order to evaluate the q-dimensional area of the part of σ_q contained inside $E[(C), x]$, we assume that $x = O$ and the axes of $E[(C), x]$ are the coordinate axes. Moreover, instead of the preceding area, we prefer to calculate the part of the q-dimensional tangent plane to σ_q in O contained in $E[(C), O]$. Let

$$\frac{(x^1)^2}{a_1^2} + \cdots + \frac{(x^n)^2}{a_n^2} = 1, \tag{6}$$

be the equation of $E[(C), O]$ and let

$$x^i \cos \alpha_i^k = 0, \qquad (k = 1, \ldots, n - q), \tag{7}$$

be the equation of the q-dimensional tangent plane to σ_q in O. Clearly $\cos k_i^k$ ($i = 1, \ldots, n$) are the directing cosines of the normal to the $(n - 1)$-dimensional plane

$$x^i \cos \alpha_i^k = 0.$$

for the corresponding k. The intersection of $E[(C), O]$ with the q-dimensional plane (7) yields a $(q - 1)$-dimensional ellipsoid E which contains inside the portion of the plane (7) we intend to estimate and which is expressed analytically by the system of equations (6) and (7). If we eliminate x^m ($m = q + 1, \ldots, n$) among the $n - p + 1$ equations (6) and (7), we obtain

$$\begin{aligned}
&(x^1)^2\left\{\frac{1}{a_1^2} + \frac{1}{\delta^2_{q+1,\ldots,n}}\left[\left(\frac{\delta_{1,q+2,\ldots,n}}{a_{q+1}}\right)^2 + \cdots + \left(\frac{\delta_{q+1,\ldots,n-1,1}}{a_n}\right)^2\right]\right\} + \cdots \\
&+ (x^q)^2\left\{\frac{1}{a_q^2} + \frac{1}{\delta^2_{q+1,\ldots,n}}\left[\left(\frac{\delta_{q,q+2,\ldots,n}}{a_{q+1}}\right)^2 + \cdots + \left(\frac{\delta_{q+1,\ldots,n-1,q}}{a_n}\right)^2\right]\right\} + \\
&+ \frac{2x^1x^2}{\delta^2_{q+1,\ldots,n}}\left(\frac{\delta_{1,q+2,\ldots,n}\,\delta_{2,q+2,\ldots,n}}{a^2_{q+1}} + \frac{\delta_{q+1,\ldots,n-1,1}\,\delta_{q+1,\ldots,n-1,2}}{a_n^2}\right) + \\
&+ \cdots + \frac{2x^{q+1}x^q}{\delta^2_{q+1,\ldots,n}}\left(\frac{\delta_{q-1,q+2,\ldots,n}\,\delta_{q,q+2,\ldots,n}}{a^2_{q+1}} + \cdots + \right. \\
&\left. + \frac{\delta_{q+1,\ldots,n-1,q-1}\,\delta_{q+1,\ldots,n-1,q}}{a_n^2}\right) = 1,
\end{aligned} \tag{8}$$

where

$$\delta_{i_1, \ldots, i_{n-q}} = \begin{vmatrix} \cos \alpha^1_{i_1} \cdots \cos \alpha^1_{i_{n-q}} \\ \cdots\cdots\cdots\cdots \\ \cos \alpha^{n-q}_{i_1} \cdots \cos \alpha^{n-q}_{i_{n-q}} \end{vmatrix},$$

i.e. the equation of the cylinder which intersected with the p-dimensional plane (7) yields just the $(q - 1)$-dimensional ellipsoid E. Let E' be the $(q - 1)$-dimensional ellipsoid obtained by the intersection of the cylinder (8) with the q-dimensional plane $x^m = 0\ (m = q + 1, \ldots, n)$. Then clearly

$$d\sigma'_q = d\sigma_q \cos \varphi, \tag{9}$$

where $d\sigma'_q$ is the q-dimensional volume element of E' and φ the angle between the two q-dimensional planes (7) and $x^m = 0\ (m = q+1, \ldots, n)$, which is defined by the formula (see E. Cartan [1], p. 9):

$$\cos \varphi = \frac{\delta_{q+1, \ldots, n}}{\sqrt{\sum\limits_{i_1, \ldots, i_{n-q}=1}^{n} \delta^2_{i_1, \ldots, i_{n-q}}}}.$$

In order to calculate $d\sigma'_q$, we notice that the equation of E' can be written also under the form

$$(\xi^1)^2 S_1 + \cdots + (\xi^q)^2 S_q = 1, \quad \xi^m = 0 \quad (m = q + 1, \ldots, n),$$

where $S_k (k = 1, \ldots, q)$ are the roots of the equation in S of the form

$$\det |a_{i_k} - S\delta_{i_k}| = 0$$

with the coefficients of the corresponding determinant of order q taken of (8). The q-dimensional volume of E' is

$$d\sigma_q = \frac{\omega_q}{\sqrt{S_1 \ldots S_q}}, \tag{10}$$

where the product $S_1 \ldots S_q$ is equal to the free term of the equation in S, i.e. to

$$\delta = \begin{vmatrix} a_{11} \cdots a_{1q} \\ \cdots\cdots \\ a_{1q} \cdots a_{qq} \end{vmatrix} = \frac{\sum\limits_{i_1, \ldots, i_{n-q}=1}^{n} a^2_{i_1} \cdots a^2_{i_{n-q}} \delta^2_{i_1, \ldots, i_{n-q}}}{a^2_1 \cdots a^2_n \delta^2_{q+1, \ldots, n}},$$

where $a_{kp}(k, p = 1, \ldots, q)$ denote the coefficients of $x^k x^p$ in (8). The preceding formula and (10) yield

$$\mathrm{d}\sigma'_q = \frac{\omega_q a_1 \cdots a_n \delta_{q+1, \ldots, n}}{\sqrt{\sum\limits_{i_1, \ldots, i_{n-q}=1}^{n} a_{i_1}^2 \cdots a_{i_{n-q}}^2 \delta_{i_1, \ldots, i_{n-q}}^2}},$$

which combined with (9) and the formula giving $\cos\varphi$ implies that

$$\mathrm{d}\sigma_q = \frac{\mathrm{d}\sigma'_q}{\cos\varphi} = \frac{\omega_q a_1 \cdots a_n \sqrt{\sum\limits_{i_1, \ldots, i_{n-q}=1}^{n} \delta_{i_1, \ldots, i_{n-q}}^2}}{\sqrt{\sum\limits_{i_1, \ldots, i_{n-q}=1}^{n} a_{i_1}^2 \cdots a_{i_{n-q}}^2 \delta_{i_1, \ldots, i_{n-q}}^2}}. \tag{11}$$

Set $f(x) = f_1(x, O) + o\,(|x|)$ and let $\mathrm{d}\sigma_q^E$ be the image by $f_1(x, O)$ of the part of σ_q contained inside the ellipsoid $E_h[(C), O]$ with $h = a_n$. Since $f(x)$ maps the infinitesimal ellipsoid $E[(C), O]$ into an infinitesimal sphere, we conclude that

$$\mathrm{d}\sigma_q^E = \omega_q r^q \ (q \text{ is not a dummy index}), \tag{12}$$

where r is the radius of the sphere corresponding to the ellipsoid $E_h[(C), O]$ by $f_1(x,O)$. Since in every A-point of $f(x)$

$$J = \frac{r^n}{a_1 \cdots {}_n},$$

(11) and (12) yield

$$\frac{\mathrm{d}\sigma_q^E}{\mathrm{d}\sigma_q} = \frac{r^q \sqrt{\sum\limits_{i_1, \ldots, i_{n-q}=1}^{n} a_{i_1}^2 \cdots a_{i_{n-q}}^2 \delta_{i_1, \ldots, i_{n-q}}^2}}{a_1 \cdots a_n \sqrt{\sum\limits_{i_1, \ldots, i_{n-q}=1}^{n} \delta_{i_1, \ldots, i_{n-q}}^2}} =$$

$$= \frac{J^{\frac{q}{n}} \sqrt{\sum\limits_{i_1, \ldots, i_{n-q}=1}^{n} a_{i_1}^2 \cdots a_{i_{n-q}}^2 \delta_{i_1, \ldots, i_{n-q}}^2}}{\sqrt{\sum\limits_{i_1, \ldots, i_{n-q}=1}^{n} \delta_{i_1, \ldots, i_{n-q}}^2}\,(a_1 \cdots a_n)^{\frac{n-q}{n}}} \geqslant \tag{13}$$

$$\geqq \frac{J^{\frac{q}{n}} a_{q+1} \cdots a_n}{(a_1 \cdots a_n)^{\frac{n-q}{n}}} = \frac{J^{\frac{q}{n}} (a_{q+1} \cdots a_n)^{\frac{q}{n}}}{(a_1 \cdots a_q)^{\frac{n-q}{n}}} \geqq \left(\frac{a_n}{a_1}\right)^{\frac{(n-q)q}{n}} J^{\frac{q}{n}}$$

a.e. in D. But, since $f(x)$ is K-QCfH in Markuševič—Pesin's sense, $\frac{a_1}{a_n} \leqq K$, hence

$$\frac{d\sigma_q^E}{d\sigma_q} \geqq J^{\frac{q}{n}} K^{\frac{-q(n-q)}{n}} \tag{14}$$

a.e. in D.

Next, let $\rho \in F(\Sigma_q')$ and set

$$\rho^*[f(x)] = \frac{\rho(x) K^{\frac{n-q}{n}}}{J(x)^{\frac{1}{n}}}. \tag{15}$$

Clearly $\rho^* \in F_{D^*}(\Sigma_q'^*)$, because (14) and (15) imply

$$\int_{\sigma_q^*} \rho^{*q} \, d\sigma_q^* \geqq \int_{\sigma_q} \rho^q \, d\sigma_q \geqq 1,$$

so that conditions 1°, 2°, 3° (Chapter 1.5) are satisfied. Hence

$$M(\Sigma_q^*) = M(\Sigma_q'^*) \leqq \inf_{\rho^*} \int_{D^*} \rho^{*n} \, d\tau \leqq \inf_{\rho} \int_D \frac{\rho^n K^{n-q}}{J} J d\tau = K^{n-q} M(\Sigma_q') = K^{n-q} M(\Sigma_q).$$

The same argument for the inverse mapping $x = f^{-1}(x^*)$ (which is again K-QCfH in Markuševič—Pesin's sense) yields

$$M(\Sigma_q) \leqq K^{n-q} M(\Sigma_q^*).$$

The last two formulae allow us to conclude that f is K-QCfH in Šabat's sense.

Corollary. *Every K-QCfH according to Gehring's metric definition is K-QCfH also in Šabat's sense.*

Equivalence of the definition of K-QCfH in Šabat's sense for $q = n-1$ and Gehring's analytic and geometric definition. In order to establish this equivalence, let us first introduce several preliminary concepts and lemmas.

A compact set σ is said *to separate the boundary components of a ring A* if $\sigma \subset A$ and each component of $C\sigma$ contains at most a component of CA.

Set

$$a(\rho) = a(A, \rho) = \inf_{\sigma} \int_{\sigma} \rho(x)^{n-1} d\sigma, \tag{16}$$

where the infimum is taken over all compact piecewise smooth surfaces σ which separate the boundary components of A.

LEMMA 4. *Let $u(x)$ be continuous and ACL in the ring $A : a < |x| < b$. Then*

$$\int_a^b (\operatorname*{osc}_S u)^n \frac{dr}{r} \leqq \frac{A_0}{2^n} \int_A |\nabla u|^n \, d\tau, \tag{17}$$

where A_0 is given by (1.5.13) *and $S = S(r) \, (a < r < b)$.*

We recall first that the *oscillation of a real function* $u(x)$ on a set $E \subset R^n$ is the quantity

$$\operatorname*{osc}_E u(x) = \sup_E u(x) - \inf_E u(x).$$

We may clearly assume that $|\nabla u|$ is L^n-integrable in A as otherwise there is nothing to prove. Now fix $a < a' < b' < b$ and let

$$v_c(x) = \frac{1}{\omega_n c^n} \int_{B(c)} u(x + y) \, d\tau,$$

where $0 < c < \min(a' - a, \; b - b')$. Then, arguing as in lemma 2.2 for w, we conclude that $v_c \in C^1$ in A, $|\nabla v_c| \in L^n$ and

$$|\nabla v_c(x)| \leqq \frac{1}{\omega_n c^n} \int_{B(c)} |\nabla u(x + y)| \, d\tau.$$

It is easy to see that

$$\sup_S \int_{B(c)} u(x + y) \, d\tau = \int_{B(c)} \sup_S u(x + y) \, d\tau,$$

hence, since $\sup_S u(x + y)$ is a function continuous with respect to y, by virtue of the proposition 1.6.9 we get

$$\lim_{c \to 0} \frac{1}{\omega_n c^n} \int_{B(c)} \sup_S u(x + y) \, d\tau = \sup_S u(x),$$

and analogously

$$\lim_{c \to 0} \inf_S \frac{1}{\omega_n c^n} \int_{B(c)} u(x + y) \, d\tau = \lim_{c \to 0} \frac{1}{\omega_n c^n} \int_{B(c)} \inf_S u(x + y) \, d\tau = \inf_S u(x),$$

whence

$$\operatorname*{osc}_S u(x) = \sup_S u(x) - \inf_S u(x) = \lim_{c\to 0} \frac{1}{\omega_n c^n}\int_{B(c)} \sup_S u(x+y)\,d\tau -$$

$$- \lim_{c\to 0} \frac{1}{\omega_n c^n}\int_{B(c)} \inf_S u(x+y)\,d\tau = \lim_{c\to 0}\sup_S \frac{1}{\omega_n c^n}\int_{B(c)} u(x+y)\,d\tau - \tag{18}$$

$$- \lim_{c\to 0}\inf_S \frac{1}{\omega_n c^n}\int_{B(c)} u(x+y)\,d\tau = \lim_{c\to 0}\left[\sup_S v_c(x) - \inf_S v_c(x)\right] = \lim_{c\to 0}\operatorname*{osc}_S v_c(x).$$

If in the corollary of proposition 1.5.10 we set $\rho(x) = |\nabla v_c(x)|$ and we choose x_1, x_2 so that $v_c(x_1) = \sup_S v_c(x)$, $v_c(x_2) = \inf_S v_c(x)$, then on account of lemma 2.3, we obtain

$$(\operatorname*{osc}_S v_c)^n \leqq \left(\int_\alpha |\nabla v_c|\,ds\right)^n \leqq \frac{A_0 r}{2^n}\int_S |\nabla v_c|^n\,d\sigma\,, \tag{19}$$

where α is a circular arc joining x_1 and x_2.

Since $\lim_{c\to 0} f_c(x) = f(x)$, where $f_c(x)$ are continuous on $[a, b]$, implies $\lim_{c\to 0} \int_a^b f_c(x)\,dx = \int_a^b f(x)\,dx$ (Lebesgue's theorem, see Natanson [1], p. 169), then (18) and (19) combined with Tonelli's theorem (Chapter 1.4) yield

$$\int_{a'}^{b'} (\operatorname*{osc}_S u)^n \frac{dr}{r} = \int_{a'}^{b'} \lim_{c\to 0} (\operatorname*{osc}_S v_c)^n \frac{dr}{r} = \lim_{c\to 0}\int_{a'}^{b'} (\operatorname*{osc}_S v_c)^n \frac{dr}{r} \leqq$$

$$\leqq \frac{A_0}{2^n}\lim_{c\to 0}\int_{a'}^{b'}\left(\int_S |\nabla v_c|^n\,d\sigma\right)dr = \frac{A_0}{2^n}\lim_{c\to 0}\int_{a'<|x|<b'} |\nabla v_c|^n\,d\tau.$$

Finally, with Minkowski's inequality (see G. H. Hardy, J. E. Littlewood and G. Polya [1], p. 148) it is easy to show that

$$\int_{a'}^{b'} (\operatorname*{osc}_S u)^n \frac{dr}{r} \leqq \frac{A_0}{2^n}\lim_{c\to 0}\int_{a'<|x|<b'} |\nabla v_c|^n\,d\tau =$$

$$= \frac{A_0}{2^n}\lim_{c\to 0}\int_{a'<|x|<b'} \left|\frac{1}{\omega_n c^n}\int_{B(c)} \nabla u(x+y)\,d\tau\right|^n d\tau \leqq$$

$$\leqq \frac{A_0}{2^n}\lim_{c\to 0}\left\{\frac{1}{\omega_n c^n}\int_{B(c)}\left[\int_{a'<|x|<b'} |\nabla u(x+y)|^n\,d\tau\right]^{\frac{1}{n}} d\tau\right\}^n,$$

hence, by proposition 1.6.9,

$$\int_{a'}^{b'} (\operatorname*{osc}_S u)^n \frac{\mathrm{d}r}{r} \leqq \frac{A_0}{2^n} \lim_{c\to 0} \left\{ \frac{1}{\omega_n c^n} \int_{B(c)} \left[\int_{a'<|x|<b'} |\nabla u(x+y)|^n \,\mathrm{d}\tau \right]^{\frac{1}{n}} \mathrm{d}\tau \right\}^n =$$

$$= \frac{A_0}{2^n} \int_{a'<|x|<b'} |\nabla u(x)|^n \,\mathrm{d}\tau \leqq \frac{A_0}{2^n} \int_{a<|x|<b} |\nabla u(x)|^n \mathrm{d}\tau$$

and letting $a' \to a$, $b' \to b$, yields (17) as desired.

LEMMA 5. *Let $u(x)$ be an admissible function for the ring A, let $|\nabla u| \in L^n$ and extend u to be 0 on c_0 and 1 on C_1. Then $u(x)$ is continuous and ACL everywhere and $|\nabla u| = 0$ a.e. in CA.*

The continuity everywhere is immediate. We next show that u is ACL in R^n. For this let X denote any line parallel to the x^1-axis. Then by Fubini's theorem,

$$\int \Big(\int_{X\cap A} |\nabla u|^n \,\mathrm{d}x^1 \Big) \,\mathrm{d}x^2 \cdots \mathrm{d}x^n = \int_A |\nabla u|^n \,\mathrm{d}\tau < \infty.$$

From this and the *ACL* property in A it follows that almost all lines X have the following property: $u(x)$ is *AC* on each compact interval $E \subset X \cap A$ and

$$\int_{X\cap A} |\nabla u|^n \,\mathrm{d}x^1 < \infty.$$

Fix such a line X. Then using the fact that u is 0 on C_0 and 1 on C_1 it is not difficult to show that

$$|u(x) - u(y)| = \left| \int_{u(y)}^{u(x)} \mathrm{d}u \right| \leqq \int_E \left| \frac{\partial u}{\partial x^1} \right| \mathrm{d}x^1 \leqq \int_E |\nabla u| \,\mathrm{d}x^1 = \int_{E\cap A} |\nabla u| \,\mathrm{d}x^1,$$

for each compact interval $E \subset X$ with endpoints x and y. From this and Hölder's inequality we obtain

$$[\Sigma_m |u(x_m) - u(y_m)|]^n \leqq \Big(\int_{E\cap A} |\nabla u| \,\mathrm{d}x^1 \Big)^n \leqq (\Sigma_m |x_m - y_m|)^{n-1} \int_{X\cap A} |\nabla u|^n \mathrm{d}x^1,$$

where x_m, y_m are the endpoints of a sequence of non-overlapping compact intervals in X. Hence $u(x)$ is *AC* on X and, by symmetry, on almost all lines parallel to the coordinate axes.

Finally let $x_0 \in CA$ be a point of linear density for CA in the directions of the coordinates axes (see p. 99) and at which ∇u exists. Then $\nabla u(x_0) = 0$ and, since almost all points of a set are points of linear density in the directions of the coordinate axes (see S. Saks [2], p. 298), it follows that almost all the points of CA satisfy the above requirements, and then $|\nabla u(x)| = 0$ a.e. in CA.

Let $\{f(x)\}$ be a family of functions continuous on a domain D. They are said to be *equicontinuous* if given any $\varepsilon > 0$, there exists a $\delta > 0$, depending only on ε, such that $|x'' - x'| < \delta$ imply $|f(x'') - f(x')| < \varepsilon$ for every function of the family.

The functions of $\{f(x)\}$ are called *uniformly bounded* if there exists a constant K such that $|f(x)| < K$ for every function of the family.

Theorem of Arzela–Ascoli. *If $\{f(x)\}$ is a family of functions equicontinuous and uniformly bounded in a domain D, then it is possible to pick out a uniformly convergent sequence of functions.* (For the proof see for instance Natanson [1], theorem 3, p. 552.)

Let $u(x)$ be continuous in a domain D and let $u(x)$ have boundary values at each point ∂D. Following Lebesgue, we say that $u(x)$ is *monotone* in D if

$$\sup_{\partial\Delta} u = \sup_{\Delta} u, \quad \inf_{\partial\Delta} u = \inf_{\Delta} u$$

for each domain $\Delta \subset D$. Alternatively, u is monotone in D if and only if there exists no domain $\Delta \subset D$ such that u is constant on $\partial\Delta$ without being constant in Δ itself (Lebesgue [1]).

Now let u be an admissible function for a ring A (Chapter 2), and let $\{r_m\}$ be an ordering of rationals. We define a sequence of functions $\{u_m\}$ as follows: Fix $x \in A$. If x lies in a domain $\Delta \subset A$ on whose boundary $u(x) = r_1$, we set $u_1(x) = r_1$; otherwise we set $u_1(x) = u(x)$. We obtain u_2 from u_1 in the same way, using r_2 in place of r_1, and the sequence is defined by induction.

Lebesgue's theorem for monotone functions. (Lebesgue [1]). *The sequence $\{u_m\}$ from above converges uniformly in A to a function v which is monotone in A and which has the same boundary values as u.*

Let us show first that $\{u_m\}$ converges to v. Fix $x_1 \in A$ and let u^1 be the first u_m for which $u_m(x_1) \neq u_{m-1}(x_1)$ holds. Then $u^1(x) = r^1$ in a certain domain Δ^1. Let u' be the first function following u^1. Compare $u^1(x_1)$ and $u'(x_1)$. Suppose first $u^1(x_1) = u'(x_1)$. Then in whole Δ^1, $u^1(x) = u'(x)$. Indeed, if there were a point $x_0 \in \Delta^1$ such that $u^1(x_0) \neq u'(x_0)$, then x_0 would be contained in a domain Δ such that $u^1(x) \not\equiv$ const. in Δ and $u^1(x) \equiv$ const. on $\partial\Delta$. Hence $\Delta \supset \Delta^1$ and then $u'(x_1) = u'(x_0) \neq u^1(x_0) = u^1(x_1)$ contradicting so $u^1(x_1) = u'(x_1)$. In other words $u^1(x_0) \neq u'(x_0)$ implies $u^1(x) \neq u'(x)$ for all $x \in \Delta^1$. If u^2 denotes the first function following u^1 and for which $u^2(x_1) \neq u^1(x_1)$, then $u^2 = r^2$ in a whole domain $\Delta^2 \supset \Delta^1$. We obtain u^3, r^3, Δ^3 in the same way and the sequence is defined by induction. If there is only a finite number of such functions u^m then $v(x_1)$ will be equal to $u^m(x_1)$, where u^m is the last function of the sequence. Suppose now that there exists an infinite sequence of such functions u^m and let $\overline{x_1x_0}$ be a segment with an end point at x_1, the other in ∂A and all its other points in A. Let x_0^m denote the nearest point from x_0 among all the points of the intersection of $\overline{x_1 x_0}$ with $\partial\Delta^m$. Then $u^m(x_1) = u^m(x_0^m) = u(x_0^m)$. The first equality is obvious and the second follows from the fact that every x such that $u^m(x) \neq u(x)$ is an inner point of a domain in which $u^m \not\equiv u$. But $x_0^m \to x_0'$, hence

$$v(x_1) = \lim_{m\to\infty} u^m(x_1) = \lim_{m\to\infty} u(x_0^m) = u(x_0').$$

If $x_0' = x_0$, then $v(x_0) = v(x_1) = u(x_0)$., but also in general $v(x_0) = u(x_0)$.

The convergence $u^m \to v$ is uniform. Indeed, u^m $(m=1, 2, \ldots)$ are uniformly continuous in A and the oscillation of u^m in any interval does not increase by passing from u^m to u^{m+k}, hence $\{u^m\}$ is a family of equicontinuous functions. Since $u(x)$ is bounded in A, it follows that $\{u^m\}$ is a family of uniformly bounded functions in A. Hence, by Arzela—Ascoli's theorem (quoted above), we conclude that the sequence $\{u^m\}$ (and then also the sequence $\{u_m\}$) converges uniformly to v.

Now let us prove that the function v obtained in this way is monotone. Let $\Delta \subset D$. The difference between the supremum (infimum) of v in Δ and $\partial\Delta$ is the limit of the corresponding difference relatively to u_m. Suppose that $\sup_{x\in\Delta} u_m(x) > \sup_{x\in\partial\Delta} u_m(x)$ or $\inf_{x\in\Delta} u_m(x) < \inf_{x\in\partial\Delta} u_m(x)$. Then there exists a point $x_0 \in \Delta$ such that $u_m(x_0) > \sup_{x\in\partial\Delta} u_m(x)$. The continuity of u_m implies the existence of a rational number r, $u_m(x_0) > r > \sup_{x\in\partial\Delta} u_m(x)$, among $r_{m+1}, r_{m+2}, \ldots$ and of a domain Δ_r such that $x_0 \in \Delta_r \subset \Delta$ and $u_{m|\partial\Delta_r} = r$. Since x_0 may be chosen so that $u_m(x_0)$ be as nearly equal to $\sup_{x\in\Delta} u_m(x)$ as one pleases and the same may be made for r with respect to $\sup_{x\in\partial\Delta} u_m(x)$, it follows that if $r = r_p$, then the quantity $\sup_{x\in\Delta} |u_p(x) - u_m(x)|$ will be as nearly equal to $\sup_{x\in\Delta} u_m(x) - \sup_{x\in\partial\Delta} u_m(x)$ as one pleases. But $\sup_{x\in\Delta} |u_p(x) - u_m(x)| \to 0$ as $m \to \infty$ (it is sufficient to have $m \geqq p$), and since the same is also true with respect to the infimum, it follows that

$$\sup_{x\in\Delta} v(x) = \sup_{x\in\partial\Delta} v(x), \qquad \inf_{x\in\Delta} v(x) = \inf_{x\in\partial\Delta} v(x)$$

i.e. $v(x)$ is monotone.

The function v in the preceding Lebesgue's theorem, obtained from an admissible function for the ring A is called a *monotone admissible function*. It is also easy to verify that $\operatorname{osc}_E v \leqq \operatorname{osc}_E u$ for each segment $E \subset A$. Hence v is *ACL* in A, $|\nabla v| \leqq |\nabla u|$ a.e. there, and we conclude that

$$\operatorname{cap} A = \inf_v \int_A |\nabla v|^n \, d\tau,$$

where the infimum is taken over all the class of monotone admissible functions for A.

LEMMA 6. *Let A be a ring with non-degenerate boundary components, let $u(x)$ be a monotone admissible function for A, let*

$$\int_A |\nabla u|^n \, d\tau \leqq K < \infty. \tag{20}$$

and extend u so that u is 0 on C_0 and 1 on C_1. If $d(C_0) = b$, then

$$|u(x) - u(y)|^n \leqq \frac{A_0 K}{2^n} \left(\log \frac{b}{a} \right)^{-1} \tag{21}$$

whenever $|x - y| \leqq a < b$. If $d(C_0, C_1) = c$, then

$$|u(x) - u(y)|^n \leqq \frac{A_0 K}{2^n} \left(\log \frac{a}{c} \right)^{-1} \tag{22}$$

whenever $d(x, C_0)$, $d(y, C_0) \geqq a > c$. Here A_0 is given by (1.5.13).

We begin with the proof for (21). Fix x and y so that $|x - y| \leqq a < b$. Since $u(x)$ is monotone admissible it follows $0 \leqq u(x), u(y) \leqq 1$ and then we may assume that

$$\frac{A_0 K}{2^n} \left(\log \frac{b}{a} \right)^{-1} < 1, \tag{23}$$

for otherwise (21) follows trivially. Lemma 5 implies that $u(x)$ is continuous and ACL everywhere and that $|\nabla u| = 0$ a.e. in CA. Hence we can apply lemma 4 to conclude there exists an $r \left(\frac{a}{2} < r < \frac{b}{2} \right)$, for which

$$(\operatorname*{osc}_S u)^n \leqq \frac{A_0 K}{2^n} \left(\log \frac{b}{a} \right)^{-1}, \tag{24}$$

where $S = S\left(\frac{x+y}{2}, r \right)$, because if for all $r \in \left(\frac{a}{2}, \frac{b}{2} \right)$ we had

$$(\operatorname*{osc}_S u)^n > \frac{A_0 K}{2^n} \left(\log \frac{b}{a} \right)^{-1},$$

then, integrating with respect to r, we would obtain

$$\int_{\frac{a}{2}}^{\frac{b}{2}} (\underset{S}{\operatorname{osc}} u)^n \frac{dr}{r} > \frac{A_0 K}{2^n} \left(\log \frac{b}{a}\right)^{-1} \int_{\frac{a}{2}}^{\frac{b}{2}} \frac{dr}{r} = \frac{A_0 K}{2^n},$$

which would contradict the inequalities

$$\int_{\frac{a}{2}}^{\frac{b}{2}} (\underset{S}{\operatorname{osc}} u)^n \frac{dr}{r} \leqq \frac{A_0}{2^n} \int_{\frac{a}{2} < \left|\xi - \frac{x+y}{2}\right| < \frac{b}{2}} |\nabla u|^n \, d\tau \leqq \frac{A_0}{2^n} \int_A |\nabla u|^n \, d\tau \leqq \frac{A_0 K}{2^n}.$$

which hold as a consequence of lemma 4 and of inequality (20).

Suppose first that $S \subset A$. Since the $d(S) < d(C_0)$, then $B\left(\frac{x+y}{2}, r\right) \subset A$ and the monotone admissible function $u(x)$ for the ring A satisfies the maximum and minimum principles in $B\left(\frac{x+y}{2}, r\right)$ and

$$|u(x) - u(y)|^n \leqq (\underset{S}{\operatorname{osc}} u)^n \leqq \frac{A_0 K}{2^n} \left(\log \frac{b}{a}\right)^{-1},$$

hence (21) is satisfied in this case.

Suppose next that $S \cap A$, $S \cap C_0 \neq \emptyset$. Then (23) and (24) imply that

$$\underset{S}{\operatorname{osc}} u < 1, \tag{25}$$

and then $S \cap C_1 = \emptyset$. Hence the boundary of each component of $B\left(\frac{x+y}{2}, r\right) \cap A$ is contained in $C_0 \cup S$ and, since $u(x) = 0$ on C_0, we conclude that

$$|u(x) - u(y)|^n \leqq (\sup_S u)^n = (\underset{S}{\operatorname{osc}} u)^n \leqq \frac{A_0 K}{2^n} \left(\log \frac{b}{a}\right)^{-1}.$$

Suppose now that $S \cap C_1$, $S \cap A \neq \emptyset$. Then (25) yields $S \cap C_0 = \emptyset$, and since the components of $B\left(\frac{x+y}{2}, r\right) \cap A$ are contained in $C_1 \cup S$ and $u(x) = 1$ on C_1, it follows that

$$|u(x) - u(y)|^n \leqq (1 - \inf_S u)^n = (\underset{S}{\operatorname{osc}} u)^n \leqq \frac{A_0 K}{2^n} \left(\log \frac{b}{a}\right)^{-1}.$$

Suppose finally that S lies in a component of CA. Then $B\left(\frac{x+y}{2}, r\right)$ lies in the same component, $u(x) = u(y)$, and the proof for the first part of this lemma is complete.

For the second part fix x and y so that $d(x, C_0), d(y, C_0) \geqq a > c > 0$ and assume that

$$\frac{A_0 K}{2^n}\left(\log \frac{a}{c}\right)^{-1} < 1, \tag{26}$$

for otherwise (22) follows trivially. Next let $x_0 \in C_0$ for which $d(x_0, C_1) = c$. Then applying lemma 4 as above we can find $r_1 \in (c, a)$ for which

$$(\operatorname*{osc}_{S_1} u)^n \leqq \frac{A_0 K}{2^n}\left(\log \frac{a}{c}\right)^{-1},$$

where $S_1 = S(x_0, r_1)$. (The same argument as in the preceding case.) Since $r_1 > c$ it follows $S_1 \cap C_1 \neq \emptyset$ and then (26) implies that $S_1 \cap C_0 = \emptyset$. The points x and y lie outside S_1. Fix $k > 1$ such that $kc > \max[d(x_0, x), d(x_0, y)]$, then by lemma 4 there exists at least a number $r_2 \in (kc, ka)$, such that the sphere S_2 with the radius r_2 and concentric with S_1 contain inside the points x, y and

$$(\operatorname*{osc}_{S_2} u)^n \leqq \frac{A_0 K}{2^n}\left(\log \frac{a}{c}\right)^{-1}, \quad S_2 \cap C_1 \neq \emptyset, \quad S_2 \cap C_0 = \emptyset.$$

But then C_0 does not meet the spherical ring U bounded by S_1 and S_2. Thus the boundary of each component of $U \cap A$ lies in $S_1 \cup S_2 \cup C_1$ and it is easy to verify that

$$|u(x) - u(y)|^n \leqq (1 - \inf_{S_1 \cup S_2} u)^n = (\operatorname*{osc}_{S_1 \cup S_2} u)^n \leqq \frac{A_0 K}{2^n}\left(\log \frac{a}{c}\right)^{-1},$$

as desired.

LEMMA 7. (Monotoneity and super-additivity properties for moduli of rings.) *If A' is a ring which separates the boundary components of A, then*

$$\operatorname{mod} A \geqq \operatorname{mod} A'. \tag{27}$$

If $A_1, \ldots, A_m$ are disjoint rings each of them separating the boundary components of A, then

$$\operatorname{mod} A \geqq \sum_{k=1}^{m} \operatorname{mod} A_k. \tag{28}$$

We consider only the proof for (28), since the first part of this lemma is a consequence of the second part. For each ring A_k let $u_k \in C^1$, $u_k(x) = 0$ on C_{0k} and $u_k(x) = 1$ on C_{1k}, where C_{0k}, C_{1k} are the components of CA_k and let $\nabla u_k = 0$ off a compact subset of A_k. Next set

$$u = \sum_{k=1}^{m} a_k u_k, \quad \sum_{k=1}^{m} a_k = 1,$$

where $a_k \geqq 0$ $(k = 1, \ldots, m)$. Then

$$\int_A |\nabla u|^n \, d\tau = \sum_{k=2}^{m} a_k^n \int_{A_k} |\nabla u_k|^n \, d\tau.$$

Since $C_0 \subset C_{0k}$ and $C_1 \subset C_{1k}$ for all k, u is admissible for A and taking infimums over all such u_k gives

$$\text{cap}\, A \leqq \sum_{k=1}^{m} a_k^n \,\text{cap}\, A_k. \tag{29}$$

If $\text{cap}\, A_k > 0$ for all k, setting

$$a_k = (\text{cap}\, A_k)^{\frac{-1}{n-1}} \left[\sum_{p=1}^{m} (\text{cap}\, A_p)^{\frac{-1}{n-1}} \right]^{-1}$$

n (29) and taking into account that mod A is given by (2.1), we obtain (28).

If some $\text{cap}\, A_k = 0$, then setting $a_k = 1$ and $a_p = 0$ for $k \neq p$ again yields (28).

LEMMA 8. *Let A be a ring. Then*

$$\frac{s(A)^n}{(mA)^{n-1}} \leqq \text{cap}\, A \leqq \frac{mA}{d(C_0, C_1)^n},$$

where $s(A) = \inf_E H^{n-1}(E)$ and the infimum is taken over all the sets $E \subset A$ which separate C_0 and C_1, and where $H^{n-1}(E)$ is the area of E (in the sense of Hausdorff).

Let us prove first the second part of the double inequality. Set

$$u(x) = \min\left[1, \frac{d(x, C_0)}{d(C_0, C_1)}\right].$$

Then $u(x)$ is admissible for A and

$$|\nabla u| \leqq \frac{1}{d(C_0, C_1)} \tag{30}$$

a.e. in A. Indeed, by proposition 1.6.7, ∇u exists a.e. in A. Fix $x \in A$ at which ∇u exists and $d(x, C_0) < d(C_0, C_1)$. The definition of $d(x, C_0)$ implies $d(x, y) \geqq d(x, C_0)$ for all $y \in C_0$ and also that given a sequence $\{\varepsilon_m\}$, $\varepsilon_m > 0$, $\lim_{m \to \infty} \varepsilon_m = 0$, there exists a sequence $\{y_m\}$, $y_m \in C_0$ $(m = 1, 2, \ldots)$ such that $d(x, y_m) < d(x, C_0) + \varepsilon_m$. Hence, for each natural m and for Δx^i such that $d[(x^1, \ldots, x^{i-1}, x^i + \Delta x^i, x^{i+1}, \ldots, x^n), C_0] < d(C_0, C_1)$, we have

$$0 \leqslant \left| \frac{u(x^1, \ldots, x^{i-1}, x^i + \Delta x^i, x^{i+1}, \ldots, x^n) - u(x)}{\Delta x^i} \right| =$$

$$= \frac{1}{d(C_0, C_1)} \left| \frac{d\,[(x^1, \ldots, x^{i-1}, x^i + \Delta x^i, x^{i+1}, \ldots, x^n), C_0] - d(x, C_0)}{\Delta x^i} \right| <$$

$$< \frac{1}{d(C_0, C_1)} \left| \frac{d\,[(x^1, \ldots, x^{i-1}, x^i + \Delta x^i, x^{i+1}, \ldots, x^n), y_m] - d(x, y_m) + \varepsilon_m}{\Delta x^i} \right| =$$

$$= \frac{1}{d(C_0, C_1)} \left| \frac{2(x^i - y_m^i) + \Delta x^i}{d[(x^1, \ldots, x^{i-1}, x^i + \Delta x^i, x^{i+1}, \ldots, x^n), y_m] + d(x, y_m)} + \frac{\varepsilon_m}{\Delta x^i} \right|,$$

and letting $m \to \infty$, we get

$$\left| \frac{u(x^1, \ldots, x^{i-1}, x^i + \Delta x^i, x^{i+1}, \ldots, x^n) - u(x)}{\Delta x^i} \right| \leqq$$

$$\leqq \frac{1}{d(C_0, C_1)} \left| \frac{2(x^i - y_0^i) + \Delta x^i}{d[(x^1, \ldots, x^{i-1}, x^i + \Delta x^i, x^{i+1}, \ldots, x^n), y_0] + d(x, y_0)} \right|,$$

where $\lim_{m \to \infty} y_m = y_0 \in C_0$ since C_0 is compact. And then letting $\Delta x^i \to 0$, we obtain

$$\left| \frac{\partial u\,(x)}{\partial x^i} \right| \leqq \left| \frac{x^i - y_0^i}{d(C_0, C_1)\, d(x, x_0)} \right|,$$

hence (30) holds for almost all $x \in A$ for which $d(x, C_0) < \mathrm{d}(C_0, C_1)$. But, since almost all the points of the set $A_0 \subset A$ where $d(x, C_0) \geqq d(C_0, C_1)$ are points of linear density for A_0 with respect to the coordinate axes (see S. Saks [2], p. 298), it follows that $|\nabla u| = 0$ a.e. in A_0. Thus (30) holds a.e. in A and, since $u(x)$ is admissible for A,

$$\operatorname{cap} A \leqq \int_A |\nabla u|^n \, d\tau \leqq \frac{mA}{d(C_0, C_1)^n}.$$

For the first inequality let $u \in C^1$ and admissible for A. For each $0 < a < 1$, let E_a denote the set where $u = a$. Since $E_a \subset A$ and separates C_0 and C_1

$$s(A) \leqq H^{n-1}(E_a)$$

for $0 < a < 1$. Since $u \in C^1$ it follows that u satisfies a uniform Lipschitz condition on each compact subset of A and a recent result due to Federer—Young (Chapter 1.3) gives

$$s(A) \leqq \int_0^1 H^{n-1}(E_a)\, da \leqq \int_A |\nabla u|\, d\tau,$$

Hölder's inequality implies that

$$s(A)^n \leqq (mA)^{n-1} \int_A |\nabla u|^n\, d\tau,$$

and taking the infimum over all such u yields then also the first inequality.

LEMMA 9. *cap* $A < \infty$ *for every ring* $A \subset R^n$.

Let $c = d(C_0, C_1) \leqq \infty$ and $d(C_0) = b < \infty$. Set $0 < \varepsilon < \frac{c}{3}$, and let A' be a ring, which separates the boundary components of A, with $d(A') < b + 4\varepsilon < \infty$ and $c' = d(C_0', C_1')$, where C_0', C_1' are the components of CA'. Then we conclude from lemmas 7 and 8 that

$$\operatorname{cap} A \leqq \operatorname{cap} A' \leqq \frac{mA'}{c'^n} < \frac{\omega_n (b + 4\varepsilon)^n}{c'^n} < \infty.$$

LEMMA 10. *If* A *has non-degenerate boundary components, then there exists a unique admissible function* u *for which*

$$\operatorname{cap} A = \int_A |\nabla u|^n\, d\tau.$$

We call u *the extremal function for* A.

Let $\{u_m\}$ be any sequence of monotone admissible functions for A such that

$$\operatorname{cap} A = \lim_{m \to \infty} \int_A |\nabla u_m|^n d\tau.$$

The existence of such a sequence follows from the definition of the monotone admissible functions. We will show that these functions converge uniformly on A to the extremal function u. The preceding lemma implies that cap $A < \infty$. Hence we may suppose $\{u_m\}$ be so that

$$\int_A |\nabla u_m|^n\, d\tau \leqq K < \infty.$$

for all m. We begin by showing that the gradients ∇u_m converge in L^n. For this we apply the right inequality in (1.2.4) (Clarkson's theorem) for $x = \frac{\nabla u_m}{2}$, $y = \frac{\nabla u_k}{2}$ and $p = n$, which shows the space of functions L^n-integrable over A is uniformly convex. In particular we obtain

$$\int_A \left| \frac{\nabla u_m - \nabla u_k}{2} \right|^n d\tau + \int_A \left| \frac{\nabla u_m + \nabla u_k}{2} \right|^n d\tau \leqq \frac{1}{2} \int_A \left| \nabla u_m \right|^n d\tau + \\ + \frac{1}{2} \int_A \left| \nabla u_k \right|^n d\tau,$$

for all m, k. Then, since $\frac{u_m + u_k}{2}$ is admissible for A, this inequality yields

$$\overline{\lim_{m,k\to\infty}} \int_A \left| \frac{\nabla u_m - \nabla u_k}{2} \right|^n d\tau = \operatorname{cap} A - \underline{\lim_{m,k\to\infty}} \int_A \left| \frac{\nabla u_m + \nabla u_k}{2} \right| d\tau = 0.$$

Hence $\{\nabla u_m\}$ is a Cauchy sequence. Since, by proposition 1.2.2, $L^n(A, R^n)$ is a Banach space, it follows that the ∇u_m converge in L^n to a vector function $f = (f_1, \ldots, f_n)$ and

$$\operatorname{cap} A = \int_A |f|^n d\tau < \infty.$$

Next let $b = d(C_0) > 0$ (since A is with non-degenerate components by hypothesis) and $c = d(C_0, C_1)$. Extend u_m to be 0 on C_0 and 1 on C_1. Lemma 6 implies that

$$|u_m(x) - u_m(y)|^n \leqq \frac{A_0 K}{2^n} \left(\log \frac{b}{a} \right)^{-1}$$

for all m whenever $d(x, C_0)$, $d(y, C_0) \geqq a > c$. Thus $\{u_m\}$ is a sequence of equal continuous functions, and since u_m are also monotone admissible, it follows that $|u_m| < 1$. Arzela—Ascoli's theorem then implies that a subsequence $\{u_{m_k}\}$ converges uniformly for all x to a function u in R^n.

We show that u is admissible for A and that $\nabla u = f$ a.e. First it is clear that u is continuous in A since u_m are continuous in A. Next, since u_m have values 0 on C_0 and 1 on C_1, we deduce that u has the values 0 on C_0 and 1 on C_1. Next appealling to Hölder's inequality, it is not difficult to verify that if $F \subset A$ is a compact set, then

$$\left| \int_F \left(f_1 - \frac{\partial u_{m_k}}{\partial x^1} \right) d\tau \right| \leqq \int_F \left| f_1 - \frac{\partial u_{m_k}}{\partial x^1} \right| d\tau \leqq \int_F \left| f - \nabla u_{m_k} \right| d\tau \leqq \\ \leqq \left(\int_F \left| f - \nabla u_{m_k} \right|^n d\tau \right)^{\frac{1}{n}} (mF)^{\frac{n-1}{n}},$$

hence

$$\lim_{k\to\infty} \int_F \left|f - \nabla u_{m_k}\right|^n d\tau = 0$$

implies that

$$\int_F f_1 \, d\tau = \lim_{k\to\infty} \int_F \frac{\partial u_{m_k}}{\partial x^1} d\tau < \infty,$$

and appealing to Fubini's theorem, we obtain on almost all the lines X parallel to the x^1-axis

$$\int_E f^1 \, dx^1 = \lim_{k\to\infty} \int_E \frac{\partial u_{m_k}}{\partial x^1} dx^1,$$

where $E \subset X \cap A$ is a compact interval with endpoints x, y. Hence, by the ACL property and the uniform convergence of the functions u_{m_k}, we obtain

$$\int_E f_1 \, dx^1 = \lim_{k\to\infty} \int_E \frac{\partial u_{m_k}}{\partial x^1} dx^1 = \lim_{k\to\infty} \left[u_{m_k}(x) - u_{m_k}(y)\right] = u(x) - u(y).$$

From this it follows that u is AC on $X \cap A$ and that $\dfrac{\partial u}{\partial x^1} = f_1$ a.e. in $X \cap A$. Thus, by symmetry, u is ACL in A, $\nabla u = f$ a.e. there and hence u has the desired extremal property.

To complete the proof of the theorem we show that u is unique. In particular we prove that if u and v are admissible functions for A and if

$$\operatorname{cap} A = \int_A \left|\nabla u\right|^n d\tau = \int_A \left|\nabla v\right|^n d\tau, \tag{31}$$

then $u = v$ in A. For this let $w = u - v$ and consider for real t the integral

$$W(t) = \int_A \left|\nabla u + t\nabla w\right|^n d\tau.$$

Since $u + tw$ is admissible for A,

$$(0) \leqq W(t) \tag{32}$$

for all t. We show that we may differentiate $W(t)$ with respect to t under the integral sign. For this it suffices to show that

$$\left|\frac{\left|\nabla u + (t + \Delta t)\nabla w\right|^n - \left|\nabla u + t\nabla w)\right|^n}{\Delta t}\right| \tag{33}$$

is majorized by an integrable function not depending on Δt. Suppose first that

$$|\nabla u + (t + \Delta t)\nabla w| \geqslant |\nabla u + t\nabla w|.$$

Then

$$0 \leqq \frac{|\nabla u + (t + \Delta t)\nabla w|^n - |\nabla u + t\nabla w|^n}{|\Delta u|} \leqq$$

$$\leqq \frac{(|\nabla u + \nabla w| + |\Delta t|\,|\nabla w|)^n - |\nabla u + t\nabla w|^n}{\Delta t} =$$

$$= \sum_{m-1}^{n} C_n^m |\Delta t|^{m-1} |\Delta w|^m |\nabla u + t\nabla w|^{n-m}.$$

Now suppose

$$|\nabla u + (t + \Delta t)\nabla w| < |\nabla u + t\nabla w\,|.$$

Then

$$0 < \left| \frac{|\nabla u + (t + \Delta t)\nabla w|^n - |\nabla u + t\nabla w|^n}{|\Delta t|} \right| \leqq$$

$$\leqslant \left| \frac{(|\nabla u + t\nabla w| - |\Delta t|\,|\nabla w|)^n - |\nabla u + t\nabla w|^n}{|\Delta t|} \right| =$$

$$= \left| \sum_{m=1}^{n} C_n^m (-1)^m |\Delta t|^{m-1} |\nabla w|^m |\nabla u + t\nabla w|^{n-m} \right| \leqslant$$

$$\leqslant \sum_{m=1}^{n} C_n^m |\Delta t|^{m-1} |\nabla w|^m |\nabla u + t\nabla w|^{n-m}.$$

Thus (33) is majorized in both hypotheses by

$$\sum_{m=1}^{n} C_n^m |\nabla w|^m |\nabla u + t\nabla w|^{n-m} \tau_0^{m-1}$$

for $|\Delta t| < \tau_0$. We show that this function is integrable. Indeed

$$|\nabla w|^m|\nabla u + t\nabla w|^{n-m} \leqq (|\nabla u| + |\nabla v|)^m[(1 + |t|)|\nabla u| + |t|\,|\nabla v|]^{n-m} =$$

$$= \sum_{p=0}^{m} C_m^p|\nabla u|^p|\nabla v|^{m-p} \sum_{q=0}^{n-m} C_{n-m}^q(1 + |t|)^q|\nabla u|^q|t|^{n-m-q}|\nabla v|^{n-m-q} =$$

$$\sum_{p=0}^{m}\sum_{q=0}^{n-m} C_m^p C_{n-m}^q(1 + |t|)^q|t|^{n-m-q}|\nabla u|^{p+q}|\nabla v|^{n-(p+q)}.$$

But with Hölder's inequality we see that

$$\int_A |\nabla u|^{p+q}|\nabla v|^{n-(p+q)}\,d\tau \leqq \left(\int_A |\nabla u|^n\,d\tau\right)^{\frac{p+q}{n}}\left(\int_A |\nabla v|^n\,d\tau\right)^{\frac{n-(p+q)}{n}} < \infty.$$

Hence $W(t) < \infty$ for all t and (33) is majorized by an integrable function not depending on Δt. $W(t)$ verifies then the conditions which allow us to differentiate it with respect to t under the integral sign and (33) yields

$$\frac{\partial W(t)}{\partial t} = \int_A n|\nabla u + t\nabla w|^{n-2}(\nabla u + t\nabla w)\nabla w\,d\tau.$$

Then setting $t = 0$, since by (32) 0 is an extremum for $W(t)$, we see that

$$\left[\frac{\partial W(t)}{\partial t}\right]_{t=0} = n\int_A |\nabla u|^{n-2}\nabla u\cdot\nabla w\,d\tau = 0, \tag{34}$$

where $\nabla u.\nabla v$ is the scalar product of ∇u and ∇v. Thus, with Hölder's inequality and taking into account that $w = u - v$,

$$\operatorname{cap} A = \int_A |\nabla u|^n\,d\tau = \int_A |\nabla u|^{n-2}\nabla u\cdot\nabla v\,d\tau \leqq \int_A |\nabla u|^{n-1}|\nabla v|\,d\tau \leqq$$

$$\leqq \left(\int_A |\nabla u|^n\,d\tau\right)^{\frac{n-1}{n}}\left(\int_A |\nabla v|^n\,d\tau\right)^{\frac{1}{n}} = \operatorname{cap} A.$$

Hence we have equality throughout. But

$$\int_A |\nabla u|^{n-1}|\nabla v|\,d\tau = \left(\int_A |\nabla u|^n\,d\tau\right)^{\frac{n-1}{n}}\left(\int_A |\nabla v|^n\,d\tau\right)^{\frac{1}{n}}$$

implies (see G. H. Hardy, J. E. Littlewood and G. Polya [1], theorem 189, p. 140) that $|\nabla u|$ and $|\nabla v|$ are proportional a.e. in A and hence, by (31), that

$$|\nabla u| = |\nabla v| \tag{35}$$

a.e. in A. Next

$$\int_A |\nabla u|^{n-2}\nabla u \cdot \nabla v \, d\tau = \int_A |\nabla u|^{n-1}|\nabla v| \, d\tau$$

implies that $\nabla u . \nabla v = |\nabla u|\,|\nabla v|$ a.e. in A. Thus from (35) we obtain that $\nabla(u-v)=0$ a.e. in A, and since u and v are both admissible functions and then ACL, we have $w(x) = u(x) - v(x) =$ constant in A (see for instance P. I. Natanson [1], theorem 2, p. 322). Next $u(x) = v(x) = 0$ on F_0, $u(x) = v(x) = 1$ on F_1, so that $w(x) = 0$ on A and then $u = v$ everywhere in A, as desired.

COROLLARY. *Let u be the extremal function for A and let w be a function which is continuous and ACL in A, which has boundary value* 0 *on ∂A and for which $|\nabla w| \in L^n(A)$. Then*

$$\int_A |\nabla u|^{n-2}\nabla u \nabla w \, d\tau = 0.$$

The corollary is a consequence of the formula (34).

LEMMA 11. *Let A be a ring with non-degenerate boundary components, let u be the extremal function for A, and let σ be a compact set that separates the boundary components of A. Then*

$$\int_{\sigma(b)} |\nabla u|^{n-1} \, d\tau \geqq 2b \operatorname{cap} A$$

or $0 < b < d(\sigma, \partial A)$, *where*

$$\sigma(b) = \{x;\, d(x, \sigma) < b\}. \tag{36}$$

Fix $0 < b < d(\sigma, \partial A)$. Next, for $k = 0, 1$, let D_k be the component of $C\sigma$ that contains C_k, and let

$$E_k = \{x;\, 0 < d(x, CD_k) < b\} \qquad (k = 0, 1).$$

Then $E_k \subset D_k$ and $E_0 \cup E_1 \subset \sigma(b)$. Hence it is sufficient to show that

$$\int_{E_k} |\nabla u|^{n-1} \, d\tau \geqq b \operatorname{cap} A \qquad (k = 0, 1).$$

We consider first the case where $k = 1$. For this set $w = v - bu$, where $v(x) = \min[b, d(x, CD_1)]$. Then w is clearly continuous and ACL in A. Next, it is easy to see that v has boundary values 0 on F_0 and b on F_1, and hence that w has boundary value 0 on ∂A. Finally, arguing as for the inequality (30) it follows that $|\nabla v| \leqq 1$ a.e. in E_1 and $|\nabla v| = 0$ a.e. in $A - E_1$. Since E_1 is bounded, $|\nabla w| \in L^n(A)$ and we can apply corollary of the preceding lemma to conclude that

$$\int_{E_1} |\nabla u|^{n-1} d\tau \geqq \int_A |\nabla u|^{n-2} \nabla u \nabla v \, d\tau = b \int_A |\nabla u|^n d\tau = b \operatorname{cap} A,$$

as desired. A trivial modification of the argument above for $k = 0$ completes the proof.

We introduce now, following Gehring [6], the concept of *simple admissible functions*, i.e. admissible functions whose level surfaces are particularly well behaved. Let $w \in C^1$, $w = 0$ on C_0 and $w = 1$ on C_1, $0 \leqq w \leqq 1$ in A and $\nabla w = 0$ outside a compact subset of A. The set where $0 < w < 1$ is bounded and lies at a distance b from CA. Now consider a decomposition of the space into congruent simplices $\{T\}$ with diameter $c < b$. Then define a new function u so that u is a linear function of the coordinate variables in each simplex and $u = w$ on the vertices of each simplex. Then u is admissible for A and

$$\lim_{c \to 0} \int_A |\nabla u|^n d\tau = \int_A |\nabla w|^n d\tau. \tag{37}$$

Clearly $0 \leqq u \leqq 1$. Now fix a, $0 \leqq a < 1$, and let F be the set where $u \leqq a$ and σ the set where $u = a$. Then F is a closed polyhedron, that is the union of a finite number of closed (possible degenerate) simplices, and $\partial F \subset \sigma$. If, in addition, we choose a to be different from the finite set of values assumed by w, and hence by u, on the vertices of the simplices $\{T\}$, then it is easy to see that each point of σ is a boundary point of F, whence $\partial F = \sigma$. If $w = a$ (and then also $u = a$) in some vertices of simplices, then σ consists of ∂F and those vertices. We say that any such a function is a simply admissible function for A.

Remark. Lemma 2.2 implies that

$$\operatorname{cap} A = \inf_w \int_A |\nabla w|^n d\tau,$$

since the functions w used there satisfy the properties of the functions w defined above. Hence and by (37), we obtain

$$\operatorname{cap} A = \inf_u \int_A |\nabla u|^n d\tau,$$

where the infimum is taken over all simple admissible functions u.

LEMMA 12. *If A is a ring, then*

$$\operatorname{cap} A = \sup_{\rho} \frac{a(\rho)^n}{v(\rho)^{n-1}}, \tag{38}$$

where

$$v(\rho) = v(A, \rho) = \int_A \rho^n \, d\tau,$$

and the supremum is taken over all functions $\rho \geqq 0$, Borel measurable and for which $a(\rho)$ and $v(\rho)$ are not simultaneously 0 or ∞.

Fix $\alpha > 0$. From above it follows that there exists an admissible function $u(x)$ with the following properties: $u(x)$ is piecewise linear in A,

$$\int_A |\nabla u|^n \, d\tau < \operatorname{cap} A + \alpha$$

and, for all but a finite set of b in $0 < b < 1$, the points where $u = b$ form a polyhedral surface σ_0 that separates the boundary components of A. Thus

$$a(\rho) \leqq \int_{\sigma_0} \rho^{n-1} \, d\rho,$$

and integrating over all such b, on account of lemma 2.3, we find that

$$a(\rho) \leqq \int_0^1 \left(\int_{\sigma_0} \rho^{n-1} \, d\sigma \right) db \leqq \int_A \rho^{n-1} |\nabla u| \, d\tau.$$

Hölder's inequality yields the result

$$a(\rho)^n \leqq \left(\int_A \rho^n \, d\tau \right)^{n-1} \left(\int_A |\nabla u|^n \, d\tau \right) < v(\rho)^{n-1} (\operatorname{cap} A + \alpha)$$

and, letting $a \to 0$, we conclude that

$$\frac{a(\rho)^n}{v(\rho)^{n-1}} \leqq \operatorname{cap} A \tag{39}$$

for all ρ with $a(\rho)$ and $v(\rho)$ not simultaneously 0 or ∞.

To complete the proof of (38), we must show that

$$\operatorname{cap} A \leqq \sup_{\rho} \frac{a(\rho)^n}{v(\rho)^{n-1}}. \tag{40}$$

This is clear so when cap $A = 0$, and hence we need only consider the case where A has non-degenerate boundary components.

Let u be the extremal function for A, and, for each $r > 0$, let

$$\rho(x, r) = \left[\frac{1}{mB(r)} \int_{B(r)} |\nabla u(x + y)|^{n-1} \, d\tau\right]^{\frac{1}{n-1}},$$

where $|\nabla u|$ is taken as 0 in CA. Next, for each $\alpha > 0$, let

$$g(x) = \sup_{0<r<\alpha} \rho(x, r), \quad h(x) = \sup_{0<r<\infty} \rho(x, r).$$

Both g and h are non-negative Borel measurable functions, and the theorem of Hardy — Littlewood — Smith (Chapter 1.3) (which can also be proved in the same way in our case) implies $h \in L^n(A)$. Then, since by proposition 1.6.9

$$\lim_{r \to 0} \rho(x, r) = |\nabla u(x)|$$

a.e. in A, we conclude from Lebegues's dominated convergence theorem (see for instance I. P. Natanson [1], theorem 1, p. 200) that

$$\lim_{\alpha \to 0} \nu(g) = \int_A \left\{\lim_{\alpha \to 0} \sup_{0<r<\alpha} \left[\frac{1}{mB(r)} \int_{B(r)} |\nabla u(x + y)|^{n-1} \, d\tau\right]^{\frac{1}{n-1}}\right\}^n d\tau = \tag{41}$$

$$= \int_A |\nabla u|^n \, d\tau = \operatorname{cap} A.$$

Now fix $\alpha > 0$, let σ be a compact piecewise smooth surface that separates the boundary components of A, and choose $b > 0$ and $r > 0$ so that $r < \alpha$ and $b + r < d(\sigma, \partial A)$. By Fubini's theorem

$$\int_{\sigma(b)} \rho(x, r)^{n-1} \, d\tau = \int_{\sigma(b)} \left[\frac{1}{\omega_n r^n} \int_{B(r)} |\nabla u(x + y)|^{n-1} \, d\tau\right] d\tau = \tag{42}$$

$$= \frac{1}{\omega_n r^n} \int_{B(r)} \left[\int_{\sigma_y(b)} |\nabla u(x)|^{n-1} \, d\tau\right] d\tau,$$

where $\sigma(b)$ is given by (36), σ_y is the translation of σ through the vector y, and

$$\sigma_y(b) = \{x; d(x, \sigma_y) < b\}.$$

The preceding lemma implies that

$$\int_{\sigma_y(b)} |\nabla u|^{n-1}\, d\tau \geqq 2b \operatorname{cap} A$$

for all $y \in B(r)$, and on account of (42), that

$$\int_{\sigma(b)} \rho(x, r)^{n-1}\, d\tau \geqq 2b \operatorname{cap} A.$$

Since ρ is continuous and σ piecewise smooth,

$$\int_\sigma g(x)^{n-1}\, d\sigma \geqq \int_\sigma \rho(x, r)^{n-1}\, d\sigma = \lim_{b\to 0} \frac{1}{2b} \int_{\sigma(b)} \rho(x, r)^{n-1}\, d\tau \geqq \operatorname{cap} A,$$

and we conclude that

$$a(g) \geqq \operatorname{cap} A. \tag{43}$$

This, together with (41), implies that

$$\operatorname{cap} A \leqq \lim_{\alpha\to 0} \frac{a(g)^n}{v(g)^{n-1}} \leqq \sup_{0<\alpha<\infty} \frac{a(g)^n}{v(g)^{n-1}}, \tag{44}$$

and we obtain (40), which, combined with (39), yields (38), as desired.

COROLLARY 1.

$$\operatorname{cap} A = M(\Sigma_A)^{1-n} \tag{45}$$

where Σ_A *is the family of surfaces which separate the boundary components of the ring* A.

Suppose first that $\operatorname{cap} A > 0$. The preceding lemma implies $\sup_\rho a(\rho) > 0$, where the supremum is taken over all the Borel measurable functions $\rho(x) \geqq 0$. Let ρ be such a function with $a(\rho) > 0$ and let $\rho_1(x) = \dfrac{\rho(x)}{a(\rho)^{\frac{1}{n-1}}}$. Then

$$a(\rho_1) = \inf_{\Sigma_A} \int_\sigma \rho_1^{n-1}\, d\sigma = \inf_{\Sigma_A} \int_\sigma \frac{\rho^{n-1}}{a(\rho)}\, d\sigma = 1$$

and, by the preceding lemma,

$$\frac{1}{\operatorname{cap} A} = \inf_\rho \frac{v(\rho)^{n-1}}{a(\rho)^n} = \inf_{\rho_1} v(\rho_1)^{n-1} \geqq \inf_{\rho'} v(\rho')^{n-1} = M(\Sigma_A)^{n-1}, \tag{46}$$

where the last minimum is taken over all Borel measurable functions $\rho'(x) \geqq 0$ with $a(\rho') \geqq 1$. On the other hand,

$$\inf_{\rho'} v(\rho')^{n-1} \geqq \inf_{\rho'} \frac{v(\rho')^{n-1}}{a(\rho')^n} \geqq \inf_{\rho} \frac{v(\rho)^{n-1}}{a(\rho)^n}.$$

Hence, the inequality (46) reduces to an equality, i.e. (45) holds.

If cap $A = 0$, then by the preceding lemma,

$$0 \leqq \frac{a(\rho)^n}{v(\rho)^{n-1}} \leqq 0,$$

i.e.

$$\frac{a(\rho)^n}{v(\rho)^{n-1}} = 0$$

for all $\rho \geqq 0$ and Borel measurable. Hence, if $\rho \in F(\Sigma_A)$, and then, $a(\rho) \geqq 1$ (condition 3°), we must have $v(\rho) = \infty$. But such an admissible ρ exists (for instance $\rho \equiv \infty$) and we conclude that $M(\Sigma_A) = \infty$ (see the definition of the modulus of a surface family in Chapter 1.5).

COROLLARY 2.

$$M(A) = n\omega_n M(\Sigma_A)^{n-1}. \tag{47}$$

Indeed, from the corollary of lemma 2.6 and the preceding one, we deduce

$$M(\Gamma_A) = M(\Sigma_A)^{1-n} = \text{cap } A,$$

hence

$$M(A) = \frac{n\omega_n}{M(\Gamma_A)} = n\omega_n M(\Sigma_A)^{n-1},$$

as desired.

LEMMA 13. *A differentiable mapping $f: D \rightleftarrows D^*$ with $J \neq 0$ and the characteristic parameters $p_k = \dfrac{a_k}{a_n}$ $(k = 1, \ldots, n-1)$ transforms the infinitesimal spheres into infinitesimal ellipsoids with the characteristic parameters*

$$p'_1 = p_1, \quad p'_m = \frac{p_1}{p_{n+1-m}} \qquad (m = 2, \ldots, n-1). \tag{48}$$

The mapping f tranforms in particular an infinitesimal ellipsoid $E[(C), x_0]$ into an infinitesimal sphere. Let $\xi = A(x)$ be an affine mapping which transforms the ellipsoids $E_h[(C), x_0]$ into spheres $S[f(x_0), r]$. Evidently, $A(x)$ will transform the spheres centered at x_0 into ellipsoids with the characteristics (C'). Let us prove that the corresponding characteristic parameters $p_k'(k = 1, \ldots, n-1)$ are given by (48).

Indeed, clearly

$$\frac{|A(x') - A(x_0)|_{s'}}{|A(x'') - A(x_0)|_{s''}} = 1,$$

where $x', x'' \in E_h$ and $x' - x_0$, $x'' - x_0$ are vectors having the directions s' and s'', respectively. Hence and from

$$\frac{A^k(x'') - A^k(x_0)}{|x'' - x_0|_{s''}} = \frac{a_i^k(x_0)(x''^i - x_0^i)}{|x'' - x_0|_{s''}} = a_i^k(x_0) \cos(s'', x^i) =$$

$$= \frac{A^k(x_1) - A^k(x_0)}{|x_1 - x_0|_{s''}} \qquad (k = 1, \ldots, n),$$

where $x_1 - x_0$ is an arbitrary vector of direction s'', we deduce

$$\frac{|x'' - x_0|_{s''}}{|x' - x_0|_{s'}} = \frac{\dfrac{|A(x') - A(x_0)|}{|x' - x_0|_{s'}}}{\dfrac{|A(x'') - A(x_0)|}{|x'' - x_0|_{s''}}} = \frac{\dfrac{|A(x') - A(x_0)|}{|x' - x_0|_{s'}}}{\dfrac{|A(x_1) - A(x_0)|}{|x_1 - x_0|_{s''}}} = \frac{|A(x') - A(x_0)|_{s'}}{|A(x_1) - A(x_0)|_{s''}}, \tag{49}$$

where we choose x_1 so that $|x_1 - x_0| = |x' - x_0|$. In particular, if s_1 and s_n are the directions of the maximal and minimal axis, respectively, then

$$\frac{|\Delta A(x_0)|_{s_n}}{|\Delta A(x_0)|_{s_1}} = p_1,$$

where $\Delta A(x_0) = A(x_0 + \Delta x) - A(x_0)$. If we denote by $\lambda_i'(i = 1, \ldots, n)$ the dilatations in the directions of the axes, then (49) implies

$$\lambda_1' \leqq \cdots \leqq \lambda_n'.$$

According to E. Hedrick and L. Ingold [3], the principal directions of a mapping differentiable at a point x_0 are the directions corresponding to the extremal values of the dilatation

$$R = \frac{\delta_{ik} x^{*i}_{m} x^{*k}_{p} \mathrm{d}x^m \mathrm{d}x^p}{\delta_{ik} \mathrm{d}x^i \mathrm{d}x^k}.$$

They are mutually orthogonal and are the solutions of the system

$$\delta_{ik} x^{*i}_{m} x^{*k}_{p} \mathrm{d}x^m = R\, \mathrm{d}x^p \quad (p = 1, \ldots, n).$$

Clearly, these principal directions are the directions of the axes of the infinitesimal ellipsoid $E[(C), x_0]$.

Next, if $\mathrm{d}x$, δx are two principal directions, then the preceding system and the orthogonality of the directions yield for the corresponding directions $\mathrm{d}x^*$, δx^*,

$$\mathrm{d}x^* \delta x^* = \delta_{ik} x^{*i}_{m} x^{*k}_{p} \mathrm{d}x^m \delta x^p = R \delta_{mp} \mathrm{d}x^m \delta x^p = R\, \mathrm{d}x\, \delta x = 0,$$

which allows us to conclude that the directions $\mathrm{d}x^*$, δx^* are orthogonal too, i.e. a differentiable mapping transforms the principal directions into mutually orthogonal directions.

Then, from (49), we deduce

$$p_k' = \frac{\lambda_{n+1-k}'}{\lambda_1'} = \frac{|\Delta x|_{s_1}}{|\Delta x|_{s_{n+1-k}}} = \frac{a_1}{a_{n+1-k}} = \frac{\dfrac{a_1}{a_n}}{\dfrac{a_{n+1-k}}{a_n}} = \frac{p_1}{p_{n+1-k}} \qquad (k = 2, \ldots, n-1).$$

Remark. The characteristic parameters $p_k'(x)$ $(k = 1, \ldots, n-1)$ of the infinitesimal ellipsoids in which a differentiable homeomorphism maps infinitesimal spheres, clearly coincide with the characteristic parameters $p_k^*(x^*)$ $(k = 1, \ldots, n-1)$ of the infinitesimal ellipsoids transformed by the inverse mapping $x = f^{-1}(x^*)$ into infinitesimal spheres.

LEMMA 14. *Let $\rho \geqq 0$ be a Borel measurable and L^n-integrable function, such that $\int_\gamma \rho \mathrm{d}s \geqq 1$ for every arc joining the bases B_0 and B_1 of Z. For each $a > 0$ there exists a $b > 0$ with the following properties: If x_0 and x_1 are endpoints of a rectifiable arc $\beta \subset Z$ and if $d(x_i, B_i) < k \leqq b$ and $d(x_i, \partial Z - B_i) > 2k$ $(i = 0, 1)$ for an arbitrary k, then*

$$\int_\beta \rho\, \mathrm{d}s \geqq 1 - a.$$

The proof is the same as for lemma 2.1.

LEMMA 15.

$$M(Z) = \operatorname{cap} Z. \tag{50}$$

Let us prove first the inequality

$$M(Z) \leqq \operatorname{cap} Z.$$

If $|\nabla u| \notin L^n$, the inequality is trivial. Let us suppose $|\nabla u| \in L^n$. But then u is ACL_n and, by corollary of lemma 2.3,

$$\int_\gamma |\nabla u| \, ds \geqq 1,$$

for almost all $\gamma \in \Gamma_Z$. By proposition 1.6.7, there exists a Borel measurable function ρ such that $\rho(x) = |\nabla u(x)|$ a.e. in Z, hence and by proposition 1.5.5,

$$\int_\gamma \rho \, ds = \int_\gamma |\nabla u| \, ds$$

for almost all $\gamma \in \Gamma_Z$. If we denote the two exceptional arc families by Γ_1, Γ_2 then clearly

$$M[\Gamma_Z - (\Gamma_1 \cup \Gamma_2)] \leqq M(\Gamma_Z)$$

and

$$M(\Gamma_Z) \leqq M[\Gamma_Z - (\Gamma_1 \cup \Gamma_2)] + M(\Gamma_1) + M(\Gamma_2) = M[\Gamma_Z - (\Gamma_1 \cup \Gamma_2)],$$

hence

$$M(\Gamma_Z) = M[\Gamma_Z - (\Gamma_1 \cup \Gamma_2)],$$

and then

$$M(Z) = M(\Gamma_Z) = M[\Gamma_Z - (\Gamma_1 \cup \Gamma_2)] \leqq \operatorname{cap} Z.$$

Now, let us prove the inverse inequality. Let $\rho \geqq 0$ be a Borel function such that

$$\int_\gamma \rho \, ds \geqslant 1 \tag{51}$$

for all $\gamma \in \Gamma_Z$. By Rengel's inequality (1.5.40), $\rho \in L^n(Z)$.

$$\rho_1(x) = \min\left[1, \frac{\delta(x)}{2}\right],$$

where $\delta(x) = \mathrm{d}(x, \partial Z)$ is Lipschitzian, since

$$|\rho_1(x) - \rho_1(x')| \leqq \frac{1}{2}|x - x'|. \tag{52}$$

Indeed, if $x_0, x_0' \in \partial Z$ are such that $\delta(x) = |x - x_0|$ and $\delta(x') = |x' - x_0'|$, and if $|x - x_0| \geqq |x' - x_0'|$, then

$$|\delta(x) - \delta(x')| = \big||x - x_0| - |x' - x_0'|\big| \leqq \big||x - x_0'| - |x' - x_0'|\big| \leqq |x - x'|. \tag{53}$$

If $|x - x_0| \leqq |x' - x_0'|$, then

$$|\delta(x) - \delta(x')| = \big||x - x_0| - |x' - x_0'|\big| \leqq \big||x - x_0| - |x' - x_0|\big| \leqq |x - x'|. \tag{54}$$

Thus, if $\delta(x), \delta(x') \geqq 2$, clearly $\rho_1(x) - \rho_1(x') = 0$, if $\delta(x), \delta(x') \leqq 2$, then (53) and (54) imply that

$$|\rho_1(x) - \rho_1(x')| = \frac{1}{2}|\delta(x) - \delta(x')| \leqq \frac{1}{2}|x - x'|,$$

and if for instance $\delta(x) \geqq 2$ and $\delta(x') \leqq 2$, then

$$|\rho_1(x) - \rho_1(x')| = \left|1 - \frac{\delta(x')}{2}\right| \leqq \frac{1}{2}|\delta(x) - \delta(x')| \leqq |x - x'|,$$

and $\rho_1(x)$ is Lipschitzian as desired.

Next, let

$$h(x) = \begin{cases} 0 & \text{if } \mathrm{d}(x, B_0) < k \leqq b \text{ and } \mathrm{d}(x, \partial Z - B_0) > 2k, \\ 0 & \text{if } \mathrm{d}(x, B_1) < k \leqq b \text{ and } \mathrm{d}(x, \partial Z - B_1) > 2k, \\ \dfrac{\rho(x)}{1 - a} & \text{otherwise.} \end{cases}$$

For $0 < \varepsilon \leqq 1$ and each $\xi \in B(\varepsilon)$, $y = y(x; \xi) = x + \xi\rho_1(x)$ is a one-to-one mapping of Z onto itself. The function

$$g(x, \varepsilon) = \frac{1}{\omega_n \varepsilon^n} \int_{B(\varepsilon)} h[y(x; \xi)]\, \mathrm{d}\tau$$

is continuous in x, since h is integrable and $\rho_1(x)$ is continuous (the same argument as for the continuity of $g(x)$ in the proof of lemma 2.6). We point out that $g(x, \varepsilon)$ is continuous even on the bases B_0 and B_1 of Z. But then, for all $\gamma \in \Gamma_Z$, on account of the preceding lemma,

$$\int_\gamma h\mathrm{d}s = \int_\beta h\mathrm{d}s = \frac{1}{1-a}\int_\beta \rho\,\mathrm{d}s \geqq 1, \tag{55}$$

and, by Fubini theorem,

$$\begin{aligned}\int_\gamma g(x, \varepsilon)\,\mathrm{d}s(x) &= \int_\gamma \left\{\frac{1}{\omega_n\varepsilon^n}\int_{B(\varepsilon)} h[x+\xi\rho_1(x)]\,\mathrm{d}\tau\right\}\mathrm{d}s(x) = \\ &= \frac{1}{\omega_n\varepsilon^n}\int_{B(\varepsilon)}\left[\int_{y(\gamma;\,\xi)} h(y)\frac{\mathrm{d}s(x)}{\mathrm{d}s(y)}\,\mathrm{d}s(y)\right]\mathrm{d}\tau\end{aligned} \tag{56}$$

for every $\gamma \in \Gamma_Z$, where the linear elements $\mathrm{d}s(x)$ and $\mathrm{d}s(y)$ (in the space of the variable x and y, respectively) satisfy, on account of (52), the inequalities

$$\begin{aligned}\mathrm{d}s(y) &\leqq |\mathrm{d}x + \xi\mathrm{d}\rho_1(x)| = \left|\mathrm{d}x + \xi\frac{\partial\rho_1}{\partial x^i}\,\mathrm{d}x^i\right| \leqq \mathrm{d}s(x) + |\xi|\left|\frac{\partial\rho_1}{\partial x^i}\right||\mathrm{d}x^i| \leqq \\ &\leqq \mathrm{d}s(x) + |\xi|\,|\nabla\rho_1|\,\mathrm{d}s(x) = \\ &(1+|\xi|\,|\nabla\rho_1|)\,\mathrm{d}s(x) \leqq \left(1+\frac{|\xi|}{2}\right)\mathrm{d}s(x) \leqq \left(1+\frac{\varepsilon}{2}\right)\mathrm{d}s(x).\end{aligned}$$

(55) and (56) imply that

$$\begin{aligned}\int_\gamma g(x, \varepsilon)\,\mathrm{d}s(x) &= \frac{1}{\omega_n\varepsilon^n}\int_{B(\varepsilon)}\left[\int_{y(\gamma;\,\xi)} h(y)\frac{\mathrm{d}s(x)}{\mathrm{d}s(y)}\,\mathrm{d}s(y)\right]\mathrm{d}\tau \geqq \\ &\geqq \frac{1}{\omega_n\varepsilon^n}\int_{B(\varepsilon)}\left[\int_{y(\gamma;\,\xi)}\frac{h(y)}{1+\frac{\varepsilon}{2}}\,\mathrm{d}s(y)\right]\mathrm{d}\tau \geqq \frac{1}{\omega_n\varepsilon^n\left(1+\frac{\varepsilon}{2}\right)}\int_{B(\varepsilon)}\mathrm{d}\tau = \frac{1}{1+\frac{\varepsilon}{2}}.\end{aligned} \tag{57}$$

Since, on account of Hölder's inequality,

$$\int_Z |g(x, \varepsilon) - h(x)|^n \, d\tau = \int_Z \left| \frac{1}{\omega_n \varepsilon^n} \int_{B(\varepsilon)} [h(x + \xi\rho_1) - h(x)] \, d\tau \right|^n d\tau \leqq$$

$$\leqq \int_Z \left[\frac{1}{\omega_n \varepsilon^n} \int_{B(\varepsilon)} |h(x + \xi\rho_1) - h(x)|^n \, d\tau \right] d\tau = \frac{1}{\omega_n \varepsilon^n} \int_{B(\varepsilon)} \left[\int_Z |h(x + \xi\rho_1) - h(x)|^n \, d\tau \right] d\tau,$$

it follows that if h is approximated in L^n by means of continuous transformations φ with compact support, then

$$\int_Z |g(x, \varepsilon) - h(x)|^n \, d\tau \leqq \frac{1}{\omega_n \varepsilon^n} \int_{B(\varepsilon)} \left[\int_Z |h(x + \xi\rho_1) - h(x)|^n \, d\tau \right] d\tau \leqq$$

$$\leqq \frac{1}{\omega_n \varepsilon^n} \int_{B(\varepsilon)} \left[\int_Z |h(x + \xi\rho_1) - \varphi(x + \xi\rho_1)|^n \, d\tau + \int_Z |\varphi(x + \xi\rho_1) - \varphi(x)|^n \, d\tau + \int_Z |\varphi(x) - h(x)|^n \, d\tau \right] d\tau,$$

hence, it is clear that $\int_Z |g(x, \varepsilon) - h(x)|^n \, d\tau \to 0$ as $\varepsilon \to 0$, and even $\int_Z |v(x, \varepsilon) - h(x)|^n d\tau \to 0$ as $\varepsilon \to 0$, where

$$v(x, \varepsilon) = \left(1 + \frac{\varepsilon}{2}\right) g(x, \varepsilon),$$

i.e. $v(x, \varepsilon)$ converges in L^n to $h(x)$, and since h converges to ρ in Z, for $a \to 0$, we conclude that $v(x, \varepsilon)$ converges to ρ in L^n. The inequality (57) yields $\int_\gamma v(x, \varepsilon) ds \geqq 1$ for all $\gamma \in \Gamma_Z$. Set

$$u(x, \varepsilon) = \min \left[1, \inf_{\gamma'} \int_{\gamma'} v(x, \varepsilon) \, ds\right], \tag{58}$$

where the infimum is taken over all rectifiable curves $\gamma' \subset Z$, joining B_0 with x. By the same argument as for the function $u(x)$ given by (2.30), it follows that the continuity of $v(x, \varepsilon)$ implies that $u(x, \varepsilon)$ satisfies a local Lipschitz condition, and then it is *ACL* and (arguing again as in lemma 2.6 for u), from (58) we deduce that $|\nabla u(x, \varepsilon)| \leqq \nu(x, \varepsilon)$. Thus, for each Borel measurable function $\rho(x) \geqq 0$, satisfying (51) for all $\gamma \in \Gamma_Z$ and for every $\delta > 0$, it is possible to associate a function $u(x, \varepsilon)$, ACL in Z and which admits a continuous extension at the bases so that $u(x, \varepsilon)_{|B_0} = 0$, $u(x, \varepsilon)_{|B_1} = 1$, and which satisfies the inequalities

$$\operatorname{cap} Z \leqq \int_Z |\nabla u|^n \, d\tau \leqq \int_Z \rho^n \, d\tau + \delta.$$

Hence, taking the infimum over admissible ρ and letting $\delta \to 0$, we get

$$\operatorname{cap} Z \leqq M(Z),$$

which together with the converse inequality, gives (50).

LEMMA 16.

$$\operatorname{cap} Z \geqq \frac{s(Z)^n}{(mZ)^{n-1}}, \tag{59}$$

where $s(Z) = \inf_E H^{n-1}(E)$, *and the infimum is taken over all sets which separate* B_0 *from* B_1 *in* Z.

Let $u = u(x)$ be an admissible function for Z. We may suppose, without loss of generality, that $0 \leqq u \leqq 1$, because if it were not so, we should consider the function

$$\nu(x) = \begin{cases} 0 & \text{if } u(x) < 0, \\ u(x) & \text{if } 0 \leqq u(x) \leqq 1, \\ 1 & \text{if } u(x) > 1, \end{cases}$$

which is admissible too and clearly

$$\int_Z |\nabla \nu|^n \, d\tau \leqq \int_Z |\nabla u|^n \, d\tau,$$

so that the fact of considering only those admissible functions which satisfy the condition that $0 \leqq u \leqq 1$, does not influence on the infimum, i.e. on cap Z.

For each $a \in (0, 1)$, let $E_a = \{x;\ u(x) = a\}$. Since $E_a \subset Z$ and separates B_0 from B_1 in Z, we have for every $a \in (0, 1)$:

$$s(Z) \leqq H^{n-1}(E_a). \tag{60}$$

Since u is ACL in Z, it follows, (see W. Ziemer [4]), that

$$\int_Z |\nabla u| \, d\tau = \int_0^1 H^{n-1}(E_a) \, da,$$

hence, and taking into account (60), we obtain

$$s(Z) \leqq \int_0^1 H^{n-1}(E_a) \, da = \int_Z |\nabla u| \, d\tau$$

and, applying Hölder inequality,

$$s(Z)^n \leqq (\int_Z |\nabla u| \, d\tau)^n \leqq \int_Z |\nabla u|^n \, d\tau (mZ)^{n-1},$$

which, taking the infimum over all admissible u, yields (59) as desired.

THEOREM 2. *The definition of K-QCfH in Šabat's sense (with $q = n - 1$) is equivalent to Gehring's analytic and geometric definition.*

Let f be a K-QCfH in Šabat's sense in D, i.e. satisfying (1) with $q = n - 1$ and set $A^* = \{x^*; \ a^* < |x^*| < b^*\}$ and $\Sigma^* = \{S(r^*); \ (a^* < r^* < b^*)\}$. We may suppose, without loss of generality, that $A^* \subset D^* = f(D)$. If Σ_{A*} is the family of all surfaces separating the boundary components of A^*, then, clearly, $M(\Sigma_{A*}) = = M(\Sigma^*)$, hence and by (1),

$$M(\Sigma_{A*}) = M(\Sigma^*) \leqq KM(\Sigma) \leqq KM(\Sigma_A),$$

because $\Sigma \subset \Sigma_A$ (see corollary 2 of proposition 1.5.3). Since by (47),

$$M(A^*) = n\omega_n M(\Sigma_{A*})^{n-1}, \quad M(A) = n\omega_n M(\Sigma_A)^{n-1},$$

it follows that

$$M(A^*) \leqq K^{n-1} M(A),$$

which implies the QCfH of f according to Gehring's metric definition (see lemma 1.2), and then also according to Gehring's analytic and geometric definition, hence we deduce its differentiability with $J \neq 0$ a.e. in D.

In the remainder of the proof, we determine the constant K of the quasiconformality of the mappings which satisfy (1).

Since $M(Z) = \operatorname{cap} Z$, by (50), and

$$\operatorname{cap} Z = M(\Sigma_Z)^{1-n}$$

(see W. Ziemer [1]), it follows that

$$M(Z) = M(\Sigma_Z)^{1-n},$$

and since a cube Q and its image Q^* are particular cases of cylinders, the preceding relation comes to

$$1 = M(Q) = M(\Sigma_Q)^{1-n}, \quad M(Q^*) = M(\Sigma_{Q^*})^{1-n}.$$

For the family Σ of $(n-1)$-dimensional cubes which separate the bases of the cube Q, the double inequality (1) holds and then, since $\Sigma \subset \Sigma_Q$,

$$1 = M(\Sigma_Q) = M(\Sigma) \leqq KM(\Sigma^*) \leqq KM(\Sigma_{Q^*}),$$

hence

$$M(Q^*) = M(\Sigma_{Q^*})^{1-n} \leqq K^{n-1}. \tag{61}$$

Similarly, if Q_1^* is a cube, $\Sigma^{*\prime}$ is the family of $(n-1)$-dimensional cubes that separates its bases and $\Sigma' = f^{-1}(\Sigma^{*\prime})$, we have

$$1 = M(Q_1^*) = M(\Sigma^{*\prime}) \leqq KM(\Sigma') \leqq KM(\Sigma_{Q_1}),$$

from which

$$M(Q_1) = M(\Sigma_{Q_1})^{1-n} \leqq K^{n-1}. \tag{62}$$

Next, let x_0 be an A-point of f. We show that (61) and (62) imply

$$\frac{\max |f'(x_0)|^n}{K^{n-1}} \leqq |J(x_0)| \leqq K^{n-1} \min |f'(x_0)|^n. \tag{63}$$

Since translations, rotations and reflexions have no influence on (61), (62) and (63), we may assume that $x_0 = f(x_0) = O$ and that $f'(O)$ is given by (1.22). The differentiability of f implies that

$$f(x) = f'(O)\,x + \varepsilon(O, x)\,|x|,$$

where $\varepsilon(O, x) \to O$ as $x \to O$ and (1.23) and

$$\min |f'(O)| = a_n \tag{64}$$

hold. Let the faces $x^n = 0$, h be the bases of Q. Then (1.22) and (1.24) imply that

$$|f(x_1, \dots, x^{k-1}, h, x^{k+1}, \dots, x^n) - f(x_1, \dots, x^{k-1}, 0, x^{k+1}, \dots, x^n)| > (a_k - 2\varepsilon)h$$

$$(k = 1, \dots, n).$$

On the other hand, (1.22) and (1.24) yield

$$mQ^* \leqq h^n(a_1 + 2\varepsilon) \cdots (a_n + 2\varepsilon).$$

Then, by (50), (59), (61) and (62)

$$\frac{(a_1 - 2\varepsilon)^n \cdots (a_{n-1} - 2\varepsilon)^n}{(a_1 + 2\varepsilon)^{n-1} \cdots (a_n + 2\varepsilon)^{n-1}} < \frac{s(Q^*)^n}{(mQ^*)^{n-1}} \leqq M(Q^*) \leqq K^{n-1},$$

$$\frac{(a_1^* - 2\varepsilon)^n \cdots (a_{n-1}^* - 2\varepsilon)^n}{(a_1^* + 2\varepsilon)^{n-1} \cdots (a_n^* + 2\varepsilon)^{n-1}} < \frac{s(Q_1)^n}{(mQ_1)^{n-1}} \leqq M(Q_1) \leqq K^{n-1},$$

where $a_1^*, \dots, a_n^*$ $(a_1^* \geqq \dots \geqq a_n^*)$ with respect to f^{-1} have the same meaning as $a_1, \dots, a_n$ with respect to f. Letting $\varepsilon \to 0$, we obtain

$$\frac{(a_1 \cdots a_{n-1})^n}{(a_1 \cdots a_n)^{n-1}} \leqq K^{n-1}, \quad \frac{(a_1^* \cdots a_{n-1}^*)^n}{(a_1^* \cdots a_n^*)^{n-1}} \leqq K^{n-1},$$

hence

$$\frac{a_1 \cdots a_n}{a_n^n} \leqq K^{n-1}, \quad \frac{a_1^* \cdots a_n^*}{a_n^{*n}} \leqq K^{n-1}. \tag{65}$$

The preceding inequality, on account of (48), yields

$$\frac{a_1^* \cdots a_n^*}{a_n^{*n}} = p_1^* \cdots p_{n-1}^* = \frac{p_1^{n-1}}{p_2 \cdots p_{n-1}} = \frac{\dfrac{a_1^{n-1}}{a_n^{n-1}}}{\dfrac{a_2 \cdots a_{n-1}}{a_n^{n-1}}} = \frac{a_1^n}{a_1 \cdots a_n} \leqq K^{n-1} \tag{66}$$

Thus, (65) and (66), by taking into account (1.23) and (64), yield (63), which allows us to conclude that f is K^{n-1}-QCfH according to Väisälä's definitions, and then, K-QCfH according to Gehring's geometric and analytic definitions.

For the second part of the proof, let $f\colon D \rightleftarrows D^*$ be a K-QCfH according to Gehring's geometric definition. But then f is QCfH also according to his metric definition (theorem 3.1), and then also to Markuševič—Pesin's definition (theorem 3.13), and we are in the hypotheses of theorem 1. Arguing as in that theorem, we obtain in this case too, but only for $q = n - 1$, the inequality (13), hence

$$\frac{d\sigma^*_{n-1}}{d\sigma_{n-1}} \geqq J^{\frac{n-1}{n}} \frac{a_n}{(a_1 \cdots a_n)^{\frac{1}{n}}} = J^{\frac{n-1}{n}} \left(\frac{a_n^n}{a_1 \cdots a_n}\right)^{\frac{1}{n}} = J^{\frac{n-1}{n}} \left(\frac{a'_1 \cdots a'_n}{a_1'^n}\right)^{\frac{1}{n}}, \tag{67}$$

a.e. in D, where $a'_1, \ldots, a'_n$ $(a'_1 \geqq \ldots \geqq a'_n)$ are the semi-axes of the ellipsoid onto which f(x) maps a sphere of radius r. But since f is K-QCfH according to Gehring's geometric definition, it is K^{n-1}-QCfH according to Väisälä's definitions, which, by (1.12), yields

$$\frac{a_1'^n}{a'_1 \cdots a'_n} \leqq K^{n-1},$$

hence, by (67),

$$\frac{d\sigma^*_{n-1}}{d\sigma_{n-1}} \geqq J^{\frac{n-1}{n}} \left(\frac{a'_1 \cdots a'_n}{a_1'^n}\right)^{\frac{1}{n}} \geqq \left(\frac{J}{K}\right)^{\frac{n-1}{n}}$$

a.e. in D. The last part of the proof is the same as in the preceding theorem (with $q = n - 1$), and finally, we obtain (1) with $q = n - 1$.

Remarks. 1. A similar theorem is not yet proved for Šabat's definition with $q = 2, \ldots, n - 2$. It is an open problem also the equivalence of Šabat's definitions for diferent q.

2. S. Agard [2] and H. M. Reimann [4] established that if f is a QCfH in D and Σ_q is a family of q-dimensional QCf surfaces (i.e. the image of a domain $\Delta \subset R^q$ by a q-dimensional QCfH), then there exists a subfamily Σ'_q of Σ_q, such that $M(\Sigma_q - \Sigma'_q) = 0$ and (1) holds for Σ'_q. But, as it is easy to see, the image of a QCf surface by an n-dimensional QCfH may, in general, not be a QCf surface, and since Agard and Reimann characterize the modulus only for families of QCf surfaces, it follows that generally it may be meaningless to speak of modulus of $\Sigma^*_q = f(\Sigma_q)$, where Σ_q is a family of q-dimensional QCf surfaces. Thus, so much the less, we can assert that Σ^*_q, where $M(\Sigma_q) = 0$, is an exceptional family, i.e. that $M(\Sigma^*_q) = 0$. And moreover, given a surface family $\Sigma_q \subset D$, the subfamily $\Sigma^0_q(f)$, which must be removed, in order that the family $\Sigma'_q = \Sigma_q - \Sigma^0_q$ may satisfy (1), depends on f, or more exactly on the subset of D, where f is not differentiable with $J \neq 0$.

3. Cabiria Andreian [2] proved that a QCfH verifies (1) for all the families $\{\sigma^\zeta_q\}$ (considered at the begining of this chapter). She uses the more general concept of weighted modulus, but if the weight is equal to 1, one obtains a con-

cept of the modulus of the kind considered in this book, the admissible functions $\rho(x)$ being defined only a.e. in D. Hence we deduce that $\rho(x)$ is defined only a.e. in almost all the surfaces of the family $\{\sigma_q^\zeta\}$, and since an exceptional family of surfaces σ_q^ζ are mapped by an n-dimensional QCfH again into an exceptional family of the same kind, we deduce that the double inequality (1) is verified for all the families $\{\sigma_q^\zeta\}$, and not only for subfamilieas which differ from them by an exceptional subfamily. However, in the characterization of the quasiconformality given by us at the beginning of this chapter, we supposed the double inequality (1) is verified also by the families of concentric spheres, because this was necessary in order to be able to prove, in the case $q = n - 1$, the equivalence of this definition to the other.

Kühnau's definitions. R. Kühnau [1] gives the following more precise form of Šabat's definitions for $q = n - 1$:

A homeomorphism $f(x)$ is said to be K-QCfH ($1 \leqq K < \infty$) in D if

$$\frac{M(\Gamma)}{K_1} \leqq M(\Gamma^*) \leqq K_O M(\Gamma) \tag{68}$$

for every curve $\Gamma \subset D$, where $K = \max(K_I, K_O)$.

A homeomorphism $f(x)$ is said to be K-QCfH ($1 \leqq K < \infty$) in D if

$$\frac{M(\Sigma)}{K_1} \leqq M(\Sigma^*) \leqq K_O M(\Sigma) \tag{69}$$

for every piecewise smooth surface family $\Sigma \subset D$, where $K = \max(K_I, K_O)$.

Finally R. Kühnau gives also the following more precise form of J. Väisälä's metric definition:

A homeomorphism $f(x)$ is said to be K-QCfH ($1 \leqq K < \infty$) in D if

$$\bar{\delta}(x) \leqq K_I, \quad \underline{\delta}(x) \leqq K_O$$

a.e. in D, where $K = \max(K_I, K_O)$.

Remark. In fact, R. Kühnau [1] establishes only that the homeomorphisms $f \in C^1$ with $J \neq 0$ in $D \subset R^3$, with outer local dilatation $\bar{\delta}(x)$ and inner local dilatation $\delta(x)$ satisfy the inequalities (68) and (69), where $K_I = \sup_{x \in D} \bar{\delta}(x)$, $K_0 = \sup_{x \in D} \delta(x)$.

Equivalence between Kühnau's definitions and Šabat's and Väisälä's definitions.

THEOREM 3. *The first two Kühnau's definitions are equivalent to the corresponding Šabat's definitions, for $q = 1, n - 1$, and third Kühnau's definition is equivalent to Väisälä's metric definition.*

Remark. Clearly, Kühnau's definitions would be no more equivalent to the corresponding Šabat's and Väisälä's definitions if K_I, K_0 were given by the formulae of the preceding remarks instead of being arbitrary constants.

CHAPTER 6

Gehring's analytic definitions and their equivalence to his geometric definition

Gehring's first analytic definition. (Gehring [5]). A homeomorphism $x^* = f(x)$ is said to be K-QCfH ($1 \leqq K < \infty$) in D if $x^* = f(x)$ and $x = f^{-1}(x^*)$ are ACL in D and D^*, respectively and satisfy the inequalities

$$\Lambda_f(x)^n \leqq K^{n-1} J_G(x), \quad \Lambda_{f^{-1}}(x^*)^n \leqq K^{n-1} J_G^*(x^*) \tag{1}$$

a.e. in D and a.e. in D^*, respectively, where

$$J_G^*(x^*) = \overline{\lim_{r \to 0}} \frac{mf^{-1}[B(x^*, r)]}{mB(x^*, r)}, \quad \Lambda_{f^{-1}}(x^*) = \overline{\lim_{|\Delta x^*| \to 0}} \frac{|f^{-1}(x^* + \Delta x^*) - f^{-1}(x_*)|}{|\Delta x^*|}$$

Its equivalence to his geometric definition. We begin with some preliminary results (Gehring [5]).

LEMMA 1. *Let $x^* = f(x)$ be a topological mapping of a domain D and let $x_0 \in D$. If $\delta_L(x_0) < \infty$, then*

$$\Lambda_f(x_0)^n \leqq \delta_L(x_0)^n J_G(x_0), \quad \Lambda_{f^{-1}}(x_0^*)^n \leqq \delta_L(x_0)^n J_G^*(x_0^*), \tag{2}$$

where $x^* = f(x_0)$. If, in addition, $f(x)$ is differentiable at x_0, then

$$\Lambda_f(x_0)^n \leqq \delta_L(x_0)^{n-1} |Jx_0|, \quad \Lambda_{f^{-1}}(x_0^*)^n \leqq \delta_L(x_0)^{n-1} |Jx_0^*|. \tag{3}$$

First inequality (2) is a consequence of the inequalities (1.27) and (1.38). For the second inequality let $t = L(x_0, r)$, $s = l(x_0, r)$, for small r, where $L(x_0, r)$ and $l(x_0, r)$ are given by (1.5) and (1.6). Then $r = L^*(x^*, s) = l^*(x^*, t)$ and we have

$$\Lambda_{f^{-1}}(x_0^*)^n = \left(\overline{\lim_{s \to 0}} \frac{r}{s} \right)^n \leqq \left(\overline{\lim_{s \to 0}} \frac{t}{s} \right)^n \left(\overline{\lim_{s \to 0}} \frac{r}{t} \right)^n \leqq \delta_L(x_0)^n J_G^*(x_0^*). \tag{4}$$

as desired. (1.27) and (1.15) yield the first inequality (3); the second is obtained as follows

$$\Lambda_{f^{-1}}(x_0^*)^n = \left(\overline{\lim_{s \to 0}} \frac{r}{s} \right)^n \leqq \overline{\lim_{r \to 0}} \frac{(x_n, r)}{\omega_n l(x_0, r)^n} \overline{\lim_{r \to 0}} \frac{\omega_n r^n}{T(x_0, r)} = \underline{\delta}(x_0) | \ (x_0)|^{-1} \leqq$$

$$\leqq \delta_L(x_0)^{n-1} |J^*(x_0^*)|.$$

A sequence of sets $\{E_m\}$ is said *to converge uniformly to a set* E, if for each $\varepsilon > 0$, there exists an N such that $n > N$ implies that each point of E_m lies within distance ε of E and each point of E lies within distance ε of E_m.

We establish next the continuity property for the modulus of a ring.

LEMMA 2. *Let* $\{A_m\}$ *be a sequence of rings and let* A *be a bounded ring. If each of the components of* ∂A_m *converges uniformly to the corresponding component of* ∂A, *then*

$$\operatorname{mod} A = \lim_{m\to\infty} \operatorname{mod} A_m. \tag{5}$$

Suppose first that C_0 reduces to a point x_0 and fix $0 < a < b < \mathrm{d}(x_0, C_1)$. Then the ring $a < |x - x_0| < b$ separates the boundary components of A_m for large m. Hence, by (2.3) and lemma 5.7,

$$\varliminf_{m\to\infty} \operatorname{mod} A_m \geqslant \log \frac{b}{a}.$$

and, since $\operatorname{mod} A = \infty$, holding b fix and letting $a \to 0$ yields (5) as desired.

Now suppose that C_0 is non-degenerate, let u be the extremal function for A and extend u so that u is 0 on C_0 and 1 on C_1. Next fix $0 < a < \frac{1}{2}$ and pick $b > 0$ so that $|u(x) - u(y)| \leqq a$ whenever $|x - y| \leqq b$. If we define v by (2.17) then v is continuous and ACL everywhere. Now v is 0 and 1 at points within distance b of C_0 and C_1, respectively, and hence v is admissible for A_m when m is large. Thus

$$\operatorname{cap} A_m \leqq \int_{A_m} |\nabla v|^n \,\mathrm{d}\tau \leqq (1 - 2a)^{-n} \int_A |\nabla u|^n \,\mathrm{d}\tau = \frac{\operatorname{cap} A}{(1 - 2a)^n}$$

for $m > m(a)$, and letting $a \to 0$ yields

$$\varlimsup_{m\to\infty} \operatorname{cap} A_m \leqq \operatorname{cap} A. \tag{6}$$

To complete the proof for (5) we show that

$$\varliminf_{m\to\infty} \operatorname{cap} A_m \geqq \operatorname{cap} A. \tag{7}$$

Since F_{m0} converges uniformly to F_0, we may assume that $d(C_{m0}) \geqq c > 0$ for all m. Next let u_m be the extremal function for A_m and extend u_m to be 0 on C_{m0} and 1 on C_{m1}. Inequality (6) and lemma 5.9 imply that we can find a $K < \infty$ such that

$$\operatorname{cap} A_m = \int_{A_m} |\nabla u|^n \,\mathrm{d}\tau \leqq K$$

for all m. Fix $a \in \left(0, \dfrac{1}{2}\right)$. Then lemma 5.6 implies that $|u_m(x) - u_m(y)| \leqq a$ whenever $|x - y| \leqq b$, where

$$b = ce^{-\frac{A_0 K}{2^n a^n}}.$$

If we define v_m as in (2.17), with u_m in place of u, then v_m will be admissible for A when m is large. Thus

$$\operatorname{cap} A \leqq \int_A |\nabla v_m|^n \, d\tau \leqq (1 - 2a)^{-n} \int_{A_m} |\nabla u_m|^n \, d\tau = \frac{\operatorname{cap} A_m}{(1 - 2a)^n}$$

for $m > m(a)$, and letting $a \to 0$ yields (7), completing the proof of lemma 2.

LEMMA 3. *Let $x^* = f(x)$ be a topological mapping of a domain D and let*

$$\operatorname{mod} A^* \leqq K \operatorname{mod} A \tag{8}$$

hold for all bounded rings $A \subset\subset D$. Then

$$\Lambda_f(x)^n \leqq K^{n-1} |J(x)| \tag{9}$$

whenever $f(x)$ is differentiable and

$$\Lambda_{f^{-1}}(x^*)^n \leqq K^{(n-1)^2} |J^*(x^*)| \tag{10}$$

whenever $f^{-1}(x^*)$ is differentiable.

We begin by establishing (9) at each point $x_0 \in D$ where $f(x)$ is differentiable. If $J(x_0) = 0$, then (9) follows from (3), since condition (8) implies that $\delta_L(x)$ is bounded in A (lemma 1.2). Thus, we may assume $J(x_0) > 0$. Next, by performing preliminary translations, rotations and reflections we can reduce our problem to the case where $x_0 = x_0^* = O$ and $f(x)$ is of the form

$$x^{*i} = a_i x^i + 0(|x|) \qquad (i = 1, \ldots, n) \text{ (do not sum with respect to } i\text{)}, \tag{11}$$

with $0 < a_1 \leqq \ldots \leqq a_n$.

For each $b > 0$ let A_0^* be the ring bounded by the $(n-1)$-ball $\{|x^*| \leqq b, x^{*n} = 0\}$ and by the surface of the cylinder $\{(x^{*1})^2 + \ldots + (x^{*n-1})^2 \leqq (b+1)^2, |x^{*n}| \leqq 1\}$. Then let A be the preimage of A_0^* under the affine mapping

$$x^{*i} = a_i x^i \quad (i = 1, \ldots, n) \text{ (do not sum with respect to } i\text{)},$$

and let A_m^* be the image of A under $mf\left(\frac{x}{m}\right)$. The A_m^* are defined for large m and (8) yields

$$\operatorname{mod} A_m^* \leqq K \operatorname{mod} A.$$

since mod A is conformally invariant with respect to homotheties, because $|\nabla u(x)| = |\nabla u(m\xi)| = \frac{1}{m}|\nabla v(\xi)|$ and $J(\xi) = m^n$, hence $\int_A |\nabla u(x)|^n d\tau = \int_{A_1} |\nabla v(\xi)|^n \, d\tau$ and then cap $A =$ cap A_1, where A_1 is the homothetical image of A. Moreover, since $f(x)$ is as in (11), each component of ∂A_m^* converges uniformly to the corresponding component of ∂A_0^* and the preceding lemma gives

$$\operatorname{mod} A_0^* = \lim_{m \to \infty} A_m^* \leqq K \operatorname{mod} A.$$

Now lemma 5.8 and the above imply that

$$\frac{s(A)^n}{(mA)^{n-1}} \leqq \operatorname{cap} A \leqq K^{n-1} \operatorname{cap} A_0^* \leqq K^{n-1} \frac{mA_0^*}{d(C_0^*, C_1^*)^n}. \tag{12}$$

If C_0^*, C_1^* denote the components of CA^*, then

$$d(C_0^*, C_1^*) = 1, \; s(A_0^*) = a_1 \cdots a_{n-1} \, s(A) = 2\omega_{n-1} \, b^{n-1},$$

$$mA_0^* = a_1 \cdots a_n \, mA = 2\omega_{n-1} (b+1)^{n-1}.$$

Hence and by (12) we obtain

$$\frac{\left(\dfrac{2\omega_{n-1} \, b^{n-1}}{a_1 \cdots a_{n-1}}\right)^n}{\left[2 \dfrac{\omega_{n-1} (b+1)^{n-1}}{a_1 \cdots a_n}\right]^{n-1}} \leqq 2K^{n-1} \, \omega_{n-1} (b+1)^{n-1}$$

and then

$$\frac{a_n^{n-1}}{a_1 \cdots a_{n-1}} \leqq K^{n-1} \left(1 + \frac{1}{b}\right)^{n(n-1)}.$$

Finally letting $b \to \infty$ yields

$$a_n^{n-1} \leqq K^{n-1} \, a_1 \cdots a_{n-1}, \tag{13}$$

whence, since $\Lambda_f(O) = a_n^n$ and $|J(O)| = a_1 \cdots a_n$,

$$\Lambda_f(O)^n = a_n^n \leqq K^{n-1} \, a_1 \cdots a_n = K^{n-1} |J(O)|,$$

as desired.

Next fix $x_0^* \in f(D)$ where $f^{-1}(x^*)$ is differentiable. Again (10) follows from (3) when $J^*(x_0^*) = 0$ and we may assume that $J^*(x_0^*) > 0$. In this case $f(x)$ will have a differential at $x_0 = f^{-1}(x_0^*)$ and, arguing as above, we may further assume that $x_0 = f(x_0) = O$ and that $f(x)$ is as in (11). Again (13) holds and we obtain

$$a_i a_n^{n-2} \leqq a_n^{n-1} \leqq K^{n-1} a_1 \cdots a_{n-1} \leqq K^{n-1} a_1 a_n^{n-2} \quad (i = 2, \ldots, n),$$

hence

$$(a_1)^{-1} \leqq K^{n-1}(a_i)^{-1} \quad (i = 2, \ldots, n)$$

and the product of these $n - 1$ inequalities yields

$$(a_1)^{-n} \leqq K^{(n-1)^2}(a_1 \cdots a_n)^{-1},$$

hence

$$\Lambda_{f-1}(O)^n = (a_1)^{-n} \leqq K^{(n-1)^2}(a_1 \cdots a_n)^{-1} = K^{(n-1)^2}|J^*(O)|,$$

i.e. the inequality (10) for $x^* = O$, completing the proof.

LEMMA 4. *Let $x^* = f(x)$ be a topological mapping of a domain D. Then $f(x)$ satisfies the modulus condition* (8) *for all bounded rings $A \subset\subset D$ if and only if $f(x)$ is ACL in D with*

$$\Lambda_f(x)^n \leqq K^{n-1} {}_G(x) \tag{14}$$

a.e. in D.

The necessity is a consequence of the preceding lemma, because (1.38), lemma 1.2 and theorem 1.2 imply that $f(x)$, and then also $f^{-1}(x^*)$ are QCfH according to Gehring's geometric definition, and then, since all the definitions are equivalent, they are QCfH also according to all Gehring's and Väisälä's definitions. Hence $f(x)$ is differentiable a.e. in D, and thus, the conditions of the preceding lemma are fulfilled, implying a.e. in D (14) which reduces to (9) whenever $f(x)$ is differentiable. On the other hand, since $f(x)$ is QCfH also according to Väisälä's analytic definition, it follows that $f(x)$ is ACL in D as desired.

For the sufficiency we shall establish first that $f(x)$ is differentiable a.e. in D. Indeed, $\varphi(E) = m[f(E)]$ is a function of set completely additive in D and with finite values on every compact subset of D. Hence, by (1.40) and theorem of Lebesgue, $J_G(x) < \infty$ a.e. in D and, by (14), $\Lambda_f(x) < \infty$ a.e. in D. Then, by Rademacher—Stepanov's theorem (Chapter 1.4), $f(x)$ is differentiable a.e. in D.

Next, let A be a bounded ring with $\overline{A} \subset D$, let v be a continuously differentiable admissible function for A^* and set $u(x)$ equal to $v[f(x)]$ or to $1 - v[f(x)]$, depending on whether F_0^* or F_1^* is the boundary of the bounded component of CA^*. Then u is admissible for A (proposition 1.6.4) and

$$| \ u(x) \leqq |\nabla v[f(x)]| \Lambda_f(x). \tag{15}$$

a.e. in A. Indeed,

$$\left| \left| \frac{\partial u(x)}{\partial x^i} \cos(s, x^i) \right| - |\varepsilon(x, |\Delta x|)| \right| \leq \left| \frac{\partial u(x)}{\partial x^i} \cdot \frac{\Delta x^i}{|\Delta x|} + \varepsilon(x, |\Delta x|) \right| =$$

$$= \frac{|\Delta u(x)|}{|\Delta x|} = \frac{|\Delta v[f(x)]|}{|\Delta x|} = \frac{|v(x^* + \Delta x^*) - v(x^*)|}{|\Delta x^*|} \frac{|\Delta x^*|}{|\Delta x|} =$$

$$= \left| \frac{\partial v(x^*)}{\partial x^{*i}} \frac{\Delta x^{*i}}{|\Delta x^*|} + \varepsilon(x^*, |\Delta x^*|) \right| \frac{|\Delta x^*|}{|\Delta x|} \leq \left[\left| \frac{\partial v(x^*)}{\partial x^{*i}} \cos(s^*, x^{*i}) \right| + \right.$$

$$\left. + |\varepsilon(x^*, |\Delta x^*|)| \right] \frac{|\Delta x^*|}{|\Delta x|},$$

where s, s^* are the directions of the vectors Δx and $\Delta x^* = f(x + \Delta x) - f(x)$, respectively and hence, by (2.19),

$$\sup_s \left| \frac{\partial u(x)}{\partial x^i} \cos(s, x^i) \right| - \inf_{|\Delta x| = r} |\varepsilon(x, |\Delta x|)| \leq |\nabla u(x)| - \inf_{|\Delta x| = r} |\varepsilon(x, |\Delta x|)| \leq$$

$$\leq \left[\sup_s \left| \frac{\partial v(x^*)}{\partial x^{*i}} \cos(s^*, x^{*i}) \right| + \sup_{|\Delta x| = r} |\varepsilon(x^*, |\Delta x^*|)| \right] \sup_{|\Delta x| = r} \frac{|\Delta x^*|}{r} =$$

$$= \left[|\nabla v(x^*)| + \sup_{|\Delta x| = r} |\varepsilon(x^*, |\Delta x^*|)| \right] \sup_{|\Delta x| = r} \frac{|\Delta x^*|}{r},$$

and letting $r \to 0$ yields just (15). Then appealing to the theorem of change of measure (Chapter 1.3), we obtain

$$\operatorname{cap} A \leq \int_A |\nabla u|^n \, d\tau \leq \int_A |\nabla v|^n \Lambda_f^n \, d\tau \leq K^{n-1} \int_A |\nabla v|^n J d\tau \leq$$

$$\leq K^{n-1} \int_{A^*} |\nabla v|^n \, d\tau$$

and taking the infimum over all such v we conclude that

$$\operatorname{cap} A \leq K^{n-1} \operatorname{cap} A^*,$$

which implies (8), as desired.

Remark. We wish to point out that in his second analytic definition Väisälä [3] renounces the condition of the differentiability of $f(x)$ a.e. in D because it was a consequence of the condition ACL_n, whereas, in Gehring's analytic definition (as it follows from the preceding lemma), the differentiability a.e. in D is a consequence of the inequality (14) and this allowed him to replace the condition ACL_n by the less restrictive condition ACL. The inequality (14) is stronger than (1.37) and reduces to it at the points of differentiability, since at such a point the formulae (1.40), (1.10) and (3.8) hold. If (14) holds a.e. in D, then (1.37) holds a.e. in D, too (it seems that the converse is no more true).

THEOREM 1. *Gehring's analytic and geometric definitions are equivalent.*

This is an immediate consequence of the preceding lemma applied both to $x^* = f(x)$ and to $x = f^{-1}(x^*)$.

THEOREM 2. *Let* $x^* = f(x)$ *be a homeomorphism and let* (8) *hold for all bounded rings* $A \subset\subset D$. *Then* $f(x)$ *is* K^{n-1}-QCfH *according to Gehring's analytic and geometric definitions. The bound* K^{n-1} *is the best possible.*

By the preceding lemma, $f(x)$ is ACL in D and (14) holds a.e. in D, hence, by the theorem 1.2, $f(x)$ is QCfH in D^*, and then also $x = f^{-1}(x)$ is QCfH in D^*. This allows us to conclude that $f(x)$ is differentiable a.e. in D and $f^{-1}(x^*)$ a.e. in D^* and that $f^{-1}(x^*)$ is ACL in D^*. Hence and by lemma 3, it follows that the inequalities (9) and (10) hold a.e. in D and D^*, respectively. Thus $f(x)$ is K^{n-1}-QCfH in D according to Gehring's analytic definition.

Let us show that the bound K^{n-1} cannot be improved. Indeed, it is easy to verify that the affine mapping $f(x) = (x^1, K^{n-1}x^2, \ldots, K^{n-1}x^n)$ has the maximal dilatation $\Lambda_f(x) = K^{n-1}$ and then (15) takes the form

$$\left|\nabla u(x)\right| \leqq K^{n-1}\left|\nabla v(x^*)\right|.$$

Since $J(x) = K^{(n-1)^2}$, it follows that

$$\operatorname{cap} A \leqq \int_A \left|\nabla u(x)\right|^n d\tau \leqq K^{n(n-1)} \int_A \left|\nabla v[f(x)]\right|^n d\tau =$$

$$= K^{n-1} \int_A \left|\nabla v[f(x)]\right|^n J(x)\, d\tau = K^{n-1} \int_{A^*} \left|\nabla v(x^*)\right|^n d\tau,$$

which implies (8) and then also (9). Taking into account that $J^*(x^*) = K^{-(n-1)^2}$ and $\Lambda_{f-1}(x^*) = 1$, we infer that the inequality (10) reduces to an identity, whereas the second inequality (1) is not satisfied. The two inequalities (1) are satisfied if we put K^{n-1} in place of K. Thus the affine mapping from above satisfies the one side modulus condition (8) and is K^{n-1}-QCfH according to Gehring's analytic and geometric definitions.

Remark. The preceding theorem yields that if $n = 2$, then, in contrast with the case $n > 2$, the inequality (8) is sufficient to characterize the K-QCfH according to Gehring's analytic and geometric definitions, since, for $n = 2$, $K^{n-1} = K$.

Gehring's second analytic definition. Gehring's analytic definition characterized by the condition (1) has the disadvantage to involve both the homeomorphism $x^* = f(x)$ and its inverse $x = f^{-1}(x^*)$. That is why we shall give the following equivalent definition which involves only the homeomorphism $x^* = f(x)$.

A homeomorphism $f(x)$ is said to be K-QCfH ($1 \leqq K < \infty$) in D if it is ACL in D and the inequality

$$\frac{\Lambda_f(x)^n}{K^{n-1}} \leqq J_G(x) \leqq K^{n-1}\, \lambda_f(x)^n \tag{16}$$

holds a.e. in D.

The equivalence of the two Gehring's analytic definitions.

THEOREM 3. *The two Gehring's analytic definitions are equivalent.*

Suppose $x^* = f(x)$ is a K-QCfH according to Gehring's first analytic definition. Then $f(x)$ is ACL in D and the first inequality of (16) holds a.e. in D. From the equivalence of all Gehring's and Väisälä's definitions of the QCfH (without specifying K), it follows that $f(x)$ is QCfH in D also according to Väisälä's definitions, hence $f(x)$ is differentiable with $J(x) \neq 0$ a.e. in D. Then, the second inequality (1), taking into account that

$$J^*(x_0^*) = \frac{1}{J(x_0)} \tag{17}$$

holds whenever x_0 is an A-point of D and that

$$\Lambda_{f^{-1}}(x_0^*) = \overline{\lim_{x^* \to x_0^*}} \frac{|f^{-1}(x^*) - f^{-1}(x_0^*)|}{|x^* - x_0^*|} = \overline{\lim_{x \to x_0}} \frac{|x - x_0|}{|f(x) - f(x_0)|} =$$

$$= \frac{1}{\underline{\lim}_{x \to x_0} \dfrac{|f(x) - f(x_0)|}{|x - x_0|}} = \frac{1}{\lambda_f(x_0)}, \tag{18}$$

implies the second part of the double inequality (16), which allows us to conclude that $f(x)$ is K-QCfH also according to the second Gehring's analytic definition.

Conversely, suppose $f(x)$ is K-QCfH according to the second Gehring's analytic definition. Then, $f(x)$ is ACL in D and the first inequality (1) holds a.e. in D. The second part of (16), on account of (17) and (18), implies that $f^{-1}(x^*)$ satisfies the second inequality (1) a.e. in D. On the other hand, since $f(x)$ is ACL in D and satisfies the first inequality (1), it follows, by lemma 4, that (8) holds, and then, by theorem 2, $f(x)$ is QCfH in D also according to the first Gehring's analytic definition hence $f^{-1}(x^*)$ is ACL in D^*. Thus $f(x)$ and $f^{-1}(x^*)$ are ACL in D and D^*, respectively and (1) hold a.e. in D and D^*, respectively, i.e. $f(x)$ is K-QCfH also according to Gehring's first analytic definition.

Church's definition. In the particular case of the diffeomorphisms, Gehring's analytic definitions are equivalent to the following Church's [1] definition:

A diffeomorphism $f(x)$ is said to be K-QCf ($1 \leq K < \infty$) in D, if for every pair of directions s, s'

$$\left|\frac{\partial f(x)}{\partial s}\right| \leq K \left|\frac{\partial f(x)}{\partial s'}\right| \tag{19}$$

a.e. in D.

Equivalence between Church's definition of QCf diffeomorphisms and the others.

THEOREM 4. *Church's definition of* QCf *diffeomorphisms is equivalent to Gehring's analytic definition of the* QCf *diffeomorphisms.*

Let $f(x)$ be a K-QCf diffeomorphism according to Gehring's analytic definitions. Then lemma 3.3 implies

$$\frac{\left|\dfrac{\partial f(x)}{\partial s}\right|}{\left|\dfrac{\partial f(x)}{\partial s'}\right|} \leq \frac{\Lambda_f(x)}{\lambda_f(x)} \leq K^{\frac{2(n-1)}{n}}$$

everywhere in D, and we conclude that $f(x)$ is $K^{2\left(\frac{n}{n}-1\right)}$-QCf according to Church's definition.

Conversely, let $f(x)$ be a K-QCf diffeomorphism according to Church's definition. Then (19), (3.3) and (3.4) imply

$$\frac{\Lambda_f(x)}{\lambda_f(x)} = \frac{\sup\limits_{s}\left|\dfrac{\partial f(x)}{\partial s}\right|}{\inf\limits_{s}\left|\dfrac{\partial f(x)}{\partial s}\right|} = \sup_{s,\, s'} \frac{\left|\dfrac{\partial f(x)}{\partial s}\right|}{\left|\dfrac{\partial f(x)}{\partial s'}\right|} \leqslant K,$$

hence, by (3.10), (3.8) and (3.9), we get (16), i.e. $f(x)$ is K-QCf according to Gehring's second analytic definition.

Remark. Since all the definitions are equivalent (if we do not specify the constant K), the preceding theorem allows us to conclude the equivalence of Church's definition to all the other definitions of QCf diffeomorphisms.

CHAPTER 7

A more general form for Gehring's metric definition of K-QCfH equivalent to his preceding metric definition

The new form for Gehring's metric definition (Gehring [5]). A homeomorphism $f(x)$ is said to be K-QCfH ($1 \leqq K < \infty$) in D, if $\delta_L(x) < \infty$ except in a set of Σ-finite area and

$$\delta_L(x) \leqq K$$

a.e. in D.

Equivalence to his preceding metric definition. We will require some lemmas to obtain the equivalence of the two last definitions. (Gehring [5] and [20]).

For $p \geqq 1$ and $0 < q < \frac{1}{2} d(D)$, let Δ_{pq} denote those points $x \in D$ such that $L(x, r) \leqq pl(x, r)$ whenever $0 < r \leqq q$, where $L(x, r)$ and $l(x, r)$ are given by (1.5), (1.6), and set $\Delta_p = \bigcup\limits_{q>0} \Delta_{pq}$. It is easy to verify that $\Delta_{p,q}$ is closed with respect to D and Δ_p contains the points $x \in D$ at which $\delta_L(x) < p$.

LEMMA 1. *Suppose $f(x)$ is a homeomorphism in D, $y_0 \in I_n$. Then for each F_σ-Borel set $E \subset Z_{y_0}$,*

$$[\Lambda(E^* \cap \Delta_p^*)]^n \leqq \frac{2^{n+1} p^n \omega_{n-1}}{\omega_n} D_{sym}\varphi(y_0)[m_1(E \cap \Delta_p)]^{n-1}, \tag{1}$$

where $\Delta_p^* = f(\Delta_p)$.

Since $\Delta_{p,q_1} \supset \Delta_{p,q_2}$ for $0 < q_1 < q_2$, it follows (see C. Carathéodory [3], theorem 8, p. 254) that

$$\lim_{q \to 0} \Delta_{p,q} = \Delta_p, \quad \lim_{q \to 0} \Delta_{p,q}^* = \Delta_p^*$$

imply that

$$\lim_{q \to 0} m_1(E \cap \Delta_{p,q}) = m_1(E \cap \Delta_p), \quad \lim_{q \to 0} \Lambda(E^* \cap \Delta_{p,q}^*) = \Lambda(E^* \cap \Delta_p^*),$$

and then it is sufficient to prove that

$$[\Lambda(E^* \cap \Delta_{p,q}^*)]^n \leqq \frac{2^{n+1} p^n \omega_{n-1}}{\omega_n} D_{sym}\varphi(y_0)[m_1(E \cap \Delta_{p,q})]^{n-1} \tag{2}$$

for all $q > 0$. Next we can express E as the union of an expanding sequence of compact sets and then, by the same limiting argument as in lemma 1.5, it suffices to establish (2) in the hypothesis that E is a compact set. Finally, we assume, for convenience of notations, that $y_0 = O$.

Since $f(x)$ is continuous on the compact set E, we can pick a $\rho_1 > 0$ such that $B(x, r) \subset I$ and $L(x, r) < \frac{\varepsilon}{2}$ for $x \in E$, $0 < r < \rho_1$. If we apply lemma 1.4 to the compact set $F = E \cap \Delta_{p,q}$, we can find ρ_2, $0 < \rho_2 < \rho_1$, q such that for each r, $0 < r < \rho_2$, there exist points $x_1, \ldots, x_N \in E$ such that the conditions (1.28), (1.29) and (1.30) of lemma 1.4 hold. Fix such an r, and let $B_m = B(x_m, r)$, $L_m = L(x_m, r)$, $l_m = l(x_m, r)$. Since $x_m \in \Delta_{p,q}$ and $r < q$, we see that $L_m < pl_m$. Now each $B_m^* = f(B_m)$ is contained in a ball of diameter $2L_m < \varepsilon$, and since these balls cover $F^* = E^* \cap \Delta_{p,q}^*$, taking into account also the definition of the length of a set (Chapter 1.3), we obtain

$$\Lambda^{\varepsilon}(F^*) \leqq \sum_{m=1}^{N} 2L_m \leqq 2p \sum_{m=1}^{N} l_m.$$

But each B_m^* also contains a ball of radius l_m, and hence by Hölder's inequality

$$\Lambda^{\varepsilon}(F^*)^n \leqq (2p)^n \left(\sum_{m=1}^{N} l_m \right)^n \leqq (2p)^n N^{n-1} \sum_{m=1}^{N} l_m^n = \frac{(2p)^n}{\omega_n} N^{n-1} \sum_{m=1}^{N} \omega_n l_m^n \leqq$$

$$\leqq \frac{(2p)^n}{\omega_n} N^{n-1} \sum_{m=1}^{N} mB^*.$$

Arguing as in lemma 1.5, this inequality implies the inequality (1), as desired.

A topological mapping $f\colon R^1 \to R^n$ is said *to satisfy the Lusin condition N, on a linear interval* (a, b) if $\Lambda(E^*) = 0$ whenever $E \subset (a, b)$, $m_1 E = 0$.

LEMMA 2. *Suppose $f(x)$ is a homeomorphism in D, I an interval with $I \subset D$, I_n its base, $y_0 \in I_n$, $D_{sym}\varphi(y_0) < \infty$ and $\delta_L(y_0 + \xi e_n) = \infty$ for at most an enumerable set of $\xi \in (\alpha^n, \beta^n)$. Then $f(y_0 + \xi e_n)$ satisfies the Lusin condition N in (α^n, β^n).*

Let E be a G_δ-Borel set in Z_{y_0} with $m_1 E = 0$ and let p run through the positive integers. Then

$$E^* = \bigcup_{p=1}^{\infty} (E^* \cap \Delta_p^*) \cup (E^* \cap \Delta_\infty^*),$$

where $\Delta_\infty = \{x;\ \delta_L(x) = \infty\}$ and $\Delta_\infty^* = f(\Delta_\infty)$. By hypothesis, $E^* \cap \Delta_\infty^*$ is at most enumerable and, since Z_{y_0} is an F_σ-Borel set, we conclude from (1) that

$$\Lambda(E^* \cap \Delta_p^*) \leqq \Lambda(Z_{y_0}^* \cap \Delta_p^*) < \infty$$

for all p. Thus E^* is a G_δ-Borel set of Σ-finite length. Now pick any compact set $F^* \subset E^*$. If F is the pre-image of F^* then F is compact, $F \subset E$ and $\Lambda(F) \leqq \leqq \Lambda(E) = 0$. Furthermore

$$F^* = \left[\bigcup_{p=1}^{\infty} (F^* \cap \Delta_p^*) \right] \cup (\ ^* \cap \Delta_\infty^*).$$

Again $F^* \cap \Delta_\infty^*$ is at most enumerable and, by (1), $\Lambda(F^* \cap \Delta_p^*) = 0$ for all p. Hence $\Lambda(F^*) = 0$ and, with lemma 5.1 we conclude that

$$\Lambda(E^*) = \sup\{\Lambda(F^*);\quad F^* \text{ compact},\quad ^* \subset E^*\} = 0. \tag{3}$$

Finally pick any $E \subset Z_{y_0}$ with $m_1 E = 0$. We can find a G_δ-Borel set E_0 such that $E \subset E_0 \subset Z_{y_0}$ and $m_1 E_0 = 0$. By formula (3) we see that $\Lambda(E^*) \leqq \Lambda(E_0^*) = 0$ and the proof is complete.

LEMMA 3. *Suppose $f(x)$ is a homeomorphism in D, $y_0 \in I_n$, $D_{sym}\varphi(y_0) < \infty$. If $f(y_0 + \xi e_n)$ satisfies the Lusin condition N in Z_{y_0} and $\delta_L(y_0 + \xi e_n) \leqq K$ a.e. in Z_{y_0}, then $f(y_0 + \xi e_n)$ is AC in Z_{y_0}.*

Set $E = \bigcup_{m=1}^{N} \{y_0 + a_m e_n, y_0 + y_0 + b_m e_n\}$, where the intervals are disjoint. If $p > K$, then $\delta_L(y_0 + \xi e_n) < p$ a.e. in Z_{y_0}. Hence $m_1(E - \Delta_p) = 0$. Because $f(y_0 + \xi e_n)$ satisfies the Lusin condition N in Z_{y_0}, $\Lambda(E^* - \Delta_p^*) = 0$ and by (1) we conclude that

$$\Lambda(E^*)^n = \Lambda(E^* \cap \Delta_p^*)^n \leqq \frac{2^{n+1}p^n\omega_{n-1}}{\omega_n} D_{sym}\varphi(y_0)[m_1(E \cap \Delta_p)]^{n-1} =$$

$$= \frac{2^{n+1}p^n\omega_{n-1}}{\omega_n} D_{sym}\varphi(y_0)(m_1 E)^{n-1},$$

whence, on account of the definition of E, we obtain (1.35) (with p in place of K_1), i.e. $f(y_0 + \xi e_n)$ is AC on Z_{y_0}.

LEMMA 4. *Suppose $f(x)$ is a homeomorphism in D with $\delta_L(x) < \infty$ in D except possibly a set of Σ-finite area and with $\delta_L(x) \leqq K$ a.e. in D. Then $f(x)$ is ACL in D.*

We want to show that $f(x)$ is AC on Z_y for almost all $y \in I_n$. Let D_1 be the subset of D where $\delta_L(x) \leqq K$ and D_f the subset where $\delta_L(x) < \infty$. By Lebesgue's theorem (Chapter 1), $D_{sym}\varphi(y) < \infty$ a.e. in I_n. On the other hand, $\delta_L(x) \leqq K$ a.e. in D implies $\delta_L(x) \leqq K$ on Z_y except possibly on a set of linear measure zero and for almost all $y \in I_n$, since, by Fubini's theorem,

$$m(I - I \cap D_k) = \int_{I - D_k} dx = \int \cdots \int dx^1 \cdots dx^{n-1} \int dx^n = \int_{I_n - E_k} m_1(Z_y - D_k)\, d\sigma = 0.$$

But then, the subset of I_n where $m_1(Z_y - D_k) \neq 0$ is a set of Lebesgue's $(n-1)$-dimensional measure zero. Moreover, by corollary 2 of Gross' theorem (Chapter 1.3), since $D - D_f$ is a set of Σ-finite area, it follows that $f(x)$ satisfies the Lusin condition N in Z_y for almost all $y \in I_n$. Hence, by the preceding lemma and taking into account that $\delta_L x) \leqq K$ in Z_y except possibly a set of linear measure zero, then it follows that $f(x)$ is AC in Z_y for almost all $y \in I_n$, i.e. $f(x)$ is ACL in I, and, since I is an arbitrary interval in D, we conclude that $f(x)$ is ACL in D.

Now we shall prove that the conditions of the preceding lemma imply the quasiconformality of $f(x)$ according to Gehring's metric definition.

THEOREM 1. *Let $f(x)$ be a homeomorphism of D, $\delta_L(x) \leqq K$ a. e. in D and $\delta_L(x) < \infty$ in D except possibly on a set of Σ-finite area. Then $f(x)$ is K-QCfH according to Gehring's metric definition.*

First $f(x)$ is *ACL* in D by the preceding lemma. Next lemma 6.1 implies $\Lambda_f(x)^n \leqq K^n J_G(x)$ a.e. in D. Thus

$$\operatorname{mod} A^* \leqq K^{\overline{n-1}}{}^{n} \operatorname{mod} A$$

by lemma 6.4 and theorem 6.2 allows us to conclude that $f(x)$ is QCfH according to Gehring's analytic and geometric definitions, hence, since all the definitions are equivalent, it follows that $f(x)$ is QCfH also according to Gehring's metric definition. Then $\delta_L(x)$ is bounded in D, and since $\delta_L(x) \leqq K$ a.e. in D (from above) we conclude that $f(x)$ is K-QCfH in D according to Gehring's metric definition, as desired.

This theorem points out that in the definition of a K-QCf according to Gehring's metric definition, it is possible to neglect a set of Σ-finite measure, as follows.

COROLLARY 1. *Let $f(x)$ be a homeomorphism in D, K-QCf according to Gehring's metric definition in $D-E$, where E is a set of Σ-finite area in D. Then $f(x)$ is K-QCfH also in D.*

Indeed, $f(x)$ has $\delta_L(x) < \infty$ except possibly on a set E (of Σ-finite area) and $\delta_L(x) \leqq K$ a.e. in D.

A more general form for the preceding Gehring's and Väisälä's definitions of K-QCfH and the equivalence to the corresponding definitions.

COROLLARY 2. *Let E be a set of Σ-finite area of D and let $f(x)$ be a homeomorphism in D, K-QCfH according to Gehring's analytic or geometric definition in a neighbourhood of every point of $D-E$. Then $f(x)$ is K-QCfH in D according to Gehring's analytic, respectively geometric, definition.*

$f(x)$ K-QCfH in D according to Gehring's analytic (geometric) definition in every neighbourhood of every point of $D-E$ implies, by theorem 3.3, that $f(x)$ is $K^{\frac{2(n-1)}{n}}$-QCfH according to his metric definition in a neighbourhood of every point of $D-E$, and then in $D-E$, hence, by the preceding corollary, also in D, and then, on account of the equivalence of all definitions of QCfH, we conclude that $f(x)$ is QCfH in D according to Gehring's analytic and geometric definitions. Next, $f(x)$ is K-QCfH in a neighbourhood of every point of $D-E$. Then (6.16) holds a.e. in $D-E$ and then a.e. in D. Hence, we conclude that $f(x)$ is K-QCfH in D according to Gehring's analytic definition.

Finally, if $f(x)$ is K-QCfH according to Gehring's geometric definition in a neighbourhood of every point of $D-E$, then by theorem 6.1, $f(x)$ is K-QCfH according to his analytic definition in a neighbourhood of every point of $D-E$, hence, by what was just proved, it is K-QCfH in D according to his analytic definition and by theorem 6.1, $f(x)$ is K-QCfH in D also according to his geometric definition.

Remark. As a consequence of this corollary it follows that in Gehring's geometric definition of K-QCfH it is not necessary to consider all the rings

$A \subset\subset D$, but it is sufficient to deal only with the rings of a neighbourhood as small as one pleases of every point of D (or of $D-E$). Thus, Gehring's geometric definition of the K-QCfH can be given in the following more general form:

A homeomorphism $f(x)$ in D is said to be K-QCfH ($1 \leqq K < \infty$) in D, if the double inequality (2.32) holds for all the rings of a neighbourhood of every point of D, except possibly a set of Σ-finite area.

In the same way, Gehring's second analytic definition can be given as follows:

A homeomorphism $f(x)$ in D is called K-QCfH ($1 \leqq K < \infty$) in D, if it is ACL in the neighbourhood of every point of D, except possibly a set of Σ-finite area, and (6.16) holds a.e. in D.

And for Gehring's first analytic definition, we have:

A homeomorphism $f(x)$ in D is K-QCfH in D, if and only if $f(x)$ and $f^{-1}(x^*)$ are ACL in the neighbourhood of every point of D and of D^*, respectively, except possibly a set of Σ-finite area and (6.1) holds a.e. in D, respectively in D^*.

It is easy to see that Väisälä's definitions can be given in the following more general form:

A K-QCfH ($1 \leqq K < \infty$) in D according to *Väisälä's first geometric definition* is said to be a homeomorphism which satisfies (1.18) for every curve family contained in the neighbourhood of every point of D except possibly a set of Σ-finite area.

A K-QCfH ($1 \leqq K < \infty$) in D according to *Väisälä's second geometric definition* is said to be a homeomorphism which satisfies (1.56) for all rings contained with their closure in a neighbourhood of every point of D, except possibly a set of Σ-finite area.

A K-QCfH ($1 \leqq K < \infty$) in D according to *Väisälä's third geometric definition* is said to be a homeomorphism which satisfies (1.62) for all cylinders contained with their closure in a neighbourhood of every point of D, except possibly a set of Σ-finite area.

A K-QCfH ($1 \leqq K < \infty$) in D according to *Väisälä's first analytic definition* is said to be a homeomorphism ACL in a neighbourhood of every point of D except possibly a set of Σ-finite area and which is differentiable and satisfies (1.19) a.e. in D.

A K-QCfH ($1 \leqq K < \infty$) in D according to *Väisälä's second analytic definition* is said to be a homeomorphism ACL_n in a neighbourhood of every point of D except possibly a set of Σ-finite area and which satisfies (1.52) a.e. in D.

COROLLARY 3. *Let $f(x)$ be a homeomorphism in D and $E \subset D$ a closed set of Σ-finite area. If $f(x)$ is K-QCfH in $D - E$ according to one of the preceding definitions, then $f(x)$ is K-QCfH in D.*

CHAPTER 8

Conformal mappings and the corresponding elliptic system

Conformal mappings. In this chapter we define the conformal mappings and establish the elliptic system which characterizes them.

A conformal mapping $f(x)$ in D is a 1-QCfH in D.

The theorems establishing the connexion between the different definitions of the K-QCfH allow us to conclude that all the classes of 1-QCfH coincide, i.e. one obtains the same class of conformal mappings if we put $K = 1$ in any definition of K-QCfH. Hence conformal mappings are differentiable with $J(x) \neq 0$ a.e. in D. (It will be proved in chapter 16 that conformal mappings are differentiable with $J \neq 0$ everywhere in D.) Thus almost all the points of D are A-points for $f(x)$.

The corresponding elliptic system.

THEOREM 1. *A conformal mapping $x^* = f(x)$ satisfies in an A-point the following first order partial derivative system*:

$$\delta_{ik} x_p^{*i} x_q^{*k} = \delta_{pq} \sqrt[n]{J^2} \quad (p, q = 1, \ldots, n). \tag{1}$$

Since conformal mappings are 1-QCfH also according to Kopylov's definition, it follows that they map infinitesimal spheres into infinitesimal spheres. It is possible to express this property at an A-point by the proportionality of linear elements, i.e. analytically by the relation

$$\delta_{ik}\, \mathrm{d}x^{*i}\, \mathrm{d}x^{*k} = \mu^2 \delta_{pq}\, \mathrm{d}x^p\, \mathrm{d}x^q,$$

which implies

$$\delta_{ik} x_p^{*i} x_q^{*k} = \delta_{pq} \mu^2 \quad (p, q = 1, \ldots, n), \tag{2}$$

hence $J^2 = \mu^{2n}$, completing the proof.

COROLLARY. *The conformal mapping* $\vartheta_k^* = \vartheta_k^*(\vartheta_1, \ldots, \vartheta_{n-1})$ $(k = 1, \ldots, n-1)$ *of the spherical (n — 1)-space satisfies in an A-point the following system*:

$$\vartheta_{1p}^* \vartheta_{1q}^* + \sin^2 \vartheta_1^* \vartheta_{2p}^* \vartheta_{2q}^* + \cdots + \sin^2 \vartheta_1^* \cdots \sin^2 \vartheta_{n-2}^* \vartheta_{n-1p}^* \vartheta_{n-1q}^* =$$

$$= \delta_{pq} \mu^2 \sin^2 \vartheta_1 \cdots \sin^2 \vartheta_{p-1} \quad (p, q = 1, ..., n-1), \tag{3}$$

where $\vartheta_{pq}^* = \dfrac{\partial \vartheta_p^*}{\vartheta \partial_q^*} (p, q = 1, \ldots, n-1)$ *and* $ds^{*2} = \mu^2 ds^2$.

This is an immediate consequence of the preceding theorem, taking into account that the square of *the linear element of the sphere of radius r, expressed in the spherical coordinates* $\vartheta_1, \ldots, \vartheta_{n-1}$ is of the form

$$ds^2 = r^2 (d\vartheta_1^2 + \sin^2 \vartheta_1 \, d\vartheta_2^2 + \cdots + \sin^2 \vartheta_1 \cdots \sin^2 \vartheta_{n-2} \, d\vartheta_{n-1}^2).$$

CHAPTER 9

The definition of Θ-mappings and its equivalence to the definition of QCfH

Most results of this chapter are from F. Gehring ([6], [8] and [5]).

Θ-mappings. (F. Gehring [8]). A homeomorphism $x^* = f(x)$ is said to be a *Θ-mapping*, if there exists a continuous and strictly increasing real function $\Theta(t)$ in $[0, 1)$ with $\Theta(0) = 0$ and such that the following are true:

(i) If $x \in D$ and $|x' - x| < d(x, D)$, then

$$\frac{|f(y) - f(x)|}{d[f(x), \partial D^*]} \leqslant \Theta\left[\frac{|y - x|}{d(x, \partial D)}\right];$$

(ii) The restriction of $f(x)$ to any subdomain Δ satisfies (i).

In this chapter, it will be proved (following Gehring) that the Θ-mappings are K-QCfH according to Gehring's metric definition with $K = [\Theta^{-1}(1)]^{-1}$, but this requires a preliminary study of some geometrical properties of spherically symmetrized sets.

Spherical symmetrization of sets. Given an open set D we define (following Gehring [6]) a second set D°, the *spherical symmetrization of D*, as follows. For each $r \geqq 0$, set $S(r) \cap D^\circ = \emptyset$ if and only if $S(r) \cap D = \emptyset$. Next, $S(r) \subset D^\circ$ if and only if $S(r) \subset D$. For the remaining case let D meet $S(r)$ in a set whose area is a. Then $0 < a \leqq n\omega_n r^{n-1}$ and D° is to meet $S(r)$ in a single open spherical cap of area a with centre at $(-r, 0, \ldots, 0)$; when $a = n\omega_n r^{n-1}$, this cap will consist of $S(r)$ minus the point $(r, 0, \ldots, 0)$. It is easy to see that D° itself is an open set and that CD° is connected whenever CD is. Next given a closed set F we define F° exactly as above except in the last case. Here $0 \leqq a < n\omega_n r^{n-1}$ and F° is to meet $S(r)$ in a closed spherical cap of area a with its centre at $(-r, 0, \ldots, 0)$; when $a = 0$, this cap will consist only of the centre point $(-r, 0, \ldots, 0)$. Then F° is closed and F° is connected whenever F is.

It is easy to see that the measure of a closed set F is preserved under spherical symmetrization. For if $a(r)$ denotes the area of $S(r) \cap F$, then

$$mF = \int_0^\infty a(r)\,\mathrm{d}r = mF^\circ.$$

A property of the spherical symmetrization of a polyhedron. *The closed convex hull of a set E* is the intersection of all closed half spaces which contain E.

A simplex $\tau \subset R^n$ is the closed convex hull of a set consisting of $n+1$ points called *the vertices of the simplex*. A simplex is said *to be degenerated* if its vertices lie in an $(n-1)$-dimensional plane.

A closed polyhedron is a finite union of simplices (possibly degenerated).

The following result shows that in certain cases we can say something about what happens to the area of ∂F, where F is a certain closed set, under spherical symmetrization.

LEMMA 1. *If F is a closed polyhedron and if F° is the spherical symmetrization of F, then ∂F° is a surface of revolution whose area does not exceed that of ∂F.*

Let $\sigma(r)$ and $\sigma^\circ(r)$ denote the area of the parts of ∂F and ∂F° contained in $\overline{B(r)}$ for $r \geqq 0$. We shall show that

$$\sigma^\circ(r_2) - \sigma^\circ(r_1) \leqq \sigma(r_2) - \sigma(r_1), \tag{1}$$

for all $0 \leqq r_1 < r_2 < \infty$.

Now let $a(r)$ be the area of $S(r) \cap F$. Then $a(r)$ is continuous and satisfies a uniform Lipschitz condition. We show it only when F is a closed n-dimensional simplex and hence the result follows when F is a closed polyhedron. Suppose first that $S(r)$ meets only one $(n-1)$-dimensional face of the simplex and let $\psi(r)$ denote the acute positive angle between the radius to a point x, with $|x| > 0$, of this intersection and the normal to the corresponding face. Since the area of the corresponding cap is

$$a(r) = (n-1)\omega_{n-1} r^{n-1} \int_0^{\psi} \sin^{n-2} \vartheta \, d\vartheta, \tag{2}$$

where ϑ is the positive acute angle between the radius to a point of the cap and the normal to the corresponding face of the simplex, then

$$a(r+\Delta r) - a(r) = (n-1)\omega_{n-1}\Big[(r+\Delta r)^{n-1} \int_0^{\psi+\Delta\psi} \sin^{n-2}\vartheta \, d\vartheta -$$

$$- r^{n-1}\int_0^{\psi} \sin^{n-2}\vartheta \, d\vartheta\Big] < (n-1)\omega_{n-1}\Big(|\Delta r| K + r^{n-1}\Big|\int_{\psi}^{\psi+\Delta\psi} \sin^{n-2}\vartheta \, d\vartheta\Big|\Big) \leqq \tag{3}$$

$$\leqq (n-1)\omega_{n-1}\Big(|\Delta r| K + r^{n-1}\Big|\int_{\psi}^{\psi+\Delta\psi} \sin\vartheta \, d\vartheta\Big|\Big) \leqq$$

$$\leqq (n-1)\omega_{n-1}\big[|\Delta r| K + r^{n-1}|\cos\psi - \cos(\psi+\Delta\psi)|\big].$$

Suppose now that $S(r + \Delta r)$ meets also only the same $(n - 1)$-dimensional face of the simplex. Then

$$(r + \Delta r) \cos(\psi + \Delta\psi) = r \cos\psi$$

yields

$$\cos(\psi + \Delta\psi) = \frac{r}{r + \Delta r} \cos\psi, \tag{4}$$

and (3) implies that

$$a(r + \Delta r) - a(r) \leqq (n - 1)\omega_{n-1} \left[|\Delta r| K + r^{n-1} \left(1 - \frac{r}{r + \Delta r} \right) \cos\psi \right] \leqq$$

$$\leqq (n - 1)\omega_{n-1} |\Delta r| (K + r^{n-2} \cos\psi) \leqq |\Delta r| (K + R^{n-2}), \tag{5}$$

where R is the radius of a sphere concentric to $S(r)$ and containing inside the considered simplex. This allows us to conclude that $a(r)$ satisfies a uniform Lipschitz condition in this particular case.

Next, if one vertex of the simplex is in the centre of the sphere $S(r)$ and all the others are outside it, then the area $a(r)$ of the $(n - 1)$-dimensional spherical simplex is

$$a(r) = r^{n-1} \int_{\vartheta_1^0}^{\vartheta_1^1} \cdots \int_{\vartheta_{n-1}^0}^{\vartheta_{n-1}^1} \sin^{n-2} \vartheta_1 \cdots \sin \vartheta_{n-2} \, d\vartheta_1 \cdots d\vartheta_{n-1},$$

where ϑ_k^0, ϑ_k^1 $(k = 1, \ldots, n - 1)$ are independent of r. This generalizes the bidimensional area of a spherical triangle: $a_3(r) = r^2\delta$ (δ the geodetic excess). Clearly $a(r)$ satisfies a uniform Lipschitz condition in this case also.

Now consider the case of a simplex with at least two vertices inside $S(r)$ one of them being at the centre of the sphere, with the only restrictive hypothesis that $S(r)$ and $S(r + \Delta r)$ meet the same number of $(n - 1)$-dimensional faces of the simplex. Then, if $S(r) \cap F \neq \emptyset$, the part of $S(r)$ inside the simplex is contained between the $(n - 1)$-dimensional plane corresponding to the $(n - 1)$-dimensional face of the simplex which does not contain the centre of the sphere and the $(n - 1)$-dimensional plane parallel to it through the centre. The angle between the radius of a point of $S(r)$ contained in the former plane and the normal to this plane is just the angle ψ corresponding to this face of the simplex. Thus the area of the part of $S(r)$ inside the simplex is

$$a(r) = r^{n-1} \int_{\psi^0}^{\psi} \int_{\vartheta_1^0}^{\vartheta_1^1} \cdots \int_{\vartheta_{n-2}^0}^{\vartheta_{n-2}^1} \sin^{n-2} \varphi \sin^{n-3} \vartheta_1 \cdots \sin \vartheta_{n-3} \, d\varphi \, d\vartheta_1 \cdots d\vartheta_{n-2},$$

where ψ^0 is independent of r and $\vartheta_k^0, \vartheta_k^1$ $(k = 1, \ldots, n-2)$, also independent of r, are obtained from the relations

$$A_1^k \sin\varphi \sin\vartheta_1 \cdots \sin\vartheta_{n-2} + A_2^k \sin\varphi \sin\vartheta_1 \cdots \sin\vartheta_{n-3} \cos\vartheta_{n-2} + \cdots +$$

$$+ A_p^k \sin\varphi \sin\vartheta_1 \cdots \sin\vartheta_{n-p-1} \cos\vartheta_{n-p} + \cdots + A_n^k \cos\varphi = 0 \quad (k = 1, \cdots, n),$$

representing the equations of the $(n-2)$-dimensional spheres intersected from $S(r)$ by the $(n-1)$-dimensional faces of the simplex containing O, i.e. by $A_i^k x^i = 0$ $(k = 1, \ldots, n)$. Then, by (4)

$$|a(r+\Delta r) - a(r)| \leqq |\Delta r| K + 2r^{n-1}\pi^{n-2} \left| \int_{\psi}^{\psi+\Delta\psi} \sin\varphi \, d\varphi \right| < |\Delta r| K +$$

$$+ 2r^{n-1}\pi^{n-2}[\cos\psi - \cos(\psi + \Delta\psi)] \leqq |\Delta r| K + 2r^{n-2}\pi^{n-2} \cos\psi |\Delta r| \leqq$$

$$\leqq |\Delta r|(K + 2R^{n-2}\pi^{n-2}),$$

and we conclude that $a(r)$ satisfies a uniform Lipschitz condition in this case too.

However we may dispose of even the restrictive hypothesis made above, because if $S(r+\Delta r)$ meets a face which $S(r)$ does not intersect, then the quantity $a(r+\Delta r)$, which is to be added or subtracted to obtain the difference between the areas of $S(r+\Delta r) \cap F$ and $S(r) \cap F$, is majorized using the inequalities (3) and (5) with $\psi = 0$.

The general case in which the simplex has no vertex in the centre of $S(r)$ reduces to the preceding one because such a simplex may be obtained from the sum of some disjoint simplices with a vertex in the centre of $S(r)$ by subtracting another sum of disjoint simplices with a vertex in the centre of $S(r)$.

Now as int F (the interior of F, i.e. the set of the inner points of F) and CF have a finite number of components, the sets $\{r;\ a(r) > 0\}$ and $\{r;\ a(r) < (n-1)\omega_{n-1} r^{n-1}\}$ each consists of a finite number of open intervals. Hence the set $\{r;\ 0 < a(r) < (n-1)\omega_{n-1} r^{n-1}\}$ is the union of open disjoint intervals I. Since $\sigma^\circ(r)$ is constant in each complementary interval and since both σ and σ° are continuous in r, it will be sufficient to establish (1) for the case where r_1 and r_2 belong to one of the intervals I.

Let I_0 denote the closed interval $r_1 \leqq r \leqq r_2$ and for each x with $|x| > 0$ let φ denote the angle between the radius to x and the negative part of the x^1-axis. The area of the cap $S(r) \cap F$ has the representation

$$a(r) = (n-1)\omega_{n-1} r^{n-1} \int_0^{\varphi} \sin^{n-2}\vartheta \, d\vartheta \tag{6}$$

for $r \in I_0$. Since $a(r)$ satisfies a uniform Lipschitz condition there, so too does $\varphi(r)$. Indeed, the recurrence formula

$$\int_0^{\varphi} \sin^{n-2} \vartheta \, d\vartheta = -\frac{1}{n-2} \cos \varphi \sin^{n-3} \varphi + \frac{n-3}{n-2} \int_0^{\varphi} \sin^{n-4} \vartheta \, d\vartheta$$

and (6) yield

$$\cos \varphi \sin^{n-3} \varphi = \frac{a_{n-3}(r)}{\omega_{n-3} r^{n-3}} - \frac{(n-2)a(r)}{(n-1)\omega_{n-1} r^{n-1}}, \tag{7}$$

where

$$a_{n-3}(r) = (n-3)\omega_{n-3} r^{n-3} \int_0^{\varphi} \sin^{n-4} \vartheta \, d\vartheta.$$

For $n = 3$, (7) reduces to

$$\cos \varphi = 1 - \frac{a_2(r)}{2\pi r^2},$$

[see F. Gehring [7], formula (14)] and for $n = 2$, $\varphi = \dfrac{a_1(r)}{2r}$. Since $a(r)$ and $a_{n-3}(r)$ satisfy a uniform Lipschitz condition, (7) implies the same for $\varphi(r)$.

Next, the area of an $(n-1)$-dimensional surface obtained by the rotation of a meridian curve C of the half plane x^1Ox^2, $x^2 > 0$, around the x^1-axis, has the representation

$$\sigma^{\circ} = (n-1)\omega_{n-1} \int_C (x^2)^{n-2} \, ds.$$

It is now easy to see that

$$\left| r \frac{d\varphi}{dr} \right| = \cot \psi^{\circ}, \tag{8}$$

where $\psi^0(r)$ denotes the positive acute angle between the radius to $x \in \partial F^0$ and the normal to ∂F^0 at x, whenever the latter exists. Clearly $\psi^{\circ}(r)$ exists a.e. in I_0, because $\varphi(r)$ in the preceding relation satisfies a uniform Lipschitz condition and then has a derivative a.e. in I_0 (see for example Natanson [1],p. 283, and corollary 1, p. 287). (8) and

$$ds^2 = dr^2 + r^2 \, d\varphi^2 = \left[1 + r^2 \left(\frac{d\varphi}{dr}\right)^2\right] dr^2,$$

where ds is the arc element of C, yield

$$\sigma^\circ(r_2) - \sigma^\circ(r_1) = (n-1)\omega_{n-1}\int_{r_2}^{r_1} r^{n-2}\sin^{n-2}\varphi \sqrt{\left| r\frac{d\varphi}{dr}\right|^2 + 1}\, dr =$$

$$= \int_{r_1}^{r_2} \Sigma^\circ \operatorname{cosec} \psi^\circ\, dr, \tag{9}$$

where

$$\Sigma^\circ = (n-1)\omega_{n-1} r^{n-2} \sin^{n-2}\varphi \tag{10}$$

denotes the area of the $(n-2)$-dimensional sphere $S(r) \cap \partial F^\circ$ of radius $r \sin \varphi$.

Next, for each $x \in \partial F$ with $|x| > 0$, let $\psi = \psi(x)$ be the corresponding angle between the radius to x and the normal to ∂F at x, whenever the latter exists, i.e. a.e. because F is a polyhedron. Clearly, for $|x| = r$, $\psi(x) = \psi(r)$ defined above. Then because ∂F is a polyhedral surface, it follows that $S(r) \cap \partial F$ consists of portions of $(n-2)$-dimensional spheres contained in $(n-1)$-dimensional faces of F. The radius of such an $(n-2)$-dimensional sphere is $\rho = r \sin \psi$. It is easy to see that

$$r \cos \psi(x) = \text{const},$$

on one $(n-1)$-dimensional face, hence

$$d\psi = \frac{dr \cot \psi}{r}. \tag{11}$$

Then from the preceding relation it follows that

$$d\rho^2 = dr^2 + r^2\, d\psi^2 = \frac{dr^2}{\sin^2 \psi},$$

whence

$$d\rho = \frac{dr}{\sin \psi}.$$

Thus the area of the polyhedral surface ∂F contained between $S(r_1)$ and $S(r_2)$ is

$$\sigma(r_2) - \sigma(r_1) = \int_{\partial F} d\sigma = \int_{\rho_1}^{\rho_2} \int_{S(r)\cap\partial F} d\rho\, d\sigma_{n-2} = \int_{r_1}^{r_2} \left(\int_{S(r)\cap\partial F} \operatorname{cosec}\psi\, d\sigma_{n-2} \right) dr, \tag{12}$$

where $d\sigma_{n-2}$ is the area element of the $(n-2)$-dimensional sphere corresponding to the angle ψ. For $r = \text{const.}$ this angle is constant on the same $(n-1)$-dimensional face of F, but is generally different from one face to the other.

Now fix r, $r + h \in I_0$ with $h > 0$ and let E be the central projection on $S(r)$ of that part of ∂F which lies in $r \leqq |x| \leqq r + \Delta r$. If $x \in S(r)$ and if the radius through x meets just one of the sets $S(r) \cap F$, $S(r + h) \cap F$, then $x \in E$. Letting $\alpha = \alpha(r, \Delta r)$ denote the area of E we thus obtain

$$\alpha \geqq \left| a(r) - \left(\frac{r}{r + \Delta r}\right)^{n-1} a(r + \Delta r) \right| = \alpha^\circ, \tag{13}$$

where α^0 denotes the $(n - 1)$-dimensional area of the projection on $S(r)$ of the part of ∂F^0 which lies between $S(r)$ and $S(r + \Delta r)$. Now suppose that $S(r)$ meets only one $(n-1)$-dimensional face of the polyhedron. Then by (2), we shall have the following central projection

$$\alpha = (n - 1)\omega_{n-1} r^{n-1} \int_{\psi}^{\psi + \Delta\psi} \sin^{n-2} \vartheta \, d\vartheta.$$

Taking the ratio $\frac{\alpha}{\Delta r}$, letting $r \to 0$, applying the formula of differentiation under the integral sign and taking into account (11), we obtain

$$\lim_{\Delta r \to 0} \frac{\alpha}{\Delta r} = \lim_{\Delta r \to 0} (n - 1)\omega_{n-1} r^{n-1} \frac{\int_0^{\psi + \Delta\psi} \sin^{n-2} \vartheta \, d\vartheta - \int_0^{\psi} \sin^{n-2} \vartheta \, d\vartheta}{\Delta r} = \tag{14}$$

$$= (n - 1)\omega_{n-1} r^{n-1} \sin^{n-2} \psi \frac{d\psi}{dr} = (n - 1)\omega_{n-1} r^{n-2} \cot \psi \sin^{n-2} \psi =$$

$$= \int_\theta^{2\pi} \int_0^{\pi} \cdots \int_0^{\pi} \cot \psi \rho^{n-2} \sin^{n-3} \vartheta_1 \cdots \sin \vartheta_{n-3} \, d\vartheta_1 \cdots d\vartheta_{n-2} = \int_{S(r) \cap \partial F} \cot \psi \, d\sigma_{n-2}.$$

In the general case, when $S(r)$ meets some edges of F, then, for every $(n - 1)$-dimensional face, $\frac{\alpha}{\Delta r}$ has the same representation as above. We only observe that instead of

$$\int_0^{2\pi} \int_0^{\pi} \cdots \int_0^{\pi} \cot \psi \rho^{n-2} \sin^{n-3} \vartheta_1 \cdots \sin \vartheta_{n-3} \, d\vartheta_1 \cdots d\vartheta_{n-2}$$

we shall have

$$\int_{\vartheta_1^0}^{\vartheta_1^1} \cdots \int_{\vartheta_{n-2}^0}^{\vartheta_{n-2}^1} \cot \psi \rho^{n-2} \sin^{n-3} \vartheta_1 \cdots \sin \vartheta_{n-3} \, d\vartheta_1 \cdots d\vartheta_{n-2}$$

and also same terms which approach zero as $\Delta r \to 0$.

Next from (6) it follows that

$$\lim_{r\to 0}\frac{1}{\Delta r}\left[\frac{a(r+\Delta r)}{(r+\Delta r)^{n-1}}-\frac{a(r)}{r^{n-1}}\right]=(n-1)\omega_{n-1}\sin^{n-2}\varphi\,\frac{d\varphi}{dr},$$

and from (13), (10) and (8), we obtain

$$\lim_{\Delta r\to 0}\frac{\alpha^\circ}{\Delta r}=\lim_{\Delta r\to 0}\frac{r^{n-1}}{\Delta r}\left|\frac{a(r)}{r^{n-1}}-\frac{a(r+\Delta r)}{(r+\Delta r)^{n-1}}\right|=\Sigma^\circ\left|r\,\frac{d\varphi}{dr}\right|=\Sigma^\circ\cot\psi^\circ.$$

(13) and (14) then imply that

$$\Sigma^\circ\cot\psi^\circ\leqq\int_{S(r)\cap\partial F}\cot\psi\,d\sigma_{n-2}. \tag{15}$$

Finally, for each $r\in I_0$, $S(r)\cap\partial F$ bounds $S(r)\cap F$, a set of area $a(r)$. Hence, applying the isoperimetric inequality for the $(n-1)$-dimensional spherical space (Chapter 1.3) on $S(r)$, we conclude that the $(n-2)$-dimensional area of $S(r)\cap\partial F$ is not less than the area of the $(n-2)$-dimensional boundary of the equivalent spherical cap $S(r)\cap F^0$, that is

$$\Sigma^\circ\leqslant\int_{S(r)\cap\partial F}d\sigma_{n-2}.$$

Hence, by (15) and applying Minkowski's inequality {see G. H. Hardy, J. E. Littlewood and G. Polya [1], (6.13.8), p. 148}, yields

$$\Sigma^\circ\operatorname{cosec}\psi^\circ=[(\Sigma^\circ\cot\psi^\circ)^2+\Sigma^{\circ 2}]^{\frac{1}{2}}\leqq\left[\left(\int_{S(r)\cap\partial F}\cot\psi\,d\sigma_{n-2}\right)^2+\right.$$

$$\left.+\left(\int_{S(r)\cap\partial F}d\sigma_{n-2}\right)^2\right]^{\frac{1}{2}}\leqq\int_{S(r)\cap\partial F}(\cot^2\psi+1)^{\frac{1}{2}}\,d\sigma_{n-2}=\int_{S(r)\cap\partial F}\operatorname{cosec}\psi\,d\sigma_{n-2}$$

a.e. in I_0, and integrating with respect to r we obtain, on account of (9) and (12), the inequality (1), as desired.

This lemma was proved for $n = 3$ by F. Gehring [6].

Spherical symmetrization of a function. The equivalence of the definition of the Θ-mappings to that of QCfH requires also the introduction (following F. Gehring [6]) of the definition of the spherically symmetrized functions.

Let $u(x)$ be everywhere continuous. We symmetrize u to obtain a new function as follows. For each a, let G_a and F_a be the open and closed sets where $u < a$

and $u \leqq a$, respectively, and let G_a^0 and F_a^0 denote the spherical symmetrizations of these sets. Then, given any point x, we see that $x \in F_a^0$ for sufficiently large a and we define

$$u^\circ(x) = \inf\{a;\ x \in F_a^\circ\}.$$

Properties of spherical symmetrization of functions.

LEMMA 2. *For each a, G_a^0 and F_a^0 are precisely the sets where $u^0 < a$ and $u^\circ \leqq a$, respectively.*

Indeed, since $G_a^0 \subset G_b^0$ and $F_a^0 \subset F_b^0$ for $a \leqq b$, we conclude that for every x_0 with $u^0(x_0) = b \leqq c$, we have $x_0 \in F_b^0 \subset F_c^0$. The converse is also true. Let $x_1 \in F_c^0$ and suppose that $u^0(x_1) = b > c$. But this contradicts the hypothesis that b is by definition the infimum of the numbers a such that $x_1 \in F_a^0$, because $x_1 \in F_c^0$ would imply $b \leqq c$. The same argument may be used to obtain the assertion on G_a^0.

LEMMA 3. *$u^0(x)$ is continuous everywhere.*

Indeed let $u^0(x_0) = a$ and let $\{x_m\}$ be a sequence of points such that $\lim\limits_{m\to\infty} x_m = x_0$. Let the sequence $\{x_m\}$ be decomposed into two subsequences $\{x_m'\}$ and $\{x_m''\}$ (where one of them may be a null set) such that $u^\circ(x_m') \leqq a$, $u^0(x_m'') > a'$. Then

$$\lim_{m\to\infty} u^\circ(x_m') = a,$$

for if this lower limit were equal to $a - \varepsilon$, $\varepsilon > 0$, we could find a subsequence $\{x_{m_k}'\}$ such that $u^0(x_{m_k}') \leqq a - \dfrac{\varepsilon}{2}$, which would imply $x_{m_k}' \in F^0_{a-\frac{\varepsilon}{2}}$ $(k = 1, 2, \ldots)$, where $d(x_0, F_{a-\frac{\varepsilon}{2}}) > 0$, contradicting thus that $\lim\limits_{k\to\infty} x_{m_k}' = x_0$.

By the same argument we show that $\lim\limits_{m\to\infty} u^0(x_m'') = a$.

LEMMA 4. *If u^0 is the spherical symmetrization of u and if*

$$|u(x_1) - u(x_2)| < K|x_1 - x_2| \tag{16}$$

for all points x_1, x_2, then

$$|u^\circ(y_1) - u^\circ(y_2)| < K|y_1 - y_2| \tag{17}$$

for all y_1, y_2.

Fix two points y_1, y_2 with $u^0(y_1) \leqq u^\circ(y_2)$ and let $a_1 = u^\circ(y_1)$ and $d = d(y_1, y_2)$. For (17) it is sufficient to prove that

$$u^\circ(y_2) < a_2 = a_1 + Kd.$$

Let $r_k = |y_k|$ $(k = 1, 2)$, let E_1 be the set of points on $S_1 = S(r_1)$, where $u(x) \leqq a_1$ and let E_2 be the points on $S_2 = S(r_2)$ whose distance from E_1 does not exceed d. Since $|r_2 - r_1| \leqq d$, E_2 is clearly non-empty. E_2 is closed and for each $x_2 \in E_2$ there exists an $x_1 \in E_1$ such that $|x_1 - x_2| \leqq d$. Thus, by (16),

$$u(x_2) \leqq u(x_1) + Kd \leqq a_2,$$

and $u(x) \leqq a_2$ at every point of E_2.

Next, let E_1^0, E_2^0 be the spherical symmetrizations of E_1 and E_2. Then E_1^0 is just the set of points on S_1 where $u^0 \leqq a_1$ while $u^0 \leqq a_2$ everywhere in E_2^0. Hence $y_1 \in E_1^0$ and to obtain (17) we only need to show that $y_2 \in E_2^0$. We do this by appealing to the Brunn—Minkowski inequality for spherical geometry (Chapter 1.3) in the following manner.

Let $\alpha_k r_k$ denote the radii of the closed spherical caps E_k^0 measured along the spherical surfaces S_k ($k = 1, 2$). If H is the central projection of E_1 on S_2 and if we choose α so that

$$0 \leqq \alpha \leqq \pi, \quad d^2 = r_1^2 + r_2^2 - 2r_1 r_2 \cos \alpha,$$

then E_2 is just the union of the closed spherical caps with centres in H and radii αr_2 measured along S_2. Since H has the same area as the spherical cap of radius $\alpha_1 r_2$ on S_2, it follows from the aforementioned inequality of Brunn—Minkowski, that

$$\alpha_2 r_2 \geqq r_2 \min(\alpha_1 + \alpha, \pi). \tag{18}$$

In fact, Brunn—Minkowski's inequality asserts that for α_1, $\alpha_2 \leqq \pi$, we have

$$\alpha_2 r_2 \geqq \alpha_1 r_2 + \alpha r_2. \tag{19}$$

But the spherical radius of a spherical cap is the length of an arc of a great circle joining the centre of the cap with one of its boundary points. Thus, for a sphere of radius r_2, a cap cannot have a spherical radius greater than πr_2, and is equal to πr_2 only if the cap is the whole sphere. And then (19) and $\alpha_2 \leqq \pi$, imply (18),which yields

$$\alpha_2 \geqq \min(\alpha_1 + \alpha, \pi).$$

Hence either $\alpha_2 - \alpha_1 \geqq \alpha$ or $\alpha_2 = \pi$. In both cases, taking into account that the spherical distance between y_2 and the centre of the spherical cap of S_2 is $\leqq (\alpha_1 + \alpha) r_2$, it follows that $y_2 \in E_2^0$ and the proof is complete.

Let u^0 be the spherical symmetrization of u, let D and D^0 be the sets where $a_1 < u < a_2$ and $a_1 < u^0 < a_2$, respectively, and let f and f^0 be a pair of functions related to u, u^0 as follows. For each $r \geqq 0$ and each $a_1 < a < a_2$, f and f^0 are constant and have the same value at the points of S_r where $u = a$ and $u^0 = a$, respectively.

And now we shall establish the equimeasurability of the functions f and f^0 by

LEMMA 5. *If f° is continuous in D°, then f is continuous in D and*

$$\int_D |f|^q \, d\tau = \int_{D^\circ} |f^\circ|^q \, d\tau \tag{20}$$

for all $q > 0$.

If f is not continuous in D, we can find an $x_0 \in D$ and a sequence $\{x_m\} \subset D$ converging to x_0 such that

$$\lim_{m \to \infty} f(x_m) = a \neq f(x_0). \tag{21}$$

Since u and u° assume exactly the same values on $S_r \cap D$ and $S_r \cap D^\circ$, respectively, we can find a sequence $\{y_m\} \subset D^\circ$ such that $|y_m| = |x_m|$ and $u^\circ(y_m) = u(x_m)$. Then a subsequence $\{y_{m_k}\}$ will converge to a point y_0 for which $|y_0| = |x_0|$ and $u^\circ(y_0) = u(x_0)$. Hence, $y_0 \in D^\circ$ and

$$f(x_0) = f^\circ(y_0) = \lim_{k \to \infty} f^\circ(y_{m_k}) = \lim_{k \to \infty} f(x_{m_k}) = a.$$

This contradicts (21) and we conclude that f is continuous in D.

We turn to the proof of (20). Fix b, let G and G° be the sets where $f < b$ and $f^\circ < b$, respectively, for each $r \geqq 0$ let E_r be the set of values assumed by u° on $S_r \cap G^\circ$. Then, since $S_r \cap G^\circ$ is open on S_r and u° is continuous, E_r is the countable union of disjoint (possibly degenerate) linear intervals I. Next, since u° is the spherical symmetrization of u, the sets of points of S_r for which $u \in I$ and $u^\circ \in I$ have equal area. Now $S_r \cap G$ and $S_r \cap G^\circ$ are just sets on S_r which assign to u and u°, respectively, values in E_r and, summing over the linear intervals I, we conclude that $S_r \cap G$ and $S_r \cap G^\circ$ have equal area.

Finally fix $b_1 < b_2$ and let G and G° be the sets where $b_1 \leqq f < b_2$ and $b_1 \leqq f^\circ < b_2$. From the above we see that $S_r \cap G$ and $S_r \cap G^\circ$ have equal areas and integration yields $mG = mG^\circ$. Thus f and f° are equimeasurable functions and (20) follows directly (see for example G. H. Hardy, J. E. Littlewood and G. Polya [1], p. 277).

Spherical symmetrization of rings. Let A be a ring. Then $A \cup C_0$ is open, C_0 is closed and we define the spherical symmetrization of A as

$$A^\circ = (A \cup C_0)^\circ - C^\circ.$$

It is easy to verify that A° is again a ring.

Properties of spherical symmetrization of rings.

LEMMA 6. mod $A \leqq$ mod A°.

We want to prove that mod $A \leqq$ mod A° or alternatively, that

$$\text{Cap}\, A^\circ \leqq \text{Cap}\, A. \tag{22}$$

To establish (22) let u be one of the simple admissible functions for A and let u° be the spherical symmetrization of u. Since $\nabla u = 0$ in all but a finite set of the simplices T where it is linear with respect to each variable, it follows that $u(x)$ satisfies a uniform Lipschitz condition. Hence, by lemma 4, the same is true of u°. Now, since $u(x) = 0$ on C_0 and $u(x) = 1$ on C_1, it follows that C_0 contains those points where $u(x) \leqq 0$, while $A \cup C_0$ contains those points where $u(x) < 1$. This together with the fact that $0 \leqq u^\circ \leqq 1$ implies that u° is zero on C_0° and 1 on C_1°. In particular we conclude that u° is admissible for the ring A°.

The remainder of the argument is devoted to showing that

$$\int_{A^\circ} |\nabla u^\circ|^n \, d\tau \leqq \int_A |\nabla u|^n \, d\tau, \tag{23}$$

for with (23) we obtain

$$\operatorname{Cap} A^\circ \leqq \int_A |\nabla u|^n \, d\tau,$$

and taking the infimum over all simple admissible functions u yields (22) as desired.

Now let $0 = a_1 < a_2 < \ldots < a_N = 1$ be the finite set of values assumed by u on the vertices of the simplices T and let G_k and G_k° be the sets where $a_k < u < a_{k+1}$ and $a_k < u^\circ < a_{k+1}$, respectively. If the set where $u^\circ = a_k$ has positive measure, then almost all of its points are points of linear density in the directions of the coordinate axes. Since ∇u will vanish at almost all such density points, we conclude that the integral of $|\nabla u^\circ|^n$ over this set will vanish. Hence, to establish (23), it suffices to show that

$$\int_{G_k^\circ} |\nabla u^\circ|^n \, d\tau \leqq \int_{G_k} |\nabla u|^n \, d\tau \qquad (k = 1, \ldots, N). \tag{24}$$

Fix such a k, let $f^\circ \geqq 0$ be defined on G_k° and let f° be continuous and symmetric in the x^1-axis. That is, the value which f° assumes at x depends only on $|x|$ and φ, the angle the radius to x makes with the negative half of the x^1-axis. Next, for each $a_k < a < a_{k+1}$, let F, F° denote the sets where $u \leqq a$, $u^\circ \leqq a$ and σ, σ° the sets where $u = a$, $u^\circ = a$. Then, as observed by the definition of the simple admissible functions (chapter 5), F is a closed polyhedron and since $a \neq a_k$ ($k = 1, \ldots, N$), it follows (as it was observed on that occasion) that $\sigma = \partial F$. Hence $\sigma^\circ \subset F^\circ$ and $\partial F^\circ \subset \sigma^\circ$. In order to prove that conversely $\sigma^\circ \subset \partial F^\circ$, it is sufficient to show that every $x_0 \in \sigma^\circ$, with $x_0 \bar{\in} \partial F^\circ$, is either an inner point of σ°, or every sufficiently small spherical neighbourhood of x_0 contains inner points of σ°; but this would imply $u^\circ = \text{const.}$ on a set of strictly positive measure lying inside a simplex, so contradicting the hypothesis that u is simple admissible and then a linear mapping, which is not constant because $a_1 < \ldots < a_N$ and $a \neq a_k$ ($k = 1, \ldots, N$). Thus, $S_r \cap \sigma^\circ$, where $S_r = S(r)$, is an $(n-2)$-dimensional sphere with the centre on the x^1-axis, and then $|x| = \text{const.}$, $\varphi = \text{const.}$ on S_r, implying $f^\circ = \text{const.}$ on $S_r \cap \sigma^\circ$. And now define a second function f on G_k by requiring that f take on $S_r \cap \sigma$ the same value as f° on the corresponding intersection $S_r \cap \sigma^\circ$.

The functions f and f° defined in this way are just those of the preceding lemma, so that, on account of this lemma, the continuity of f° on G_k° implies the continuity of f on G_k. We shall further show that

$$\int_{G_k^\circ} f^\circ |\nabla u^\circ| \, d\tau \leqq \int_{G_k} f |\nabla u| \, d\tau. \tag{25}$$

First, let $\sigma = \sigma(a, r)$ and $\sigma^\circ = \sigma^\circ(a, r)$ denote the area of the part of σ and σ° contained in $B(r)$. Then, by lemma 1,

$$\sigma^\circ(a, r_2) - \sigma^\circ(a, r_1) \leqq \sigma(a, r_2) - \sigma(a, r_1)$$

for $0 \leqq r_1 < r_2$ and, since f and f° are equal on the corresponding intersections $S_r \cap \sigma$ and $S_r \cap \sigma^\circ$, we obtain

$$\int_{\Sigma^\circ} f^\circ d\sigma^\circ \leqq \int_{\Sigma} f \, d\sigma.$$

Now this inequality holds for $a \in (a_k, a_{k+1})$ and, since u, u° satisfy the uniform Lipschitz condition, we can apply Federer—Young's theorem (Chapter 1.3) to conclude that

$$\int_{G_k^\circ} f^\circ |\nabla u^\circ| \, d\tau = \int_{a_k}^{a_{k+1}} \left(\int_{\Sigma^\circ} f^\circ d\sigma^\circ \right) da \leqq \int_{a_k}^{a_{k+1}} \left(\int_{\Sigma} f \, d\sigma \right) da = \int_{G_k} f |\nabla u| \, d\tau,$$

which establishes the inequality (25).

Now the function $|\nabla u^\circ|^{n-1}$ is bounded, measurable and symmetric in the x^1-axis. Hence, appealing also to a theorem of Fréchet (see for instance P. Natanson [1], theorem 3, p. 144), we can find a sequence of functions $\{f_m^\circ\}$, $f_m^\circ \geqq 0$, which are continuous in G_k°, symmetric in the x^1-axis and which converge boundedly to $|\nabla u^\circ|^{n-1}$ a.e. in D°. Let $\{f_m\}$ be the corresponding sequence of functions defined on G_k as above. Then Lebesgue's theorem on the limit under the integral sign (see for instance P. Natanson [1], p. 169) and (25) yield

$$\int_{G_k^\circ} |\nabla u^\circ|^n \, d\tau = \lim_{m \to \infty} \int_{G_k^\circ} f_m^\circ |\nabla u^\circ| \, d\tau \leqq \underline{\lim}_{m \to \infty} \int_{G_k} f_m |\nabla u| \, d\tau. \tag{26}$$

Applying Hölder's inequality and (20) of the preceding lemma we obtain

$$\int_{G_k} f_m \, |\nabla u| \, d\tau \leqq \left(\int_{G_k} f_m^{\frac{n}{n-1}} \, d\tau \right)^{\frac{n-1}{n}} \left(\int_{G_k} |\nabla u|^n \, d\tau \right)^{\frac{1}{n}} =$$

$$= \left[\int_{G_k^\circ} (f_m^\circ)^{\frac{n}{n-1}} \, d\tau \right]^{\frac{n-1}{n}} \left(\int_{G_k} |\nabla u|^n \, d\tau \right)^{\frac{1}{n}},$$

and hence, since $f_m^\circ \to |\nabla u^\circ|^{n-1}$, we conclude that

$$\lim_{m\to\infty} \int_{G_k} f_m |\nabla u| \, d\tau \leqq \left(\int_{G_k^\circ} |\nabla u^\circ|^n \, d\tau\right)^{\frac{n-1}{n}} \left(\int_{G_k} |\nabla u|^n \, d\tau\right)^{\frac{1}{n}}.$$

Hence and by (26) we obtain (24) which implies (22) and the proof for the lemma 6 is complete.

The Grötzsch ring. For each $a > 1$, the *Grötzsch ring* $A_G = A_G(a)$ is the ring whose complementary components consist of the ball B and the ray $a \leqq x^1 < \infty$, $x^2 = \ldots = x^n = 0$.

The Teichmüller ring. Similarly for each $b > 0$, the *Teichmüller ring* $A_T = A_T(b)$ is the ring bounded by the segment $-1 \leqq x^1 \leqq 0$, $x^2 = \ldots = x^n = 0$ and the ray $b \leqq x^1 < \infty$, $x^2 = \ldots = x^n = 0$.

Properties of the moduli of these rings. Following O. Teichmüller [2] we set

$$\text{mod}\, A_G = \log \Phi(a), \ \text{mod}\, A_T = \log \Psi(b). \tag{27}$$

LEMMA 7. $\dfrac{\Phi(a)}{a}$ *is non-decreasing in* $(1, \infty)$ *and*

$$\Psi(b) = \Phi[(b+1)^{\frac{1}{2}}]^2 \tag{28}$$

for $b > 0$.

For the first part fix $1 < a < b$, let $A = A_G(b)$ and let A' and A'' be two rings into which A is split by $S\left(\dfrac{b}{a}\right)$. Then lemma 5.7 and (2.7) yield

$$\log \Phi(b) = \text{mod}\, A \geqq \text{mod}\, A' + \text{mod}\, A'' = \log \frac{b}{a} + \log \Phi(a),$$

whence

$$\frac{\Phi(b)}{b} \geqq \frac{\Phi(a)}{a}$$

as desired.

For the second part fix $b > 0$, set $a = \sqrt{b+1}$ and let A be the ring bounded by the segment $0 \leqq x^1 \leqq \dfrac{1}{a}$, $x^2 = \cdots = x^n = 0$ and by the ray $a \leqq x^1 < \infty$, $x^2 = \cdots = x^n = 0$. Next let A' and A'' be the part of A contained in B and $C\overline{B}$, respectively. The rings A' and A'' may be obtained one from the other by the inversion $x^* = \dfrac{x}{|x|^2}$ which is conformal (i.e. an 1-QCfH according to whatsoever definition) (see Chapter 8), hence, by Gehring's geometric definition,

$$\operatorname{mod} A' = \operatorname{mod} A''.$$

Then, lemma 5.7, taking also into account that $A_T(b)$ is obtained from A by the similarity $y = ax - e_1$, yields

$$\log \Psi(b) = \operatorname{mod} A \geqq 2 \operatorname{mod} A' = 2 \log \Phi(a).$$

Hence to complete the proof of (28) it is sufficient to show that

$$\operatorname{mod} A \leqq 2 \operatorname{mod} A',$$

or that

$$\operatorname{cap} A' \leqq 2^{n-1} \operatorname{cap} A. \tag{29}$$

Let $u \in C^1$ be an admissible function for A and let $w = u + v$, where

$$v = v(x) = 1 - u\left(\frac{x}{|x|^2}\right).$$

Then w is admissible for A' and, since

$$|\nabla w(x)| = \frac{\left|\nabla w\left(\dfrac{x}{|x|^2}\right)\right|}{|x|^2},$$

and the Jacobian of the inversion is $J = \dfrac{1}{|x|^{2n}}$, we obtain

$$\int_{A'} |\nabla w|^n \, d\tau = \int_{A''} |\nabla w|^n \, d\tau = \frac{1}{2} \int_A |\nabla w|^n \, d\tau. \tag{30}$$

Minkowski's inequality (1.73), taking into account that the ring A is preserved by the inversion $y = \dfrac{x}{|x|^2}$, yields

$$\left(\int_A |\nabla w|^n \, d\tau\right)^{\frac{1}{n}} \leqq \left(\int_A |\nabla u|^n \, d\tau\right)^{\frac{1}{n}} + \left(\int_A |\nabla v|^n \, d\tau\right)^{\frac{1}{n}} = 2\left(\int_A |\nabla u|^n \, d\tau\right)^{\frac{1}{n}},$$

and we conclude from (30) that

$$\operatorname{cap} A' \leqq 2^{n-1} \int_A |\nabla u|^n \, d\tau.$$

Taking the infimum over all admissible u gives (29) and the proof is complete.

LEMMA 8. *Let A be the ring $1 < |x| < b$, let $x^* = f(x)$ be a topological mapping of A onto a second ring A^* and let $f \in C^1$ with $J(x) \neq 0$ in D. Then*

$$\operatorname{mod} A^* \leqq \operatorname{mod} A + \int_1^b [D(r) - 1] \frac{dr}{r}, \tag{31}$$

where for $1 < r < b$

$$D(r) = \max_{|x|=r} \left[\frac{\Lambda_f(x)^n}{|J(x)|} \right]^{\frac{1}{n-1}}. \tag{32}$$

Let $v = v(y)$, $v \in C^1$, be an admissible function for A^* and let $u(x) = v[f(x)]$. Then integrating along a fixed radius and taking into account (6.15) we obain

$$1 \leqq \int_1^b |\nabla u| \, dr \leqq \int_1^b |\nabla v| \Lambda_f \, dr \leqq \int_1^b |\nabla v| \, D(r)^{\frac{n-1}{n}} \, J^{\frac{1}{n}} \, dr,$$

and with Hölder's inequality, we obtain

$$1 \leqq \left(\int_1^b |\nabla v|^n J r^{n-1} \, dr \right) \left[\int_1^b D(r) \frac{dr}{r} \right]^{n-1}.$$

Since this holds for all radii, we have

$$n\omega_n \left[\int_1^b D(r) \frac{dr}{r} \right]^{1-n} \leqq \int_A |\nabla v|^n |J| \, d\tau = \int_{A^*} |\nabla v|^n \, d\tau,$$

and taking the infimum over all such functions v gives

$$n\omega_n \left(\int_1^b D(r) \frac{dr}{r} \right)^{1-n} \leqq \operatorname{cap} A^*.$$

Hence, and by (2.1), (2.3),

$$\operatorname{mod} A^* \leqq \int_1^b D(r) \frac{dr}{r} = \operatorname{mod} A + \int_1^b [D(r) - 1] \frac{dr}{r}$$

and the proof is complete.

LEMMA 9. *$\Phi(a)$ satisfies the inequality*

$$a \leqq \Phi(a) \leqq \lambda a \tag{33}$$

where $1 < a < \infty$ and λ is a finite constant.

The ring $1 < |x| < a$ separates the boundary components of $A_G(a)$. Hence, and by (2.3) and lemma 5.7, mod $A_G \geqq \log a$ and by (27),

$$\Phi(a) \geqq a.$$

Next $\dfrac{\Phi(a)}{a}$ is nondecreasing by lemma 7 and approaches a limit λ as $a \to \infty$. We will show that $\lambda < \infty$. This then gives the upper bound

$$\Phi(a) \leqq \lambda a.$$

Let $A_E = A_E(a)$ denote the ring bounded by the segment $-1 \leqq x^1 \leqq 1$, $x^2 = \dots$ $\dots = x^n = 0$ and by the ellipsoid

$$\frac{(x^1)^2}{a^2+1} + \frac{(x^2)^2}{a^2} + \dots + \frac{(x^n)^2}{a^2} = 1.$$

Next for $a > 4$ let A' and A'' denote the rings bounded by the above segment and by the spherical surfaces with centres at $(-1, 0, \dots, 0)$ and radii $a-2$ and $a+2$, respectively. Then A' separates the boundary components of A_E while A_E separates those of A''. Hence

$$\text{mod } A' \leqq \text{mod } A_E \leqq \text{mod } A'', \tag{34}$$

where

$$\text{mod } A' = \log \Phi\left(\frac{a}{2} - 1\right),$$

since the inversion

$$y = \frac{(a-2)(x+e_1)}{|x+e_1|^2},$$

carries A' into $A_G\left(\dfrac{a}{2} - 1\right)$, and where

$$\text{mod } A'' = \log \Phi\left(\frac{a}{2} + 1\right),$$

since the inversion

$$y = \frac{(a+2)(x+e_1)}{|x+e_1|^2}$$

carries A'' into $A_G\left(\frac{a}{2}+1\right)$. Hence, by (34) and taking into account that $\lambda = \lim\limits_{a\to\infty} \frac{\Phi(a)}{a}$, we conclude that

$$\log\lambda = \lim_{a\to\infty} \log \frac{\Phi\left(\frac{a}{2}-1\right)}{\frac{a}{2}-1} \leq \lim_{a\to\infty}\left[\operatorname{mod} A_E - \log\left(\frac{a}{2}-1\right)\right] =$$

$$= \lim_{a\to\infty}\left[\operatorname{mod} A_E - \log\frac{a}{2}\right] = \lim_{a\to\infty}\left[\operatorname{mod} A_E -\right.$$

$$\left. - \log\left(\frac{a}{2}+1\right)\right] \leq \lim_{a\to\infty} \log \frac{\Phi\left(\frac{a}{2}+1\right)}{\frac{a}{2}+1} = \log\lambda,$$

and then

$$\log\lambda = \lim_{a\to\infty}\left(\operatorname{mod} A_E - \log\frac{a}{2}\right). \tag{35}$$

Hence the problem is reduced to considering the asymptotic behaviour of $\operatorname{mod} A_E$ as $a \to \infty$.

In the case $n = 2$ we know that, when $b = a + \sqrt{a^2+1}$, the transformation

$$y^1 + iy^2 = \frac{1}{2}\left(x^1 + ix^2 + \frac{1}{x^1+ix^2}\right) \quad (i^2 = -1) \tag{36}$$

maps the plane ring $1 < |x| < b$ conformally onto the ring bounded by the segment $-1 \leq y^1 \leq 1$, $y^2 = 0$ and by the ellipse

$$\frac{(y^1)^2}{a^2+1} + \frac{(y^2)^2}{a^2} = 1. \tag{37}$$

We thus obtain the modulus of the plane analogue of the ring A_E.

The situation is more complicated in n-space ($n > 2$). Here (as it will be proved in Chapter 16) a topological mapping preserves the moduli of rings if and only if it is conformal and the only such mappings are the Möbius transformations. Hence for no number b can we map the ring $A: 1 < |x| < b$ conformally onto the ring A_E bounded by $-1 \leqq y^1 \leqq 1$, $y^2 = \ldots = y^n = 0$ and by the ellipsoid

$$\frac{(y^1)^2}{a^2+1} + \frac{(y^2)^2}{a^1} + \cdots + \frac{(y^n)^2}{a^2} = 1.$$

On the other hand, when a is large, there will exist a number b and mappings $f(x)$ of A onto A_E which are nearly conformal for large $|x|$. The lemma yields an upper bound for mod A_E in terms of mod A for such mappings $f(x)$.

Introduce first polar coordinates $\rho, \alpha_1, \ldots, \alpha_{n-2}$ and $\rho^*, \alpha_1^*, \ldots, \alpha_{n-2}^*$ in the planes $x^1 = 0$ and $x^{1*} = 0$, respectively. Next let $b = a + \sqrt{a^2+1}$, let A be the ring $1 < |x| < b$ and define the mapping $x^* = f(x)$ as follows

$$x^{*1} + i\rho^* = \frac{1}{2}\left(x^1 + i\rho + \frac{1}{x^1 + i\rho}\right) (i^2 = -1),\ \alpha_k^* = a_k\,(k = 1, \ldots, n-2). \tag{38}$$

Since (36) maps the plane ring $1 < |x| < b$ conformally onto the ring bounded by the segment $-1 \leqq y^1 \leqq 1$, $y^2 = 0$ and by the ellipse (37), then $f(x)$ maps A onto A_E. If we introduce again the Cartesian coordinates by

$$\begin{cases} x^1 = x^1, \\ x^2 = \rho \sin\alpha_1 \cdots \sin\alpha_{n-2}, \\ \cdots\cdots\cdots\cdots \\ x^k = \rho \sin\alpha_1 \cdots \sin\alpha_{n-k} \cos\alpha_{n-k+1}, \\ \cdots\cdots\cdots\cdots \\ x^n = \rho\cos\alpha_1, \end{cases}$$

(38) becomes

$$\begin{cases} x^{*1} = \dfrac{x^1}{2}\left(1 + \dfrac{1}{r^2}\right), \\ x^{*2} = \dfrac{\rho}{2}\left(1 - \dfrac{1}{r^2}\right) \sin\alpha_1 \cdots \sin\alpha_{n-2}, \\ \cdots\cdots\cdots\cdots \\ x^{*k} = \dfrac{\rho}{2}\left(1 - \dfrac{1}{r^2}\right) \sin\alpha_1 \cdots \sin\alpha_{n-k}\cos\alpha_{n-k+1}, \\ \cdots\cdots\cdots\cdots \\ x^{*n} = \dfrac{\rho}{2}\left(1 - \dfrac{1}{r^2}\right) \cos\alpha_1, \end{cases}$$

where $r = |x|$. It is a simple technical matter to verify that

$$\Lambda_f(x_0) = \frac{1}{2} \overline{\lim_{x \to x_0}} \left\{ 1 + \frac{1}{r^2 r_0^2} - \frac{1}{r^2} - \frac{1}{r_0^2} + \right.$$

$$\left. + \frac{(r^2 + r_0^2 - 2x^1 x_0^1)^2 - 4\rho^2 \rho_0^2}{r^2 r_0^2[(x^1 - x_0^1)^2 + \rho^2 + \rho_0^2 - 2\rho\rho_0 \cos(\alpha - \alpha_0)]} \right\}^{\frac{1}{2}}, \tag{39}$$

where $r_0 = |x_0|$ and

$$\cos(\alpha - \alpha_0) = \sin\alpha_1 \cdots \sin\alpha_{n-2} \sin\alpha_1^0 \cdots \sin\alpha_{n-2}^0 + \cdots +$$

$$+ \sin\alpha_1 \cdots \sin\alpha_{n-k} \cos\alpha_{n-k+1} \sin\alpha_1^0 \cdots \sin\alpha_{n-k}^0 \cos\alpha_{n-k+1}^0 + \cdots + \cos\alpha_1 \cos\alpha_1^0.$$

Since

$$|x - x_0|^2 = (x^1 - x_0^1)^2 + \rho^2 + \rho_0^2 - 2\rho\rho_0 \cos(\alpha - \alpha_0) =$$

$$= r^2 + r_0^2 - 2[x^1 x_0^1 + \rho\rho_0 \cos(\alpha - \alpha_0)] \geqq 0$$

for every α, α_0, then for $\alpha = \alpha_0$,

$$(r^2 + r_0^2 - 2x^1 x_0^1) - 2\rho\rho_0 \geqq 0$$

and for $\alpha - \alpha_0 = \pi$,

$$(r^2 + r_0^2 - 2x^1 x_0^1) + 2\rho\rho_0 \geqq 0,$$

hence, by multiplying them,

$$(r^2 + r_0^2 - 2x^1 x_0^1)^2 - 4\rho^2 \rho_0^2 \geqq 0,$$

which allows us to conclude that we have $\cos(\alpha - \alpha_0) = 1$ in (39), i.e. $x \to x_0$ in the direction given by α_0. Then (39) yields

$$\Lambda_f(x_0) = \frac{1}{2} \overline{\lim_{x \to x_0}} \sqrt{1 + \frac{1}{r^2 r_0^2} - \frac{1}{r^2} - \frac{1}{r_0^2} + \frac{r^2 + r_0^2 - 2(x^1 x_0^1 - \rho\rho_0)}{r^2 r_0^2}} =$$

$$= \frac{1}{2r_0^2} \sqrt{r_0^4 + 1 - 2[(x_0^1)^2 - \rho_0^2]}$$

and since $(x_0^1)^2 + \rho_0^2 = r_0^2$, $\Lambda_f(x_0)$ may be written also as follows

$$\Lambda_f(x_0) = \frac{1}{2r_0^2}\sqrt{(r_0^2+1)^2 - 4(x_0^1)^2} \tag{40}$$

or

$$\Lambda_f(x_0) = \frac{1}{2r_0^2}\sqrt{(r_0^2-1) + 4\rho_0^2}. \tag{41}$$

Then (32), on account of (40), (41) and

$$J(x) = \frac{(r^2-1)^{n-2}[(r^2-1)^2+4\rho]}{2^n r^{2n}} \tag{42}$$

yields

$$D(r)^{n-1} = \max_{|x|=r}\frac{\Lambda_f(x)^n}{|J(x)|} = \max_{|x|=r}\frac{[(r^2+1)^2-4(x^1)^2]^{\frac{n-2}{2}}[(r^2-1)^2+4\rho^2]}{(r^2-1)^{n-2}[(r^2-1)+4\rho^2]} =$$

$$= \max_{|x|=r}\frac{[(r^2+1)^2-4(x^1)^2]^{\frac{n-2}{2}}}{(r^2-1)^{n-2}} = \left(\frac{r^2+1}{r^2-1}\right)^{n-2}.$$

Hence, and by (31) and (2.3), we obtain

$$\operatorname{mod} A_E \leqq \operatorname{mod} A + \int_1^\infty\left[\left(\frac{r^2+1}{r^2-1}\right)^{\frac{n-2}{n-1}} - 1\right]\frac{dr}{r} = \log \lambda' b. \tag{43}$$

In order to calculate λ', we make in the last integral the following change of the variable of integration:

$$t = \left(\frac{r^2+1}{r^2-1}\right)^{\frac{1}{n-1}},$$

which transforms it into

$$(n-1)\int_1^\infty \frac{[t^{2(n-2)} - t^{n-2}]\,dt}{t^{2(n-1)}-1}. \tag{44}$$

Since mod $A = \log b$ and

$$2(n-1)\int \frac{x^{m-1}dx}{x^{2(n-1)}-1} = \log(x-1) + (-1)^m \log(x+1) +$$

$$+\sum_{k=1}^{n-2}\left[\cos\frac{km\pi}{n-1}\log\left(x^2 - 2x\cos\frac{k\pi}{n-1}+1\right) - \right.$$

$$\left. -2\sin\frac{km\pi}{n-1}\arctan\frac{x-\cos\dfrac{k\pi}{n-1}}{\sin\dfrac{k\pi}{n-1}}\right] + C,$$

(see E. Brahy [1], p. 71), then from (44) and (43) we deduce for even n

$$\log\lambda' = \sum_{k=1}^{\frac{n-1}{2}}\left\{\left[\cos\frac{k\pi}{n-1} - (-1)^k\right]\log\cot\frac{k\pi}{(2n-1)} + \frac{\pi}{2}\sin\frac{k\pi}{n-1}\right\} < \infty, \tag{45}$$

and for odd n

$$\log\lambda' = \log\frac{1}{\sin\dfrac{\pi}{2(n-1)}} + \frac{\pi}{4} - \sum_{k=1}^{\frac{n-3}{2}}\left[\log\frac{\sin\dfrac{(2k+1)\pi}{2(n-1)}}{\sin\dfrac{k\pi}{n-1}} + \right. \tag{46}$$

$$\left. + \cos\frac{k\pi}{n-1}\log\tan\frac{k\pi}{2(n-1)} - \frac{\pi}{2}\sin\frac{k\pi}{n-1}\right] < \infty.$$

In both cases $0 < \lambda' < \infty$, since for $n \geq 4$

$$0 < \tan\frac{\pi}{2(n-1)} \leqq \tan\frac{k\pi}{2(n-1)}, \quad \cot\frac{k\pi}{2(n-1)} \leqq \cot\frac{\pi}{2(n-1)} < \infty,$$

$$\frac{(n-2)\pi}{4(n-1)} < \frac{\pi}{2} \quad \left(k = 1, \ldots, \frac{n-2}{2}\right),$$

and for $n \geqq 3$

$$\sin\frac{k\pi}{n-1},\quad \sin\frac{(2k+1)\pi}{2(n-1)} \geqq \sin\frac{\pi}{n-1} > 0 \qquad \left(k = 1,\dots,\frac{n-3}{2}\right),$$

$$0 < \tan\frac{\pi}{2(n-1)} \leqq \tan\frac{k\pi}{2(n-1)} \leqq \tan\frac{(n-3)\pi}{4(n-1)} < \infty \left(k = 1,\dots,\frac{n-3}{2}\right).$$

For $n = 3$, $\lambda' = \sqrt{2}\, e^{\frac{\pi}{4}}$ and for $n = 4$, $\lambda' = 3^{\frac{3}{4}} e^{\frac{\pi\sqrt{3}}{4}}$.

Finally (35) and (43), taking into account that $b = a + \sqrt{a^2 + 1}$, yield

$$\log \lambda = \lim_{a\to\infty}\left(\operatorname{mod} A_E - \log\frac{a}{2}\right) \leqq \lim_{n\to\infty} \log\frac{2\lambda' b}{a} = \log 4\lambda',$$

and we obtain

$$\lambda \leqq 4\lambda' < \infty, \tag{47}$$

where λ' is given by (45) and (46).

For $n = 3$, this lemma is proved by F. Gehring [6].

Thus we obtain an upper bound for mod A_E which depends on b and n.

And now let us get an upper bound for the modulus of an arbitrary ring.

LEMMA 10. *Let A be a ring and let $x_0 \in C_0$, and $0 < a \leqq b$. If C_0 and C_1 contain points which lie at distances a and b from x_0, then*

$$\operatorname{mod} A \leqq \log \Psi\left(\frac{b}{a}\right) \leqq \log \lambda^2\left(\frac{b}{a} + 1\right), \tag{48}$$

where λ is the constant in lemma 9.

By performing a translation we may assume that $x_0 = O$. Next let A° be the spherical symmetrization of A. C_0° will contain the segment $-a \leqq x^1 \leqq 0$, $x^2 = \cdots = x^n = 0$ while C_1° will contain the ray $b \leqq x^1 < \infty$, $x^2 = \cdots = x^n = 0$. Hence A° separates the boundary components of the ring bounded by the above segment and ray. We conclude from lemmas 5.7 and 6 that

$$\operatorname{mod} A \leqq \operatorname{mod} A^\circ \leqq \operatorname{mod} A_T\left(\frac{b}{a}\right) = \log \Psi\left(\frac{b}{a}\right),$$

and (48) follows from (28) and (33).

Lemmas 7 and 9 allow us to conclude that $\Phi(a)$ is a strictly increasing function, continuous in $(1, \infty)$ and for which

$$\lim_{a\to\infty} \frac{\Phi(a)}{a} = \lambda < \infty \tag{49}$$

and (28) hold.

Next for $1 \leqq K < \infty$ and $0 < t < 1$ we define

$$\Theta_K(t) = \left\{ \Psi^{-1}\left[\Phi\left(\frac{1}{t}\right)^{\frac{1}{K}}\right]\right\}^{-1}, \tag{50}$$

where Ψ^{-1} is the inverse function for Ψ. Then $\Theta_K(t)$ is strictly increasing and continuous in $0 < t < 1$, and then, by (28) and (33),

$$\Psi^{-1}\left[\Phi\left(\frac{1}{t}\right)^{\frac{1}{K}}\right] \leqq \Psi^{-1}\left[\Phi\left(\frac{1}{t}\right)^{2}\right] = \frac{1}{t^2} - 1,$$

hence

$$\lim_{t\to 1} \Theta_K(t) = \infty.$$

On the other hand, (28) yields

$$\lim_{b\to\infty} \frac{\Psi(b)}{b} = \lim_{b\to\infty} \left\{\left[\frac{\Phi(\sqrt{b+1})}{\sqrt{b+1}}\right]^2 \frac{b+1}{b}\right\} = \lambda^2,$$

hence

$$\lim_{c\to\infty} \frac{\Psi^{-1}(c)}{c} = \frac{1}{\lambda^2},$$

and then

$$\lim_{t\to 0} \frac{\Theta_K(t)}{t^{\frac{1}{K}}} = \frac{\Phi\left(\frac{1}{t}\right)^{\frac{1}{K}}}{\Psi^{-1}\left[\Phi\left(\frac{1}{t}\right)^{\frac{1}{K}}\right]} \left[\frac{\frac{1}{t}}{\Phi\left(\frac{1}{t}\right)}\right]^{\frac{1}{K}} = \lambda^{2-\frac{1}{K}}. \tag{51}$$

Loewner's lemma [1]. Cap $A > 0$ *if and only if neither component degenerates into a point.*

If at least one of the boundary components is degenerated into a point, then cap $A = 0$ by lemma 2.5.

Suppose now that none of the boundary components consists only of one point. Then we can evidently fix a finite point $O \in R^n$ such that two concentric spheres with centre O and radii r_1, r_2 $(0 < r_1 < r_2)$ intersect the two components C_0 and C_1 of CA. Because C_0 and C_1 are two continua, every sphere $S(r)$ with a radius $r \in (r_1, r_2)$ will intersect both C_0 and C_1. Set $A_r = A \cap S(r)$. Then, by Fubini's theorem,

$$\int_A |\nabla u|^n \, d\tau = \int_{r_1}^{r_2} \left(\int_{A_r} |\nabla u|^n \, d\sigma \right) dr. \tag{52}$$

In order to estimate $\int_{A_r} |Vu|^n d\sigma$ let us assume first that $S(r)$ meets C_0 and C_1 in two diametrally opposed points, which we make the north and south poles of a spherical coordinate system on $S(r)$. Each half meridian of $S(r)$ contains an arc $\gamma \in A_r$ in whose end points $u(x)$ has the values 0 and 1, respectively. Let $\vartheta_1, \ldots$ $\ldots, \vartheta_{n-1}$ be the spherical coordinates on $S(r)$. Then the meridians are given by $\vartheta_2 = \ldots = \vartheta_{n-1} = \text{const.}$ Calling ϑ_1^0, ϑ_1^1 the values of ϑ_1 at the end points and $\nabla' u$ the gradient of u on $S(r)$, we have by lemma 2.3

$$\left| \frac{\partial u}{\partial \vartheta_1} \right| = \left| \frac{du}{ds} \right| \left| \frac{ds}{d\vartheta_1} \right| \leqq |\nabla' u| r,$$

where $s = r\vartheta_1$ is the arc-length on the considered meridian, hence

$$\int_{\vartheta_1^0}^{\theta_1^1} |\nabla' u|^n \sin^{n-2} \vartheta_1 \, d\vartheta_1 \geqq \frac{1}{r^n} \int_{\vartheta_1^0}^{\vartheta_1^1} \left| \frac{\partial u}{\partial \vartheta_1} \right|^n \sin^{n-2} \vartheta_1 \, d\vartheta_1 \cdot \tag{53}$$

Use of Hölder's inequality gives further

$$1 \leqq \int_{\vartheta_1^0}^{\vartheta_1^1} \left| \frac{\partial u}{\partial \vartheta_1} \right| d\vartheta_1 \leqq \left(\int_{\vartheta_1^0}^{\vartheta_1^1} \left| \frac{\partial u}{\partial \vartheta_1} \right|^n \sin^{n-2} \vartheta_1 \, d\vartheta_1 \right)^{\frac{1}{n}} \left(\int_{\vartheta_1^0}^{\vartheta_1^1} \frac{d\vartheta_1}{\sin^{\frac{n-2}{n-1}} \vartheta_1} \right)^{\frac{n-1}{n}},$$

and therefore

$$\int_{\vartheta_1^0}^{\vartheta_1^1} \left| \frac{\partial u}{\partial \vartheta_1} \right|^n \sin^{n-2} \vartheta_1 \, d\vartheta_1 \geqq k = \frac{1}{\left(\int_0^{\pi} \frac{d\vartheta_1}{\sin^{\frac{n-2}{n-1}} \vartheta_1} \right)^{n-1}} > 0, \tag{54}$$

since

$$\lim_{x \to 0} x^{\frac{n-2}{n-1}} \cdot \frac{1}{\sin^{\frac{n-2}{n-1}} x} = 1, \quad \lim_{x \to \pi} (\pi - x)^{\frac{n-2}{n-1}} \cdot \frac{1}{\sin^{\frac{n-2}{n-1}} (\pi - x)} = 1,$$

whence (see V. I. Smirnov [1], p. 256):

$$\int_0^{\pi} \frac{d\vartheta_1}{\sin^{\frac{n-2}{n-1}} \vartheta_1} < \infty.$$

Combination of (53) and (54) and subsequent integration with respect to $\vartheta_2, \ldots$ $\ldots, \vartheta_{n-1}$ gives finally

$$\int_{A_r} \left|\nabla' u^n\right| d\sigma \geqq \frac{(n-1)\,\omega_{n-1} k}{r}. \tag{55}$$

Suppose now $S(r)$ meets C_0, C_1 in two points x_0, x_1 with the spherical distance (the length of an arc of a great circle) equal to $r\omega (0 < \omega \leqq \pi)$. We perform a conformal mapping $\tilde{\vartheta} = \tilde{\vartheta}(\vartheta)$ of $S(r)$ onto itself, so that these points are mapped into two diametrally opposed points and call the function into which $u(\vartheta_1, \ldots, \vartheta_{n-1})$ is transformed $\tilde{u}(\tilde{\vartheta}_1, \ldots, \tilde{\vartheta}_{n-1})$. The relation

$$u_p = \sum_{q=1}^{n-1} \tilde{u}_q \tilde{\vartheta}_{q,p} \qquad (p = 1, \ldots, n-1),$$

where $u_p = \dfrac{\partial u}{\partial \vartheta_p}$, $\tilde{u}_q = \dfrac{\partial \tilde{u}}{\partial \tilde{\vartheta}_q}$, $\tilde{\vartheta}_{q,p} = \dfrac{\partial \tilde{\vartheta}_q}{\partial \vartheta_p}$, taking into account the formula of the gradient in spherical coordinates, yields

$$|\nabla' u|^2 = \frac{1}{r^2}\left(u_1^2 + \frac{u_2^2}{\sin^2 \vartheta_1} + \cdots + \frac{u_{n-1}^2}{\sin^2 \vartheta_1 \cdots \sin^2 \vartheta_{n-2}}\right) =$$

$$= \left[\frac{1}{r^2} \sum_{k=1}^{n-1} \tilde{u}_k^2 \left(\tilde{\vartheta}_{k,1}^2 + \frac{\tilde{\vartheta}_{k,2}^2}{\sin^2 \vartheta_1} + \cdots + \frac{\tilde{\vartheta}_{k,n-1}^2}{\sin^2 \vartheta_1 \cdots \sin^2 \vartheta_{n-2}}\right) + \right.$$

$$\left. + 2 \sum_{\substack{p,q=1 \\ p \neq q}}^{n-1} \tilde{u}_p \tilde{u}_q \left(\vartheta_{p,1} \vartheta_{q,1} + \frac{\tilde{\vartheta}_{p,2} \tilde{\vartheta}_{q,2}}{\sin^2 \vartheta_1} + \cdots + \frac{\tilde{\vartheta}_{p,n-1} \tilde{\vartheta}_{q,n-1}}{\sin^2 \vartheta_1 \cdots \sin^2 \vartheta_{n-2}}\right)\right] = \tag{56}$$

$$= \frac{1}{r^2}\left(\tilde{\vartheta}_{1,1}^2 + \cdots + \frac{\tilde{\vartheta}_{1,n-1}^2}{\sin^2 \vartheta_1 \cdots \sin^2 \vartheta_{n-2}}\right) \sum_{k=1}^{n-1} \frac{\tilde{u}_k^2 \left(\tilde{\vartheta}_{k,1}^2 + \cdots + \dfrac{\tilde{\vartheta}_{k,n-1}^2}{\sin^2 \vartheta_1 \cdots \sin^2 \vartheta_{n-2}}\right)}{\tilde{\vartheta}_{1,1}^2 + \cdots + \dfrac{\tilde{\vartheta}_{1,n-1}^2}{\sin^2 \vartheta_1 \cdots \sin^2 \vartheta_{n-2}}} +$$

$$+ \frac{2}{r^2} \sum_{\substack{p,q=1 \\ p \neq q}}^{n-1} \tilde{u}_p \tilde{u}_q \left(\tilde{\vartheta}_{p,1} \tilde{\vartheta}_{q,1} + \cdots + \frac{\tilde{\vartheta}_{p,n-1} \tilde{\vartheta}_{q,n-1}}{\sin^2 \vartheta_1 \cdots \sin^2 \vartheta_{n-2}}\right),$$

Since $\widetilde{\vartheta}(\vartheta)$ is a conformal mapping, (8.3) holds, i.e. in our notations

$$\widetilde{\vartheta}_{p,1}\widetilde{\vartheta}_{q,1} + \cdots + \frac{\widetilde{\vartheta}_{p,n-1}\widetilde{\vartheta}_{q,n-1}}{\sin^2 \vartheta_1 \cdots \sin^2 \vartheta_{n-2}} = 0 \quad (p \neq q;\ q, p = 1,\ldots,n-1),$$

$$\frac{\widetilde{\theta}^2_{k,1} + \cdots + \dfrac{\widetilde{\theta}^2_{k,n-1}}{\sin^2 \vartheta_1 \cdots \sin^2 \vartheta_{n-2}}}{\widetilde{\theta}^2_{1,1} + \cdots + \dfrac{\widetilde{\theta}^2_{1,n-1}}{\sin^2 \vartheta_1 \cdots \sin^2 \vartheta_{n-2}}} = \frac{1}{\sin^2 \widetilde{\theta}_1 \cdots \sin^2 \widetilde{\theta}_{k-1}}, \quad (k = 2, \ldots, n-1)$$

and then (56) becomes of the form

$$|\nabla' u|^2 = \left(\widetilde{\theta}^2_{1,1} + \cdots + \frac{\widetilde{\vartheta}^2_{1,n-1}}{\sin^2 \vartheta_1 \cdots \sin^2 \vartheta_{n-2}}\right) |\nabla' u_1|^2. \tag{57}$$

where

$$\widetilde{\vartheta}^2_{1,1} + \cdots + \frac{\widetilde{\vartheta}^2_{1,n-1}}{\sin^2 \vartheta_1 \cdots \sin^2 \vartheta_{n-2}} = \mu^2 \tag{58}$$

is the square of the factor of proportionality of the two linear elements $d\widetilde{s}$ and ds obtained from one another by the conformal mapping $\widetilde{\vartheta} = \widetilde{\vartheta}(\vartheta)$ from above (as it follows by corollary of theorem 8.1). Hence

$$d^{\sim} = \mu^{n-1}\, d\sigma, \tag{59}$$

where $d\widetilde{\sigma}$ and $d\sigma$ are the two elements of $(n-1)$-dimensional area which correspond to one another by $\widetilde{\vartheta} = \widetilde{\vartheta}(\vartheta)$. Next, rising (57) to $\dfrac{n-1}{2}$, multiplying by $d\sigma$ and taking into account (58) and (59), we get

$$|\nabla' u|^{n-1}\, d\sigma = |\nabla' u_1|^{n-1}\, d\widetilde{\sigma},$$

which means that $|\nabla' u|^{n-1}$ is invariant under a conformal mapping of the sphere onto itself. Hence

$$|\nabla u|^n d\sigma \geqq |\nabla' u|^n\, d\sigma = |\nabla' u_1|^n \left|\frac{d\widetilde{\sigma}}{d\sigma}\right|^{\frac{1}{n-1}} d\widetilde{\sigma}, \tag{60}$$

holds in every A-point of $S(r)$, i.e. a.e. in $S(r)$ for a. e. $r \in (r_1, r_2)$.

For the sake of simplicity, we suppose that one of the two points, say x_0, is held fix and the other, x_1, is mapped into a point diametrally opposed to x_0. Choose the origin of the arc γ in x_0, i.e. set $\vartheta_1^0 = 0$. Next, consider the mapping

$$\begin{cases} \widetilde{\vartheta}_1 = 2\arctan \dfrac{\tan\dfrac{\omega}{2}\tan\dfrac{\vartheta_1}{2}}{\left(\tan^2\dfrac{\vartheta_1}{2}+\tan^2\dfrac{\omega}{2}-2\tan\dfrac{\omega}{2}\tan\dfrac{\vartheta_1}{2}\cos\vartheta_2\right)^{\frac{1}{2}}}, \\ \widetilde{\vartheta}_2=\arctan\dfrac{\tan\dfrac{\omega}{2}\sin\vartheta_2}{\tan\dfrac{\vartheta_1}{2}-\tan\dfrac{\omega}{2}\cos\vartheta_2}, \\ \widetilde{\vartheta}_k = \vartheta_k \quad (k = 3, \dots, n-1) \end{cases}$$

and show that it is a conformal mapping of $S(r)$ onto itself. It is sufficient for this to prove that the linear elements are proportional. Clearly

$$\begin{aligned} d\widetilde{s}^2 &= r^2(d\widetilde{\vartheta}_1^2 + \sin^2\widetilde{\vartheta}_1 d\widetilde{\vartheta}_2^2 + \sin^2\widetilde{\vartheta}_1\sin^2\widetilde{\vartheta}_2 d\widetilde{\vartheta}_3 + \cdots + \sin^2\widetilde{\vartheta}_1 \cdots \sin^2\widetilde{\vartheta}_{n-2} d\widetilde{\vartheta}_{n-1}^2) = \\ &= r^2[d\widetilde{\vartheta}_1^2 + \sin^2\widetilde{\vartheta}_1 d\widetilde{\vartheta}_2^2 + \sin^2\widetilde{\vartheta}_1\sin^2\widetilde{\vartheta}_2 (d\vartheta_3^2 + \cdots + \sin^2\vartheta_3 \cdots \sin^2\vartheta_{n-2} d\vartheta_{n-1}^2)] = \\ &= r^2(d\widetilde{\vartheta}_1^2 + \sin^2\widetilde{\vartheta}_1 d\widetilde{\vartheta}_2^2 + \sin^2\widetilde{\vartheta}_1\sin^2\widetilde{\vartheta}_2 d\alpha^2) = r^2[(\widetilde{\vartheta}_{1,1}^2 + \sin^2\widetilde{\vartheta}_1\widetilde{\vartheta}_{2,1}^2)d\vartheta_1^2 + (\widetilde{\vartheta}_{1,2}^2 + \\ &+ \sin^2\widetilde{\vartheta}_1\widetilde{\vartheta}_{2,2}^2)d\vartheta_2^2 + \sin^2\widetilde{\vartheta}_1\sin^2\widetilde{\vartheta}_2 d\alpha^2] = r^2(\widetilde{\vartheta}_{1,1}^2 + \sin^2\widetilde{\vartheta}_1\widetilde{\vartheta}_{2,1}^2)\Bigg(d\vartheta_1^2 + \\ &+ \frac{\widetilde{\vartheta}_{1,2}^2 + \sin^2\widetilde{\vartheta}_1\,\widetilde{\vartheta}_{2,2}^2}{\widetilde{\vartheta}_{1,1}^2 + \sin^2\widetilde{\vartheta}_1\,\widetilde{\vartheta}_{2,1}^2} d\vartheta_2^2 + \frac{\sin^2\widetilde{\vartheta}_1\sin^2\widetilde{\vartheta}_2}{\widetilde{\vartheta}_{1,1}^2 + \sin^2\widetilde{\vartheta}_1\,\widetilde{\vartheta}_{2,1}^2} d\alpha^2\Bigg) \end{aligned}$$

where

$$d\alpha^2 = d\vartheta_3^2 + \cdots + \sin^2\vartheta_3 \cdots \sin^2\vartheta_{n-2} d\vartheta_{n-1}^2,$$

is the part of ds^2 which is unaffected by the mapping. It is a simple technical matter to verify that

$$\frac{\widetilde{\vartheta}_{1,2}^2 + \sin^2\widetilde{\vartheta}_1\widetilde{\vartheta}_{2,2}^2}{\widetilde{\vartheta}_{1,1}^2 + \sin^2\widetilde{\vartheta}_1\widetilde{\vartheta}_{2,1}^2} = \sin^2\vartheta_1, \quad \frac{\sin^2\widetilde{\vartheta}_1\sin^2\widetilde{\vartheta}_2}{\widetilde{\vartheta}_{1,1}^2 + \sin^2\widetilde{\vartheta}_1\widetilde{\vartheta}_{2,1}^2} = \sin^2\vartheta_1\sin^2\vartheta_2,$$

and therefore

$$\begin{aligned} d\widetilde{s}^2 &= r^2(\widetilde{\vartheta}_{1,1}^2 + \sin^2\widetilde{\vartheta}_1\widetilde{\vartheta}_{2,1}^2)(d\vartheta_1^2 + \sin^2\vartheta_1 d\vartheta_2^2 + \sin^2\vartheta_1\sin^2\vartheta_2 d\vartheta_3^2 + \\ &+ \cdots + \sin^2\vartheta_1 \cdots \sin^2\vartheta_{n-2} d\vartheta_{n-1}^2) = K^2 ds^2, \end{aligned}$$

.e. the linear elements are proportional in every point of $S(r)$. (8.3) yields

$$\tilde{\vartheta}_{1,1}^2 + \sin^2\tilde{\vartheta}_1\tilde{\vartheta}_{2,1}^2 = \mu^2,$$

and in this case

$$\mu^2 = \tilde{\vartheta}_{1,1}^2 + \sin^2\tilde{\vartheta}_1\tilde{\vartheta}_{2,1}^2 = \frac{\tan^4\frac{\omega}{2}\left(1+\tan^2\frac{\vartheta_1}{2}\right)^2}{\left[\left(1+\tan^2\frac{\omega}{2}\right)\tan^2\frac{\vartheta_1}{2}+\tan^2\frac{\omega}{2}-2\tan\frac{\omega}{2}\tan\frac{\vartheta_1}{2}\cos\vartheta_2\right]^2}.$$

We remark now that $\omega > 0$ since $d(C_0, C_1) > 0$, hence the minimum of μ, obtained for $\vartheta_2=\pi$, and $\tan\vartheta_1= -2\tan\frac{\omega}{2}$, is a number $\beta^2 > 0$. Then (59) implies

$$\left(\frac{d\tilde{\sigma}}{d\sigma}\right)^{\frac{1}{n-1}} = \mu \geqq \beta^2 > 0,$$

and (60) yields $\quad |\nabla u|^n d\sigma \geqq |\nabla' u_1|^n \beta^2 d\tilde{\sigma} > 0.$

Using the inequality (55) for u_1 we obtain

$$\int_{A_r} |\nabla u|^n d\sigma \geqq \beta^2 \int_{A_r} |\nabla' u_1|^n d\tilde{\sigma} \geqq \frac{(n-1)\omega_{n-1}k\beta^2}{r} > 0,$$

or, setting $\beta_0 = \min\limits_{r_1\leqq r\leqq r_2} \beta$,

$$\int_{A_r} |\nabla u|^n d\sigma \geqq \frac{(n-1)\omega_{n-1}k\beta_0^2}{r} > 0$$

and (52) gives finally

$$\int_A |\nabla u|^n d\sigma = \int_{r_1}^{r_2}\left(\int_{A_r} |\nabla u|^n d\sigma\right)dr \geqq \log\frac{r_2}{r_1}(n-1)\,\omega_{n-1}k\beta_0^2 > 0$$

where $\log\frac{r_2}{r_1}(n-1)\omega_{n-1}K\beta_0^2$ is a constant which depends only on the geometrical structure of A and not on u. Hence, it represents a lower bound for cap A.

Loewner's theorem [1], *A* QCfH f: $R^n \to R^n$ *is a homeomorphism* f: $R^n \rightleftarrows R^n$.

Let $f(x)$ be a K-QCfH according to Gehring's geometric definition. Then

$$\operatorname{mod} A^* \geqq \frac{1}{K} \operatorname{mod} A, \tag{61}$$

hence

$$\operatorname{cap} A \geqq \frac{\operatorname{cap} A^*}{K^{n-1}}.$$

Let us apply this inequality to $f(x)$ by choosing for the ring A the exterior of a sphere S. Then one boundary component is the point ∞ and hence, by the preceding lemma, $\operatorname{cap} A = 0$. But then, we conclude from the preceding inequality that also $\operatorname{cap} A^* = 0$ and A^* must, according to the preceding lemma, also have a degenerate component (i.e. which reduce to a point). But $x^* = f(x)$ maps S onto an $(n-1)$-dimensional Jordan surface, the ball bounded by S onto a Jordan domain bounded by $f(S)$, and the ring A onto a ring $A^* \subset R^n$ (since the topological image of a ring is again a ring), with the boundary components $f(S)$ and the point ∞, because if the degenerated component were a finite point, then ∂A^* would have three components, since $f(x)$ is a homeomorphism and then $f(x) < \infty$ whenever $x < \infty$, hence $\infty = f(\infty)$.

Remark. This theorem generalizes the corresponding results from the theory of functions of a complex variable concerning the invariance of the type (hyperbolic, parabolic) of a Riemann surface under a QCfH (see S. Kakutani [1], O. Teichmüller [1]). Later on, it was established also for $n = 3$ by B. V. Šabat (see Ju. G. Rešetnjak and B. V. Šabat [1], p. 676) on the basis of another of his results ([3], theorem 2).

Equivalence of the definition of Θ-mappings to the definitions of QCfH.

THEOREM 1. *Let $x^* = f(x)$ be a K-QCfH according to Gehring's metric definition in a domain D. If $\partial D \neq \emptyset$, then $\partial D^* \neq \emptyset$ and*

$$\frac{|x_0^* - y_0^*|}{d(x_0^*, \partial D^*)} \leqq \Theta_K\left[\frac{|x_0 - y_0|}{d(x_0, \partial D)}\right] \tag{62}$$

for all $x_0, y_0 \in D$ with $|x_0 - y_0| < d(x_0, \partial D)$.

By the preceding theorem, $D^* = R^n$ if and only if $D = R^n$, hence $\partial D^* = \emptyset$ if and only if $\partial D = \emptyset$.

Now fix $x_0, y_0 \in D$, so that $|x_0 - y_0| < d(x_0, \partial D)$, let $|x_0 - y_0| < a < d(x_0, \partial D)$ and let A be the ring bounded by the segment $\overline{x_0 y_0}$ and by the sphere $S(x_0, a)$. Then, clearly, $\bar{A} \subset D$ and

$$\operatorname{mod} A = \log \Phi\left(\frac{a}{|x_0 - y_0|}\right), \tag{63}$$

since Gröztsch's ring corresponding to $\Phi\left(\frac{a}{|x_0 - y_0|}\right)$ is obtained from A by the inversion

$$\tilde{x} = \frac{a(x - x_0)}{|x - x_0|^2},$$

which is a conformal mapping, and hence conserves the moduli of rings. Evidently, $x_0^* = f(x_0)$ and $y_0^* = f(y_0)$ are contained in the component C_0^* of $\complement A^*$, and $F_1^* = \partial C_1^*$ contains a point at a distance

$$a^* < d(x_0^*, \partial D^*) \tag{64}$$

from x_0^*. Hence lemma 10 yields

$$\text{mod}^* A \leqq \log \Psi\left(\frac{a^*}{|x_0^* - y_0^*|}\right). \tag{65}$$

Then, because $f(x)$ is K-QCfH according to Gehring's metric definition, it is K-QCfH also according to his geometric definition by theorem 3.1 and then satisfies (61), which, on account of (63) and (65), implies

$$\Psi\left(\frac{a^*}{|x_0^* - y_0^*|}\right) \geqq \Phi\left(\frac{a}{|x_0 - y_0|}\right)^{\frac{1}{K}}.$$

Thus, by (64) and (50) and since $\Psi(t)$ is increasing, we have

$$\frac{|x_0^* - y_0^*|}{d(x_0^*, \partial D^*)} < \frac{|x_0^* - y_0^*|}{a^*} = \frac{1}{\Psi^{-1}\left[\Psi\left(\frac{a^*}{|x_0^* - y_0^*|}\right)\right]} \leqq$$

$$\leqq \frac{1}{\Psi^{-1}\left[\Phi\left(\frac{a}{|x_0 - y_0|}\right)^{\frac{1}{K}}\right]} = \Theta_K\left(\frac{|x_0 - y_0|}{a}\right)$$

and letting $a \to d(x_0, \partial D)$ and taking into account that x_0, y_0 are arbitrary points of D with $|x_0 - y_0| < d(x_0, \partial D)$ yields the inequality (62).

Remarks. 1. $\Theta_K(t)$ is a strictly increasing function which depends only on t and K and not on the considered K-QCfH or on the domain D. Moreover $\Theta_K(O) = 0$.

2. An analogous theorem was established by J. Väisälä [1] for K-QCfH (according to his definitions) of the unit ball onto itself.

THEOREM 2. *A* Θ*-mapping is a* K-QCfH *according to Gehring's metric definition, where* $K = \dfrac{1}{\Theta^{-1}(1)}$.

Since the inverse mapping of a K-QCfH according to Gehring's metric definition is again a K-QCfH according to the same definition, it follows that it will be sufficient to show that $f(x)$ is a K-QCfH according to Gehring's metric definition under the assumption that the inverse mapping $f^{-1}(x^*)$ is a Θ-mapping. For this let $x = f^{-1}(x^*)$ be a Θ-mapping, fix a point $x \in D$ and choose a, $0 < a < d(x_0, \partial D)$, so that

$$|f(x) - f(x_0)| < d[f(x_0), \partial D^*] \tag{66}$$

whenever $|x - x_0| < a$. Next for each $r \in (0, a)$, choose x_1, x_2 so that $|x_1 - x_0| = |x_2 - x_0| = r$ and so that

$$L(x_0, r) = |f(x_1) - f(x_0)|, \; l(x_0, r) = |f(x_2) - f(x_0)|, \tag{67}$$

where $L(x_0, r)$, $l(x_0, r)$ are given by (1.5) and (1.6).

Now suppose that

$$l(x_0, r) < L(x_0, r), \tag{68}$$

let $\Delta = D - x_1$ and let $\Delta^* = f(\Delta)$. Then $\Delta^* = D^* - f(x_1)$, and we see from (66), (67) and (68) that

$$|f(x_2) - f(x_0)| = l(x_0, r) < L(x_0, r) = d[f(x_0), \partial\Delta^*].$$

Since $f^{-1}(x^*)$ was assumed to be a Θ-mapping, we have

$$1 = \frac{|f^{-1}[f(x_2)] - f^{-1}[f(x_0)]|}{d\{f^{-1}[f(x_0)], \partial\Delta\}} \leqq \Theta\left\{\frac{|f(x_2) - f(x_0)|}{d[f(x_0), \partial\Delta^*]}\right\} = \Theta\left[\frac{l(x_0, r)}{L(x_0, r)}\right],$$

and hence it follows that

$$\frac{L(x_0, r)}{l(x_0, r)} \leqq \frac{1}{\Theta^{-1}(1)} \tag{69}$$

and taking into account that $\Theta(t)$ is a strictly increasing function with $\Theta(O) = 0$ (as it follows from the definition), we conclude that $0 < \Theta(t) < \infty$ whenever $0 < t < 1$, whence

$$0 < \frac{1}{\Theta^{-1}(T)} < 1 \text{ for } 0 < T < \infty,$$

and then

$$\frac{1}{\Theta^{-1}(1)} = K, \; 1 < K < \infty,$$

hence, by (69), we have

$$\frac{L(x_0, r)}{l(x_0, r)} \leqq K. \tag{70}$$

If (68) does not hold, $l(x_0, r) = L(x_0, r)$, and so in either case we obtain (70) for $0 < r < a$. Thus

$$\delta_L(x_0) = \overline{\lim_{r\to 0}} \frac{L(x, r)}{l(x_0, r)} \leqq K$$

for each $x_0 \in D$ and $f(x)$ is K-QCfH according to Gehring's metric definition.

COROLLARY. *The definition of Θ-mappings is equivalent to each of the definitions of* QCfH.

The equivalence of Θ-mappings to QCfH according to Gehring's metric definition (without specifying K) is a consequence of the theorems 1 and 2.

Remark. In spite of the last two theorems, we cannot assert that the definition of the Θ-mappings is equivalent to Gehring's metric definition of the K-QCfH, since although a K-QCfH according to Gehring's metric definition is a Θ_K-mapping, i.e. a particular Θ-mapping defined by (50), however (50), (28) and (33) yield

$$\frac{2^K}{\lambda} \leqq [\Theta_K^{-1}(1)]^{-1} \leqq 2^K \lambda^{2K},$$

so that, setting $[\Theta_K^{-1}(1)]^{-1} = K'$, it follows that the Θ_K-mappings are K'-QCfH according to Gehring's metric definition, where, for a sufficiently large K, we have $K < K'$.

Generalization of Koebe's theorem. And now, in order to obtain a ε lower bound for the constant λ in lemma 9, we give, following F. Gehring [5], a generalization for K-QCfH of an important result ("Koebe Viertelsatz") on bidimensional conformal mappings, which was established for bidimensional QCfH by A. Pfluger [3].

THEOREM 3. *Let* $x^* = f(x)$ *be a* K-QCfH *in* D *according to Gehring's geometric definition and let* $\partial D \neq \emptyset$. *Then*

$$\overline{\lim_{r\to 0}} \frac{L(x, r)}{r^{\frac{1}{K}}} \leqq \lambda^{2-\frac{1}{K}} \frac{d(x^*, \partial D^*)}{d(x, \partial D)^{\frac{1}{K}}} \tag{71}$$

and

$$\overline{\lim_{r\to 0}} \frac{l(x, r)}{r^{\frac{1}{K}}} \leqq \lambda \frac{d(x^*, \partial D^*)}{d(x, \partial D)^{\frac{1}{K}}} \tag{72}$$

for each $x \in D$, *where* $L(x, r)$ *and* $l(x, r)$ *are given by* (1.5) *and* (1.6), *respectively, and where* λ *is the constant in lemma* 9.

For the first part fix $0 < r < d(x, \partial D)$. Then theorem 1 implies that

$$\frac{L(x, r)}{r^{\frac{1}{K}}} \leqq \frac{d(x^*, \partial D^*)}{d(x, \partial D)^{\frac{1}{K}}} \cdot \frac{\Theta_K(t)}{t^{\frac{1}{K}}}, \qquad t = \frac{r}{d(x, \partial D)}$$

and we obtain (71) from (51) by letting $r \to 0$.

For the second part fix $0 < r < a < d(x_0, \partial D)$ and let A be the ring $r < |x - x_0| < a$. Then, by (2.3),

$$\operatorname{mod} A = \log \frac{a}{r}. \tag{73}$$

Consider next the ring A^*. C_0^* contains the ball

$$|x^* - x_0^*| \leqq r^* = l(x_0, r), \tag{74}$$

while F_1^* contains a point which lies at distance

$$a^* < d(x_0^*, \partial D^*) \tag{75}$$

from x_0^*. By performing a translation we may assume that x_0^* is the origin. Let $A^{*\circ}$ be the spherical symmetrization of A^*. C_0^* contains the ball $B(r^*)$ and C_1^* contains the ray $a^* \leqq x^{*_1} < \infty$, $x^{*_2} = \dots = x^{*_n} = 0$. Hence, the ring $A^{*\circ}$ separates the boundary components of the ring bounded by the sphere $S(r^*)$ and the preceding ray, which is homothetic to Grötzsch's ring $A_G = A_G\left(\frac{a^*}{r^*}\right)$. Thus, by lemmas 6, 5.7 and 9, and since a homothety preserves the modulus of a ring, it follows that

$$\operatorname{mod} A^* \leqq \operatorname{mod} A^{*\circ} \leqq \operatorname{mod} A_G\left(\frac{a^*}{r^*}\right) = \log \Phi\left(\frac{a^*}{r^*}\right) \leqq \log\left(\lambda \frac{a^*}{r^*}\right). \tag{76}$$

The modulus condition (61), on account of (73) and (76), yields

$$\frac{1}{K}\log\frac{a}{r} \leqq \log\left(\lambda\,\frac{a^*}{r^*}\right),$$

whence, by (74) and (75), we obtain

$$\frac{l(x_0, r)}{r^{\frac{1}{K}}} \leqq \lambda\,\frac{d(x_0^*, \partial D)}{a^{\frac{1}{K}}}$$

and we get (72) from first letting $r \to 0$ and then $a \to d(x_0, \partial D)$.

Remark. This theorem allows us to find a lower bound for λ. Indeed, fix $x_0 = (1, 0, \ldots, 0)$, $D = B(x_0, 1)$ and let D^* be the half space $x^* > 2$. Then the inversion $x^* = \dfrac{4x}{|x|^2}$ maps D conformally onto D^*, hence, since

$$\Lambda_f(x_0) = \overline{\lim_{r\to 0}}\,\frac{\max\limits_{|x-x_0|=r}\left|\dfrac{4x}{|x|^2} - 4e_1\right|}{r} = 4\,\overline{\lim_{r\to 0}}\,\frac{\left|\dfrac{e_1}{1-r} - e_1\right|}{r} = 4\,\overline{\lim_{r\to 0}}\,\frac{1}{1-r} = 4,$$

and on account of the inequality (71) of the preceding theorem, where $K = 1$, we obtain

$$4 = \Lambda_f(x_0) = \overline{\lim_{r\to 0}}\,\frac{L(x_0, r)}{r} \leqq \lambda\,\frac{d(x_0^*, \partial D^*)}{d(x_0, \partial D)} = 2\lambda,$$

because $d(x_0, \partial D) = 1$ and $d(x_0^*, \partial D^*) = 2$, and then $\lambda \geqq 2$.

CHAPTER 10

Agard – Gehring's definition of *K*-QCfH

In this chapter we generalize for arbitrary n the bidimensional characterization of K-QCfH by means of an angle distortion given by S. B. Agard and F. Gehring [1].

We say (following Agard and Gehring [1]) that *two arcs* γ_1 *and* γ_2 *form a topological angle at a point* x_0 if both γ_1 and γ_2 have x_0 as an end point and if x_0 is the only point γ_1 and γ_2 have in common. We then define *the inner measure* $A(\gamma_1, \gamma_2)$ *of this topological angle* as follows:

$$A(\gamma_1, \gamma_2) = \varliminf_{x_1, x_2 \to x_0} 2 \arcsin \frac{|x_1 - x_2|}{|x_1 - x_0| + |x_2 - x_0|}, \quad x_k \in \gamma_k \quad (k = 1, 2).$$

We see that $0 \leqq A(\gamma_1, \gamma_2) \leqq \pi$, that $A(\gamma_1, \gamma_2)$ does not depend upon the behaviour of γ_1 and γ_2 outside a neighbourhood of x_0 and that

$$A[f(\gamma_1), f(\gamma_2)] = A(\gamma_1, \gamma_2),$$

when $f(x)$ is a similarity mapping or a reflexion. To see how this inner measure is related to the usual unsigned measure of an angle, given two distinct points $x_1, x_2 \neq x_0$, let $\vartheta = \vartheta(x_1, x_0, x_2)$ denote the radian measure of the angle at x_0 in the triangle whose vertices are x_1, x_0, x_2. Then by the law of cosines

$$\left(\frac{|x_1 - x_2|}{|x_1 - x_0| + |x_2 - x_0|}\right)^2 = 1 - \frac{4|x_1 - x_0|\,|x_2 - x_0|}{(|x_1 - x_0| + |x_2 - x_0|)^2} \cos^2 \frac{\vartheta}{2},$$

and we obtain

$$\varliminf_{x_1, x_2 \to x_0} \vartheta \leqq A(\gamma_1, \gamma_2) \leqq \varliminf_{\substack{x_1, x_2 \to x_0 \\ |x_1 - x_0| = |x_2 - x_0|}} \vartheta,$$

hence

$$A(\gamma_1, \gamma_2) = \varliminf_{x_1, x_2 \to x_0} \vartheta(x_1, x_2, x_0),$$

provided that γ_1 or γ_2 has a tangent at x_0. In particular, if both γ_1 and γ_2 have tangents λ_1 and λ_2 at x_0, then $A(\gamma_1, \gamma_2)$ gives the radian measure of the smaller of the two angles determined by λ_1 and λ_2 at x_0.

Remark. Clearly

$$\overline{\lim_{x_1, x_2 \to x_0}} \; 2 \arcsin \frac{|x_1 - x_2|}{|x_1 - x_0| + |x_2 - x_0|} = \pi, x_k \in \gamma_k \quad (k = 1, 2).$$

This allows us to conclude the impossibility of the definition of an outer measure of a topological angle in the same way starting from $\overline{\lim}$.

We are able to give now, following S. Agard and F. Gehring [1], the definition of K-QCfH by angle distortion:

A homeomorphism $f(x)$ is said to be K-QCfH ($1 \leqq K < \infty$) in D, if

(i) for a fixed angle $\alpha_0 \left(0 < \alpha_0 < \dfrac{\pi}{2}\right)$ and for each $x \in D - E$, where E is a set of Σ-finite area, there exists at least an angle α ($0 < \alpha \leqq \alpha_0$) so that for all pairs γ_1, γ_2, consisting of a segment γ_1 and an arc γ_2 which form a topological angle of inner measure α at x,

$$A[f(\gamma_1), f(\gamma_2)] > 0;$$

(ii) for almost all $x_0 \in D$ and for all pairs γ_1, γ_2, consisting of a segment γ_1 and an arc γ_2 which form a topological angle in D at x_0,

$$\frac{A(\gamma_1, \gamma_2)}{K} \leqq A[f(\gamma_1), f(\gamma_2)]. \tag{1}$$

Remarks 1. For $n = 2$, S. Agard and F. Gehring [1] established the equivalence between this definition of K-QCfH and Gehring's analytic definition (with the same K). In this chapter, we shall prove for a general n the equivalence of this definition of K-QCfH to Gehring's metric definition (with the same K) (see our joint paper with Monica Corduneanu [1]). This result may seem strange, because we proved that for a general n, Gehring's metric definition of K-QCfH is equivalent neither to Väisälä's nor to Gehring's analytic definitions (with the same K). However, for $n = 2$, all Gehring's and Väisälä's definitions are equivalent (with the same K).

2. In Agard–Gehring's bidimensional definition of K-QCfH, they prove that (i) and (ii) imply the condition ACL, whence $f(x)$ has first order partial derivatives a.e. in D, and then, appealing to a theorem of F. Gehring and O. Lehto [1], they obtain the differentiability of $f(x)$ a.e. in D. Since Gehring–Lehto's theorem holds no more for $n > 2$, we are obliged to establish the ACL_n condition, which implies the differentiability a.e. in D.

3. Condition (i) has in Agard–Gehring's definition the following weaker form:

"For all x_0 in D, and for all segments γ_1 and γ_2 which form an angle in D at x_0, $A[f(\gamma_1), f(\gamma_2)] > 0$".

The more restrictive form of condition (i) in our definition is connected with our proof. It remains an open question if conditions (i) and (ii) cannot be given in the more general form met in Agard and Gehring's. Following the argument which establishes the equivalence between Agard—Gehring's definition and Gehring's metric definition of the K-QCfH, it follows that it suffices that condition (i) hold only for segments γ_1 parallel to coordinate axes.

We study first how the inner measure is changed under a homeomorphism which is differentiable at the vertex of the angle. We require two preliminary results.

LEMMA 1. *Suppose that $f(x)$ is a homeomorphism of a neighbourhood U of the origin, that*

$$f(x) = x + \varepsilon(O, |x|)|x|,$$

near the origin, where $\varepsilon(O, |x|) \to 0$ as $|x| \to 0$, and that γ_1 and γ_2 are two arcs in U which form a topological angle at O. Then $f(\gamma_1)$ and $f(\gamma_2)$ form a topological angle and

$$A[f(\gamma_1), f(\gamma_2)] = A(\gamma_1, \gamma_2). \tag{2}$$

Given that $0 < \varepsilon_0 < 1$, we may choose $\delta > 0$ such that $\varepsilon(O, |x|) < \varepsilon_0$ for $|x| < \delta$. Choose $x_k \in \gamma_k$ so that $0 < |x_k| < \delta$ $(k = 1, 2)$. Then

$$|f(x_1)| + |f(x_2)| \geqq (1 - \varepsilon_0)(|x_1| + |x_2|),$$

$$|f(x_1) - f(x_2)| \leqq |x_1 - x_2| + \varepsilon_0(|x_1| + |x_2|),$$

hence

$$\frac{|f(x_1) - f(x_2)|}{|f(x_1)| + |f(x_2)|} \leqq \frac{1}{1 - \varepsilon_0} \cdot \frac{|x_1 - x_2|}{|x_1| + |x_2|} + \frac{\varepsilon_0}{1 - \varepsilon_0}$$

and letting $x_1, x_2 \to O$, yields

$$\sin \frac{1}{2} A[f(\gamma_1), f(\gamma_2)] \leqq \frac{1}{1 - \varepsilon_0} \sin \frac{1}{2} A(\gamma_1, \gamma_2) + \frac{\varepsilon_0}{\varepsilon_0 - 1} .$$

Since ε_0 is arbitrary, we obtain

$$A[f(\gamma_1), g(\gamma_2)] \leqq \frac{1}{K} A(\gamma_1, \gamma_2). \tag{3}$$

On the other hand, from

$$\frac{|f(x_1) - f(x_2)|}{|f(x_1)| + |f(x_2)|} \geqq \frac{1}{1 + \varepsilon_0} \cdot \frac{|x_1 - x_2|}{|x_1| + |x_2|} - \frac{\varepsilon_0}{1 + \varepsilon_0},$$

follows the reverse inequality too

$$A[f(\gamma_1), f(\gamma_2)] \geqq A(\gamma_1, \gamma_2),$$

which, together with (3), implies (2).

LEMMA 2. *Suppose that*

$$g(x) = a_i x^i e_i, \quad a_1 \geqq \cdots \geqq a_n = 1. \tag{4}$$

If $a_1 \leqq K$, *then*

$$A[g(\gamma_1), g(\gamma_2)] \geqq \frac{1}{K} A(\gamma_1, \gamma_2) \tag{5}$$

for each pair of arcs γ_1, γ_2 *which form a topological angle in D at O. Conversely, if* (5) *holds for each pair of segments* γ_1, γ_2 *which form an angle at O, then* $a_1 \leqq K$.

Choose $x_1 \in \gamma_1$, $x_2 \in \gamma_2$, $x_1, x_2 \neq O$, and set

$$\varphi = \arcsin \frac{|x_1| - |x_2|}{|x_1| + |x_2|}, \quad \varphi^* = \arcsin \frac{|g(x_1) - g(x_2)|}{|g(x_1)| + |g(x_2)|}.$$

Then (4) yields

$$\tan^2 \varphi^* = \frac{1}{\cos^2 \arcsin \dfrac{|g(x_1) - g(x_2)|}{|g(x_1)| + |g(x_2)|}} - 1 = \frac{a\,(x_1^i - x_2^i)^2}{2\sqrt{a_i^2 (x_1^i)^2 a_j^2 (x_2^j)^2} + 2a_i^2 x_1^i x_2^i}. \tag{6}$$

Next, we shall prove that

$$a_i^2 x_1^i x_2^i + \sqrt{a_i^2 (x_1^i)^2 a_i^2 (x_2^i)^2} \leqq a_1^2 \left[\sum_{i=1}^n x_1^i x_2^i + \sqrt{\sum_{i=1}^n (x_1^i)^2 \sum_{i=1}^n (x_2^i)^2} \right]. \tag{7}$$

We verify it first for $n = 2$, i.e.

$$a_1^2 x_1 x_2 + y_1 y_2 + \sqrt{(a_1^2 x_1^2 + y_1^2)(a_1^2 x_2^2 + y_2^2)} \leqq a_1^2 [x_1 x_2 + y_1 y_2 + $$
$$+ \sqrt{(x_1^2 + y_1^2)(x_2^2 + y_2^2)}\,], \tag{8}$$

where we set $x_1^1 = x_1$, $x_1^2 = y_1$, $x_2^1 = x_2$, $x_2^2 = y_2$. Hence

$$y_1 y_2 + \sqrt{(a_1^2 x_1^2 + y_1^2)(a_1^2 x_2^2 + y_2^2)} \leqq a_1^2 [y_1 y_2 + \sqrt{(x_1^2 + y_1^2)(x_2^2 + y_2^2)}],$$

which is obvious if $y_1y_2 \geqq 0$. Thus we may assume $y_1y_2 < 0$. Then

$$0 \leqq (1 - a_1^2)y_1y_2 \leqq a_1^2\sqrt{(x_1^2 + y_1^2)(x_2^2 + y_2^2)} - \sqrt{(a_1^2x_1^2 + y_1^2)(a_1^2x_2^2 + y_2^2)},$$

hence, squaring both sides,

$$\begin{aligned}(1 - a_1^2)^2y_1^2y_2^2 \leqq {} & a_1^4(x_1^2x_2^2 + y_1^2y_2^2 + x_1^2y_2^2 + x_2^2y_1^2) + a_1^4x_1^2x_2^2 + a_1^2(x_1^2y_2^2 + x_2^2y_1^2) + \\ & + y_1^2y_2^2 - 2a_1^2\sqrt{(x_1^2 + y_1^2)(x_2^2 + y_2^2)(a_1^2x_1^2 + y_1^2)(a_1^2x_2^2 + y_2^2)},\end{aligned}$$

whence

$$\begin{aligned}2\sqrt{(x_1^2 + y_1^2)(x_2^2 + y_2^2)(a_1^2x_1^2 + y_1^2)(a_1^2x_2^2 + y_2^2)} \leqq {} & a_1^2(2x_1^2x_2^2 + x_1^2y_2^2 + x_2^2y_1^2) + \\ & + (x_1^2y_2^2 + x_2^2y_1^2 - 2y_1^2y_2^2)\end{aligned}$$

and squaring again

$$\begin{aligned}4(x_1^2 + y_1^2)(x_2^2 + y_2^2)(a_1^2x_1^2 + y_1^2)(a_1^2x_2^2 + y_2^2) \leqq {} & [a_1^2(2x_1^2x_2^2 + x_1^2y_2^2 + x_2^2y_1^2) + \\ & + (x_1^2y_2^2 + x_2^2y_1^2 - 2y_1^2y_2^2)]^2\end{aligned}$$

which implies

$$(x_1^2y_2^2 - x_2^2y_1^2)^2(a_1^2 - 1)^2 \geqq 0,$$

and this inequality is trivial. Thus (8) is proved in general.

Next, in order to establish (7), let us prove that

$$\begin{aligned}& a_1^2(x_1^1x_2^1 + \cdots + x_1^kx_2^k) + a_{k+1}^2x_1^{k+1}x_2^{k+1} + \cdots + a_n^2x_1^nx_2^n + \\ & + \{a_1^2[(x_1^1)^2 + \cdots + (x_1^k)^2] + a_{k+1}^2(x_1^{k+1})^2 + \cdots + a_n^2(x_1^n)^2\}^{\frac{1}{2}}\,\{a_1[(x_2^1)^2 + \cdots + \\ & + (x_2^k)^2] + a_{k+1}^2(x_2^{k+1})^2 + \cdots + a_n^2(x_2^n)^2\}^{\frac{1}{2}} \leqq a_1^2(x_1^1x_2^1 + \cdots + x_1^{k+1}x_2^{k+1}) + \qquad (9) \\ & + a_{k+2}^2x_1^{k+2}x_2^{k+2} + \ldots + a_n^2x_1^nx_2^n + \{a_1^2[(x_1^1)^2 + \cdots + (x_1^{k+1})^2] + a_{k+2}^2(x_1^{k+2})^2 + \\ & + \cdots + a_n^2(x_1^n)^2\}^{\frac{1}{2}}\{a_1^2[(x_2^1)^2 + \cdots + (x_2^{k+1})^2] + a_{k+2}^2(x_2^{k+2})^2 + \cdots + a_n^2(x_2^n)^2\}^{\frac{1}{2}}.\end{aligned}$$

Hence, setting

$$a_1^2[(x_1^1)^2 + \cdots + (x_1^k)^2] + a_{k+2}^2(x_1^{k+2})^2 + \cdots + a_n^2(x_1^n)^2 = a_1^2 t_1^2,$$

$$a_1^2[(x_2^1)^2 + \cdots + (x_2^k)^2] + a_{k+2}^2(x_2^{k+2})^2 + \cdots + a_n^2(x_2^n)^2 = a_1^2 t_2^2,$$

we obtain

$$a_{k+1}^2 x_1^{k+1} x_2^{k+1} + \sqrt{[a_1^2 t_1^2 + a_{k+1}^2(x_1^{k+1})^2][a_1^2 t_2^2 + a_{k+1}^2(x_2^{k+1})^2]} \leqq$$

$$a_1^2\{x_1^{k+1} x_2^{k+1} + \sqrt{[t_1^2 + (x_1^{k+1})^2][t_2^2 + (x_2^{k+1})^2]}$$

and setting $\widetilde{a} = \dfrac{a_1}{a_{k+1}}$,

$$x_1^{k+1} x_2^{k+1} + \sqrt{[\widetilde{a}_1^2 t_1^2 + (x_1^{k+1})^2][\widetilde{a}_1^2 t_2^2 + (x_2^{k+1})^2]} \leqq$$

$$\leqq \widetilde{a}_1^2\{x_1^{k+1} x_2^{k+1} + \sqrt{[t_1^2 + (x_1^{k+1})^2][t_2^2 + (x_2^{k+1})^2]}\}$$

which is satisfied by (8). Since (9) holds for $k = 1, \ldots, n-1$, it follows

$$a_i^2 x_1^i x_2^i + \sqrt{a_i^2 (x_1^i)^2 a_i^2 (x_2^i)^2} \leqq \cdots \leqq a_1^2(x_1^1 x_2^1 + \cdots + x_1^k x_2^k) + a_{k+1}^2 x_1^{k+1} x_2^{k+1} +$$

$$+ \cdots + a_n^2 x_1^n x_2^n + \{a_1^2[(x_1^1)^2 + \cdots + (x_1^k)^2] + a_{k+1}^2(x_1^{k+1})^2 + \cdots +$$

$$+ a_n^2(x_1^n)^2\}^{\frac{1}{2}} \{a_1^2[(x_2^1)^2 + \cdots + (x_2^k)^2] + a_{k+1}^2(x_2^{k+1})^2 + \cdots + a_n^2(x_2^n)^2\}^{\frac{1}{2}} \leqq \cdots$$

$$\leqq a_1^2\left[\sum_{i=1}^n x_1^i x_i^2 + \sqrt{\sum_{i=1}^n (x_1^i)^2 \sum_{i=1}^n (x_2^i)^2}\right].$$

Hence and by (6),

$$\tan\varphi^* \geqq \frac{a_i^2(x_1^i - x_2^i)^2}{2\sqrt{a_i^2(x_1^i)^2 a_i^2(x_2^i)^2} + 2a_i^2 x_1^i x_2^i} \geqq \frac{\sum_{i=1}^n (x_1^i - x_2^i)^2}{2a_1^2\left[\sqrt{\sum_{i=1}^n (x_1^i)^2 \sum_{i=1}^n (x_2^i)^2} + \sum_{i=1}^n x_1^i x_2^i\right]} =$$

$$= \frac{1}{a_1^2} \frac{|x_1 - x_2|^2}{(|x_1| + |x_2|)^2 - |x_1 - x_2|^2} = \frac{1}{a_1^2} \tan^2\varphi,$$

which implies that

$$\varphi^* \geqq \arctan\left(\frac{1}{a_1}\tan\varphi\right) \geqq \frac{1}{a_1}\varphi$$

and we obtain

$$A[g(\gamma_1), g(\gamma_2)] = \varliminf_{x_1, x_2 \to O} 2\varphi^* \geqq \frac{1}{a_1} \varliminf_{x_1, x_2 \to O} 2\varphi = \frac{1}{a_1} A(\gamma_1, \gamma_2).$$

Thus (5) holds if $a_1 \leqq K$.

Conversely, suppose (5) holds. Then, for $\vartheta > 0$ choose two points on S which lie in the x^1x^n-plane and are symmetrical with respect to the x^n-axis, so that the segments γ_1, γ_2 joining them to the origin form an angle ϑ. Then the inequality (5) and an easy computation show that

$$A[g(\gamma_1), g(\gamma_2)] = 2\arctan\left(\frac{1}{a_1}\tan\theta\right) \geqq \frac{A(\gamma_1, \gamma_2)}{K} = \frac{2\vartheta}{K}.$$

Letting $\vartheta \to 0$ and taking into account that

$$\lim_{\vartheta \to 0} \frac{\arctan\left(\dfrac{\tan\vartheta}{a_1}\right)}{\vartheta} = \frac{1}{a_1},$$

it follows that $a_1 \leqq K$.

Lemma 3. *Suppose that $f(x)$ is a homeomorphism of a domain D, that $f(x)$ has a differential at x_0 and that* (1) *holds for each pair of segments which form an angle in D at x_0, then*

$$\max |f'(x_0)|^n \leqq K^{n-1} |J(x_0)| \tag{10}$$

holds.

Suppose, as we may, $\max|f'(x_0)| > 0$, because otherwise inequality (10) is trivial. Consider first the case $J(x_0) \neq 0$. Hence, by performing preliminary similarity mappings and reflexions, we may assume that $x_0 = f(x_0) = O$ and that near $x_0 = O$

$$f(x) = g(x) + O[|g(x)|] = g(x) + O(|x|), \tag{11}$$

where g is as in (4). Then lemma 1 and (1) imply that

$$A[g(\gamma_1), g(\gamma_2)] = A[f(\gamma_1), f(\gamma_2)] \geqq \frac{1}{K} A(\gamma_1, \gamma_2)$$

for each pair of segments γ_1 and γ_2 which form an angle in D at O. Hence $a_1 \leqq K$ by lemma 2, and, on account of (1.13), we obtain

$$\frac{\max |f'(x_0)|^n}{|J(x_0)|} = \frac{a_1^n}{a_1 \cdots a_{n-1}} \leqq a_1^{n-1} \leqq K^{n-1},$$

which implies (10).

Finally, to complete the proof, we observe that either $\max|f'(x_0)| = 0$ and then (10) is trivial, or $\max|f'(x_0)| > 0$ and then also $|J(x_0)| > 0$. Indeed, in the last case, if $J(x_0) = 0$, then by performing preliminary similarity mappings, we may assume that $x_0 = f(x_0) = O$ and near $x_0 = O$

$$f(x) = a_p x^p + o(|x|)$$

where p takes on the values $1, \ldots, q < n$. Next, for $0 < \vartheta < \frac{\pi}{2}$, let γ_1, γ_2 be (as in the preceding lemma) two segments lying in the x^1x^n-plane with an endpoint at O, symmetrical with respect to x^n-axis and forming with it an angle ϑ. Then the same computation as in the preceding lemma in the corresponding case, yields $A[f(\gamma_1), f(\gamma_2)] = 0$, $A(\gamma_1, \gamma_2) = 2\vartheta$, contradicting so (1), and this completes the proof.

THEOREM 1. *A K-QCfH according to Agard—Gehring's definition in D is ACL in D.*

Using the notations of lemma 1.5, we shall have $D_{\text{sym}}\varphi(y) < \infty$ a.e. in I_n. Let y_0 be such a point. By performing a preliminary similarity transformation, we may assume that I is the interval $I = \{x;\ 0 < x^i < c < 1\ (i = 1, \ldots, n)\}$. Next, fix r so that $0 < r < d(\bar{I}, \partial D)$, and for each $x_0 \in I - E$, where E is the set involved in condition (i) (p. 255), let us consider a cone C with the vertex at x_0, the generatrices forming with the axis an angle $\alpha \leqq \alpha_0$ for which (i) holds (this angle depends on the vertex x_0). Let γ_1 be a segment which lies on the axis of the cone, has an end point in x_0 and the length less than r. By condition (i), we have

$$\overline{\lim_{x_1, x_2 \to x_0}} \frac{|f(x_1) - f(x_0)| + |f(x_2) - f(x_0)|}{|f(x_1) - f(x_2)|} < \infty, \quad x_1 \in \gamma_1, x_2 \in \partial C, \tag{12}$$

since if there were two sequences of points $\{x_{1m}\}$, $\{x_{2m}\}$, so that $x_{1m} \to x_0$, $x_{2m} \to$ $\to x_0$, $x_{1m} \in \gamma_1$, $x_{2m} \in \gamma_2$, $|x_{2m} - x_0| > |x_{2m+1} - x_0|$ $(m = 1, 2, \ldots)$ and

$$\lim_{m \to \infty} \frac{|f(x_{1m}) - f(x_0)| + |f(x_{2m}) - f(x_0)|}{|f(x_{1m}) - f(x_{2m})|} = \infty,$$

then, joining all the points x_{2m}, x_{2m+1} by an arc $\tilde{\gamma}_m$ so that two different arcs have at most an end point in common, we should obtain an arc $\gamma = \bigcup_{m=1}^{\infty} \tilde{\gamma}_m \cup x_0$, with $A(\gamma_1, \gamma) = \alpha > 0$, and $A|f(\gamma_1), f(\gamma)| = 0$, contradicting so (i).

Now for each pair of integers p and q, with $p > 0$ and $0 < \frac{1}{q} < r$, let E_{pq} denote the set of x_0 in I such that

$$|f(x_1) - f(x_0)| + |f(x_2) - f(x_0)| \leqq p\,|f(x_1) - f(x_2)|, \tag{13}$$

whenever $|x_k - x_0| \leqq \frac{1}{q}\,(k = 1, 2)$ and $x_1 \in \gamma_1$, $x_2 \in \partial C$. Then E_{pq} is compact and, by (12),

$$I - E = \bigcup_{p,q} E_{pq} \tag{14}$$

where the sum is taken over relevant p and q.

Next, if $\tilde{E}$ denotes the set where (1) does not hold, it follows that, for every point $x_0 \in D - \tilde{E}$, we can choose for instance $\alpha = \frac{\pi}{4}$, and then (ii) implies that $A[f(\gamma_1), f(\gamma_2)] \geqq \frac{\pi}{8K}$ for all pairs γ_1, γ_2, consisting of a segment γ_1 and an arc γ_2, which form a topological angle $\frac{\pi}{4}$ in D at x_0. Hence, for every $x_0 \in D - \tilde{E}$

$$\overline{\lim_{x_1, x_2 \to x_0}} \frac{|f(x_1) - f(x_0)| + |f(x_2) - f(x_0)|}{|f(x_1) - f(x_2)|} < p_0, \qquad x_1 \in \gamma_1, x_2 \in \gamma_2, \tag{15}$$

where γ_1 is the axis of a cone C with a vertex at x_0, γ_2 is an arc of ∂C, the angle between the axis and the generatrices is $\frac{\pi}{4}$ and $p_0 > \operatorname{cosec} \frac{\pi}{8K}$.

For each integer $p > 0$, set

$$E_p = \bigcup_{p,\,q} E_{pq},$$

where the sum is taken over all relevant q. Then (15) yields $I - \tilde{E} \subset D - \tilde{E} \subset E_{p_0}$, hence $I - E_{p_0} \subset \tilde{E}$ and then $m(I - E_{p_0}) = 0$. By Fubini's theorem,

$$m_1(Z_{y_0} - E_p) = 0 \tag{16}$$

for almost all $y_0 \in I_n$. Thus $D_{\text{sym}}\varphi(y) < \infty$ and (16) hold a.e. in I_n. Let y_0 be such a point, $J_0 = Z_{y_0}$ and $F \subset J_0 \cap E_{pq}$ a compact set. We shall prove first that

$$\Lambda[f(\ )]^n \leqq 2^{n-2} n p\, D_{\text{sym}}\varphi(y_0)\,(m_1\ \ )^{n-1}, \tag{17}$$

where $\Lambda(E)$ means the length of the set E (see Chapter 1.3). Suppose that $J \subset J_0$ is a closed interval with the end points $a, b \in F$, where $b - a > 0$ and

$$b - a < \min\left[\frac{1}{q}, \mathrm{d}(F, \partial I) \cot \alpha_0\right], \tag{18}$$

and let $\widetilde{C}$ be the open set obtained as the intersection of the two cones with the axis J, the vertices a and b, respectively, and the base of one passing through the vertex of the other. We say that $\widetilde{C}$ is *the conical figure* associated with the interval J. Clearly $\widetilde{C} \subset I$, by (18). By performing a change of variables, we may assume that $f(a) = O$ and $f(b) = l^* e_n$ $(l^* > 0)$. We shall show that

$$|f(a) - f(b)|^n = l^{*n} \leqq \frac{2^{n-2} np}{\omega_{n-1}} mf(C). \tag{19}$$

For each u_0^*, $0 < u_0^* < l^*$, on each ray with the origin at a point $\tilde{x}_1^* \in f(J)$ and which lies in the plane $x^{*n} = u_0^*$, there is a pair of points $x_1^* \in f(J)$, $x_2^* \in f(\partial \widetilde{C})$ such that $\overline{x_1^* x_2^*} \subset f(\widetilde{C})$. Let be $x_1 = f^{-1}(x_1^*)$, $x_2 = f^{-1}(x_2^*)$ and let C_a and C_b be the parts of the conical surface of $\widetilde{C}$ corresponding to the cones with the vertices a and b, respectively. Suppose that $x_2 \in C_a$. Then $|x_1 - a|$, $|x_2 - a| \leqq b - a < \dfrac{1}{q}$ by (18), and since $a \in E_{pq}$, (13) yields

$$2u_0^* \leqq |x_1^*| + |x_2^*| \leqq p\,|x_1^* - x_2^*|.$$

If $x_2 \in C_b$, then $|x_1 - b|$, $|x_2 - b| \leqq b - a < \dfrac{1}{q}$ and since $b \in E_{pq}$, (13) yields

$$2(l^* - u_0^*) \leqq |x_1^* - l^*| + |x_2^* - l^*| \leqq p\,|x_1^* - x_2^*|.$$

Hence, for $0 < u_0^* < 1^*$, every ray with the origin at a point $x_1^* \in f(J)$ and lying in the plane $x^{*n} = u_0^*$ contains an open interval which lies in $f(\widetilde{C})$ and has length not less than $\dfrac{2}{p} \min(u_0^*, l^* - u_0^*)$. By Fubini's theorem,

$$mf(C) \geqq \frac{2}{p} \int_0^{l^*} \omega_{n-1} \min(u^*, l^* - u^*)^{n-1} \, du^* = \frac{\omega_{n-1}\, l^{*n}}{2^{n-2}\, np},$$

from which follows (19). Since F is closed, $F = F_1 \cup F_2$, where F_1 is countable and F_2 is perfect (one of them may be empty). Obviously $m_1 F = m_1 F_2$, $\Lambda[f(F)] = \Lambda[f(F_2)]$ and hence, for the proof of (17), we may assume that F is a perfect set.

Fix $\varepsilon > 0$, choose the corresponding ρ of lemma 1.4 and fix $r_0 \in (0, \rho)$ so that

$$r_0 \leqslant \frac{1}{2} \min \left[\frac{1}{q}, d(F, \partial I) \operatorname{cotg} \alpha_0 \right]. \tag{20}$$

Next, let $B(x_1, r_0), \ldots, B(x_N, r_0)$ be the covering of F described in lemma 1.4 and let $\widetilde{C}_m$ be the conical figure associated with the interval $J_0 \cap B(x_m, r)$. Then each pair of points $a, b \in F \cap B(x_m, r)$ with $b - a > 0$, bounds a closed interval J whose associated conical figure $\widetilde{C}$ lies in $B_m = B(x_m, r_0)$ (since $\alpha \leqq \alpha_0$). Since $b - a \leqq 2 r_0$, (20) implies (18), and hence

$$|f(a) - f(b)|^n \leqq \frac{2^{n-2} np}{\omega_{n-1}} mf(C) \leqq \frac{2^{n-2} np}{\omega_{n-1}} mf(C_m),$$

by (19). From this it follows that

$$d[f(E_m)]^n = d_m^n \leqq \frac{2^{n-2} np}{\omega_{n-1}} mf(C_m) \tag{21}$$

where $E_m = F \cap B_m$. Let $d = \max(d_1, \ldots, d_N)$. Then the sets $f(E_m)$ form a covering of $f(F)$, $d[f(E_m)] \leqq d$, and hence, by (21), (1.30) and Hölder's inequality,

$$\Lambda^d[f(F)]^n \leqq \left\{ \sum_{m=1}^{N} d[f(E_m)] \right\}^n \leqq \frac{2^{n-2} npN^{n-1}}{\omega_{n-1}} \sum_{m=1}^{N} mf(C_m) \leqq$$

$$\leqq 2^{n-2} np(Nr)^{n-1} \frac{\varphi(|y| < r)}{\omega_{n-1} r^{n-1}} \leqq 2^{n-2} np \frac{\varphi(|y| < r)}{\omega_{n-1} r^{n-1}} (m_1 F + \varepsilon)^{n-1},$$

where $\Lambda^d(E) = \Xi(\mathfrak{M}, \chi_n^q)$ (see Chapter 1.3). If we now let $r_0 \to 0$, then $d \to 0$ (by the continuity of f), and we obtain

$$\Lambda[f(\ )]^n \leqq 2^{n-2} npD_{\text{sym}} \varphi(y_0)(m_1 F + \varepsilon)^{n-1}.$$

Letting $\varepsilon \to 0$, yields (17).

We shall show now that if $E' \subset J_0$ with $m_1 E' = 0$, then $\Lambda[f(E')] = 0$, provided that $J_0 - E$ is countable.

Suppose first that E' is compact. Then $F = E' \cap E_{pq}$ is compact for relevant p and q, and from (14) and (17), we conclude that

$$\Lambda[f(E)] \leqq \sum_{p,q} \Lambda[f(E \cap E_{pq})] = 0.$$

Suppose next that E' is a G_δ-Borel set. Then, since $F = J_0 \cap E_{pq}$ is a compact and the length of J_0 is $c < 1$,

$$\Lambda[f'(E' \cap E_{pq})]^n \leqq 2^{n-2} n p D_{\text{sym}} \varphi(y_0) m_1(J_0 \cap E_{pq})^{n-1} \leqq$$

$$\leqq 2^{n-2} n p D_{\text{sym}} \varphi(y_0) < \infty$$

by (17). Hence, by (14), $f(E')$ is of Σ-finite length (see Chapter. 1.3). Since $f(E')$ is itself a G_δ-Borel set, lemma 5.1 implies that

$$\Lambda[f(E)] = \sup\{\Lambda(F^*);\ F^* \text{ compact},\ F^* \subset f(E)\}. \tag{22}$$

Now, let F^* be any compact subset of $f(E')$ and set $F = f^{-1}(F^*)$. Then F is compact and $F \subset E'$. Hence $m_1 F = 0$ and $\Lambda(F^*) = 0$ by what was proved above. Thus $\Lambda[f(E')] = 0$ by (22).

Finally, in the general case, we find a G_δ-Borel set E'' such that $E' \subset E'' \subset J_0$ and $m_1 E'' = m_1 E' = 0$. Then, clearly $\Lambda[f(E')] \leqq \Lambda[f(E'')] = 0$. Thus $m_1 E' = 0$ implies $\Lambda[f(E')] = 0$ for an arbitrary set $E' \subset J_0$.

Now, let us prove that $J_0 - E$ is countable. Suppose γ_1 is parallel to one o the coordinate axes. Condition (i) implies that $I - E$ is of Σ-finite area. Then by corollary of Gross' theorem (Chapter 1.3), $J_0 - E$ is at most countable for almost all $y_0 \in I_n$.

With the help of what was established above, we can now complete the proof of ACL property of f as follows. Let E be any compact set in J_0. Then $E = (E \cap E_p) \cup (E - E_p)$, where $m_1(E - E_p) = 0$ by (16). Hence, by (17) and since $m_1 E = 0$ implies $\Lambda[f(E)] = 0$,

$$\Lambda[f(E)]^n = \Lambda[f(E \cap E_p)]^n = \lim_{q \to \infty} \Lambda[f(E \cap E_{pq})]^n \leqq$$

$$\leqq 2^{n-2} n p D_{\text{sym}} \varphi(y_0) \lim_{q \to \infty} m_1(E \cap E_{pq})^{n-1} = 2^{n-2} n p D_{\text{sym}} \varphi(y_0)(m_1 E)^{n-1},$$

and it follows that $f(x)$ is AC in $J_0 = Z_{y_0}$. Since the preceding inequality holds a.e. in I_n, $f(x)$ has the desired ACL property in I. But $I \subset\subset D$, is an arbitrary interval with the edges parallel to the coordinate axes, hence $f(x)$ is ACL in D and the proof is complete.

THEOREM 2. *A-QCfH according to Agard—Gehring's definition in D is ACL_n in D.*

Clearly, it is enough to prove that

$$\int_I \left| \frac{\partial f(x)}{\partial x^n} \right|^n d\tau < \infty \tag{23}$$

for all intervals $I \subset\subset D$, with the axes parallel to the coordinate axes. Given such an interval, $f(x)$ is AC on almost every segment parallel to Ox^n, and, by proposition 1.6.6, the image of such a segment J under f has the length $\Lambda[f(J)] = \int_J \left|\frac{\partial f(x)}{\partial x^n}\right| dx^n$.

Next, define the measurable functions

$$g_m(x) = \frac{m}{2} \int\limits_{J_m(x^n)} \left|\frac{\partial f(x)}{\partial x^n}\right| dx^n \quad (m = 1, 2, \dots),$$

where $J_m(x^n) = \left\{\xi;\ x^n - \frac{1}{m} \leqq \xi \leqq x^n + \frac{1}{m}\right\}$. Then,

$$g(x) = \lim_{m\to\infty} g_m(x) = \left|\frac{\partial f(x)}{\partial x^n}\right| \tag{24}$$

a.e. in J, and since this happens for almost every J in I, we conclude, by Fubini's theorem, that (24) holds a.e. in I. Now, let us fix x^n and p. By (17) with $F = J_m(x^n) = E_m(y)$ and $\varphi(y) = \varphi[y, J(x^n)] = mf(Z_y)$, $Z_y = \left\{x;\ x = y + \xi e_n,\ x^n - \frac{1}{m} \leqq \xi \leqq x^n + \frac{1}{m}\right\}$, we have

$$g_m(x)^n = \left(\frac{\Lambda\{f[E_m(y, x^n)]\}}{\Lambda[E_m(y, x^n)]}\right)^n \leqq \frac{2^{n-2} npD_{\text{sym}}\,\varphi[y, J_m(x^n)]}{\Lambda[E_m(y, x^n)]}.$$

Integrating over I_n, we find

$$\int\limits_{I_n} g_m(x)^n d\sigma < m2^{n-3}\, np \int\limits_{I_n} D_{\text{sym}}\,\varphi[y, J_m(x^n)]\, d\sigma \leqq 2^{n-3}\, mnp\varphi[I_n, J_m(x^n)],$$

and letting $m \to \infty$, by Fatou's theorem: "If $\{u_m(x)\}$ is a sequence of functions integrable on a set E so that $|u_m(x)| \leqq S(x)$, where $\int_E S(x)\, d\tau < \infty$ and $\underline{\lim}_{m\to\infty} \int_E u_m(x)\, d\tau < < \infty$, then $\int_E \underline{\lim}_{m\to\infty} u_m(x)\, d\tau \leqq \underline{\lim}_{m\to\infty} \int_E u_m(x)\, d\tau$" (see C. Carathéodory [3], p. 443), we conclude

$$\int_{I_n} g(x)^n d\sigma \leqq \underline{\lim}_{m\to\infty} 2^{n-3} mnp\varphi[I_n, J_m(x^n)] = 2^{n-3} np \lim_{m\to\infty} \frac{\varphi[I_n, J_m(x^n)]}{\frac{2}{m}} =$$

$$= 2^{n-3} np\varphi'(I_n, x^n),$$

where $\varphi'(I^n, x^n)$ is measurable and $\int_J \varphi'(I^n, x^n)\,dx^n \leqq \varphi(I_n, J)$. Hence, by Fubini's theorem

$$\int_I g(x)^n\,d\tau = \int_I \left|\frac{\partial f(x)}{\partial x^n}\right|^n d\tau \leqq 2^{n-3} np\varphi(I_n, J) < \infty.$$

The proof is completed.

REMARK. This proof is following S. Agard [1]. The same thing was proved also by O. Taari [1].

COROLLARY. *A K-*QCfH *according to Agard—Gehring's definition in D is differentiable a.e. in D.*

This is an immediate consequence of lemma 1.10 and of the preceding theorem.

THEOREM 3. *A* QCfH *in D satisfies conditions* (ii).

Clearly, $f^{-1}(x^*)$ is also a QCfH in D, which, by theorem 9.1, satisfies the inequality

$$\frac{|f^{-1}(y^*) - f^{-1}(x^*)|}{d[f^{-1}(x^*), \partial\Delta]} \leqq \Theta\left[\frac{|y^* - x^*|}{d(x^*, \partial\Delta^*)}\right],$$

where the domain $\Delta \subset D$ and where $\Theta(t)$ is continuous and strictly increasing in $[0, 1)$ with $\Theta(0) = 0$. Hence

$$\frac{|y^* - x^*|}{d(x^*, \partial\Delta^*)} \geqq \Theta^{-1}\left\{\frac{|f^{-1}(y^*) - f^{-1}(x^*)|}{d[f^{-1}(x^*), \partial\Delta]}\right\},$$

where Θ^{-1} is the inverse of $\Theta(t)$ and then is also strictly increasing in [0,1) with $\Theta^{-1}(0) = 0$. If we set $\Delta = D - \{x_0\}$, $y = x_1$, $x = x_2$, where x_1, x_2 are in a sufficiently small neighbourhood of x_0, then

$$\frac{|f(x_1) - f(x_2)|}{|f(x_2) - f(x_0)|} \geqq \Theta^{-1}\left(\frac{|x_1 - x_2|}{|x_2 - x_0|}\right) \geqq \Theta^{-1}\left(\frac{|x_1 - x_2|}{|x_1 - x_0| + |x_2 - x_0|}\right) > 0,$$

and if we set $\Delta = D - \{x_0\}$, $y = x_1$, $x = x_2$, then

$$\frac{|f(x_1) - f(x_2)|}{|f(x_1) - f(x_0)|} \geqq \Theta^{-1}\left(\frac{|x_1 - x_2|}{|x_1 - x_0| + |x_2 - x_0|}\right) > 0.$$

Combining these inequalities

$$\frac{|\ (x_1) - f(x_2)|}{|f(x_1) - f(x_0)| + |f(x_2) - f(x_0)|} \geqq$$

$$\geqq \frac{|f(x_1) - f(x_2)|}{2\max\left[|f(x_1) - f(x_0)|, |f(x_2) - f(x_0)|\right]} \geqq$$

$$\frac{1}{2}\Theta^{-1}\left(\frac{|x_1 - x_2|}{|x_1 - x_0| + |x_2 - x_0|}\right) > 0,$$

which implies (ii), as desired.

THEOREM 4. *The definition of the K-*QCfH *according to Agard and Gehring is equivalent to Gehring's metric definition.*

Suppose $f(x)$ is K-QCfH according to Gehring's metric definition in D. Then $f(x)$ is differentiable a.e. in D, and condition (i) follows as an immediate consequence of the preceding theorem. The corollary of theorem 1.1 implies $J(x) \neq 0$ a.e. in D. Since at an A-point, by (1.14), $\delta_L(x) = p_1(x) = \dfrac{a_1(x)}{a_n(x)}$ from the definition of $f(x)$ it follows $p_1(x) = \delta_L(x) \leqq K$ a.e. in D, hence, lemma 2 allows us to conclude that also condition (ii) holds.

Now suppose that $f(x)$ is a K-QCfH according to Agard—Gehring's definition, i.e. is characterized by the conditions (i), (ii). Then, by theorems 1 and 2, $f(x)$ is ACL_n in D. Moreover, $f(x)$ is differentiable a.e. in D, by corollary of theorem 2, and at the points of differentiability lemma 3 implies (10). This allows us to conclude that $f(x)$ is QCfH according to Väisälä's definitions, and then, since all the definitions are equivalent, $f(x)$ is QCfH according to Gehring's metric definition. On the other hand, by lemma 2, $p_1(x) \leqq K$ in almost all points of differentiability of $f(x)$, i.e. a.e. in D. Since, by corollary of theorem 1, 1, $J(x) \neq 0$ a.e. in D, it follows by (1.14) that $\delta_L(x) = p_1(x) \leqq K$ in almost all the A-points of $f(x)$ in D, i.e. a.e. in D, hence, taking into account that $f(x)$ is QCfH according to Gehring's metric definition in D, we conclude that $f(x)$ is K-QCfH in D according to Gehring's metric definition, and thus the proof of the equivalence is complete.

Remark. Another bidimensional definition of the K-QCfH by means of the angle distortion was given by O. Taari [1]. His definition of the K-QCfH cannot be generalized to n-space ($n > 2$) since his definition of the measure of an angle involves Riemann mapping theorem. We remark also that even for $n = 2$, Taari's definition of the measure of an angle does not coincide with Agard and Gehring's.

Recently S. Agard [1] and O. Taari [2] characterized the QCfH in a way similar to that given by S. Agard and F. Gehring [1] for $n = 2$. S. Agard [1] proved the equivalence of the K-QCfH with Gehring's metric definition (with the same K), while O. Taari [2] succeeded only in proving its equivalence to J. Väisälä's analytic definition in [3], but with different constants K.

CHAPTER 11

Definitions of Grötzsch's, of Lavrent'ev's, of Teichmüller's and of Andreian's type for the QCfH

A general formulation for the definitions of Grötzsch's type. The different definitions of the K-QCfH in n-space have a common essential characteristic: *a pair of functions is bounded in D.* These functions are point functions, or set functions. The point functions involved in the different definitions are: the maximal local dilatation $\delta(x) = \max[\bar{\delta}(x), \underline{\delta}(x)]$ of Väisälä's metric definition, the linear local dilatation of Gehring's metric definition, the outer and inner local dilatations $\bar{\delta}(x)$ and $\underline{\delta}(x)$, respectively, of Kühnau's definitions, the characteristic $q(x)$ of Markuševič—Pesin's definition, the principal characteristic parameters $p_1(x)$, $p_1'(x)$ of the definition of K-QCfH with two sets of characteristics, the ratios $\dfrac{|J(x)|}{\min|f'(x)|^n}$, $\dfrac{\max|f'(x)|^n}{|J(x)|}$ of Väisälä's definition and $\dfrac{\Lambda_f(x)^n}{J_G(x)}$, $\dfrac{J_G(x)}{\lambda_f(X)^n}$ of Gehring's analytic definition, $\dfrac{|\nabla f|^n}{|J(x)|}$ of Callencer's definition and the function $\sup\limits_{s,s'}\dfrac{\left|\dfrac{\partial f(x)}{\partial s}\right|}{\left|\dfrac{\partial f(x)}{\partial s'}\right|}$ of Church's definition. The set functions involved in the different definitions are the moduli of rings: $M(A)$ and mod A of Väisälä's and of Gehring's geometric definitions, respectively, the conformal capacity of a ring: cap A of Loewner's definition, the modulus of an arc family or of a cylinder of Väisälä's definitions, the modulus of a q-dimensional surface family $(1 \leqq q \leqq n-1)$ of Šabat's definition and the inner measure of a topological angle of Agard—Gehring's definition. In fact, the pair of set functions involved in the different definitions is formed by the ratio between one of the set functions from above and the set function of the image under a QCfH of the corresponding geometrical configuration and the inverse of this ratio.

It is interesting that for each definition characterized by means of set functions, it is possible to give an equivalent definition characterized by means of point functions and involving the same geometrical configuration. All the results of this chapter were established by us in [24, 26].

Starting from the two characteristics mentioned above, i.e. the condition that a pair of functions is bounded in D (involved in all the definitions of the K-QCfH) and the possibility to associate to each definition characterized by means of set functions, an equivalent definition characterized by means of point functions, we succeeded in obtaining for the K-QCfH a general formulation, which, by a corresponding particularization of the terms involved in it, reduces to a definition equivalent to each of the definitions of the K-QCfH, with the same K and expressed by means of the same terms. Such a definition is the following:

I. A homeomorphism $f(x)$ in a domain $D \subset R^n$ is said to be a K-QCfH $(1 \leqq K < \infty)$ in D if a certain pair of point functions $K_I^f(x)$, $K_O^f(x)$ is bounded in D and

$$1 \leqq K_I^f(x), \quad K_O^f(x) \leqq K$$

a.e. in D.

By specifying the point functions we obtain definitions equivalent (with the same K) to each of the other (excepting the analytical ones) and expressed by means of the same terms.

We observe that the preceding general definition may be given in the following more general but equivalent form:

I′. A homeomorphism $f(x)$ is said to be a K-QCfH $(1 \leqq K < \infty)$ in D if a certain pair of point functions $K_I^f(x)$, $K_O^f(x)$ is defined in D, such that

(i) $1 \leqq K_I^f(x)$, $K_O^f(x) < \infty$ except possibly on a set of Σ-finite area.

(ii) $1 \leqq K_I^f(x)$, $K_O^f(x) \leqq K$ a.e. in D.

Now we shall show that by a suitable particularization of the point functions $K_I^f(x)$, $K_O^f(x)$ we obtain a definition equivalent to each of the metric and geometric definitions. In order to include also the analytic definitions, we shall give later on a more general (but also a little more complicated) formulation.

Definition I reduces to one equivalent to Väisälä's first geometric definition if we particularize $K_I^f(x)$, $K_0^f(x)$ as follows:

For every subdomain $\Delta \subset D$, set

$$K_I^\Delta = \sup_\Gamma \frac{M(\Gamma)}{M(\Gamma^*)}, \quad K_O^\Delta = \sup_\Gamma \frac{M(\Gamma^*)}{M(\Gamma)}, \tag{1}$$

where $M(\Gamma)$ and $M(\Gamma^*)$ are not both 0 or ∞, and

$$K_I^f(x) = \inf_{B \supset x} K_I^B, \; K_O^f(x) = \inf_{B \supset x} K_0^B, \tag{2}$$

where the infimum is taken over all the balls $B = B(x, r)$ contained in D.

THEOREM 1. *The general definition I, with $K_I^f(x)$, $K_O^f(x)$ defined as above, is equivalent to Väisälä's first geometric definition.*

Let $x^* = f(x)$ be a K-QCfH according to Väisälä's first geometric definition. It is clear that $K_I^\Delta, K_O^\Delta \leqq K$ for all $\Delta \subset D$, and *a fortiori* $K_I^f(x)$, $K_O^f(x) \leqq K$ in D. Thus $f(x)$ is K-QCfH according to I.

Now let $x^* = f(x)$ be K-QCfH according to I with $K_I^f(x)$, $K_O^f(x)$ defined as above and $x \in D$. Then I implies $K_I^f(x)$, $K_O^f(x) < K'$ and hence there exists a ball $B(x, r)$ such that

$$\frac{M(\Gamma^*)}{M(\Gamma)}, \; \frac{M(\Gamma)}{M(\Gamma^*)} \leqq 2K' < \infty \tag{3}$$

for all $\Gamma \subset B(x, r)$. But the preceding inequality may be written also

$$\frac{M(\Gamma)}{2K'} \leqq M(\Gamma^*) \leqq 2K'M(\Gamma) \tag{4}$$

for all $\Gamma \subset B(X, r)$. Then the equivalence of Väisälä's definitions implies that $f(x)$ is $2K'$-QCfH according to Väisälä's metric definition in the neighbourhood of every point of D, hence $f(x)$ is QCfH in D according to Väisälä's metric definition, and then according to Väisälä's geometric definition.

Now let $D_K = E\{x;\ x \in D,\ K_I^f(x),\ K_O^f(x) \leqq K\}$. For every $\varepsilon > 0$ and $x \in D_K$ there exists by (2) a ball such that K_I^B, $K_O^B < K + \varepsilon$. Then (1) implies (3) and (4) with $K + \varepsilon$ in place of $2K'$. Hence $f(x)$ is $(K + \varepsilon)$-QCfH according to Väisälä's first geometric definition, and then also according to his metric definition in the neighbourhood of every point of D_K. Since the radii r aproach zero as $\varepsilon \to 0$, it follows that letting $\varepsilon \to 0$, we obtain that $f(x)$ is K-QCfH in D according to Väisälä's first geometric definition.

If we put in (1), in place of the modulus of a curve family, the modulus of a ring (as defined by Väisälä's or Gehring), of a cylinder, of a q-dimensional $(q = 1, n - 1)$ surface family or the conformal capacity of a ring and maintain (2), then arguing as in the preceding theorem definition I reduces to one equivalent to Väisälä's second geometric definition (or Gehring's geometric definition) to Väisälä's third definition, to Šabat's definitions, or to Loewner's definition. The conclusion still holds also for Kühnau's definitions.

For Väisälä's and for Gehring's metric definition, it is enough to choose $K_I^f(x) = K_O^f(x) = \delta(x)$ and $K_I^f(x) = K_O^f(x) = \delta_L(x)$ in order that I reduce even to Väisälä's and to Gehring's metric definitions, respectively (and not only to definitions equivalent to them as above).

THEOREM 2. *The definition I is equivalent to I′ for all particular cases considered above.*

Obviously I is a particular case of I′.

Now consider I′. It implies I in the particular case of Gehring's metric definition by theorem 7.1. Hence I and I′ are equivalent in the case of Gehring's metric, of Markuševič—Pesin's and of Kopylov's definitions, and also of definitions of K-QCfH with one or two principal characteristics.

The equivalence between I and I′ in the case of Gehring's geometric definition is a consequence of corollary 2 of theorem 7.1. Hence it follows the equivalence between I and I′ in the case of Väisälä's metric and geometric definitions, and also Loewner's.

We have been obliged to except for the moment the analytical definitions. That is because their structure is somewhat different from that of the others. Indeed, $f(x)$ is no more only a homeomorphism with the property that certain quantities defined by means of f verify a double inequality, but in the analytical definitions f verifies some additional regularity conditions as for instance the differentiability a.e. in D, the ACL or ACL_n condition (see Väisälä's definitions), the existence of Sobolev's first order derivatives L^n-integrable (Kreines' definition), etc.; in exchange, the corresponding point functions $K_I^f(x)$, $K_O^f(x)$ are defined only a.e. in D. In order to include also the analytic definitions in the general definitions I, I′, we shall formulate them in the following more general (but also more complicated) form:

I_1. A homeomorphism $f(x)$ (verifying some regularity conditions) is said to be a K-QCfH ($1 \leqq K < \infty$) in D if a certain pair of functions $K_I^f(x)$, $K_O^f(x)$ is defined in D such that $K_I^f(x)$, $K_O^f(x)$ be uniformly bounded in D and $K_I^f(x)$, $K_O^f(x) \leqq K$ a.e. in D (or only the last condition).

I_1'. A homeomorphism $f(x)$ (verifying some regularity conditions) is said to be a K-QCfH ($1 \leqq K < \infty$) in D if a certain pair of functions $K_I^f(x)$, $K_O^f(x)$ is defined in D (or a.e. in D) such that (i) and (ii) [or only (i)] hold.

The parts of the definitions contained in the parantheses are relevant to the analytic definitions; in this case I_1 and I_1' coincide.

The analytic definitions are obtained directly as a particular case of I by specifying the regularity conditions and the point functions.

For Väisälä's first analytic definition,

$$K_I^f(x) = \frac{|\ (x)|}{\min\limits_{|\Delta x|=1} |f'(x)\Delta x|^n}, \qquad K_O^f(x) = \frac{\min\limits_{|\Delta x|=1} |f'(x)\Delta x|^n}{|J(x)|}$$

for Gehring's second analytic definition,

$$K_I^f(x) = \left[\frac{J_G(x)}{\lambda_f(x)^n}\right]^{\frac{1}{n-1}}, \qquad K_O^f(x) = \left[\frac{\Lambda_f(x)^n}{J_G(x)}\right]^{\frac{1}{n-1}}$$

for Kreines' definition

$$K_I^f(x) = K_O^f(x) = \frac{|\nabla f(x)|^n}{n^{\frac{n}{2}}|J(x)|},$$

for Callender's definition

$$K_I^f(x) = K_O^f(x) = \frac{|\nabla f(x)|^2}{n|J(x)|^{\frac{2}{n}}},$$

and so on.

Remark. The general definition I_1 (or I_1') has not only the advantage to contain (as particular cases) definitions of K-QCfH equivalent to each other and expressed by the same terms and with the same K, but also the advantage to allow us to obtain three new classes of definitions characterized again by a general definition.

We shall call the first class characterized by I′ *the class of definitions of* QCfH *of Grötzsch's type* (because they involve inequalities of Grötzsch's [1] type).

The class of definitions of Lavrent'ev's type. The general formulation of the definitions of this second class of definitions is obtained from I_1 if, instead of supposing that the two functions $K_I^f(x)$, $K_O^f(x)$ are bounded, we suppose they are continuous in D, i.e.

II. A homeomorphism $f(x)$ (satisfying some regularity conditions) is said to be a QCfH in D if a certain pair of functions $K_I^f(x)$, $K_O^f(x)$ is continuous in D.

We call this new class of definitions of QCfH *of Lavrent'ev's type* (because M. A. Lavrent'ev [14] gave a plane definition of this type).

Remark. As we can see, in this case we have not K-QCfH. However, it is easy to see that to each definition of QCfH of Grötzsch's type corresponds a definition of Lavrent'ev's type. Between every pair of corresponding definitions of these two types there is also this very strong connexion: each QCfH in D according to one of the definitions of Lavrent'ev's type is K_Δ-QCfH of Grötzsch's type in every bounded domain $\Delta, \subset\subset D$. (Clearly the converse is not true.)

The class of definitions of Teichmüller's type. The third class of definitions contains the two preceding ones and also a lot of new possible definitions and is obtained from I_1 if instead of the constant K which bounds $K_I^f(x)$, $K_O^f(x)$ a.e. in D, we have a pair of functions $K_I(x)$, $K_O(x)$ continuous in D. i.e.

III. A homeomorphism $f(x)$ (which satisfies some regularity conditions) is said to be a QCfH in D if a certain pair of functions satisfies the inequalities

$$1 \leqq K_I^f(x) \leqq K_I(x), \quad 1 \leqq K_O^f(x) \leqq K_O(x),$$

where $K_I(x)$, $K_O(x)$ are arbitrary functions continuous in D.

We call this new class of definitions *of Teichmüller's type* (because O. Teichmüller [3], p. 15, suggested a plane definition of this type).

If $K_I^f(x) = K_O^f(x) = K = \text{const}$, we obtain I_1, and if $K_I^f(x) = K_I(x)$, $K_O^f(x) = K_O(x)$, we obtain II.

The class of definitions of Andreian's type. The fourth class of definitions which obviously contains the three preceding ones is the following:

IV. A homeomorphism $f(x)$ is said to be QCfH in D if it is K_Δ-QCfH according to I_1 in every bounded domain $\Delta \subset\subset D$.

We call this new class of definitions *of Andreian's type* (because Cabiria Andreian-Cazacu [15] gave a plane definition of this type).

Clearly, every homeomorphism of one of the first two types is a K_Δ-QCfH in each bounded subdomain $\Delta \subset\subset D$. As for QCfH of Teichmüller's type, the two continuous functions $K_I(x)$, $K_O(x)$ are bounded in every bounded domain $\Delta \subset\subset D$ by a constant K_Δ, hence the two functions $K_I^f(x)$, $K_O^f(x)$ are bounded in Δ by the greater of the two constants, and this for any bounded domain $\Delta\subset\subset D$. Clearly, in all cases the constant K_Δ depends on the subdomain Δ.

CHAPTER 12

Characterization of QCfH by the Hurwitz property

All the definitions of QCfH from above give an individual characterization of them. Now we shall give following Gehring [8] a global characterization of the QCfH by means of Hurwitz property. The disadvantage of such a definition is that, given a homeomorphism $f(x)$, it is not possible to settle if it is QCfH according to this definition or not without including it into a family of QCfH according to this definition.

Normal families of homeomorphisms. A family H of homeomorphisms defined in D is said to be *a normal family* if each sequence of homeomorphisms in H, which are bounded at a pair of points in D contains a subsequence, which converges uniformly on each compact subset of D.

Hurwitz property. A family H of homeomorphisms is said *to have the Hurwitz property* if each finite function, which is the limit of homeomorphisms in H, is either a homeomorphism or a constant.

Families of homeomorphisms complete with respect to the similarity mappings. A family H of homeomorphisms is said *to be complete with respect to the similarity mappings* if given any pair of similarity mappings $T^*(x^*)$, $T(x)$, such that $T: D \to D$, then the composite homeomorphism $T^*\{f[T(x)]\}$ is in H whenever $f \in H$.

QCfH characterized by the Hurwitz property. A homeomorphism $f(x)$ in a bounded domain D is a QCfH in D if it belongs to a normal family, complete with respect to the similarity mappings and having the Hurwitz property.

Equivalence with the other definitions. In order to establish the equivalence of this definition to Gehring's geometric definition (and then also to the other), we shall establish (following Gehring [8]) some preliminary results. We begin by establishing some equicontinuity properties of QCfH.

THEOREM 1. *Suppose that $\{f_m(x)\}$ is a sequence of K-QCfH of D which are uniformly bounded on each compact subset of D. Then the $f_m(x)$ are equicontinuous on each compact subset of D.*

Let F be a compact subset of D, $x_0 \in D - F$ and $\Delta = D - x_0$. By hypothesis there exists a finite constant $A_1 = A(F, x_0) < \infty$ such that

$$|f_m(x) - f_m(x_0)| \leqq |f_m(x)| + |f_m(x_0)| \leqq A_1$$

for $x \in F$ and all m. Hence

$$d[f_m(x), \partial\Delta_m^*] \leqq |f_m(x) - f_m(x_0)| \leqq A_1 \tag{1}$$

for $x \in F$ and all m, where $\Delta^* = f_m(\Delta)$. Since F is compact

$$d(x, \partial\Delta) \geqq a > 0 \tag{2}$$

for $x \in F$. Now fix $x \in F$ and choose y so that $|y - x| < a$. Then theorem 9.1 applied to the restriction of $f_m(x)$ to Δ, yields

$$\frac{|f_m(y) - f_m(x)|}{d[f_m(x), \partial\Delta_m^*]} \leqq \Theta\left[\frac{|y - x|}{d(x, \partial\Delta)}\right]$$

and combining (1) and (2) we obtain

$$|f_m(y) - f_m(x)| \leqq A_1\Theta\left[\frac{|y - x|}{a}\right]. \tag{3}$$

Since by (9.51)

$$\lim_{t \to 0} \Theta(t) = 0,$$

then (3) implies the desired equicontinuity.

Remark. An analogous theorem was established by M. Kreines [1].

THEOREM 2. *Suppose that* $\{f_m(x)\}$ *is a sequence of* K-QCfH, $f_m\colon D \rightleftarrows D^*$, *that*

$$\sup_m |f_m(x_0)| < \infty$$

for some fix point $x_0 \in D$, *and that*

$$\sup_m d(O, \partial D_m^*) < \infty,$$

Then the $f_m(x)$ *are uniformly bounded and equicontinuous on each compact subset of* D.

Fix $a \in (0, 1)$. Then if we choose $x \in D$ and y so that $|y - x| < ad(x, \partial D)$, theorem 9.1 implies that

$$|f_m(y) - f_m(x)| \leqq \Theta(a)d[f_m(x), \partial D_m^*]$$

for all m. Since

$$d[f_m(x), \partial D_m^*] \leqq |f_m(x)| + d(O, \partial D_m^*),$$

we thus obtain

$$|f_m(y)| \leqq A_1 |f_m(x)| + A_2, \tag{4}$$

where A_1, A_2, are constants,

$$A_1 = 1 + \Theta(a), \; A_2 = \Theta(a) \sup d(O, \partial D_m^*) < \infty.$$

In particular we conclude that each point $x \in D$ has a neighbourhood $U = U_x \in D$ such that (4) holds for all $y \in U$.

Next if we choose $y \in D$ and x so that $|x - y| < \frac{a}{2} d(y, \partial D)$, then it is easy to show that

$$|y - x| \leqq \frac{a}{2} d(y, \partial D) \leqq \frac{a}{2} |y - x| + \frac{a}{2} d(x, \partial D),$$

which implies that

$$\frac{1}{2} |y - x| \leqq \left(1 - \frac{a}{2}\right) |y - x| \leqq \frac{a}{2} d(x, \partial D),$$

whence

$$|y - x| \leqq a d(x, \partial D).$$

Hence we see that each point $y \in D$ has a neighbourhood $V = V_y \subset D$ such that (4) holds for all $x \in V$.

Now let E denote the set of points $x \in D$ for which

$$\sup_m |f_m(x)| = c(x) < \infty. \tag{5}$$

If $x \in E$ and if U is the neighbourhood described above, then (4) implies that

$$\sup_m |f_m(y)| \leqq A_1 \sup |f_m(x)| + A_2 < \infty$$

for all $y \in U$. Hence $U \subset E$ and E is open. Similarly if $y \in D - E$ and if V is the neigbourhood described above, then (4) implies that

$$\infty = \sup_m |f_m(y)| \leqq A_1 \sup |f_m(x)| + A_2,$$

for all $x \in V$, hence $V \subset D - E$ and then $D - E$ is open too. Since D is connected and $x_0 \in E$, we conclude that (5) holds for all $x \in D$.

Finally, suppose that F is a compact subset of D. Then the neighbourhoods $U(x)$ described above cover F as x ranges through F, and we can choose $x_1, \ldots, x_p$ so that

$$F \subset \bigcup_{k=1}^{p} U(x_k).$$

It then follows from (4) that

$$|f_m(y)| \leqq A_1 \max[c(x_1), \ldots, c(x_p)] + A_2 < \infty$$

for $y \in F$, and hence $f_m(x)$ are uniformly bounded on F. The equicontinuity is now a consequence of the preceding theorem.

THEOREM 3. *Suppose* $\{f_m(x)\}$ *is a sequence of* K-QCfH *of* D *and that*

$$\sup_m |f_m(x_0)| < \infty, \quad \sup_m |f_m(x_1)| < \infty$$

for a pair of distinct points $x_0, x_1 \in D$. *Then the* $f_m(x)$ *are uniformly bounded and equicontinuous on each compact subset of* D.

Set $\Delta = D - x_1$ and $\Delta_m^* = f_m(\Delta)$. Then $f_m(x_1) \in \partial\Delta_m^*$, and hence

$$\sup_m d(O, \partial\Delta_m^*) \leqq \sup_m |f_m(x_1)| < \infty,$$

The preceding theorem now implies the desired conclusions on each compact subset of Δ. Interchanging the roles of x_0 and x_1 then yields these results on each compact subset of D.

And now, combining theorem 9.1 and the preceding one, we generalize for QCfH Hurwitz's theorem (see A. I. Markuševič [2], p. 317) about the limit functions of some normal families of analytic functions. A particular case of this theorem was established by P. P. Belinskiĭ [1] before F. Gehring [8].

THEOREM 4. *Suppose that* $\{f_m(x)\}$ *is a sequence of* K-QCfH *of* D, *that*

$$\lim_{m\to\infty} f_m(x) = {}_J(x), \quad |f(x)| < \infty, \tag{6}$$

in D, *that* $f_m(x) \neq x_m^{*\prime}$ *in* D, *and that*

$$\lim_{m\to\infty} x_m^{*\prime} = x^{*\prime}.$$

Then either $f(x) \neq x^{*\prime}$ *in* D *or* $f(x) \equiv x^{*\prime}$ *in* D.

Let E be the set of points $x \in D$ for which $f(x) \equiv x^{*\prime}$. The preceding theorem implies that the $f_m(x)$ are equicontinuous on each compact subset of D. Hence, by Artzela—Ascoli's theorem (Chapter 5) and by (6), it follows that $f(x)$ is continuous and E is closed in D. Now suppose $x \in E$ and let U be the set of points y for which $|y - x| < ad(x, \partial D)$, where $a \in (0, 1)$. Then theorem 9.1 implies that

$$|f_m(y) - f_m(x)| \leqq \Theta(a)\, \mathrm{d}[f_m(x), \partial D_m^*]$$

for all $y \in U$, where $D_m^* = f_m(D)$. Since $x_m^{*\prime} \in D_m^*$, we see that

$$d[f_m(x), \partial D_m^*] \leqq |f_m(x) - x_m^{*\prime}|$$

and hence that

$$|f(y) - f(x)| = \lim_{m \to \infty} |f_m(y) - f_m(x)| \leqq \Theta(a) \lim_{m \to \infty} |f_m(x) - x_m^{*\prime}| = 0$$

for all $y \in U$. Hence $U \subset E$ and E is open. Since D is connected, and E is closed in D, we conclude that either $E = \emptyset$ or that $E = D$. Thus either $f(x) \neq x^{*\prime}$ in D or else $f(x) \equiv x^{*\prime}$ in D as desired.

THEOREM 5. *Suppose that* $\{f_m(x)\}$ *is a sequence of* K-QCfH *of* D *and that* (6) *holds in* D. *Then* $f(x)$ *is either a homeomorphism or a constant.*

Theorem 3 implies that the $f_m(x)$ are equicontinuous on each compact subset of D, hence by Arzela—Ascoli's theorem (Chapter 5) and by (6), $f(x)$ is continuous in D. If $f(x)$ is not one to one, we can find a pair of distict points $x', y' \in D$ such that $f(x') = f(y') = x^{*\prime}$. Let $\Delta = D - y'$ and $f_m(y') = x_m^{*\prime}$. Clearly, $f_m(x) \neq x_m^{*\prime}$ in Δ and $\lim_{m\to\infty} x_m^{*\prime} = x^{*\prime}$. Since $x' \in \Delta$ and $f(x') = x^{*\prime}$, the preceding theorem implies $f(x) = x^{*\prime}$ in Δ and hence, $f(x)$ is constant in D. The desired conclusion now follows from the well known theorem of invariance of domains under homeomorphisms of Brouwer [1] (see M. H. A. Newman [1], theorem 21.4, p. 137 and also our Notes [3], [5]).

THEOREM 6. *Suppose that* H *is a family of homeomorphisms of a bounded domain* D *and that* H *is normal, complete with respect to the similarity mappings and has the Hurwitz property, then each homeomorphism in* H *is* K-QCfH *according to Gehring's metric and geometric definition for some fixed* K.

We may assume, by performing a preliminary change of variables (for instance a similarity mapping which preserves the family H), that $D \supset B$. Then for each homeomorphism $f \in H$ we set

$$K(f) = \frac{\max\limits_{S} |f(x) - f(O)|}{\min\limits_{S} |f(x) - f(O)|}.$$

Now the fact that H has the Hurwitz property implies that

$$K = \sup_{f \in H} K(f) < \infty. \tag{7}$$

For if (7) does not hold, we can find a sequence of homeomorphisms $f_m \in H$ such that

$$ml_m = m \min_S |f_m(x) - f_m(O)| \leqq \max_S |f_m(x) - f_m(O)| = L_m,$$

Then, since H is complete with respect to the similarity mappings,

$$\varphi_m(x) = \frac{f_m(x) - f_m(O)}{L_m} \in H.$$

Now $|\varphi_m(x)| \leqq 1$ for $|x| \leqq 1$, and because H is a normal family, we can find a subsequence $\{m_p\}$ such that

$$\lim_{p \to \infty} \varphi_{m_p}(x) = \varphi(x), \ |\varphi(x)| < \infty,$$

uniformly on each compact subset of D. Next for each m, there exist points $x_m, y_m \in S$ such that

$$|\varphi_m(x_m)| = 1, \ |\varphi_m(y_m)| = \frac{l_m}{L_m} \leqq \frac{1}{m}. \tag{8}$$

Because S is compact, we may assume, by choosing a second subsequence and then relabelling, that

$$\lim_{p \to \infty} x_{m_p} = x_0, \ \lim_{p \to \infty} y_{m_p} = y_0$$

and by virtue of the uniform convergence of the sequence $\{\varphi_{m_p}(x)\}$, we conclude from (8) that

$$|\varphi(x_0)| = \lim_{p \to \infty} |\varphi_{m_p}(x_{m_p})| = 1, \ |\varphi(y_0)| = \lim_{p \to \infty} |\varphi_{m_p}(y_{m_p})| = 0.$$

Now $\varphi(0) = 0$ and $y_0 \neq 0$. Thus $\varphi(x)$ is neither a homeomorphism nor a constant, and then H does not have the Hurwitz property. This contradiction implies (7).

We complete the proof of theorem 6 by showing that each homeomorphism in H is K-QCfH according to Gehring's metric and geometric definitions, where K is the constant given by (7). For this, fix $f \in H$ and $x_0 \in D$. Since D is bounded, we can choose $a < \infty$ so that $D \subset B(a)$. Then for $0 < ar < d(x_0, \partial D)$, the similarity mapping $T(x) = x_0 + rx$ maps D into itself, and hence, since H is complete with respect to similarity mappings,

$$\psi_r(x) = f(x_0 + rx) \in H.$$

If we now apply (7) to $\psi_r(x)$, we obtain

$$L(x_0, r) = \max_{|x-x_0|=r} |f(x)-f(x_0)| = \max_{|x|=1} |\psi_r(x) - \psi_r(O)| \leqq$$

$$\leqq K \min_{|x|=1} |\psi_r(x) - \psi_r(O)| = K \min_{|x-x_0|=r} |f(x)-f(x_0)| = K l(x_0, r) \tag{9}$$

for $0 < ar < d(x_0, \partial D)$. We conclude that

$$\delta_L(x_0) = \overline{\lim_{r\to 0}} \frac{L(x_0, r)}{l(x_0, r)} \leqq K$$

for all $x_0 \in D$, and hence $f(x)$ is K-QCfH according to Gehring's metric definition. Thus, by theorem 3.1, $f(x)$ is K-QCfH also according to his geometric definition.

Remark. (7) and (9) allow us to conclude that for each family of homeomorphisms H which is normal, complete with respect to the similarity mappings and has the Hurwitz property, there exists a constant K ($1 \leqq K < \infty$) such that

$$\frac{\max\limits_{|\Delta x|=r} |f(x+\Delta x) - f(x)|}{\min\limits_{|\Delta x|=r} |f(x+\Delta x) - f(x)|} \leqq K,$$

for each $f \in H$, $x \in D$ and where $ar < d(x, \partial D)$ and $\dfrac{d(D)}{2} < a < \infty$.

Hence we obtain a definition of K-QCfH characterized by Hurwitz property as follows:

A homeomorphism $f(x)$ of a bounded domain D is a K-QCfH ($1 \leqq K < \infty$) in D, if it is contained in a normal family, complete with respect to the similarity mappings, having the Hurwitz property and with the constant corresponding to (7) equal to K.

THEOREM 7. *The preceding definition of* QCfH *is equivalent to Gehring's metric definition.*

The fact that each K-QCfH according to the preceding definition is a K-QCfH according to Gehring's metric definition is a consequence of the preceding theorem.

Now let H be the class of K-QCfH according to Gehring's metric definition. This class is a normal family of homeomorphisms (theorem 3 and Arzela—Ascoli's theorem), having the Hurwitz property (theorem 5) and complete with respect to the similarity mappings since the similarity mappings are conformal mappings, and then preserves the class of K-QCfH (no matter according to which definition). Hence, we conclude that this class is of QCfH according also to the preceding definition.

CHAPTER 13

A compactness characterization for QCfH

Some properties of the sequences of QCfH (Gehring [5]).

THEOREM 1. *Suppose that* $\{f_m(x)\}$ *is a sequence of* K-QCfH *(according to Gehring's geometric definition) of* D, *that*

$$\lim_{m\to\infty} f_m(x) = f(x)$$

uniformly on each compact subset of D, *and that* $f(x)$ *is a homeomorphism. The* $f(x)$ *is* K-QCfH *in* D *according to Gehring's geometric definition.*

Let A be a bounded ring with $\bar{A} \subset D$ and let A_m^* and A^* be the images of A under $f_m(x)$ and $f(x)$, respectively. Then A^* is a bounded ring and each component of ∂A_m^* converges uniformly to the corresponding component of ∂A^*. But then, by lemma 6.2,

$$\operatorname{mod} A^* = \lim_{m\to\infty} \operatorname{mod} A_m^*,$$

and, since

$$\frac{1}{K} \operatorname{mod} A \leqq \operatorname{mod} A_m^* \leqq K \operatorname{mod} A$$

for all m, we conclude that $f(x)$ itself is K-QCfH according to Gehring's geometric definition.

THEOREM 2. *Let* ∂D *be non-empty and let* $\{f_m(x)\}$ *be a sequence of* K-QCfH, $f_m ; D \rightleftharpoons D^*$. *Then either*

$$\lim_{m\to\infty} |f_m(x)| = \infty \tag{1}$$

everywhere in D, *or there exists a subsequence which converges uniformly on each compact set in* D *to a function* $f(x)$. *This limit function is either a constant vector in* ∂D^* *or a* K-QCfH *(according to Gehring's geometric definition)* $f : D \rightleftharpoons D^*$.

Let E be the set of $x \in D$ for which

$$\overline{\lim_{m\to\infty}} |f_m(x)| < \infty. \tag{2}$$

We show first that either $E = \emptyset$ or $E = D$. For this fix $x_0 \in E$ and pick x_1 so that $|x_0 - x_1| < \dfrac{d(x_0, \partial D)}{2}$. Then theorem 9.1 yields

$$|f_m(x_1)| \leqq |f_m(x_0)| + d[f_m(x_0), \partial D^*]\Theta_k\left(\frac{1}{2}\right), \tag{3}$$

and since the quantities on the right are bounded, we conclude that $x_1 \in E$. Hence E is open. Similarly, if $x_1 \in D - E$ and x_0 is chosen so that $|x_0 - x_1| < \dfrac{d(x_0, \partial D)}{3}$, then $|x_0 - x_1| < \dfrac{d(x_0, \partial D)}{3} \leqq \dfrac{|x_0 - x_1|}{3} + \dfrac{d(x_0, \partial D)}{3}$, hence $|x_0 - x_1| < \dfrac{d(x_0, \partial D)}{2}$ and (3) implies $|f_m(x_0)| = \infty$, i.e. $x_0 \in D - E$. Hence $D - E$ is also open and, since D is connected, it follows $E = \emptyset$ or $E = D$.

From the above we now see that either (1) holds for all $x \in D$ or there exists a subsequence $\{f_{m_k}\}$ such that

$$\overline{\lim_{k\to\infty}} |f_{m_k}(x)| < \infty$$

for all $x \in D$. Theorem 12.3 then shows that the $f_{m_k}(x)$ are equicontinuous in D and we can apply to Arzela—Ascoli's theorem to find a second subsequence which converges uniformly on each compact set in D to a function $f(x)$. To complete the proof we must show that $f(x)$ is either a constant vector in ∂D^* or a K-QCfH according to Gehring's geometric definition of D onto D^*.

By relabelling we may assume that the original sequence $\{f_m(x)\}$ converges to $f(x)$. Now fix a point $x_0^* \in \partial D^*$ and let E_1 be the set of x in D for which $f(x) = x_0^*$. Since $f(x)$ is continuous, E_1 is closed in D. On the other hand if $x_0 \in E_1$ and if $|x_0 - x_1| < \dfrac{d(x_0, \partial D)}{2}$, then theorem 9.1 yields

$$|f(x_1) - x_0^*| = \lim_{m\to\infty} |f_m(x_1) - f_m(x_0)| \leqq \underline{\lim}_{m\to\infty}\, d[f_m(x_0), \partial D^*]\Theta_k\left(\frac{1}{2}\right) = 0,$$

Hence $x_1 \in E_1$ and E_1 is open. We conclude that either $E_1 = D$ and $f(x)$ is a constant vector in ∂D^* for x running over D, or that $E_1 = \emptyset$ and then $f(x) \in D^*$ for all $x \in D$.

We must prove that in this case $f(x)$ is a K-QCfH, $f: D \rightleftarrows D^*$ (according to Gehring's geometric definition). We begin by showing that

$$\lim_{m\to\infty} f_m^{-1}[f(x)] = x \tag{4}$$

for all $x \in D$, where $f_m^{-1}(x^*)$ is the inverse of $f_m(x)$. For this fix $x_0 \in D$. Then $f(x_0) \in D^*$ and theorem 9.1 applied to $f_m^{-1}(x^*)$ gives

$$\left|f_m^{-1}[f(x_0)] - x_0\right| < d(x_0, \partial D)\Theta_k\left\{\frac{|f(x_0) - f_m(x_0)|}{d[f_m(x_0), \partial D^*]}\right\}$$

for sufficiently large m in order that $\dfrac{|f(x_0) - f_m(x_0)|}{d[f_m(x_0), \partial D^*]} < 1$, and then letting $m \to \infty$ yields (4).

Next all of the above arguments can be applied to the sequence $f_m^{-1}(x^*)$. Thus

$$\lim_{m \to \infty} \left|f_m^{-1}(x^*)\right| = \infty$$

for all $x^* \in D^*$, or

$$\overline{\lim_{m \to \infty}} \left|f_m^{-1}(x^*)\right| < \infty$$

for all $x^* \in D^*$. But (4) implies clearly the preceding inequality for $x^* \in D^*$, and then $f_m^{-1}(x^*)$ are equicontinuous in D^* and uniformly bounded on every compact subset of D^*. Then, by Arzela—Ascoli's theorem, a subsequence converges uniformly on each compact subset of D^* to a continuous function $f^{-1}(x^*)$. Relabelling allows us to replace this subsequence by the sequence $\{f_m^{-1}(x^*)\}$. Then (4) implies that $f^{-1}f[(x)] = x$ for all $x \in D$. It then follows (taking into account also what was proved above), that $f^{-1}(x^*) \in D$ for all $x^* \in D^*$ and arguing as above we conclude that $f[f^{-1}[(x^*)] = x^*$ for all $x^* \in D^*$. Thus $x^* = f(x)$ is a homeomorphism of D onto D^*. The preceding theorem implies that $f(x)$ is a K-QCfH according to Gehring's geometric definition and the proof of the theorem 2 is complete.

Remark. This Gehring's [5] theorem was established in the particular case $D = B$ by Väisälä [1].

COROLLARY. *Let D and D^* be domains, let ∂D be non-empty and let x_0, x_0^* be fixed points in D and D^*, respectively. Then the K-QCfH (according to Gehring's geometric definition) of D onto D^* which map x_0 onto x_0^* form a closed normal family.*

Compactness condition (A). Let H be a family of homeomorphisms $f(x)$ of the whole space (i.e. $D = R^n$). Let $O_1 = (1, 0, \ldots, 0)$. We say that a homeomorphism $f(x)$ *is normalized* if $f(O) = O$ and $f(O_1) = O_1$. Then the family is said to *satisfy the compactness condition* (A) if every infinite set of normalized homeomorphisms in H contains a sequence which converges uniformly on compact sets to a homeomorphism.

QCfH characterized by a compactness condition. A homeomorphism $f(x)$ is a QCfH in R^n, if the family of all the mappings of the form $T^* \circ f \circ T$, where T and T^* are similarity mappings, satisfies *the compactness condition* (A).

The equivalence to other definitions.

THEOREM 3. *Let H be a family of homeomorphisms of the space which is complete with respect to the similarity mappings. Then H satisfies the compactness condition (A) if and only if each mapping in H is K-*QCfH *according to Gehring's geometric definition for some fixed K.*

Suppose first that the homeomorphisms in H are K-QCfH. The preceding corollary implies that the K-QCfH of the space which keep O and O_1 constitute a closed normal family. Hence, each sequence of normalized homeomorphisms in H contains a subsequence which converges uniformly on each compact set to a K-QCfH of the space and H clearly satisfies the compactness condition (A).

For the converse let H be a family of homeomorphisms of R^n, complete with respect to the similarity mappings and satisfying the compactness condition (A), and let

$$K = \sup \left[\max_S |f(x)|\right], \tag{5}$$

where the supremum is taken over all normalized mappings $f \in H$. Then there exists a sequence of normalized mappings $f_m \in H$ for which

$$K = \lim_{m\to\infty} \left[\max_S |f_m(x)|\right],$$

and since, by the compactness condition (A), which is satisfied by $\{f_m(x)\}$, there exists a subsequence converging uniformly on the compact set S to a homeomorphism, it follows that $1 \leqq K < \infty$.

Now let $f(x)$ be any mapping in H, fix x_0 and $r > 0$ and choose $x_1 \in S(x_0, r)$ so that $|x_0^* - x_1^*| = l(x_0, r)$. Next let T and T^* be the similarity mappings which map S onto $S(x_0, r)$ and $|x^* - x_0^*| = l(x_0, r)$ onto $|x^*| = 1$, so that

$$T(O) = x_0,\ T(O_1) = x_1,\ T^*(x_0^*) = O,\ T^*(x_1^*) = O_1.$$

Then $T^* \circ f \circ T$ is a normalized mapping in H and we thus obtain

$$L(x_0, r) = l(x_0, r) \max_S \left|T^*\{f[T(x)]\}\right| \leqq K l(x_0, r),$$

hence

$$\frac{L(x_0, r)}{l(x_0, r)} \leqq K,$$

letting $r \to 0$, yields $\delta_L(x_0) \leqq K$, and we conclude that $f(x)$ is K-QCfH according to Gehring's metric definition in R^n. As f was an arbitrary homeomorphism in H, the proof is complete.

Remarks. 1. The definition of the QCfH by means of the compactness condition (A), for a family defined not in the whole space, but only in a bounded domain, is equivalent (as it is easy to see) to the definition of the QCfH characterized by the Hurwitz property.

2. The preceding definition which characterizes QCfH by a compactness condition still holds for $n = 1$ (see L. Ahlfors and A. Beurling [1]). They established there also a theorem similar to the preceding one: "The homeomorphism of a family closed with respect to the linear mappings satisfy the compactness condition (A) if and only if they satisfy the ρ-condition (3.1), with the same ρ for all the homeomorphisms of the family".

As in the case of the preceding chapter, we can give also in this case a definition for the K-QCfH:

A homeomorphism $f(x)$ is said to be K-QCfH $(1 \leqq K < \infty)$ in R^n if the family of all the homeomorphisms of the form $T^* \circ f \circ T$, where T, T^* are similarity mappings, satisfy the compactness condition (A), and the constant corresponding to the formula (5) is equal to K.

Remarks. 1. It is easy to see that when $f(x)$ is defined in a bounded domain, the preceding definition of the K-QCfH is equivalent to that of the preceding chapter.

2. The first 13 chapters of the second part allow us to conclude that all the definitions of Grötzsch's type, the Θ-mappings and also the definitions of the last two chapters are all equivalent if we do not specify the constant K involved in the corresponding definition.

CHAPTER 14

Markuševič's definition of the K-QCf mappings

In this chapter we shall mention just in passing a definition of QCfH where one gets rid of the restriction for the function to be one to one.

Interior transformations in the sense of Stoïlow [1]. A mapping $f\colon R^n \to R_u$ is said to *be open* if it maps open sets into open sets.

A mapping is called *zerodimensional* if it maps every point into a discrete set.

A mapping $f\colon R^n \to R^n$ is said to be *an interior mapping in Stoïlow's sense* if

1° $f(x)$ is continuous;

2° $f(x)$ is open;

3° $f^{-1}(x^*)$ is zerodimensional.

Markuševič's definition. A K-QCf mapping ($1 \leq K < \infty$) in D is an interior transformation $f(x)$ in D, which is a K-QCfH in a neighbourhood of every point where it is univalent.

Remarks. 1. A. I. Markuševič [1] (1940) and Marie-Hélène Schwartz [1]—[4] (1950) adopt for QCfH the definition of K-QCfH with a system of characteristics.

2. Also in the bidimensional case, even in the first geometric definition given by H. Grötzsch [1] (1928), the QCf mappings are considered only locally one to one.

3. A. I. Markuševič [1] considered also the class of continuous mappings $f(x)$ in a domain $D \subset R^n$, so that for every $x \in D - E$, where $mE = 0$,

(a) $f(x)$ is one to one in a neighbourhood U_x of x and

(b) in U_x there is a sequence $\{\sigma_m(x)\}$ of surfaces that are homeomorphic to spheres and

$$\text{(A)}\qquad \begin{gathered} \lim_{m\to\infty} \frac{r_{m+1}(x)}{r_m(x)} > 0, \qquad \varliminf_{m\to\infty} \frac{r_m(x)}{R_m(x)} = \frac{1}{k} > 0, \\ \lim_{m\to\infty} \frac{r_m^*(x^*)}{R_m^*(x^*)} = \frac{1}{k^*} > 0, \qquad \lim_{m\to\infty} R_m(x) = 0, \end{gathered}$$

where $r_m(x)$, $R_m(x)$ denote the minimum, respective the maximum, of the distances from x to $\sigma_m(x)$, $\sigma_m^*(x)$ is the image of $\sigma_m(x)$, and $r_m^*(x)$, $R_m^*(x)$ the corresponding distances from x^* to $\sigma_m^*(x)$. He observed that one can substitute each sequence $\{\sigma_m(x)\}$ by the family $\{\sigma_\lambda(x)\}$, which fills a neighbourhood of x and then the first condition of (A) becomes unnecessary.

We remark that if we impose to this class of mappings the additional conditions to be one to one everywhere in D (i.e. $E = \emptyset$) and to have inf $k(x)\,k'(x)$ bounded in D and inf $k(x)k'(x) \leqq K$ a.e. in D, where the infimum is taken over all the families $\{\sigma_\lambda(x)\}$ of surfaces that are homeomorphic to spheres, then the class of Markuševič's continuous locally one to one mappings reduces to the class of K-QCfH in Markuševič—Pesin's sense. The main result Markuševič proved about his class from above is its differentiability a.e. in D.

n-Dimensional pseudoconformal transformation (PCT) and $2n$-dimensional QCf diffeomorphisms

PCT. Let be $z = (z^1, \ldots, z^n)$, $z^* = (z^{*1}, \ldots, z^{*n})$, $z^k = x^{2k-1} + ix^{2k}$, $z^{*k} = x^{*2k-1} + ix^{*2k}$($i^2 = -1$) and C^n the space of complex variables z^k ($k=1, \ldots, n$).

A *PCT* $z^* = f(z)$ of a domain $\Omega \subset C^n$ onto a domain $\Omega^* \subset C^n$ is a diffeomorphism in Ω whose components $z^{*k}(z)$ ($k = 1, \ldots, n$) are holomorphic in each variable z^k ($k = 1, \ldots, n$). This term is used by S. Bergman [1], we meet also the term of holomorphic mappings (S. Hitotumatu [1]), biholomorphic mappings (B. A. Fuks [2]), analytic transformations (H. Cartan [1]) and analytic isomorphisms (for instance M. Jurchescu in the joint book with Cabiria Andreian-Cazacu and C. Constantinescu [1]).

First order elliptic systems. In order to see one of the aspects of the connexions between QCf diffeomorphisms and *PCT*, we shall consider the first order partial differential equations satisfied by QCf diffeomorphisms and by *PCT*.

Since all the definitions of the QCfH of Grötzsch's type are equivalent (without specifying the constant K) and since also the QCfH of other types are K-QCfH in every bounded subdomain $\Delta \subset\subset D$, lemma 1.3 and corollary of theorem 1.1 imply that almost all points of D are A-points, and then $f(x)$ maps the infinitesimal ellipsoids

$$\alpha_{ik}\, dx^i\, dx^k = d\rho^2, \quad \det|\alpha_{ik}| = 1, \tag{1}$$

centered at an A-point into infinitesimal spheres (corollary of lemma 3.7).

And now, let us prove theorem 1.

THEOREM 1. *At a point of differentiability an n-dimensional* QCfH $f(x)$ *satisfies the system*

$$\frac{\delta_{ik} x_p^{*i} x_q^{*k}}{\alpha_{pq}} = J^{\frac{2}{n}} \qquad (p, q = 1, \ldots, n), \tag{2}$$

where α_{pq} are the coefficients of the equation of the infinitesimal ellipsoids (1), *which are mapped onto infinitesimal spheres by $f(x)$.*

If x_0 is a point of differentiability in D where $J(x_0) = 0$, the system (2) is satisfied by theorem 3.5. Suppose $J(x_0) \not\equiv 0$. Then

$$d\rho^2 = \delta_{ik} dx^{*i} dx^{*k} = \delta_{ik} x_p^{*i} x_q^{*k} dx^p dx^q;$$

in other words, the infinitesimal ellipsoids

$$\delta_{ik} x_p^{*i} x_q^{*k} \, dx^p \, dx^q = d\rho^2 \tag{3}$$

are mapped by $f(x)$ onto the infinitesimal spheres

$$\delta_{ik} \, dx^{*i} \, dx^{*k} = d\rho^2.$$

Since (1) and (3) represent the same equation, it follows that their coefficients are proportional, hence we obtain the system (2). As a matter of fact, the equality to $J^{\frac{2}{n}}$ is a consequence of the other $n^2 + n - 2$ equations representing the equality between the left-hand sides in (2).

THEOREM 2. *Let $f(x)$ be a PCT of $D \subset R^{2n}$. Then the coefficients α'_{km} of the corresponding infinitesimal ellipsoid* (1) *satisfy the relations*

$$\alpha_{2m-1,\,2k-1} = \alpha_{2m,\,2k} \qquad (m, k = 1, \ldots, n), \tag{4}$$

$$\alpha_{2m-1,\,2k} = -\,\alpha_{2m,\,2k-1} \qquad (m, k = 1, \ldots, n). \tag{5}$$

We firstly remark that the diffeomorphism $f(x)$ may be written also in the form $z^* = \varphi(z)$, where $z = z^k e_k$, $z^* = z^{*k} e_k$, $z^k = x^{2k-1} + ix^{2k}$, $z^{*k} = x^{*2k-1} + ix^{*2k}$, $i^2 = -1$. If we suppose that this diffeomorphism is a *PCT*, then $J \neq 0$ and $z^{*k} = z^k(z)$ $(k = 1, \ldots, n)$ will be an analytic function of z, i.e. an analytic function in each variable z^k $(k = 1, \ldots, n)$ separately. But this means that the system of functions $x^{*p} = x^{*p}(x^1, \ldots, x^{2n})$ $(p = 1, \ldots, 2n)$ satisfies in D the first order differential system

$$x_{2k-1}^{*2m-1} = x_{2k}^{*2m}, \quad x_{2k}^{*2m-1} = -\,x_{2k-1}^{*2m} \qquad (m, k = 1, \ldots, n), \tag{6}$$

which represents the Cauchy—Riemann systems for all couples of indices m, k. But the system (6) implies that

$$\begin{aligned} \delta_{pq} x_{2m-1}^{*p} x_{2k-1}^{*q} &= \delta_{pq} x_{2k}^{*p} x_{2k}^{*q} \qquad (m, k = 1, \ldots, n), \\ \delta_{pq} x_{2m-1}^{*p} x_{2k}^{*q} &= -\,\delta_{pq} x_{2m}^{*p} x_{2k-1}^{*q} \qquad (m, k = 1, \ldots, n), \end{aligned} \tag{7}$$

where the dummy indices run over $1, \ldots, 2n$ and (2) changes as

$$\frac{\delta_{ab} x_p^{*a} x_q^{*b}}{\alpha_{pq}} = |J|^{\frac{1}{n}} \qquad (p, q = 1, \ldots, 2n), \tag{8}$$

where the dummy indices a, b take on the values $1, \ldots, 2n$. Hence a *PCT* $f(x)$ satisfies the systems (7) and (8), and then also the systems (4) and (5).

THEOREM 3. *Let $f(x)$ be a PCT and $a_p(x)$ $(p = 1, \ldots, 2n)$ the semi-axes of the corresponding infinitesimal ellipsoids* (1). *Then* $a_{2k-1} = a_{2k}$ $(k = 1, \ldots, n)$.

The system (5) comes for $m = k$, to

$$\alpha_{2k-I,\, 2k} = 0 \qquad (k = 1, \ldots, n)$$

so that the equation (1) of the corresponding infinitesimal ellipsoid (i.e. where the indices run over the values $1, \ldots, 2n$) comes to the more particular form:

$$\alpha_{2k,\, 2k}[(dx^{2k-1})^2 + (dx^{2k})^2] + 2\alpha_{2m,\, 2k}[dx^{2m-1}\, dx^{2k-1} + dx^{2m}\, dx^{2k}) +$$
$$+ 2\alpha_{2m-1,\, 2k}(dx^{2m-1}\, dx^{2k} - dx^{2m}\, dx^{2k-1}) = d\rho^2,$$

where $k \neq m$. If we consider the differentials dx^p $(p = 1, \ldots, 2n)$ as variables and we intersect these infinitesimal ellipsoids with the bidimensional plane

$$dx^m = 0 \qquad (m = 1, \ldots, 2k - 2, 2k + 1, \ldots, 2n),$$

then we obtain

$$(dx^{2k-1})^2 + (dx^{2k})^2 = \frac{d\rho^2}{\alpha_{2k,\, 2k}},$$

which is the equation of an infinitesimal circle. Hence, if $a_p(x)$ $(p = 1, \ldots, 2n)$ denote the semi-axes of the preceding ellipsoid, we conclude that $a_{2k-1}(x) = a_{2k}(x)$.

Remarks. 1. A similar result has been obtained for two complex variables by S. Bergman [1,2] and for analytical quaternions by G. Haefeli [1].

2. From theorems 2 and 3 it follows that it is possible a *PCT* not to be a QCfH of Grötzsch type, because, for instance, the principal characteristic $p_1(x)$ in the definition of the QCfH with one or two sets of characteristics can not be bounded in the domain of definition, as for the *PCT*

$$z^{*1} = z^1, \quad z^{*2} = (z^2 - 1)^2$$

of the dicylinder $\{|z^1| < 1, |z^2| < 1\}$, where the ratio of the extreme semi-axes of the infinitesimal ellipsoid (1) corresponding to the preceding *PCT* approaches ∞ as $z^2 \to 1$.

3. However, the *PCT* are QCf of Lavrent'ev's type because $p_1(x)$ is continuous in D since it is a continuous function of the first order partial derivatives of the components of a *PCT*, which is of class C^∞ (see F. Gehring [5]).

4. The *PCT* are QCfH also according to definition of Teichmüller's and of Andreian's type.

Discwise QCf diffeomorphisms (Hitotumatu). Remark 2 allows us to conclude that *PCT* are not in general 1-QCfH. S. Hitotumatu [1] introduces the so called *K-discwise QCf mappings* which generalize the *K*-QCfH from the theory of functions of a complex variable to several complex variables in such a way that a 1-discwise-QCf mapping be a *PCT.*

A homeomorphism $z^* = f(x)$ is said to be *discwise* QCf *with* dilatation K if

(i) $z^{*i} \in C^1$ in D $(i = 1, \ldots, n)$, and

(ii) all z^{*i} $(i = 1, \ldots, n)$ are K-QCf on each holomorphic plane.

More precisely, the condition (ii) means the following: as far as we have a linear mapping

$$z^p = a^p\zeta + b^p \qquad (p = 1, \ldots, n) \tag{9}$$

(a^p, b^p: complex constants, ζ complex variable) defined on the unit disc B^2 whose image lies completely in D, the composite functions

$$w^p(\zeta) = w^p(a^1\zeta + b^1, \ldots, a^n\zeta + b^n) \qquad (p = 1, \ldots, n)$$

are always K-QCf in B^2.

A mapping $f(x)$ satisfying conditions (i), (ii) is called a *K-discwise* QCf *mapping.* If $n = 1$, we obtain a definition for the K-QCf mappings, but where condition (i) is more restrictive than in the corresponding definitions of this book. The K-discwise QCf mapping f: $C^n \to C^2$ is termed a *K-discwise* QCf *function.*

Remark. This class of homeomorphisms, in contradistinction to that of the QCfH, does not form a group (the inverse of a discwise QCfH or the product of two such homeomorphisms can be no more a discwise QCfH). In order that the composition mapping of two discwise QCfH $w = f(z)$ and $z = \varphi(\zeta)$ be again a discwise QCfH, it is necessary that each component of $f(z)$, in which we put $\varphi(\zeta)$ in place of z, satisfy condition (i), (ii) of the preceding definition. For example, if $f(z)$ is a 1-discoidal QCfH in $D \subset C^2$, with the component $w^1(z) = z^1 + z^2$, and if $z^1 = 2\zeta^1 + 2\zeta^2 + \bar{\zeta}^1 + \bar{\zeta}^2$, $z^2 = -\zeta^1 - \zeta^2$, where $\zeta = (\zeta^1, \zeta^2)$, then $\psi^1(\zeta) = w^1[\varphi(\zeta)] = 2(\xi^1 + \xi^2)$, where $\zeta^k = \xi_k + i\eta_k$, $\bar{\zeta}^k = \xi^k - i\eta^k$ $(k = 1, 2)$. But $\psi^1(\zeta)$ does no more satisfy condition (ii), because, on the holomorphic plane $\zeta^2 = \text{const}$ the function $\psi^1 = 2(\xi^1 + \xi^2)$ assumes only real values, hence it is not a QCfH of z^1.

QPCT (Bergman) [1], [2]. A homeomorphism $x^* = f(x)$ of $D \subset R^{2n}$ is said to be a *QPCT* (*quasi-pseudoconformal transformation*) in D if

$$\left| \frac{\partial(z^{*1}, \ldots, z^{*n})}{\partial(x^1, \ldots, x^n)} \right|^2 \leq C \frac{\partial(x^{*1}, y^{*1}, \ldots, x^{*n}, y^{*n})}{\partial(x^1, y^1, \ldots, x^n, y^n)}, \tag{10}$$

where $C = C(x^1, y^1, \ldots, x^n, y^n) > 0$, $z^k = x^k + iy^k$ and $z^{*k} = x^{*k} + iy^{*k}$ $(k = 1, \ldots, n)$.

This definition generalizes that of *PCT* because the inequality (10) generalizes the relation

$$\frac{\partial(x^{*1}, y^{*1}, \ldots, x^{*n}, y^{*n})}{\partial(x^1, y^1, \ldots, x^n, y^n)} = \left| \frac{\partial(z^{*1}, \ldots, z^{*n})}{\partial(x^1, \ldots, x^n)} \right|^2,$$

satisfied by the *PCT*.

Connection between discwise QCf diffeomorphisms and PCT. A point z_0 is said to be *ordinary* with respect to a discwise QCf function $f(z)$ if at least one of the derivatives $\dfrac{\partial f(z_0)}{\partial z^k}$ $(k = 1, \ldots, n)$ does not vanish at z_0.

THEOREM 4. (*Hitotumatu*). *At every ordinary point z_0 a K-discwise* QCf *function satisfies the differential equations*

$$\frac{\partial \bar{f}}{\partial z^p} = \chi \frac{\partial f}{\partial z^p} \tag{11}$$

where $\bar{f}$ is the conjugate of f and

$$|\chi| \leqq k_0 = \frac{K-1}{K+1} < 1. \tag{12}$$

We may assume, by changing the coordinates if necessary, that $z_0 = O$, $\dfrac{\partial f(O)}{\partial z^1} \neq 0$ and that the linear mapping (9) maps B^2 into D, provided that

$$|a^p| < \delta \quad (p = 1, \ldots, n), \tag{13}$$

where $\delta > 0$ is sufficiently small. By definitions of *K*-discwise QCf functions,

$$\tilde{f}(\zeta) = f(a^1\zeta, \ldots, a^n\zeta) \tag{14}$$

is a *K*-QCfH in B^2. But for $n = 1$, *K*-QCf mappings satisfy the Beltrami system

$$\frac{\partial f}{\partial \bar{z}} = \chi(z) \frac{\partial f}{\partial z}$$

where $\chi(z)$ satisfies (12), as it follows from formulae (1.25), p. 236 and (5.6), p. 115 in Künzi's book [3]. The coefficient $\chi(z)$ is called the *complex dilatation* and was introduced by O. Lehto [1]. For every set of constants satisfying (13), the relations

$$\sum_{p=1}^{n} \bar{a}^p \left(\frac{\partial f}{\partial \bar{z}^p} \right)_0 = \left(\frac{\partial \tilde{f}}{\partial \bar{\zeta}} \right)_0 = \chi \left(\frac{\partial \tilde{f}}{\partial \zeta} \right)_0 = \chi \sum_{p=1}^{n} a^p \left(\frac{\partial f}{\partial z^p} \right)_0 \tag{15}$$

hold. Here $\chi = \chi(a^1, \ldots, a^n)$ depends upon the parameters, but it always satisfies (12), i.e. $|\chi|$ is bounded by k_0. For each $q = 2, \ldots, n$, we put

$$S_q(w) = \frac{\left(\dfrac{\partial \bar{f}}{\partial z^1} \right)_0 w + \left(\dfrac{\partial \bar{f}}{\partial z^q} \right)_0}{\left(\dfrac{\partial f}{\partial z^1} \right)_0 w + \left(\dfrac{\partial f}{\partial z^q} \right)_0} \qquad (q = 2, \ldots, n),$$

where w is a complex variable on the whole complex sphere $\{|w| \leqq \infty\}$. We are going to prove that $S_q(w) = \text{const.}$ $(q = 2, \ldots, n)$. In fact, for every w, we can choose the n numbers a^p $(p = 1, \ldots, n)$ in (13) in such a way that the conditions

$$a^2 = \cdots = a^{q-1} = a^{q+1} = \cdots = a^n = 0, \qquad w = \frac{a^1}{a^q} \tag{16}$$

are satisfied. $w = \infty$ corresponds to $a^q = 0$. Inserting (16) in (15), we obtain the boundedness of $S_q(w)$, namely

$$|S_q(w)| = \left| \frac{\left(\dfrac{\partial f}{\partial \bar{z}^1} \right)_0 \bar{w} + \left(\dfrac{\partial f}{\partial \bar{z}^q} \right)_0}{\left(\dfrac{\partial f}{\partial z^1} \right)_0 w + \left(\dfrac{\partial f}{\partial z^q} \right)_0} \right| = \left| \frac{\left(\dfrac{\partial f}{\partial \bar{z}^1} \right)_0 \bar{a}^1 + \left(\dfrac{\partial f}{\partial \bar{z}^q} \right)_0 \bar{a}^q}{\left(\dfrac{\partial f}{\partial z^1} \right)_0 a^1 + \left(\dfrac{\partial f}{\partial z} \right)_0 a^q} \right| =$$

$$= |\chi(a^1, 0, \ldots, 0, a^q, 0, \ldots, 0)| \leqq k_0.$$

Hence, by virtue of Liouville's theorem (see for instance A. I. Markuševič [2], p. 227), $S_q(w)$ reduces to a constant $\chi_q = \chi(a^1, 0, \ldots, 0, a^q, 0, \ldots, 0)$, which implies the relations

$$\left(\frac{\partial \bar{f}}{\partial z^1} \right)_0 = \chi_q \left(\frac{\partial f}{\partial z^1} \right)_0, \qquad \left(\frac{\partial \bar{f}}{\partial z^q} \right)_0 = \chi_q \left(\frac{\partial f}{\partial z^q} \right)_0 \qquad (q = 2, \ldots, n).$$

Since we have assumed $\dfrac{\partial f(O)}{\partial z^1} \neq 0$, the value of χ_q is the same for each $q = 2, \ldots, n$. Denoting this common value by χ, we have the relations (11).

THEOREM 5. (*Hitotumatu* [1]). *The K-discwise* QCfH *in* $D \subset C^n$ *satisfy in D the inequality*

$$\left|\frac{\partial(z^{*1}, \ldots, z^{*n})}{\partial(x^1, \ldots, x^n)}\right|^2 \leq \left[\frac{(K+1)^2}{4K}\right]^n \frac{\partial(x^{*1}, y^{*1}, \ldots, x^{*n}, y^{*n})}{\partial(x^1, y^1, \ldots, x^n, y^n)}.$$

It follows by an easy computation and taking into account the relations

$$\frac{\partial \bar{z}^{*q}}{\partial z^p} = \zeta^q \frac{\partial z^{*q}}{\partial z^p} \qquad (p, q = 1, \ldots, n), \tag{17}$$

which represent the set of all the systems (11) established in the preceding theorem. which are satisfied by the n components of the considered homeomorphism.

We conclude that Bergman's *PCT* are 1-discwise QCfH.

Another definition of discwise QCf diffeomorphisms (Hitotumatu) [1]). By *K-discwise* QCf *diffeomorphism* we mean any solution $w = f(z)$ of the system (17) in $D \subset C^n$, where $\chi(z)$ is a continuous function.

THEOREM 6. (*Hitotumatu* [1]). *Suppose* $\chi(x^1, y^1, \ldots, x^n, y^n)$ *is a given continuous function in a domain* $D \subset R^{2n}$. *We assume that* χ *satisfies the inequalities* (12). *If* $f \in C^1$ *is a solution of differential equations* (17) *in D, then f is a K-discwise* QCf *diffeomorphism in D* (*according to the first definition*).

Let (9) be a linear mapping from B^2 into D. Then the composite function (14) satisfies the condition

$$\overline{\left(\frac{\partial \tilde{f}}{\partial \bar{\zeta}}\right)} = \frac{\partial \bar{\tilde{f}}}{\partial \zeta} = \sum_{p=1}^{n} a^p \frac{\partial \bar{f}}{\partial z^p} = \chi \sum_{p=1}^{n} a^p \frac{\partial f}{\partial z^p} = \chi \frac{\partial \tilde{f}}{\partial \zeta},$$

where χ satisfies (12), and we conclude that $\tilde{f}(\zeta)$ is K-QCf in B^2, as desired.

COROLLARY. *The two definitions of the K-discwise* QCf *diffeomorphisms are equivalent.*

The theory of *PCT* has been developed in its proper way and is a vast and independent domain. That is why we shall not deal with the *PCT* any more in our book.

CHAPTER 16

Different definitions of the conformal mappings

Characterization of the conformal mappings by means of the modulus of a ring. Since the different definitions of Grötzsch's type are equivalent, it follows that the conformal mappings can be defined as 1-QCfH with respect to whatsoever definition. Thus, for instance, setting $K = 1$ in Gehring's geometric definitions, the inequality (2.32) reduces to

$$\operatorname{mod} A \leqq \operatorname{mod} A^* \leqq \operatorname{mod} A,$$

for all rings $A \subset\subset D$, i.e.

$$\operatorname{mod} A^* = \operatorname{mod} A, \tag{1}$$

that is the modulus of a ring is invariant under conformal mappings. However, in order to characterize the conformal mappings, it suffices that only the corresponding inequality holds, as it follows

THEOREM 1. *Let $x^* = f(x)$ be a homeomorphism in D. Then $x^* = f(x)$ is conformal in D if and only if*

$$\operatorname{mod} A^* \leqq \operatorname{mod} A \tag{2}$$

for all rings $A \subset\subset D$.

The necessity is obvious and the sufficiency is an immediate consequence of theorem 6.2.

The conformal mappings form a group.

THEOREM 2. *The class of conformal mappings form a group under the composition (i.e. the product in functional sense).*

This is an immediate consequence of (1).

In what follows, we shall show that an 1-QCfH in a domain D is the restriction of a Möbius transformation to D. (We recall that a Möbius transformation is a finite product of inversions.) We begin with some preliminary results.

Hölder's continuity of QCfH.

THEOREM 3. *Let* $x^* = f(x)$ *be a K-QCfH according to Gehring's geometric definition in D. Then, for each compact set* $F \subset D$, *there exists a constant C (depending on D, f and F) such that*

$$|x^* - y^*| \leqq C|x - y|^{\frac{1}{k}} \tag{3}$$

for all $x, y \in F$.

Suppose that $\partial D \neq \emptyset$. Let

$$a = \mathrm{d}(F, \partial D), \quad a^* = \mathrm{d}(F^*, \partial D^*), \quad b^* = \mathrm{d}(F^*), 0 < c < 1.$$

Then by virtue of (9.51) we can find a constant α such that

$$\Theta_k(t) \leqq \alpha t^{\frac{1}{k}}$$

for $0 < t \leqq c$, and applying theorem 9.1 yields

$$|x^* - y^*| \leqq C_1|x - y|^{\frac{1}{k}}, \; C_1 = \frac{a^* + b^*}{a^{\frac{1}{k}}}\alpha, \tag{4}$$

for all $x, y \in F$ with $|x - y| \leqq ac$. But clearly

$$\frac{|x^* - y^*|}{b^*} < 1 < \left(\frac{|x - y|}{ac}\right)^{\frac{1}{k}}$$

for all $x, y \in F$ with $|x-y| \geqq ac$ and we obtain in these conditions for x, y the inequality

$$|x^* - y^*| \leqq C_2|x - y|^{\frac{1}{k}}, \; C_2 = \frac{b^*}{(ac)^{\frac{1}{k}}} \tag{5}$$

(4) and (5) yield the desired inequality (3) for all $x, y \in F$, where $C = \max(C_1, C_2)$, and $\partial D \neq \emptyset$. Thus $C = C(a, a^*, b^*)$.

Finally the case $\partial D = \emptyset$ can be handled as above by first choosing F and then replacing D by any open sphere containing F. (This is possible because a compact set of the Euclidean space cannot coincide with the whole space.)

COROLLARY 1. *Every K-QCfH according to Gehring's geometric definition in D is continuous Hölder with exponent* $\frac{1}{K}$ *everywhere in D.*

COROLLARY 2. *Every K-QCfH in D according to whatsoever definition of Grötzsch's type is continuous Hölder with exponent* $K^{-e(n)}$, *where we infer* $e(n)$ *from the corresponding theorem establishing the connexion between the considered definition of Grötzsch's type and Gehring's geometric definition.*

Remark. This result is generalized by D. Callender [2] for the almost *K*-QCfH, when D^* is the unit ball.

COROLLARY 3. *Every* QCfH *in D according to whatsoever definition of Lavrent'ev's type is continuous Hölder in every compact set* $F \subset D$, *with an exponent depending on D, f and F.*

Conformal mappings are restrictions of Möbius transformations. The proof is divided into several steps. First we use an important recent result due to de Giorgi [1] to show that the extremal function u for a ring A is continuously differentiable. Next, a theorem of Morrey [1] allows us to pass from C^1 to C^n. The analyticity finally follows from a theorem due to E. Hopf [1].

THEOREM OF DE GIORGI. *Let the functions* a_{ik} $(i, k = 1, \ldots, n)$ *be measurable in D, let* $a_{ik} = a_{ki}$ *and let constants* $K_1, K_2 > 0$ *exist such that for each vector* $b = (b^1, \ldots, b^n)$

$$K_1 |b|^2 \leqq a_{ik} b^i b^k \leqq K_2 |b|^2 \tag{6}$$

a.e. in. D. Next let v be continuous and ACL in D, let

$$\int_D v^2 d\tau \leqq K_3 < \infty, \tag{7}$$

let $|\nabla v| \in L^2(D)$ *and let*

$$\int_D \sum_{i,k=1}^{n} a_{ik} \frac{\partial v}{\partial x^i} \frac{\partial w}{\partial x^k} d\tau = 0 \tag{8}$$

for all functions $w \in C^1$ *with compact support in D. Then, given a compact set* $F \subset D$ *there exist constants C and* α, *which depend only on* K_1, K_2 *and F, such that*

$$|v(x) - v(y)| \leqq C |x - y|^{\alpha}$$

for all $x, y \in F$.

This particular form of de Giorgi's theorem is given by J. Moser [1].

Before stating Morrey's theorem, we begin with some notations and definitions involved in it.

First let $u \in C^1$ in $\bar{D}$ and set $\nabla u(x) = p(x)$. A function $g(x, u, p)$ is said to be *the integrant of a regular variational problem near* $u(x)$ if and only if $g, g_p \in C^1$ and g satisfy

$$g_{p_i p_k}(x, u, p) f^i f^k > 0, \quad |f| \neq 0$$

for all $(x, u, p) \in \bar{G}$, G being a bounded domain of the form

$$G = \{x, u, p;\ x \in D,\ |p - p(x)| < h,\ |u - u(x)| < h,\ h > 0\},$$

and

$$g_{p_i p_k} = \frac{\partial^2 g}{\partial u_i \partial u_k}, \quad u_i = \frac{\partial u}{\partial x^i}.$$

In what follows, we shall assume, without loss of generality, that f satisfies

$$g_{p_i p_k}(x, u, p) f^i f^k \geqq \frac{|f|^2}{K^{n-1}}, \quad |g_{p_i p_k}(x, u, p)| \leqq K_1, \quad (x, u, p) \in G. \tag{9}$$

A function $u \in C^1$ on $\bar{D}$ is said to *furnish a stationary value to the integral*

$$I(u, D) = \int_D g[x, u(x), \nabla u(x)]\, dx,$$

if and only if the first variation

$$J(u, w; D) = \int_D \{g_p[x, u(x), \nabla u(x)] \nabla w(x) +$$

$$+ g_u[x, u(x), \nabla u(x)] w(x)\}\, dx = 0 \tag{10}$$

for every $w \in C^1$ on $\bar{D}$ which vanishes on and near ∂D.

A function $u(x)$ will be said of *class* C^p_μ, $0 < \mu < 1$, $p \geqq 0$, on $\bar{D}$ if $u \in C^p$ on $\bar{D}$ (i.e. the derivatives tend to limits on ∂D) and $\nabla^p u$ satisfies a uniform Lipschitz condition with exponent μ on $\bar{D}$, where $\nabla^p u$ is the vector whose components are the set of the partial derivatives of the form $\dfrac{\partial^p u}{\partial x^{i_1} \ldots \partial x^{i_p}}$

THEOREM OF MORREY. *Suppose* $g(x, u, p)$, $g_p(x, u, p) \in C^p_\mu$, $p \geqq 1$, $0 < \mu \leqq 1$ *on* $\bar{D}$. *If* $g(x, u, p)$ *satisfies* (9), *and* $u \in C^1$ *satisfies* (10), *then* $u \in C^{p+1}_\mu$ *on* D.

Before enunciating the theorem concerning the analyticity of the solution of an elliptic equation, we recall that a partial differential equation of second order in n variables

$$\Phi\left[\frac{\partial^2 u}{\partial (x^1)^2}, \frac{\partial^2 u}{\partial x^1 \partial x^2}, \ldots, \frac{\partial^2 u}{\partial (x^n)^2}; \frac{\partial u}{\partial x^1}, \ldots, \frac{\partial u}{\partial x^n}, u; x^1, \ldots, x^n\right] = 0 \tag{11}$$

is said to be *elliptic* with respect to a given solution $u(x)$ at a point x_0 if the quadratic form

$$\sum_{\nu \leqslant \mu} \frac{\partial \Phi}{\dfrac{\partial^2 u}{\partial x^\mu \partial x^\nu}} \xi^\mu \xi^\nu$$

is definite. Without loss of generality we may suppose it positive definite.

GIRAUD—HOPF'S THEOREM. *Suppose* (11) *is an elliptic equation with respect to the given solution* $u(x)$ *and analytic at the origin with* $\frac{\partial^2 u}{\partial x^2}$ *satisfying a Hölder condition. Then* $u(x)$ *is analytic in a neighbourhood of the origin where these conditions still hold.* (For the proof see G. Giraud [1], E. Hopf [1].)

LEMMA 1. *Let* u *be the extremal function for a ring* A, *and for each compact set* $F \subset A$, *let a constant* $K > 0$ *exist such that*

$$\frac{1}{K} \leqq |\nabla u| \leqq K \tag{12}$$

a.e. in F. *Then* $u(x)$ *is analytic and satisfies the differential equation*

$$\operatorname{div}(|\nabla u|^{n-2} \nabla u) = 0 \tag{13}$$

everywhere in A.

The proof is divided in several steps. First we use de Giorgi's theorem to show that $u \in C^1$. Next Morrey's theorem allows us to pass from C^1 to C^n. Then we can apply corollary of lemma 5.10 and Gauss theorem relatively to the change of a volume integral into a surface integral to obtain (13). The analyticity finally follows from Giraud-Hopf's theorem.

We begin with the proof of the continuous differentiability of $u(x)$. Let D denote any bounded domain with $\overline{D} \subset A$, let $b = d(D, \partial A)$ and fix $0 < a < b$. The set $\{x; d(x, D) \leqq a\}$ is a compact subset of A and hence we can find a constant $K > 0$ such that (12) holds a.e. in this set. Now let h denote a vector parallel to the positive half of the x^1-axis with $0 < |h| \leqq a$. Then

$$\frac{1}{K} \leqq |\nabla u(x)|, |\nabla u(x+h)| \leqq K \tag{14}$$

a.e. in D. Next let

$$v_h(x) = \frac{u(x+h) - u(x)}{|h|}. \tag{15}$$

Then $v_h(x)$ is continuous and ACL in D. We now use (14) to show that $|v_h(x)| \leqq K$ a.e. in D. In fact, $u(x)$ has first order partial derivatives with respect to all the variables, and from (14) we infer that ∇u satisfies this inequality in almost all points of almost all line segments in D parallel to the coordinate axes. Fix x_0 on such a line segment with the property that $\nabla u(x_0)$ exists and satisfies (14). Clearly $u(x)$ is derivable at x_0 and $|\nabla u| \leqq K$ a.e. in $[x_0, x_0 + h]$. Let y_0 be such a point. Then, for every $\varepsilon > 0$, there exists $\delta > 0$ such that for each y, which belongs to the straight line containing the segment $[x_0, x_0 + h]$ and for which $|y - y_0| < \delta$, to have

$$\left| \frac{u(y) - u(y_0)}{y^1 - y_0^1} - \frac{\partial u(y_0)}{\partial x^1} \right| < \varepsilon.$$

Let y_1, y_2 be two such points y lying on each side of y_0. Then

$$\frac{\partial u(y_0)}{\partial x^1} - \varepsilon < \frac{u(y_1) - u(y_0)}{y_1^1 - y_0^1} < \frac{\partial u(y_0)}{\partial x^1} + \varepsilon,$$

$$\frac{\partial u(y_0)}{\partial x^1} - \varepsilon < \frac{u(y_0) - u(y_2)}{y_0^1 - y_2^1} < \frac{\partial u(y_0)}{\partial x^1} + \varepsilon$$

hold. On the other hand, if we multiply the first inequality by $\dfrac{y_1^1 - y_0^1}{y_1^1 - y_2^1}$ and the second by $\dfrac{y_0^1 - y_2^1}{y_1^1 - y_2^1}$ and then we add them, taking also into account that

$$\frac{u(y_1) - u(y_2)}{y_1^1 - y_2^1} = \frac{u(y_1) - u(y_0)}{y_1^1 - y_0^1} \frac{y_1^1 - y_0^1}{y_1^1 - y_2^1} + \frac{u(y_0) - u(y_2)}{y_0^1 - y_2^1} \frac{y_0^1 - y_2^1}{y_1^1 - y_2^1},$$

and that

$$\frac{y_1^1 - y_0^1}{y_1^1 - y_2^1} + \frac{y_0^1 - y_2^1}{y_1^1 - y_2^1} = 1,$$

it follows that

$$\frac{\partial u(y_0)}{\partial x^1} - \varepsilon < \frac{u(y_2) - u(y_1)}{y_2^1 - y_1^1} < \frac{\partial u(y_0)}{\partial x^1} + \varepsilon.$$

And now, let us associate to each $y \in [x_0, x_0 + h]$, which satisfies (14), an interval (y_1, y_2) such that $y \in (y_1, y_2)$ and

$$\frac{\partial u(y)}{\partial x^1} - \varepsilon < \frac{u(y_2) - u(y_1)}{y_2^1 - y_1^1} < \frac{\partial u(y)}{\partial y^1} + \varepsilon. \tag{16}$$

Since this happens for almost all $y \in [x_0, x_0 + h]$, it follows that the set of the intervals form a covering of $[x_0, x_0+h]$, and then, by Heine—Borel—Lebesgue covering theorem we can pick out a finite covering of $[x_0, x_0 + h]$ and to this covering it is possible to associate a finite system of closed intervals having in common only the endpoints, covering $[x_0, x_0 + h]$ satisfying (16). Let y_k $(k = 1, \ldots, m)$ be the endpoints of these intervals, where $y_1 = x_0$, $y_m = x_0 + h$ and let $x_p \in (y_p, y_{p+1})$ $(p = 1, \ldots, m - 1)$ be the points where (16) holds. Then (16), which in this case is of the form

$$\frac{\partial u(x_{p-1})}{\partial x^1} - \varepsilon < \frac{u(y_p) - u(y_{p-1})}{y_p^1 - y_{p-1}^1} < \frac{\partial u(x_{p-1})}{\partial x^1} + \varepsilon,$$

and

$$\left| \frac{\partial u(x_{p-1})}{\partial x^1} \right| \leqq \left| \nabla u(x_{p-1}) \right| \leqq K \quad (p = 1, \ldots, m-1)$$

yield

$$\frac{u(x_0 + h) - u(x_0)}{|h|} = \sum_{p=1}^{m-1} \frac{u(y_{p+1}) - u(y_{p-1})}{y_{p+1}^1 - y_p^1} \frac{y_{p+1}^1 - y_p^1}{|h|} < \tag{17}$$

$$< \sum_{p=1}^{m-1} \left[\frac{\partial u)x_p)}{\partial x^1} + \varepsilon \right] \frac{y_{p+1}^1 - y_p^1}{|h|} \leqq K + \varepsilon.$$

Arguing similarly, we obtain also the inequality

$$\frac{u(x_0 + h) - u(x_0)}{|h|} > -(K + \varepsilon).$$

Since this holds for all small $\varepsilon > 0$, we obtain

$$\left| v_h(x_0) \right| \leqq K, \tag{18}$$

as desired.

Next, let us show that $v_h(x)$ satisfies the hypotheses of de Giorgi's theorem with respect to $v(x)$. First let $w \in C^1$ and with compact support in D. Then $w(x)$ and $w(x - h)$ vanish on ∂A and corollary of lemma 5.10 yields

$$\int_D [|\nabla u(x + h)|^{n-2} \nabla u(x + h) - |\nabla u(x)|^{n-2} \nabla u(x)] \nabla w(x) d\tau = 0. \tag{19}$$

Now let

$$g = \frac{1}{n} |y|^n, \quad g_i = \frac{\partial g}{\partial y^i} = |y|^{n-2} y^i, \quad g_{ki} = \frac{\partial^2 g}{\partial y^i \partial y^k} =$$

$$= |y|^{n-2} \left[\delta_{ki} + \frac{(n-2)\, y^i y^k}{|y|^2} \right] \tag{20}$$

and

$$y = t \nabla u(x + h) + (1 - t) \nabla u(x), \quad 0 \leqq t \leqq 1. \tag{21}$$

Then (19) yields

$$\int_D \sum_{i=1}^n \{g_i[y(1)] - g_i[y(0)]\} \frac{\partial w}{\partial x^i} d\tau = 0, \tag{22}$$

and on account of (15), (21) may be written as follows

$$y = \nabla v_h |h| t + \nabla u, \quad 0 \leqq t \leqq 1,$$

and

$$y^i = \frac{\partial v_h}{\partial x^i} |h| t + \frac{\partial u}{\partial x^i} \qquad (i = 1, \ldots, n).$$

Hence, it is easy to verify, on account of (20), that

$$\frac{\mathrm{d}g_k}{\mathrm{d}t} = \sum_{i=1}^{n} \frac{\partial g_k}{\partial y^i} \frac{\mathrm{d}y^i}{\mathrm{d}t} = \sum_{i=1}^{n} g_{ik} \frac{\partial v_h}{\partial x^i} |h|,$$

and we conclude that

$$g_k[y(1)] - g_k[y(0)] = |h| \sum_{i=1}^{n} \left[\int_0^1 g_{ik}(y)\, \mathrm{d}t \right] \frac{\partial v_h}{\partial x^i}$$

and then, by (22), relation (8) in de Giorgi's theorem holds with

$$a_{ik} = \int_0^1 g_{ik}(y)\, \mathrm{d}t \quad (i, k = 1, \ldots, n). \tag{23}$$

The a_{ik} $(i, k = 1, \ldots, n)$ are clearly measurable in D and $a_{ik} = a_{ki}$ $(i, k = 1, \ldots, n)$. Next, by (23) and (20), for each vector b we have

$$a_{ik} b^i b^k = \int_0^1 g_{ik}(y) b^i b^k \mathrm{d}t = \int_0^1 |y|^{n-2} \left[|b|^2 + \frac{(n-2)(y \cdot b)^2}{|y|^2} \right] \mathrm{d}t,$$

where (y, b) means the scalar product of the vectors y and b (chapter 1.2), and since $(y . b)^2 \leqq |y|^2 |b|^2$, it follows

$$|b|^2 \int_0^1 |y|^{n-2}\, \mathrm{d}t \leqq a_{ik} b^i b^k \leqq (n-1) |b|^2 \int_0^1 |y|^{n-2} \mathrm{d}t \tag{24}$$

and from (21) we deduce that, for every x satisfying (14), we have

$$|y| \leqq t |\nabla u(x+h)| + (1-t) |\nabla u(x)| \leqq tK + (1-t)K = K,$$

which implies that

$$\int_0^1 |y|^{n-2}\, \mathrm{d}t \leqq K^{n-2}. \tag{25}$$

And now, in order to establish the inequality

$$\int_0^1 |y|^{n-2}\, dt \geqq \frac{1}{(n-1)(2K)^{n-2}}\,. \tag{26}$$

we consider $\nabla u(x)$ and $\nabla u(x+h)$ as two points in R joined by the segment (21), and $|y(t)|$ as the distance between the origin of the segment (21) corresponding to the value t of the parameter. Consider first the particular case in which the origin lies on the half of this segment corresponding to the endpoint $\nabla u(x)$. The points $\nabla u(x)$ and $-\nabla u(x)$ are symmetric with respect to the origin and the part of the segment (21) joining them is given by

$$y(\tau) = -\tau\nabla u(x) + (1-\tau)\,\nabla u(x) = (1-2\tau)\,\nabla u(x), \quad 0 \leqq \tau \leqq 1.$$

Then, by (14),

$$\int_0^1 |y(t)|^{n-2}\, dt \geqq \int_0^1 |y(\tau)|^{n-2}\, d\tau = |\nabla u(x)|^{n-2}\left[\int_0^{\frac{1}{2}} (1-2\tau)^{n-2}\, d\tau + \right.$$

$$\left. + \int_{\frac{1}{2}}^1 (2\tau-1)^{n-2} d\tau\right] \geqq$$

$$\geqq \frac{1}{K^{n-2}}\left[\frac{-(1-2\tau)^{n-1}}{2(n-1)}\Bigg|_0^{\frac{1}{2}} + \frac{(2\tau-1)^{n-1}}{2(n-1)}\Bigg|_{\frac{1}{2}}^1\right] = \frac{1}{(n-1)K^{n-2}}\,.$$

Arguing similarly for the case in which the origin lies on the half of the segment (21) corresponding to $\nabla u(x+h)$, we obtain

$$y(\tau) = \tau\nabla u(x+h) + (\tau-1)\nabla u(x+h) = (2\tau-1)\nabla u(x+h), \quad 0 \leqq \tau \leqq 1$$

and then

$$\int_0^1 |y(t)|^{n-2}\, dt \geqq \int_0^1 |y(\tau)|^{n-2}\, dt = \frac{|\nabla u(x+h)|^{n-2}}{n-1} \geqq \frac{1}{(n-1)K^{n-2}}\,.$$

Finally we have only to consider the case in which the origin does not lie on the segment (21). If this segment is outside the sphere $S\left(\dfrac{1}{2K}\right)$, then

$$\int_0^1 |y(t)|^{n-2}\, dt \geqq \frac{1}{2^{n-2}K^{n-2}}\,.$$

If the segment (21) meets the sphere $S\left(\frac{1}{2K}\right)$, then, by (14), their intersection consists of two different points and the distance between the origin and the endpoints of the projection of the segment (21) on a line through the origin and parallel to it is at least $\frac{\sqrt{3}}{2K} > \frac{1}{2K}$. And then, arguing as above, but only taking into account that in this case the distance between the origin and the endpoints is greater than $\frac{1}{2K}$ $\left(\text{instead of } \frac{1}{K}\right)$, we obtain in the general case the inequality

$$\int_0^1 |y(t)|^{n-2}\, dt \geqq \frac{1}{(n-1)(2K)^{n-2}}\,.$$

Hence, and by (24), (25), (26), it follows that the inequality (6) in de Giorgi's theorem holds a.e. in D with

$$K_1 = \frac{1}{(n-1)(2K)^{n-2}}\,, \quad K_2 = (n-1)K^{n-2}.$$

Finally, from (18) and taking into account that D is bounded, we infer (7), and from (22) and (23), we deduce (8) and we can apply de Giorgi's theorem to conclude that the functions $v_h(x)$ satisfy a Hölder condition at each point of D, uniformly in h.

Now, by Arzela—Ascoli's theorem (Chapter 5), from the equicontinuity and the uniformly boundedness of the family $\{v_h(x)\}$, it follows the existence of a sequence $\{h_m\}$, $|h_m| \to 0$, such that

$$\frac{u(x+h_m) - u(x)}{|h_m|}$$

converge uniformly on each compact subset of D. It then follows that $\frac{\partial u}{\partial x^1}$ exists and is continuous everywhere in D; by symmetry $u \in C^1$ in D.

In order to prove that $u \in C^\infty$, we show that $g(y)$ given by (20) satisfies the hypothesis of Morrey's theorem. Again, let $w \in C^1$ be with compact support in D. Then corollary of lemma 5.10 yields

$$\int_D \nabla g(\nabla u) \nabla w \, d\tau = 0,$$

where g is as in (20) and then $g(\nabla u) = g[y(0)]$ and $\nabla g(\nabla u) = |\nabla u|^{n-2}\nabla u$. Moreover (14) and the continuity of ∇u, arguing as for (24), imply that, for each b,

$$\frac{1}{K^{n-2}}|b|^2 \leqq g_{ik}(\nabla u) b^i b^k \leqq (n-1)|b|^2 K^{n-2}$$

everywhere in D. Since $g(y)$ is analytic in $|y| > 0$ we can apply Morrey's theorem (quoted above) to conclude that $u(x)$ has continuous derivatives of all orders, i.e. $u \in C^\infty$.

Next, the Gauss theorem concerning the change of a volume integral into a surface integral in an n-space (see for instance E. Hopf [1]) yields

$$\int\limits_{B(x,\,r)} \operatorname{div}(w|\nabla u|^{n-2}\nabla u)\,d\tau = \int\limits_{S(x,\,r)} w|\nabla u|^{n-2}\frac{\partial u}{\partial n}\,d\sigma, \tag{27}$$

where n is the outer normal to $S(x, r)$, $\dfrac{\partial}{\partial n} = \dfrac{\partial}{\partial x^i}\cos\alpha^i$, $\cos\alpha^i$ $(i = 1, \ldots, n)$ are the directing cosines of n at the considered point and $w \in C^1$. If w has, in addition, compact support in $B(x, r)$, then the surface integral vanishes, and (27) yields

$$\int\limits_{B(x,\,r)} w \operatorname{div}(|\nabla u|^{n-2}\nabla u)\,d\tau + \int\limits_{B(x,\,r)} |\nabla u|^{n-2}\nabla u \nabla w\,d\tau = 0,$$

which, on account of corollary of lemma 5.10, comes to

$$\int\limits_{B(x,\,r)} w \operatorname{div}(|\nabla u|^{n-2}\nabla u)\,d\tau = 0.$$

Since w is an arbitrary continuous differentiable function with compact support in $B(x, r)$ we conclude that (13) holds a.e. in $B(x, r)$, and since $u \in C^\infty$, we deduce that $\operatorname{div}(|\nabla u|^{n-2}\nabla u)$ is continuous and vanishes everywhere in each $B(x, r) \subset D$, and then everywhere in D. From (20) it follows that the equation (13) is of the form

$$\sum_{i,\,k=1}^{n}\left[|\nabla u|^2\delta_{ik} + (n-2)\frac{\partial u}{\partial x^i}\,\frac{\partial u}{\partial x^k}\right]\frac{\partial^2 u}{\partial x^i \partial x^k} = 0;$$

i.e. $u(x)$ is the solution of a quasilinear elliptic equation in D. Since the coefficients of $\dfrac{\partial^2 u}{\partial x^i \partial x^k}$ $(i, k = 1, \ldots, n)$ in this equation are analytic functions of the partial derivatives $\dfrac{\partial u}{\partial x^i}$ $(i=1,\ldots,n)$, we can appeal to Giraud—Hopf's theorem (quoted above) to conclude that u is real analytic in D. Now D was chosen to be any bounded domain with closure in A. Thus the above results hold throughout A and the proof of lemma 1 is complete.

LIOUVILLE'S THEOREM. *Let $f(x)$ be a topological mapping of D and let $f \in C^4$, $J(x) > 0$ and $\delta_L(x) = 1$ everywhere in D. Then $f(x)$ is the restriction of Möbius transformation in D.* (For the proof see R. Nevanlinna [1]).

THEOREM 4. *A mapping $f(x)$ conformal in D has $0 < \Lambda_f(x) < \infty$ in D.*

Indeed, since $f(x)$ satisfies a Hölder condition with exponent 1 (theorem 3), i.e. is Lipschitzian on every compact subset of D, it follows $\Lambda_f(x) < \infty$ in D. Next, since the inverse of a conformal mapping is again a conformal mapping, theorem 3 implies

$$|f^{-1}(x_1^*)-f^{-1}(x^*)| \leqq C|x_1^*-x^*|,$$

hence

$$0<\frac{1}{C} \leqq \frac{|x_1^*-x^*|}{|f^{-1}(x_1^*)-f^{-1}(x^*)|}=\frac{|f(x_1)-f(x)|}{|x_1-x^*|},$$

and we conclude that $\Lambda_f(x)>0$ on every compact subset of D, and then also everywhere in D.

THEOREM 5. *A topological mapping $f(x)$ of D is the restriction of a Möbius transformation in D if and only if the inequality* (2) *holds for all rings $A \subset\subset D$.*

The necessity is obvious. We shall prove the sufficiency. Theorem 1 implies that $f(x)$ is 1-QCfH (i.e. conformal) in D. Next, the preceding theorem and the inequality (3.12) imply

$$0<\Lambda_f^n(x) \leqq |J(x)|, \tag{28}$$

in every point of differentiability of D, and, from the equivalence of the definitions of the QCfH of Grötzsch's type, it follows $\delta_L(x)=1$ a.e. in D (theorem 3.14).

The remainder of the proof is devoted to show that $f(x)$ satisfies the hypotheses of Liouville's theorem. It is clear we need only show that $f \in C^4$ in D, for then, $|J(x)|>0$ at every point of differentiability means $|J(x)|>0$ everywhere in D, and since at a point of differentiability $p_1(x)$ is a continuous function of the first order partial derivatives of $f(x)$ it follows that $p_1(x)$ is a continuous function of the variable x in D, where $p_1(x)$ denotes the ratio of the extremal semi-axes of the ellipsoid which is mapped onto a sphere by the affine mapping (3.2) corresponding to $f(x)$, and then, by lemma 3.1, $p_1(x)=q(x)$, where $q(x)$ is the characteristic involved in Markuševič—Pesin's definition, hence *a fortiori* $p_1(x)=\delta_L(x)$ everywhere in D, and since $\delta_L(x)=1$ a.e. in D, it follows $\delta_L(x)=1$ everywhere in D by continuity.

Moreover for $f \in C^4$ it is sufficient to show there exists an absolute constant $a>0$ with the following property: for each $x_0 \in D$ the function $|f(x)-f(x_0)|$ is real analytic in the punctured sphere $0<|x-x_0|<ad(x_0, \partial D)$. Let $\Theta_1(t)$ be the distortion function (9.50) with $K=1$ and pick $0<a<1$ so that $\Theta_1(a)<1$. Now fix $x_0 \in D$ and let A^* be the ring $c<|x^*-f(x_0)|<d$ where $0<c<d=$ $=L(x_0, b)$ and set $b=ad(x_0, \partial D)$. (9.62) implies

$$\max_{|x-x_0|=b} |f(x)-f(x_0)| \leqq d(x_0^*, \partial D^*)\Theta_1(a)<d(x_0^*, \partial D^*),$$

and hence $A^* \subset D^*$. Next let $v(x^*)$ be the extremal function for A^* and set $u(x)=$ $=v[f(x)]$. Then we see that

$$v(x^*)=\frac{\log\left|\dfrac{x^*-f(x_0)}{c}\right|}{\log\dfrac{d}{c}},$$

by (2.21). Besides, $u(x)$ is also admissible for A (proposition 1.6.4) and (6.15) holds a.e. in A. Hence, by (28),

$$\operatorname{cap} A \leqq \int_A |\nabla u|^n \, d\tau \leqq \int_A |\nabla v|^n |J| \, d\tau \leqq \int_{A^*} |\nabla v|^n \, d\tau = \operatorname{cap} A^* = \operatorname{cap} A,$$

and u is the extremal function for A.

Theorem 3 implies that we can find a constant $c > 0$ such that $\Lambda_f(x) \leqq c$ in A and that $\Lambda_{f^{-1}}(x^*) \leqq c$ in Λ^*. Hence, taking into account that $|x^* - f(x_0)| > c$, it follows that

$$\overline{\lim_{h\to 0}} \frac{|u(x+h) - u(x)|}{|h|} =$$

$$= \frac{1}{\log \frac{d}{c}} \overline{\lim_{h\to 0}} \left[\frac{\log \frac{|f(x+h) - f(x_0)|}{c} - \log \frac{|f(x) - f(x_0)|}{c}}{\frac{|f(x+h) - f(x_0)| - |f(x) - f(x_0)|}{c}} \times \right.$$

$$\left. \times \frac{|f(x+h) - f(x_0)| - |f(x) - f(x_0)|}{c|h|} \right] \leqq \tag{29}$$

$$\leqq \frac{1}{|x^* - f(x_0)| \log \frac{d}{c}} \overline{\lim_{h\to 0}} \frac{|f(x+h) - f(x)|}{|h|} < \frac{\Lambda_f(x)}{c \log \frac{d}{c}} \leqq \frac{C}{c \log \frac{d}{c}}.$$

On the other hand, the inequality $\Lambda_{f^{-1}}(x^*) \leqq c$ implies that

$$C \geqq \Lambda_{f^{-1}}(x^*) = \overline{\lim_{k\to 0}} \frac{|f^{-1}(x^* + k) - f^{-1}(x^*)|}{|k|} =$$

$$= \overline{\lim_{k\to 0}} \frac{|h|}{|f(x+h) - f(x)|} = \frac{1}{\underline{\lim_{h\to 0}} \frac{|f(x+h) - f(x)|}{|h|}},$$

whence

$$\overline{\lim_{h\to 0}} \frac{|f(x+h) - f(x)|}{|h|} \geqq \underline{\lim_{h\to 0}} \frac{|f(x+h) - f(x)|}{|h|} \geqq \frac{1}{C},$$

and then, since $|x^* - f(x_0)| < d$,

$$\overline{\lim_{h\to 0}} \frac{|u(x+h) - u(x)|}{|h|} =$$

$$= \frac{1}{\log \frac{d}{c}} \overline{\lim_{h\to 0}} \left\{ \frac{\log \frac{|f(x+h) - f(x_0)|}{c} - \log \frac{|f(x) - f(x_0)|}{c}}{\frac{|[f(x+h) - f(x_0)] - [f(x) - f(x_0)]|}{c}} \cdot \right.$$

$$\left. \cdot \frac{|f(x+h) - f(x)|}{c|h|} \right\} \geq \tag{30}$$

$$\geq \frac{1}{Cd \log \frac{d}{c}} \overline{\lim_{h\to 0}} \frac{|f(x+h) - f(x_0)| - |f(x) - f(x_0)|}{|f(x+h) - f(x_0)|} =$$

$$= \frac{1}{Cd \log \frac{d}{c}} \overline{\lim_{h\to 0}} \frac{|x^* + k| - |x^*|}{|k|},$$

where $\overline{\lim_{k\to 0}} \frac{|x^* - k| + |x^*|}{|k|} = 1$, as it follows by the same argument as in the proof of lemma 2.3. which establishes the inequalitity (2.19) (this upper limit is obtained letting $k \to 0$ in the direction Oy). (29) and (30) yield

$$\frac{1}{Cd \log \frac{d}{c}} < \overline{\lim_{h\to 0}} \frac{|u(x+h) - u(x)|}{|h|} < \frac{C}{c \log \frac{d}{c}}$$

everywhere in A, or since

$$\overline{\lim_{h\to 0}} \frac{|u(x+h) - u(x)|}{|h|} = |\nabla u(x)|,$$

(where the upper limit is reached letting $h \to 0$ in the direction ∇u) we conclude that the extremal function $u(x)$ satisfies an inequality of the form (12). Thus, by the preceding lemma, $u(x)$ is real analytic in A. This means that

$$|f(x) - f(x_0)| = c\left(\frac{d}{c}\right)^{u(x)}$$

is real analytic in A and letting $c \to 0$, we conclude that $|f(x) - f(x_0)|$ is real analytic in $0 < |x - x_0| < ad(x, \partial D)$, hence $f \in C^\infty$ in a punctured neighbourhood of x_0. Since x_0 is an arbitrary point of D it follows that the hypotheses of Liouville's theorem are satisfied and this completes the proof of theorem 5.

COROLLARY 1. *A conformal mapping of D is differentiable with* $0 < |J(x)| < \infty$ *everywhere in D.*

The preceding theorem implies the differentiability of $f(x)$ and then (28) holds everywhere in D.

Other characterizations of the conformal mappings.

COROLLARY 2. *A topological mapping $f(x)$ of a domain D is the restriction of a Möbius transformation to the domain D if and only if the conditions* (i), (ii) (chapter 10) *hold with* $K = 1$.

COROLLARY 3. *A topological mapping $f(x)$ of a domain D is the restriction of a Möbius transformation to the domain D if and only if* (i) *holds and $f(x)$ preserves angles a.e. in D.*

Remark. This result was obtained by A. P. Kopylov [3] in the more particular case in which all the segments γ_1, γ_2 which form an angle in D at an arbitrary point $x \in D$ are mapped by $f(x)$ in two arcs which possess tangent at $f(x)$.

We obtained in a joint paper with M. Corduneanu [1] the following Looman—Menšov's theorem generalizing the preceding corollary and which was proved for $n = 2$ by S. Agard and F. Gehring [1]:

THEOREM 6. *Let $f(x)$ be a topological mapping of a domain D. Then $f(x)$ is the restriction of a Möbius transformation to D if and only if the following conditions are satisfied*:

(i') *for each $x_0 \in D - E$, where E is of Σ-finite area, there is at least an* α, $0 < \alpha \leqq \alpha_0 < \frac{\pi}{2}$, *such that for every segment γ_1 parallel to one of the edges of a fixed simplex τ and every arc γ_2 which form a topological angle in D at x_0 of inner measure α,*

$$A[f(\gamma_1), f(\gamma_2)] > 0;$$

(ii') *for almost all $x_0 \in D$* and *for all pairs γ_1, γ_2 consisting of a segment γ_1 and an arc γ_2 which form an angle in D at x_0 equal to an angle of τ,*

$$A[f(\gamma_1), f(\gamma_2)] \geqq A(\gamma_1, \gamma_2).$$

We begin by showing that $f(x)$ is ACL in D and differentiable a.e. in D, under the assumption that τ has its vertices $O, e_1, \ldots, e_n$. For this, let $\varphi(y)$, J_0, γ_1, γ_2, E_{pq} and E_p be as in the proof of theorem 10.1 and set $E = \bigcup_p E_p$. If $\varphi(y_0) < \infty$, then, arguing as in theorem 10.1, we conclude that inequality (10.17) holds for every compact set $F \subset J_0 \cap E_{pq}$, and $E' \subset J_0$ such that $m_1 E' = 0$ implies $\Lambda[f(E')] = 0$, and condition (ii) implies $m(I - E_p) = 0$ for $p > \operatorname{cosec} \frac{\alpha}{2}$, where α

is an angle equal to one of the angles of τ. Fix such a p. By Fubini's theorem, (10.16) holds with this p a.e. in I_n, hence (10.17) holds for every compact set $E \subset J_0$. Thus $f(x)$ is AC on J_0, and then on all segments of I parallel to γ_1. The same argument shows that $f(x)$ is AC also on almost all the segments parallel to each of the coordinate axes. Since $I \subset \subset D$ is an arbitrary interval of D, it follows that $f(x)$ is ACL in D, and arguing as in theorem 10.2, $f(x)$ is ACL_n in D, and then differentiable *a.e.* in D.

We prove next that $f(x)$ is conformal or what comes to the same thing, we show that $f(x)$ is 1-QCfH according to Väisälä's first analytic definition under the assumption that f is differentiable a.e. in D and ACL and τ is an arbitrary simplex. The first two conditions of this definition are satisfied, and then it is sufficient to show that

$$\max |f'(x)| = |J(x)| = \min |f'(x)| \tag{31}$$

on each point $x_0 \in D$ where (ii′) holds and where $f(x)$ is differentiable, i.e. a.e. in D. Let x_0 be such a point. By performing preliminary similarity transformations (which preserves angles), we may assume that $x_0 = f(x_0) = O$ and, near $x_0 = O$, $f(x)$ is of the form

$$f(x) = a_k x^k e_k + x^n e_n + o(|x|),$$

where the dummy index k takes on the values $1, \ldots, n-1$. Suppose $\max |f'(O)| > 0$. This does not restrict the generality because if $\max |f'(0)| = O$, then $J(O) = O$ and (31) holds also in this particular case. Next, choose the similarity transformation from above so that $a_k \geqq 0$ $(k = 1, \ldots, n-1)$. But condition (ii′) implies $a_k > 0$ $(k = 1, \ldots, n-1)$, since if $a_k = 0$ for at least one index k, then it is easy to find two segments γ_1, γ_2 which form an angle parallel to an angle in τ and such that $A[f(\gamma_1), f(\gamma_2)] = 0$, contradicting so (ii′).

In order to prove that (31) holds, it suffices to show that $a_k = 1$ $(k = 1, \ldots$ $\ldots, n-1)$. By lemma 10.1 we may assume that

$$f(x) = a_k x^k e_k + x^n e_n \qquad (a_k > 0, k = 1, \ldots, n-1).$$

We first prove that $f(x)$ preserves the angles of τ. Let γ_1, γ_2 be two edges in τ which form an angle, i.e. belong to a triangle $\Delta \subset \partial\tau$, which is mapped by $f(x)$ onto a triangle Δ^*. Then, by (ii′), $A[f(\gamma_1), f(\gamma_2)] \geqq A(\gamma_1, \gamma_2)$. But, since the sum of the interior angles of any triangle is constant ($= 180°$), if $A[f(\gamma_1), f(\gamma_2)] > A(\gamma_1, \gamma_2)$, then it would follow that at least an angle of Δ must decrease, contradicting thus condition (ii′), which holds for all angles formed by pairs of segments with the vertex in O and parallel to angles in τ. Thus

$$A[f(\gamma_1), f(\gamma_2)] = A(\gamma_2, \gamma_2).$$

In order to prove $a_k = 1$ $(k = 1, \ldots, n-1)$, consider first the equations

$$\frac{x^1}{l_1^1} = \cdots = \frac{x^n}{l_n^1} \ (\delta^{ik} l_i^1 l_k^1 = 1), \quad \frac{x^1}{l_1^2} = \cdots = \frac{x^n}{l_n^2} \ (\delta^{ik} l_i^2 l_k^2 = 1),$$

of the two lines through O and corresponding to γ_1 and γ_2. Thus setting $\alpha = A(\gamma_1, \gamma_2)$ we get $\cos\alpha = \delta^{ik}e^1_i e^2_k$. Since $f(x)$ preserves the angle α, it follows that

$$a_1^2 l_1^1 l_1^2 + \cdots + a_{n-1}^2 l_{n-1}^1 l_{n-1}^2 + l_n^1 l_n^2 = \delta^{ik} l_i^1 l_k^2,$$

hence

$$(1 - a_1^2) l_1^1 l_1^2 + \cdots + (1 - a_{n-1}^2) l_{n-1}^1 l_{n-1}^2 = 0. \tag{32}$$

Next, an easy computation shows that, for $n \geqq 3$, an $(n-1)$-dimensional face contains at least n edges. The equations of the lines through O corresponding to the n edges of the same $(n-1)$-dimensional face are

$$\frac{x^1}{l_1^i} = \cdots = \frac{x^n}{l_n^i} \qquad (\delta^{pq} l_p^i l_q^i = 1) \qquad (i = 1, \ldots, n).$$

The condition of the preservation of angles leads to the following equations, analogous to (32):

$$(1 - a_1^2) l_1^i l_2^k + \cdots + (1 - a_{n-1}^2) l_{n-1}^i l_{n-1}^k = 0 \qquad (i \neq k;\ i, k = 1, \ldots, n).$$

If we consider the system obtained from the preceding one for an index i fix and set

$$(1 - a_i^2) l_k^i = \lambda_k^i \qquad (k = 1, \ldots, n-1), \tag{33}$$

then the new system comes to

$$\lambda_1^i l_1^k + \cdots + \lambda_{n-1}^i l_{n-1}^k = 0 \qquad (k = 1, \ldots, n),$$

which implies either $\lambda_k^i = 0$ $(k = 1, \ldots, n-1)$ or the condition that the n edges lie in the same $(n-1)$-dimensional plane $\lambda_k^i x^k = 0$. But, since clearly n edges of an n-dimensional non-degenerate simplex cannot lie in the same $(n-2)$-dimensional plane, it follows that the n edges determine the $(n-1)$-dimensional plane $\lambda_k^i x^k = 0$, which is the plane corresponding to the $(n-1)$-dimensional face containing these edges. But each of the planes of at least two $(n-1)$-dimensional faces in an n-dimensional simplex meets the x^n-axis only in one point. Suppose that the n edges from above belong to one of these two faces. Hence all n cannot belong to the plane $\lambda_k^i = 0$ $(k = 1, \ldots, n-1)$, which on account of the notations (33) yields

$$(1 - a_k^2) l_k^i = 0 \qquad (k = 1, \ldots, n-1),$$

and this for $i = 1, \ldots, n$. Next, consider the system

$$(1 - a_k^2) l_k^i = 0 \qquad (i = 1, \ldots, n),$$

obtained from the preceding one for a fixed index k. Suppose to prove it is false, that $a_k \neq 1$. Then $l_k^i = 0$ $(i = 1, \ldots, n)$, i.e. the n considered edges would lie in the plane $x^k = 0$ (containing the x^n-axis), which is not possible, since the $(n-1)$-dimensional plane of the considered face meets the x^n-axis only in a point. Hence $a_k = 1$, and this is true for $k = 1, \ldots, n-1$. We conclude that if $f(x)$ is ACL in D and (i'), (ii') hold for a fix but arbitrary simplex τ then $f(x)$ is conformal in D.

Finally, to complete the proof, we must show that f is differentiable a.e. in D and ACL_n under the assumption that τ is an arbitrary simplex. For this let $g(\xi)$ be an affine mapping with the dilatation K (i.e. representing a K-QCfH), which carries the vertices of τ into the points $O, e_1, \ldots, e_n$, and set $h = f \circ g^{-1}$. Then h satisfies conditions (i'), (ii') with τ, D, E replaced by $g(\tau)$, $g(D)$, $g(E)$ and $A(\gamma_1, \gamma_2)$ by $\dfrac{A(\gamma_1, \gamma_2)}{K}$ in (ii'). Since $g(x)$ is a non-degenerate affine mapping, also $g(E)$ is of Σ-finite area. We establish that $h(\xi)$ is ACL_n in $g(D)$ by the same argument used above to prove $f(x)$ is ACL_n in D, the only difference consisting in the fact that in this case we have no more $p > \operatorname{cosec}\dfrac{\pi}{8}$, but $p > \operatorname{cosec}\dfrac{\pi}{8K}$. Hence $h(\xi)$ is differentiable a.e. in $g(D)$. Since $g(x)$ is an affine mapping, it follows that $f = h \circ g$ is differentiable a.e.in D and ACL_n.

Remark. For some time it has been an open question as to what *a priori* differentiability hypotheses are necessary in Liouville's theorem on the conformal mappings in space. Liouville's theorem stated above asks that the mapping have continuous fourth order partial derivatives. The following corollary shows that no such differentiability hypotheses are necessary. The first to prove it is Ju. G. Rešetnjak [2].

THEOREM 7. *A topological mapping $f(x)$ of a domain D is the restriction of a Möbius transformation to the domain D if and only if $\delta_L(x) = 1$ a.e. in D and $\delta_L(x) < \infty$ in D except possibly on a set of Σ-finite area.*

This theorem is an immediate consequence of theorem 5 and of the equivalence between Gehring's geometric and metric definitions for $K = 1$ (theorem 3.14).

Remark. For the moment it is an open question if the set of $(n-1)$-dimensional Σ-finite area which is involved in the definitions of Grötzsch's type of the form I' or I'_1 is the most general exceptional set which does not affect the quasiconformality (and in particular the conformality) of a topological mapping. However, there are example of larger exceptional sets where the mapping is no more conformal.

Example. Let X^i $(i = 1, \ldots, n)$ denote the *ith* coordinate axis, let F be any nonenumerable closed set in X^1 and let $u(x^1)$ be a continuous non-decreasing singular function which is constant in each interval of $X^1 - F$. Then the mapping $(x^{*1}, \ldots, x^{*n}) = [x^1 + u(x^1), x^2, \ldots, x^n]$ is a topological mapping of the space R^n onto itself for which $\delta_L(x) = 1$ in CE where $E = F \times X^1 \times \cdots \times X^n$. But $x^* = f(x)$ is clearly not a Möbius transformation.

The preceding corollary leads to the following definition of the conformal mappings (clearly equivalent to the other):

A *conformal mapping* of D is a diffeomorphism in D, which for each point $x \in D$, except possibly a set of $(n-1)$-dimensional Σ-finite area, carries infinitesimal spheres onto infinitesimal spheres.

This definition and the property of the conformal mappings in D to be differentiable everywhere in D yield the following new definition.

A *conformal mapping* $f(x)$ of D is a diffeomorphism with the property that the linear element and its image under f are proportional everywhere in D.

And now, in view of a new definition, let us prove:

LEMMA 2. *The extremal dilatations of a homeomorphism $f(x)$ which carries infinitesimal ellipsoids centered at x_0 onto infinitesimal ellipsoids and has a modulus of dilatation $\left|\frac{\partial f(x_0)}{\partial s}\right|$ in a direction s, satisfy the inequality*

$$\frac{\Lambda_f(x_0)}{\lambda_f(x_0)} \leqq p_1^2(x_0)p_1'^2(x_0) < \infty, \tag{34}$$

where $p_1(x_0)$, $p_1'(x_0)$ are the principal characteristics of the two families of infinitesimal ellipsoids.

Let $\{x_k\}$, $\{x_m\}$ be two sequences of points such that

$$\lim_{k\to\infty} x_k = \lim_{m\to\infty} x_m = x_0,$$

$$\lim_{k\to\infty}\frac{|f(x_k)-f(x_0)|}{|x_k - x_0|} = \varliminf_{x\to x_0}\frac{|f(x)-f(x_0)|}{|x-x_0|} = \lambda_f(x_0),$$

$$\lim_{m\to\infty}\frac{|f(x_m)-f(x_0)|}{|x_m - x_0|} = \varlimsup_{x\to x_0}\frac{|f(x)-f(x_0)|}{|x-x_0|} = \Lambda_f(x_0),$$

Every x_k and x_m lie on an ellipsoid E_{h_k} and E_{h_m}, respectively centred at x_0 (all these ellipsoids are supposed to be homothetical). Let us establish between the two sequences a one-to-one correspondence. Then, if the first sequence is $\{x_k\}$, the second one may be written as $\{\widetilde{x}_k\}$.

Set $\widetilde{x}_k'' \in E_{h_k}$ and $x_k'' \in E_{h_k}$ so that $\widetilde{x}_k'' - x_0$ and $x_k'' - x_0$ be vectors in the direction s. Then, since

$$\frac{1}{p_1(x_0)} \leqq \frac{|\widetilde{x}_k'' - x_0|}{|\widetilde{x}_k - x_0|}, \quad \frac{|x_k'' - x_0|}{|x_k - x_0|} \leqq p_1(x_0),$$

$$\frac{|f(\widetilde{x}_k)-f(x_0)|}{|f(\widetilde{x}_k'')-f(x_0)|} \leqq p_1'(x_0)\frac{(\widetilde{h}_1')_k}{(\widetilde{h}_2')_k}, \quad \frac{1}{p_1'(x_0)}\cdot\frac{(h_2')_k}{(h_1')_k} \leqq \frac{|f(x_k)-f(x_0)|}{|f(x_k'')-f(x_0)|},$$

– where $(\widetilde{h}_1')_k$, $(\widetilde{h}_2')_k$ are semi-axes of the ellipsoids circumscribing $f(E_{\widetilde{h}_k})$, respectively inscribed in $f(E_{\widetilde{h}_k})$, and $(h_1')_k$, $(h_2')_k$ are the semi-axes of the ellipsoids circumscribing $f(E_{h_k})$, respectively inscribed in $f(E_{h_k})$, we obtain

$$\frac{\Lambda_f(x_0)}{\lambda_f(x_0)} = \frac{\overline{\lim_{x\to x_0}} \frac{|f(x)-f(x_0)|}{|x-x_0|}}{\underline{\lim_{x\to x_0}} \frac{|f(x)-f(x_0)|}{|x-x_0|}} = \frac{\lim_{k\to\infty} \frac{|f(\widetilde{x}_k)-f(x_0)|}{|\widetilde{x}_k - x_0|}}{\lim_{k\to\infty} \frac{|f(x_k)-f(x_0)|}{|x-x_0|}} \leqq$$

$$\leqq \lim_{k\to\infty} \frac{p_1(x_0)p_1'(x_0)\dfrac{(\widetilde{h}_1')_k}{(\widetilde{h}_2')_k}}{\dfrac{(h_2')_k}{p_1(x_0)p_1'(x_0)(h_1')_k}} \cdot \frac{\dfrac{|f(\widetilde{x}_k'')-f(x_0)|}{|\widetilde{x}_k''-x_0|}}{\dfrac{|f(x_k'')-f(x_0)|}{|x_k''-x_0|}} =$$

$$= p_1^2(x_0)p_1'^2(x_0) \lim_{k\to\infty} \frac{(\widetilde{h}_1')_k}{(\widetilde{h}_2')_k} \lim_{k\to\infty} \frac{(h_1')_k}{(h_2')_k} \cdot \frac{\lim_{k\to\infty} \dfrac{|f(\widetilde{x}_k'')-f(x_0)|}{|\widetilde{x}_k''-x_0|}}{\lim_{k\to\infty} \dfrac{|f(x_k'')-f(x_0)|}{|x_k''-x_0|}} =$$

$$= p_1^2(x_0)p_1'^2(x_0),$$

on account of

$$\lim_{k\to\infty} \frac{(\widetilde{h}_1')_k}{(\widetilde{h}_2')_k} = \lim_{k\to\infty} \frac{(h_1')_k}{(h_2')_k} = 1,$$

as it follows from the definition of the homeomorphisms carrying infinitesimal ellipsoids onto infinitesimal ellipsoids, and of

$$\lim_{k\to\infty} \frac{|f(x_k'')-f(x_0)|}{|x_k''-x_0|} = \lim_{k\to\infty} \frac{|f(\widetilde{x}_k''-f(x_0)|}{|\widetilde{x}_k''-x_0|} = \left| \frac{\partial f(x_0)}{\partial s} \right|,$$

by hypothesis.

THEOREM 8. *Let $f(x)$ be a homeomorphism of D. Then $f(x)$ is a conformal mapping if and only if* $\dfrac{\Lambda_f(x)}{\lambda_f(x)} < \infty$ *in D, except possibly a set of Σ-finite area and one of the following additional conditions is satisfied:*

(i) $\Lambda_f(x) = \lambda_f(x) \neq 0$ *a.e. in D;*

(ii) $\lim_{x'\to x} \dfrac{|f(x')-f(x)|}{|x'-x|} = |J(x)|^{\frac{1}{n}}$ *with* $0 < |J(x)| < \infty$ *a.e. in D.*

For the necessity we take into account corollary of lemma 3.4 and observe that Gehring's second analytic definition and corollary 1 of theorem 5 imply $\Lambda_f(x) = \lambda_f(x)$ everywhere in D, $\Lambda_f(x)^n = |J(x)| = \lambda_f(x)^n$ a.e. in D and $0 < |J(x)| < < \infty$ in D, i.e. the conditions (i) and (ii).

We next show the sufficiency. Condition (i), on account of (3.31), yields $\delta_L(x) = 1$ a.e. in D and $\delta_L(x) < \infty$ in D except possibly on a set of Σ-finite area, hence $f(x)$ is conformal in D (according to Gehring's metric definition). As (ii) implies (i), the proof is complete.

And now we give, following F. and R. Nevanlinna ([1] p. 27) the following new definition of conformal mappings:

A diffeomorphism $x^* = f(x)$ of D is said to be a *conformal mapping* in D if it has the functional matrix

$$\mathfrak{J}(x) = \|x_k^{*i}\| = \mu(x) \|\alpha_k^i(x)\|, \ \mu(x) \neq 0, \tag{35}$$

where $\|\alpha_k^i(x)\|$ is an orthogonal matrix and $\mu(x)$ is a real factor.

We remind that a square matrix $\|\alpha_k^i\|$ is called *orthogonal* if

$$\delta_{ik}\alpha_p^i\alpha_q^k = \delta_{pq} \qquad (p, q = 1, \ldots, n).$$

The function $|\mu(x)|$ is termed the *norm of the matrix* $\mathfrak{J}(x)$ *or of the function* $f(x)$. If $\mu(x) \neq 0$, then $f(x)$ is called *non-degenerate*. Clearly, $\mu(x)^2 = \sqrt[n]{J(x)^2}$. (We find again in this way a result furnished by the theorem 8.1.)

THEOREM 9. *The preceding definition of conformal mappings is equivalent to the others.*

The system (8.1) satisfied by the conformal mappings implies (35). Conversely, the preceding definition implies (8.1) in D, hence

$$\left|\frac{\partial f}{\partial s}\right|^2 = \frac{|dx^*|^2}{|dx|^2} = \frac{\delta_{ik}x_p^{*i}x_q^{*k}\,dx^p\,dx^q}{\delta_{ik}\,dx^i\,dx^k} = \sqrt[n]{J^2} \neq 0, \infty,$$

and then, on account of the differentiability of $f(x)$ and combining (3.3) and (3.4), we obtain condition (*i*) of the preceding theorem and the inequality $\dfrac{\Lambda_f(x)}{\lambda_f(x)} < \infty$, as desired.

Now we give, according to I. Cristea [1], the following definition of conformal mappings:

A diffeomorphism $f: R^n \to R^n$, is said to be *conformal* in R^n if every surface $\sigma \subset R^n$ is mapped conformally (in the sense of differential geometry) onto its image.

As it is well known, two surfaces σ, σ^* are said to be *conformally equivalent* if their linear elements are proportional and the factor of proportionality is not 0 or ∞.

THEOREM 10. *Cristea's definition is equivalent to the others.*

Let

$$x^i = x^i(u^1, \ldots, u^{n-1}) \qquad (i = 1, \ldots, n-1) \tag{36}$$

be the parametrical equations of the surface σ, let

$$\tilde{x}^1 = x^{*i}[x^1(u), \ldots, x^n(u)] = \tilde{x}^i(u), \qquad (i = 1, \ldots, n), \tag{37}$$

where $u = (u^1, \ldots, u^n)$, be the parametrical equations of its image σ^* and let

$$e_{pq} = \delta_{ik} x^i_p x^k_q, \; e^*_{pq} = \delta_{ik} \tilde{x}^i_p \tilde{x}^k_q \quad (p, q = 1, \ldots, n-1) \tag{38}$$

be the fundamental coefficients of the first order of the surfaces σ, σ^*, respectively.

According to Cristea's definition, $f(x)$ is a conformal mapping if for any surface σ we have proportionality of the fundamental coefficients of the first order i.e.

$$\frac{e^*_{pq}}{e_{pq}} = \mu^2. \tag{39}$$

On the other hand, by (36), (37), (38) and on account of the differentiability of (x), we obtain

$$e^*_{pq} = \delta_{ik}(x^{*i}_m x^m_p)(x^{*k}_c x^c_q) = (\delta_{ik} x^{*i}_m x^{*k}_c) x^m_p x^c_q = \tag{40}$$

$$= A_{mc} x^m_p x^c_q \qquad (p, q = 1, \ldots, n-1),$$

where $A_{mc} = \delta_{ik} x^{*i}_m x^{*k}_c$ $(m, c = 1, \ldots, n)$. Then, (39) and (40) yield

$$e^*_{pq} - \mu^2 e_{pq} = (A_{mc} - \mu^2 \delta_{mc}) x^m_p x^c_q = 0 \; (p, q = 1, \ldots, n-1).$$

Since this system holds for any surface σ, i.e. for any equation (36), it follows

$$A_{mc} - \mu^2 \delta_{mc} = 0 \qquad (m, c = 1, \ldots, n),$$

which are just the equations (8.2) expressing that the Jacobian is of the form (35).

Conversely, starting with the definition of the brothers Nevanlinna and following the argument in a reversing sense we obtain (39), which expresses the proportionality of the fundamental coefficients of the first order; and this for whatever surface $\sigma \subset R^n$. Hence, any surface σ is conformally equivalent to its image under f.

Conformal-conjugate functions. In view of the definition of the conformal-conjugate functions (for $n = 3$, see I. Cristea [1]) we begin with some preliminary remarks.

Let $x^* = f(x)$ be a conformal mapping. Then formula (35) implies that

$$\delta^{ik} x^{*p}_i x^{*q}_k = \delta^{pq} \mu^2 \qquad (p, q = 1, \ldots, n), \tag{41}$$

which yields

$$\frac{x_i^{*p}}{\begin{vmatrix} x_1^{*1} \cdots & x_{i-1}^{*1} & x_{i+1}^{*1} \cdots & x_n^{*1} \\ \cdots & \cdots & \cdots & \cdots \\ x_1^{*p-1} \cdots & x_{i-1}^{*p-1} & x_{i+1}^{*p-1} \cdots & x_n^{*p-1} \\ x_1^{*p+1} \cdots & x_{i-1}^{*p+1} & x_{i+1}^{*p+1} \cdots & x_n^{*p+1} \\ \cdots & \cdots & \cdots & \cdots \\ x_1^{*n} \cdots & x_{i-1}^{*n} & x_{i+1}^{*n} \cdots & x_n^{*n} \end{vmatrix}} = \frac{1}{\mu^{n-2}} \qquad (i, p = 1, \ldots, n),$$

hence

$$\mu^{n-2} x_i^{*p} = \begin{vmatrix} x_1^{*1} \cdots & x_{i-1}^{*1} & x_{i+1}^{*1} \cdots & x_n^{*1} \\ \cdots & \cdots & \cdots & \cdots \\ x_1^{*p-1} \cdots & x_{i-1}^{*p-1} & x_{i+1}^{*p-1} \cdots & x_n^{*p-1} \\ x_1^{*p+1} \cdots & x_{i-1}^{*p+1} & x_{i+1}^{*p+1} \cdots & x_n^{*p+1} \\ \cdots & \cdots & \cdots & \cdots \\ x_1^{*n} \cdots & x_{i-1}^{*n} & x_{i+1}^{*n} \cdots & x_n^{*n} \end{vmatrix} \qquad (i, p = 1, \ldots, n).$$

And now fix the index p and let us differentiate with respect to x^i each equation of the preceding system. Since

$$\sum_{i=1}^{n} \frac{\partial}{\partial x^i} \begin{vmatrix} x_1^{*1} \cdots & x_{i-1}^{*1} & x_{i+1}^{*1} \cdots & x_n^{*1} \\ \cdots & \cdots & \cdots & \cdots \\ x_1^{*p-1} \cdots & x_{i-1}^{*p-1} & x_{i+1}^{*p-1} \cdots & x_n^{*p-1} \\ x_1^{*p+1} \cdots & x_{i-1}^{*p+1} & x_{i+1}^{*p+1} \cdots & x_n^{*p+1} \\ \cdots & \cdots & \cdots & \cdots \\ x_1^{*n} \cdots & x_{i-1}^{*n} & x_{i+1}^{*n} \cdots & x_n^{*n} \end{vmatrix} = 0,$$

we get

$$\sum_{i=1}^{n} \frac{\partial(\mu^{n-2} x_i^{*p})}{\partial x^i} = 0 \qquad (p = 1, \ldots, n),$$

whence

$$\sum_{i=1}^{n} \left[\mu^2 x_{ii}^{*p} + (n-2) x_i^{*p} \mu \frac{\partial \mu}{\partial x_i} \right] = 0 \qquad (p = 1, \ldots, n),$$

which, since, on account of (41),

$$\mu \frac{\partial \mu}{\partial x^m} = \delta^{ik} x_i^{*p} x_{km}^{*p},$$

yields

$$\mu^2 \Delta x^{*p} + (n-2)\delta^{qm}\delta^{ik} x_i^{*p} x_q^{*p} x_{km}^{*p} = 0 \quad (p = 1, ..., n).$$

We define, following Cristea [1], operator $\diamond$ as

$$\diamond g = \delta^{ik} \frac{\partial g}{\partial x^i} \frac{\partial g}{\partial x^k} \Delta g + (n-2)\delta^{ik}\delta^{qm} \frac{\partial g}{\partial x^i} \frac{\partial g}{\partial x^q} \frac{\partial^2 g}{\partial x^k \partial x^m}, \tag{42}$$

where

$$\mu(x) = \sqrt{\delta^{in} \frac{\partial g}{\partial x^i} \frac{\partial g}{\partial x^k}}$$

is called the *norm of the function*. We see that the n components of a conformal mapping $x^* = f(x)$ satisfy the condition $\diamond x^{*i} = 0\ (i = 1, \ldots, n)$, justifying the following definition:

The operator $\diamond$ defined by (42) is said to be *conformal* and any function $g \in C^2$ of D satisfying in D the equation $\diamond g = 0$ is called a *conformal function in D.*

Thus the components of a conformal mapping are conformal functions. Hence we obtain the following new definition:

n conformal functions $x^{*i}(x)\ (i = 1, \ldots, n)$ of $D \subset R^n$ considered as the components of a mapping of D onto a domain $D^* \subset R^n$ are called *conformal conjugate* in D if the corresponding mapping is a conformal mapping $f : D \rightleftarrows D^*$.

As a consequence of the definition of the conformal mappings given by F. and R. Nevanlinna, we have

THEOREM 11. *n conformal functions $x^{*i}(x)\ (i = 1, \ldots, n)$ are conformal conjugate if and only if their functional matrix is orthogonal up to a real factor.*

Remark. This theorem was proved in a different way by N. Ciorănescu [1].

And now given a conformal function, we establish, following Cristea [1], the degree of freedom of a set of $n - 1$ functions conformal conjugate to the first.

THEOREM 12. *Given a conformal function $x^{*n}(x)$ of D and two sets of $n-1$ functions $\{x^{*k}\}$ and $\{\tilde{x}^k\}$ $(k = 1, \ldots, n-1)$, conformal conjugate to x^{*n} in D, then*

$$\tilde{x}^k = \alpha_p^k x^{*p} + \alpha_0^k \qquad (k = 1, ..., n-1), \tag{43}$$

where $\|\alpha_p^k\|$ is an orthogonal matrix.

Since the n functions $\{x^{*i}\}$ $(i = 1, \ldots, n)$ are conformal conjugate, i.e. $x^* = f(x)$ is a conformal mapping, it follows that $J(x) \neq 0$ in D and then we can choose $\{x^{*i}\}$ $(i = 1, \ldots, n)$ as independent vectors, set then

$$\tilde{x}^k = \tilde{x}^k(x) = x'^k(x^*) \qquad (k = 1, ..., n-1).$$

Since these functions are differentiable, it follows

$$\tilde{x}^k_p = x'^k_i x^{*i}_p \qquad (p = 1, \ldots, n; \; k = 1, \ldots, n-1). \tag{44}$$

The functions $\tilde{x}^1, \ldots, \tilde{x}^{n-1}, x^{*n}$ are conformal conjugate, too, hence

$$\begin{cases} \delta^{pq}\tilde{x}^k_p\tilde{x}^k_q = \delta^{pq}x^{*n}_p x^{*n}_q = \mu^2 & (k = 1, \ldots, n-1), \\ \delta^{pq}\tilde{x}^k_p\tilde{x}^m_q = \delta^{pq}\tilde{x}^k_p x^{*n}_q = 0 & (m \neq k; m, k = 1, \ldots, n-1), \end{cases}$$

which, on account of (44), may be written as

$$\begin{cases} x'^k_i x'^k_c(\delta^{pq}x^{*i}_p x^{*c}_q) = \mu^2 & (k = 1, \ldots, n-1), \\ x'^k_i x'^m_c(\delta^{pq}x^{*i}_p x^{*c}_q) = x'^k_i(\delta^{pq}x^{*i}_p x^{*n}_q) = 0 & (m \neq k, m, k = 1, \ldots, n-1). \end{cases} \tag{45}$$

But, since $\|x^{*_k}_i\|$ is orthogonal (up to a real factor), we have

$$\delta^{pq}x^{*i}_p x^{*k}_q = \delta^{ik}\mu^2 \qquad (i, k = 1, \ldots, n),$$

hence, by (45),

$$\delta^{ic}x'^k_i x'^k_c = 1, \; \delta^{ic}x'^k_i x'^m_c = 0, \; x'^k_n = 0 \quad (m \neq k; \; m, k = 1, \ldots, n-1).$$

From the last $n-1$ equations, we conclude that x'^k $(k = 1, \ldots, n-1)$ does not depend on x^{*_n}, and then the other equations reduce to the system (8.1) for the $(n-1)$-dimensional case. Thus, the $n-1$ functions $x'^k(x^{*1}, \ldots, x^{*_{n-1}})$ constitute a conformal mapping of the $(n-1)$-dimensional domain $D \cap \{x^{*_n} = 0\}$ onto itself with $|J(x)| = 1$. Thus, $\{\tilde{x}^k\}$ $(k = 1, \ldots, n-1)$ express in terms of $\{x^{*_k}\}$ $(k = 1, \ldots, n-1)$, by (43).

In other words, a conformal mapping is uniquely determined by a conformal function up to a rotation about the x^{*_n}-axis and a translation in the plane $x^{*_n} = 0$.

Connection between conformal mappings and harmonic functions. A function $u(x)$ is called harmonic in D if it satisfies the equation

$$\Delta u(x) = 0,$$

where $\Delta = \delta^{ik}\dfrac{\partial^2}{\partial x^i \partial x^k}$.

THEOREM 13. *In order that the functions* $x^{*_i} \in C^2$ $(i = 1, \ldots, n)$ *conformal conjugate in* D *be harmonic in* D, *it is necesssary and sufficient that* $J(x) \equiv$ *const. in* D.

Since $f(x)$ is a conformal mapping it follows that $J(x) \neq 0$ in D, hence we can choose $\{x^{*i}\}$ $(i = 1,\ldots, n)$ as new independent variables. If $\mu(x)$ denotes the norm of these functions, then $\mu = \mu(x) = \tilde{\mu}(x^*)$, and $x^{*i} \in C^2$ $(i = 1,\ldots, n)$ imply $\tilde{\mu} \in C^1$, hence

$$\mu_i = \tilde{\mu}_k x_i^{*k} \qquad (i = 1, \ldots, n), \tag{46}$$

where $\mu_i = \dfrac{\partial \mu}{\partial x^i}$, $\tilde{\mu}_k = \dfrac{\partial \tilde{\mu}}{\partial x^{*k}}$, and, by the preceding definition of the conformal mappings

$$\Diamond x^{*i} = \mu^2 \Delta x^{*i} + (n-2)\, \mu \delta^{pq} \mu_p x_q^{*i} = \mu^2 \Delta x^{*i} +$$

$$+ (n-2)\, \mu \tilde{\mu}_k (\delta^{pq} x_p^{*k} x_q^{*i}) = 0 \qquad (i = 1, \ldots, n),$$

whence, by (8.1),

$$\Diamond x^{*i} = \mu^2 \Delta x^{*i} + (n-2)\, \mu^3 \mu_i = 0 \qquad (i = 1,\ldots, n),$$

which implies that

$$\Delta x^{*i} + (n-2)\, \mu \tilde{\mu}_i = 0 \qquad (i = 1, \ldots, n). \tag{47}$$

Suppose now that $\{x^{*i}\}$ $(i = 1,\ldots, n)$ are harmonic in D. Then the preceding equations yield $\tilde{\mu}_i = 0$ $(i = 1,\ldots, n)$, and then, by (46), $\mu_i = 0$ $(i = 1,\ldots, n)$, and we conclude that $\mu(x) \equiv$ const. But since from the definition of the brothers Nevanlinna $\mu(x) = \sqrt[n]{J(x)^2}$, it follows that $\mu(x) \equiv$ const. implies $J(x) \equiv$ const. as desired.

Suppose now conversely that $J(x) \equiv$ const. Then clearly $\mu(x) =$ const. and (47) yields $\Delta x^{*i} = 0$ $(i = 1,\ldots, n)$, i.e. $\{x^{*i}\}$ $(i = 1,\ldots, n)$ are harmonic functions in D, as desired.

THEOREM 14. *If $\{x^{*i}\}$ are conformal conjugate in D, then the norm of these unctions is a constant if and only if the corresponding conformal mapping is given by*

$$x^{*i} = \alpha_k^i x^k + \alpha^i \qquad (i = 1, \ldots, n), \tag{48}$$

where α_k^i, $\alpha^i (i, k = 1,\ldots, n)$ are constant and the matrix of the coefficients is orthogonal.

Since $x^{*i} (i = 1,\ldots, n)$ are conformal conjugate, theorem 11 implies (41). Differentiating with respect to $x^m (m = 1,\ldots, n)$ the equations of this system which correspond to $\delta^{pq} = 1$, i.e. the equations

$$\delta^{ik} x_i^{*p} x_k^{*p} = \mu^2 \qquad (p = 1, \ldots, n),$$

where $\mu(x) \equiv$ const. we get

$$\delta^{ik} x^{*p}_{im} x^{*p}_{k} = 0 \qquad (p = 1, \ldots, n).$$

Hence, and from the equations of (41), which correspond to $\delta^{pq} = 0$, we obtain for fix m and p

$$\begin{cases} \delta^{ik} x^{*p}_{im} x^{*p}_{k} = 0, \\ \delta^{ik} x^{*q}_{i} x^{*p}_{k} = 0 \qquad (q = 1, \ldots, p - 1, p + 1, \ldots, n). \end{cases}$$

But since $x^* = f(x)$ is conformal by hypothesis, and then $J(x) \neq 0$ in D, it follows that the preceding system has not the trivial solution $x^{*p}_{k} = 0$ $(k = 1, \ldots, n)$, which would imply $J(x) = 0$. Thus the determinant of the coefficients is zero, and then

$$x^{*p}_{im} = \nu^{p}_{mq} x^{*q}_{i} \qquad (i, m = 1, \ldots, n), \tag{49}$$

where the dummy index q takes on the values $1, \ldots, p - 1, p + 1, \ldots n$. Differentiating the equation

$$\delta^{ik} x^{*p}_{i} x^{*c}_{k} = 0 \qquad (p \neq c;\, p, c = 1, \ldots, n) \tag{50}$$

with respect to x^m yields

$$\delta^{ik}(x^{*p}_{im} x^{*c}_{k} + x^{*p}_{i} x^{*c}_{km}) = 0 \qquad (p \neq c;\; p, c = 1, \ldots, n),$$

which, on account of (49) and (50), comes to

$$\nu^{p}_{mc} + \nu^{c}_{mp} = 0 \qquad (c, m, p = 1, \ldots, n). \tag{51}$$

Next, from (49), and since $x^{*p}_{im} = x^{*p}_{mi} (i, m = 1, \ldots, n)$, we obtain

$$\nu^{p}_{mq} x^{*q}_{i} = \nu^{p}_{iq} x^{*q}_{m} \qquad (i, m = 1, \ldots, n).$$

Fix $c, b \neq p$; adding the equations of the preceding system previously multiplied by $x^{*p}_{i} x^{*c}_{m}$, $x^{*c}_{i} x^{*b}_{m}$ $(i, m = 1, \ldots, n)$, we obtain

$$\delta^{am} \nu^{p}_{mc} x^{*p}_{a} = 0 \quad (c = 1, \ldots, p - 1,\;\; p + 1, \ldots, n;\;\; p = 1, \ldots, n), \tag{52}$$

$$\delta^{am} \nu^{p}_{mc} x^{*b}_{a} - \delta^{am} \nu^{p}_{mb} x^{*c}_{a} = 0,$$

$$(b, c = 1, \ldots, p - 1,\;\; p + 1, \ldots, n;\;\; p = 1, \ldots, n). \tag{53}$$

Since $b \neq c$ in (53) (otherwise they would reduce to an identity), we may change there p and c between them, getting

$$\delta^{am}\nu^{c}_{mp}x^{*b}_{a} - \delta^{am}\nu^{c}_{mb}x^{*p}_{a} = 0. \tag{54}$$

Here, changing also b and c between them, we obtain

$$\delta^{am}\nu^{b}_{mp}x^{*c}_{a} - \delta^{am}\nu^{b}_{mc}x^{*p}_{a} = 0,$$

which, by (51), imply that

$$\delta^{am}\nu^{p}_{mb}x^{*c}_{a} - \delta^{am}\nu^{c}_{mb}x^{*p}_{a} = 0. \tag{55}$$

Then, adding the equations (55) to the differences between the equations (53) and the corresponding equation (54) and taking into account (51), we obtain

$$\delta^{am}\nu^{p}_{mc}x^{*b}_{a} = 0 \quad (b, c, p = 1,..., n;\ b, c, p \text{ different}). \tag{56}$$

But (51) implies that $\nu^{p}_{mp} = 0$ $(m, p = 1,..., n)$, hence

$$\delta^{am}\nu^{p}_{mp}x^{*b}_{a} = 0 \quad (b, p = 1,..., n),$$

where it is no more necessary to assume $b \neq p$. Then, (56), (52), (51) and the preceding equations imply (56), where this time there is no need to suppose b, c, p different. Finally

$$\delta^{am}\nu^{p}_{mc}x^{*b}_{a} = 0 \quad (b = 1,..., n),$$

with c, p fix, form a homogeneous system of n equations with n unknowns ν^{p}_{mc} $(m = 1,\ldots, n)$. Since the determinant of the coefficients is just $J \neq 0$, it follows that this system has only the trivial solution $\nu^{p}_{mc} = 0$ $(m = 1,\ldots, n)$ and since c and p are arbitrary, we conclude that $\nu^{p}_{mc} = 0$ $(c, m, p = 1,\ldots, n)$, which, by (49) yield $x^{*p}_{im} = 0$ $(i, m, p = 1,\ldots, n)$, and then all the functions x^{*p}_{i} $(i, p = 1,\ldots, n)$ reduce to constants, hence (48) follows, as desired.

The converse is obvious.

From the two preceding theorems we infer the following

COROLLARY. *The n components of a conformal mapping f(x) are harmonic functions if and only if f(x) is a similarity mapping.*

Remark. Since the inversions are conformal mappings, but not similarity mappings, we conclude that the class of conformal mappings is strictly richer than the class of conformal mappings whose components are harmonic functions, and which were studied by N. Ciorănescu [1]. Three harmonic functions $u(x, y, z)$, $v(x, y, z)$, $w(x, y, z)$ are called conjugate if

$$\psi(x, y, z) = \frac{1}{\sqrt{(u + \lambda)^2 + (v + \mu)^2 + (w + \nu)^2}}$$

is harmonic function whatsoever would be the constants λ, μ, ν. Ciorănescu [1] proved that the Jacobian of the corresponding mapping is orthogonal and constant.

Theorem of Reade [1]. *A conformal mapping $f\colon B \rightleftarrows B$ with $f(O) = O$ is of the form*

$$x^{*i} = \alpha^i_k x^k \quad (i = 1,\dots, n),$$

(i.e. a rotation), where $\alpha^i_k (i, k = 1, \dots, n)$ are constant and the corresponding matrix is orthogonal.

Theorem 8.1 implies

$$(\mathrm{d}x^*)^2 = \mu^2\, \mathrm{d}x^2, \mu^2 = \sqrt[n]{J^2}, \tag{57}$$

so that

$$T(x_0, r) = \int_{B(x_0, r)} \mu^n\, \mathrm{d}\tau, \quad A(x_0, r) = \int_{S(x_0, r)} \mu^{n-1}\, \mathrm{d}\sigma, \tag{58}$$

where $T(x_0, r)$ is as in (1.7), and

$$A(x_0, r) = \int_{f[S(x_0, r)]} \mathrm{d}\sigma. \tag{59}$$

The isoperimetric inequality (see T. Rado [2] or H. Hadwiger [1], (174), p. 268) implies

$$n^n \omega_n T(x_0, r)^{n-1} \leqq A(x_0, r)^n,$$

hence

$$\left[\frac{1}{\omega_n} \int_{B(x_0, r)} \mu^n\, \mathrm{d}\tau\right]^{n-1} \leqq \left[\frac{1}{n\omega_n} \int_{S(x_0, r)} \mu^{n-1}\, \mathrm{d}\sigma\right]^n.$$

Applying Hölder's inequality to the right-hand side, yields

$$\left[\frac{1}{\omega_n} \int_{B(x_0, r)} \mu^n\, \mathrm{d}\sigma\right]^{n-1} \leqq \left[\frac{1}{n\omega_n} \int_{S(x_0, r)} \mu^n\, \mathrm{d}\sigma\right]^{n-1},$$

which gives us, on account of (1.7.1) and (1.7.2), the following inequality between the volume and the spherical averages:

$$V(\mu^n, x_0, r) \leqq S(\mu^n, x_0, r)$$

for all $\overline{B(x_0, r)} \subset B$. Since $f(x)$ is conformal, it follows that $f \in C^1$ and $|J(x)| \neq 0$ (corollary 1 of theorem 5), and then $\mu(x)$ is continuous in D. We conclude, by proposition 1.7.4, that μ^n is subharmonic in B, and then proposition 1.7.4 implies

$$\mu(O)^n \leqq V(\mu^n, O, r), \tag{60}$$

where $V(\mu^n, 0, r)$ is non-decreasing with r (proposition 1.7.5). Hence, since we have $f : B \rightleftarrows B$, then from (1.7.1), (58) and (60) where $x_0 = O$ and $r \to 1$, it follows

$$\mu(O)^n \leqq V(\mu^n, O, 1) \leqq 1. \tag{61}$$

But we could also have considered the inverse mapping, which by theorem 2 is conformal, too, with $dx^2 = \frac{1}{\mu^2}(dx^*)^2$ by (57), and, arguing as above, we should have found $\frac{1}{\mu(O)^n} \leqq 1$. Hence we conclude that $|\mu(O)| = 1$, whence, by (60) and (61) and taking into account that $V(\mu^n, O, r)$ is a non-decreasing function of r (proposition 1.7.5), it follows that

$$1 = \mu(O)^n \leqq V(\mu^n, O, r) \leqq 1,$$

i.e. $\mu(O)^n = V(\mu^n, O, r)$. Because $\mu(x)$ is subharmonic in B, corollary of proposition 1.7.6 implies that $\mu(x)^n$ is harmonic in B.

If we again apply Hölder's inequality to the isoperimetric inequality and we make use of the harmonic character of $\mu(x)^n$ in B, then, combining corollaries 2 of propositions 1.7.2 and 1.7.3, we obtain

$$\mu(x_0)^{n(n-1)} = \left[\frac{1}{\omega_n} \int_{B(x_0, r)} \mu^n \, d\tau\right]^{n-1} \leqq \left[\frac{1}{n\omega_n} \int_{S(x_0, r)} \mu^{n-1} \, d\sigma\right]^n \leqq$$

$$\leqq \left[\frac{1}{n\omega_n} \int_{S(x_0, r)} \mu^n \, d\sigma\right]^{n-1} = \mu(x_0)^{n(n-1)},$$

which clearly implies

$$\mu(x_0)^{n-1} = \frac{1}{n\omega_n} \int_{S(x_0, r)} \mu^{n-1} \, d\sigma$$

for all balls $B(x_0, r) \subset B$, hence $\mu^{n-1}(x)$ is harmonic in B (corollary 2 of proposition 1.7.3). Since $\mu^{n-1}(x)$ and $\mu^n(x)$ are both harmonic, proposition 1.7.7 allows us to conclude that $\mu(x) \equiv$ const., and since $\mu(O) = 1$, it follows $\mu(x) \equiv 1$ in B, i.e. $|J(x)| \equiv 1$ in B, hence, since $f : B \rightleftarrows B$, Reade's theorem is a consequence of the preceding one.

Areolar derivatives and conformal mappings. Some generalizations of plane conformal mappings to more than two dimensions and where the components are still harmonic functions are given by M. Nicolescu [1]—[3], Gr. C. Moisil and N. Teodorescu [1]. These definitions are based on the concept of areolar derivatives introduced by D. Pompeiu [1], [2].

We define, following D. Pompeiu [1], [2], the *areolar derivative* of $f(z)$ at z_0 as

$$\frac{Df(z_0)}{Dz} = \lim \frac{1}{2i} \frac{\int_C f(z)\,dz}{\int_\Omega d\sigma},$$

where $f(z)$ is a continuous function in a domain Ω with boundary a rectifiable curve C and where the limit is taken as C shrinks to the point z_0.

Two functions $F(z, z')$, $G(z, z')$ of the complex variables z, z' are said to be *areolar-conjugate* (M. Nicolescu [1]) if

$$\frac{DF}{Dz} = \overline{\frac{DG}{Dz'}}, \quad \frac{DF}{Dz'} = -\overline{\frac{DG}{Dz}},$$

where $\frac{DF}{Dz}, \frac{DF}{Dz'}, \frac{DG}{Dz}, \frac{DG}{Dz'}$ are the areolar partial derivatives of F and G with respect to z and z', and where $\frac{DG}{Dz}, \overline{\frac{DG}{Dz}}$ and $\frac{DG}{Dz'}, \overline{\frac{DG}{Dz'}}$, respectively, are complex conjugate numbers.

Two areolar-conjugate functions satisfy the same equation

$$\frac{D}{Dz}\left(\overline{\frac{D\vartheta}{Dz}}\right) + \frac{D}{Dz'}\left(\overline{\frac{D\vartheta}{Dz'}}\right) = 0.$$

A function satisfying this equation is called *α-harmonic.*

M. Nicolescu [1] proved that: "A function $Z(z, z')$ is a harmonic α function if and only if its real and imaginary parts are harmonic functions in R^4".

Now let us say some words about G. C. Moisil's [1] generalization. Let $\psi_\alpha \in C^1$ be a system of 2 functions and let $\gamma_i (i = 1, \ldots, n)$ be a system of n square matrices with r rows and r columns and which satisfy Dirac's equations

$$\gamma_i \gamma_k + \gamma_k \gamma_i = 2\delta_{ik} e \quad (i, k = 1, \ldots, n), \tag{62}$$

where e is an identity matrix. One considers the linear partial differential system

$$F\psi_\alpha = \psi_\alpha \qquad (\alpha = 1, \ldots, r), \tag{63}$$

where $F = \sum_{i=1}^{n} \gamma_i \frac{\partial}{\partial x^i}\cdot$ The equations (62) yield $F^2 = e\Delta$, hence, applying the operator F to the equation (63), we obtain $\Delta\psi_\alpha = F\Psi_\alpha$ $(\alpha = 1,\ldots, r)$. If $r = 2$, Ψ is the areolar derivative of ψ. If the system (63) is homogeneous i.e. $\Psi_\alpha = 0$ $(\alpha = 1,\ldots, r)$ then ψ_α $(\alpha = 1,\ldots, r)$ are the solutions of the equation $\Delta\psi_\alpha = 0$, i.e. harmonic functions.

N. Teodorescu generalized in his course (1957—1958): "Special chapters of the theory of the equations of the mathematical physics" the definition of the areolar derivative for n-space as follows:

Given a field of matrices $\varphi = \{\varphi^i_k\}$, the elements of each matrix being continuous functions in D and given a regular family (Δ, σ) of subdomains Δ, with $\sigma = \partial\Delta$, shrinking to a point $x \in D$, *this field* is said to *possess a spatial derivative* at x with respect to the field of constant matrices $\gamma(\gamma_1,\ldots, \gamma_n)$, if

$$\Phi(x) = -\lim_{m\Delta\to 0} \frac{\int_\sigma (\vec{n}_y\cdot\gamma)\varphi(y)\,d\sigma}{m\Delta}$$

where $\vec{n}_y$ is the vector normal to σ, exists and is independent of the family (Δ, σ). Then the numerical matrix $\Phi(x) = D\varphi$ is called the *spatial derivative of the field φ with respect to γ at the point x*. The fields φ the matrices elements of which are continuous matrices in D, differentiable (γ) in a domain $\Omega \subset D$ and with the spatial derivatives contained in Ω, are termed *functions of the class $C^1_\gamma(\Omega)$*.

A monogenic function at $x \in R^n$ is a matrix $\varphi \in C^1$ in a neighbourhood of x, satisfying at x the conditions $D\varphi = 0$ and

$$\bar{\gamma}_i\gamma_k + \bar{\gamma}_k\gamma_i = \delta_{ik}e \quad (i, k = 1, \ldots, n),$$

where

$$\gamma_i\bar{\gamma}_i = e \qquad (i = 1, \ldots, n).$$

Since $\bar{D}D = \Delta e$, where $\bar{D} = \sum_{i=1}^{n} \bar{\gamma}_i \frac{\partial}{\partial x^i}$, it follows that the monogeneity of φ implies $\Delta\varphi = 0$, hence φ is harmonic.

We see that the monogenic vector functions, according to all preceding definitions, have harmonic components, and then, by corollary of theorem 14 a conformal mapping is not, in general, monogenic according to these definitions (as for instance the inversions). That is why we shall not deal with these extensions any more. Let us mention just in passing, an analogous generalization given by Gr. C. Moisil [2], by means of hypercomplex functions of a hypercomplex variable and also different results obtained by Gr. Moisil [1], [2], N. Teodorescu [1]—[5], D. Pascali [1]—[6], A. V. Bicadze [1], I. Elianu [1], Vinogradov [1], M. Roşculeţ[1]—[2], A. I. Ward [1], D. Ivanenko and K. Nikolski [1], V. S. Fedorov [1], I. H. Michael [1], etc.

Finally, we also mention the *analytic vectorial functions* (Si-ping Cheo [1], Rainich [1], etc.) characterized by the conditions

$$\operatorname{div} f = 0, \ \operatorname{rotor} f = 0.$$

Si-ping Cheo shows that a vectorial function $f=\operatorname{grad} h$ if and only if $\Delta h = 0$.

Points of analyticity. We close this chapter with a necessary and sufficient condition for an A-point to be a point of analyticity (we established this in [13], [17]). By a *point of analyticity of a homeomorphism* $f(x)$ we mean a point at which $f(x)$ carries infinitesimal spheres onto infinitesimal sphere (in the sense of the definition of mappings carrying infinitesimal ellipsoids onto infinitesimal ellipsoids: chapter 3). We begin with some preliminary results.

LEMMA 3. *If a homeomorphism $f(x)$ carries infinitesimal ellipsoids centred at x_0 onto infinitesimal ellipsoids and $\Lambda_f(x_0) = 0$, then $f(x)$ is differentiable at x_0 and the rank of the functional matrix at x_0 is* 0.

Indeed, $\Lambda_f(x_0) = 0$ implies

$$\left|x_k^{*i}\, e_i\right| = 0 \qquad (k = 1, \ldots, n),$$

hence $x_k^{*i} = 0$ $(i, k = 1,\ldots, n)$, and then the functional matrix has the rank zero. But $\Lambda_f(x_0) = 0$ yields also

$$f(x) - f(x_0) = 0 \qquad (\left|x - x_0\right|),$$

and then $f(x)$ is differentiable at x_0.

LEMMA 4. *A homeomorphism $f(x)$, which carries infinitesimal ellipsoids centred at x_0 onto infinitesimal ellipsoids and has $J(x_0) = 0$ is differentiable at x_0 if and only if the rank of the functional matrix is zero.*

For the necessity we observe that a homeomorphism which carries infinitesimal ellipsoids centred at x_0 onto infinitesimal ellipsoids is a homeomorphism with two sets of characteristics at x_0 and then *a fortiori* a QCfH in Markuševič—Pesin's sense at x_0, so that, by theorem 3.5, the functional matrix has the rank zero.

The sufficiency is a consequence of the fact that if the functional matrix is of rank zero at x_0, then $\Lambda_f(x_0) = 0$ and by preceding theorem, $f(x)$ is differentiable at x_0, as desired.

THEOREM 15. *In order that a homeomorphism $f(x)$ with $\Lambda_f(x) > 0$ be analytical at x_0 it is necessary and sufficient that*

$$\Lambda_f(x_0) = \lambda_f(x_0) < \infty. \tag{64}$$

The necessity is a consequence of (34), where $p_1(x_0) = p_1'(x_0) = 1$.

For the sufficiency we show that (64) implies

$$\lim_{r\to 0} \frac{\max\limits_{x\in S_r} |\ (x) - f(x_0)|}{\min\limits_{x\in S_r} |f(x) - \ (x_0)|} = 1, \tag{65}$$

which expresses analytically that infinitesimal spheres centred at x_0 are carried by $f(x)$ onto infinitesimal spheres. Set $S_r = S(x_0, r)$ and let $x', x'' \in S_r$ be so that

$$|f(x') - f(x_0)| = \max_{x \in S_r} |f(x) - f(x_0)|,$$

$$|f(x'') - f(x_0)| = \min_{x \in S_r} |f(x) - f(x_0)|.$$

Letting $r \to 0$ in the obvious formula

$$\frac{|f(x') - f(x_0)|}{|x' - x_0|} = \frac{|f(x') - f(x_0)|}{|f(x'') - f(x_0)|} \frac{|f(x'') - f(x_0)|}{|x'' - x_0|}$$

and taking into account that $\Lambda_f(x_0) = \lambda_f(x_0) \neq 0$, we obtain (64), as desired.

Remarks. 1. The restrictive condition $\Lambda_f(x_0) \neq 0$ is essential, as follows from the example

$$\begin{cases} u = x\sqrt{x^2 + a^2 y^2}, \\ v = ay\sqrt{x^2 + a^2 y^2}, \end{cases} \quad a^2 > 1, \tag{66}$$

because this mapping is a homeomorphism in the neighbourhood of O with $\Lambda_f(O) = \lambda_f(O)$ and nevertheless

$$\lim_{r \to 0} \frac{\max\limits_{x \in S_r} |\Delta f(O)|}{\min\limits_{z \in S_r} |\Delta f(O)|} = a^2 > 1 \quad (z = x + iy).$$

2. The preceding theorem is no more true if we replace (64) by the condition

$$\left| \frac{\partial f(x)}{\partial s} \right| = K(x), \tag{67}$$

as it follows by considering the example (3.32), because this homeomorphism in the neighbourhood of the origin satisfies in O the condition (67) with $K(O) = K$ and has also $\Lambda_f(O) \neq 0$. But if we choose a sequence $\{z_m\}$, satisfying the conditions

$$\lim_{m \to \infty} z_m = O, \ |z_m| = \arg z_m,$$

then

$$\Lambda_f(O) \geqq \lim_{m \to \infty} \frac{|\ (z_m) - \ (O)|}{|z_m|} = K + 1 > K \geqq \lambda_f(O),$$

hence $\Lambda_f(O)=\lambda_f(O)$, and then, by the preceding theorem, O is not a point of analyticity for $f(x)$.

THEOREM 16. *A point of differentiability x_0 of a homeomorphism $f(x)$ with a functional matrix of a rank different of zero is a point of analyticity if and only if $f(x)$ satisfies the conditions*

$$\left|\frac{\partial f(x_0)}{\partial s_q}\right| = K(x_0) \quad \left[q = 1, \ldots, \frac{1}{2}n(n+1)\right], \tag{68}$$

where s_q are $\frac{n(n+1)}{2}$ different directions such that the $\frac{n(n+1)}{2}$ corresponding lines passing through x_0 do not lie on the same right circular cone.

For the necessity suppose first that $J(x_0) = 0$. Then, by the preceding lemma, in the particular case in which the ellipsoids reduce to spheres, the functional matrix has the rank 0, hence $\Lambda_f(x_0) = 0$ and then $\left|\frac{\partial f(x_0)}{\partial s}\right| = 0$ in all directions s. We assumed however that the rank of the functional matrix does not vanish because in this case the condition (68) is no more sufficient.

If $J(x_0) \neq 0$, then at least a partial derivative $x^{*i}_k(x_0) \neq 0$, hence $\left|\frac{\partial f(x_0)}{\partial x^k}\right| \neq 0$, which implies $\Lambda_f(x_0) \neq 0$, and then, by the preceding theorem, (64) holds, which, by (3.3), and (3.4), yields

$$\left|\frac{\partial f(x_0)}{\partial s}\right| = K(x_0)$$

for all s.

For the sufficiency, suppose that $J(x_0) = 0$ and that the functional matrix has the rank $r (1 < r < n-1)$. Since at a point of differentiability x_0 of $f(x)$

$$\left|\frac{\partial f(x_0)}{\partial s}\right| = \left|\frac{\partial f_1(x, x_0)}{\partial s}\right|,$$

where $f_1(x, x_0)$ is the corresponding linear transformation (3.2), it follows that if $f_1(x, x_0)$ does not satisfy (67), neither $f(x)$ does.

Let us assume, as we may, $x_0 = f(x_0) = O$. Then $f_1(x, x_0)$ becomes of the form

$$\begin{cases} x^{*c} = a^c_i x^i & (c = 1, \ldots, r), \\ x^{*m} = v^m_c a^c_i x^i & (m = r+1, \ldots, n), \end{cases} \tag{69}$$

where the matrix $\|a^c_i\|$ has the rank r, $\delta^{cp} v^m_c v^m_p \neq 0$ $(m = r+1, \ldots, n)$ and the indices c, p run over $1, \ldots, r$. Then from

$$\delta_{ik} x^{*i} x^{*k} = \delta_{cp}(a^c_i x^i)(a^p_k x^k) + \delta_{mq} v^m_c v^q_p (a^c_i x^i)(a^p_k x^k) =$$
$$= (\delta_{cp} + \delta_{mq} v^m_c v^q_p)(a^c_i x^i)(a^p_k x^k),$$

where c, p take on the values $1, \ldots, r$; m, q the values $r+1, \ldots, n$ and i, k the values $1, \ldots, n$, it follows that the affine mapping (69) carries the cylinder

$$(\delta_{cp} + \delta_{mq} v_c^m v_p^q)(a_i^c x^i)(a_k^p x^k) = R^2 \tag{70}$$

in the $(r-1)$-dimensional sphere

$$\delta_{ik} x^{*i} x^{*k} = R^2, \; x^{*m} = v_c^m x^{*c} \quad (m = r+1, \ldots, n).$$

The cylinder (70) has as directrix the $(r-1)$-dimensional ellipsoid given by the system of equation of the form

$$(\delta_{cp} + \delta_{mq} v_c^m v_p^q) \xi^c \xi^p = R^2, \; \xi^m = 0 \quad (m = r+1, \ldots, n)$$

and as element the $(n-r)$-planes given by

$$a_i^c x^i = \text{const.} \quad (c = 1, \ldots, r).$$

In the particular case $r = 1$, both the directrix of (70) and the sphere which is the image of (70) under (69), reduce to two points, and the cylinder (70) to two parallel planes.

Next we observe that

$$\left| \frac{\partial f_1(x, x_0)}{\partial s} \right| = \frac{|f_1(x, x_0) - f(x_0)|}{|x - x_0|_s},$$

where $|x - x_0|_s$ is the modulus of a vector $x - x_0$ having the direction s. In order to find the directions in which

$$\left| \frac{\partial f_1(x, O)}{\partial s} \right| = K(O),$$

it is sufficient to try to find out the points of the cylinder (70) lying on the sphere S_r, since they are carried into the sphere $|f_1(x, x_0)| = K(O) R$. These points lie then on the $(n-2)$-dimensional surface

$$(\delta_{cp} + \delta_{mq} v_c^m v_p^q) a_i^c a_k^p x^i x^k = R^2, \; \delta_{ik} x^i x^k = R^2,$$

which may be written also as

$$(\delta_{cp} + \delta_{mq} v_c^m v_p^q)(a_i^c a_k^p - \delta_{ik}) x^i x^k = 0, \; \delta_{ik} x^i x^k = R^2,$$

i.e. lie on the intersection of a right circular cone with a sphere. Hence it follows that it is impossible that (68) holds for $\frac{n(n+1)}{2}$ directions specified in the theorem if $f_1(x, x_0)$ is a singular linear transformation and the same is true for the corresponding $f(x)$.

And now let us consider the case $J(x_0) \neq 0$. It is well known that a non-singular linear transformation carries certain ellipsoids into spheres and lemma 3.7 shows that if x_0 is an A-point of a homeomorphism $f(x)$, then a necessary and sufficient condition for $f(x)$ to carry the infinitesimal ellipsoids $E[(C), x_0]$ onto infinitesimal spheres is that $f_1(x, x_0)$ carries the ellipsoids $E_h[(C), x_0]$ onto spheres. Thus, we shall prove that if the condition (68) holds for $f_1(x, x_0)$, then the ellipsoid $E_h[(C), x_0]$ reduces to a sphere and by lemma 3.7, x_0 will be a point of analyticity of $f(x)$. Next, arguing as in the preceding case (the singular one), the directions s in which

$$\left|\frac{\partial f(x_0)}{\partial s}\right| = K(x_0),$$

are those corresponding to the points where $E_h[(C), x_0]$ meets $S(x_0, r)$, i.e. the points of the surface

$$\delta_{ik} x_p^{*i}(x_0) x_q^{*k}(x_0)(x^p - x_0^p)(x^q - x_0^q) = R^2,$$

$$\delta_{pq}(x^p - x_0^p)(x^q - x_0^q) = R^2,$$

which may be written also as

$$[\delta_{ik} x_p^{*i}(x_0) x_q^{*k}(x_0) - \delta_{pq}](x^p - x_0^p)(x^q - x_0^q) = 0,$$

$$\delta_{pq}(x^p - x_0^p)(x^q - x_0^q) = R^2.$$

Thus an ellipsoid has in common with the concentric sphere only the points lying on the right circular cone

$$[\delta_{ik} x_p^{*i}(x_0) x_q^{*k}(x_0) - \delta_{pq}](x^p - x_0^p)(x^q - x_0^q) = 0,$$

and if we suppose a common point which is not on the cone, then, necessarily the ellipsoids must coincide with the sphere.

We have only to specify that $\frac{n(n+1)}{2} - 1$ is the maximal number of arbitrary concurent lines such that there exists a corresponding right circular cone containing all of them and is given by the number of the coefficients of the variables in the general equation of a cone.

Remarks. 1. This theorem represents our generalization [17] of the corresponding Menšov's [1] bidimensional theorem which asserts that the condition (68) (which holds for three different directions) is sufficient to assure the monogeneity at a point of a differentiable function. In the enunciation of our theorem we were obliged to put also the additional condition that the functional matrix be of rank different from zero (this condition does not appear in Menšov's theorem) since, for $n = 2$, the condition (68) implies the monogeneity of $f(z)$ in z_0 also in the hypothesis that the rank of the functional matrix is zero (as we saw in the proof of our theorem for a general n), however (68) is no more sufficient to imply the analyticity as it follows from the example (66), where $\left|\frac{\partial f(O)}{\partial s}\right| = 0$ in all directions s, and $f(z)$ is monogenic with $f'(O) = 0$, but the origin is not a point of analyticity because $f(z)$ carries the ellipses $x^2 + a^2y^2 = \text{const.}$ onto the circles $u^2 + v^2 = \text{const.}$, and not infinitesimal circles onto infinitesimal circles.

We recall at this point that the function (Menšov [1])

$$f(z) = \begin{cases} |z|\, e^{2i \arg z} & \text{if } 0 \leq \arg z < \frac{\pi}{2}, \\ |z|\, e^{\frac{2}{3} i(\arg z + \pi)} & \text{if } \frac{\pi}{2} \leqslant \arg z < 2\pi, \end{cases}$$

is an example of mapping analytic at the origin without being monogenic there. However, theorem 1 in Menšov's book ([1], p. 34) asserts that, for $n = 2$, if $f(z)$ is a conformal mapping in D, except possibly in a countable set of points, then $f(z)$ or $\overline{f(z)}$ is monogenic in D.

2. The preceding theorem is more general than the theorem 15, since instead of (64) we have (68), however it is more restrictive since it assumes the differentiability of $f(x)$ at x_0.

3. For a global result of the same kind see theorem 6.

THEOREM 17. *If $f(x)$ is a conformal mapping in $D \subset R_n$ $(n > 2)$ then the vectors $\frac{\partial f}{\partial x^i}$ $(i = 1, \ldots, n)$ have the same length $\left|\frac{\partial f}{\partial x^i}\right| = \sqrt[n]{|J|}$ and are mutually orthogonal in D.*

This is an immediate consequence of the system (8.1).

THEOREM 18. *A homeomorphism preserves the angles with the vertex at an A-point of analyticity x_0.*

The cosine of an angle formed by the unit vectors e_s and e_{s_1} is given by the scalar product

$$e_s e_{s_1} = \frac{x - x_0}{|x - x_0|_s} \frac{x - x_0}{|x - x_0|_{s_1}} = \left(\frac{x_i - x_0^i}{|x - x_0|_s} e_i\right)\left(\frac{x^k - x_0^k}{|x - x_0|_{s_1}} e_k\right) =$$

$$= [\cos(s, x^i)\, e_i][\cos(s_1, x^k)\, e_k] = \delta_{ik} \cos(s, x^i) \cos(s_1, x^k) = \cos(s, s_1).$$

On the other side

$$\frac{\partial f(x_0)}{\partial s} = \frac{\partial f(x_0)}{\partial x^i} \cos (s, x)^i = x_i^{*k}(x_0) \cos (s, x^i)\, e_k^i,$$

where, by theorem 15 and system (8.1),

$$\left| \frac{\partial f(x_0)}{\partial s} \right| = \sqrt[n]{J(x_0)^2}$$

but then, the cosine of the angle formed by the vectors $\frac{\partial f(x_0)}{\partial s}$ and $\frac{\partial f(x_0)}{\partial s_1}$ is

$$\frac{\frac{\partial f(x_0)}{\partial s} \frac{\partial f(x_0)}{\partial s_1}}{\left| \frac{\partial f(x_0)}{\partial s} \right| \left| \frac{\partial f(x_0)}{\partial s_1} \right|} = \frac{\delta_{ik} x_m^{*i}(x_0) x_p^{*k}(x_0) \cos (s, x_0) \cos (s_1, x_0)}{\sqrt[n]{J(x_0)^2}},$$

hence by (8.1)

$$\frac{\frac{\partial f(x_0)}{\partial s} \frac{\partial f(x_0)}{\partial s_1}}{\left| \frac{\partial f(x_0)}{\partial s} \right| \left| \frac{\partial f(x_0)}{\partial s_1} \right|} = \cos (s, s_1) = e_s e_{s_1}.$$

CHAPTER 17

K-QCfH between surfaces

In this chapter we generalize for $(n-1)$-dimensional surfaces Gehring—Väisälä's [1] corresponding results obtained for bidimensional surfaces.

Inner, outer and maximal dilatations of homeomorphisms. We define the *inner dilatation* $K_I(f)$ and the *outer dilatation* $K_O(f)$ of the homeomorphism f as

$$K_I(f) = \sup_{A} \frac{M(A)}{M(A^*)}, \; K_O(f) = \sup_{A} \frac{M(A^*)}{M(A)},$$

where the suprema are taken over all rings $A \subset\subset D$ for which $M(A)$ and $M(A^*)$ are not both zero or infinite.

Remark. F. Gehring and J. Väisälä [1] denote $K_I(f) = \sup_A \frac{\operatorname{mod} A}{\operatorname{mod} A^*}$ and $K_O(f) = \sup \frac{\operatorname{mod} A^*}{\operatorname{mod} A}$; we preferred the notation from above for the simplifications implied in what follows.

We call

$$K(f) = \max\left[K_I(f),\ K_O(f)\right]. \tag{1}$$

the *maximal dilatation* of f.

Properties of the dilatations. Obviously

$$K_I(f) = K_O(f^{-1}),\ K_O(f) = K_I(f^{-1}),\ K(f) = K(f^{-1}), \tag{2}$$

Moreover, it follows from theorem 6.2 and taking into account that $M(A) = (\operatorname{mod} A)^{n-1}$ (corollary of lemma 2.6), that

$$K_I(f) \leqq K_O(f)^{n-1},\ K_O(f) \leqq K_I(f)^{n-1}. \tag{3}$$

Thus the three dilatations of a homeomorphism f are simultaneously finite or infinite. In the former case $f(x)$ is said to be QCfH.

A homeomorphism $f(x)$ is said to be K-QCfH $(1 \leqq K < \infty)$ in D if $K(f) \leqq K$.

LEMMA 1. *Suppose that $f(x)$ is a homeomorphism of D. If f is not differentiable with $J(x) > 0$ a.e. in D or if f is not ACL in D, then*

$$K_I(f) = K_O(f) = K(f) = \infty.$$

If f is differentiable with $J > 0$ a.e. in D and if f is ACL in D, then

$$K_I(f) = \operatorname*{ess\,sup}_{x \in D} \frac{J_G(x)}{\lambda_f(x)^n}, \quad K_O(f) = \operatorname*{ess\,sup}_{x \in D} \frac{\Lambda_f(x)^n}{J_G(x)}.$$

We recall that

$$\operatorname*{ess\,sup}_{x \in D} g(x) = \inf_E \big[\sup_{x \in D-E} g(x) \big], \quad \operatorname*{ess\,inf}_{x \in D} g(x) = \sup_E \big[\inf_{x \in D-E} g(x) \big],$$

where $\inf_E$ and $\sup_E$ are taken over all the subsets of D of measure zero. Some authors use the notation vrai max and vrai min, respectively (Ju. G. Rešetnjak [6], I. N. Pesin [3]).

This lemma is a consequence of Väisälä's definitions, of lemmas 1.3 and 1.5, of the corollary of theorem 1.1, of the relation $M(A) = (\text{mod } A)^{n-1}$, of lemma 6.4 applied to f and to f^{-1} and of the equivalence of both Gehring's analytic definitions (theorem 6.3).

If we apply the preceding lemma to the transformation $f(x) = (Kx^1, \dots, Kx^{n-1}, x^n)$, where $K > 1$, we obtain $K_I(f) = K^{n-1}$ and $K_O(f) = K$, and we conclude that the inequalities (3) are the best possible.

COROLLARY. *A K-QCfH $(1 \leqq K < \infty)$ $f(x)$ in D satisfies*

$$K_I(f) = \operatorname*{ess\,sup}_{x \in D} \underline{\delta}(x), \quad K_O(f) = \operatorname*{ess\,sup}_{x \in D} \overline{\delta}(x), \quad K(f) = \operatorname*{ess\,sup}_{x \in D} \delta(x).$$

This is an immediate consequence of (1.1), (1.12) and (1.13) which hold at an A-point of $f(x)$.

Since the modulus of a ring can be expressed by means of the arc family joining its boundary components, it follows that it is possible to define also the dilatations of f in terms of what happens to the moduli of the arc families joining the boundary components of rings with closure in D under $f(x)$. Surprisingly enough, if we know what happens to the moduli of this particular class of arc families under f, we know what happens to the moduli of all arc families under f, i.e. we have the following

LEMMA 2. *Suppose that f is a homeomorphism of D. Then*

$$K_I(f) = \sup_\Gamma \frac{M(\Gamma^*)}{M(\Gamma)}, \quad K_O(f) = \sup_\Gamma \frac{M(\Gamma)}{M(\Gamma^*)},$$

where the suprema are taken over all families $\Gamma \subset D$ *for which* $M(\Gamma)$ *and* $M(\Gamma^*)$ *are not simultaneously equal to* 0 *or* ∞.

This is a consequence of the equivalence of Väisälä's definitions.

Quasi-isometries. We introduce the notion of quasi-isometry in order to describe a certain class of surfaces.

We say that a homeomorphism of D is a *C-isometry* $(1 \leqq C < \infty)$, if

$$\frac{|x_1 - x_2|}{C} \leqq |f(x_1) - f(x_2)| \leqq C|x_1 - x_2|$$

for all $x_1, x_2 \in D$. A homeomorphism is a *quasi-isometry* if it is a C-isometry for some C. We define $C(f)$, the *maximal distortion* of f in D, as the smallest constant C for which the preceding inequality holds for all $x_1, x_2 \in D$.

Some properties of the quasi-isometries.

THEOREM 1. *A C-isometry in D is a QCfH in D.*

Indeed, lemma 1 implies

$$K_I(f) \leqq C(f)^{2n}, \; K_O(f) \leqq C(f)^{2n}.$$

Thus f is QCfH.

On the other hand, it is clear that a QCfH need not be a quasi-isometry. (For information about the connexions between QCfH and quasi-isometries see our Note [25].)

COROLLARY. *A C-isometry in D satisfies the inequalities*

$$K_I(f) \leqq C(f)^{2(n-1)}, \quad K_O(f) \leqq C(f)^{2(n-1)}. \tag{4}$$

In order to prove the preceding inequalities it suffices, on account of lemma 1, to show that the inequalities

$$\frac{J_G(x)}{\lambda_f(x)^n} \leqq C(f)^{2(n-1)}, \quad \frac{\Lambda_f(x)^n}{J_G(x)} \leqq C(f)^{2(n-1)}$$

hold a.e. in D. Since f is QCfH, by the preceding theorem, then it is differentiable with $J \neq 0$ a.e. in D, and it is sufficient to prove the preceding inequalities at every A-point $x_0 \in D$. Suppose, as we may, $x_0 = f(x_0) = O$ and

$$x^{*i} = a^i x^i + o(|x|), \; a^i > 0 \qquad (i = 1, \dots, n),$$

which is possible to obtain by performing a preliminary translation and rotation (which comes to choosing a suitable system of coordinates and then does not affect the behaviour of the mapping), followed by symmetries corresponding to the coefficients $a^i < 0$ (which preserve $|x^*|$). Next, since at an A-point, we have $J_G(x) = |J(x)|$ and since from the fact that f is a C-isometry it follows $\frac{1}{C(f)} \leqq a^i \leqq C(f)$ $(i = 1, \ldots, n)$, we obtain

$$\frac{|J(O)|}{\lambda_f(O)^n} = \frac{a^1 \ldots a^n}{(a^n)^n} \leqq \frac{C(f)^{n-1}}{\dfrac{1}{C(f)^{n-1}}} = C(f)^{2(n-1)},$$

$$\frac{\Lambda_f(O)^n}{J_G(O)} = \frac{(a^1)^n}{a^1 \ldots a^n} \leqq \frac{C(f)^{n-1}}{\dfrac{1}{C(f)^{n-1}}} = C(f)^{2(n-1)}.$$

Admissible surfaces. A connected set $\sigma \subset R^n$ is said to be an *admissible surface* if to each point $\xi \in \sigma$ there corresponds a quasi-isometry i_ξ with the following properties: for each $\varepsilon > 0$ there exists a neighbourhood U_ξ of ξ, in which i_ξ is defined, such that i_ξ maps $\sigma \cap U_\xi$ onto a plane domain Π_ξ, and such that the maximal distortion $C(i_\xi)$ of i_ξ in U_ξ satisfies the inequalities

$$\sup_{\xi \in \sigma} C(i_\xi) < \infty, \quad \operatorname*{ess\,sup}_{\xi \in \sigma} C(i_\xi) \leqq 1 + \varepsilon, \tag{5}$$

where $\operatorname*{ess\,sup}_{\xi \in \sigma} g(\xi) = \inf_{\sigma_0} [\sup_{\xi \in \sigma - \sigma_0} g(\xi)]$ and $\inf_{\sigma_0}$ is taken over all $\sigma_0 \subset \sigma$ with $H^{n-1}(\sigma_0) = 0$.

Projection mapping. Suppose that $\sigma \subset R^n$ is homeomorphic to a plane domain and that, for all $\xi_1, \xi_2 \in \sigma$, the acute angle which the segment $\overline{\xi_1 \xi_2}$ makes with the basis vector e_n is never less than $\alpha > 0$. Next let Π denote the projection of σ onto the plane $x^n = 0$ and let D denote the set of all points x of the form

$$x = \xi + ue_n,$$

where $\xi \in \sigma$ and u is real. Then for each $x \in D$, ξ and u are uniquely determined and we may define

$$f(x) = f(\xi) + aue_n, \tag{6}$$

where $f(\xi)$ denotes the *projection* of ξ onto $x_n = 0$ and a is some fixed strictly positive number.

Some properties of the projections.

LEMMA 3. *The mapping f from above is a homeomorphism of D onto itself which maps σ onto Π and*

$$\frac{a}{A_1} |x_1 - x_2| \leqq |f(x_1) - f(x_2)| \leqq A_1 |x_1 - x_2|,$$

for all $x_1, x_2 \in D$, *where*

$$A_1 = \frac{1}{2}(a^2 \operatorname{cosec}^2 \alpha + 2a + 1)^{\frac{1}{2}} + \frac{1}{2}(a^2 \operatorname{cosec}^2 \alpha - 2a + 1)^{\frac{1}{2}}. \tag{8}$$

Fix points $x_1 = \xi_1 + u_1 e_n$, $x_2 = \xi_2 + u_2 e_n \in D$ and let β and γ denote the acute angles which the segments $\overline{x_1 x_2}$ and $\overline{\xi_1 \xi_2}$ make with e_n. From (6) it follows that

$$\begin{aligned} |f(x_1) - f(x_2)|^2 &= |f(\xi_1) - f(\xi_2)|^2 + a^2(u_1 - u_2)^2 = \\ &= |x_1 - x_2|\sin^2\beta + \|x_1 - x_2|\cos\beta \pm |\xi_1 - \xi_2|\cos\gamma|^2 a^2 = \\ &= [\sin^2\beta + a^2(\cos\beta \pm \cot\gamma \sin\beta)^2]|x_1 - x_2|^2 = K^2|x_1 - x_2|^2, \ K > 0. \end{aligned} \tag{9}$$

Next, in order to verify that

$$\frac{a}{A_2} \leqq K \leqq A_2, \tag{10}$$

where A_2 is equal to the right-hand side of (8) with γ in place of α, it is sufficient to show that the extrema of K with respect to β satisfy this double inequality. But the condition that the derivative of K with respect to β vanish, yields

$$\tan 2\beta = \mp \frac{2a^2 \cot\gamma}{1 - a^2 + a^2\cot^2\gamma}, \tag{11}$$

where the upper sign corresponds to the maximum and the lower sign to the minimum. If we substitute $\tan 2\beta$ by its expression given by (11), in K^2, we obtain A_2 and $\dfrac{a^2}{A_2}$, respectively. The definition of γ and of the projection f imply $\alpha \leqq \gamma \leqq \dfrac{\pi}{2}$, hence $C_1 \leqq A_1$, and then (9) and (10) yield (7), as desired.

COROLLARY 1. *If* $a = 1$ *in* (6), *then the projection* f *is a quasi-isometry with maximal distortion*

$$C(f) \leqq \cot\alpha + 1 \tag{12}$$

and a QCfH *with*

$$K(f) \leqq \left[\frac{1}{2}(\cot^2\alpha + 4)^{\frac{1}{2}} + \frac{1}{2}\cot\alpha\right]^n \leqq (\cot\alpha + 1).^n \tag{13}$$

If we set $a = 1$ in (8), we get

$$A_1 = \frac{1}{2}(\cot^2\alpha + 4)^{\frac{1}{2}} + \frac{1}{2}\cot\alpha \leqq \cot\alpha + 1,$$

which implies (12), and since (7) and

$$x = \xi^1 e_1 + \cdots + \xi^{n-1}e_{n-1} + (\xi^n + u)e_n,\ f(x) = \xi^1 e_1 + \cdots + \xi^{n-1}e_{n-1} + aue_n,$$

imply

$$\Lambda_f(x) \leqq A_1,\ \lambda_f(x) \geqq \frac{a}{A_1},\ J(x) = a$$

for all $x \in D$, the preceding lemma in the particular case $a = 1$ yields (13), as desired.

COROLLARY 2. *If* $a = \sin\alpha$ *in* (6), *then the projection* f *is a* QCfH *with*

$$K_I(f) \leqq 2^{-\frac{n-2}{2}} \cot\frac{\alpha}{2} \operatorname{cosec}^{n-2}\frac{\alpha}{2},\quad K_O(f) \leqq 2^{\frac{n-2}{2}} \cot\frac{\alpha}{2} \cos^{n-2}\frac{\alpha}{2}. \tag{14}$$

The same argument as in the preceding corollary, taking only into account that in this case

$$A_1 = \frac{1}{2}(2 + 2\sin\alpha)^{\frac{1}{2}} + \frac{1}{2}(2 - 2\sin\alpha)^{\frac{1}{2}} = \sqrt{2}\cos\frac{\alpha}{2}.$$

Now, by means of corollary 1 we shall give, following Gehring and Väisälä [1], a simple geometric condition which implies that a connected set σ is an admissible surface. Suppose that a point $\xi_0 \in \sigma$ has a neighbourhood V such that $\sigma \cap V$ is homeomorphic to an $(n-1)$-ball (if $n = 3$ to a disc), suppose that e is a fixed unit vector, and suppose that, for each pair of points $\xi_1, \xi_2 \in \sigma \cap V$, the acute angle which the segment $\overline{\xi_1\xi_2}$ makes with e is never less than $\alpha > 0$. Then there exists a neighbourhood U of ξ_0 such that $U \subset V$ and each point $x \in U$ has a unique representation of the form $x = \xi + ue$, where $\xi \in \sigma \cap U$ and u is real. For each such x we let $i(x) = i(\xi) + ue$, where $i(\xi)$ is the projection of ξ onto the plane through ξ_0 which has e as its normal. Then i is a projection according to the above definition and is obtained from (6) for $a = 1$. Hence (as we have seen) i maps $\sigma \cap U$ onto a plane domain Π, and it follows from corollary 1 that i is a quasi-isometry of U with the maximal distortion $C(i) \leqq \cot\alpha + 1$. Thus a connected set $\sigma \subset R^n$ is an admissible surface if to each point $\xi \in \sigma$ there corresponds a unit vector e_ξ with the following property. For each $\varepsilon > 0$ there exists a neighbourhood U_ξ of ξ such that $\sigma \cap U_\xi$ is homeomorphic to an $(n-1)$-ball and such that for each pair of points $\xi_1, \xi_2 \in \sigma \cap U_\xi$, the acute angle between the segment $\overline{\xi_1\xi_2}$ and the vector e_ξ is never less than α_ξ, where

$$\inf_{\xi \in \sigma} \alpha_\xi > 0, \quad \operatorname{ess\,inf}_{\xi \in \sigma} \alpha_\xi \geqq \frac{\pi}{2} - \varepsilon.$$

For example, a two-dimensional manifold $\sigma \subset R^n$ is an admissible surface if it has a well defined continuously turning tangent plane at each point $\xi \in \sigma$.

Geometric definition of K-QCfH between two admissible surfaces by means of quasi-isometries. Suppose that σ and σ^* are admissible surfaces and that $f: \sigma \rightleftarrows \sigma^*$ is a homeomorphism. Next for each $\xi \in \sigma$ let $\xi^* = f(\xi)$ and let i_ξ and i_{ξ^*} be the quasi-isometries associated with ξ and ξ^*. We say that $f(\xi)$ is K-QCfH ($1 \leqq K < \infty$), if for each $\varepsilon > 0$ there exist neighbourhoods U_ξ of ξ and U_{ξ^*} of ξ^* with the following properties. The quasi-isometries i_ξ and i_{ξ^*} map $\sigma \cap U_\xi$ and $\sigma^* \cap U_{\xi^*}$ onto plane domains Π_ξ and Π_{ξ^*} respectively, f maps $\sigma \cap U_\xi$ into $\sigma^* \cap U_{\xi^*}$, and

$$\sup_{\xi \in \sigma} K(g_\xi) < \infty, \quad \operatorname{ess\,sup}_{\xi \in \sigma} K(g_\xi) \leqq K + \varepsilon,$$

where $K(g_\xi)$ denotes the maximal dilatation of the plane homeomorphism

$$g_\xi = i_{\xi^*} \circ f \circ i_\xi^{-1}. \tag{15}$$

We say that $f(\xi)$ is QCfH if it is K-QCfH for some K.

We define $K(f)$, the *maximal dilatation* of f, as the smallest number K for which f is K-QCfH.

From the symmetry of the definition and since g_ξ is K-QCfH, it follows that if $f: \sigma \rightleftarrows \sigma^*$ is K-QCfH then also $f^{-1}: \sigma^* \rightleftarrows \sigma$ is K-QCfH.

Measurability of QCfH between surfaces.

THEOREM 2. *Suppose that σ, σ^* are admissible surfaces and that $f: \sigma \rightleftarrows \sigma^*$ is a* QCfH. *If $E \subset \sigma$ and $a(E) = 0$, then $a(E^*) = 0$.*

Suppose f is K-QCfH, and for $\varepsilon = 1$ and each $\xi \in \sigma$, let U_ξ and U_{ξ^*} be the neighbourhoods of the above definition of the QCfH between two admissible surfaces. By Lindelöf's covering theorem, (see C. Carathéodory [3] p. 46) we can choose a sequence of points $\{\xi_m\}$ so that $E \subset \bigcup_m U_{\xi_m}$. Set $E_m = E \cap U_{\xi_m}$. Then $E^* = \bigcup_m E_m^*$ and $a(E_m) = 0$. Since i_{ξ_m} and $i_{\xi_m^*}^{-1}$ are isometries and then ACL_n, and since g_{ξ_m} is a QCfH in R^{n-1}, then arguing as in theorem 1.1, it follows that $a(E_m^*) = 0$, hence, on account of the complete additivity of Hausdorff's measure, $a(E^*) = 0$, as desired.

Surface modulus of an arc family. The above definition for QCfH is awkward since it involves the quasi-isometries i_ξ and i_{ξ^*}. We shall give, following Gehring and Väisälä [1], two other equivalent definitions, but first we must introduce the notion of the surface modulus of an arc family.

The *surface modulus*

$$M^\sigma(\Gamma) = \inf_{\rho \in F(\Gamma)} \int_\sigma \rho^{n-1} d\sigma$$

of a family of arcs in an admissible surface σ is obtained from (1.5.1) for $q=1$ and $p=n-1$ and behaves like the familiar plane modulus $M^\pi(\Gamma) = \inf_{\rho \in F(\Gamma)} \int_{R^{n-1}} \rho^{n-1} d\sigma$, corresponding to $M(\Gamma)$ in R^n. In particular, $M^\sigma(\Gamma)$ reduces to $M^\pi(\Gamma)$ if σ is a plane domain. It is easy to see that all the assertions in proposition 1.5.3 and in corollaries 1 and 2 hold with M_p replaced by M^σ. Finally we can argue as in the proof of proposition 1.5.9 to show that the surface modulus of the family of all compact non-rectifiable arcs in an admissible surface σ is equal to zero. This means that the arcs of a family Γ, which are not locally rectifiable, have no influence on $M^\sigma(\Gamma)$. That is, if Γ_1 is the subfamily of locally rectifiable arcs in Γ, then $M^\sigma(\Gamma) = M^\sigma(\Gamma_1)$. But this follows also from

LEMMA 4. *Suppose that σ is an admissible surface, that i is a C-isometry of U which maps $\sigma \cap U$ onto a domain $\Pi \subset R^{n-1}$, that Γ is a family of arcs in $\sigma \cap U$ and that $\Gamma^* = i(\Gamma)$. Then*

$$\frac{M^\sigma(\Gamma)}{C^{2(n-1)}} \leqq M^\pi(\Gamma)^* \leqq C^{2(n-1)} M^\sigma(\Gamma). \tag{16}$$

If $\rho \in F(\Gamma)$, then $C\rho^* \in F(\Gamma^*)$, where $\rho^* = \rho \circ i^{-1}$, since

$$\int_{\gamma^*} C\rho^* \, ds^* = \int_\gamma C\rho \frac{ds^*}{ds} ds \geqq \int_\gamma \rho \, ds \geqq 1,$$

and then

$$M^\pi(\Gamma^*) \leqq \int_\pi C^{n-1} \rho^{n-1} d\sigma^* \leqq C^{n-1} \int_\sigma \rho^{n-1} \frac{d\sigma^*}{d\sigma} d\sigma \leqq C^{2(n-1)} M^\sigma(\Gamma).$$

This yields the second half of (16). The first half follows similarly if we change M^σ and M^π between them.

COROLLARY. *Suppose σ is an admissible surface. The surface modulus of the family Γ of all arcs in σ is equal to the surface modulus of the subfamily Γ_1 of all locally rectifiable arcs in σ.*

Let U be a neighbourhood of a point $\xi \in \sigma$. Consider first the subfamilies of Γ and Γ_1 contained in $U \cap \sigma$. (For the sake of simplicity we denote them again by Γ and Γ_1.) The preceding lemma then implies that

$$\frac{M^{\sigma}(\Gamma)}{C^{2(n-1)}} \leqq M^{\pi}(\Gamma^*) \leqq C^{2(n-1)} M^{\sigma}(\Gamma),$$

$$\frac{M^{\sigma}(\Gamma_1)}{C^{2(n-1)}} \leqq M^{\pi}(\Gamma_1^*) \leqq C^{2(n-1)} M^{\sigma}(\Gamma_1).$$

Subtracting the second inequality from the first one, we obtain

$$\frac{1}{C^{2(n-1)}} [M^{\sigma}(\Gamma) - M^{\sigma}(\Gamma_1)] \leqq M^{\pi}(\Gamma^*) - M^{\pi}(\Gamma_1^*) \leqq C^{2(n-1)} [M^{\sigma}(\Gamma) - M^{\sigma}(\Gamma_1)].$$

Arguing as in proposition 1.5.9, we obtain, for the families Γ^* and Γ^*, of all bounded arcs and of all rectifiable arcs, respectively, the relation $M^{\pi}(\Gamma^*) = M^{\pi}(\Gamma_1^*)$, which substituted in the preceding double inequality yields

$$M^{\sigma}(\Gamma) \leqq M^{\sigma}(\Gamma_1), \quad M^{\sigma}(\Gamma_1) \leqq M^{\sigma}(\Gamma),$$

hence

$$M^{\sigma}(\Gamma) = M^{\sigma}(\Gamma_1), \tag{17}$$

i.e.

$$M^{\sigma}(\Gamma') = 0,$$

where $\Gamma' = \Gamma - \Gamma_1$ is the family of all bounded non-rectifiable arcs in $U \cap \sigma$. Since, σ, by Lindelöf's theorem, can be covered by a countable set of neighbourhoods U, then, by proposition 1.5.3 (which still holds for the surface modulus), it follows that the surface modulus of all arcs which are not locally rectifiable vanishes, and then, on account of proposition 1.5.3, we obtain (17), where this time Γ and Γ_1 are given not only in $U \cap \sigma$, but even in the whole σ and the proof of the corollary is complete.

Analytic definition of K-QCfH between admissible surfaces. Suppose that $f(\xi)$ is a homeomorphism of an admissible surface σ. For each $\xi \in \sigma$ we let

$$J^{\sigma}(\xi) = \lim_{r \to 0} \frac{a[(\sigma \cap U)^*]}{a(\sigma \cap U)},$$

where $U = B(\xi, r)$ and $(\sigma \cap U)^* = f(\sigma \cap U)$.

A homeomorphism $f: \sigma \rightleftarrows \sigma^*$, where σ, σ^* are admissible surfaces, is said to be K-QCfH ($1 \leqq K < \infty$), if it is ACA in σ and

$$\frac{\Lambda_f(\xi)^{n-1}}{K} \leqq J^\sigma(\xi) \leqq K\lambda_f(\xi)^{n-1} \tag{18}$$

H^{n-1}-a.e. in σ, i.e. everywhere in σ except possibly a set o area (in Hausdorff's sense) zero.

Equivalence between analytic and geometric definitions.

THEOREM 3. *The analytic and the geometric definitions of* K-QCfH *between two admissible surfaces* σ, σ^*, *are equivalent.*

Suppose that $f: \sigma \rightleftarrows \sigma^*$ is K-QCfH according to geometric definition. We must show that f is ACA in σ and that (18) holds for $\xi \in \sigma - E$, where $a(E) = 0$. Fix $\varepsilon > 0$. Because σ, σ^* are admissible surfaces we can choose for each $\xi \in \sigma$ and $\xi^* = f(\xi)$ neighbourhoods U_ξ and U_{ξ^*} such that

$$\sup_{\xi \in \sigma} C(i_\xi) < \infty, \quad \operatorname*{ess\,sup}_{\xi \in \sigma} C(i_\xi) \leqq 1 + \varepsilon,$$
$$\sup_{\xi^* \in \sigma^*} C(i_{\xi^*}) < \infty, \quad \operatorname*{ess\,sup}_{\xi^* \in \sigma^*} C(i_{\xi^*}) \leqq 1 + \varepsilon, \tag{19}$$

where $C(i_\xi)$ and $C(i_{\xi^*})$ denote the maximal distortions of i_ξ and i_{ξ^*} in U_ξ and U_{ξ^*}, respectively. Next because f is K-QCfH, according to the geometric definition, we can choose these neighbourhoods so that, in addition, f maps $\sigma \cap U_\xi$ into $\sigma^* \cap U_{\xi^*}$ and

$$\sup_{\xi \in \sigma} K(g_\xi) < \infty, \quad \operatorname*{ess\,sup}_{\xi \in \sigma} K(g_\xi) \leqq (1 + \varepsilon)K, \tag{20}$$

where $K(g_\xi)$ is the maximal dilatation of the $(n-1)$-dimensional K-QCfH g_ξ given in (15).

We show first that f is ACA. Let Γ denote the family of all locally rectifiable arcs in σ which contain compact subarcs on which f is not AC. Next choose a sequence of points $\xi_m \in \sigma$ so that the corresponding neighbourhoods U_{ξ_m} cover σ. Each $\gamma \in \Gamma$ has a compact subarc β on which f is not AC. Then, there exists an index m such that β have a compact subarc $\alpha \subset \sigma \cap U_{\xi_m}$ on which f is not AC; and this for each $\gamma \in \Gamma$. These arcs α form a family Γ_0 which minorizes Γ, i.e. $\Gamma_0 < \Gamma$ (Chapter. 1.5), and hence, by corollary 2 of proposition 1.5.3 (which, as we pointed out above, still holds for the surface modulus),

$$M^\sigma(\Gamma) \leqq M^\sigma(\Gamma_0). \tag{21}$$

Let Γ_m be the subfamily of arcs of Γ_0 which lie in $\sigma \cap U_{\xi_m}$, and let $\tilde{\Gamma}_m = i_{\xi_m}(\Gamma_m)$. The analytic definition of $(n-1)$-dimensional QCfH (in this case the dimension has no influence) combined with theorem 6.1 and corollary of Fuglede's theorem implies that g_{ξ_m} is ACA in $\Pi_{\xi_m} = i_{\xi_m}(U_{\xi_m} \cap \sigma)$ and since g_{ξ_m} is not AC on any arc of Γ_0, it follows that g_{ξ_m} is not AC on any arc of $\tilde{\Gamma}_m$, and then $M^{\pi_m}(\tilde{\Gamma}_m) = 0$. Hence $M^\sigma(\Gamma_m) = 0$, by lemma 4 and we conclude that $M^\sigma(\tilde{\Gamma}_0) \leqq \sum_m M^\sigma(\tilde{\Gamma}_m) = 0$. Thus $M^\sigma(\Gamma) = 0$, by (21), i.e. f is ACA in σ.

Next we prove the left-hand side of (18). The preceding lemma and (19), (20) imply there exists a set $E_1 \subset \sigma$ such that $a(E_1) = 0$ and

$$C(i_\xi) \leqq 1 + \varepsilon, \quad C(i_{\xi^*}) \leqq 1 + \varepsilon, \quad K(g_\xi) \leqq (1 + \varepsilon) K \tag{22}$$

for $\xi \in \sigma - E_1$. Fix such a point ξ_0 and let $\xi^* = i_{\xi_0}(\xi)$ for $\xi \in \sigma \cap U_{\xi_0}$. Then, clearly, $g = g_{\xi_0}$ is a $(1 + \varepsilon)K$-QCfH of the domain $\Pi = \Pi_{\xi_0} \subset R^{n-1}$ and then, by lemma 1 and since i_{ξ_0} and $i_{\xi_0^*}$ are $C(i_{\xi_0})$-and $C(i_{\xi_0^*})$-isometries, respectively, with $C(i_{\xi_0})$ and $C(i_{\xi_0^*})$ satisfying inequalities (5), we have

$$\Lambda_f(\xi)^{n-1} = \left[\overline{\lim_{\xi' \to \xi}} \frac{|f(\xi') - f(\xi)|}{|\xi' - \xi|}\right]^{n-1} = \tag{23}$$

$$\overline{\lim_{\xi' \to \xi}} \left[\frac{|i_{\xi_0^*}^{-1}(\xi^{*\prime}) - i_{\xi_0^*}^{-1}(\xi^*)|}{|\xi^{*\prime} - \xi^*|} \cdot \frac{|g(y') - g(y)|}{|y' - y|} \cdot \frac{|i_{\xi_0}(\xi') - i_{\xi_0}(\xi)|}{|\xi' - \xi|}\right]^{n-1} \leqq$$

$$\leqq (1 + \varepsilon)^{2(n-1)} \Lambda_g(y)^{n-1} \leqq K(1 + \varepsilon)^{2n-1} J_g^\pi(y)$$

H^{n-1}-a.e. in Π_{ξ_0} and, by (22),

$$J_g^\pi(y) = \overline{\lim_{r \to 0}} \frac{m_{n-1} g[B^{n-1}(y, r)]}{m_{n-1} B^{n-1}(y, r)} = \overline{\lim_{r \to 0}} \left[\frac{m_{n-1} i_{\xi_0^*}(\sigma^* \cap U_{\xi_0^*})}{a(\sigma^* \cap U_{\xi_0^*})}\right.\cdot$$

$$\left.\cdot \frac{a(\sigma^* \cap U_{\xi_0^*})}{a(\sigma \cap U_{\xi_0})} \cdot \frac{a(\sigma \cap U_{\xi_0})}{m_{n-1} i_{\xi_0}(\sigma \cap U_{\xi_0})}\right] \leqq$$

$$\leqq (1 + \varepsilon)^{2(n-1)} \overline{\lim_{r \to 0}} \frac{a(\sigma^* \cap U_{\xi_0^*})}{a(\sigma \cap U_{\xi_0})} = (1 + \varepsilon)^{2(n-1)} J^\sigma(\xi)$$

H^{n-1}-a.e. in $\sigma \cap U_{\xi_0}$. And then, the preceding inequality and (23) imply that

$$\Lambda_f(\xi)^{n-1} \leqq K(1 + \varepsilon)^{4n-3} J^\sigma(\xi) \tag{24}$$

H^{n-1}-a.e. in $\sigma \cap U_{\xi_0}$. Since there exists a sequence of points $\xi_m \in \sigma - E_1$ whose neighbourhoods U_{ξ_m} cover $\sigma - E_1$, we can find a set E_2 such that $a(E_2) = 0$ and that (24) hold everywhere in $\sigma - E_2$. Finally, let E be the union of the exceptional sets E_2 for $\varepsilon = \dfrac{1}{m}$, $m = 1, 2, \ldots$. Then, clearly, $a(E) = 0$ and the left-hand side of (18) holds for $\xi \in \sigma - E$.

By the same argument as above, we have

$$\lambda_f(\xi)^{n-1} = \left[\lim_{\xi' \to \xi} \frac{|f(\xi') - f(\xi)|}{|\xi' - \xi|}\right]^{n-1} =$$

$$= \lim_{\xi' \to \xi} \left[\frac{|i_{\xi_0^*}^{-1}(\xi^{*\prime}) - i_{\xi_0^*}^{-1}(\xi^*)|}{|\xi^{*\prime} - \xi^*|} \cdot \frac{|g(y') - g(y)|}{|y' - y|} \cdot \frac{|i_{\xi_0}(\xi') - i_{\xi_0}(\xi)|}{|\xi' - \xi|}\right]^{n-1} \geqq$$

$$\geqq \frac{\lambda_g(y)^{n-1}}{(1+\varepsilon)^{2(n-1)}} > \frac{J_g^{\pi}(y)}{K(1+\varepsilon)^{2n-1}} =$$

$$= \frac{1}{K(1+\varepsilon)^{2n-1}} \varliminf_{r \to 0} \left[\frac{m_{n-1} i_{\xi_0^*}(\sigma^* \cap U_{\xi_0^*})}{a(\sigma^* \cap U_{\xi^*})} \frac{a(\sigma^* \cap U_{\xi_0^*})}{a(\sigma \cap U_{\xi_0})} \frac{a(\sigma \cap U_{\xi_0})}{m_{n-1} i_{\xi_0}(\sigma \cap U_{\xi_0})}\right] >$$

$$> \frac{1}{(1+\varepsilon)^{4n-3} K} \varliminf_{r \to 0} \frac{a(\sigma^* \cap U_{\xi_0^*})}{a(\sigma \cap U_{\xi_0})} = \frac{J^{\sigma}(\xi)}{(1+\varepsilon)^{4n-3} K}$$

H^{n-1}-a.e. in σ and the double inequality (18) holds H^{n-1}-a.e. in σ.

For the sufficiency part suppose $f: \sigma \rightleftarrows \sigma^*$ is a K-QCfH according to the analytic definition and fix $\varepsilon > 0$. Then for $\xi_0 \in \sigma$, $\xi_0^* = f(\xi_0)$, choose neighbourhoods U_ξ and U_{ξ^*} so that (19) holds and so that f maps $\sigma \cap U_\xi$ into $\sigma^* \cap U_{\xi_0^*}$. We show first that the homeomorphism g_ξ given by (15) is an $(n-1)$-dimensional QCfH with maximal dilatation

$$K(g_{\xi_0}) \leqq K C(i_{\xi_0})^{2(n-1)} C(i_{\xi_0^*})^{2(n-1)}. \tag{25}$$

Fix $\xi \in \sigma$, let $\tilde{\Gamma}$ be the family of all locally rectifiable arcs in Π_{ξ_0} which contain a compact subarc on which g_ξ is not AC, and let $\Gamma = i_{\xi_0}^{-1}(\tilde{\Gamma})$. Since f is by hypothesis ACA in σ, $M^{\sigma}(\Gamma) = 0$ and hence $M^{\pi_{\xi_0}}(\tilde{\Gamma}') = 0$ by lemma 4, i.e.

g_{ξ_0} is *ACA* and, *a fortiori*, *ACL* in Π_{ξ_0}. Next arguing as in (23) we see from (18) that

$$\Lambda_g(y)^{n-1} = \left[\overline{\lim_{y'\to y}} \frac{|g(y') - g(y)|}{|y' - y|}\right]^{n-1} =$$

$$= \overline{\lim_{y'\to y}} \left[\frac{|i_{\xi_0^*}(\xi^{*\prime}) - i_{\xi_0^*}(\xi^*)|}{|\xi^{*\prime} - \xi^*|} \cdot \frac{|f(\xi') - f(\xi)|}{|\xi' - \xi|} \cdot \frac{|i^{-1}(y') - i^{-1}(y)|}{|y' - y|}\right]^{n-1} \leqq$$

$$\leqq C(i_{\xi*})^{n-1} C(i_{\xi_0})^{n-1} \Lambda_f(\xi)^{n-1} \leqq C(i_{\xi_0^*})^{n-1} C(i_{\xi_0})^{n-1} K J^{\sigma}(\xi) =$$

$$= C(i_{\xi*})^{n-1} C(i_{\xi_0})^{n-1} K \overline{\lim_{r\to 0}} \frac{a(\sigma^* \cap U_{\xi_0^*})}{a(\sigma \cap U_{\xi_0})} = \tag{26}$$

$$= C(i_{\xi_0^*})^{n-1} C(i_{\xi_0})^{n-1} K \overline{\lim_{r\to 0}} \left[\frac{a(\sigma^* \cap U_{\xi_0^*})}{a(\Pi_{\xi_0^*})} \cdot \frac{a(\Pi_{\xi_0^*})}{a(\Pi_{\xi_0})} \cdot \frac{a(\Pi_{\xi_0})}{a(\sigma \cap U_{\xi_0})}\right] \leqq$$

$$\leqq C(i_{\xi_0^*})^{2(n-1)} C(i_{\xi_0})^{2(n-1)} K J^{\pi}(y)$$

H^{n-1} is a.e. in Π_{ξ_0}, where $\Pi_{\xi_0} = i_{\xi_0}(U_{\xi_0})$, $\Pi_{\xi_0^*} = i_{\xi_0^*}(U_{\xi_0^*})$, $g = g_{\xi_0}$, $\Pi = \Pi_{\xi_0}$ and where we choose U_{ξ_0} and $U_{\xi_0^*}$ so that $\Pi_{\xi_0} = B^{n-1}(\xi_0, r)$, $\Pi_{\xi_0^*} = B^{n-1}(\xi_0^*, r)$. Analogously,

$$\lambda_g(y)^{n-1} = \left[\underline{\lim_{y'\to y}} \frac{|g(y') - g(y)|}{|y' - y|}\right]^{n-1} =$$

$$= \left\{\underline{\lim_{y'\to y}} \left[\frac{|i_{\xi_0^*}(\xi^{*\prime}) - i_{\xi_0^*}(\xi^*)|}{|\xi^{*\prime} - \xi^*|} \cdot \frac{|f(\xi') - f(\xi)|}{|\xi' - \xi|} \cdot \frac{|i^{-1}(y') - i^{-1}(y)|}{|y' - y|}\right]\right\}^{n-1} \geqq$$

$$\geqq \frac{\lambda_f(\xi)^{n-1}}{C(i_{\xi_0^*})^{n-1} C(i_{\xi_0})^{n-1}} \geqq \frac{J^{\sigma}(\xi)}{C(i_{\xi_0^*})^{n-1} C(i_{\xi_0})^{n-1} K} = \tag{27}$$

$$= \frac{1}{C(i_{\xi_0^*})^{n-1} C(i_{\xi_0})^{n-1} K} \lim_{r\to 0} \left[\frac{a(\sigma^* \cap U_{\xi_0^*})}{a(\Pi_{\xi_0^*})} \cdot \frac{a(\Pi_{\xi_0^*})}{a(\Pi_{\xi_0})} \cdot \frac{a(\Pi_{\xi_0})}{a(\sigma \cap U_{\xi_0})}\right] \geqq$$

$$\geqq \frac{J^{\pi}(y)}{C(i_{\xi_0^*})^{2(n-1)} C(i_{\xi_0})^{2(n-1)} K}$$

H^{n-1}-a.e. in Π_{ξ_0}. (26) and (27), taking into account that g is *ACL* in σ by the second Gehring's analytic definition characterized by the double inequality (6.16), imply that the maximal dilatation $K(g_{\xi_0})$ satisfies the inequality (25). Next (19) and (25) yield $\sup_{\xi\in\sigma} K(g_\xi) < \infty$, hence $f: \sigma \rightleftarrows \sigma^*$ is QCfH. Then (25), (19) and preceding theorem imply that

$$\operatorname*{ess\,sup}_{\xi\in\sigma} K(g_\xi) \leqq K(1+\varepsilon)^{4(n-1)}$$

Thus f is K-QCfH according to the geometric definition and the proof is complete.

Geometric definition of K-QCfH between two admissible surfaces by means of the surface modulus of a curve family. A homeomorphism $f: \sigma \rightleftarrows \sigma^*$, where σ, σ^* are admissible surfaces, is said to be K-QCfH ($1 \leqq K < \infty$) if

$$\frac{M^\sigma(\Gamma)}{K} \leqq M^{\sigma^*}(\Gamma^*) \leqq KM^\sigma(\Gamma) \tag{28}$$

for all families Γ of arcs of σ.

Equivalence to the preceding definitions.

THEOREM 4. *The definition of the K-QCfH between two admissible surfaces by means of the surface modulus is equivalent to the other two definitions.*

Suppose that (28) holds for all arc families Γ in σ, fix $\varepsilon > 0$, and choose U_{ξ_0}, $U_{\xi_0^*}$ and g_{ξ_0} as in the last part of the proof of the preceding theorem. Then lemma 4 and (28) imply that

$$M^{\pi_{\xi_0^*}}(\Gamma_1^*) \leqq C(i_{\xi_0^*})^{2(n-1)} M^{\sigma^*}(\Gamma^*) \leqq KC(i_{\xi_0^*})^{2(n-1)} M^\sigma(\Gamma) \leqq$$

$$\leqq KC(i_{\xi_0})^{2(n-1)} C(i_{\xi_0^*})^{2(n-1)} M^{\pi_{\xi_0}}(\Gamma_1),$$

$$M^{\pi_{\xi^*}}(\Gamma_1) \geqq \frac{M^{\sigma^*}(\Gamma^*)}{C(i_{\xi_0^*})^{2(n-1)}} \geqq \frac{M^\sigma(\Gamma)}{KC(i_{\xi_0^*})^{2(n-1)}} \geqq \frac{M^{\pi_{\xi_0}}(\Gamma_1)}{KC(i_{\xi_0})^{2(n-1)} C(i_{\xi^*})^{2(n-1)}}$$

for each arc family $\Gamma_1 \subset \Pi_{\xi_0}$, where $\Gamma_1^* = g_{\xi_0}(\Gamma_1)$, which allows us to conclude that g_{ξ_0} is QCfH. Hence we obtain (25) and the proof that f is K-QCfH is concluded as in the last part of the proof of the preceding theorem.

Suppose now, conversely, that $f: \sigma \rightleftarrows \sigma^*$ is K-QCfH according to the other two definitions. Let us show that in this case (28) holds. We begin by establishing the left-hand side of the inequality. Let Γ_1 be the family of arcs in Γ which are locally rectifiable (where Γ is as above the family of all arcs in σ), and let Γ_2 be the family of arcs in Γ_1 on each compact subarc of which f is AC. Then corollary of lemma 4 and the fact that $f: \sigma \rightleftarrows \sigma^*$ is K-QCfH according to the analytic definition, and then ACA in σ, imply that

$$M^\sigma(\Gamma) = M^\sigma(\Gamma_1) = M^\sigma(\Gamma_2). \tag{29}$$

Choose $\rho^* \in F(\Gamma^*)$, set $\rho(\xi) = \rho^*[f(\xi)]\Lambda_f(\xi)$ for all $\xi \in \sigma$, and pick $\gamma \in \Gamma_2$. If β is any compact subarc of γ, then β is rectifiable, f is AC on β, and, arguing as in lemma 1.6 for $\rho(x)$ given by (1.42), we obtain

$$\int_\gamma \rho \, ds \geqq \int_\beta \rho \, ds = \int_\beta \rho^* \Lambda_f ds \geqq \int_{\beta^*} \rho^* ds.$$

Since this inequality holds for all such β,

$$\int_\gamma \rho \, ds \geqq \sup_{\beta^*} \int_{\beta^*} \rho^* \, ds = \int_{\gamma^*} \rho^* \, ds \geqq 1,$$

Because ρ is Borel measurable, we conclude that $\rho \in F(\Gamma_2)$. Thus, since $f: \sigma \rightleftarrows \rho^*$ as a K-QCfH according to the analytic definition (for surfaces) satisfies (18), we obtain

$$M^\sigma(\Gamma_2) \leqq \int_\sigma \rho^{n-1} \, d\sigma = \int_\sigma (\rho^* \Lambda_f)^{n-1} \, d\sigma \leqq K \int_\sigma \rho^{*n-1} \, J^\sigma d\sigma = K \int_{\sigma^*} \rho^{*n-1} d\sigma,$$

which holds for any $\rho^* \in F(\Gamma^*)$, hence

$$M^\sigma(\Gamma_2) \leqq K M^{\sigma^*}(\Gamma^*),$$

This together with (29) yield the left-hand side of (28). Since $f^{-1}: \sigma^* \rightleftarrows \sigma$ is K-QCfH, too, according to the first two definitions, the same argument applied to f^{-1} yields also the right-hand side of (28), and the proof of the theorem is complete.

Part 3

Some properties of the QCfH

CHAPTER 1

The Carathéodory convergence theorem

In this chapter we present, according to F. Gehring [8], certain results in order to give an analogue of Carathéodory's [1] theorem on the convergence of conformal mappings of variable domains, for QCfH in n-space.

Some properties of the sequences of homeomorphisms.

LEMMA 1. *Suppose that* $\{f_m(x)\}$ *is a sequence of homeomorphisms* $f_m: D_m \rightleftarrows D_m^*$, *that each compact subset of a domain* D *is contained in all but a finite number of* D_m, *that*

$$\lim_{m\to\infty} f_m(x) = f(x) \tag{1}$$

uniformly on each compact subset of D, *and that* $f(x)$ *is a homeomorphism of* D *onto* D^*. *Then, for each compact set* $E^* \subset D^*$ *we can find a compact set* F *and an integer* m_1 *such that* $F \subset D_m$ *and* $E^* \subset F_m^*$ *for* $m \geqq m_1$, *where* $F_m^* = f_m(F)$.

Let U^*, V^* be two balls with $\overline{U^*}, \overline{V^*} \subset D^*$ and

$$\overline{U^*} \subset V^* \tag{2}$$

and let $\overline{U}$, $\overline{V}$ be the pre-images of $\overline{U^*}$, $\overline{V^*}$ under $f(x)$. Then $\overline{V}$ is a compact subset of D and there exists an integer m_0 such that $\overline{V} \subset D_m$ for $m \geqq m_0$. We shall show, by *reductio ad absurdum*, that there exists an $m_1 \geqq m_0$ such that $\overline{U_m^*} \subset \overline{V_m^*}$ for $m \geqq m_1$, where $\overline{V_m^*}$ is the image of V under $f_m(x)$.

If this were not the case, we could find a subsequence $\{m_k\}$, $m_k \geqq m_0$, such that $\overline{U^*} - \overline{V_{m_k}^*} \neq \emptyset$ for all k. Let x_0 be the point which $f(x)$ maps onto the center of U^* and let $r > 0$ be the radius of U^*, i.e. $U^* = B[f(x_0), r]$. Then there exists a k_0 such that

$$|f_{m_k}(x_0) - f(x_0)| < r$$

for $k \geqq k_0$, and, since $\overline{U^*} \subset V^*$ implies that $\overline{U} \subset V$ and then $f_{m_k}(x_0) \in f_{m_k}(V) = V_{m_k}^*$, it follows that $\overline{U^*} \cap \overline{V_{m_k}^*} \neq \emptyset$ for $k \geqq k_0$. Since $\overline{U^*}$ is connected, we can find a sequence of points $\{x_k^*\}$ such that

$$x_k^* \in \overline{U^*} \cap \partial V_{m_k}^*. \tag{3}$$

for $k \geqq k_0$. Because $f_{m_k}^{-1}(x_k^*) \in \partial V$ and ∂V is compact, we assume, by choosing a second subsequence and then relabelling, that

$$\lim_{k\to\infty} f_{m_k}^{-1}(x_k^*) = x_1 \in \partial V. \tag{4}$$

Since $f_{m_k}(x)$ converges uniformly on ∂V, it is easy to see, from (1) that

$$x_1^* = f(x_1) = \lim_{k\to\infty} f[f_{m_k}^{-1}(x_k^*)] = \lim_{k\to\infty} f_{m_k}[f_{m_k}^{-1}(x_k^*)] = \lim_{k\to\infty} x_k^*. \tag{5}$$

Now (3) implies that $x_1^* \in \overline{U^*}$ while (4) and (5) imply that $x_1^* \in \partial V^*$. Thus $\overline{U^*} \cap \cap \partial V^* \neq \emptyset$, and this contradicts (2), establishing the assertion from the beginning of the proof.

Now suppose that E^* is a compact subset of D^*. Then we can cover E^* by a finite number of balls $U_1^*, \ldots, U_p^*$ contained with their closure in D^* (by Heine—Borel—Lebesgue's covering theorem). Choose the balls $V_1^*, \ldots, V_p^*$ so that $\overline{U_q^*} \subset V_q^*$, $\overline{V_q^*} \subset D^*$ $(q = 1, \ldots, p)$, and let F be the pre-image of

$$F^* = \bigcup_{q=1}^{p} \overline{V_q^*}$$

under $f(x)$. F is then a compact subset of D, and if what was proved above is applied to each U_q^*, it follows we can find an integer m_1 such that $F \subset D_m$ and $E^* \subset F_m^*$ for $m \geqq m_1$. This completes the proof of lemma 1.

LEMMA 2. *Under the hypotheses of the preceding lemma, each compact subset of D^* is contained in all but a finite number of D_m^* and*

$$\lim_{m\to\infty} f_m^{-1}(x^*) = f^{-1}(x^*),$$

uniformly on each compact subset of D^*, where $f_m^{-1}(x^*)$ and $f^{-1}(x^*)$ are the inverses of $f_m(x)$ and $f(x)$.

We see that the first assertion in the conclusion of this lemma is contained in the preceding lemma, because for each compact set $E^* \subset D^*$, we have $F \subset D_m$ and $E^* \subset F_m^* \subset D_m^*$ for $m \geqq m_1$.

We must also prove the second assertion in the conclusion of lemma 2. If this were not the case, we could find an $\varepsilon > 0$, a subsequence $\{m_k\}$ with $m_k \geqq m_1$, and a sequence of points $\{x_k^*\}$ in E^* such that

$$|f_{m_k}^{-1}(x_k^*) - f^{-1}(x_k^*)| \geqq \varepsilon \tag{6}$$

for all k. Since $f_{m_k}^{-1}(x_k^*) \in F$ and F is compact, we may assume as in the proof of the preceding lemma that

$$\lim_{k\to\infty} f_{m_k}^{-1}(x_k^*) = x_1 \in F, \tag{7}$$

and arguing as in (5), we get

$$x_1^* = f(x_1) = \lim_{k\to\infty} x_k^*.$$

But $f^{-1}(x^*)$ is continuous at $x^* \in E^*$, so that

$$\lim_{k\to\infty} f^{-1}(x_k^*) = f^{-1}(x_1^*) = x_1,$$

and we see that the preceding relation and (7) contradict (6), this completing the proof of lemma 2.

Carathéodory convergence of a sequence of domains. Suppose that $\{D_m\}$ is a sequence of domains which contain a fixed point x_0. We define the *kernel* D *of the sequence* $\{D_m\}$ *at* x_0 as follows (F. Gehring [8]):

(i) If there exists no fixed neighbourhood U if x_0 which is contained in all of the D_m, then D consists only of x_0.

(ii) If there exists a fixed neighbourhood U of x_0 which is contained in all of the D_m, then D is the domain with the following three properties:

(a) $x_0 \in D$,

(b) each compact set $E \subset D$ lies in all but a finite number of D_m,

(c) if Δ is a domain satisfying (*a*) and (*b*), then $\Delta \subset D$.

In the case (i), the kernel D is said to be *degenerate*. In the case (ii), it is not *a priori* obvious that any such domain exists. However, we may, for example, set $D = \bigcup_\alpha \Delta_\alpha$, where the union is taken over the collection of all domains $\{\Delta_\alpha\}$ which satisfy (*a*) and (*b*). This collection is not empty. Then D is clearly a domain which satisfies (*a*) and (*c*); but it has also the property (*b*). Indeed, suppose $F \subset D$ is compact. Then $\{\Delta_\alpha\}$ represents a covering of F, and, by Heine—Borel—Lebesgue covering theorem, it is possible to extract a finite covering $\Delta_1, \ldots, \Delta_p$. Since $F - \bigcup_{k=2}^{p} \Delta_k \subset \Delta_1$ is closed, we can find a neighbourhood U of it such that $U \subset \Delta_1$. Clearly, $F - U \subset \bigcup_{k=2}^{p} \Delta_k$ and $F = F \cap \bar{U} \cup (F - U)$. By the same argument applied to $F - U$, we finally obtain p closed sets $F_1, \ldots, F_p$ such that $F = \bigcup_{q=1}^{p} F_q$ and $F_q \subset \Delta_q$, and since each F_q is contained in all D_m, except possibly a finite number, we conclude that also their union F is contained in all D_m except possibly a finite number, and D also has the property (*b*). If $x_0 = O$, D is called the *kernel of the sequence* $\{D_m\}$, without adding "at the origin".

For an interesting alternative characterization of a kernel D in the case (ii), see MacLane [1].

Finally the D_m are said to *converge to their kernel* D *at* x_0 if every subsequence of domains $\{D_{m_k}\}$ also has D as its kernel. Again, if $x_0 = O$ we say only that the D_m *converge to their kernel* (without adding "at the origin").

Carathéodory convergence theorem.

THEOREM 1. *Suppose that* $\{f_m(x)\}$ *is a sequence of* K-QCfH, $f_m : D_m \rightleftharpoons D_m^*$, *that each compact subset of* D *is contained in all but a finite number of* D_m, *and that*

$$\lim_{m\to\infty} f_m(x) = f(x), \qquad |f(x)| < \infty, \tag{8}$$

in D. *Then the convergence is uniform on each compact subset of* D *and* $f(x)$ *is either a constant or a* K-QCfH *of* D *onto* D^*. *In this case, each compact subset of* D^* *is contained in all but a finite number of* D_m^* *and*

$$\lim_{m\to\infty} f_m^{-1}(x^*) = f^{-1}(x^*)$$

uniformly on each compact subset of D^*.

Let Δ be any domain with compact closure $\bar{\Delta} \subset D$. Then $\Delta \subset D_m$ for $m \geqq m_0$, and we can apply theorem 2.12.3 to conclude that the $f_m(x)$ are equicontinuous, and uniformly bounded, and hence converge uniformly, by (8) and by Arzela–Ascoli's theorem, on each compact subset of Δ. Theorem 2.12.5 further implies that $f(x)$ is either constant in Δ or a homeomorphism of Δ. In the latter case, we see from theorem 2.13.1 that $f(x)$ is a K-QCfH of Δ.

Now Δ may be chosen arbitrarily, with $\bar{\Delta} \subset D$. Hence the convergence in (8) is uniform on each compact subset of D and $f(x)$ is either a constant in D or a K-QCfH $f: D \rightleftharpoons D^*$. Finally in this last case, the remaining conclusion follows from the preceding lemma.

Next we give the following n-space form of the Carathéodory convergence theorem (Gehring [8]):

THEOREM 2. *Suppose that* $\{D_m\}$ *is a sequence of domains which contain the origin, that the* D_m *converge to their kernel* D, *and that* D *is a domain with a finite boundary point. Suppose further that* $\{f_m(x)\}$ *is a sequence of* K-QCfH, $f_m : D_m \rightleftharpoons D_m^*$, *and that* $f_m(O) = O$. *If*

$$\lim_{m\to\infty} f_m(x) = f(x), \qquad |f(x)| < \infty \tag{9}$$

in D, *then the* D_m^* *converge to their kernel* D^* *and* D^* *has a finite boundary point.*

Conversely if the D_m^* *converge to their kernel* D^* *and if* D^* *has a finite boundary point, then there exists a subsequence* $\{m_k\}$ *such that*

$$\lim_{k\to\infty} f_{m_k}(x) = f(x), \qquad |f(x)| < \infty \tag{10}$$

in D. *In each case,* $D^* = f(D)$ *and* $f(x)$ *is either a constant or a* K-QCfH, *depending on whether or not* D^* *is degenerate.*

Suppose that $f_m(x)$ converge to a finite limit $f(x)$ in D and let $D' = f(D)$. We prove first that $D' = D^*$. Now the preceding theorem implies that $D' \subset D^*$. For if $f(x)$ is constant, then D' consists only of the origin. Otherwise $f(x)$ is a K-QCfH $f: D \rightleftharpoons D'$ and each compact subset of D' is contained in all but a finite number of D_m^*. Since D' contains the origin, it follows from (c) that $D' \subset D^*$.

In order to show that $D^* \subset D'$, fix $x_1^* \in D^*$ and let Δ^* be any bounded domain, containing the origin and x_1^*, such that $\overline{\Delta^*} \subset D^*$. Then $\overline{\Delta^*}$ is a compact subset of D^* and hence $\Delta^* \subset D_m^*$ for $m \geqq m_0$. Let $\Delta_m = f_m^{-1}(\Delta^*)$ for $m \geqq m_0$. Then $\Delta_m \subset$ $\subset D_m$ for $m \geqq m_0$ and hence

$$\sup_m d(O, \partial\Delta_m) \leqq \sup_m d(O, \partial D_m). \tag{11}$$

Now the fact that the D_m converge to D, a domain with a finite boundary point, implies that

$$\sup_m d(O, \partial D_m) < \infty. \tag{12}$$

For otherwise we could find a subsequence $\{m_k\}$ such that

$$\lim_{k\to\infty} d(O, \partial D_{m_k}) = \infty,$$

and the kernel for the subsequence of domains $\{D_{m_k}\}$ would be the finite space.

Since $f_m^{-1}(O) = O$, we conclude from (10), (11) and theorem 2.12.2 that the $f_m^{-1}(x^*)$ are uniformly bounded and equicontinuous on each compact subset of Δ^*. Hence by Arzela—Ascoli's theorem, we may pick a subsequence $\{m_k\}$ such that

$$\lim_{k\to\infty} f_{m_k}^{-1}(x^*) = f_0^{-1}(x^*)$$

in Δ^*.

Let Δ be the kernel of the sequence of domains $\{\Delta_{m_k}\}$, and let $\Delta' = f_0^{-1}(\Delta^*)$. Then the argument from above applied to the $f_{m_k}^{-1}(x^*)$ on Δ^*, shows that $\Delta' \subset \Delta$. Moreover, since D is the kernel of the sequence $\{D_{m_k}\}$, we see that $\Delta \subset D$. Hence $\Delta' \subset D$ and since $x_1^* \in \Delta^*$, we have $f_0^{-1}(x_1^*) \in D$ and, by virtue of the equicontinuity of the $f_m(x)$, we conclude that

$$f[f_0^{-1}(x_1^*)] = \lim_{k\to\infty} f_{m_k}[f_0^{-1}(x_1^*)] = \lim_{k\to\infty} f_{m_k}[f_{m_k}^{-1}(x_1^*)] = x_1^*.$$

Thus $x_1^* \in D'$. Since x_1^* was chosen as any point in D^*, we conclude that $D^* \subset D'$ and hence with the above that $D' = D^*$.

Now let $\{m_k\}$ be any subsequence. By virtue of the hypothesis that (9) holds in D,

$$\lim_{k\to\infty} f_{m_k}(x) = f(x)$$

in D. Then, since the D_{m_k} converge to D, we can apply what was proved above to conclude that $D^* = D'$ is the kernel of the sequence $\{D_{m_k}^*\}$. Hence D^* converge to D^*, as the sequence of indices $\{m_k\}$ is arbitrary.

It is also clear that D^* has a finite boundary point, or, in other words, that $\partial D^* \neq \emptyset$, since D^* either consists of the origin or is the image of D under the K-QCfH $f(x)$, and then, by Loewner's theorem (Chapter 2.9), $\partial D^* \neq \emptyset$. This, therefore completes the proof of the first part of the theorem.

Suppose now that the domains D_m^* converge to their kernel D^* and that D^* has a finite boundary point. Next, let Δ be any bounded domain containing the origin such that $\bar{\Delta} \subset D$. Then $\Delta \subset D_m$ for $m \geqq m_0$. Set $\Delta_m^* = f_m(\Delta)$ for $m \geqq m_0$. Then $\Delta_m^* \subset D_m^*$ and, arguing as in (11) and (12), we have

$$\sup_m d(O, \partial\Delta_m^*) \leqq \sup_m d(O, \partial D_m^*) < \infty.$$

Since $f_m(O) = O$, we can use theorem 2.12.2 and Arzela—Ascoli's theorem to obtain a subsequence $\{f_{m_k}(x)\}$ which converges to a finite limit in Δ. Now D can be express-ed as the union of an expanding sequence of such domains Δ, and by means of a well-known diagonal process, we can find a subsequence $\{m_k\}$ such that (10) holds in D. Finally, since the D_{m_k} converge to D, the first half of the theorem implies that from the convergence of the D_m^* to D^* the convergence of the D_m to D, follows and then, *a fortiori*, the convergence of the D_{m_k} to D. Hence the kernel D^* of the domains D_m^* is the image of D under $f(x)$ and $f(x)$, by the preceding theorem, is either a constant or a K-QCfH. The proof of the second half of the theorem 2 is now complete.

Remarks. 1. In the preceding theorem we established a relation between the convergence of a sequence of K-QCfH $\{f_m(x)\}$ to a finite function $f(x)$ and the convergence of the domains D_m^* to a kernel D^* which has a finite boundary point. The connection established here, between these two kinds of convergence is not as close as in the usual form of Carathéodory's convergence theorem. For though the convergence of the $f_m(x)$ implies convergence of the D_m^*, conver-gence of D_m^* only implies convergence of some subsequence of the $f_m(x)$. To obtain the stronger conclusion that the convergence of the D^* implies convergence of $f_m(x)$, we should have to know that all limit functions of the $f_m(x)$ are identical. This is clear so in the case where D^* is degenerate. However in the case where D^* is a domain, we should have to include some additional normalization for the $f_m(x)$ which would guarantee that at most one normalized K-QCfH $f\colon D \rightleftharpoons D^*$ with $f(0) = O$ exists.

2. A similar, but more general theorem was proved by G. D. Suvorov [2] for a sequence of equicontinuous and equiopen homeomorphisms $f_m\, D_m \rightleftharpoons D_m^*$ in a compact metric space. We recall that a sequence of homeomorphisms $\{f_m(x)\}$ is called *equiopen in the domain D*, if for every $x \in D$ and $\varepsilon > 0$, there exists a $\delta(x, \varepsilon) > 0$ such that $f_m[B(x, \varepsilon)] \subset B[f_m(x), \delta(x, \varepsilon)]$ $(m = 1, 2, \ldots)$.

CHAPTER 2

The stability of Liouville's theorem

As we have seen in Chapter 2.16, by Liouville's theorem (proved later by Ju. G. Rešetnjak [4, 11]), a homeomorphism which carries infinitesimal spheres onto infinitesimal spheres is a finite product of inversions (product here means composition of mappings). M. A. Lavrent'ev [3] raised for the first time the problem (and even solved it in one particular case) if a small local deviation of conformality implies a small global deviation, i.e. arose the question of the stability of Liouville's theorem. The general form in which this problem was stated by Lavrent'ev is the following:

Does a real function of $\varepsilon: \lambda(\varepsilon)$, with $\lambda(\varepsilon) \to 0$ as $\varepsilon \to 0$ exist and such that for every $(1+\varepsilon)$-QCfH $f(x)$ in B we may find a conformal mapping $L(x)$ satisfying the inequality

$$|f(x) - L(x)| < \lambda(\varepsilon)$$

everywhere in B?

The stability theorem was stated in its general form, but without proof, by P. P. Belinskiĭ in [1] and then at the International Congress of Moscow (1966) [2], at the Symposium on QCf mappings at Indiana University (1967) [3] and in [5], [6].

We shall prove the theorem under more restrictive conditions following Rešetnjak [3].

THEOREM 1. *There exists a function* $\mu(\varepsilon, r)$, $0 < r < 1$, $\mu(\varepsilon, r) \to 0$ *as* $\varepsilon \to 0$, *so that for every* $(1+\varepsilon)$-QCfH $f: B \rightleftarrows B$ *we may find a conformal mapping* $L(x)$ *so that*

$$|f(x) - L(x)| < \mu(\varepsilon, r) \tag{1}$$

for all $x \in B(r)$.

Fix a sequence $\{\varepsilon_m\}$, $\varepsilon_m > 0$, $\varepsilon_m \to 0$ as $m \to \infty$, and choose a sequence $\{f_m\}$ of $(1+\varepsilon_m)$-QCfH in B such that $\lim_{m\to\infty} f_m(x) = L(x)$, $|L(x)| < \infty$. By theorem 1.1, $L(x)$ is a $(1+\varepsilon_1)$-QCfH in B or a constant and $\{f_m(x)\}$ converges uniformly at $L(x)$ on every compact subset of B. Let $A \subset\subset B(r)$ be a ring and let be $A_m^* = f_m(A)$, $A^* = L(A)$. Then the uniform convergence of $\{f_m(x)\}$ on $\bar{A}$ implies the uniform convergence of each component of ∂A_m^* to the corresponding component of ∂A^*, so that, by lemma 2.6.2, mod A_m^* converges to mod A^*. But since

$$\frac{\operatorname{mod} A}{1+\varepsilon_m} \leqq \operatorname{mod} A_m^* \leqq (1+\varepsilon_m) \operatorname{mod} A, \quad (m = 1, 2, \ldots)$$

for every ring A, $\subset \bar{A} \subset B$, letting $m \to \infty$, we obtain

$$\operatorname{mod} A = \lim_{m\to\infty} \operatorname{mod} A_m^* = \operatorname{mod} A^*,$$

and then $x^* = f(x)$ is either a conformal mapping in B or a constant.

From the above we may now derive the theorem by *reductio ad absurdum.* Suppose, to prove it is false, that given $r \in (0, 1)$, there is a sequence $\{f_m(x)\}$ of $(1+\varepsilon_m)$-QCfH, with $\varepsilon_m \to 0$ as $m \to \infty$, and so that, for every Möbius transformation $L(x)$,

$$|f_m(x) - L(x)| \geqq \varepsilon_0 > 0, \ \varepsilon_0 = \text{const.} \qquad (m = 1, 2, \ldots) \tag{2}$$

at least at a point $x \in B(r)$. However, by theorem 2.12.1 and by Arzela–Ascoli's theorem, it would be possible to obtain a subsequence $\{f_{m_k}(x)\}$, which converges uniformly on $B(r)$, hence, and from the first part of the proof, it would follow that $\lim_{k\to\infty} f_{m_k}(x) = \tilde{L}(x)$, where $\tilde{L}(x)$ is either a conformal mapping contradicting thus the preceding inequality, or a constant a; but in this case, a sequence of conformal mappings $\{L_p(x)\}$ could be found with $\lim_{p\to\infty} L_p(x) = a$, and then, if k and p are sufficiently large, we should obtain $|f_{m_k}(x) - L_p(x)| < \varepsilon_0$, which contradicts also in this case the inequality (2), and the proof is complete.

CHAPTER 3

Cluster sets of QCfH

Cluster sets. Let $f: D \to R^n$ and $\xi \in \partial D$. By *cluster set* $C(f, \xi)$ of the mapping f at the boundary point ξ we mean the set of points $x^* \in S^n$ such that there exists a sequence $x_m \to \xi$, $x_m \in D$, with $f(x_m) \to x^*$.

We recall that S^n is the n-dimensional unit sphere, or the compactification of the Euclidean space R^n obtained by adjoining a single point (usually designated by the symbol ∞) and defining the neighbourhood of ∞ to be sets containing ∞ and the complement of compact subsets of R^n.

An arc $\gamma \subset D$ with an endpoint in $\xi \in \partial D$ is called an *endcut* of D from ξ.

We denote by $C_\gamma(f, \xi)$ the set of all $x^* \in S^n$ for which there exists a sequence of points $\{x_m\}$ converging to ξ along γ such that $f(x_m) \to x^*$. The set $C_\gamma(f, \xi)$ is termed *cluster set of f with respect to* γ. We set

$$\Pi(f, \xi) = \bigcap_{\gamma} C_\gamma(f, \xi),$$

where the intersection is taken over all endcuts γ of D from ξ.

Now suppose that $f: D \rightleftarrows D^*$ is a homeomorphism and $\xi \in \partial D$. We say that a sequence of points $\{x_m\}$, $x_m \in D$, *converges in a cone* C to ξ if $x_m \to \xi$ and there exists a constant a, $1 \leqq a < \infty$, such that

$$|x_m - \xi| \leqq a d(x_m, \partial D)$$

for all m.

The set $C_c(f, \xi)$ of all points $x^* \in S^n$ for which a sequence $\{x_m\}$ converging in a cone to ξ such that $f(x_m) \to x^*$ exists is said to be a *cluster set with respect to the cone* C.

If in a proof only one function is involved we shall use, for sake of simplicity, $C(\xi)$, $C_\gamma(\xi)$, $\Pi(\xi)$ and $C_c(\xi)$ in place of $C(f, \xi)$, $C_\gamma(f, \xi)$, $\Pi(f, \xi)$ and $C_c(f, \xi)$, respectively.

Existence theorem for locally homeomorphic QCf mappings with a given cluster set (Church [1]). In order to establish this theorem we must first define some QCfH involved in the proof.

Given a positive real-valued map u on (a, b) or $[a, b)$ (both a and b or one of them may be infinite), set

$$S_u(a, b) = \left\{x; a < x^1 < b, \left[\sum_{k=2}^{n} (x^k)^2\right]^{\frac{1}{2}} < u(x^1)\right\}, \tag{1}$$

$S_u[c, d]$ denotes the set of points of $S_u(a, b)$ which satisfy $c \leqq x^1 \leqq d$.

LEMMA 1. *Let $a > 0$, $b > 0$, and let $u: (-a, b) \to (0, \infty)$ be a differentiable (C^k, C^∞) mapping such that $\frac{du}{dt} \leqq 0$ and bounded below and $u(t) \to 0$ as $t \to b$. Let $v: (-a, \infty) \to (0, \infty)$ be a differentiable (C^k, C^∞) map with $\frac{dv}{dt}$ bounded above and below. Then there exists a differentiable (C^k, C^∞) QCfH $q: S_u[0, b) \rightleftarrows S_v[0, \infty)$.*

We recall first that there exists a unique real-valued function $\tau(t)$ defined on a neighbourhood of 0 satisfying

$$\frac{d\tau}{dt} = p(t, \tau) \tag{2}$$

$\tau(0) = 0$ and also the Lipschitz condition

$$\left|p(t, \tau_1) - p(t, \tau_0)\right| < K\left|\tau_1 - \tau_0\right| \tag{3}$$

in the considered neighbourhood (see E. L. Ince [1] pp. 62, 63).

If we set $p(t, \tau) = \frac{v[\tau(t)]}{u(t)}$, with u and v given in lemma 1, then (2) comes to

$$\frac{d\tau}{dt} = \frac{v[\tau(t)]}{u(t)}, \tag{4}$$

and Lipschitz condition (3) follows from the mean value theorem, as $\frac{dv}{d\tau}$ and $\frac{1}{u(t)}$ are bounded — the latter for $t \leqq d < b$. Thus the equation (4) has a unique continuous real solution $\tau = \tau(t)$ in the neighbourhood of 0. Suppose now that $\tau(t)$ is defined on $[0, \varepsilon)$, where $0 < \varepsilon < b$, and ε is maximal. (4) implies that

$$\int_0^{\tau(t)} \frac{ds}{v(s)} = \int_0^t \frac{ds}{u(s)}, \tag{5}$$

where the right-hand integral has a finite limit as $t \to \infty$. By hypothesis $\frac{dv}{d\tau} \leqq C$, where $v(\tau) \leqq C\tau + v(0)$ and therefore

$$\int_0^{\tau(t)} \frac{ds}{v(s)} \geqq \int_0^{\tau(t)} \frac{ds}{Cs + v(0)} = C^{-1} \log\left[C\tau(t) + v(0)\right].$$

Thus $\tau(t)$ and $\frac{dv[\tau(t)]}{dt}$ have finite limits as $t \to \varepsilon$. By the theorem quoted above (Ince [1]), in a neighbourhood of $x = \varepsilon$ the differential equation (4) has a unique solution, call it $\tau_1(t)$, such that $\tau_1(\varepsilon)$ is the limit of $\tau(t)$ as $t \to \varepsilon$. Then τ may be extended past ε as τ_1 and the maximality of ε is thus contradicted, unless $\varepsilon = b$. Hence $\tau(t)$ is definite on $[0, b)$.

By hypothesis there exists $C_1 > 0$ such that $-C_1 < \frac{du}{dt} \leqq 0$, so that

$$\int_t^b du > -C_1 \int_t^b dt,$$

hence, since $u(b) = 0$, we obtain $-u(t) > C_1(t - b)$, which implies

$$\int_0^t \frac{ds}{u(s)} > \int_0^t \frac{Cds}{{}_1(b - s)} = -\frac{1}{C_1} \log(b - t) + \frac{\log b}{C_1},$$

hence the left-hand integral approaches ∞ as $t \to b$, and then also the left-hand integral of (5) approaches ∞ as $t \to b$. And now, let us prove that this implies $\tau(t) \to \infty$ as $t \to b$.

Indeed, by hypothesis, $\frac{dv}{d\tau} \geqq C_2$, and integrating we obtain

$$v(\tau) \geqq C_2\tau + v(0), \tag{6}$$

with $0 \leqq v(0) < \infty$, because $v(0) \geqq 0$ by hypothesis, and $v(0) = \infty$ would imply, by (6) that $v(\tau) = \infty$, contradicting the hypothesis. Next, suppose first $C_2 \geqq 0$. Then, by integrating (6), we obtain

$$\int_0^{\tau(t)} \frac{ds}{v(s)} \leqq \int_0^{\tau(t)} \frac{ds}{C_2 s + v(0)}. \tag{7}$$

If $C_2 > 0$, (7) implies

$$\int_0^{\tau(t)} \frac{ds}{v(s)} \leqq \frac{1}{C_2} \log\left[\tau + v(0)\right].$$

Since the left-hand side approaches ∞ as $t \to b$, it follows that the right-hand side approaches ∞, too, and then $\tau(t) \to \infty$ as $t \to b$.

If $C_2 = 0$, (7) yields

$$\int_0^{\tau(t)} \frac{ds}{v(s)} \leqq \frac{\tau(t)}{v(0)},$$

and again $\tau(t) \to \infty$ as $t \to b$.

If $C_2 < 0$, from (6) we obtain

$$\frac{C_2}{v(\tau)} \geqq \frac{C_2}{v(0) + C_2\tau},$$

and by integration we see that

$$\int_0^{\tau(t)} \frac{ds}{v(s)} \leqq \frac{1}{C_2} \log\left[C_2\tau + v(0)\right],$$

which implies also in this case $\tau(t) \to \infty$ as $t \to b$.

Next, since $u(t)$ and $v(\tau)$ are differentiable or u, $v \in (C^k, C^\infty)$, it follows that also $\dfrac{d\tau}{dt}$ is differentiable or (C^k, C^∞), respectively, and then $\tau \in C^1$ or $\tau \in (C^k, C^\infty)$, respectively. Moreover, since u, $v > 0$, we have also $\dfrac{d\tau}{dt} > 0$ and we conclude that $\tau(t)$ is strictly increasing. Since $\tau(0) = 0$ and $\tau(b) = \infty$, we get $\tau\colon [0, b) \to [0, \infty)$.

Let $q : x \to y$, where

$$\begin{cases} y^1 = y^1(x^1), \\ y^k = \dfrac{v[y^1(x^1)]x^k}{u(x^1)} \qquad (k = 2, \ldots, n) \end{cases}$$

and $y^1(x^1)$ is given by (4). Clearly $q(x)$ is a differentiable (C^1, C^∞) homeomorphism mapping $S_u[0, b)$ onto $S_v[0, \infty)$.

It remains only to prove that $q(x)$ is a QCfH. The modulus of dilatation (i.e. the absolute values of the directional derivative) $\left|\dfrac{\partial q}{\partial s}\right|$ of q in the direction

s with direction cosines $(\alpha^1, \ldots, \alpha^n)$, where $\alpha^i = \cos(s, x^i)$ $(i = 1, \ldots, n)$, is $\left|\dfrac{\partial q}{\partial s}\right| = r\dfrac{dy}{dx}$, where

$$r = \left\{(\alpha^1)^2 + \sum_{k=2}^{n}\left[\frac{\alpha^1 x^k}{u(x^1)}\left(\frac{dv}{dy^1} - \frac{du}{dx^1}\right) + \alpha^k\right]^2\right\}^{\frac{1}{2}}.$$

Since r is bounded above and below by positive numbers, $q(x)$ is QCfH according to Church's definition, and then, by theorem 6.4, also according to all the other definitions.

Let $u: (-1, 1) \to (0, 1)$, $u \in C^\infty$, be such that $u(t) = \sqrt{1 - t^2}$ for $t \leqq 0.3$, $u(t) = 4(1 - t)$ in some neighbourhood of 1,

$$4(1 - t) \leqq u(t) \leqq \sqrt{1 - t^2} \tag{8}$$

for $t \geqq 0.9$, and $u'(t) < 0$ for $t > 0$.

We shall now construct a homeomorphism $f: S_u(-1, 1) \to \bar{B}$, whose restriction to $S_u(-1, 1)$ is a QCfH C^∞. Let $\tau = \left[\sum_{k=2}^{n} (x^k)^2\right]^{\frac{1}{2}}$; let $g(\tau)$ be the inverse of $u(x^1)$ for $x^1 > 0$. Thus $g(\tau) = 1 - \dfrac{\tau}{4}$ in some neighbourhood of $\tau = 0$. There is a C^∞ function λ: $(-\infty, \infty) \to [0, 1]$, such that $\lambda(s) = 0$ for $s \leqq 0$, $\lambda(s) = 1$ for $s \geqq 1$ and $0 < \lambda'(s) \leqq K$ for $0 < s < 1$. Let

$$s(x^1, v) = \frac{x^1 - 1 + \left(\frac{1}{2} + \varepsilon\right)\tau}{\left(\frac{1}{6} + \varepsilon\right)\tau},$$

where ε is to be specified later, $0 < \varepsilon \leqq \dfrac{1}{2}$; and let

$$\psi(x^1, \tau) = 1 + \lambda[s(x^1, \tau)]\,\frac{\sqrt{1 - \tau^2} - g(\tau)}{g(\tau) - 1 + \frac{\tau}{2}}.$$

The homeomorphism $x^* = f(x)$ we are constructing is given by $x^{*k} = x^k$ $(k = 2, \ldots, n)$, $x^{*1} = x^1$ for $\tau \geqq 0.5$ and for $\tau = 0$, and $x^{*1} = 1 - \dfrac{\tau}{2} + \psi(x^1, \tau)\left(x^1 - 1 + \dfrac{\tau}{2}\right)$

for $0 < \tau < 0.6$. Observe that the expression for x^{*1} yields x^1 for $0.5 \leqq \tau \leqq 0{,}6$ or $x^1 \leqq 1 - \left(\frac{1}{2} + \varepsilon\right)\tau$ (in particular for $\tau = 0$). For $0 < \tau < 0.6$, x^{*1} is defined by composition, sum, product, and reciprocal of C^∞ maps, and thus is C^∞ itself. Now

$$\frac{\partial x^{*1}}{\partial x^1} = \psi + \left(x^1 - 1 + \frac{\tau}{2}\right)\frac{\partial \psi}{\partial x^1} \tag{9}$$

for $\tau < 0{,}6$. Next $\psi \geqq 1$, since (8) implies

$$\sqrt{(1-(x^1)^2} \geqq \tau = u(x^1) \geqq 4(1 - x^1) \geqq 2(1 - x^1),$$

hence

$$\sqrt{1-\tau^2} \geqq x^1 = g(\tau) \geqq 1 - \frac{\tau}{4} \geqq 1 - \frac{\tau}{2}. \tag{10}$$

and then

$$\frac{\sqrt{1-\tau^2} - g(\tau)}{g(\tau) - 1 + \dfrac{\tau}{2}} \geqq 0$$

and $0 \leqq \lambda(s) \leqq 1$, whence $\psi \geqq 1$. On the other hand,

$$\frac{\partial \lambda[s(x^1, \tau)]}{\partial x^1} = \frac{\lambda'(s)}{\left(\dfrac{1}{6} + \varepsilon\right)\tau} \geqq 0,$$

so that for $x^1 \geqq 1 - \dfrac{\tau}{2}$, the second term in (9) is non-negative, and, since $\psi \geqq 1$ we conclude that $\dfrac{\partial x^{*1}}{\partial x^1} \geqq 1$ for $x^1 \geqq 1 - \dfrac{\tau}{2}$. For $x^1 < 1 - \left(\dfrac{1}{2} + \varepsilon\right)\tau$, the preceding remark implies $x^{*1} = x^1$ and then $\dfrac{\partial x^{*1}}{\partial x^1} = 1$. For

$$1 - \left(\frac{1}{2} + \varepsilon\right)\tau < x^1 < 1 - \frac{\tau}{2}, \tag{11}$$

$$\left|\left(x^1 - 1 + \frac{\tau}{2}\right)\frac{\partial \psi}{\partial x^1}\right| \leqq \frac{1 - \dfrac{\tau}{2} - x^1}{\left(\dfrac{1}{6} + \varepsilon\right)\tau} \cdot \frac{\mathrm{d}\lambda}{\mathrm{d}s} \cdot \frac{\sqrt{1-\tau^2} - g(\tau)}{g(\tau) - 1 + \dfrac{\tau}{2}}. \tag{12}$$

And now we show first that the last factor of the left-hand side of (12) is bounded. Indeed, from (10) we deduce

$$\sqrt{1-\tau^2} \geqq x^1 = g(\tau) \geqq 1 - \frac{\tau}{4},$$

so that, for $0 \leqq v \leqq 0.5$,

$$\frac{\sqrt{1-\tau^2} - g(\tau)}{g(\tau) - 1 + \dfrac{\tau}{2}} \leqq \frac{4\sqrt{1-\tau^2} - 4 + \tau}{\tau} \leqq 1.$$

Hence, on account of the definition of $\lambda(s)$, the product of the last two factors of the right-hand side of the inequality (1) is bounded say by K, and then, by (11) the whole right-hand side of (12) is bounded by $6K\varepsilon$. Thus, for $\varepsilon < \dfrac{1}{12K}$ $\left(0 < \varepsilon \leqq \dfrac{1}{2}\right)$,

$$\left|\left(x^1 - 1 + \frac{\tau}{2}\right)\frac{\partial \psi}{\partial x^1}\right| \leqq \frac{1}{2} \tag{13}$$

and then $\dfrac{\partial x^{*1}}{\partial x^1} \geqq \dfrac{1}{2}$ for $0 \leqq x^1 \leqq 1$ and $0 \leqq \tau \leqq 1$, and f is a homeomorphism.

Next, we prove that $\dfrac{\partial x^{*1}}{\partial x^1}$ is bounded also above. From the definition of ψ and the inequality (10), it follows that

$$\psi = 1 + \lambda(s)\frac{\sqrt{1-\tau^2} - g(v)}{g(\tau) - 1 + \dfrac{\tau}{2}} \leqq 1 + \frac{\sqrt{1-\tau^2} - 1 + \dfrac{\tau}{4}}{\dfrac{\tau}{4}} \leqq 2$$

and arguing as above, for $x^1 > 1 - \left(\dfrac{1}{2} + \varepsilon\right)\tau$, we obtain (13), which, on account of (9), implies $\dfrac{\partial x^{*1}}{\partial x^1} \leqq \dfrac{5}{2}$, and if $x^1 \leqq 1 - \left(\dfrac{1}{2} + \varepsilon\right)\tau$, we obtain $\dfrac{\partial x^{*1}}{\partial x^1} = 1$. Thus, $\dfrac{\partial x^{*1}}{\partial x^1}$ is bounded above, which implies the boundedness of the modulus of dilatation

$$\left|\frac{\partial f}{\partial s}\right| = \left[\left(\sum_{i=1}^{n} \alpha^i \frac{\partial x^{*1}}{\partial x^i}\right)^2 + \sum_{k=2}^{p} (\alpha^k)^2\right]^{\frac{1}{2}},$$

where s is a direction with direction cosines $(\alpha^1, \ldots, \alpha^n)$, and we conclude that $f(x)$ is QCfH according to Church's definition and, on account of theorem 2.6.4, $f(x)$ is QCfH also according to the other definitions.

Let $v(\tau)$ be any C^∞ homeomorphism such that $\frac{d^i v(0)}{(d\tau)^i} = \frac{d^i u(0)}{(dt)^i}$ $(i=1, 2, \ldots)$, and v satisfies the hypotheses of the preceding lemma, and let $u: [-1, 1] \rightleftharpoons [0, 1]$ be as above. If v is extended to $[-1, 1]$ by $v \equiv u$ on $[-1, 0]$, then there is, by the preceding lemma, a QCfH $q: S_u(-1, 1) \rightleftharpoons S_v(-1, \infty)$; let F be the natural extension of $q \circ f^{-1}$ to $\bar{B} - \{(1, 0, \ldots, 0)\}$. Thus we constructed a homeomorphism $F: \bar{B} - \{(1, 0, \ldots, 0)\} \rightleftharpoons \overline{S_v(-1, \infty)}$ (with its closure in R^n), and with the restriction to B: $F/B \in C^\infty$ and QCfH.

LEMMA 2. *Given any continuum* $C \subset S^n$, *there exists a* C^∞ *map* $h(t)$, $h: [0, \infty) \to R^n$ *such that*

(i) *on some neighbourhood of* 0, $h^1 = t$ *and* $h^k = const.$ $(k = 2, \ldots, n)$;

(ii) $\sum_{i=1}^{n} \left(\frac{dh^i}{dt}\right)^2 > 0$ *and*

(iii) *as* $t \to \infty$, *the set of limit points of* $h(t)$ *is* C.

First suppose that $C \subset R^n$. Let

$$U_m = \left\{x^*; x^* \in R^n, d(x^*, C) < \frac{1}{m}\right\},$$

and let $\{x^*_{mp}\}$ $(p = 1, \ldots, p_m)$ be a set of points $\frac{1}{m}$-dense in U_m. Recall that a set E is said to be $\frac{1}{m}$-*dense* in a set E', if for every point $x' \in E'$, there exists at least a point $x \in E$ such that $d(x, x') < \frac{1}{m}$.

Define a mapping $\varphi: [0, a_1] \to R^n$ $(0 < a_1 < \infty)$ so that $x^*_{1p} \in \varphi([0, a_1])$ $(p = 1, \ldots, p_1)$, $\varphi([0, a_1]) \subset U_1$, $\varphi(a_1) \in U_2$ and $\varphi/[0, a_1]$ satisfies (i) and (ii). Extend φ to $[0, a_2]$ $(a_1 < a_2)$, so that $x^*_{2p} \in \varphi([a_1, a_2])$ $(p = 1, \ldots, p_2)$, $\varphi([a_1, a_2]) \subset U_2$, $\varphi(a_2) \subset U_3$ and $\varphi/[0, a_2]$ satisfies (i) and (ii), and so on. If $\lim_{m\to\infty} a_m = b < \infty$, let $h = \varphi \circ \gamma$, where $\gamma(t) = \frac{bt}{1+t}$, otherwise let $h = \varphi$. This mapping $h(t)$ satisfies all the conditions in lemma 2.

If C is not contained in R^n, and $C \neq \{\infty\}$, replace U_m by

$$[B(x_0, m) - B(x_0, m-1)] \cap [B(x_0, m) \cup U_m] \quad (m = 1, 2, \ldots),$$

where $x_0 \in C - \{\infty\}$, and proceed as above.

Finally, if $C = \{\infty\}$, let $h^1 = t$, $h^k = 0$ $(k = 2, \ldots, n)$.

THEOREM 1. *Given any continuum* $C \subseteq S^n$ *and* $\xi \in S = \partial B$, *there exists a* C^∞ QCf *local homeomorphism* $f: B \to R^n$ *such that* $C(f, \xi) = C$. *Moreover, each arc cluster set* $C_\gamma(\xi) = C$.

If C is a single point in R^n, f is a rigid motion of $\bar{B}$ in R^n, sending ξ into that point. Suppose, for simplicity's sake, $\xi = (1, 0, \ldots, 0)$. If $C = \{\infty\}$, then $f = F$, where $F: \bar{B} - \{(1, 0, \ldots, 0)\} \rightleftharpoons S_v(-1, \infty)$ is the homeomorphism from above obtained by means of the QCfH $q(x)$ of lemma 1.

If C is an arbitrary nondegenerate continuum, let h be the map in the preceding lemma. We may suppose that h is reparametrized so that t (call it x^1 now) is arc length; then (ii) comes to

$$\sum_{i=1}^{n} \left(\frac{dh^i}{dx^1}\right)^2 = 1. \tag{14}$$

For each x^1 choose a normal $(n-1)$-frame β_k $(k = 2, \ldots, n)$ to $h([0, \infty))$ at $h(x^1)$, so that each coordinate function β_k^i $(i = 1, \ldots, n;\ k = 2, \ldots, n)$ is C^∞ on $[0, \infty)$ and, by condition (i) of the preceding lemma, $\beta_k^k = 1$, $\beta_k^i = 0$ for $i \neq k$ on a neighbourhood of 0, for it is sufficient to choose β_k^i to be the derivatives of some functions satisfying condition (i) of the preceding lemma. Define $\psi: \{x,\ x \in R^n,\ x^1 \geqq 0\} \to R^n$ by

$$\psi^i(x) = h^i(x^1) + x^k \beta_k^i(x^1) \quad (i = 1, \ldots, n). \tag{15}$$

On the line $L = \{x;\ x^k = 0\ (k = 2, \ldots, n)\}$ the Jacobian of $\psi(x)$ has the entries β_k^i $(i, k = 1, \ldots, n)$, where $\beta_1^i(x^1) = \dfrac{dh^i(x^1)}{dx^1}$, and then, by (14) and taking into account the conditions satisfied by β_k^i $(i = 1, \ldots, n;\ k = 2, \ldots, n)$, we obtain $J = \pm 1$. Thus there exists a decreasing function $\lambda: [0, \infty) \to (0, \infty)$ such that $\psi / S_\lambda[0, \infty)$ has $J \neq 0$, and hence it is a local homeomorphism. On the line L, $\left|\dfrac{\partial \psi}{\partial s}\right| = \left[\sum_{k=1}^{n} (\alpha_i \beta_k^i)^2\right]^{\frac{1}{2}}$, where $(\alpha_1, \ldots, \alpha_n)$ are the direction cosines of the direction s. Since the right-hand side may be interpreted as the length of a vector $\beta(\alpha)$ which is the image of the unit vector $\alpha = (\alpha_1, \ldots, \alpha_n)$ under a linear transformation with the matrix of the coefficients $\beta = \|\beta_k^i\|$, which is an orthogonal matrix, it follows that $\left|\dfrac{\partial \psi}{\partial s}\right| \equiv 1$ on L. Hence and since $\beta_k^i \in C^\infty$ $(i, k = 1, \ldots, n)$, we conclude that given an $\varepsilon > 0$, there exists a sufficiently small neighbourhood of L such that $\left|\frac{\partial \psi}{\partial s}\right| < 1 + \varepsilon$ and then there exists a decreasing function $\nu: [0, \infty) \to (0, \infty)$ such that $K - 1 < \varepsilon$, where K is the constant of the K-QCf restriction $\psi(x)/S_\nu[0, \infty)$ according to Church's definition.

Next, let $\mu: [0, \infty) \to (0, \infty)$, $\mu \in C^\infty$, be a function which satisfies the conditions

$$\mu(x^1) < \min\left[\lambda(x^1), \nu(x^1)\right], \tag{16}$$

and set $\mu(x^1) = [\mu(0)^2 - (x^1)^2]^{\frac{1}{2}}$ in a neighbourhood of 0 and on the interval $[-\mu(0), 0]$. Thus, by (1), $S_\mu[-\mu(0), \infty) = T\{S_v[-1, \infty)\}$, where v is as above and T is the transformation of similitude given by $T^i(x) = ax^i$, $a>0$ $(i=1, \ldots, n)$. If $\delta > 0$ is in the neighbourhood of conclusion (i) of lemma 2, since $\psi(x)$ is the identity on $S_\mu[0, \delta]$, extend $\psi(x)$ to $S_\mu[-\mu(0), 0]$ as the identity map, too.

The desired local homeomorphism $f \in C^\infty$ is

$$\{\psi|S_\mu[-\mu(0), \infty)\}[T(F|B)], \tag{17}$$

where F is the homeomorphism from above, because if a point of B approaches $(1, 0, \ldots, 0)$, the corresponding point of $S_v(-1, \infty)$ approaches $(\infty, 0, \ldots, 0)$, and then (15) implies $\lim\limits_{x^1 \to \infty} \psi(x^1, 0, \ldots, 0) = \lim\limits_{x^1 \to \infty} h(x^1) = C$.

To prove that for each endcut γ of B from ξ, $C_\gamma(f, \xi) = C$, it suffices to prove the analogous result for the homeomorphism ψ, the domain $S_\mu[-\mu(0), \infty)$ and the ray $0 < x^1 < \infty$. In $\psi(S_\mu)$ the distance from each point $\psi(x)$ to $\psi(x^1, 0, \ldots, 0)$ is bounded by $\mu(x^1)$, as it follows from (15) and (1). From the definition of $\mu(x^1)$ it follows that $x^1 \to \infty$ as $\mu(x^1) \to 0$ and the set of limit points of $\psi(x^1, \ldots, 0)$ is C as we conclude from (15).

Existence theorem for QCfH with a given cluster set.

LEMMA 3. *Given any continuum $C \subset S^n$ so that $S^n - C$ has a component E whose boundary is C, then h in lemma 2 may be chosen so that, besides the conditions* (i), (ii), (iii), *to satisfy also the conditions*

(iv) $h\{[0, \infty)\} \subset E$,

(v) $h(t)$ *is one-to-one.*

We may as well suppose $O \in E$ and $d(O, C) > 1$. We consider three cases:

(a) $n = 2$. Let $\Phi: (0, \infty) \to B^2$ be the spiral map given in polar coordinates by:

$$r(t) = \frac{t}{\pi + t}, \theta(t) = t \qquad (0 \leqq t < \infty).$$

Let $\Psi: B^2 \rightleftharpoons E$ be a conformal mapping (its existence is assured by Riemann mapping theorem). Since Ψ is one-to-one, it follows that if $\Psi\{\Phi[0, \infty)\}$ meets ∞, it meets it for at most one t, so that there exists $\delta > 0$ such that $\infty \bar{\in} \Psi\{\Phi[\delta, \infty)\}$; with a suitable reparametrization sending δ into 0 and a change of $\Psi \circ \Phi$ on a neighbourhood of δ (now 0) to satisfy condition (i) in the preceding lemma, $\Psi \circ \Phi / [\delta, \infty)$ is the desired map h, i. e satisfies (i)—(v).

(b) $n \geqq 3$, $\infty \bar{\in} C$. Let $\{Q_m\}$ be the family of all closed n-cubes which have edges parallel to coordinate axes, side $\frac{1}{2^m}$, and vertices $\left(\frac{m_1}{2^m}, \ldots, \frac{m_n}{2^m}\right)$, where m_i $(i = 1, \ldots, n)$ are integers. Let $E_m = \bigcup_{Q_m \subset E} Q_m$. If $\infty \in E$, let F_m be the unbounded component of E_m, together with ∞; otherwise, suppose, as we may, $O \in E$, and let F_m be the component of E_m containing O. Then $F_{m+1} \supset F_m$ and $\bigcup_{m=1}^{\infty} \operatorname{int} F_m = E$ ($\operatorname{int} E =$ the set of inner points of E).

F_m is simply connected, for if it were not so, then $S^n - F_m$ would contain at least two components C_1 and C_2, each of them containing at least by a point of $S^n - E$. Since $C = \partial E$, it follows that every arc joining a point of $F_m \subset E$ with a point of $S^n - \bar{E}$ meets the set C. Hence $C_1 \cap C$, $C_2 \cap C \neq \emptyset$ thus contradicting the fact that on the one hand C is a continuum of $S^n - F_m$ and on the other hand each arc joining $C_1 \cap C$ and $C_2 \cap C$ meets F_m. Thus F_m is simply connected, and then its boundary is connected. We show next that the open set $E - F_m$ is connected, too. Set $\delta = d(C, F_m)$. For a number p sufficiently large $\partial F_m \cap \partial F_{m+p} = \emptyset$ $\Bigl($it is enough to take $p > \frac{\sqrt{n}}{\delta \log 2} - m$, since the cubes with the edges of length $\frac{1}{2^{m+p}}$ have the diagonal of length $\frac{\sqrt{n}}{2^{m+p}} < \delta$, and if such a cube meets F_m it is contained in F_{m+1}, hence $F_m \subset \operatorname{int} F_{m+p}\Bigr)$. Choose now x, $x' \in E - F_m$ and let $\Lambda \subset E$ be a polygonal line joining them (this is possible since E is connected). If $\Lambda \subset E - F_m$, then x and x' may be joined by a polygonal line contained in $E - F_m$. But in general $\Lambda \not\subset E - F_m$. In this case $\Lambda \cap \partial F_m$ consists of some polygonal lines L_p $(p = 1, \ldots, q)$ and of some points x_k, x'_k $(k=1, \ldots, a)$, where x_k, x'_k are the endpoints of the polygonal line $L'_k \subset F_m$. Let L_{q+k} $(k = 1, \ldots, a)$, $L_{q+k} \subset \partial F_m$, be polygonal lines joining x_k and x'_k $(k=1, \ldots, a)$. Let $Q_{b_1}, \ldots, Q_{b_c}$ be the cubes of $F_{m+p} - \operatorname{int} F_m$ which meet L_b $(1 \leqq b \leqq q + a)$. These cubes form clearly a chain, i.e. two successive cubes have a common $(n-1)$-dimensional face. Set $D_b = \bigcup_{j=1}^{c} Q_{b_j}$. We replace each L_b $(b = 1, \ldots, q + a)$ by a polygonal line $L'_b \subset \operatorname{int} D_b$ (of course the endpoints of L'_b do not coincide with the endpoints of L_b, but are other two points of $\Lambda \cap D_b$). The broken line Λ', consisting of parts of Λ and of the polygonal lines L'_b $(b = 1, \ldots, q + a)$, joins x and x' and is contained completely in $E - F_m$. Since x, x' is an arbitrary pair of points of $E - F_m$, we conclude that $E - F_m$ is connected.

For each m, choose a set of points x_{ma} $(a = 1, ..., a_m)$, $\frac{1}{m}$-dense in $E-F_m$ and distinct for distinct indices. As in the previous lemma, define a one-to-one mapping $h: [0, \infty) \to E$ such that (i) and (ii) are satisfied and $h[0, \infty)$ joins the points x_{m_a} in lexicographic order. Since $E - F_m$ is connected, it follows that two points with the first index m may be joined by a broken line contained in $E - F_m$, so that there exist $\delta_m > 0$, $\delta_{m+1} > \delta_m$, with $h[\delta_m, \infty) \subset E - F_m$. Since $n \geqq 3$, $h(t)$ may be chosen one-to-one. Finally, since $\bigcap_{m=1}^{\infty} \overline{E - F_m} = C$, the set of limit points of $h(t)$ is C as $t \to \infty$.

(c) $n \geqq 3$, and $\infty \in C$. Suppose, as we may, $O \in E$ and let E_m be the union of the cubes of $\{Q_m\}$ contained in $E \cap B(O, m+1)$, and F_m the component of E_m containing the origin. The argument proceeds as above, except that the points x_{ma} are defined to be $\frac{1}{m}$-dense in $(E-F_m) \cap B(O, m+2)$.

THEOREM 2. *Given any continuum* $C \subset S^n$ *and* $\xi \in S$, *there exists a* C^∞ QCfH $f: B \to R^n$, *as in the preceding theorem, if and only if* $S^n - C$ *has a component whose boundary is* C.

If C is a single point, the existence of f follows from the preceding theorem, because the condition that $S^n - C$ has a component whose boundary is C is obvious in this case.

In the general case, arguing as in the preceding theorem, we obtain ψ given by (15), where in this case $h(x^1)$ satisfies the conditions (i)—(v) of lemmas 2 and 3. We may construct, as in the preceding theorem, the local homeomorphism $\psi/S_\mu[0, \infty)$, which this time may be so that $\psi\{S_\mu[0, \infty)\} \subset E$. Set $R_m = \overline{S_{\frac{1}{m}}[0, 1]}$ and let $L = \{x; x^k = 0\ (k = 2, ..., n)\}$. We shall show that, for some natural m, $\psi/(L \cup R_m)$ is one-to-one. If not, then there exist distinct points $x_m, y_m \in L \cup R_m$ such that $\psi(x_m) = \psi(y_m)$ $(m = 1, 2, ...)$. For m sufficiently large, $\psi(R_m) \subset E$ and then $d[\psi(R_m), C] \geqq \delta > 0$. Since ψ/L is one-to-one, it follows that for any number m it is impossible to have $x_m, y_m \in L$, so that we may suppose that $x_m \in R_m$ $(m = 1, 2,...)$; thus, by the definition of R_m, $\{x_m\}$ has a limit point $x \in L$. If $\{y_m\}$ has no limit point in L, then this sequence contains a subsequence contained in L and which approach $(\infty, 0, ..., 0)$ as $m \to \infty$; then the set of limit points of $\{\psi(y_m)\}$ is contained in C, contradicting the fact that $\psi(x_m) = \psi(y_m)$ and $d[\psi(R_m), C] \geqq \delta > 0$. Therefore, we may suppose that $\{y_m\}$ has a limit point $y \in L$. Since ψ/L is one-to-one, and since $\psi(x) = \psi(y)$, $x = y$. Since ψ is locally one-to-one in a neighbourhood of L, it follows that for a sufficiently large m, also $\psi(x_m) = \psi(y_m)$ implies $x_m = y_m$, contradicting thus the hypothesis $x_m \neq y_m$.

Similarly, there exists a natural number k such that

$$\psi|(L \cup \overline{S_{\frac{1}{m}}[0, 1]} \cup \overline{S_{\frac{1}{k}}[1, 2]})$$

is one-to-one. By an inductive argument there exists a strictly decreasing function $\chi : [0, \infty) \to (0, \infty)$ such that $\psi/S_\chi[0, \infty)$ is one-to-one.

The desired homeomorphism is given by (17), where in this case μ satisfies the inequality $\mu(x^1) < \min[\chi(x), \nu(x)]$ instead of (16), as in the preceding theorem.

Conversely, suppose that $f: B \to S^n$ is a homeomorphism such that $C = C(f, \xi)$, $\xi \in S$. To complete the proof it suffices to prove that $S^n - C$ has a component whose boundary is C. We shall show first that $f(B) \cap C = \emptyset$. If $x^* \in f(B)$, then there exists a ball $B(x, r)$ centered at $x = f^{-1}(x^*)$ and so that $\overline{B(x, r)} \subset B$, hence $x^* \in f[B(x, r)]$. Suppose that x^* is also in C. Then there exists a sequence $\{x_m\}$, $x_m \in B$ $(m = 1, 2, \dots)$, $x_m \to \xi$ with $f(x_m) \to x^*$. The sequence $\{x_m\}$ has a subsequence $\{x_{m_k}\}$, in $B - \overline{B(x, r)}$ such that $x_{m_k} \to \xi$ as $k \to \infty$. However, there exists a sequence $\{x'_{m_k}\}$ with $f(x'_{m_k}) = f(x_{m_k})$ and $x'_{m_k} \in B(x, r)$ for a k sufficiently large. Then $f[B(x, r)] \cap f[B - B(x, r)] \neq \emptyset$. Since f is a homeomorphism we have a contradiction.

Since $f(B)$ is connected, it must be in one component of $S^n - C$. Thus C is the boundary of this component, and the proof of theorem 2 is complete.

Space analogue of Lindelöf's theorem. In order to obtain, following Gehring [8], an n-space analogous for QCfH of a theorem due to Lindelöf on the boundary behaviour of conformal mappings of a disc we require first two preliminary results.

LEMMA 4. *Let $u(x)$ be continuous and ACL in $B(c) \cap \{x^n > 0\}$, let $S_r = S(r) \cap \{x^n > 0\}$. Then*

$$\int_a^b (\operatorname{osc}_{S_r} u)^n \frac{dr}{r} \leqq \frac{A_0}{2^{n-1}} \int_\Delta |\nabla u|^n \, d\tau, \tag{18}$$

where A_0 is as in (1.5.13).

For each $\alpha > 0$ let $v_\alpha = u$ for $x^n \geqq \alpha$ and let v_α be symmetric to u with respect to the plane $x^n = \alpha$ for $x^n < \alpha$, that is

$$v_\alpha(x) = u(x^1, \dots, x^{n-1}, \alpha + |x^n - \alpha|).$$

Then v_α is continuous and ACL in $B(r)$ $(r < c - 2\alpha)$ and lemma 2.5.4 gives for $\alpha < \dfrac{c-b}{2}$

$$\int_a^b (\operatorname{osc}_{S(r)} v_\alpha)^n \frac{dr}{r} \leqq \frac{A_0}{2^n} \int_{a<|x|<b} |\nabla v_\alpha|^n \, d\tau \leqq \frac{A_0}{2^n} \int_{\Delta_1} |\nabla u|^n \, d\tau + \frac{A_0}{2^n} \int_{\Delta_2} |\nabla u|^n \, d\tau, \tag{19}$$

where $\Delta_1 = \Delta \cap (x^n \geqq \alpha)$ and Δ_2 is a domain symmetrical with the domain $\{a < |x| < b\} \cap \{x^n \geqq \alpha\}$ with respect to the plane $x^n = \alpha$. On the other hand,

$$\operatorname{osc}_{S_r} u \leqq \varliminf_{\alpha \to 0} \operatorname{osc}_{S(r)} v_\alpha. \tag{20}$$

whenever $r > 0$.

Indeed, let $\{x_m\}$, $\{y_m\}$ be two sequences of points such that $x_m, y_m \in S_r$ $(m = 1, 2, \ldots)$ and

$$\lim_{m\to\infty} \left|u(x_m) - u(y_m)\right| = \operatorname*{osc}_{S_r} u(x). \tag{21}$$

Since $u(x) = v(x)$ for $x^n \geqq \alpha$, it follows that for each natural number m,

$$\left|u(x_m) - u(y_m)\right| \leqq \operatorname*{osc}_{S(r)} v_\alpha(x), \ \alpha \leqq \alpha_m = \min(x_m^n, y_m^n)$$

hence

$$\left|u(x_m) - u(y_m)\right| \leqq \inf_{\alpha \leqq \alpha_m} \operatorname*{osc}_{S(r)} v_\alpha(x) \leqq \varliminf_{\alpha\to 0} \operatorname*{osc}_{S(r)} v_\alpha(x)$$

and letting $m \to \infty$

$$\operatorname*{osc}_{S_r} u = \lim_{m\to\infty} \left|u(x_m) - u(y_m)\right| \leqq \varliminf_{\alpha\to 0} \operatorname*{osc}_{S(r)} v_\alpha(x)$$

by (21). And now, we establish the inequality

$$\int_a^b \varliminf_{\alpha\to 0}\left[\operatorname*{osc}_{S(r)} v_\alpha\right]^n \frac{dr}{r} \leqq \varliminf_{\alpha\to 0} \int_a^b \left[\operatorname*{osc}_{S(r)} v_\alpha\right]^n \frac{dr}{r}.$$

Clearly $v_\alpha(x)$ is continuous with respect to the $n+1$ variables $x^1, \ldots, x^n$, α for $a < |x| < b$ and $0 \leqq \alpha \leqq \alpha_0 < \dfrac{c-b}{2}$, and then it reaches its maximum and its minimum in two points of this $(n+1)$-dimensional domain. We conclude the existence of a constant K such that

$$\operatorname*{osc}_{S(r)} v_\alpha(x) = \sup_{S(r)} v_\alpha(x) - \inf_{S(r)} v_\alpha(x) < K < \infty,$$

whence

$$\int_a^b \left[\operatorname*{osc}_{S(r)} v_\alpha\right]^n \frac{dr}{r} < K^n \log\frac{b}{a}.$$

Next, let $\{\alpha_m\}$, $\alpha_m \geqq 0$, be a sequence satisfying the condition

$$\lim_{m\to\infty} \int_a^b \left[\operatorname*{osc}_{S(r)} v_{\alpha_m}(x)\right]^n \frac{dr}{r} = \varliminf_{\alpha\to 0} \int_a^b \left[\operatorname*{osc}_{S(r)} v_\alpha(x)\right]^n \frac{dr}{r}.$$

Then, applying the following Fatou's lemma (see C. Carathéodory [3] p. 443): "If $\{u_m(x)\}$ is a sequence of summable functions over the set E and such that $|u_m(x)| \leqq S(x)$, where $\int_E S(x)\,d\tau < \infty$ and $\varliminf_{m\to\infty} \int_E u_m(x)\,d\tau < \infty$, then $\int_E \varliminf_{m\to\infty} u_m(x)\,d\tau \leqq$ $\leqq \varliminf_{m\to\infty} \int_E u_m(x)\,d\tau$," yields

$$\int_a^b \varliminf_{\alpha\to 0} \left[\operatorname{osc}_{S(r)} v_\alpha\right]^n \frac{dr}{r} = \int_a^b \varliminf_{m\to\infty} \left[\operatorname{osc}_{S(r)} v_{\alpha_m}\right]^n \frac{dr}{r} \leqq$$

$$\leqq \varliminf_{m\to\infty} \int_a^b \left[\operatorname{osc}_{S(r)} v_{\alpha_m}\right]^n \frac{dr}{r} = \varliminf_{\alpha\to 0} \int_a^b \left[\operatorname{osc}_{S(r)} v_\alpha\right]^n \frac{dr}{r},$$

which combined with (20) and (19) and taking into account that $\lim_{\alpha\to 0} \Delta_1 = \lim_{\alpha\to 0} \Delta_2 = \Delta$, yields (18), as desired.

THEOREM 3. *Suppose that* $x^* = f(x)$ *is a* K-QCfH *of* $B(c) \cap \{x^n > 0\}$, *that* $\Delta = \{a < |x| < b < c,\ x^n > 0\}$ *and that* $\Delta^* = f(\Delta)$. *Then*

$$\int_a^b \left[\operatorname{osc}_{S_r} f(x)\right]^n \frac{dr}{r} \leqq \frac{A_0 n^{\frac{n}{2}} K^{n-1}}{2^{n-1}} m\,\Delta^*,$$

where $S_r = S(r) \cap \{x^n > 0\}$ *and* $\operatorname{osc}_{S_r} f(x) = \sup_{y, z \in S_r} |f(y) - f(z)|$.

Since $x^{*i}(x)$ $(i = 1, \ldots, n)$ are continuous and ACL, then the preceding lemma implies

$$\int_a^b \left[\operatorname{osc}_{S_r} x^{*i}(x)\right]^n \frac{dr}{r} \leqq \frac{A_0}{2^{n-1}} \int_\Delta |\nabla x^{*i}(x)|^n\,d\tau, \tag{22}$$

where A_0 is as in (1.5.13). Since $f(x)$ is a K-QCfH, $f(x)$ is differentiable with

$$|\nabla x^{*i}(x)|^n \leqq K^{n-1} |J(x)|, \tag{23}$$

a.e. in Δ, as it follows from Gehring's analytic definition characterized by (2.6.16) and from the inequality

$$|\nabla x^{*i}(x)| \leqq \Lambda_{x^{*i}}(x) \leqq \Lambda_f(x),$$

which holds in every point of differentiability. Now

$$\left[\operatorname{osc}_{S_r} f(x)\right]^2 \leqq \sum_{i=1}^n \left[\operatorname{osc}_{S_r} x^{*i}(x)\right]^2$$

and applying Minkowski's inequality we obtain

$$\int_a^b \left[\operatorname{osc} f(x)\right]^n \frac{dr}{r} = \int_a^b \left\{\left[\operatorname{osc} f(x)\right]^2\right\}^{\frac{n}{2}} \frac{dr}{r} \leqq \int_a^b \left\{\sum_{i=1}^n \left[\operatorname{osc} x^{*i}(x)\right]^2 \frac{1}{r^{\frac{2}{n}}}\right\}^{\frac{n}{2}} dr \leqq$$

$$\leqq \left(\sum_{i=1}^n \left\{\int_a^b \frac{[\operatorname{osc} x^{*i}(x)]^n}{r} dr\right\}^{\frac{2}{n}}\right)^{\frac{n}{2}} \leqq \frac{A_0 n^{\frac{n}{2}} K^{n-1}}{2^{n-1}} \int_\Delta |J(x)| \, d\tau = \frac{A_0 n^{\frac{n}{2}} K^{n-1}}{2^{n-1}} m\Delta^*.$$

by (22) and (23), as desired.

Now, combining the preceding theorem and theorem 2.9.1, we obtain, following Gehring [9], an n-space analogue of a theorem due to Lindelöf [1] which gives a relation between the sets $\Pi(\xi)$ and $C_c(\xi)$.

THEOREM 4. *If* $f(x)$ *is a* QCfH *of* $B(x, r)$, *then*

$$\Pi(f, \xi) = C_c(f, \xi) \tag{24}$$

for all $\xi \in S(x, r)$.

Fix $\xi_0 \in S(x, r)$. We perform first a Möbius transformation $x = \varphi(y)$ in R^n mapping $B(x, r)$ onto the half space $y^n > 0$ and sending ξ_0 into O. This transformation may be decomposed into a rigid motion, which carries $B(x, r)$ onto a ball tangent at O to the plane $y^n = 0$ and lying in the half space $y^n < 0$, and an inversion with respect to a sphere with the centre $(0, ..., 0, -2r)$ and the radius $2r$. Since $\Pi(f, \xi)$ and $C_c(f, \xi)$ are unaffected by a rigid motion, we may assume, without loss of generality, that $\xi_0 = 0$, that $B(x, r)$ is the ball centered at $x = (0, \ldots, 0, -2r)$ and that $x = \varphi(y)$ is the inversion from above. Next we show the invariance of $\Pi(f, \xi)$ and of $C_c(f, \xi)$ under the inversion $x = \varphi(y)$. But, on the one hand, every endcut γ of $B(x, r)$ from O is mapped onto an endcut of $y^n > 0$ from O, and conversely, every endcut γ' of $y^n > 0$ from O is mapped onto an endcut of $B(x, r)$ from O. On the other hand, since $f(x)$ is a homeomorphism, it follows that $C_\gamma(f, O) = C_{\gamma'}(F, O)$, where $F(y) = f[\varphi(y)]$. Hence $\Pi(f, \xi_0) = \Pi(F, O)$. Next, let $\{x_m\}$ be a sequence of points in $B(x, r)$, which converges in a cone to O. But this means (according to the definition) that there exists a constant $a \in [1, \infty)$ such that $|x_m| \leqq a \, d\,[x_m, S(x, r)]$, and then, clearly, the sequence $\{x_m\}$ is contained in a cone of $x^n < 0$ with the vertex at O, the x^n-axis as axis and with the angle between it and the generatrices equal to α, where $\cot \alpha = a$. The planes tangent to this cone are mapped by $x = \varphi(y)$ onto spheres which do not contain inside points of the sequence $\{x_m\}$ and are tangent to the cone symmetrical with the preceding one with respect to the plane $y^n = 0$ (see for instance theorem 2.16.18). If we suppose that the points of the sequence $\{x_m\}$ lie in a neighbourhood sufficiently small of O, then there exist numbers $a' > a$ such that $|y_m| < a' \, d\,(y_m, y^n = 0)$ and then $\{y_m\}$ converges in a cone (of $y^n > 0$) to O.

Conversely, suppose $\{y_m\}$ converges in a cone contained in $y^n > 0$ to O. Then the spheres lying outside the cone and tangent to it do not contain inside points of $\{y_m\}$ and are mapped by $x = \varphi(y)$ onto a family of planes enveloping a cone symmetrical with the preceding one with respect to the plane $y^n = 0$ and which contains inside all the points $x_m = \varphi(y_m)$, i.e. the sequence $\{x_m\}$ converges in a cone contained in $B(x, r)$. Hence, and since the conformal mappings are homeomorphisms, it follows that $C_c(F, O) = C_c(f, O)$. Thus it is sufficient to prove that

$$\Pi(F, O) = C_c(F, O). \tag{25}$$

Let γ be a segment joining O to a point of $y^n > 0$. Clearly γ is an endcut of $y^n > 0$ from O which lies in a cone, and then $\Pi(F, O) \subset C_\gamma(F, O) \subset C_c(F, O)$.

To complete the proof for (24) we must show that given $x_0^* \in C_c(F, O)$ and any endcut γ of $y^n > 0$ from O, there exists a sequence of points $\{y_k\}$, $y_k \in \gamma \cap \{y^n > O\}$ $(k = 1, 2, \ldots)$, $\lim\limits_{k\to\infty} y_k = O$, such that

$$\lim_{k\to\infty} F(y_k) = x_0^*. \tag{26}$$

Choose $x_0^* \in C_c(F, O)$ and let γ be an endcut of $y^n > 0$ from O. Then there exists a sequence of points $\{y_m'\}$ in $y^n > 0$ which converge to O and a constant $a \in [1, \infty)$, such that

$$\lim_{k\to\infty} F(y_k') = x_0^*, \quad |y_k'| \leqq a d(y_k', y^n > 0) \quad (k = 1, 2, \cdots). \tag{27}$$

Pick $c > 0$ so that the hemisphere $\Delta : B(c) \cap \{y^n > 0\}$ contains the y_k', and assume for the moment that $F(y)$ is a K-QCfH of Δ onto the bounded domain Δ^*. Next set

$$b = 1 - \frac{1}{2a}, \tag{28}$$

et Δ_k be the half spherical annulus $\{b|y_k'| < |y| < |y_k'|, y^n > 0\}$, and set $\Delta_k^* = F(\Delta_k)$. Since $\Delta_k^* \subset \Delta^*$ and Δ^* is bounded, hence $m\Delta^* < \infty$, it follows that

$$\lim_{k\to\infty} m\,\Delta_k^* = 0. \tag{29}$$

The preceding theorem implies that for each k

$$\int_{b|y_k'|}^{|y_k'|} \left[\operatorname*{osc}_{S_r} F(y)\right]^n \frac{dr}{r} \leqq \frac{A_0 n^{\frac{n}{2}} K^{n-1}}{2^{n-1}}\, m\Delta^*,$$

hence, for each k there exists an $r_k \in (b|y_k'|, |y_k'|)$ such that

$$\operatorname*{osc}_{S_{r_k}} F(y) \leqq \left(\frac{A_0 n^{\frac{n}{2}} K^{n-1} m\Delta^*}{-2^{n-1}\log b}\right). \tag{30}$$

Now $\gamma \cap S_{r_k} \neq \emptyset$ for $k \geqq k_0$; choose $y_k \in \gamma \cap S_{r_k}$ and let y_k'' denote the point where the radius Oy_k' meets S_{r_k}. Then (29) and (30) imply that

$$\lim_{k \to \infty} |F(y_k) - F(y_k'')| = 0. \tag{31}$$

On the other hand, we see from (27) and (28) that

$$|y_k'' - y_k'| \leqq |y_k'| - b|y_k'| \leqq \frac{1}{2} d(y_k', y^n > 0) \quad (k = 1, 2, \ldots),$$

and hence theorem 2.9.1 yields

$$|F(y_k'') - F(y_k')| \leqq \Theta_K\left(\frac{1}{2}\right) d[F(y_k'), F(y^n = 0)], \tag{32}$$

where K is the constant of K-QCf (quasiconformality) of the homeomorphism $f(x)$. Now let $F(y_k') \in \Delta^*$ $(k = 1, 2, \ldots)$, where Δ^* was supposed to be bounded. Hence the set $\{F(y_k')\}$ is bounded, too, and then $y_k' \to O \in \{y^n = 0\}$ implies

$$\lim_{k \to \infty} d[F(y_k'), F(y^n = 0)] = 0, \tag{33}$$

since on the one hand, clearly, $\{F(y_k')\}$ cannot have as limit point a point of the complement of $F(y^n > 0)$ and, on the other hand, if $\lim_{k \to \infty} F(y_k') = x_0^* \in F(y^n > 0)$, then, taking into account that F is a homeomorphism, $x_0^* = F(y_0)$, where y_0 is an inner point of $y^n > 0$, and then, there exists a closed ball $\overline{B(y_0, r)} \subset \{y^n > 0\}$ with d $[B(y_0, r), \{y^n = 0\}] = \delta > 0$ so that for sufficiently large indices $y_k' \in B(y_0, r)$, contradicting so the hypothesis that $y_k' \to O$. Then

$$\lim_{k \to \infty} F(y_k) = \lim_{k \to \infty} [F(y_k) - F(y_k'')] + \lim_{k \to \infty} [F(y_k'') - F(y_k')] + \lim_{k \to \infty} F(y_k') = x_0^*$$

follows from (27), (31), (32) and (33); but this, only under the assumption that Δ^* is a bounded domain. Suppose that this is not the case. Since $F(y)$ is a homeomorphism, there exists a ball $B(x^*, r)$ such that $\Delta^* \cap B(x^*, r) = \emptyset$. Let $\tilde{x} = \Phi(x^*)$ denote the inversion with respect to $B(x^*, r)$. Then $\Phi[F(y)]$ is a K-QCfH of Δ onto a bounded domain, and arguing as above, we can find a sequence of points $\{y_k\}$, $y_k \in \gamma \cap \{y^n > 0\}$, $y_k \to O$ such that

$$\lim_{k \to \infty} \Phi[F(y_k)] = \Phi(x_0^*). \tag{34}$$

Since $\Phi(x^*)$ is an inversion, (34) implies (26) also in this case. But from (26), we obtain $C_c(F, O) \subset \Pi(F, O)$, and on account of the opposite inclusion, we get (25), which — taking into account that $C_c(F, O)$ and $\Pi(F, O)$ are invariant under conformal mappings — yields (24) and the proof of the theorem 4 is complete.

The following immediate consequence of this theorem is an analogue of a very well known theorem due to Lindelöf [1] on bounded analytic functions.

COROLLARY. *If $f(x)$ is a* QCfH *of a ball $B(x, r)$ and if $f(x) \to x_1^*$ as $x \to \xi_1 \in S(x_0, r)$ along some endcut γ of $B(x_0, r)$, then $f(x) \to x_1^*$ as $x \to \xi_1$ in a cone.*

Indeed, the preceding theorem implies $\Pi(\xi_1) = C_\gamma(\xi_1) = C_c(\xi_1)$, and, since from the convergence of $f(x)$ to x_1^* along γ we conclude that $C_\gamma(\xi_1)$ reduces to a single point, it follows that also $C_c(\xi_1)$ reduces to a single point (which coincides with the preceding one), i.e. $f(x) \to x_1^*$ in a cone.

CHAPTER 4

Boundary correspondence of two domains QCf (quasiconformally) equivalent

Boundary correspondence induced by QCfH. In this paragraph we deal with some theorems (J. Väisälä [2]) about the extension of QCfH from a ball or a Jordan domain to the closed ball or closed Jordan domain, respectively. But we begin with some definitions.

A domain D is called *locally connected at a boundary point* ξ if for every neighbourhood U of ξ there exists a neighbourhood V of ξ such that any two distinct points in $V \subset D$ can be joined by an arc $\gamma \subset U \cap D$. If D is locally connected at each boundary point, it is said to be *locally connected on the boundary.*

THEOREM 1. *Let* $x^* = f(x)$ *be a* QCfH *of* B *onto a domain* D^* *and let* D^* *be locally connected on the boundary. Then* $f(x)$ *can be extended to a homeomorphism of* $\overline{B}$ *onto* $\overline{D^*}$.

We first prove, by *reductio ad absurdum*, that $f(x)$ can be extended to a continuous mapping $f_1(x)$ of $\overline{B}$ onto $\overline{D^*}$. If this were not the case, there would exist two sequences $\{a_m\}$, $\{b_m\}$ in B converging to the boundary point $a \in S$ such that their image sequences $\{a_m^*\}$, $\{b_m^*\}$ converge to two distinct boundary points a^*, b^* of D^*. Because D^* is locally connected at a^*, there exists a sequence $U_1^* \supset U_2^* \supset \ldots.$ of spherical neighbourhoods of a^* which converges to a^* and such that any two points in $U_{m+1}^* \cap D^*$ can be joined by an arc in $U_m^* \cap D^*$. In the same way, we can obtain a similar sequence $V_1^* \supset V_2^* \supset \ldots$ of spherical neighbourhoods of b^*. We may assume that

$$d(U_1^*), d(V_1^*) < \frac{2d(a^*, b^*)}{3}, \quad \overline{U_1^*} \cap \overline{V_1^*} = \emptyset, \; a_m^* \in U_{m+1}^*, \; b^* \in V_{m+1}^*.$$

Let γ^* be an arc which joins a_m^* and a_{m+1}^* in $U_m^* \subset D^*$. Denote

$$F_1^* = \bigcup_{m-1}^{\infty} \gamma_m^*.$$

The same procedure applied to b^* gives a set F_2^*. Let Γ^* be the family of all arcs which join F_1^* and F_2^* in D^* and let Γ be the pre-image of Γ^*. We shall prove that, in this case, $M(\Gamma) = \infty$ while $M(\Gamma^*) < \infty$, which will contradict the QCf of $f(x)$.

Let C_r be the spherical cap $B \cap S(a, r)$. For all $r < \min(|a_1 - a|, |b_1 - b|)$, C_r meets both F_1 and F_2, where F_1, F_2 are the pre-images of F_1^* and F_2^*, respectively, under $f(x)$. If $\rho \in F(\Gamma)$, then, by proposition 1.5.10,

$$\int_{C_r} \rho^n \, d\sigma > \frac{1}{A_0 r},$$

where A_0 is as in (1.5.13); integrating with respect to r yields

$$\int_B \rho^n \, d\tau = \infty,$$

hence $M(\Gamma) = \infty$. But f is a K-QCfH, hence

$$M(\Gamma) \leqq KM(\Gamma^*), \tag{1}$$

which implies $M(\Gamma^*) = \infty$. And now we shall prove in another way that $M(\Gamma^*) < \infty$. First suppose $a^*, b^* \neq \infty$ as we may, for if at least one of these points would be infinite, then the condition $a^*, b^* \neq \infty$ may be obtained by an inversion $X = \varphi(x^*)$, which is a conformal mapping in R^n and the continuity of $X = \psi(x) = \varphi[f(x)]$ at a would imply the continuity of $f(x) = \varphi^{-1}[\varphi(x)]$ at a. Next, let $\delta = d(U^*, V^*)$, where $d(U, V) = \inf\limits_{\substack{x \in U \\ y \in V}} d(x, y)$, and let $B(r^*)$ be a ball containing U_1^* and V_1^*, and set

$$\rho^*(x^*) = \begin{cases} \dfrac{1}{\delta} & \text{whenever } |x^*| \leqq r^*, \\ 0 & \text{in the rest.} \end{cases}$$

Clearly, $\int_{\gamma^*} \frac{ds}{\delta} \geqq 1$ whenever $\gamma^* \in \Gamma^*$, hence $\rho^* \in F(\Gamma^*)$. Then

$$M(\Gamma^*) \leqq \int_{B(r^*)} \rho^{*n} \, d\tau = \frac{mB(r^*)}{\delta^n} < \infty,$$

contradicting (1) and we conclude that f is QCf in B.

Next we show, again by *reductio ad absurdum*, that $f: \bar{B} \rightleftharpoons \overline{D^*}$. If not, there would exist two sequences $\{a_m\}$, $\{b_m\}$ in B converging to two distinct boundary points a, b, such that the image sequences $\{a_m^*\}$, $\{b_m^*\}$ converge to a single boundary point $a^* \in D^*$. We may assume that $a^* \neq \infty$ and that a and b lie in the planes $x^n = \pm h$, $h > 0$, respectively. Let F be the segment $x^k = 0$,

$(k=1,\ldots,n-1)$ $-\frac{h}{2} \leqq x^n \leqq \frac{h}{2}$ and let $F^* = f(F)$. Because D^* is locally connected at a^*, we can for each $r>0$ find an integer $m(r)$ such that $a^*_{m(r)}$ and $b^*_{m(r)}$ can be joined by an arc $\gamma^*_r \subset D^* \cap B(a^*, r)$. Obviously, $\lim_{r\to 0} m(r) = \infty$. Let Γ^*_r be the family of all arcs which join F^* and γ^*_r in D^* and let Γ_r be its pre-image in B. We will prove that as r tends to zero, $M(\Gamma^*_r) \to 0$, while $M(\Gamma_r)$ does not approach 0. This contradicts (1), which is a consequence of the QCf of $f(x)$, and proves the theorem.

Choose $r < d(a^*, F^*)$ and let A^*_r be the spherical ring $r < |x^* - a^*| < d(a^*, F^*)$. Then Γ^*_r is minorized (see Chapter 1.5) by the family Γ''_r of all arcs which join the boundary components of A^*_r in A^*_r. Hence, by corollary 1 of proposition 1.5.3 and (1.5.9),

$$M(\Gamma^*_r) \leqq M(\Gamma''_r) = \frac{n\omega_n}{\left[\log \frac{d(a^*, F^*)}{r}\right]^{n-1}} \to 0$$

as $r \to 0$.

In order to estimate $M(\Gamma_r)$, we choose r so small that a_m, b_m lie in the half spaces $x^n > \frac{h}{2}$, $x^n < -\frac{h}{2}$, respectively, for $m \geqq m(r)$. Suppose $\rho \in F(\Gamma_r)$. For $-\frac{h}{2} \leqq \alpha \leqq \frac{h}{2}$, the plane $\Pi_\alpha : x^n = \alpha$ meets both F and γ_r (the pre-image of γ^*_r). By proposition 1.5.10,

$$\int_{\Pi_\alpha \cap B} \rho^n d\sigma > \frac{4}{A_0}.$$

Integrating with respect to α yields

$$\int_B \rho^n \, d\tau > \frac{4h}{A_0}.$$

Thus, $M(\Gamma_r) \geqq \frac{4h}{A_0}$, whence $M(\Gamma_r)$ does not approach 0. The theorem is thus completely proved.

The reflection principle for QCfH.

THEOREM 2. *Let $x^* = f(x)$ be a K-QCfH of a ball $B(a_1, r)$ onto a ball $B(a^*_2, r^*)$. Let b_2, b^*_1 be the inverses of $f^{-1}(a^*_2)$ and $f(a_1)$ with respect to $S(a_1, r)$, $S(a^*_2, r^*)$, respectively. Then the mapping $x^* = F(x)$, obtained from $f(x)$ by continuous extension to the boundaries $S(a_1, r)$, $S(a^*_2, r^*)$, and by reflection with respect to $S(a_1, r)$ and $S(a^*_2, r^*)$, is a K-QCfH $F : R^n - b_2 \rightleftharpoons R^n - b^*_1$.*

According to the preceding theorem, the construction of $F(x)$ is always possible and $F(x)$ is a homeomorphism. $F(x)$ is even a K-QCfH in $\overline{B(a_1, r)}$, as it follows from corollaries 1 and 2 of theorem 2.7.1. Moreover, since $F(x)$ in the complement of $B(a_1, r)$ is of the form $\varphi^*\{f[\varphi(y)]\}$, where $y = \varphi^{-1}(x)$ is the inversion with respect to $S(a_1, r)$, $y^* = \varphi^*(x^*)$, the inversion with respect to $S(a_2^*, r^*)$, $b_1^* = \varphi^*[f(a_1)]$, $b_2 = \varphi^{-1}[f^{-1}(a_2^*)]$, $\varphi^{-1}(a_1) = \varphi^*(a_2^*) = \infty$, it follows that *$F(x)$ is a K-QCfH of $R^n - b_2$* onto *$R^n - b_1^*$.*

Remark. In the preceding theorem the K-QCfH $f(x)$ induces a K-QCfH $F: S^n \rightleftharpoons S^n$.

And now applying the preceding theorem (J. Väisälä [1]) we establish following F. Gehring [5], the Hölder condition of QCfH.

COROLLARY. *Let $f: B(R) \rightleftharpoons B(R)$, $f(O) = O$, be a K-QCfH according to Gehring's geometric definition. Then $x^* = f(x)$ can be extended to a K-QCfH of the whole space onto itself, such that there exists a constant $C = C(R, K)$ for which*

$$C^{\frac{1}{K}}|x_1 - x_0|^K \leqq |x_1^* - x_0^*| \leqq C|x_1 - x_0|^{\frac{1}{K}} \tag{2}$$

for all $x_0, x_1 \in S^n$.

We can extend $f(x)$ to a K-QCfH of the whole space onto itself by means of the preceding theorem. Next pick $x_0, x_1 \in B(R)$ and let A be the spherical annulus $|x_1 - x_0| < |x - x_0| < R + |x_0|$. Then clearly by (2.2.3)

$$\operatorname{mod} A = \log \frac{R + |x_0|}{|x_1 - x_0|} \geqq \log \frac{R}{|x_1 - x_0|}. \tag{3}$$

For an estimate of mod A^*, we observe that $x_0^*, x_1^* \in C_0^*$ and $C_1^* \cap S(R) \neq \emptyset$, because $C_1 \cap S(R) \neq \emptyset$. Thus C_0^* and C_1^* contain points which lie at distances $|x_1^* - x_0^*|$ and $R + |x_0^*|$ from x_0^* and we can apply lemma 2.9.10 to obtain

$$\operatorname{mod} A^* \leqq \log \psi\left(\frac{R + |x_0^*|}{|x_1^* - x_0^*|}\right) \leqq \log \frac{\lambda^2(R + |x_0^*| + |x_1^* - x_0^*|)}{|x_1^* - x_0^*|} \leqq \log \frac{4\lambda^2 R}{|x_1^* - x_0^*|}.$$

The inequality (3) and the preceding one, taking also into account that $f(x)$ is a K-QCfH, yield

$$\log \frac{4\lambda^2 R}{|x_1^* - x_0^*|} \geqq \operatorname{mod} A^* \geqq \frac{\operatorname{mod} A}{K} \geqq \frac{1}{K} \log \frac{R}{|x_1 - x_0|},$$

which implies the right-hand side of the inequality (2) with $C = 4\lambda^2 R^{\frac{k-1}{k}}$, where λ is as in lemma 2.9.10.

The left-hand side of (2) is obtained by the same argument applied to the homeomorphism $x = f^{-1}(x^*)$ which is a K-QCfH, too.

Remark. This result was previously obtained in another way by Ju. G. Rešetnjak [1] and by B. V. Šabat [3].

Boundary correspondence of two Jordan domains. By a *Jordan domain* we mean a domain with the boundary homeomorphic to a sphere.

The following two theorems (J. Väisälä [2]) are consequences of theorem 1.

THEOREM 3. *Let D and D^* be two Jordan domains and suppose that D can be mapped* QCf *onto a ball. Then every* QCfH $f : D \rightleftharpoons D^*$ *can be extended to a homeomorphism* $f : \overline{D} \to \overline{D^*}$.

Let $\varphi(y)$ be a QCfH of B onto D. The homeomorphism $x^* = F(y) = f[\varphi(y)]$ is a QCfH as a composition of two QCfH and maps B onto D^*. Then from theorem 1, taking also into account that a Jordan domain is locally connected on the boundary (see R. L. Wilder [1], theorem 5.35, p. 66), it follows that the homeomorphisms $x = \varphi(y)$ and $x^* = F(y)$ can be extended to homeomorphisms $\varphi : \overline{B} \rightleftharpoons \overline{D}$ and $F : \overline{B} \rightleftharpoons \overline{D^*}$. But since $y = \varphi^{-1}(x)$ is a homeomorphism in $\overline{D}$, we conclude that also $x^* = f(x) = F[\varphi^{-1}(x)]$ is a homeomorphism of $\overline{D}$ onto $\overline{D^*}$.

COROLLARY. *A* QCfH $f : B \rightleftharpoons B$ *can be extended to a homeomorphism* $f : \overline{B} \rightleftharpoons \overline{B}$.

THEOREM 4. *Let D be a domain which is locally connected on the boundary without being a Jordan domain. Then D cannot be mapped* QCf *onto a ball.*

If there existed a QCfH f of D onto a ball $B(x^*, r)$, then, from theorem 1 it would follow that f can be extended to a homeomorphism $f : \overline{D} \rightleftharpoons \overline{B(x^*, r)}$, where $\partial D \rightleftharpoons S(x^*, r)$ would be a homeomorphism, too, and then D would be a Jordan domain, contradicting the hypotheses of the theorem.

We give some examples of domains which are not QCf equivalent to the ball:

(a) *The finite space R^n* (see Loewner's theorem, chapter 2.9).

(b) *The complement of any closed set of topological dimension at most $n - 2$.* This follows from the theorem of W. Hurewicz and H. Wallman ([1] theorem IV.4, p. 48), which asserts that such a set is connected and then cannot be the homeomorphic image of the complement of a sphere, which is disconnected (since it consists of the interior and the exterior of the sphere).

We recall that the *topological dimension of a set* is defined by an inductive method by W. Hurewicz and H. Wallman ([1], p. 24), as follows:

The empty set and only the empty set has dimension -1.

A space X has *dimension* $\leqq n$ $(n \geqq 0)$ *at a point* x if x has arbitrarily small neighbourhoods whose boundaries have dimension $\leqq n - 1$.

X has *dimension* $\leqq n$, $\dim X \leqq n$, if X has dimension $\leqq n$ at each of its points.

X has *dimension* n if $\dim X \leqq n$ is true and $\dim X \leqq n - 1$ is false.

We recall also (W. Hurewicz and H. Wallman ([1], theorem A, p. 27) that a set $E \subset X$ has $\dim E \leqq n$ if and only if every point of E has arbitrarily small neighbourhoods in X, whose boundaries have intersections with E of dimensions $\leqq n - 1$. It is obvious that the property of having dimension n is topologically invariant.

(c) *The domain $D-\gamma$ where D is a Jordan domain and γ is an arc lying in D except for one endpoint on ∂D.* Let x_0 be the endpoint of γ lying in D and set $\delta = d(x_0, \partial D)$ and $S(x_0, r)$, $0 < r \leqq \delta$. Suppose, to prove it is false, there would exist a QCfH of $D-\gamma$ onto B. But, on account of the preceding theorem, and of the fact that $D-\gamma$ is locally connected on its boundary $\partial D \bigcup \gamma$, this would be possible only if $D-\gamma$ were a Jordan domain. But this would imply, by theorem 3, that $\bar{D}$ is topologically equivalent to $\bar{B}$ and $\partial D \bigcup \gamma$ to S. But this homeomorphism maps $S(x_0, r)$ onto a Jordan surface $S^* \subset \bar{B}$ and containing inside the image of x_0. This contradicts the fact that the image of x_0 belongs to S, and then, proves there exists no QCfH of $D-\gamma$ onto B.

(d) *The domain $D = S^n - \overline{B(x_1, r_1)} \bigcup \overline{B(x_2, r_2)}$, where $n > 2$, and $B(x_1, r_1)$; $B(x_2, r_2)$ are disjoint balls whose boundaries have exactly one point x_0 in common.* The local connectedness of D at x_0 (as even on the whole boundary) is obvious. On the other hand, D is not a Jordan surface (i.e. is not the topological image of sphere), as follows from Jordan—Brower's separation theorem (see for instance R. L. Wilder [1], theorem 5.23, p. 63), which asserts that if σ is a Jordan surface, then $S^n - \sigma$ consists of just two disjoint domains of which σ is the common boundary, while ∂D separates S^n into three disjoint domains : $B(x_1, r_1)$, $B(x_2, r_2)$ and D.

(e) *The domain between two parallel planes.* This is a special case of (*d*).

Space analogue of Koebe's theorem. In this paragraph, we shall give following V. Zorič [1], [2], an n-space analogous for QCfH of a theorem due to P. Koebe [1] on the boundary correspondence of admissible points of conformal mappings of a disc. We begin with some preliminary definitions and results.

Let δ be the greatest lower bound of the lengths of the curves which join in D the sets $E_1 \subset D$ and $E_2 \subset D$, and let D', E_1', E_2' be the images of D, E_1, E_2, under an inversion with respect to a non-degenerate sphere of finite radius; let δ' be the greatest lower bound of the lengths of the curves joining in D' the sets E_1', E_2'. Then the *distance between the sets E_1, E_2 with respect to the domain D* is

$$d_D(E_1, E_2) = \begin{cases} \delta \text{ whenever } \delta' > 0, \\ 0 \text{ whenever } \delta' = 0. \end{cases}$$

Two endcuts γ_1, γ_2 of D from the same boundary point ξ are said to be *D-equivalent* if in every neighbourhood U_ξ of ξ we have $d_D(\gamma_1 \cap U_\xi, \gamma_2 \cap U_\xi) = 0$.

Remark. Since the distance $d_D(E_1, E_2)$ was defined by means of an arbitrary inversion, it follows that also two endcuts of D from ∞ may be D-equivalent. Thus the point ∞ will no more play a special role.

THEOREM 5. *If $f(x)$ is a K-*QCfH of $B(x.r)$, *then in $B(x, R)$ there is no sequence of arcs $\{\gamma_m\}$ with the endpoints converging to two different points of $S(x, R)$ and such that their images $\gamma_m^* = f(\gamma_m)$ shrink into a point.*

Assume, as we may, $B(x, R) = B(R)$. The theorem will be proved by *reductio ad absurdum*. Let then $\{\gamma_m\}$ be a sequence of arcs with endpoints converging to two points $\xi_1, \xi_2 \in S(R)$, $\xi_1 \neq \xi_2$, and suppose, to prove it is false, that γ_m^* shrinks into a point ξ_0^*. We may assume $\xi_0^* \neq \infty$, since otherwise, it is enough to perform a previous inversion. Let l be a segment of length $\dfrac{d(\xi_1, \xi_2)}{2}$, parallel to $\overline{\xi_1\xi_2}$ and symmetric with respect to the origin. And now, consider the sequence of arc families $\{\widetilde{\Gamma}_m\}$ which join in $B(r)$ the segment l with the corresponding arc γ_m, of index m. Without loss of generality, we may suppose that the endpoints of the arcs γ_m are sufficiently near to the points ξ_1, ξ_2, respectively, so that each of the two planes orthogonal to l at its endpoints meet all the arcs of the sequence $\{\gamma_m\}$. Then, by proposition 1.5.14, for each family $\widetilde{\Gamma}_m$, we have

$$M(\widetilde{\Gamma}_m) \geqq \frac{d(\xi_1, \xi_2)}{2R} = \varepsilon > 0.$$

Then, by inequality (2.1.18), which characterizes the K-QCfH according to Väisälä's definitions, we obtain

$$M(\widetilde{\Gamma}_m^*) \geqq \frac{M(\widetilde{\Gamma}_m)}{K} > \frac{\varepsilon}{K} \qquad (m = 1, 2, \ldots). \tag{4}$$

If $B(\xi_0^*, r_1^*)$, $B(\xi_0^*, r_2^*)$ $(0 < r_1^* r_2^*)$ are two balls which do not meet the image of l, then, since γ_m^* shrinks into ξ_0^*, it follows that for any $r_1^* > 0$ we can find an $m(r_1^*)$ so that $\gamma_m^* \subset B(\xi_0^*, r_1^*)$ whenever $m > m(r_1^*)$. But then, for $m > m(r_1^*)$, all the arcs $\widetilde{\gamma}_m^* \in \widetilde{\Gamma}_m^*$ will meet both spheres $S(\xi_0^*, r_1^*)$ and $S(\xi_0^*, r_2^*)$. If $\widetilde{\Gamma}_m'^*$ denotes the family of subarcs of arcs of $\widetilde{\Gamma}_m^*$ joining $S(\xi_0^*, r_1^*)$ with $S(\xi_0^*, r_2^*)$, then it minorizes $\widetilde{\Gamma}_m^*$, i.e. $\widetilde{\Gamma}_m'^* < \widetilde{\Gamma}_m^*$. Hence, by corollary 1 of proposition 1.5.3,

$$M(\widetilde{\Gamma}_m^*) \leqq M(\widetilde{\Gamma}_m'^*). \tag{5}$$

Moreover, $\widetilde{\Gamma}_m'^*$ is contained in the family $\widetilde{\Gamma}^*$ of all arcs joining $S(\xi_0^*, r_1^*)$ and $S(\xi_0^*, r_2^*)$, and then, corollary 2 of proposition 1.5.3, on account of (1.5.9), implies that

$$M(\widetilde{\Gamma}_m'^*) \leqq M(\widetilde{\Gamma}^*) = \frac{n\omega_n}{\left(\log \dfrac{r_2^*}{r_1^*}\right)^{n-1}} \quad \text{whenever } m > m(r_1^*),$$

which, if

$$r_2^* = r_1^* e^{\left(\frac{2n\omega_n K}{\varepsilon}\right)^{\frac{1}{n-1}}}$$

and on account of (5), yields

$$M(\tilde{\Gamma}^*_m) \leqq \frac{\varepsilon}{2K} \text{ whenever } m > m(r^*_1),$$

contradicting so (4) and the proof is complete.

THEOREM 6. *If* $f: B(x, R) \rightleftharpoons D^*$ *is a* K-QCfH, *then* $d_B(E_1, E_2) = 0$ *implies* $d_{D^*}(E^*_1, E^*_2) = 0$, *where* $E_1, E_2 \subset B(x, R)$.

$d_B(E_1, E_2) = 0$ implies $E_1 \cap E_2 \neq \emptyset$, which, by proposition 1.5.15, implies $M(\Gamma) = \infty$, where Γ is the arc family joining E_1 and E_2 in $B(x, R) - E_1 - E_2$. The continuity of $f(x)$ implies $E^*_1 \cap E^*_2 \neq \emptyset$ and its K-QCf implies

$$M(\Gamma^*) = \infty. \tag{6}$$

Now suppose, to prove it is false, that $d_{D^*}(E^*_1, E^*_2) \neq 0$. This would mean that the greatest lower bound of the lengths of the arcs $\gamma^* \in \Gamma^*$ would be a number $s > 0$. But then, for $\rho = \frac{1}{s}$ we should have $\int_\gamma \rho ds \geqq 1$, and then $\rho \in F(\Gamma^*)$, hence

$$M(\Gamma^*) \leqq \int_{D^*} \rho^n \, d\tau = \frac{mD^*}{s^n} < \infty,$$

contradicting (6). Thus $d_{D^*}(E^*_1, E^*_2) = 0$. We assumed $mD^* < \infty$, because it is possible to obtain it by performing a previous suitable inversion, which leaves the distance unaffected.

THEOREM 7. *Let* $f: B(x, R) \rightleftharpoons D^*$ *be a homeomorphism and let* $\gamma_1, \gamma_2 \subset B(x, R)$ *be two* $B(x, R)$*-equivalent arcs. If* $f(x)$ *has a limit as* x *approaches* $S(x, R)$ *along each arc* γ_1, γ_2, *then* γ^*_1 *and* γ^*_2 *are* D^**-equivalent.*

Assume $B(x, R) = B$. By the preceding theorem, $d_{D^*}(\gamma^*_1, \gamma^*_2) = 0$. Hence γ^*_1, γ^*_2 are endcuts of D^* from the same point x^*_0, and we may suppose $x^*_0 \neq \infty$, since otherwise we obtain it by performing a previous inversion, which affects neither d_{D^*} nor the K-QCf. Let U^* be a neighbourhood of x^*_0, let $x_0 = f^{-1}(x^*_0)$ and let V' be a set defined as follows: the part of V' contained in B is given by $f^{-1}(U^* \cap D^*)$, the part of V' lying outside S is obtained by taking on each ray from the origin the symmetric with respect to S of the corresponding point of $f^{-1}(U^* \cap D^*)$ lying on this ray, and finally the common boundary point of both subsets of V' from above give the part of V' lying on S. Let V denote the interior of the set V' obtained as the union of the three sets from above. V is a neighbourhood of x_0, since otherwise, it would exist a sequence $x'_m \to x_0$, $x'_m \bar{\in} V$ and $x'_m \in B$ $(m = 1, 2, \ldots)$. Let $\{x''_m\}$ be a sequence such that $x''_m \to x_0$, $x''_m \in \gamma_1$. Clearly $|x'_m - x''_m| \to 0$, which, since $f(x)$ is continuous, implies $|f(x'_m) - f(x''_m)| \to 0$, where, by hypothesis, $f(x'_m) \to x^*_0$. But this would mean that, for a number m sufficiently large, $f(x') \in U^* \cap D^*$, contradicting so the hypothesis $x'_m \bar{\in} U \cap B = f^{-1}(U^* \cap D^*)$. Hence, V is a neighbourhood of $x_0 = f^{-1}(x^*_0)$.

Let us show that $d_{D^*}(U^*\cap\gamma_1^*, U^*\cap\gamma_2^*) = 0$. From above we now see that V is a neighbourhood of x_0 and that $V\cap\gamma_1 = f^{-1}(U^*\cap\gamma_1^*)$, $V\cap\gamma_2 = f^{-1}(U^*\cap\gamma_2^*)$. The B-equivalence of γ_1, γ_2 then implies $d_{D^*}(U^*\cap\gamma_1^*, U^*\cap\gamma_2^*) = 0$.

The pair (ξ, γ) of a boundary point ξ of a domain D and an endcut γ of D from ξ is called an *accessible boundary point of D*. Two accessible boundary points (ξ_1, γ_1) and (ξ_2, γ_2) of a domain D are considered as identical if and only if γ_1 and γ_2 are D-equivalent.

From the preceding definition, it follows that it is possible to identify an accessible point (ξ, γ) with the class of the D-equivalent endcuts of D from ξ. Thus, an accessible point (γ, ξ) as well as the corresponding class of D-equivalent arcs is completely determined by an arc of the class.

THEOREM 8. *By a K-QCfH $f: B(x, R) \rightleftharpoons D^*$, to each accessible point $(\xi_0^*, \gamma_0^*) \in \partial D^*$ it corresponds a unique point $\xi_0 \in S(x, R)$ such that if $x^* \to \xi_0^*$ along every arc γ_0^* which defines the accessible point (ξ_0^*, γ_0^*), then the corresponding point $x = f^{-1}(x^*)$ approaches ξ_0 and for different accessible points of ∂D^* correspond different points of $S(x, R)$. The subset of $S(x, R)$ corresponding to the set of the accessible boundary points of ∂D^* is everywhere dense in every continuum $F \subset S(x_1, R)$.*

Assume, as we may, $B(x_0, R) = B$ and set $\gamma_0 = f^{-1}(\gamma_0^*)$. The arc γ_0 has on S only one cluster point ξ_0, since otherwise it would be possible to find a sequence of arcs contained in B and with the endpoints converging to two points of S and such that their images under K-QCfH $f(x)$ shrink into a point, contradicting thus theorem 5. Clearly, if $x^* \to \xi_0^*$ along γ_0^*, then $x = f^{-1}(x^*) \to \xi_0$ along γ_0. We associate in this way to each accessible point (x_0^*, γ_0^*) a point $\xi_0 \in S$.

We must prove that this correspondence is the same no matter how we choose the endcut γ^* to determine the accessible point (ξ_0^*, γ_0^*). Indeed, let γ_1^*, γ_2^* be two endcuts corresponding to the same accessible point (ξ_0^*, γ_0^*) and let $\gamma_1 = f^{-1}(\gamma_1^*)$, $\gamma_2 = f^{-1}(\gamma_2^*)$. If γ_1, γ_2 were two endcuts of B from two different boundary points ξ_1, ξ_2, then, since $\gamma_1^*\cap U^*$ and $\gamma_2^*\cap U^*$ may be joined by arcs in $D^* - \gamma_1^* - \gamma_2^*$ of length as small as one pleases, it would follow that there exists a sequence of such arcs which shrink to ξ_0^* and their images under $x = f^{-1}(x^*)$ form a sequence of arcs with the endpoints converging to boundary points $\xi_1 \neq \xi_2$, thus contradicting theorem 5.

Now, if γ_1^*, γ_2^* were two endcuts of D^* from two different accessible boundary points, then clearly, $d_{D^*}(\gamma_1^*, \gamma_2^*) > 0$ and then, by theorem 7, the corresponding points $\xi_1, \xi_2 \in S$, would be different.

Finally, let us show that the set E of those points of S which correspond to the accessible points of ∂D^* is everywhere dense in every continuum of the sphere. Suppose, to prove it is false, it would exist a continuum $F \subset S$, which would contain no point of E. This would mean that the homeomorphism $f(x)$ would have no limit along any endcut of B from a boundary point of F, i.e. to such an endcut γ of B from $\xi_0 \in F$, there would correspond an arc of infinite length. But then, fix $\overline{xy} \subset B$, and let Γ be the arc family joining this segment with F in $B - \overline{xy}$. By proposition 1.5.9, $M(\Gamma^*) = 0$, contradicting so that $f(x)$ is K-QCfH and the proof of the n-space analogous of Koebe's theorem is complete.

Other results on boundary correspondence induced by QCfH. Before establishing, following V. Zorič [2], new results on the boundary correspondence of two domains QCf equivalent, we give some definitions and preliminary results.

A sequence of domains $\{U_m\}$, $U_m \subset D(m = 1, 2, \ldots)$ is said to be *regular* if

a) $U_{m+1} \subset U_m \quad (m = 1, 2, \ldots)$;

b) $\left(\bigcap_{m=1}^{\infty} \overline{U_m}\right) \subset \partial D$;

c) $\sigma_m = \partial U_m \cap D$ (the relative boundary of U_m in D) is a connected set;

d) $d_D(\sigma_m, \sigma_{m+1}) > 0$;

e) there exists at most an accessible boundary point of D which is an accessible boundary point for each domain of the sequence $\{U_m\}$.

Two sequences of domains $\{U'_m\}$, $\{U''_m\}$ are said to be *equivalent* if every term of each sequence contains all the terms of the other from a sufficiently large index on.

LEMMA 1. *Let* $\{U_m\}$ *be a sequence of subdomains of* $B(x, R)$ *satisfying* a) *and* b) (*from the above definition*) *and let* $\{x_k\}$ *be a sequence of* $B(x, R)$ *which converges to a point of the continuum* $F = \bigcap_{m=1}^{\infty} \overline{U_m}$; *then each domain* $U_m \in \{U_m\}$ *contains all the points* $x_p \in \{x_k\}$ *from a sufficiently large index* $p(m)$ *on.*

Assume $B(x, R) = B$. We prove the theorem by *reductio ad absurdum.* Suppose, to prove it is false, that there would exist a sequence $\{\tilde{x}_p\}$, $\tilde{x}_p \in B$, $\tilde{x}_p \bar{\in} U_m$ $(p = 1, 2, \ldots)$ for a certain m and that this sequence would converge to a point $\xi_0 \in F$. But then, since $\xi_0 \in \partial U_m$ would be an accumulation point of points $\tilde{x}_p \bar{\in} U_m$, it would follow that $\xi_0 \in \sigma_m$, and, by the same argument, $\xi_0 \in \sigma_{m+1}$, hence $d_B(\sigma_m, \sigma_{m+1}) = 0$, thus contradicting condition d).

COROLLARY 1. *If* $\{U_m\}$ *is a sequence of subdomains of* $B(x_0, R)$ *satisfying a), b), d), then all the points of* $F = \bigcap_{m=1}^{\infty} \overline{U_m}$ *are accessible boundary points for each of the subdomains of the sequence.*

Indeed, from the preceding lemma, we conclude that for every endcut $x = \varphi(t)$ $(0 \leqq t < 1)$ of $B(x_0, R)$ from $\xi_0 \in F$, there is a point $x_m = \varphi(t_m)$ such that the arc $x = \varphi(t)$ $(t_m \leqq t < 1)$ lie in the corresponding domain $U_m \in \{U_m\}$, i.e. every point $\xi \in F$ is an accessible boundary point for each subdomain of the sequence.

COROLLARY 2. *A regulary sequence of subdomains of the ball shrinks into only one point, i.e.* $F = \bigcap_{m=1}^{\infty} \overline{U_m}$ *consists of only one point.*

This is an immediate consequence of the preceding corollary and of e).

COROLLARY 3. *Two regular sequences of subdomains of a ball are equivalent if and only if they shrink into the same boundary point of the ball.*

This is an immediate consequence of the preceding corollary and of the definition of the equivalence of two regular sequences.

THEOREM 9. *A* K-*QCfH* $f: B(x, R) \rightleftharpoons D^*$ *carries a regular sequence* $\{U_m\}$ *of subdomains of* $B(x, R)$ *into a regular sequence* $\{U^*_m\}$ *of subdomains of* D^*; *the* *pre-image of a regular sequence of subdomains of D^* is again a regular sequence of subdomains of $B(x, R)$ and equivalent regular sequences are carried into equivalent regular sequences.*

Suppose $B(x, R) = B$ and let $\{U_m^*\}$ be the image of a regular sequence $\{U_m\}$. In order to prove that $\{U_m^*\}$ is regular, too, it is sufficient to verify the conditions d) and e), because a), b), c) are satisfied since $\{U_m\}$ is regular and $f(x)$ is a homeomorphism.

We prove d) and e) by *reductio ad absurdum*. Suppose, to prove it is false, $d_{D^*}(\sigma_m^*, \sigma_{m+1}^*) = 0$. But then, it would be possible to find a sequence of arcs $\{\gamma_m^*\}$ which join σ_m^* and σ_{m+1}^* and shrinks into a point ξ_0^* (this point may be even ∞). As $\sigma_m^* \cap \sigma_{m+1}^* \neq \emptyset$ – for otherwise $\sigma_m \cap \sigma_{m+1} = \emptyset$ since $f(x)$ is a homeomorphism, and then $d_B(\sigma_m, \sigma_{m+1}) = 0$, contradicting thus the regularity of $\{U_m\}$ – it would follow $\xi_0^* \in \partial D^*$. Since $d_B(\sigma_m, \sigma_{m+1}) > 0$, it would be possible to find a subsequence of the sequence $\{\gamma_k\}$ joining σ_m with σ_{m+1}, such that the end points of its arcs converge to two different points of F, contradicting thus theorem 5.

Suppose now that e) would not hold, i.e. $F^* = \bigcap_{m=1}^{\infty} \overline{U_m^*}$ would contain at least two accessible boundary points. Then, theorem 8 would imply that $F = \bigcap_{m=1}^{\infty} \overline{U_m}$ would contain two different points, contradicting thus corollary 2 of the preceding lemma.

In order to prove the converse, it suffices again to show that d) and e) are satisfied by the sequence $\{U_m\}$ pre-image of the regular sequence $\{U_m^*\}$. We do it by *reductio ad absurdum*.

If d) did not hold, i.e. $d_B(\sigma_m, \sigma_{m+1}) = 0$, then, by theorem 6, $d_{D^*}(\sigma_m^*, \sigma_{m+1}^*) = 0$, contradicting so the regularity of $\{U_m^*\}$.

If e) did not hold, i.e. the continuum F did not reduce to a point, then it would contain, by theorem 8, at least two points $\xi_1, \xi_2, \xi_1 \neq \xi_2$, which correspond to two accessible points (ξ_1^*, γ_1^*), (ξ_2^*, γ_2^*) of D^*. The endcuts γ_1, γ_2 are from ξ_1, ξ_2 and since the sequence $\{U_m\}$ satisfies a), b), d), the corollary of the preceding lemma implies that on each arc

$$x_k + \varphi_k(t) \ (0 \leqq t < 1) \qquad (k = 1, 2) \tag{7}$$

there is by one point $x_{km} = \varphi_k(t_m)$ $(k = 1, 2)$, such that the arc $x_k = \varphi_k(t)$ $(t_{km} \leqq t < 1)$ $(k = 1, 2)$ lie in the corresponding domain $U_{km} \in \{U_{km}\}$. But since $x^* = f(x)$ is a homeomorphism, the images of the arcs (7) would be contained in $U_m^* = f(U_m)$, which would imply that the accessible boundary points (ξ_k^*, γ_k^*) $(k = 1, 2)$ of D^* would be accessible also for each subdomain $U_m^* \in \{U_m^*\}$, contradicting thus the condition e) for $\{U_m^*\}$. And thus the regularity of $\{U_m\}$ is established.

Finally, the assertion of the theorem that equivalent sequences are carried into equivalent sequences follows directly from the fact that $f(x)$ is a homeomorphism.

And now we introduce, following V. Zorič [2], the concept of boundary elements as follows:

A *boundary element* of a domain D is said to be the pair $(F, \{U_m\})$ which consists of a regular sequence $\{U_m\}$ of subdomains of D and the continuum $F = \bigcap_{m=1}^{\infty} \overline{U_m}$. Two *boundary elements* $(F', \{U'_m\})$ and $(F'', \{U''_m\})$ are considered as *identical* if and only if the two corresponding regular sequences $\{U'_m\}$ and $\{U''_m\}$ are equivalent.

Thus, to whatever of the equivalent regular sequences there corresponds one and the same boundary element.

Combining corollary 2 and 3 of the preceding lemma, the preceding theorem may be formulated also by means of boundary elements as follows:

COROLLARY 1. *For any* K-QCfH $f: B(x, R) \rightleftharpoons D^*$, *between the points of* $S(x, R)$ *and the boundary elements of* D^* *it is possible to establish a one-to-one correspondence such that to each boundary element* $(F^*, \{U^*_m\})$ *of* D^* *corresponds on* $S(x, R)$ *the point determined by the regular sequence* $\{U_m\} = f^{-1}(\{U^*_m\})$.

A regular sequence $\{D_m\}$ of subdomains of D is said *to be contained into the regular sequence* $\{U_m\}$ of subdomains of D, if for any integer p there exists an integer $m(p)$ such that $D_{m(p)} \subset U_p$.

A boundary element $(F, \{U_m\})$ is called *divisible* by a boundary element $(F', \{U'_m\})$ if the sequence $\{U'_m\}$ is contained in the sequence $\{U_m\}$. In this case, the boundary element $(F', \{U'_m\})$ is said to be a *divisor* of $(F, \{U_m\})$. This divisor is *proper* if the two boundary elements are not equivalent.

Remark. The concept of boundary element generalizes that of C. Carathéodory's [1] *prime end*; instead of the condition not to have proper divisors which was contained in the definition of the *prime end*, we have, in the definition of the boundary element, the condition e), i.e. the existence of at most one accessible boundary point of D which is accessible boundary point for each of the domains of the sequence $\{U_m\}$. Every prime end contains at most an accessible boundary point (C. Carathéodory [1], theorem 8, p. 353).

And now, in order to establish a necessary and sufficient condition for a domain to be QCf equivalent to a ball we introduce the minimal sequences:

A sequence $\{U_m\}$ of subdomains of D is said to be *minimal* if every sequence of subdomains of D contained in it is equivalent to it. Thus, a boundary element $(F, \{U_m\})$ with minimal $\{U_m\}$ does not possess proper divisor.

Remark. From the preceding definition it follows that the concept of boundary element is more general than that of prime end. Hence, the preceding corollary represents a generalization of C. Carathéodory's ([1], theorem 13, p. 350) theorem on the boundary correspondence under conformal mapping.

COROLLARY 2. *If a domain D is* QCf *equivalent to a ball, then every regular sequence of subdomains of D is minimal.*

Indeed, if there existed in D two non-equivalent regular sequences of subdomains contained into one another, by theorem 9, there would correspond to them in $B(x, R)$ two non-equivalent sequences contained into one another thus contradicting the corollaries 2 and 3 of the preceding lemma.

Remark. As it follows from examples, the condition of the preceding corollary is not sufficient.

And now, combining corollary 1 of the preceding theorem and lemma 1, we establish, following Zorič [2], two theorems on the boundary correspondence induced by a K-QCfH.

THEOREM 10. *A K-QCfH $f: B(x, R) \rightleftharpoons D^*$ induces a continuous mapping on $\overline{B(x, R)}$ if and only if to every regular sequence of subdomains of D^* only one point corresponds on ∂D^*.*

Suppose $B(x, R) = B$. The necessity is evident, for if this were not the case, the correspondence established by the preceding corollary would be no longer a continuous one on S if to each point of S there corresponded not a point of ∂D^*, but a whole continuum (of course nondegenerate in a point).

For the sufficiency, fix $\xi_0 \in S$ and let $(\xi_0^*, \{U_m^*\})$ be the boundary element corresponding to it by corollary 1 of the preceding theorem. The preceding lemma implies that in this case, the images of all sequences of points converging to ξ_0 will converge to a point ξ_0^* which we assume, as we may, to be different from ∞. Set $\xi_0^* = \bar{f}(\xi_0)$. In order to prove the continuity of the mapping $\bar{f}(x)$ which coincides in B with $x^* = f(x)$, clearly it suffices to prove its continuity on S. Let $\{\xi_m\}$ be an arbitrary sequence of points $\xi_m \in S$ $\xi_m \to \xi_0 \in S$. The continuity on S from inside, was proved above, hence, for $\xi_m \in \{\xi_m\}$ there exists a point $x_m \in B$ such that $|x_m - \xi_m| < \dfrac{1}{m}$ and

$$|\bar{f}(x_m) - \bar{f}(\xi_m)| < \frac{1}{m}. \tag{8}$$

The sequence $\{x_m\}$ consists of inner points of B, and clearly converges to the same point ξ_0 as $\{\xi_m\}$, and then, for every $\varepsilon > 0$, there exists an $m(\varepsilon)$ such that

$$|\bar{f}(x_m) - \bar{f}(\xi_0)| < \frac{\varepsilon}{2}$$

for all $m > m(\varepsilon)$. Hence and by (8), we obtain

$$|\bar{f}(\xi_m) - \bar{f}(\xi_0)| < \varepsilon,$$

whenever $m > \max\left[m(\varepsilon), \dfrac{2}{\varepsilon}\right]$, and this proves the continuity of $\bar{f}(x)$ in $\bar{B}$.

THEOREM 11. *A K-QCfH $f: B(x, R) \rightleftharpoons D^*$ induces a homeomorphism $\bar{f}: \overline{B(x, R)} \rightleftharpoons \overline{D^*}$ if and only if to each regular sequence of subdomains of D^* only one point corresponds on ∂D^* and non-equivalent regular sequences converge to different points of ∂D^*.*

The necessity is evident.

For the sufficiency, suppose $\overline{B(x, R)} = B$. The preceding theorem implies that $x^* = f(x)$ has a continuous extension $\bar{f} : \bar{B} \rightleftharpoons \overline{D^*}$. The univalence is a consequence of corollary 2 of theorem 9.

Remark. These results of V. Zorič [2] were generalized by him in [3] to a class of homeomorphisms more general than the K-QCfH, and which is characterized by the property that for each pair of connected sets $E_1, E_2 \subset D$, $d_D(E_1, E_2) = 0$ implies $d_{D^*}(E_1^*, E_2^*) = 0$.

In the rest of this paragraph, going deeply in the study of the problem of the boundary correspondence induced by a QCfH, we shall establish, following F. Gehring and J. Väisälä [1], the conditions a boundary must satisfy in order to obtain an extension of the considered QCfH to a QCfH of the closed domain (i.e. the domain and its boundary). We begin with some definitions and preliminary results.

We say that an $(n-1)$-dimensional manifold σ is a *free boundary surface* of D if

$$\sigma \subset \bar{D}, \sigma \cap \overline{\partial D - \sigma} = \emptyset. \tag{9}$$

As it is easy to see, such a surface is open or has no boundary.

LEMMA 2. *If $f(x)$ is continuous in D and if D is locally connected at $\xi \in \partial D$, then the cluster set $C(f, \xi)$ is a continuum.*

If for each m we let $E_m = D \cap B\left(\xi, \dfrac{1}{m}\right)$, then it follows that

$$F = C(f, \xi) = \bigcap_m \overline{f(E_m)}.$$

Clearly F is closed. Because D is locally connected at ξ, for each m we can find a p such that each pair of points in E_p can be joined in E_m. Thus each pair of points in F can be joined by a connected set in $\overline{f(E_m)}$. If F were not connected, we could find a bounded open set G such that both G and CG would contain points of F while $\partial G \cap F = \emptyset$. But $\partial G \cap \overline{f(E_m)} \neq \emptyset$ for all m, hence

$$\partial G \cap F = \bigcap_m \left[\partial G \cap \overline{f(E_m)}\right] \neq \emptyset$$

thus contradicting $\partial G \cap F = \emptyset$, whence F is connected.

LEMMA 3. *Suppose that σ is a free admissible boundary surface of D and that U is a neighbourhood of $\xi \in \sigma$. Then ξ has a neighbourhood $V \subset U$ such that the quasi-isometry i_ξ maps $D \cap V$ onto a hemiball H and $\sigma \cap V$ onto the plane part of ∂H.*

By definition ξ has a neighbourhood U_ξ in which i_ξ maps $\sigma \cap U_\xi$ onto a plane domain Π_ξ. Let $x^* = i_\xi(x)$. Assume first that $\sigma \neq \partial D$. Since σ is free, it follows that $d(\xi, \partial D - \sigma) \geqq d(\xi, \overline{\partial D - \sigma}) > 0$, and then there exists an $r > 0$ such that Π_ξ divides $B(\xi^*, r)$ into two open hemiballs H_0 and H_1, and so that

$$B(\xi^*, r) \subset i_\xi(U \cap U_\xi), \qquad B(\xi^*, r) \cap i_\xi[(\partial D - \sigma) \cap U_\xi] = \emptyset. \tag{10}$$

If $\sigma = \partial D$, then (10) holds trivially.

Let V, W_0, W_1 be the images of $B(\xi^*, r)$, H_0, H_1, respectively, under i_ξ^{-1}. Then $V \subset U$ and (9) implies that one of the sets W_0 or W_1, say W_1, contains points of $C\bar{D}$. Now (10) implies $\partial D \cap W_1 = \emptyset$ and hence $D \cap W_1 = \emptyset$. Thus $D \cap V = W_0$ and i_ξ maps $D \cap V$ onto $H = H_0$, as desired.

This lemma shows that a domain is locally connected at each point of a free admissible boundary surface.

COROLLARY. *A domain D which is not a Jordan domain and whose boundary is a free admissible surface is not QCf equivalent to a ball.*

This is a consequence of lemma 3 and of theorem 4.

And now a generalization of proposition 1.5.15.

LEMMA 4. *Suppose that σ is a free admissible boundary surface of D, that E_0 and E_1 are nondegenerate connected sets in D, so that $\bar{E}_0 \cap \bar{E}_1$ contain a point $\xi \in \sigma$. If Γ is the family of arcs which join E_0 and E_1 in $D - E_0 - E_1$, then $M(\Gamma) = \infty$.*

Let V be the neighbourhood of the preceding lemma with $U = R^n$, and let Γ_1 be the family of arcs which join $F_0 = E_0 \cap V$ and $F_1 = E_1 \cap V$ in $D \cap V - F_0 - F_1$.

Then

$$M(\Gamma) \geqq M(\Gamma_1) \geqq C(i_\xi)^{-2(n-1)} M(\Gamma_1^*), \tag{11}$$

by corollary 2 of proposition 1.5.3, (2.17.4) and lemma 2.17.2, and where $\Gamma_1^* = i_\xi(\Gamma_1)$ and $C(i_\xi)$ is the maximal distortion of the quasi-isometry i_ξ in V. Since the sets E_0 and E_1 are connected, we can find $a > 0$ such that the hemisphere $S_r = S[i_\xi(\xi), r] \cap H$ meets both $i_\xi(F_0)$ and $i_\xi(F_1)$ for $0 < r < a$. Let Γ_1^* be the family of arcs which join $i_\xi(F_0)$ and $i_\xi(F_1)$ inside S_r. Hence we can argue as in the first part of the proof of proposition 1.5.15, to conclude that $M(\Gamma_1^*) = \infty$. Thus, by (11), $M(\Gamma) = \infty$.

LEMMA 5. *For each $a > 0$, let Γ_a denote the modulus of the family of arcs which join the segments*

$$x^1 = \pm a, |x^2| \leqq 1, x^k = 0 \qquad (k = 3, \ldots, n)$$

in R^n. Then $aM(\Gamma_a)$ is monotonic increasing in a and

$$\lim_{a \to 0} aM(\Gamma_a) = K_1, \qquad 0 < K_1 < \infty. \tag{12}$$

Let us prove first that $0 < M(\Gamma_a) < \infty$. Let Γ'_a be the family of arcs which join the above segments in the ball $B(1+a)$. Then, proposition 1.5.14, taking into account that $\Gamma'_a \subset \Gamma_a$, implies

$$M(\Gamma_a) \geqq M(\Gamma'_a) \geqq \frac{2}{A_0 a} > 0. \tag{13}$$

If $\rho = \dfrac{1}{2a}$ in the cube $Q = \{x; |x^i| \leqq 1 + a\,(i = 1,\ldots, n)\}$ and $\rho = 0$ in the rest, then $\rho \in F(\Gamma_a)$, hence

$$M(\Gamma_a) \leqq \int_Q \rho^n \, d\tau = \frac{(1+a)^n}{(2a)^n} < \infty. \tag{14}$$

Next, for $a' > a$, $f(x) = \left(\dfrac{a'}{a} x^1, x^2, \dfrac{a'}{a} x^3, \ldots, \dfrac{a'}{a} x^n\right)$ is a homeomorphism of R^n onto itself. Combining lemmas 2.17.1, 2.17.2, and taking into account that $\Gamma_a^* = \Gamma_{a'}$, we obtain

$$M(\Gamma_a) \leqq K_0 M(\Gamma_a^*) = \frac{a'}{a} M(\Gamma_{a'}).$$

Thus $aM(\Gamma_a)$ is monotonic increasing in a and the limit in (12) exists; then (14) implies $K_1 \leqq M(\Gamma_1) \leqq 1 < \infty$ and (13) implies $K_1 \geqq \dfrac{2}{A_0} > 0$.

THEOREM 12. *Suppose that* $D, D^* \subset \{x^n > 0\}$, *that* Π, Π^* *are plane domains in* $x^n = 0$ *which are free boundary surfaces of* D *and* D^*, *respectively, and that* g *is a homeomorphism of* $D \cup \Pi$ *onto* $D^* \cup \Pi^*$ *which is a* QCfH *in* D. *Then the induced boundary mapping* $g_\pi : \Pi \rightleftharpoons \Pi^*$ *is a* QCfH *with maximal dilatation*

$$K(g_\pi) \leqq \min\left[K_I(g), K_O(g)\right]^{n-2}. \tag{15}$$

Since Π and Π^* are free boundary surfaces, $D_1 = D \cup \Pi \cup \tilde{D}$ and $D_1^* = D^* \cup \cup \Pi^* \cup \tilde{D}^*$ are domains, where $\tilde{D}$ and $\tilde{D}^*$ denote the symmetric images of D and D^* in $x^n = 0$. We can extend g by reflection to obtain a QCfH $g_1 : D_1 \rightleftharpoons D_1^*$ (on account of corollary 3 of theorem 2.7.1) with $K_I(g_1) = K_I(g)$, $K_O(g_1) = K_O(g)$ (by lemma 2.17.1). From the equivalence of the various definitions of QCfH, it follows that $\delta_L(x)$ is bounded in D. If $\delta_L^\pi(\xi)$ denotes the linear dilatation of $g(\xi)$ in Π, it follows that $\delta_L^\pi(\xi) \leqq \delta_L(\xi)$, hence, we conclude that $\delta_L^\pi(\xi)$ is bounded in Π and then $g(\xi)$ is a QCfH in Π.

To complete the proof of (15), it is sufficient to show that g_π has maximal dilatation

$$K(g_\pi) \leqq K_O(g)^{n-2}, \tag{16}$$

for then, from (2.17.2) and (2.17.3), it follows that

$$K(g_\pi) = K(g_\pi^{-1}) \leqq K_O(g^{-1})^{n-2} = K_I(g)^{n-2}.$$

Next by virtue of the analytic definition of QCfH between surfaces characterized by (2.17.18), where instead of the homeomorphism f and the surface σ we have the homeomorphism g_π and the plane domain Π, it follows that for (16), it is sufficient to show that

$$\frac{\Lambda_{g_\pi}(\xi)^{n-1}}{K_O(g)^{n-2}} \leqq J_G^\pi(x) \leqq K_O(g)^{n-2}\lambda_{g_\pi}(\xi)^{n-1} \tag{17}$$

at each A-point $\xi \in \Pi$. Fix such a point ξ_0. By performing preliminary similarity mappings in the plane $x^n = 0$, we may assume without loss of generality that $\xi_0 = g_\pi(\xi_0) = O$ and that

$$g_\pi(\xi) = \sum_{k=1}^{n-2} p_k \xi^k e_k + \xi^{n-1} e_{n-1} + o\left(\sum_{m=1}^{n-1} |\xi^m|\right), \tag{18}$$

where e_k is the unit vector on x^k-axis, $p_1 \geqq \ldots \geqq p_{n-2} \geqq 1$, $p_k = \dfrac{a_k}{a_{n-1}}$ and $a_1 \geqq \ldots \geqq a_{n-1}$ are proportional to the semi-axes of the $(n-2)$-dimensional infinitesimal ellipsoids mapped by g_π onto an $(n-2)$-dimensional infinitesimal sphere centered at O. Then, (17) comes to

$$\frac{p_1^{n-2}}{p_2 \ldots p_{n-2}} \leqq K_O(g)^{n-2}, \qquad p_1 \ldots p_{n-2} \leqq K_O(g)^{n-2}. \tag{19}$$

Set

$$\max\left(p_k, \frac{p_1}{p_{n-k}}\right) = q_k \quad (k = 1, \ldots, n-2), \quad p_{n-1} = 1. \tag{20}$$

To prove (19), it is sufficient to show that

$$q_k \leqq K_O(g) \quad (k = 1, \ldots, n-2). \tag{21}$$

For this, fix $\varepsilon > 0$, choose, as we may (according to the preceding lemma), a $b \in (0,1)$ so that to have for a fixed k,

$$q_k b M(\Gamma_{q_k, b}) \leqq K_1 + \varepsilon, \tag{22}$$

where $\Gamma_{q_k b}$ and K_1 are as in the preceding lemma, and choose $c > 0$ so that $B(c) \subset D_1$. Then for $u \in \left(0, \dfrac{c}{\sqrt{2}}\right)$, let E_0 and E_1 be the segments $x^1 = \pm bu$,

$|x^{n-k}| \leqq u, x^j = 0$ $(j = 3, \ldots, n - k - 1, n - k + 1, \ldots, n)$, and let Γ_1 and Γ_2 be the families of arcs which join E_0 and E_1 in D_1 and R^n, respectively. Then each arc $\gamma \in \Gamma_2 - \Gamma_1$ must contain a subarc which joins $S(\sqrt{2u})$ to $S(c)$. If Γ denotes the family of arcs joining both spheres, it follows $\Gamma < \Gamma_2 - \Gamma_1$, and then, by corollary 1 of proposition 1.5.3 and (1.5.9), we obtain

$$M(\Gamma_1 - \Gamma_1) < M(\Gamma) = \frac{n\omega_n}{\left(\log \dfrac{c}{\sqrt{2u}}\right)^{n-1}}$$

Since a homothety is a conformal mapping and since the modulus of an arc family is preserved by a conformal mapping, it follows that $M(\Gamma_b) = M(\Gamma_2)$, hence, on account of the preceding inequality, we may choose u_1 so that

$$M(\Gamma_b) = M(\Gamma_2) \leqq M(\Gamma_1) + \frac{\varepsilon}{q_k b} \tag{23}$$

for $u \in (0, u_1)$.

Next for $u > 0$ and $0 < t < \dfrac{q_k b}{2}$, let F_0 and F_1 be the sets of points which lie within a distance of tu of the segments $x^1 = \pm p_1 bu$, $|x^{n-k}| \leqq p_{n-k}u$, $x^j = 0$ $(j=3, \ldots, n - k-1, n - k +1, \ldots, n)$, and let Γ^k and Γ_3^k be the arc families which join in R^n both segments and F_0^k with F_1^k, respectively. Then by proposition 1.5.17,

$$\lim_{t \to 0} M(\Gamma_3^k) = M(\Gamma^k) = M(\Gamma_{q_k b}),$$

and we may fix $t > 0$ so that

$$M(\Gamma_3^k) \leqq M(\Gamma_{q_k b}) + \frac{\varepsilon}{q_k b} \tag{24}$$

for all $u > 0$.

Finally by (18), we may choose $u_2 > 0$ so that $E_0^* \subset F_0^k$ and $E_1^* \subset F_1^k$ for $u \in (0, u_2)$, hence $\Gamma_1^* \subset \Gamma_3^k$, and then

$$M(\Gamma_1^*) \leqq M(\Gamma_3^k) \tag{25}$$

for $u \in (0, u_2)$, where E_0^*, E_1^* and Γ_1^* are the images of E_0, E_1 and Γ_1 under the homeomorphism g_1. Fix u so that $0 < u < \min(u_1, u_2)$. Then we can combine (23), (24), (25) and lemma 2.17.2, to obtain

$$M(\Gamma_1) \leqq K_O(g) M(\Gamma_1^*)$$

and

$$M(\Gamma_b) \leqq K_O(g)M(\Gamma_{q_k b}) + \frac{\varepsilon}{q_k b}[K_O(g) + 1],$$

hence, by (22) and

$$K_1 \leqq bM(\Gamma_b),$$

we get

$$q_k K_1 \leqq K_O(g) K_1 + \varepsilon[1 + 2K_O(g)].$$

Letting $\varepsilon \to 0$ then yields (21). Multiplying these inequalities for $k = 1,\ldots, n-2$ and taking into account (20), we obtain (19), which implies (as we have seen) (17), and then (15), as desired.

THEOREM 13. *Suppose $f : D \to R^n$ is a* QCfH, *that σ and σ' are free admissible boundary surfaces of D and D^*, respectively, and that*

$$C(f, \xi) \cap \sigma' \neq \emptyset \tag{26}$$

for each $\xi \in \sigma$. Then f can be extended to be a homeomorphism of $D \cup \sigma$ onto $D^* \cup \sigma^*$, where σ^* is an admissible surface contained in σ'. The induced boundary mapping f_σ: $\sigma \to \sigma^*$ is a QCfH *of σ onto σ^* with the maximal dilatation*

$$K(f_\sigma) \leqq \min [K_I(f), K_O(f)]^{n-2}. \tag{27}$$

This bound for $K(f_\sigma)$ is sharp.

We begin by showing that $C(f, \xi)$ reduces to a single point for each $\xi \in \sigma$. Fix $\xi \in \sigma$ and suppose that $C(f, \xi)$ contains two distinct points. Then lemmas 2 and 3 imply that $C(f, \xi)$ is a continuum. Let us show that $C(f, \xi) \subset \partial D^*$. Indeed, fix $\xi_0^* \in C(f, \xi)$; then there exists a sequence of points $\{x_m\}$, $x_m \in D$, $x_m \to \xi_0 \in \partial D$ so that $f(x_m) \to \xi_0^*$, where clearly $f(x_m) \in D^*$ $(m = 1, 2,\ldots)$, and then $\xi_0^* \in \overline{D^*}$. But $\xi_0^* \bar{\in} D^*$, for otherwise $\xi_0 = f^{-1}(\xi_0^*) \in D$, and it would be possible to find a sufficiently large number N such that $x_m \in B(\xi_0, r)$, where $r = \frac{1}{2} d(\xi_0, \partial D)$, although $f(x_m) \to \xi_0^* = f(\xi_0)$, contradicting the hypothesis that f is a homeomorphism in D. We conclude that $C(f, \xi) \subset \partial D^*$ for $\xi \in \partial D$. Since $C(f, \xi)$ is a continuum, we can find a pair of distinct points $\xi_0^*, \xi_1^* \in C(f, \xi) \cap \sigma'$. Indeed, from (26) it follows that $C(f, \xi) \cap \sigma'$ contains at least a point ξ_0^*. If this were the single point of $C(f, \xi) \cap \sigma'$, we should deduce that $C(f, \xi) - \xi_0^* \subset \partial D^* - \sigma'$ and then, since σ' is a free boundary surface and $C(f, \xi)$ a continuum, (9) would imply $\sigma' \cap C(f, \xi) = \sigma' \cap \overline{C(f, \xi) - \xi_0^*} \subset \sigma' \cap \overline{\partial D^* - \sigma'} = \emptyset$, thus contradicting (26). Thus $C(f, \xi)$ must contain at least also another point ξ_1^*. Then there would be in D two sequences $\{x_{km}\}$ so that $x_{km} \to \xi$ and $x_{km}^* \to \xi_k^* (k = 0,1)$. Let U_0^* and U_1^* be two disjoint neighbourhoods of ξ_0^* and ξ_1^*, respectively. By lemma 3, there exists a neighbourhood

V_k^* of $\xi_k^*(k = 0,1)$ and a quasi-isometry i_k^* which maps $V_k^* \cap D^*$ onto the hemiball H_k and $\sigma' \cap V_k^*$ onto the plane part of H_k, sending ξ_k^* into the centre $\xi_k'(k = 0,1)$ of the ball. Clearly, the sequence $\{x_{km}^*\}$ is contained in V_k^*, except possibly a finite number of points. This sequence is mapped by i_k^* into the sequence $\{\xi_{km}'\}$, $\xi_{km}' \in H_k(m = 1, 2, \ldots)$, $\xi_{km}' \to \xi_m'$. Let $\delta_k > 0$ be the distance between the sequence $\{\xi_{km}'\}$ and the hemisphere σ_k of H_k, let r_k be the radius of H_k and let H_k' be the hemiball concentric to H_k, $H_k' \subset H_k$, and with the radius $r_k - \delta_k$. Set $E_k^* = i_k^{*-1}(H_k')$ $(k = 0,1)$ and let Γ^* be the family of arcs which join E_0^* and E_1^* in $D^* - E_0^* - E_1^*$, let Γ_H be the arc family joining the hemisphere which is the boundary of H_1 with the hemisphere which is the boundary of H_1' in $H_1 \cap CH_1'$, set $\Gamma_H^* = i_1^{*-1}(\Gamma_H)$ and let Γ_S be the family of arcs which join $S(\xi_1', r_1 - \delta_1)$ with $S(\xi_1', r_1)$ in the corresponding spherical ring. Clearly, $\Gamma_H^* < \Gamma^*$ and $\Gamma_H \subset \Gamma_S$, so that by virtue of the definition for admissible surfaces (Chapter 2.17), and combining corollary of the theorem 2.17.1, proposition 1.5.9 and corollaries 1 and 2 of proposition 1.5.3, we obtain

$$M(\Gamma^*) \leqq M(\Gamma_H^*) \leqq (1 + \varepsilon)^{2n} M(\Gamma_H) \leqq (1 + \varepsilon)^{2n} M(\Gamma_S) =$$

$$= \frac{n\omega_n (1 + \varepsilon)^{2n}}{\left(\log \dfrac{r_1}{r_1 - \delta_1}\right)^{n-1}} < \infty.$$

Let $E_k = f^{-1}(E_k^*)$ $(k = 0,1)$ and $\Gamma = f^{-1}(\Gamma^*)$. Hence, $M(\Gamma) = \infty$, by lemma 4. This contradicts the fact that f is a QCfH, and we conclude that $C(f, \xi)$ must reduce to a point $\xi^* \in \sigma'$ for each $\xi \in \sigma$.

We now extend f by setting $f(\xi) = \xi^*$ for $\xi \in \sigma$, where $\xi^* = C(f, \xi_0)$ is the single limit point of $f(\xi_0)$ as $x \to \xi_0 \in \sigma$. The extended function $f(x)$ is continuous in $D \cup \sigma$. Set $\sigma^* = f(\sigma)$. Clearly $\sigma^* \subset \sigma'$ and for each $\xi^* \in \sigma^*$ we have

$$C(f^{-1}, \sigma^*) \cap \sigma = \emptyset,$$

where $\xi \in f^{-1}(\xi^*)$. The above argument shows that also $C(f^{-1}, \xi^*)$ reduces to a single point ξ, and then f is a homeomorphism of $D \cup \sigma$ onto $D^* \cup \sigma^*$. By virtue of the definition of an admissible surface it follows that if σ' is admissible, then $\sigma^* \subset \sigma'$ is admissible, too. It is clear that σ^* is a free admissible boundary surface of D^*.

Indeed, $\sigma^* \subset \sigma' \subset \partial D^*$. It remains to prove that also the condition

$$\sigma^* \cap \overline{\partial D^* - \sigma^*} = \emptyset \tag{28}$$

holds. Suppose, to prove it is false, that $\xi_0^* \in \sigma^* \cap \overline{\partial D^* - \sigma^*}$. Then there would be a sequence of points $\{\xi_m^*\}$, $\xi_m^* \in \partial D^* - \sigma^* (m = 1, 2, \ldots)$, $\xi_m^* \to \xi_0^*$, and for each boundary point ξ_m of the sequence, there would be a sequence $\{x_{km}^*\}$, $x_{km}^* \in D^*(k = 1, 2, \ldots)$, $x_{km}^* \to \xi_m^*$. Then, from each sequence $\{x_{km}\}$, $x_{km} = f^{-1}(x_{km}^*)$

$(k = 1, 2, \ldots)$, it would be possible to pick a subsequence which converges to a boundary point of D. By relabelling we may assume that the original sequences $\{x_{km}\}$ converge to a boundary point of D, i.e. $x_{km} \to \xi_m \in \partial D$. Since $C(f, \xi)$ reduces to a single point for each $\xi \in \sigma$ and this point belongs to σ^*, it follows that $\xi_m \in \partial D - \sigma$. And now, let us show that $\{\xi_m\}$, or a subsequence of it, converges to ξ_0. If it were not so, i.e. if for instance $\xi_m \to \xi_1 \neq \xi_0$, it would be possible to pick a subsequence of the sequence $\{\xi_m\}$ and subsequences of the sequences $\{x_{km}\}$ (by relabelling we may assume that all these subsequences are the original sequences), so that

$$d\left[\left(\bigcup_{k,m} \{x_{km}\}\right), \xi_0\right] > \frac{d(\xi_0, \xi_1)}{2}, \tag{29}$$

i.e. the distance between the set of the points of this sequence of sequences and ξ_0 is larger than $\dfrac{d(\xi_0, \xi_1)}{2}$. Pick from $\{\xi^*_m\}$ the subsequence of the points corresponding to the sequence $\{\xi_m\}$ and having the property from above and then proceed analogously with the sequence $\{x^*_{km}\}$ corresponding to the new sequence $\{\xi_m\}$. (By relabelling we may assume that these conditions are fulfilled by the original sequences $\{\xi^*_m\}$, $\{x^*_{km}\}$). Then there exists a subsequence of $\{\xi^*_m\}$ (by relabelling we may assume it to be the original sequence) so that $d(\xi^*_0, \xi^*_m) < \dfrac{1}{2m}$. From each sequence $\{x^*_{km}\}$ we choose a point $x^*_{k_m m}$ such that $d(\xi^*_m, x^*_{k_m m}) < \dfrac{1}{2m}$. Clearly, $d(\xi^*_0, x^*_{k_m m}) < \dfrac{1}{m}$ and then $x^*_{k_m m} \to \xi^*_0$ as $m \to \infty$. Since $f(x)$ is a homeomorphism in $D \bigcup \sigma$ it follows that $x_{k_m m} \to \xi_0 \in \sigma$ as $m \to \infty$, thus contradicting (29). Hence, $\xi^*_0 \in \sigma^* \bigcap \overline{\partial D^* - \sigma^*}$ implies $\xi_m \to \xi_0 \in \sigma$ where $\xi_m \in \partial D - \sigma$, i.e. $\xi_0 \in \in \sigma \bigcap \overline{\partial D - \sigma} \neq \emptyset$. This contradicts the hypothesis that σ is a free boundary surface, and then proves (28), allowing us to conclude that σ^* is a free boundary surface, too.

We must now show that the induced boundary mapping is a QCfH of σ onto σ^* with maximal dilatation satisfying (27). Fix $\varepsilon > 0$, and for $\xi \in \sigma$ and $\xi^* = f_\sigma(\xi)$, choose, as we may (since σ and σ^* are admissible surfaces), neighbourhoods U_ξ and U_{ξ^*} so that (18) holds. Next let V_ξ and V_{ξ^*} be the neighbourhoods of lemma 3, chosen so that $B_\xi \subset U_\xi$, $V_{\xi^*} \subset U_{\xi^*}$ and so that $f: (D \bigcup \sigma) \bigcap V_\xi \to \to (D^* \bigcup \sigma^*) \bigcap V_{\xi^*}$. Finally, let H and H' be the hemiballs corresponding to $D \bigcap V_\xi$ and $D^* \bigcap V_{\xi^*}$, and let Π and Π' be the plane parts of ∂H and $\partial H'$. Then

$$g_\xi = i_{\xi^*} \circ f \circ i_\xi^{-1} \tag{30}$$

is a homeomorphism of $H \bigcup \Pi$ onto $H'' \bigcup \Pi'' \subset H' \bigcup \Pi'$, which is QCfH in H. Since Π and $\Pi'' \subset \Pi'$ are $(n-1)$-dimensional domains which are free boundary surfaces, we have essentially the situation in the preceding theorem. Thus the boundary mapping g_ξ^π is a QCfH of Π onto Π^* with maximal dilatation

$$K(g_\xi^\pi) \leqq \min\left[K_I(g_\xi^\pi), K_O(g_\xi^\pi)\right]^{n-2}. \tag{31}$$

But (30) implies that

$$K_I(g_\xi^\pi) \leqq K_I(i_{\xi^*})\, K_I(f)\, K_I(i_\xi^{-1}),\quad K_O(g_\xi^\pi) \leqq K_O(i_{\xi^*})\, K_O(f)\, K_O(i_\xi^{-1}),$$

hence (31) and (2.17.4) yield

$$K(g_\xi^\pi) \leqq C(i_{\xi^*})^{2(n-1)(n-2)}\, C(i_\xi^{-1})^{2(n-1)(n-2)} \min\left[K_I(f), K_O(f)\right]^{n-2}.$$

Since $f(x)$ is QCfH, from theorem 2.17.2 and (18) it follows that

$$\sup_{\xi\in\sigma} K(g_\xi^\pi) < \infty,\ \operatorname*{ess\,sup}_{\xi\in\sigma} K(g_\xi^\pi) \leqq (1+\varepsilon)^{4(n-1)(n-2)} \min\left[K_I(f), K_O(f)\right]^{n-2},$$

and hence $f_\sigma : \sigma \rightleftarrows \sigma^*$ is a K-QCfH whose maximal dilatation $K(f_\sigma)$ satisfies (27). Moreover, this bound cannot be improved as we conclude from the following example. Let $f(x) = (Kx^1, x^2, \ldots, x^n)$, $K > 1$. Then f maps $x^n > 0$ onto itself with $K_I(f) = K$ and $K_O(f) = K^{n-1}$, and if $f_\sigma : x^n = 0 \rightleftarrows x^n = 0$ denotes the restriction of $f(x)$ to the plane $x^n = 0$, we obtain $K_I(f_\sigma) = K$, $K_O(f_\sigma) = K^{n-2}$, hence the maximal dilatation of the boundary mapping f_σ is

$$K(f_\sigma) = \max\left[K_I(f_\sigma), K_O(f_\sigma)\right] = K^{n-2} = \min\left[K_g(f), K_O(f)\right]^{n-2},$$

and the proof of the theorem 13 is complete.

Extension of QCfH from two to three dimensions. This problem is the inverse of the problem considered in this chapter. First, L. Ahlfors and A. Beurling [1] solved the problem of the extension of QCfH from one to two dimensions (using for $n = 1$ what was called by us to be Gehring's metric definition of QCfH and also a characterization of QCfH by a compacity condition) [see (2.3.1)]. L. Ahlfors [1] solved it also in the case of the extension from a bidimensional plane to the half space (threedimensional). The problem of the extension from an $(n-1)$-space to an n-halfspace is *an open problem*, and both methods used in their proofs by Ahlfors and Beurling [1] and by Ahlfors [1] are not sufficient in higher dimensions.

CHAPTER 5

Existence theorems in the small for QCfH

For the bidimensional conformal mappings we have the *Riemann mapping theorem,* which shows that a plane domain D can be mapped conformally onto the unit disc if and only if D is simply connected and has at least two boundary points; such a domain D is *a fortiori* QCf equivalent to a disc. But there exists no analogue of the Riemann mapping theorem when $n > 2$. Moreover, two simply connected domains of R^n $(n > 2)$, in general, are not even QCf equivalent, as it follows also from the five examples of domains which are not QCf equivalent to a ball, i.e. the examples (*a*), (*b*), (*c*), (*d*) and (*e*) from the preceding chapter. This fact gives rise to more complicated problems connected with the existence of QCfH in higher dimensions. Thus we have two kinds of problems: in small (and in this chapter we deal with it) and in the large (the next chapter is devoted to its study). The results of this chapter were obtained by us in [10], [14].

Let $f \in C^3$ be a QCfH with $J(x) \neq 0$ in $D \subset R^n$. At every point of $D, f(x)$ carries an infinitesimal ellipsoid of the form

$$\alpha_{ik}(x)\, dx^i\, dx^k = d\rho^2$$

onto an infinitesimal sphere. The functions $\alpha_{ik}(x) = \delta_{pq} x_i^{*p}(x) x_k^{*q}(x)$, $\alpha_{ik} \in C^2$, are the *characteristics* of the QCfH $f(x)$.

We recall (see for instance G. Vrănceanu [1], p. 242) that a *Riemann space* V_n is characterized by the linear element

$$ds^2 = g_{ik}\, dx^i\, dx^k$$

where $g_{ik} = g_{ki}$ are continuous and possess first order partial derivatives. Riemann spaces generalize then the Euclidean spaces characterized by the linear element $ds^2 = \delta_{ik} dx^i dx^k$.

And now we prove

THEOREM 1. *Given a distribution of the characteristics* $\alpha_{ik} \in C^2$ *in a neighbourhood* V_0 *of a point* x_0, *there exists a* QCfH $f \in C^3$ *with the characteristics* $\alpha_{ik}(x)$ *and* $J(x) \neq 0$ *in a sufficiently small neighbourhood* U_0 *of* x_0 $(U_0 \subset V_0)$ *if and only if there exists a conformal mapping of a neighbourhood* $W_0 (W_0 \subset U_0)$ *of a Riemann space with the linear element*

$$ds^2 = \alpha_{ik}(x)\, dx^i\, dx^k \tag{1}$$

into a Euclidean n-space. The degrees of freedom of these QCfH are the same as for the corresponding conformal mappings.

The formula

$$\delta_{ik}\,dx^{*i}\,dx^{*k} = J(x)^{\frac{2}{n}}\,\alpha_{ik}(x)\,dx^i\,dx^k \tag{2}$$

characterizes the QCfH $f \in C^3$ with a set of characteristics $\alpha_{ik}(x)$ and with $J(x) \neq 0$, which transform a Euclidean neighbourhood of a point x_0 into a Euclidean neighbourhood of the point $f(x_0)$. Clearly, the characteristics (C) involved in the definition of the QCfH with a set of characteristics are uniquely determined by the characteristics $\alpha_{ik}(x)$ from above, and conversely, since both determine uniquely the corresponding infinitesimal ellipsoids.

But the same formula (2) characterizes also the conformal mappings which transform a neighbourhood of a point x_0 of a Riemann space with the metric (1) into a Euclidean neighbourhood of $f(x_0)$. And then, if there exists a homeomorphism $f \in C^3$ with $J(x) \neq 0$ and satisfying (2) in a neighbourhood U_0 of x_0, then this homeomorphism can be considered either as a QCfH $f: R^n \to R^n$, or as a conformal mapping of the neighbourhood U_0 provided with the Riemann metric given by (1), i.e. of a neighbourhood of a Riemann space, into a Euclidean one. Thus, the existence of the mentioned QCfH in U_0 implies the existence of the conformal mappings of the corresponding Riemannian neighbourhood and conversely.

The assertion on the degrees of freedom is obvious.

From the equivalence of all the definitions of QCfH $f \in C^3$, it follows that the preceding theorem still holds for whatsoever definition of QCfH.

COROLLARY. *Given a distribution of the characteristics $\alpha_{ik} \in C^2$ in a neighbourhood V_0 of a point $x_0 \in R^n$ $(n > 3)$, there exists a* QCfH *$f \in C^3$ with $J(x) \neq 0$ and the characteristics $\alpha_{ik}(x)$ in a sufficiently small neighbourhood U_0 of $x_0 (U_0 \subset V_0)$ if and only if the Riemann n-space* $(n > 3)$ *with the linear element* (1) *has the conformal curvature zero at least in a neighbourhood $W_0 (W_0 \subset U_0)$ of x_0.*

This corollary is an immediate consequence of the preceding theorem and of the following Schouten's one (see J. A. Schouten and D. J. Struik [1]):

Schouten's theorem. In order that a Riemann n-space $(n > 3)$ *be mapped conformally into a Euclidean one, it is necessary and sufficient that its conformal curvature tensor vanish.*

The necessity of the condition was proved by H. Weyl [1].

For $n = 3$, we have the following:

THEOREM 2. *Given a distribution of the characteristics $\alpha_{ik} \in C^2$ in a neighbourhood V_0 of a point $x_0 \in R^3$, a necessary and sufficient condition for the existence of a* QCfH *$f \in C^3$ with $J(x) \neq 0$ and the characteristics $\alpha_{ik}(x)$ in a sufficiently small neighbourhood U_0 of x_0 $(U_0 \subset V_0)$ is that the Riemann* 3*-space corresponding to the linear element* (1) *(where i, k take on only the values* 1, 2, 3*) have the conformal curvature zero in a neighbourhood W_0 of x_0 $(W_0 \subset U_0)$ and that*

$$\nabla_i R_{km} - \nabla_k R_{im} = \frac{1}{4}(\alpha_{im} \nabla_k R - \alpha_{km} \nabla_i R) \quad (i, k, m = 1, 2, 3), \tag{3}$$

where R_{ik} is Ricci tensor and R is the curvature of the considered Riemann space.

This is a consequence of theorem 1, on account of the following E. Cotton's ([1]–[3]) theorem:

Cotton's theorem. If $n = 3$, *then a Riemann space can be mapped conformally into a Euclidean one if and only if its conformal curvature vanishes and conditions* (3) *hold.*

Remark. Theorem 1, its corollary and theorem 2 are locally theorems of existence for QCfH $f \in C^3$ with a set of characteristics and $J(x) \neq 0$.

For QCfH with two sets of characteristics we have

THEOREM 3. *Given* $\alpha_{ik}, \alpha^*_{ik} \in C^2$ $(i, k = 1, \ldots, n)$ *in a neighbourhood of a point* x_0, *there exists a* QCfH $f \in C^3$ *with two sets of characteristics* $\alpha_{ik}(x)$, $\alpha^*_{ik}(x)$ $(i, k = 1, \ldots, n)$ *and* $J(x) \neq 0$ *in a neighbourhood* U_0 *of* x_0 $(U_0 \subset V_0)$ *if and only if there exists a conformal mapping of the neighbourhood* $W_0 (W_0 \subset U_0)$ *of a Riemann space with the linear element* (1) *into a Riemann space with the linear element*

$$ds^{*2} = \alpha^*_{ik}(x)\, dx^{*i}\, dx^{*k}, \tag{4}$$

i.e. we have

$$ds^{*2} = \sqrt[n]{J(x)^2}\, ds^2. \tag{5}$$

The degrees of freedom of these QCfH are the same as for the conformal mappings between the Riemann spaces with the metrics given by (1) *and* (4).

The theorem follows by the same argument as in theorem 1, because formula (5) characterizes on the one hand the QCfH $f \in C^3$ with two sets of characteristics $\alpha_{ik}(x)$, $\alpha^*_{ik}(x)$ $(i, k = 1, \ldots, n)$ and $J(x) \neq 0$, which transform a Euclidean neighbourhood U_0 of x_0 into a Euclidean domain and on the other hand, the same formula (5) characterizes the conformal mappings of the neighbourhood provided with the Riemannian metric given by (1) into a Riemann space with the metric given by (4). Thus the existence of the QCfH with two sets of characteristics implies the existence of the corresponding conformal mappings and conversely. But since the problem of the existence in small of the conformal mappings of a Riemann space into another one was solved by T. Y. Thomas [1] and O. Veblen [1], it follows that the preceding theorem solves also the problem of the existence in small of the QCfH $f \in C^3$ with two sets of characteristics.

Remark. We were obliged to consider only the very restrictive case $f \in C^3$ since in this case $x^{*i}_k \in C^2$, which implies $\alpha_{ik}, \alpha^*_{ik} \in C^2$ $(i, k = 1, \ldots, n)$, and these conditions of regularity are necessary for the existence of the curvature and of Ricci tensor.

CHAPTER 6

Coefficients of QCf (quasiconformality)

The problem of the existence of a QCfH between two arbitrary domains is not yet solved. However, some partial interesting results have been obtained by F. Gehring [5], J. Väisälä [2], B. V. Šabat [3], V. Zorič [2] and especially in the extensive study on this topic of F. Gehring and J. Väisälä [1]. If there exists a QCfH between two given domains (we deal only with the case when one of the domains is the unit ball), the following extremal problem arises: which is the smallest number K for which there exists a K-QCfH of the first domain onto the second. As we shall see in this chapter, the solution of this problem depends on the structure of the considered domains. The results of this chapter are an n-space analogous of F. Gehring and J. Väisälä's [1] results for $n = 3$.

Coefficients of QCf. Suppose that $D \subset R^n$ is a domain homeomorphic to the unit ball B. We set

$$K_I(D) = \inf_f K_I(f), \quad K_O(D) = \inf_f K_O(f), \quad K(D) = \inf_f K(f), \tag{1}$$

where the infima are taken over all homeomorphisms $f : D \rightleftarrows B$. We call these three numbers $K_I(D)$, $K_O(D)$ and $K(D)$ the *inner*, *outer* and *total coefficients of* QCf of D. When not necessary we shall suppress the word total in the case of $K(D)$.

From (2.17.1) and (2.17.3), we obtain

$$K_I(D) \leqq K_O(D)^{n-1}, \quad K_O(D) \leqq K_I(D)^{n-1}, \tag{2}$$

$$\max\left[K_I(D), K_O(D)\right] = K(D) \leqq \min\left[K_I(D), K_O(D)\right]^{n-1}. \tag{3}$$

If there is no K-QCfH of D onto B we put $K(D) = \infty$. The preceding inequalities imply that these three coefficients of QCf are simultaneously finite or infinite. In the former case we say that D is QCf *equivalent to a ball.*

The two problems we deal with in this chapter are the following: first determine what kinds of domains D are QCf equivalent to a ball; next given such a domain D, determine the coefficients of QCf $K_I(D)$, $K_O(D)$, $K(D)$. Concerning the second problem we shall obtain only upper and lower bounds for the coefficients of certain domains. It is easier to obtain upper bounds for a given domain, since it is only necessary to construct a suitable K-QCfH $f : D \rightleftarrows B$ and calculate $K_I(f)$, $K_O(f)$ and $K(f)$. The problem of obtaining significant lower bounds, i.e. which exceed 1 [because $1 \leqq K_I(D)$ $K_O(D)$, $K(D) < \infty$, and then 1 is always a lower bound] is much more difficult, since one must find lower bounds for the various dilatations of all $f: D \rightleftarrows B$. We do this by considering what happens to certain arc families under each homeomorphism $f: D \rightleftarrows B$ and then appealing to lemma 2.17.2.

In all this chapter by K-QCfH we mean K-QCfH according to Gehring's geometric definition.

In the last part of this chapter we prove (following Gehring and Väisälä [1]) that the space of all domains which are QCf equivalent to a ball has a natural metric, and that this metric space is complete and non-separable. In this space the distance between two domains D, D^* is defined as

$$d(D, D^*) = \inf \log K,$$

where the infimum is taken over all K-QCfH $f: B \rightleftarrows B$.

If $n = 2$, Riemann mapping theorem implies that $K_I(D) = K_O(D) = K(D) = 1$ for any simply connected domain with at least two boundary points. Let us see what are the domains $D \subset R^n$ $(n > 2)$ for which $K_I(D) = K_O(D) = K(D) = 1$. We begin by proving:

Lower semicontinuity of dilatations $K_I(f)$, $K_O(f)$, $K(f)$.

LEMMA 1. *Suppose that $\{f_m\}$ is a sequence of homeomorphisms of domains $D_m \subset R^n$, that each compact subset of a domain $D \subset R^n$ is contained in all but a finite number of D_m, and that f_m converges uniformly on each compact subset of D to a homeomorphism f of D. Then*

$$K_I(f) \leqq \varliminf_{m\to\infty} K_I(f_m), \quad K_O(f) \leqq \varliminf_{m\to\infty} K_0(f_m), \quad K(f) \leqq \varliminf_{m\to\infty} K(f_m). \tag{4}$$

Let A be a bounded ring with $\bar{A} \subset D$. Then $\bar{A} \subset D_m$ for $m \geqq m_0(A)$ and the uniform convergence of the sequence $\{f_m\}$ on each compact subset of D implies the uniform convergence of each component of ∂A_m^* to the corresponding component of ∂A_0^*, where $A^* = f(A)$, $A_m^* = f_m(A)$. Hence, by lemma 2.6.2, mod A_m^* converges to mod A^* and we have

$$\operatorname{mod} A \leqq \varliminf_{m\to\infty} \left[K_I(f_m)^{\frac{1}{n-1}} \operatorname{mod} A_m^*\right] = \operatorname{mod} A^* \varliminf_{m\to\infty} K_I(f_m)^{\frac{1}{n-1}}.$$

Since this inequality holds for all such A, we obtain the first inequality of (4). The proof for the other two is similar.

Extremal QCfH.

THEOREM 1. *If D is a domain with $K(D) < \infty$, then, for each $x_0 \in D$, there exist extremal* QCfH *$f_I, f_O, f: D \rightleftarrows B$, with $f_I(x_0) = f_O(x_0) = f(x_0) = O$ for which*

$$K_I(f_I) = K_I(D), \quad K_O(f_O) = K_O(D), \quad K(f) = K(D). \tag{5}$$

Fix K so that $K(D) < K < \infty$. Then we may choose a decreasing sequence $\{K_m\}$ converging to $K(D)$ such that for each m there exists a K_m-QCfH $f_m : D \rightleftarrows B$. If $f_m(x_0) = \alpha \neq O$, we may perform the inversion

$$y + \frac{\alpha}{|\alpha|^2} = \left(1 - \frac{1}{|\alpha|^2}\right) \frac{x - \dfrac{\alpha}{|\alpha|^2}}{\left|x - \dfrac{\alpha}{|\alpha|^2}\right|^2}.$$

of the unit sphere S onto itself sending α into O. Let us show that the preceding mapping preserves S, i.e.

$$\left| -\frac{\alpha}{|\alpha|^2} + \left(1 - \frac{1}{|\alpha|^2}\right) \frac{e - \dfrac{\alpha}{|\alpha|^2}}{\left|e - \dfrac{\alpha}{|\alpha|^2}\right|^2} \right| = 1.$$

But the point $e \in S$ and the line containing the points O, α, $\dfrac{\alpha}{|\alpha|^2}$, determine a bidimensional plane cutting S along the unit circumference. Choosing in this plane the complex variable z, a point of the unit circumference is of the form $e^{i\vartheta}$, and then the preceding formula may be written as

$$\left| -\frac{\alpha}{|\alpha|^2} + \left(1 - \frac{1}{|\alpha|^2}\right) \frac{e^{i\vartheta} - \dfrac{\alpha}{|\alpha|^2}}{\left|e^{i\vartheta} - \dfrac{\alpha}{|\alpha|^2}\right|^2} \right| = 1,$$

It is a simple technical matter to verify it. Since e was an arbitrary point of S, it follows that the preceding inversion maps S onto itself. By composing f_m with this inversion, we obtain for each m a K-QCfH $\tilde{f}_m : D \rightleftarrows B$ sending x_0 into O.

Suppose, as we may, $K_m \leqq K\,(m = 1, 2, \ldots)$, hence $f_m(m = 1, 2, \ldots)$ are K-QCfH. Then, by corollary of theorem 2.13.2, there exists a subsequence $\{f_{m_k}(x)\}$ which converges to a homeomorphism $f : D \rightleftarrows B$, uniformly on compact subsets of D, so that $f_I(x_0) = O$. And now let A be an arbitrary ring, with $\bar{A} \subset D$. Since $f_m(x)$ are K-QCfH,

$$\frac{1}{K_m} \operatorname{mod} A \leqq \operatorname{mod} A_m^* \leqq K_m \operatorname{mod} A$$

for all m. Letting $m \to \infty$, it follows that $f(x)$ is $K(D)$-QCfH, hence, by the preceding lemma,

$$K(D) \leqq K(f) \leqq \varliminf_{m\to\infty} K(f_m) \leqq \varliminf_{m\to\infty} K_m = K(D),$$

and hence f satisfies the inequality (5). The proofs for the existence of f_0 and f_I follow exactly the same lines.

Remarks. 1. Combining theorem 2.16.2 and the preceding one, we conclude that $K(D) = 1$ if and only if D is either a ball or a half space. This result contrasts strikingly with case $n = 2$, where for each simply connected domain with at least two boundary points $K(D) = 1$. As a matter of fact, the importance of this chapter consists precisely in the fact that here the problems which arise are specific for the case $n > 2$.

2. Clearly

$$K_I(D) \geqq 1, \quad K_O(D) \geqq 1, \quad K(D) \geqq 1$$

for all $D \subset R^n$ and (3) implies that there is simultaneously equality or strict inequality in these inequalities.

Lower semicontinuity of the coefficients of QCf.

THEOREM 2. *Suppose that $\{D_m\}$ is a sequence of domains in R^n which contain the point x_0, that the D_m converge to their kernel D at x_0, and that $D \neq \{x_0\}, R^n$. Then*

$$K_I(D) \leq \varliminf_{m\to\infty} K_I(D_m), \quad K_O(D) \leq \varliminf_{m\to\infty} K_O(D_m), \quad K(D) \leq \varliminf_{m\to\infty} K(D_m).$$

We establish the first inequality. (The proof of the other two follows exactly the same lines.) For this let

$$K = \varliminf_{m\to\infty} K_I(D_m). \tag{6}$$

We may assume that $K < \infty$, for otherwise there is nothing to prove. Next for each m, let f_m be one of the extremal homeomorphisms of D_m onto B for which $K_I(f_m) = K_I(D_m)$. By choosing a subsequence and relabelling, we may assume that

$$K = \lim_{m\to\infty} K_I(f_m) \tag{7}$$

and that $K_I(f_m) \leqq K + 1$ for all m. Furthermore, by composing the f_m with suitable Möbius transformations of B onto itself as in the proof of the preceding theorem, we may assume that $f_m(x_0) = O$ $(m = 1, 2, \cdots)$. From (2.17.1) and

(2.17.2), it follows that $f_m (m = 1, 2, \ldots)$ are $(K+1)^{n-1}$-QCfH in D, and $D \subset R^n$ implies that D has a finite boundary point. Hence,we may apply theorems 1.2 and 1.1 to obtain a subsequence $\{f_{m_k}\}$ which converges to a homeomorphism $f : D \rightleftarrows B$, uniformly on each compact subset of D. Then, with the preceding lemma and (6), (7), we have

$$K_I(D) \leqq K_I(f) \leqq \varliminf_{k\to\infty} K_I(f_{m_k}) = K = \varliminf_{m\to\infty} K_I(D_m).$$

Upper bounds for the coefficients of QCf of starlike domains.

A domain D is said to be *starlike at a point* $x_0 \in D$ if the closed segment $\overline{xx_0} \subset D$ lies in D whenever $x \in D$.

Suppose that D is a domain starlike at O, and that $\xi_0 \in \partial D$. For each $\xi \in \partial D$, $\xi \neq \xi_0$, we let $\alpha(\xi, \xi_0)$ denote the acute angle which the segment $\overline{\xi\xi_0}$ makes with the ray from O through ξ_0 and we define

$$\alpha(\xi_0) = \varliminf_{\xi\to\xi_0} \alpha(\xi, \xi_0), \quad 0 \leqq \alpha(\xi_0) \leqq \frac{\pi}{2}. \tag{8}$$

If D has a tangent plane at ξ_0 whose normal forms an acute angle β with the ray from O through ξ_0, then $\alpha(\xi_0) = \dfrac{\pi}{2} - \beta$.

THEOREM 3. *Suppose that D is a domain which is bounded and starlike at O, and that* $\alpha(\xi) \geqq \alpha > 0$ *for all* $\xi \in \partial D$. *Then*

$$K_I(D) \leqq K(D) \leqq 2^{\frac{n-2}{2}} \cot\frac{\alpha}{2} \operatorname{cosec}^{n-2}\frac{\alpha}{2},$$
$$K_O(D) \leqq 2^{\frac{n-2}{2}} \cot\frac{\alpha}{2} \cos^{n-2}\frac{\alpha}{2}. \tag{9}$$

Fix $a > 0$. Since D is bounded and starlike at the origin and since $\alpha(\xi) > 0$ for all $\xi \in \partial D$, each point $x \in D$, $x \neq O$, has a unique representation of the form

$$x = u\xi, \tag{10}$$

where $\xi \in \partial D$ and $u \in (0, 1)$. For each such x we define

$$x^* = f(x) = u^a f(\xi),\ f(\xi) = \frac{\xi}{|\xi|},\ f(O) = O. \tag{11}$$

Then f is a homeomorphism of R^n onto itself which carries D onto B. In order to calculate the Jacobian $J = \det \|x^{*i}_k\|$ of $f(x)$, we choose first the polar coordinates

$x = (u|\xi|, \alpha_1, \ldots, \alpha_{n-1})$ and $f(x) = (u^a, \alpha_1, \ldots, \alpha_{n-1})$, where the passage from the polar coordinates $r, \alpha_1, \ldots, \alpha_{n-1}$ to the Cartesian ones is given by

$$x^i = r \sin \alpha_1 \cdots \sin \alpha_{i-1} \cos \alpha_i \qquad (i = 1, \ldots, n), \tag{12}$$

Set $r = u|\xi|$ and $R = u^a$, and let us denote by J_{x^*x}, J_{x^*R}, J_{Rr}, J_{rx} the Jacobians of the transformations from the Cartesian coordinates x^* to the Cartesian coordinates x, from the Cartesian coordinates x^* to the corresponding polar coordinates, from them to the polar coordinates corresponding to x and from them to the Cartesian coordinates x. Then

$$J_{x^*x} = J_{x^*R} J_{Rr} J_{rx} =$$

$$= \frac{1}{u^{(n-1)a} \sin^{n-3} \alpha_1 \cdots \sin \alpha_{n-2}} \frac{|\xi|}{au^{a-1}} u^{n-1}|\xi|^{n-1} \sin^{n-3} \alpha_1 \cdots \sin \alpha_{n-2} = \frac{u^n|\xi|^n}{au^{na}},$$

and since $J = \dfrac{1}{J_{x^*x}}$, on account of the notations (10) and (11), we obtain

$$J = \frac{au^{na}}{|\xi|^n u^n} = a \frac{|f(x)|^n}{|x|^n}. \tag{13}$$

And now we shall estimate $\Lambda_f(x)$ and $\lambda_f(x)$ as follows [1]). Observe first that Pitagora's generalized theorem applied to the vectors $f(\xi_1)$, $f(\xi_2)$, where $\xi_1, \xi_2 \in \partial D$, gives

$$|f(\xi_1) - f(\xi_2)|^2 = 1 + 1 - 2.1.1. \cos \vartheta = 2(1 - \cos \vartheta), \tag{14}$$

where ϑ is the angle between the vectors $f(\xi_1)$ and $f(\xi_2)$, and then also between $f(x_1)$ and $f(x_2)$. But then, applying Pitagora's theorem to the vectors $f(x_1)$, $f(x_2)$, it follows:

$$|f(x_1) - f(x_2)|^2 = |f(x_1)|^2 + |f(x_2)|^2 - 2|f(x_1)|\,|f(x_2)| \cos \vartheta =$$

$$= |f(x_1)|^2 - 2|f(x_1)|\,|f(x_2)| + |f(x_2)|^2 + 2|f(x_1)|\,|f(x_2)|(1 - \cos \vartheta) = \tag{15}$$

$$= [|f(x_1)| - |f(x_2)|]^2 + |f(x_1)|\,|f(x_2)|\,|f(\xi_1) - f(\xi_2)|^2.$$

---

1) The corresponding calculus was communicated to me by F. Gehring in a letter.

But

$$|f(x_1)| - |f(x_2)| = u_1^a - u_2^a = \frac{|x_1|^a}{|\xi_1|^a} - \frac{|x_2|^a}{|\xi_2|^a} =$$

$$= |x_1|^a \left(\frac{1}{|\xi_1|^a} - \frac{1}{|\xi_2|^a} \right) + \frac{1}{|\xi_2|^a} (|x_1|^a - |x_2|^a) =$$

$$= \frac{|x_1|^a}{|\xi_1|^a |\xi_2|^a} (|\xi_2|^a - |\xi_1|^a) + \frac{1}{|\xi_2|^a} (|x_1|^a - |x_2|^a) =$$

$$= \frac{|x_1|^a a}{|\xi_1|^a |\xi_2|^a} (|\xi_2|^{a-1} + \varepsilon_1)(|\xi_2| - |\xi_1|) +$$

$$+ \frac{a}{|\xi_2|^a} (|x_1|^{a-1} + \varepsilon_2)(|x_1| - |x_2|) =$$

$$= \frac{|x_1|^a}{|\xi_1|^a} a \left(\frac{1}{|\xi_2|^a} + \frac{\varepsilon_1}{|\xi_2|^a} \right) (|\xi_2| - |\xi_1|) +$$

$$+ \frac{|x_1|^a}{|\xi_2|^a} a \left(\frac{1}{|x_1|} + \frac{\varepsilon_2}{|x_1|^a} \right) (|x_1| - |x_2|),$$

where $\varepsilon_1, \varepsilon_2 \to 0$ as $x_2 \to x_1$. Then choosing b, c, d, e so that

$$|f(\xi_1) - f(x_2)| = \frac{b|\xi_1 - \xi_2|}{|\xi_1|} = d\frac{|x_1 - x_2|}{|x_1|},$$
$$||\xi_1| - |\xi_2|| = c|\xi_1 - \xi_2|, \; ||x_1| - |x_2|| = e|x_1 - x_2|, \tag{16}$$

it follows

$$|f(x_1)| - |f(x_2)| = \pm \frac{|x_1|^a}{|\xi_1|^a} ac \left(\frac{1}{|\xi_2|} + \frac{\varepsilon_1}{|\xi_2|^a} \right) |\xi_1 - \xi_2| \pm$$

$$\pm \frac{|x_1|^a}{|\xi_2|^a} ae \left(\frac{1}{|x_1|^a} + \frac{\varepsilon_2}{|x_1|^a} \right) |x_1 - x_2| =$$

$$= a\frac{|x_1|^a}{|\xi_1|^a}\left[\pm c\frac{|\xi_1-\xi_2|}{|\xi_1|}\left(\frac{|\xi_1|}{|\xi_2|}+\frac{\varepsilon_1|\xi_1|}{|\xi_2|^a}\right)\pm\right.$$

$$\left.\pm e\frac{|\xi_1|^a}{|\xi_2|^a}\cdot\frac{|x_1-x_2|}{|x_1|}\left(1+\frac{\varepsilon_2}{|x_1|^{a-1}}\right)\right]=$$

$$= a\frac{|x_1|^a}{|\xi_1|^a}\left[\pm\frac{cd}{b}\left(\frac{|\xi_1|}{|\xi_2|}+\frac{\varepsilon_1|\xi_1|}{|\xi_2|^a}\right)\pm\right. \tag{17}$$

$$\left.\pm e\frac{|\xi_1|^a}{|\xi_2|^a}\left(1+\frac{\varepsilon_2}{|x_1|^{a-1}}\right)\right]\frac{|x_1-x_2|}{|x_1|} = a\frac{|f(x_1)|}{|x_1|}\left[\pm dm\left(\frac{|\xi_1|}{|\xi_2|}+\right.\right.$$

$$\left.\left.+\frac{\varepsilon_1|\xi_1|}{|\xi_2|^a}\right)\pm e\frac{|\xi_1|^a}{|\xi_2|^a}\left(1+\frac{\varepsilon_2}{|x_1|^{a-1}}\right)\right]|x_1-x_2| =$$

$$= a\frac{|f(x_1)|}{|x_1|}[\pm dm(1+\varepsilon_3)+e(1+\varepsilon_4)]|x_1-x_2|,$$

where $\varepsilon_3, \varepsilon_4 \to 0$ as $x_2 \to x_1$ and

$$m = \frac{c}{b} = \frac{\dfrac{||\xi_1|-|\xi_2||}{|\xi_1-\xi_2|}}{\dfrac{|f(\xi_1)-f(\xi_2)|}{\dfrac{|\xi_1-\xi_2|}{|\xi_1|}}} = \frac{||\xi_1|-|\xi_2||}{|\xi_1|\,|f(\xi_1)-f(\xi_2)|} = \cot\gamma, \tag{18}$$

hence, by (14)

$$d^2+e^2 = \frac{|f(\xi_1)-f(\xi_2)|^2|x_1|^2}{|x_1-x_2|^2}+\frac{||x_1|-|x_2||^2}{|x_1-x_2|^2} =$$

$$= \frac{|f(\xi_1)-f(\xi_2)|^2|x_1||x_2|+(|x_1|-|x_2|)^2}{|x_1-x_2|^2}+\varepsilon_5\frac{|f(\xi_1)-f(\xi_2)|^2|x_1||x_2|}{|x_1-x_2|^2} =$$

$$= \frac{2(1-\cos\vartheta)|x_1||x_2|+|x_1|^2+|x_2|^2-2|x_1||x_2|}{|x_1-x_2|^2}+$$

$$+\varepsilon_5\frac{|f(\xi_1)-f(\xi_2)|^2|x_1||x_2|}{|x_1-x_2|^2} = 1+\varepsilon_5\frac{2(1-\cos\vartheta)|x_1||x_2|}{|x_1-x_2|^2},$$

where $\varepsilon_5 \to 0$ as $x_2 \to x_1$. But clearly, $\dfrac{2(1-\cos\vartheta)\,|x_1|\,|x_2|}{|x_1-x_2|^2} \leqq 1$, hence $d^2+e^2 = 1+\varepsilon_6$, where $\varepsilon_6 \to 0$ as $x_2 \to x_1$. By above, it follows that $d = \sin\beta + \varepsilon_7$, $e = \cos\beta$, where $\varepsilon_7 \to 0$ as $x_2 \to x_1$. But then, by (16) and (17), formula (15) becomes of the form

$$\frac{|f(x_1)-f(x_2)|^2}{|x_1-x_2|^2} = \frac{[|f(x_1)|-|f(x_2)|]^2 + |f(x_1)||f(x_2)||f(\xi_1)-f(\xi_2)|^2}{|x_1-x_2|^2} =$$

$$= \left\{a\frac{|f(x_1)|}{|x_1|}[\pm dm(1+\varepsilon_3) \pm e(1+\varepsilon_4)]\right\}^2 + \frac{|f(x_1)||f(x_2)|d^2}{|x_1|^2} =$$

$$= \frac{|f(x_1)|^2}{|x_1|^2}\{d^2(1+\varepsilon_8) \pm a^2[e(1+\varepsilon_4) \pm dm(1+\varepsilon_3)]^2\} =$$

$$= \frac{|f(x_1)|^2}{|x_1|^2}\{\sin^2\beta(1+\varepsilon_9) + a^2[\cos\beta(1+\varepsilon_4) \pm m(\sin\beta+\varepsilon_7)]^2\} =$$

$$= \frac{|f(x_1)|^2}{|x_1|^2}[\sin^2\beta + a^2(\cos\beta \pm \cot\gamma\sin\beta)^2 \pm \varepsilon],$$

where $\varepsilon_8, \varepsilon_9, \varepsilon \to 0$ as $x_2 \to x_1$, hence, we conclude that

$$\Lambda_f(x_1) = \frac{|f(x_1)|}{|x_1|}\varlimsup_{x_2\to x_1}[\sin^2\beta + a^2(\cos\beta \pm \cot\gamma\sin\beta)],$$

$$\lambda_f(x_1) = \frac{|f(x_1)|}{|x_1|}\varliminf_{x_2\to x_1}[\sin^2\beta + a^2(\cos\beta \pm \cot\gamma\sin\beta)]. \tag{19}$$

But the same argument as in lemma 2.17.3, allows us to conclude that the maximum and the minimum of the expression in brackets are obtained for the values of β given by (2.17.11) and are equal to

$$\frac{1}{2}(a^2\operatorname{cosec}^2\gamma + 2a + 1)^{\frac{1}{2}} + \frac{1}{2}(a^2\operatorname{cosec}^2\gamma - 2a + 1)^{\frac{1}{2}},$$

and

$$\frac{a}{\frac{1}{2}(a^2\operatorname{cosec}^2\gamma + 2a + 1)^{\frac{1}{2}} + \frac{1}{2}(a^2\operatorname{cosec}^2\gamma - 2a + 1)^{\frac{1}{2}}}$$

respectively.

Next, let δ be the acute angle between the vectors $\xi_1 - \xi_2$ and $\xi_1 - \frac{\xi_2|\xi_1|}{|\xi_2|}$. Then combining (18), (11), (8) and appealling to the law of sines, we get

$$\cot \gamma = \frac{\|\xi_1| - |\xi_2\|}{\left|\xi_1 - \frac{\xi_2 |\xi_1|}{|\xi_2|}\right|} = \frac{\sin \delta}{\sin\left(\frac{\pi}{2} - \delta + \varepsilon_{10}\right)} = \frac{\cos\left(\frac{\pi}{2} - \delta\right)}{\sin\left(\frac{\pi}{2} - \delta + \varepsilon_{10}\right)} =$$

$$= \frac{\cos\left[\alpha(\xi_1, \xi_2) + \varepsilon_{11}\right]}{\sin\left[\alpha(\xi_1, \xi_2) + \varepsilon_{12}\right]} \leqq \frac{\cos\left[\alpha(\xi_1) + \varepsilon_{11}\right]}{\sin\left[\alpha(\xi_1) + \varepsilon_{12}\right]},$$

where $\varepsilon_{10}, \varepsilon_{11}, \varepsilon_{12} \to 0$ as $x_2 \to x_1$, hence, by (19),

$$\Lambda_f(x_1) \leqq \frac{|f(x_1)|}{|x_1|} A_1, \quad \lambda_f(x_1) \geqq \frac{|f(x_1)|}{|x_1|} \frac{a}{A_1}, \tag{20}$$

for all $x_1 \in D$, where A_1 is as in (2.17.8).

(2.1.4), (13) and (20) yield

$$\delta_L(x)^n = \overline{\lim_{r\to 0}} \left[\frac{L(x, r)}{l(x, r)}\right]^n \leqq \overline{\lim_{r\to 0}} \left[\frac{L(x, r)}{r}\right]^n \overline{\lim_{r\to 0}} \left[\frac{r}{l(x, r)}\right]^n =$$

$$= \frac{\Lambda_f(x)^n}{J(x)} \cdot \frac{J(x)}{\lambda_f(x)^n} \leqq \frac{A_1^n}{a} \cdot \frac{A_1^n}{a^{n-1}} = \frac{A_1^{2n}}{a^n},$$

hence (2.1.15) and lemmas 2.1.5 and 2.1.3 imply that $f(x)$ is ACL and that $f(x)$ is differentiable a.e. in D. From (13) it follows that for $0 < a < 1$, we have $J(x) > 0$ in D, hence (13), (20) and lemma 2.17.1 allow us to conclude that

$$K_I(f) \leqq \frac{A_1^n}{a^{n-1}}, \qquad K_O(f) \leqq \frac{A_1^n}{a}.$$

Finally, if we set $a = \sin \alpha$, we obtain, as in corollary 2 of lemma 2.17.3, the inequality (2.17.14), and then *a fortiori* the inequalities (9).

Upper bounds for bounded convex domains. A *convex domain* is a domain D that contains the line segment joining any two of its points.

THEOREM 4. *Suppose that* $0 < a \leqq b$, *that* D *is a convex domain, and that* $B(a) \subset D \subset B(b)$. *Then* (9) *holds with* $\alpha = \arcsin \frac{a}{b}$. *In particular,*

$$K_1(D) \leqq K(D) < n^{\frac{n}{2}} \left(\frac{b}{a}\right)^{n-1}, \qquad K_O(D) < 2^{\frac{n}{2}} \frac{b}{a}. \tag{21}$$

The hypotheses imply that D is bounded and starlike at O. Hence (9) with $\alpha = arc \sin \frac{a}{b}$, will follow if we show that $\alpha(\xi) \geqq arc \sin \frac{a}{b}$ for all $\xi \in \partial D$. For this fix $\xi \in \partial D$, let C_1 be the finite cone which consists of the union of all open segments $\overline{x\xi}$ with $x \in B(a)$, and let C_2 be the symmetric image of C_1 in ξ. Since D is convex and $B(a) \subset D$, it follows that $C_1 \subset D$ and $C_2 \subset C\overline{D}$. Thus $\partial D \cap C_1 = \partial D \cap C_2 = \emptyset$, and since $D \subset B(b)$, we conclude that

$$\alpha(\xi) \geqq \arc\sin\frac{a}{|\xi|} \geqq \arc\sin\frac{a}{b}.$$

Finally if $\alpha = arc \sin \frac{a}{b}$, then

$$2^{-\frac{n-2}{2}} \cot\frac{\alpha}{2} \operatorname{cosec}^{n-2}\frac{\alpha}{2} = \frac{2^{\frac{n}{2}} \cos^n \frac{\alpha}{2}}{\sin^{n-1}\alpha} < \frac{2^{\frac{n}{2}}}{\sin^{n-1}\alpha} = 2^{\frac{n}{2}}\left(\frac{b}{a}\right)^{n-1}$$

$$2^{\frac{n-2}{2}} \cot\frac{\alpha}{2} \cos^{n-2}\frac{\alpha}{2} = \frac{2^{\frac{n}{2}} \cos^n \frac{\alpha}{2}}{\sin\alpha} < \frac{2^{\frac{n}{2}}}{\sin\alpha} = 2^{\frac{n}{2}}\frac{b}{a},$$

hence, by (9), we obtain the less precise but simpler bounds in (21).

Lower bounds for the coefficients of QCf of certain domains. In general, it is much more difficult to obtain a significant lower bound for coefficients of QCf of a given domain D than it is to obtain an upper bound, since in order to obtain an upper bound for a given domain D, it is only necessary to calculate the dilatations of an arbitrary QCfH $f : D \rightleftarrows B$, and these dilatations will be upper bounds for the corresponding coefficients of QCf of D, whereas, for significant lower bounds one must find lower bounds for the various dilatations for all the QCfH $f : D \rightleftarrows B$. We do this by considering what happens to a suitable arc family under each QCfH and then appealing to lemma 2. Before dealing with more complicated cases, we mention first a category of domains whose coefficients of QCf have a lower bound which is obtained as an immediate consequence of theorem 4.4:

THEOREM 5. *If D is a domain locally connected at each point of its boundary, and if ∂D is not homeomorphic to the sphere, then all the coefficients of D are infinite.*

COROLLARY. *The coefficients of* QCf *of a right circular cylinder approaches ∞ as the ratio of its radius to height approaches ∞.*

Set $D = \{x;\ (x^1)^2 + \ldots + (x^{n-1})^2 < \infty,\ |x^n| < 1\}$, $\{b_m\}$, $b_m > 0$, $b_m \to \infty$ and for each m, $D_m = \{x;\ (x^1)^2 + \ldots + (x^{n-1})^2 < b_m,\ |x^n| < 1\}$. Then, from the preceding theorem, the coefficients of QCf of D are infinite. On the other hand, D_m converge to their kernel D at O, and then, since $D \subset R^n$, $D \neq \{O\}$, theorem 2 yields

$$\lim_{m\to\infty} K_I(D_m) = \lim_{m\to\infty} K_O(D_m) = \lim_{m\to\infty} K(D_m) = \infty.$$

And now let D be the half space $x^n > 0$, let E_0, E_1 be two continua in $\overline{D}$ and suppose $x_0, O \in E_0$ and $x_1, \infty \in E_1$, where $x_0 \neq O$, $x_1 \neq \infty$. Next, set $E_0' = \{x; -|x_0| \leqq x^1 < 0,\ x^2 = \ldots = x^n = 0\}$, $E_1' = \{x; |x_1| \leqq x^1 \leqq \infty,\ x^2 = \ldots = x^n = 0\}$ and let Γ' be the family of the arcs which join E_0' and E_1' in D. By means of the family Γ' we obtain a lower bound for $M(\Gamma)$, where Γ is the family of the arcs which join E_0 and E_1 in D. This extremal property is given by

LEMMA 2. $M(\Gamma) \geqq M(\Gamma')$.

We first observe that $M(\Gamma) = \infty$ whenever $E_0 \cap E_1 \neq \emptyset$, by proposition 1.5.15, and in this case the desired inequality follows trivially.

Suppose now that E_0 and E_1 are disjoint, let $\widetilde{E}_0$ and $\widetilde{E}_1$ be the symmetric images of E_0 and E_1 in ∂D, and let Γ_1 be the family of arcs which join $E_0 \cup \widetilde{E}_0$ to $E_1 \cup \widetilde{E}_1$ in R^n. Proposition 1.5.18 implies that there exists a ring A which has C_0 and C_1 as the components of the complement, where $\partial C_p \subset E_p \subset C_p (p = 0,1)$. Hence the family of arcs which join the components of ∂A in A is a subfamily of Γ_1, and we obtain

$$M(\Gamma) = \frac{1}{2} M(\Gamma_1) \geqq \frac{1}{2} \operatorname{cap} A, \tag{22}$$

from proposition 1.5.19 and corollary of lemma 2.2.6. We now apply lemma 2.9.6 to conclude that

$$\operatorname{cap} A \geqq \operatorname{cap} A^\circ, \tag{23}$$

where A° is the spherical symmetrization of A. But A° separates the components E_0', E_1' of $\partial A'$, so that (2.5.27) implies

$$\operatorname{cap} A^\circ \geqq \operatorname{cap} A'. \tag{24}$$

Because the sets E_0', E_1' are symmetric in ∂D, proposition 1.5.9 and corollary of lemma 2.2.6 imply

$$M(\Gamma') = \frac{1}{2} M(\Gamma_1') = \frac{1}{2} \operatorname{cap} A', \tag{25}$$

where Γ_1' is the family of arcs joining E_0', E_1' in R^n. Hence lemma 2 follows from (22), (23), (24) and (25).

For each $u > 0$ we let $\psi(u)$ denote the modulus of the family of arcs which join the segment $\{x;\ -1 \leqq x^1 \leqq 0,\ x^k = 0\ (k = 2, \ldots, n)\}$ to the ray $\{x;\ u \leqq x^1 \leqq \infty,\ x^k = 0\ (k = 2, \ldots, n)\}$ in the half space $x^n > 0$. From (25), (2.2.1), corollary of lemma 2.2.6 and (2.9.27), it follows that

$$\psi(u) = \frac{n\,\omega_n}{2[\log \Psi(u)]^{n-1}}, \tag{26}$$

and if we combine (2.9.28) and (2.9.33), we obtain

$$\frac{n\,\omega_n}{2[\log \lambda^2(u+1)]^{n-1}} \leqq \psi(u) \leqq \frac{n\,\omega_n}{2[\log (u+1)]^{n-1}},$$

where $\lambda \leqq 4\lambda' < \infty$ and λ' is the constant in (2.9.45) and (2.9.46).

COROLLARY 1. *Let D be the half space $x^n > 0$ and suppose that E_0 and E_1 are continua in $\bar{D}$, that E_0 separates O and ∞ in ∂D, and that $O, \infty \in E_1$. If Γ is the family of arcs which join E_0 and E_1 in D, then*

$$M(\Gamma) \geqq \psi\left(\frac{1}{2}\right) > \frac{n\,\omega_n}{(2\log 24\lambda'^2)^{n-1}}.$$

Since E_0 separates O and ∞ in ∂D, we can find a pair of points $x_0, x_1 \in E_0 \cap \partial D$ such that

$$0 < |x_0| \leqq \frac{1}{2}|x_1 - x_0|. \tag{27}$$

For example we may take x_0 as one of the points of $E_0 \cap \partial D$ nearest to O and then let x_1 be any point in the intersection of E_0 with the ray from O through $-x_0$. After the change of variables $y = x - x_0$, the hypotheses of the preceding lemma hold with O, $x_1 - x_0 \in E_0$ and ∞, $-x_0 \in E_1$, so that we obtain

$$M(\Gamma) \geqq M(\Gamma') = \psi\left(\frac{|x_0|}{|x_1 - x_0|}\right). \tag{28}$$

But lemma 2.9.7 implies that $\dfrac{\Phi(a)}{a}$ is monotonic increasing, hence $\Phi(a)$ is strictly increasing for $a \in (1, \infty)$, and then, from (2.9.28), we conclude that $\Psi(u)$ is strictly increasing in u and by (26) that $\psi(u)$ is stricly decreasing, so that (26) and (28) yield

$$M(\Gamma) \geqq \psi\left(\frac{|x_0|}{|x_1 - x_0|}\right) \geqq \psi\left(\frac{1}{2}\right).$$

COROLLARY 2. *Suppose that x_1, x_2, x_3, x_4 are distinct points in ∂D, D is the half space $x^n > 0$ and E_1, E_2, E_3, E_4 are continua in $\bar{D}$ which join x_1 to x_2, x_2 to x_3, x_3 to x_4, x_4 to x_1, respectively. If Γ_1 and Γ_2 are the families of arcs which join E_1 to E_3, and E_2 to E_4 in D, respectively, then*

$$M(\Gamma_1) \geqq \psi(1) \quad \text{or} \quad M(\Gamma_2) \geqq \psi(1).$$

By performing a preliminary Möbius transformation of D onto itself (which preserves moduli of rings) we may assume without loss of generality, that $x_2 = O$ and $x_4 = \infty$. Then since $x_1, O \in E_1$, $x_3, \infty \in E_3$, we have

$$M(\Gamma_1) \geqq \psi\left(\frac{|x_3|}{|x_1|}\right)$$

by the preceding lemma. Next, since $x_3, O \in E_2$ and $x_1, \infty \in E_4$, the preceding lemma yields

$$M(\Gamma_2) \geqq \psi\left(\frac{|x_1|}{|x_3|}\right).$$

Thus, since $\psi(u)$ is strictly decreasing (as the argument in the proof of the preceding corollary shows), it follows that $M(\Gamma_1) \geqq \psi(1)$ if $|x_1| \geqq |x_3|$ and $M(\Gamma_2) \geqq \psi(1)$ if $|x_1| \leqq |x_3|$, as desired.

THEOREM 6. *Suppose that* $0 < a < b$, *that* D *is a domain in* R^n, *and that* $CD \cap B(b)$ *has at least two components which meet* $S(a)$. *Then*

$$K_I(D) \geqq A_3 \left(\log \frac{b}{a}\right)^{n-1}, \tag{29}$$

where

$$A_3 = \frac{\psi\left(\frac{1}{2}\right)}{n\,\omega_n}, \tag{30}$$

and $\psi(u)$ *is as in* (26).

Let $f : D \rightleftarrows B$ be an arbitrary homeomorphism. We must show that

$$K_I(f) \geqq A_3 \left(\log \frac{b}{a}\right)^{n-1}. \tag{31}$$

Since the right-hand side of (31) is continuous in b, it is sufficient to establish (31) in the slight stronger hypothesis that the set $H = CD \cap \overline{B(b)}$ has at least two components which meet $S(a)$.

We consider first the special case where f can be extended to be a homeomorphism of $\bar{D}$ onto $\bar{B}$. By hypothesis there exist points $\xi_1, \xi_2 \in S(a)$ which belong to different components of H, and hence we can find disjoint compact sets H_1, H_2, such that $H = H_1 \cup H_2$ and $\xi_1 \in H_1$, $\xi_2 \in H_2$. Let F_0 be the closed segment $[\xi_1, \xi_2]$, let x_1 be the last point in $F_0 \cap H_1$ as we move from ξ_1 towards ξ_2 along F_0, let x_2 be the first point in $F_0 \cap H_2$ as we move from x_1 towards ξ_2 along F_0, and let E_0 be the closed segment $[x_1, x_2]$. Then $E_0 \subset \bar{D}$ and x_1, x_2 are points of ∂D which lie in different components of H.

Next let $F_1 = \partial D \cap CB(b)$ and let G be any connected set in ∂D which contains both x_1 and x_2. Since x_1 and x_2 belong to different components of H, $F_1 \cap G \neq \emptyset$. Hence F_1 separates x_1 and x_2 in ∂D, and because ∂D is homeomorphic to S and x_1, x_2 are separated by one of the components of F_1 (see M. H. A. Newman [1], theorem 14.3, p. 123 and the remark of p. 137), we can find a continuum $E_1 \subset F_1$ which separates x_1 and x_2 in ∂D.

Now let Γ be the family of arcs which join E_0 and E_1 in D. Since $E_0 \subset \overline{B(a)}$ and $E_1 \subset CB(b)$, Γ is minorized by the family of arcs which join $S(a)$ and $S(b)$ in R^n, and hence

$$M(\Gamma) \leqq \frac{n\,\omega_n}{\left(\log \dfrac{b}{a}\right)^{n-1}}. \tag{32}$$

On the other hand, we see that E_0^* joins x_1^* and x_2^* in B, that E_1^* separates x_1^* and x_2^* in S, and that Γ^* is the family of arcs which join E_0^* and E_1^* in B. Hence if we map B conformally onto $x^n > 0$ so that x_1^* and x_2^* map onto O and ∞, then, since conformal mappings preserve the modulus of arc families, we can apply corollary 1 of the preceding lemma to conclude that

$$M(\Gamma) \geqq \psi\left(\frac{1}{2}\right), \tag{33}$$

which, on account of (32) and lemma 2.17.2, yields (31) where A_3 is as in (30).

We consider now the general case. For each integer $m > 0$ set

$$D_m = f^{-1}\left[B\left(\frac{m}{m+1}\right)\right], \quad H_m = CD_m \cap \overline{B(b)}, \tag{34}$$

and let f_m denote the restriction of f to D_m. The hypothesis implies there exist points $\xi_1, \xi_2 \in S(a)$ which belong to different components of H. Let C and C_m denote the components of H and H_m which contain ξ_1. Then the C_m are monotonic decreasing in m,

$$C = \bigcap_m C_m$$

and since $\xi_2 \bar{\in} C$, there exists an m such that $\xi_2 \bar{\in} C_m$. Thus ξ_1 and ξ_2 lie in different components of H_m, and since $K_I(f_m)$ is an increasing function of m, we can appeal to what was proved, to conclude that

$$K_I(f) \geqq K_I(f_m) \geqq A_3\left(\log \frac{b}{a}\right)^{n-1}$$

which holds for all QCfH $f : D \rightleftarrows B$, hence on account of (1),

$$K_I(D) = \inf_f K_I(f) \geqq A_3 \left(\log \frac{b}{a}\right)^{n-1},$$

and this completes the proof of theorem 6.

COROLLARY. *If Z is a right circular cylinder with radius b and height h, then*

$$K_I(D) \geqq A_3 \left(\log \frac{\sqrt{4r^2 + h}}{h}\right)^{n-1} \geqq A_3 \left(\log \frac{2r}{h}\right)^{b-1}.$$

It is sufficient to set in the preceding theorem $a = \frac{h}{2}$, $b = \sqrt{r^2 + \frac{h^2}{4}}$.

THEOREM 7. *Suppose $0 < a < b$ and that $CD \cap CB(a)$ has at least two components which meet $S(b)$. Then* (29) *holds.*

Let f be a homeomorphism of D onto B. We want to show that (31) holds; the last argument in the proof of the preceding theorem shows we may assume that f can be extended to be a homeomorphism of $\overline{D}$ onto $\overline{B}$. By hypothesis we can find $\xi_1, \xi_2 \in S(b)$ which belong to different components of $H = CD \cap CB(a)$. Let F_0 be any closed arc in $S(b)$ joining these points. Arguing as before, we can find a closed subarc $E_0 \subset \overline{D}$ with endpoints x_1, x_2 which lie in different components of H. Let $F_1 = \partial D \cap \overline{B(a)}$. Then F_1 separates x_1 and x_2 in ∂D, and hence, a continuum $E_1 \subset F_1$ separates these points in ∂D. If we let Γ denote the family of arcs which join E_0 and E_1 in D, then arguing as in the preceding theorem we obtain (32) and (33), which, on account of lemma 2.17.2, implies (31), and then, by (1), yields (29) with A_3 as in (30).

The lower bounds of the inner coefficient of QCf in the preceding theorem represent a rather poor estimation, but this is natural on account of the large class of domains to which it applies. If these domains are also convex, we have the following:

THEOREM 8. *Suppose that $0 < a < b$, that D is a convex domain, and that $CD \cap B(b)$ has at least two components which meet $S(a)$. Then*

$$K_I(D) \geqq \frac{n\omega_n A_3}{2\omega_{n-1}} \sqrt{\left(\frac{b}{a}\right)^2 - 1}, \tag{35}$$

where A_3 is as in (30).

Let f be any homeomorphism $f : D \rightleftarrows B$. We must show that

$$K_I(f) \geqq \frac{n\omega_n A_3}{2\omega_{n-1}} \sqrt{\left(\frac{b}{a}\right)^2 - 1}. \tag{36}$$

As in the proof of theorem 6, it is sufficient to establish (36) under the hypothesis that $CD \cap \overline{B(b)}$ has at least two components which meet $S(a)$. Consider first the special case where f can be extended to be a homeomorphism of $\overline{D}$ onto $\overline{B}$. By hypothesis there exist two points $\xi_1, \xi_2 \in S(a)$ which belong to different components of H, and since D is convex, we can find planes Π_1, Π_2 such that $\xi_m \in \Pi_m \subset CD$ $(m = 1, 2)$. Let F_0 be the union of the two closed segments from O drawn perpendicular to Π_1 and Π_2. Now $\Pi_1 \cap \overline{B(b)}$ and $\Pi_2 \cap \overline{B(b)}$ must belong to different components of H. Hence F_0 has a closed subarc $E_0 \subset \overline{D}$ with endpoints x_1, x_2 which lie in different components of H. Then as in the proof of theorem 6, there exists a continuum $E_1 \subset \overline{D} \cap CB(b)$ which separates x_1 and x_2 in ∂D, and if we let Γ denote the family of arcs which join E_0 and E_1 in D, we have (33).

We need an upper bound for $M(\Gamma)$. Set $G = \{x;\ x \in D,\ d(x, E_0) \leqq \sqrt{b^2 - a^2}\}$ and

$$\rho(x) = \begin{cases} \dfrac{1}{\sqrt{b^2 - a^2}} & \text{whenever } x \in G, \\ 0 & \text{whenever } x \in CG. \end{cases}$$

Since D lies between the planes Π_1 and Π_2 and since the distance between E_0 and E_1 is not less than the distance between E_0 and $S(b)$, which in its turn is not less than the distance between the feet of the perpendiculars from O to Π_1 or to Π_2 and to $\Pi_1 \cap S(b)$ or to $\Pi_2 \cap S(b)$, respectively, it follows that $d(E_0, E_1) \leqq \sqrt{b^2 - a^2}$. Next from the definition of ρ, we have

$$\int_\gamma \rho \, ds \geqq \frac{\sqrt{b^2 - a^2}}{\sqrt{b^2 - a^2}} = 1,$$

hence $\rho \in F(\Gamma)$, and since $mG \leqq 2\omega_{u-1}\, a(b^2 - a^2)^{\frac{n-1}{2}}$, we conclude that

$$M(\Gamma) \leqq \int_{R^n} \rho^n \, d\tau \leqq \frac{2\omega_{n-1} a (b^2 - a^2)^{\frac{n-1}{2}}}{(b^2 - a^2)^{\frac{n}{2}}} = \frac{2\omega_{n-1} a}{\sqrt{b^2 - a^2}},$$

and we obtain (36), and then (35), from (33) and lemma 2.17.2.

For the general case let D_m and H_m be as in (34) and let f_m denote the restriction of f to D_m. Next pick points $\xi_1, \xi_2 \in S(a)$ which belong to different components of H. Then there exists an m such that ξ_1, ξ_2 belong to different components of H_m. Since D_m is a subdomain of the convex domain D, we can find planes Π_1, Π_2 such that $\xi_p \in \Pi_p \subset CD_m$ $(p = 1, 2)$. The above argument then shows that

$$K_I(f) \geqq K_I(f_m) \geqq \frac{n\omega_n A_3}{2\omega_{n-1}} \sqrt{\left(\frac{b}{a}\right)^2 - 1},$$

which implies (36) also in this case.

Remark. It is interesting to see the connexion between the two lower bounds for the inner coefficient of QCf established by theorems 6 and 8. If $n = 3, 4$, the lower bounds given by theorem 8 is better than that given by theorem 6. Let us prove that, for $n = 3$,

$$A_3(\log x)^2 \leqq \frac{4A_3\pi}{2\pi}\sqrt{x^2-1} \tag{37}$$

where $x = \dfrac{a}{b}$, i.e. that

$$\log x \leqq \sqrt{2}\cdot\sqrt[4]{x^2-1},$$

whenever $x \geqq 1$. If $x = 1$, the inequality (37) reduces to an equality. In the general case, differentiating both parts of the preceding inequality, we obtain

$$\frac{1}{x} \geqq \frac{x}{\sqrt{2}\,\sqrt[4]{(x^2-1)^3}},$$

which may be written as $4(y-1)^3 \leqq y^4$, where we set $x^2 = y$, and also as

$$4(y-1)^3 \leqq [(y-1)+1]^4 = (y-1)^4 + 4(y-1)^3 + 6(y-1)^2 + 4(y-1) + 1,$$

which clearly implies

$$(y-1)^4 + 6(y-1)^2 + 4(y-1) + 1 > 0 \quad (y \geqq 1).$$

Hence we conclude that $\sqrt{2}\,\sqrt[4]{x^2-1} - \log x$ is strictly increasing and vanishes at $x = 1$, so that

$$\sqrt{2}\,\sqrt[4]{x^2-1} - \log x \geqq 0 \quad (x \geqq 1)$$

and then (37) holds.

If $n = 4$, instead of (37) we have

$$A_3(\log x)^3 \geqq \frac{A_3\dfrac{4\pi^2}{2}\sqrt{x^2-1}}{2\cdot\dfrac{4}{3}\pi}$$

which may be written as $(\log y)^2 \leqq 36\pi^2(y-1)$, where $y = x^2$, and also as $t^6 \leqq$ $\leqq 36\pi^2(e^t-1)$, where $t = \log y$, and, by an expansion of e^t in a power series, we obtain

$$t^6 \leqq 36\pi^2\left(t + \frac{t^2}{2} + \frac{t^3}{3!} + \frac{t^4}{4!} + \frac{t^5}{5!} + \frac{t^6}{6!} + \frac{t^7}{7!} + \cdots\right),$$

which clearly holds for $0 \leqq t \leqq 1$, for $t \geqq \dfrac{7!}{36\pi^2} = \dfrac{140}{\pi^2} > 14$ and for $1 \leqq t \leqq 15$.

And now, let us show that if $n > 4$, then for some values of x, one of the bounds is better, and for other values of x, the other bound. Consider first the case $n = 5$. If $1 \leqq x \leqq e^2$ or $x > e^{17}$, then

$$A_3(\log x)^4 \leqq \frac{A_3 5 \cdot 2^5 \cdot 2^2\pi^2}{5!\pi^2} \sqrt{x^2 - 1},$$

however, if for instance $x = e^3$ or $x = e^4$, then

$$A_3(\log x)^4 > \frac{A_3 5 \cdot 2^7}{5!} \sqrt{x^2 - 1}.$$

And now, for $n = 6$, we shall consider first the case $n = 2k$, $k > 3$. The inequality

$$A_3(\log x)^{2k-1} \leqq \frac{A_3 2k\pi^k(2k-1)!}{2^{2k}k!(k-1)!\pi^{k-1}} \sqrt{x^2 - 1} \tag{38}$$

holds for values in the neighbourhood of $x = 1$. However the inequality

$$A_3(\log x)^{2k-1} > \frac{A_3(2k-1)!\pi^k}{2^{2k-1}(k-1)!\pi^{k-1}} \sqrt{x^2 - 1},$$

or, setting $x = e^t$, the inequality

$$t^{2k-1} > \frac{(2k-1)!\pi}{2(k-1)!} \sqrt{e^t - 1}$$

comes, for $t = 4$, to

$$1 > \frac{(2k-1)!\pi}{4^{2(k-1)}(k-1)!^2} = \frac{(2k-1)2(k-1)\cdots 3.2.1}{4(k-1)4(k-1)\cdots 4.4.} \cdot \frac{\sqrt{e^4 - 1}}{4},$$

which is satisfied for $\geqq 3$. However, for

$$x > 2^{\frac{(4k-1)!\,(2k-1)!}{2(2k-1)!\,\pi^2}},$$

the inequality (38) is satisfied again.

Finally, if $n = 2k + 1$ and $k \geqq 3$, for values of x in the neighbourhood of $x = 1$, we have

$$A_3(\log x)^{2k} \leqq A_3 \frac{(2k+1)2^{2k-1}k!^2\pi^k}{2(2k+1)!\pi^k} \sqrt{x^2 - 1}. \tag{39}$$

However

$$A_3(\log x)^{2k} > A_3 \frac{2^{2k}k!^2}{(2k)!}\sqrt{x^2 - 1},$$

or

$$t^{2k} > \frac{2^{4k}k!^2}{(2k)!}\sqrt{e^t - 1},$$

where $t = \log x$, comes, for $t = 4$, to

$$1 > \frac{k \cdot k \cdots 1.1}{2k(2k \quad 1) \cdots 2.1}\sqrt{e^4 - 1},$$

which clearly holds for $k \geqq 3$. But for

$$x > e^{\frac{(2k)!^2(4k+1)!}{2^{8k}(k!)^4}}$$

inequality (39) holds again.

THEOREM 9. *Inequality* (29) *gives sharp bounds for the order of growth of* $K_I(D)$ *as* $\frac{b}{a} \to \infty$ *for the domains corresponding to theorems* 6 *and* 7.

Fix $1 < b < \infty$ and let $\mathscr{D}(b)$ denote the class of domains $D \subset R^n$ such that $CD \cap B(b)$ has at least two components which meet S and let

$$g(b) = \inf_{D \in \mathscr{D}(b)} K_I(D). \tag{40}$$

If $1 < b' < b$, then

$$f(x) = x|x|^{c-1}, \tag{41}$$

where

$$c = \frac{\log b'}{\log b}, \tag{42}$$

is a homeomorphism of R^n onto itself such that for each domain $D \subset R^n$, $D \in \mathscr{D}(b)$ if and only if $D^* = f(D) \in \mathscr{D}(b')$, because the homeomorphism (41) carries the ball $B(b)$ onto the ball $B(b^c)$, where $b^c = b'$, as it follows from (42).

And now let us prove that

$$K_I(f) = c^{-(n-1)}. \tag{43}$$

Indeed, lemma 2.17.1 implies that

$$K_I(f) = \operatorname*{ess\,sup}_{x \in D} \frac{|J(x)|}{\lambda_f(x)^n}. \tag{44}$$

In order to calculate $\lambda_f(x_0)$, set $x_0 = r_0(\cos \alpha_1^0 e_1 + \cdots + \cos \alpha_n^0 e_n)$ and $x = r(\cos \alpha_1 e_1 + \cdots + \cos \alpha_n e_n)$. Then, on account of

$$\frac{r^{2c} + r_0^{2c} - 2r^c r_0^c \cos \alpha}{r^2 + r_0^2 - 2rr_0 \cos \alpha} \geq \frac{(r^c - r_0^c)^2}{(r - r_0)^2},$$

we obtain

$$\lambda_f(x_0) = \lim_{x \to x_0} \frac{|r^c(\cos \alpha_1 e_1 + \cdots + \cos \alpha_n e_n) - r_0^c(\cos \alpha_1^0 e_1 + \cdots + \cos \alpha_n^0 e_n)|}{|r(\cos \alpha_1 e_1 + \cdots + \cos \alpha_n e_n) - r_0(\cos \alpha_1 e_1 + \cdots + \cos \alpha_n e_n)|} =$$

$$= \lim_{x \to x_0} \sqrt{\frac{(r^c \cos \alpha_1 - r^c \cos \alpha_1^0)^2 + \cdots + (r^c \cos \alpha_n - r_0^c \cos \alpha_n^0)^2}{(r \cos \alpha_1 - r_0 \cos \alpha_1^0)^2 + \cdots + (r \cos \alpha_n - r_0 \cos \alpha_n)^2}} = \tag{45}$$

$$= \lim_{x \to x_0} \sqrt{\frac{r^{2c} + r_0^{2c} - 2r^c r_0^c \cos \alpha}{r^2 + r_0^2 - 2rr_0 \cos \alpha}} = \lim_{r \to r_0} \frac{r^c - r_0^c}{r - r_0} = cr_0^{c-1},$$

where $\cos \alpha = \cos \alpha_1 \cos \alpha_1^0 + \cdots + \cos \alpha_n \cos \alpha_n^0$. Next, the homeomorphism (41) may be written as a product (in the functional sense) of the following three homeomorphisms: (12),

$$R = r^c, \alpha_k = \alpha_k \quad (k = 1, \ldots, n-1) \tag{46}$$

and

$$x^{*i} = R \sin \alpha_1 \cdots \sin \alpha_{i-1} \cos \alpha_i \quad (i = 1, \ldots, n) \tag{47}$$

and then the Jacobian J corresponding to the homeomorphism (41) may be written as

$$J = J_{x^*R} J_{Rr} J_{rx} = \frac{R^{n-1} \sin^{n-2} \alpha_1 \cdots \sin \alpha_{n-2} cr^{c-1}}{r^{n-1} \sin^{n-2} \alpha_1 \cdots \sin \alpha_{n-2}} = cr^{n(c-1)},$$

where J_{x^*R}, J_{Rr}, J_{rx} denote the Jacobians of the homeomorphisms (47), (46) and (12), respectively. Hence, (44) and (45) yield (43). Next, to a homeomorphism $x' = \varphi(x^*)$

of D^* onto B, there corresponds the homeomorphism $x' = \psi(x) = \varphi[f(x)]$ of D onto B. Setting $A^* = f(A)$, $A' = \psi(A)$, then, for every ring A, it follows

$$\frac{\text{mod } A}{\text{mod } A'} = \frac{\text{mod } A}{\text{mod } A^*} \cdot \frac{\text{mod } A^*}{\text{mod } A'},$$

hence

$$\sup_{A \subset D} \frac{\text{mod } A}{\text{mod } A'} \leqq \sup_{A \subset D} \frac{\text{mod } A}{\text{mod } A^*} \sup_{A^* \subset D^*} \frac{\text{mod } A^*}{\text{mod } A'},$$

and then, by (43),

$$K_I(\psi) \leqq K_I(f) K_I(\varphi) = \frac{K_I(\varphi)}{c^{n-1}}.$$

But since this inequality holds for all homeomorphisms $\varphi(x^*)$, we obtain

$$K_I(D) \leqq \frac{K_I(D)^*}{c^{n-1}}$$

and, by (40) and (42), also

$$g(b) = \inf_{D \in \mathfrak{D}(b)} K_I(D) \leqq \frac{\inf\limits_{D^* \in \mathfrak{D}(b')} K_I(D^*)}{c^{n-1}} = \left(\frac{\log b}{\log b'}\right)^{n-1} g(b').$$

Thus

$$\frac{g(b)}{(\log b)^{n-1}} \leqq \frac{g(b')}{(\log b')^{n-1}}, \quad 1 < b' < b,$$

and we conclude that $\dfrac{g(b)}{(\log b)^{n-1}}$ is monotonic decreasing in $1 < b < \infty$ and, by (30),

$$\lim_{b \to \infty} \frac{g(b)}{(\log b)^{n-1}} = A_4 \geqq A_3 > 0.$$

Hence $g(b)$ approaches assymptotically $A_4(\log b)^{n-1}$ and we see that the lower bound for the order of growth of $K_I(D)$ given in theorem 6 cannot be improved. The above argument shows that the same is true of theorem 7.

The inner coefficient of QCf of a dihedral wedge of type ν. In this paragraph we shall calculate the inner coefficient of QCf for domains whose boundary is a dihedral angle and also for other generalizations in n-space of these domains.

To calculate a coefficient of QCf of a given domain D we first must obtain a lower bound for the corresponding dilatation of each homeomorphism $f: D \rightleftarrows B$. Then we must show that this bound is actually assumed by some extremal homeomorphism of D onto B. Clearly it is the sharp lower bounds which are most difficult to obtain.

We say that a domain D is a *dihedral wedge of angle* α, $0 < \alpha < 2\pi$, if it can be mapped by means of a similarity transformation φ onto the domain

$$D_\alpha = \{x; x = (r, \vartheta_1, x^3, \ldots, x^n),\ 0 < \vartheta_1 < x, |x| < \infty\},$$

where ϑ_1 is the angle which the projection of the vector x on x^1x^2-plane makes with the x^1-axis. The image of the $(n-2)$-dimensional plane $(x^3, \ldots, x^n)$ under φ^{-1} is said to be the *edge of the dihedral wedge* D_α.

We say that a domain D is a *dihedral wedge of type* ν $(1 \leqq \nu \leqq n-2)$ *and of angle* $\alpha = (\alpha_1, \ldots, \alpha_{n-\nu-1})$, $0 < \alpha_1 \leqq 2\pi$, $0 < \alpha_1, \ldots, \alpha_{u-\nu-1} \leqq \pi$, if it can be mapped by means of a similarity transformation φ onto the domain

$$D_{\alpha\upsilon} = \{x; x = (r, \vartheta_1, \ldots, \vartheta_{n-\nu-1}, x^{n-\nu+1}, \ldots, x^n);\ 0 < \vartheta_k < \alpha_k$$

$$(k = 1, \ldots, n-\nu-1),\ |x| < \infty\}.$$

The image of the ν-space $(x^{n-\nu+1}, \ldots, x^n)$ under φ^{-1} is said to be the *edge of the dihedral wedge D of type* ν. (If $\nu = 1$, the edge is the image of an axis as in the threedimensional case considered by F. Gehring and J. Väisälä [1]). If $\nu = n-2$, the definition of the dihedral wedge of type ν comes to the definition of the dihedral wedge of angle α from above.

We shall calculate here the inner coefficient of a convex dihedral wedge of type ν. But we first require the following preliminary results.

LEMMA 3. *Suppose that* $0 < \alpha_1 \leqq 2\pi$, $0 < \alpha_2, \ldots, \alpha_{u-\nu-1} \leqq \pi$, *that* $E_0 = \{x;\ r = 0,\ x^k = 0\ (k = n-\nu+1, \ldots, n-1),\ -1 \leqq x^n \leqq 0\}$ *and that* $E_1 = \{x;\ r = 0,\ x^k = 0\ (k = n-\nu+1, \ldots, n-1),\ 1 \leqq x^n \leqq \infty\}$. *If* Γ_α *is the family of arcs which join* E_0 *to* E_1 *in* $D_{\alpha\nu}$, *then*

$$M(\Gamma_\alpha) = \frac{\alpha_1 \cdots \alpha_{n-\nu-1}}{\pi^{n-\nu-1}} \psi(1), \tag{48}$$

where ψ *is as in* (26).

Suppose that $0 < \alpha_k \leqq 2\pi$ $(k = 1, \ldots, n-\nu-1)$. Then the homeomorphism

$$(r, \vartheta_1, \ldots, \vartheta_{n-\nu-1}, x^{n-\nu+1}, \ldots, x^n) =$$

$$= \left(r, \frac{\beta_1}{\alpha_1}\vartheta_1, \ldots, \frac{\beta_{n-\nu-1}}{\alpha_{n-\nu-1}}\vartheta_{n-\nu-1}, x^{n-\nu+1}, \ldots, x^n\right)$$

maps $D_{\alpha\nu}$ onto $D_{\beta\nu}$, Γ_α onto Γ_β, and since

$$J = \frac{\beta_1 \cdots \beta_{n-\nu-1}}{\alpha_1 \dots \alpha_{n-\nu-1}},$$

and

$$\lambda_f(x_0) = \lim_{x \to x_0} \Bigg[(r - r_0)^2 + \frac{\beta_1^2}{\alpha_1^2} (\vartheta_1 - \vartheta_1^0)^2 + \dots +$$

$$+ \frac{\beta_{n-\nu-1}^2}{\alpha_{n-\nu-1}^2} (\vartheta_{n-\nu-1} - \vartheta_{n-\nu-1}^0)^2 + (x^{n-\nu+1} - x_0^{n-\nu+1})^2 + \cdots$$

$$+ (x^n - x_0^n)^2 \Big] \big[(r - r_0)^2 + (\vartheta_1 - \vartheta_1^0)^2 + \cdots + (\vartheta_{n-\nu-1} - \vartheta_{n-\nu-1}^0)^2 +$$

$$+ (x^{n-\nu+1} - x_0^{n-\nu+1})^2 + \cdots + (x^n - x_0)^2 \Bigg]^{-1} = A,$$

we obtain

$$K_I(f) = \frac{\beta_1 \cdots \beta_{n-\nu-1}}{\alpha_1 \cdots \alpha_{n-\nu-1}}, \tag{49}$$

and lemma 2.17.2 implies

$$M(\Gamma_\beta) \leqq \frac{\beta_1 \dots \beta_{n-\nu-1}}{\alpha_1 \cdots \alpha_{n-\nu-1}} M(\Gamma_\alpha). \tag{50}$$

In order to prove the inequality of opposite sense, suppose that

$$\frac{\beta_k}{\alpha_k} = m_k \quad (m_k \text{ integers; } k = 1, \dots, n - \nu - 1). \tag{51}$$

and for $\beta = (\beta_1, \dots, \beta_{u-\nu-1})$, $p = (p_1, \dots, p_{u-\nu-1})$, $p_k = 1, \dots, m_k$ $(k = 1, \dots, n - \nu - 1)$, let Γ_β^p be the family of arcs which join E_0 and E_1 in the dihedral wedge of type ν

$$D_{\beta\nu}^p = \Bigg\{ x; x = (r, \vartheta_1, \dots, \vartheta_{n-\nu-1}, x^{n-\nu+1}, \dots, x^n), \frac{p_k - 1}{m_k} \beta_k <$$

$$< \vartheta_k < \frac{p_k}{m_k} \beta_k (k = 1, \dots, n - \nu - 1), |x| < \infty \Bigg\}.$$

Then clearly, $M(\Gamma_\beta^p) = M(\Gamma_a)$, $\Gamma_\beta \supset \overset{m}{\underset{p=1}{U}} \Gamma_\beta^p$, where $m = (m_1,\ldots, m_{u-\nu-1})$, $1 = (1,\ldots, 1)$ and Γ_β^p are separate. Hence by proposition 1.5.2, by corollary of proposition 1.5.3 and by (51)

$$M(\Gamma_\beta) \geqq \sum_{p=1}^{m} M(\Gamma_\beta^p) = m_1 \cdots m_{n-\nu-1} M(\Gamma_\alpha) = \frac{\beta_1 \cdots \beta_{n-\nu-1}}{\alpha_1 \cdots \alpha_{n-\nu-1}} M(\Gamma_\alpha),$$

With the aid of (50) it is now easy to show that

$$M(\Gamma_\beta) = \frac{\beta_1 \cdots \beta_{n-\nu-1}}{\alpha_1 \cdots \alpha_{n-\nu-1}} M(\Gamma_\alpha), \tag{52}$$

in this case and whenever $0 < \alpha_1,\ldots, \alpha_{u-\nu-1}, \beta_1,\ldots, \beta_{u-\nu-1} < 2\pi$.

Suppose next that $\frac{\beta_k}{\alpha_k}$ are rational, i.e. $\frac{\beta_k}{\alpha_k} = \frac{m_k}{q_k} > 0$ (m_k, q_k integers; $k = 1,\ldots$ $\ldots, n - \nu - 1$). Then setting $\beta'_k = q_k\beta_k$, the problem comes to the preceding case, hence, by (52) for $\frac{\beta_k}{\alpha_k}$ integers, we get

$$\frac{\beta'_1 \cdots \beta'_{n-\nu-1}}{\beta_1 \cdots \beta_{n-\nu-1}} M(\Gamma_\beta) = M(\Gamma'_\beta) = \frac{\beta'_1 \cdots \beta'_{n-\nu-1}}{\alpha_1 \cdots \alpha_{n-\nu-1}} M(\Gamma_\beta),$$

and then (52) still holds for $\frac{\beta_k}{\alpha_k}$ rational.

Finally, suppose that $\frac{\beta_k}{\alpha_k}$ ($k = 1,\ldots, n - \nu - 1$) are arbitrary real numbers, and choose a sequence of numbers $\beta_k^m = r_k^m\alpha_k$, with $r_k^m(k = 1,\ldots, n - \nu - 1;$ $m = 1, 2,\ldots)$ and so that $\lim_{m\to\infty} \beta_k^m = \beta_k$ ($k = 1,\ldots, n - \nu - 1$). Then, (52) for the case $\frac{\beta_k}{\alpha_k}$ rational, implies

$$M(\Gamma_{\beta m}) = \frac{\beta_1^m \cdots \beta_{n-\nu-1}^m}{\alpha_1 \cdots \alpha_{n-\nu-1}} M(\Gamma_\alpha) \quad (m = 1, 2,\ldots),$$

whence, letting $m \to \infty$,

$$M(\Gamma_\beta) = \lim_{m\to\infty} M(\Gamma_{\beta m}) = \lim_{m\to\infty} \frac{\beta_1^m \cdots \beta_{n-\nu-1}^m}{\alpha_1 \cdots \alpha_{n-\nu-1}} M(\Gamma_\alpha) = \frac{\beta_1 \cdots \beta_{n-\nu-1}}{\alpha_1 \cdots \alpha_{n-\nu-1}} M(\Gamma_\alpha),$$

and (52) holds in the general case.

If we set $\beta_k = \pi \; (k = 1, \ldots, n - \nu - 1)$ in (52), we obtain

$$M(\Gamma_\alpha) = \frac{\alpha_1 \cdots \alpha_{n-\nu-1}}{\pi^{n-\nu-1}} M(\Gamma_\pi).$$

Hence, since Γ_π is the family of arcs which join E_0 and E_1 in $x^n > 0$ we get $M(\Gamma_\pi) = \psi(1)$, whence the inequality (48) holds, as desired.

We now calculate the inner coefficient of QCf of a convex dihedral wedge of type ν:

THEOREM 10. *Suppose that D is a convex dihedral wedge of type* ν *and angle* α. *Then*

$$K_I(D) = \frac{\pi^{n-\nu-1}}{\alpha_1 \cdots \alpha_{n-\nu-1}}. \tag{53}$$

We may assume, for convenience of notation, that D is the dihedral wedge of type ν, $D_{\alpha\nu}$. Then, since $D_{\alpha\nu}$ is convex and $0 < \alpha \leqq \pi$, we see that the mapping

$$f(r, \vartheta_1, \ldots, \vartheta_{n-\nu-1}, x^{n-\nu+1}, \ldots, x^n) =$$

$$= \left(r, \frac{\pi}{\alpha_1} \vartheta_1, \ldots, \frac{\pi}{\alpha_{n-\nu-1}} \vartheta_{n-\nu-1}, x^{n-\nu+1}, \ldots, x^n\right)$$

is a homeomorphism of $D_{\alpha\nu}$ onto the half space D_π, and then, by (49),

$$K_I(f) = \frac{\pi^{n-\nu-1}}{\alpha_1 \cdots \alpha_{n-\nu-1}}.$$

Since we can map D_π onto B by means of a Möbius transformation g, we have

$$K_I(D_{\alpha\nu}) \leqq K_I(g \circ f) = K_I(f) = \frac{\pi^{n-\nu-1}}{\alpha_1 \cdots \alpha_{n-\nu-1}}. \tag{54}$$

To complete the proof of (53), it is sufficient to show that

$$K_I(f) \geqq \frac{\pi^{n-\nu-1}}{\alpha_1 \cdots \alpha_{n-\nu-1}} \tag{55}$$

for each QCfH $f: D_{\alpha\nu} \rightleftarrows D_\pi$. For this let

$$E_1 = \{x; r = 0, x^k = 0 \; (k = n - \nu + 1, \ldots, n - 1), -1 \leqq x^n \leqq 0\},$$

$$E_2 = \{x; r = 0, x^k = 0 \; (k = n - \nu + 1, \ldots, n - 1), 0 \leqq x^n \leqq 1\},$$

$$E_3 = \{x; r = 0, x^k = 0 \; (k = n - \nu + 1, \ldots, n - 1), 1 \leqq x^n \leqq \infty\},$$

$$E_4 = \{x; r = 0, x^k = 0 \; (k = n - \nu + 1, \ldots, n - 1), -\infty \leqq x^n \leqq -1\},$$

and let Γ_1 and Γ_2 be the families of arcs which join E_1 to E_3 and E_2 to E_4 in $D_{\alpha\nu}$. Then, by the preceding lemma,

$$M(\Gamma_1) = M(\Gamma_2) = \frac{\alpha_1 \cdots \alpha_{n-\nu-1}}{\pi^{n-\nu-1}} \psi(1). \tag{56}$$

Since $D_{\alpha\nu}$ is locally connected at each point of its boundary, $f(x)$ can be extended to be a homeomorphism of $D_{\alpha\nu}$ onto D_π by theorem 4.3. Then $E_1^*, E_2^*, E_3^*, E_4^*$ are continua in D_π which satisfy the hypotheses of corollary 2 of lemma 2. Hence

$$M(\Gamma_1^*) \geqq \psi(1) \text{ or } M(\Gamma_2^*) \geqq \psi(1).$$

and then, by (56) and lemma 2.17.2, we obtain (55), which combined with (54), implies (53), as desired.

In the more particular case of the dihedral domains (i.e. for $\nu = n - 2$), this theorem comes to

COROLLARY 1. *Suppose that D is a convex dihedral wedge of angle α. Then*

$$K_I(D) = \frac{\pi}{\alpha}.$$

As a consequence of this corollary, we have

COROLLARY 2. *Theorem 8 gives a sharp lower bound for the order of growth of $K_I(D)$ as $\frac{b}{a} \to \infty$.*

We exhibit a particular domain D to show that the order is right in theorem 8. For $0 < \alpha < \pi$, let $x_1 = \left(\cos\frac{\alpha}{2},\ 0, \ldots, 0\right)$, $x_2 = \left(-\cos\frac{\alpha}{2},\ 0, \ldots, 0\right)$, and let $D = B(x_1, 1) \cap B(x_2, 1)$. Then D is a lens shaped domain which can be mapped by means of an inversion onto a convex wedge D_α, bounded by two half planes which meet at an angle α. Hence from the preceding corollary, we obtain

$$K_I(D) = K_I(D_\alpha) = \frac{\pi}{\alpha}. \tag{57}$$

Next it is easy to see that D is itself convex and that $CD \cap B(b)$ has two components which meet $S(a)$, where $a = 1 - \cos\frac{\alpha}{2}$, $b = \sin\frac{\alpha}{2}$, $\frac{b}{a} = \cot\frac{\alpha}{4}$. Hence and from (57) it follows that

$$K_I(D) = \frac{\pi}{4}\left(\operatorname{arc\,cot}\frac{b}{a}\right)^{-1} \sim \frac{\pi}{4} \cdot \frac{b}{a} \text{ as } \frac{b}{a} \to \infty$$

$\left(\text{since arctan } \frac{b}{a} \text{ approaches assymptotically } \frac{b}{a}\right)$ and thus the order of the lower bound in theorem 8 cannot be improved.

Bounds for the outer and the total coefficients of QCf of the dihedral wedge of type ν. We shall not calculate the other coefficients of QCf of a convex dihedral wedge of type ν. However the following estimates are easily obtained.

THEOREM 11. *Suppose $D_{\alpha\nu}$ is a dihedral wedge of type ν and of angle α. Then*

$$\left(\frac{\pi^{n-\nu-1}}{\alpha_1 \cdots \alpha_{n-\nu-1}}\right)^{\frac{1}{n-1}} \leqq K_O(D_{\alpha\nu}) \leqq \frac{\pi\alpha_1 \cdots \alpha_{n-\nu-1}}{\min(\alpha_1, \cdots, \alpha_{n-\nu-1})^{n-\nu}},$$

$$\frac{\pi^{n-\nu-1}}{\alpha_1 \cdots \alpha_{n-\nu-1}} \leqq K(D_{\alpha\nu}) \leqq \frac{\pi^{\frac{\nu}{2}+1}\alpha_1 \cdots \alpha_{n-\nu-1}}{\min(\alpha_1, \ldots, \alpha_{n-\nu-1})^{n-\frac{\nu}{2}}}.$$

The lower bounds follow directly from (2), (3) and (53). The upper bounds result from the fact that the mappings

$$f(r, \vartheta_1, \ldots, \vartheta_{n-\nu-1}, x^{n-\nu+1}, \ldots, x^n) =$$

$$= \left(r, \frac{\pi}{\alpha_1}\vartheta_1, \ldots, \frac{\pi}{\alpha_{n-\nu-1}}\vartheta_{n-\nu-1}, \frac{\pi}{\alpha^m}x^{n-\nu+1}, \ldots, \frac{\pi}{\alpha^m}x^n\right),$$

$$g(r, \vartheta_1, \ldots, \vartheta_{n-\nu-1}, x^{n-\nu+1}, \ldots, x^n) =$$

$$= \left(r, \frac{\pi}{\alpha_1}\vartheta_1, \ldots, \frac{\pi}{\alpha_{n-\nu-1}}\vartheta_{n-\nu-1}, \sqrt{\frac{\pi}{\alpha^m}}x^{n-\nu+1}, \ldots, \sqrt{\frac{\pi}{\alpha^m}}x^n\right),$$

where $\alpha^m = \min(\alpha_1, \ldots, \alpha_{n-\nu-1})$, are homeomorphisms of D_α onto D_π with

$$\Lambda_f(x) = \overline{\lim_{x\to x_0}}\left[(r-r_0)^2 + \frac{\pi^2}{\alpha_1}(\vartheta_1 - \vartheta_1^0)^2 + \cdots + \frac{\pi^2}{\alpha_{n-\nu-1}}\,\vartheta_{n-\nu-1} - \vartheta_{n-\nu-1}^0)^2\right.$$

$$\left.\frac{\pi^2}{(\alpha^m)^2}(x^{n-\nu+1} - x_0^{n-\nu+1})^2 + \cdots + \frac{\pi^2}{(\alpha^m)}(x^n - x_0^n)^2\right][(r-r_0)^2 +$$

$$+ (\vartheta_1 - \vartheta_1^0)^2 + \cdots + (\vartheta_{n-\nu-1} - \vartheta_{n-\nu-1}^0)^2 +$$

$$+ (x^{n-\nu+1} - x_0^{n-\nu+1})^2 + \cdots + (x^n - x_0^n)^2]^{-1},$$

$$K_O(f) = \sup_{x \in D_{\alpha\nu}} \frac{\Lambda_f(x)^n}{J_G(x)} = \frac{\dfrac{\pi^n}{(\alpha^m)^n}}{\dfrac{\pi^{n-1}}{\alpha_1 \ldots \alpha_{n-\nu-1}\, \alpha^m)^\nu}} = \frac{\pi\alpha_1 \cdots \alpha_{n-\nu-1}}{(\alpha^m)^{n-\nu}},$$

$$K_O(g) = \sup_{x \in D_{\alpha\nu}} \frac{\Lambda_g(x)^n}{J_G(x)} = \frac{\dfrac{\pi^n}{(\alpha^m)^n}}{\dfrac{\pi^{n-\nu-1+\frac{\nu}{2}}}{\alpha_1 \cdots \alpha_{n-\nu-1}(\alpha^m)^{\frac{\nu}{2}}}} = \frac{\pi^{\frac{\nu}{2}+1}\alpha_1 \cdots \alpha_{n-\nu-1}}{(\alpha^m)^{n-\frac{\nu}{2}}}.$$

Then

$$K_O(D_{\alpha\nu}) = \inf_f K_O(f) \leqq K_O(f) = \frac{\pi\alpha_1 \cdots \alpha_{n-\nu-1}}{(\alpha^m)^{n-\nu}}.$$

Next

$$K_I(g) = \sup_{x \in D_{\alpha\nu}} \frac{J_G(x)}{\lambda_g(x)^n} = \frac{\pi^{n-\frac{\nu}{2}+1}}{\alpha_1 \cdots \alpha_{n-\nu-1}(\alpha^m)^{\frac{\nu}{2}}},$$

hence $K(g) = K_O(g) \geqq K_I(g)$, with equality for $\nu = n-2$, and then

$$K(D_{\alpha\nu}) = \inf_f K(f) \leqq K(g) = K_O(g) = \frac{\pi^{\frac{\nu}{2}+1}\alpha_1 \cdots \alpha_{n-\nu-1}}{(\alpha^m)^{n-\frac{\nu}{2}}}$$

COROLLARY.

$$\left(\frac{\pi}{\alpha}\right)^{\frac{1}{n-1}} \leqq K_O(D_\alpha) \leqq \frac{\pi}{\alpha}, \qquad \frac{\pi}{\alpha} \leqq K(D_\alpha) \leqq \left(\frac{\pi}{\alpha}\right)^{\frac{n}{2}},$$

hence

$$\left(\frac{\pi}{\alpha}\right)^{\frac{1}{n-1}} \leqq K_O(D_\alpha) \leqq K(D_\alpha) \leqq \left(\frac{\pi}{\alpha}\right)^{\frac{n}{2}}.$$

It is clear from above that the coefficients of QCf of a given domain depend strongly on the global nature of ∂D. We show now how one can obtain lower bounds for the coefficients of QCf by examining D in a neighbourhood of a fixed finite boundary point.

We say that a domain D is *raylike at a boundary point* ξ if for each point x, $x \in D$ if and only if $\xi + t(x - \xi) \in D$ for $0 < t < \infty$. That is D is raylike at ξ if each open ray from ξ lies either in D or in CD.

THEOREM 12. *Suppose that U is a neighbourhood of $\xi \in \partial D$, and that $D \cap U = \Delta \cap U$, where Δ is a domain that is raylike at ξ. Then*

$$K_I(D) \geqq K_I(\Delta), \quad K_O(D) \geqq K_O(\Delta), \quad K(D) \geqq K(\Delta).$$

We may assume without loss of generality that $\xi = O$. Choose $\alpha > 0$ so that $B(a) \subset U$, and for each integer $m > 0$ let $D_m = \left\{x; \frac{x}{m} \in D\right\}$. If $x \in \Delta$, and $m > \frac{|x|}{a}$, then because Δ is raylike at the origin

$$\frac{x}{m} \in \Delta \cap U \subset D, \ x \in D_m.$$

Hence if we fix $x_0 \in \Delta$ with $|x_0| < a$, we see that $x_0 \in D_m$ for all m. Now let D' denote the kernel of D_m at x_0. We show that $D' = \Delta$. Let F be a compact subset of Δ and x_1 a point of F such that $|x_1| = \sup_{x \in F} |x|$. Arguing as above it follows that $x_1 \in D_m$ for all $m > \frac{|x_1|}{a}$. Hence F lies in all D_m with $m > \frac{|x_1|}{a}$, and hence we see that $\Delta \subset D'$. If $x \in D'$, then, on account of the definition of the kernel it follows that $x \in D_m$ for $m > m_0$. Thus $m > \max\left(m_0, \frac{|x|}{a}\right)$ implies that

$$\frac{x}{m} \in D \cap U = \Delta \cap U \subset \Delta, \quad x \in \Delta,$$

since Δ is raylike at O. $\Delta = D'$, and repeating the above argument with a subsequence $\{D_{m_k}\}$ of $\{D_m\}$, we conclude that the D_m converge to their kernel Δ at x_0. A homothety is a conformal mapping, and then preserves the coefficients of QCf, hence $K_I(D_m) = K_I(D)$, $K_O(D_m) = K_O(D)$ and $K(D_m) = K(D)$. Since $O \in \partial\Delta$, it follows that $\Delta \neq \{x_0\}$, R^n, and we can apply theorem 2 to obtain

$$K_I(\Delta) \leqq \varliminf_{m \to \infty} K_I(D_m) = K_I(D),$$

$$K_O(\Delta) \leqq \varliminf_{m \to \infty} K_O(D_m) = K_O(D), \quad K(\Delta) \leqq \varliminf_{m \to \infty} K(D_m) = K(D). \tag{58}$$

We can combine theorems 10, 11, 12, corollary 1 of theorem 10 and corollary of theorem 11 to obtain the following lower bounds for the coefficients of QCf of a large class of domains.

THEOREM 13. *Suppose that* $D \subset R^n$, *that* U *is a neighbourhood of a point* $\xi \in \partial D$, *and that* $D \cap U = \Delta \cap U$, *where* Δ *is a dihedral wedge of type* ν *and of angle* α *which has* ξ *as a point of its edge. Then the coefficients of* QCf *of* D *are not less than the corresponding coefficients of* Δ. *In particular, if* Δ *is convex*

$$K_I(D) \geqq \frac{\pi^{n-\nu-1}}{\alpha_1 \cdots \alpha_{n-\nu-1}},$$

$$K_O(D) \geqq \left(\frac{\pi^{n-\nu-1}}{\alpha_1 \cdots \alpha_{n-\nu-1}}\right)^{\frac{1}{n-1}}, \quad K(D) \geqq \frac{\pi^{n-\nu-1}}{\alpha_1 \cdots \alpha_{n-\nu-1}},$$

and if Δ *is a dihedral wedge of angle* α, *then*

$$K_I(D) \geqq \frac{\pi}{\alpha}, \quad K_O(D) \geqq \left(\frac{\pi}{\alpha}\right)^{\frac{1}{n-1}}, \quad K(D) \geqq \frac{\pi}{\alpha}. \tag{59}$$

The preceding theorem yields lower bounds for the coefficients of QCf of all polyhedra.

COROLLARY 1. *If* $D \subset R^n$ *is a convex polyhedron with* m *faces, then*

$$K_I(D) \geqq \frac{m-n+2}{m-n}, \quad K_O(D) \geqq \left(\frac{m-n+2}{m-n}\right)^{\frac{1}{n-1}}, \quad K(D) \geqq \frac{m-n+2}{m-n}. \tag{60}$$

This is an immediate consequence of the preceding theorem, because if α is the angle corresponding to the maximal dihedral wedge bounded by the planes of a pair of adjacent faces of D, then the coefficients of QCf of D satisfy (59) with this α, and if $\alpha_0 = \inf \alpha$, where the infimum is taken over all polyhedra with m faces, then $0 < \alpha_0 \leqq \dfrac{(m-n)\pi}{m-n+2}$, and (59) yields inequalities (60).

COROLLARY 2. *If* $D \subset R^n$ *is a rectangular parallelipiped, then*

$$K_I(D) \geqq 2, \quad K_O(D) \geqq 2^{\frac{1}{n-1}}, \quad K(D) \geqq 2.$$

We can also use the preceding theorem to obtain lower bounds for the coefficients of QCf of a domain with piecewise smooth boundary.

COROLLARY 3. *Suppose that* $Z = \{x;\ x = (x^1, r, \vartheta_2, \ldots, \vartheta_{r-1}),\ 0 \leqq r < b, 0 < x^1 < h\}$ *then*

$$K_I(Z) \geqq 2, \quad K_O(Z) \geqq 2^{\frac{1}{n-1}}.$$

Fix $0 < a < b$, h and let

$$g(\rho) = \begin{cases} \sqrt{b^2 - \rho^2} & \text{whenever } \rho \leqq a, \\ \sqrt{b^2 - a^2} & \text{whenever } \rho > a. \end{cases}$$

From corollary 1 of lemma 2.17.3, it follows that $f(x) = [x^1, x^2 + g(\rho), x^3, \ldots, x^n]$, where $\rho^2 = (x^3)^2 + \cdots + (x^n)^2$, is a QCfH of R^n onto itself and that $K(f) \to 1$ as $a \to 0$. This QCfH maps the cylinder Z onto a domain Z^*, the point $x_0 = (0, -b, 0, \ldots, 0)$ into O and the spherical neighbourhood of x_0 obtained as the intersection $S(b) \cap B(x_0, r_0)$, with $r_0 < a$, onto an $(n-1)$-ball centreed at O and lying in the plane $x^{*2} = 0$. Then $Z^* \cap U = \Delta \cap U$, where $U = B(a)$ and Δ is the quarter space $\{x;\ x^1 > 0,\ x^2 > 0\}$, $\left(\text{i.e. a dihedral wedge of angle } \frac{\pi}{2}\right)$. Hence, by the preceding theorem,

$$K_I(Z)\,K(f) \geqq K_I(Z^*) \geqq 2, \quad K_O(Z)\,K(f) \geqq K_O(Z^*) \geqq 2^{\frac{1}{n-1}},$$

and letting $a \to 0$, yields

$$K_I(Z) \geqq 2, \quad K_O(Z) \geqq 2^{\frac{1}{n-1}}.$$

Bounds for the outer coefficients of QCf of an infinite cylinder.

We shall calculate in this paragraph the outer coefficient of QCf of an infinite circular cylinder. For this, we require previously some bounds for the inner coefficient of QCf of right circular cylinder.

THEOREM 14. *If* $Z = \left\{x;\ (x^1)^2 + \cdots + (x^{n-1})^2 < b^2,\ |x^n| < \frac{h}{2}\right\}$ *and* $h \in (0, 2b)$, *then*

$$\frac{n\omega_n A_3 b}{2\omega_{n-1} h} \leqq K_I(Z) \leqq 2^{\frac{n}{2}} \cot\frac{\pi}{8} \operatorname{cosec}^{n-2}\frac{\pi}{8}\,\frac{b}{h},$$

where A_3 *is the constant in theorem 8.*

Considering the ball $B(\sqrt{b^2+h^2})$ and the sphere $S(h)$, then the hypotheses of theorem 8 are fulfilled and inequality (35) implies the first part of the preceding inequality.

In order to establish also the second part of this inequality, we perform first the homeomorphism $f(x) = \left(\frac{x^1}{b}, \ldots, \frac{x^{n-1}}{b}, \frac{2x^n}{h}\right)$ which maps Z onto a right circular unit cylinder Z^*, so that $B \subset Z^* \subset B(\sqrt{2})$, and then, by theorem 4, it follows

$$K_I(Z^*) \leqq 2^{-\frac{n-2}{2}} \cot\frac{\pi}{8} \operatorname{cosec}^{n-2}\frac{\pi}{8},$$

hence, by lemma 2.17.1, $K_I(f) = \frac{2b}{h}$ and we obtain

$$K_I(Z) \leqq K_I(f)\, K_I(Z^*) \leqq 2^{-\frac{n-4}{2}} \cot\frac{\pi}{8} \operatorname{cosec}^{n-2}\frac{\pi}{8}\,\frac{b}{h}.$$

We say that a domain is an *infinite circular cylinder* if it can be mapped by means of a similarity transformation onto the domain

$$Z = \{x;\, x = (x^1, r, \vartheta_1, \ldots, \vartheta_{n-2}),\ 0 \leqq r < 1, |x| < \infty\}. \tag{61}$$

LEMMA 4. *Suppose D is the half space $x^n > 0$, that E_0 and E_1 are continuous in $\bar{D}$, that both E_0 and E_1 meet $S(a)$ where $a > 0$, and that $O \in E_0$ and $\infty \in E_1$. If Γ is the family of arcs which join E_0 and E_1 in D, then*

$$M(\Gamma) \geqq \psi(1).$$

Indeed, by hypothesis there exist two points $x_0 \in S(a) \cap E_0$, and $x_1 \in S(a) \cap E_1$, so that, by lemma 2, $M(\Gamma) \geqq \psi\left(\left|\frac{x_0}{x_1}\right|\right) = \psi(1)$, where $\psi(1)$ is the modulus of the family of arcs joining the segment $\{-1 \leqq x^1 \leqq 0,\ x^k = 0\ (k = 2, \ldots, n)\}$ with the ray $\{1 \leqq x^1 \leqq \infty,\ x^k = 0\ (k = 2, \ldots, n)\}$.

THEOREM 15. *Suppose that Z is the cylinder (61), Z^* is the half space $x^{*n} > 0$, that f is a homeomorphism of $\bar{Z} - \{\infty\}$ onto $Z^* - \{O\} - \{\infty\}$, and that*

$$\lim_{x^n \to -\infty} f(x) = O, \quad \lim_{x^n \to \infty} f(x) = \infty. \tag{62}$$

Then for each $a^* > 0$, $f^{-1}[S(a^*) \cap \bar{Z}^*]$ lies between two planes $x^n = a_0$ and $x^n = a_1$, where

$$0 \leqq a_1 - a_0 \leqq \left[\frac{\omega_{n-1}}{\psi(1)}\right]^{\frac{1}{n-1}} K_I(f)^{\frac{1}{n-1}}. \tag{63}$$

Fix $a^* > 0$, let $\sigma^* = S(a^*) \cap Z^*$, and set

$$a_0 = \inf_{x \in \sigma} x^n, \ a_1 = \sup_{x \in \sigma} x^n,$$

where $\sigma = f^{-1}(\sigma^*)$. Clearly we may assume that $a_0 < a_1$, for otherwise there is nothing to prove. Next let

$$E_0 = \{x; x \in \bar{Z}, -\infty < x^n \leqq a_0\}, \ E_1 = \{x; x \in \bar{Z}, a_1 \leqq x^n < \infty\},$$

and let Γ be the family of arcs which join E_0 and E_1 in Z. Then, by virtue of (1.5.41),

$$M(\Gamma) = \frac{\omega_{n-1}}{(a_1 - a_0)^{n-1}}. \tag{64}$$

Next it follows from (62), that $\overline{E_0^*} = E_0^* \cup \{O\}$ and $\overline{E_1^*} = E_1^* \cup \{\infty\}$. Hence $\Gamma^* = f(\Gamma)$ is the family of arcs which join the continua $\overline{E_0^*}$ and $\overline{E_1^*}$ in Z^*. Finally since $\overline{E_0^*} \cap S(a^*)$, $E_1^* \cap S(a^*) \neq \emptyset$,

$$M(\Gamma^*) \geqq \psi(1),$$

by virtue of the preceding lemma, hence, by (64) and lemma 2.17.2, we obtain

$$K_I(f) \geqq \frac{M(\Gamma^*)}{M(\Gamma)} \geqq \frac{\psi(1)(a_1 - a_0)}{\omega_{n-1}},$$

which implies (63).

THEOREM 16. *Suppose that Z is an infinite circular cylinder. Then*

$$\left(\frac{q}{n-1}\right)^{\frac{n-1}{2}} \leqq K_O(Z) \leqq \left(\frac{q}{n-1}\right)^{n-2}, \tag{65}$$

where $q = \left(\frac{\pi}{2}\right)$ *is as in* (1.5.35).

We may assume for convenience of notation, that Z is the cylinder in (61). Next let $(r^*, \vartheta_1^*, \ldots, \vartheta_{n-1}^*)$ be spherical coordinates in R^n, where the polar angle ϑ_1^* is measured from the positive half of the x^{*1}-axis, let Z^* be the half space $x^{*1} > 0$, and set

$$f_1(x^1, r, \vartheta_2, \ldots, \vartheta_{n-1}) = (r^*, \vartheta_1^*, \vartheta_2, \ldots, \vartheta_{n-1}), \tag{66}$$

where

$$r = \left(\frac{1}{q}\int_0^{\vartheta_1^*} \sin^{\frac{2-n}{n-1}} u\,du\right)^{n-1}, \ x^1 = \frac{n-1}{q}\log r^*. \tag{67}$$

Since

$$(dx^1)^2 + \cdots + (dx^n)^2 = (dx^1)^2 + dr^2 +$$

$$+ r^2 d\vartheta_2^2 + \cdots + r^2 \sin^2\vartheta_2 \cdots \sin^2\vartheta_{n-2} d\vartheta_{n-1}^2 =$$

$$= \frac{(n-1)^2 dr^{*2}}{q^2 r^{*2}} + \frac{(n-1)^2}{q^2 r^{*2}} \left(\frac{r}{\sin\vartheta_1^*}\right)^{2\left(\frac{n-2}{n-1}\right)} r^{*2} d\vartheta_1^{*2} +$$

$$+ \left(\frac{r}{r^* \sin\vartheta_1^*}\right)^2 r^{*2} \sin^2\vartheta_1^* d\vartheta_2^2 + \cdots +$$

$$+ \left(\frac{r}{r^* \sin\vartheta_1^*}\right)^2 r^{*2} \sin^2\vartheta_1^* \sin^2\vartheta_2 \cdots \sin^2\vartheta_{n-2} d\vartheta_{n-1}^2,$$

then $f_1 : Z \rightleftarrows Z^*$, $f_1 \in C^1$, maps each infinitesimal sphere onto an infinitesimal ellipsoid whose axes are proportional to

$$\frac{qr^*}{n-1}, \frac{qr^*}{n-1}\left(\frac{\sin\vartheta_1^*}{r}\right)^{\frac{n-2}{n-1}}, \frac{r^*\sin\vartheta_1^*}{r}, \ldots, \frac{r^*\sin\vartheta_1^*}{r}. \tag{68}$$

In order to prove that $\frac{r^* q}{n-1}\left(\frac{\sin\vartheta_1^*}{r}\right)^{\frac{n-2}{n-1}}$ is proportional to the semimajor axis, we show first that

$$\left(\frac{r}{\sin\theta_1^*}\right)^{\frac{n-2}{n-1}} = \frac{1}{q^{n-2}} (\sin\theta_1^*)^{\frac{2-n}{n-1}} \left(\int_0^{\theta_1^*} \sin^{\frac{2-n}{n-1}} u \, du\right)^{n-2} \tag{69}$$

increases from $\left(\frac{n-1}{q}\right)^{n-2}$ to 1 as ϑ_1^* increases from 0 to $\frac{\pi}{2}$. It is easy to verify by means of an elementary calculus, that (69) comes to $\left(\frac{n-1}{q}\right)^{n-2}$ for $\vartheta_1^* = 0$ and to 1 for $\vartheta_1^* = \frac{\pi}{2}$. Next, in order to prove that (69) is an increasing function, we differentiate it, which yields

$$\frac{2-n}{n-1}(\sin \vartheta_1^*)^{\frac{3-2n}{n-1}} \cos \vartheta_1^* \left(\int_0^{\vartheta_1^*} \sin^{\frac{2-n}{n-1}} u\, du\right)^{n-2} +$$

$$+ (n-2)(\sin \vartheta_1^*)^{\frac{2(2-n)}{n-1}} \left(\int_0^{\vartheta_1^*} \sin^{\frac{2-n}{n-1}} u\, du\right)^{n-3} = (n-2)\cos \vartheta_1^* \cdot$$

$$\cdot \left(\int_0^{\vartheta_1^*} \sin^{\frac{2-n}{n-1}} u\, du\right)^{n-3} (\sin \vartheta_1^*)^{\frac{2(2-n)}{n-1}} \left[\frac{1}{\cos \vartheta_1^*} - \frac{\int_0^{\vartheta_1^*} \sin^{\frac{2-n}{n-1}} u\, du}{(n-1)(\sin \vartheta_1^*)^{\frac{1}{n-1}}}\right],$$

where

$$(n-2)\cos \vartheta_1^* \left(\int_0^{\vartheta_1^*} \sin^{\frac{2-n}{n-1}} u du\right)^{n-3} (\sin \vartheta_1^*)^{\frac{2(2-n)}{n-1}} \geqq 0,$$

and where

$$\frac{1}{\cos \vartheta_1^*} - \frac{1}{(n-1)(\sin \vartheta_1^*)^{\frac{1}{n-1}}} \int_0^{\vartheta_1^*} \sin^{\frac{2-n}{n-1}} u du \tag{70}$$

is equal to zero for $\vartheta_1^* = 0$. And now, if (69) were not strictly increasing, then $-\frac{1}{n-1}(\sin \vartheta_1^*)^{\frac{1}{n-1}} \int_0^{\vartheta_1^*} \sin^{\frac{2-n}{n-1}} u du$ would be strictly increasing in $\left[0, \frac{\pi}{2}\right]$ and then (70) would be strictly increasing as the sum of two strictly increasing functions, and since (70) is equal to 0 for $\vartheta_1^* = 0$ and is strictly increasing for $\vartheta_1^* \in \left[0, \frac{\pi}{2}\right]$, it follows that (70) is non-negative in this interval, and then, (69) is strictly increasing, too, in the same interval, contradicting so the hypothesis that (69) would not be strictly increasing in $\left[0, \frac{\pi}{2}\right]$. Hence

$$\frac{r^* \sin \vartheta_1^*}{r} \leqq \frac{qr^*}{n-1}\left(\frac{\sin \vartheta_1^*}{r}\right)^{\frac{n-2}{n-1}}, \quad \frac{q^*}{n-1} \leqq \frac{qr^*}{n-1}\left(\frac{\sin \vartheta_1^*}{r}\right)^{\frac{n-2}{n-1}};$$

and then, from lemma 2.17.1, its corollary, formula (2.1.13) and taking into account that $f_1 \in C^2$, it follows that

$$K_O(f_1) = \operatorname*{ess\,sup}_{x\in Z} \frac{\Lambda_{f_1}(x)^n}{J_G(x)} = \operatorname*{ess\,sup}_{x\in Z} \frac{\dfrac{r^{*n}q^n}{(n-1)^n}\left(\dfrac{\sin \vartheta_1^*}{r}\right)^{n\left(\frac{n-2}{n-1}\right)}}{\dfrac{r^{*n}q^2}{(n-1)^2}\left(\dfrac{\sin \vartheta_1^*}{r}\right)^{n\left(\frac{n-2}{n-1}\right)}} = \left(\frac{q}{n-1}\right)^{n-2}, \tag{71}$$

and since we can map Z^* onto B by means of a Möbius transformation g and

$$K_O(g \circ f_1) \leqq K_O(g) K_O(f_1) = K_O(f_1),$$

$$K_O(f_1) = K_O(g^{-1} \circ g \circ f_1) \leqq K_O(g^{-1}) K_O(g \circ f_1) = K_O(g \circ f_1),$$

hence

$$K_O(g \circ f_1) = K_O(f_1), \tag{72}$$

we have

$$K_O(Z) \leqq K_O(g \circ f_1) = K_O(f_1) = \left(\frac{q}{n-1}\right)^{n-2},$$

i.e. the second part of the double inequality (65).

To complete the proof it is sufficient to show that

$$K_O(f) \geqq \left(\frac{q}{n-1}\right)^{\frac{n-1}{2}},$$

for every QCfH $f: Z \rightleftarrows Z^*$. Choose such a mapping f and let f_1 be the mapping given in (66) and (67). Then $f \circ f_1^{-1}$ is a QCfH of Z^* onto itself which can be extended by theorem 4.3 to be a homeomorphism of $\overline{Z^*}$ onto itself. We can next choose a Möbius transformation g such that $h = g \circ f \circ f_1^{-1}: \overline{Z^*} \rightleftarrows \overline{Z^*}$ is a homeomorphism with $h(O) = O$, $h(\infty) = \infty$. Since $g \circ f = h \circ f_1$, it follows from the properties of f_1 that we can extend $g \circ f$ to be a homeomorphism of $\overline{Z} - \{\infty\}$ onto $\overline{Z^*} - \{\infty\} - \{O\}$ such that

$$\lim_{x^1 \to -\infty} g \circ f(x) = O, \quad \lim_{x^1 \to \infty} g \circ f(x) = \infty.$$

Finally, arguing as for (72), $K_O(g \circ f) = K_O(f)$, so that, we may assume, without loss of generality, that the given mapping f satisfies the hypotheses of the preceding theorem. Now choose $0 < a^* < b^*$, and let D^*, σ^* and E^* be the parts of $x^{*1} > 0$, $x^{*1} = 0$ and the positive x^{*1}-axis bounded by $S(a^*)$ and $S(b^*)$. Next let Γ_1^*

be the family of arcs which join E^* to σ^* in D^* and let Γ_2^* be the family of arcs which join $S(a^*)$ to $S(b^*)$ in σ^*. Then by virtue of proposition 1.5.22,

$$M(\Gamma_1^*) = \frac{(n-1)\omega_{n-1}}{q^{n-1}} \log\frac{b^*}{a^*}, \tag{73}$$

while (1.5.9) implies that

$$M^{\sigma^*}(\Gamma_2^*) = (n-1)\omega_{n-1}\left(\log\frac{b^*}{a^*}\right)^{2-n}. \tag{74}$$

From the preceding theorem it follows that f^{-1} maps $S(a^*)\cap\overline{Z^*}$ and $S(b^*)\cap\overline{Z^*}$ into $a_0 \leqq x^1 \leqq a_1$ and $b_0 \leqq x^1 \leqq b_1$, respectively, where both $a_1 - a_0$ and $b_1 - b_0$ satisfy (63).

Since $\Gamma_1 = f^{-1}(\Gamma_1^*)$ is the family of arcs joining the portion of the x^1-axis and the cylindrical surface of the cylinder (61) between the planes $x^1 = a_0, b_1$ in the portion of (61) between those two planes, and since Γ_1 contains, for $b_0 > 1$, the family Γ_1' of the arcs which join the portion of the x^1-axis and of the cylindrical surface between the planes $x^1 = a_1, b_0$ in the portion of (61) between the same planes, then for $b_0 > 1$, we obtain

$$M(\Gamma_1) \geqq M(\Gamma_1') \geqq \frac{\omega_{n-1}(b_0 - a_1)}{(n-1)^{n-2}},$$

from proposition 1.5.21 and corollary 2 of proposition 1.5.3, whence, taking into account that $M(\Gamma_1) \geqq 0$, it follows that

$$M(\Gamma_1) \geqq \frac{\omega_{n-1}(b_0 - a_1)}{(n-1)^{n-2}} \tag{75}$$

whatsoever would be the relation between a_1 and b_0. Next, let Γ_2' be the family of arcs which join the two intersections of the cylindrical surface of Z with the planes $x^1 = a_0, b_1$ in the portion σ' of this cylindrical surface between the same planes, and let $\gamma \in \Gamma_2'$ be a line segment parallel to the x^1-axis. Then $\rho \in F(\Gamma_2')$ and Hölder's inequality yield

$$1 \leqq \left(\int_\gamma \rho\, ds\right)^{n-1} \leqq \int_\gamma \rho^{n-1}\, dx^1 \left(\int_{a_0}^{b_1} dx^1\right)^{n-2} = (b_1 - a_0)^{n-2}\int_\gamma \rho^{n-1}\, dx^1.$$

Integrating over S^{n-2} and taking into account that $\Gamma_2' > \Gamma_2$, we obtain

$$\frac{(n-1)\omega_{n-1}}{(b_1 - a_0)^{n-2}} \leqq \int_{\sigma'} \rho^{n-1}\, d\sigma,$$

from corollary 1 of proposition 1.5.3, hence

$$\frac{(n-1)\omega_{n-1}}{(b_1-a_0)^{n-2}} \leqq M^{\sigma'}(\Gamma_2') \leqq M^{\sigma}(\Gamma_2). \tag{76}$$

Now σ and σ^* are free admissible boundary surfaces of Z and Z^*, respectively. Hence by theorem 4.13, f_σ, the restriction of f to σ, is a QCf of σ onto σ^* with maximal dilatation $K(f_0)$ satisfying the inequality (4.27) and we have

$$M^{\sigma}(\Gamma_2) \leqq K(f_\sigma) M^{\sigma^*}(\Gamma_2^*) \leqq K_O(f)^{n-2} M^{\sigma^*}(\Gamma_2^*). \tag{77}$$

by theorem 2.17.4. If we combine lemma 2.17.2 with the inequalities (73), (74), (75) and (76), we obtain

$$\frac{\omega_{n-1}^{n-1}(b_0-a_1)^{n-2}}{(n-1)^{(n-2)^2-1}(b_1-a_0)^{n-2}} \leqq M(\Gamma_1)^{n-2} M^{\sigma}(\Gamma_2) \leqq$$

$$\leqq K_O(f)^{2(n-2)} M(\Gamma_1^*)^{n-2} M^{\sigma^*}(\Gamma_2^*) = \frac{(n-1)^{n-1}\omega_{n-1}^{n-1}}{q^{(n-1)(n-2)}} K_O(f)^{2(n-2)}.$$

Now this inequality holds for all $0 < a^* < b^*$, while (62) and the fact that $a_1 - a_0$ and $b_1 - b_0$ satisfy (63), imply that

$$\lim_{\substack{a^*\to 0\\ b^*\to\infty}} \frac{b_0-a_1}{b_1-a_0} = 1, \tag{78}$$

yielding the left-hand side of the inequality (65), and the proof for theorem 16 is complete.

Remark. 1. If $n=3$, then $K_0(Z) = \dfrac{q}{2}$ (F. Gehring and J. Väisälä [1]).

2. While this English version of the book was in print, Kari Hag (in her Ph. D. Thesis) improved the preceding theorem, establishing that

$$K_O(Z) = \left(\frac{q}{n-1}\right)^{n-2}.$$

This was done as follows: first she proved that

$$K_O(f_\sigma) \leqq K_O(f)$$

and

$$M^{\sigma}(\Gamma_2) \leqq K_O(f_\sigma) M^{\sigma^*}(\Gamma_2^*),$$

improving (4.27) and (2.17.28). Hence, she obtains

$$M^{\sigma}(\Gamma_2) \leqq K_0(f)M^{\sigma*}(\Gamma_2^*)$$

instead of (77), and arguing as in the preceding theorem, she gets

$$\frac{\omega_{n-1}^{n-1}(b_0 - a_1)^{n-2}}{(n-1)^{(n-2)2-1}(b_1 - a_0)^{n-2}} \leqq M(\Gamma_1)^{n-2}M^{\sigma}(\Gamma_2) \leqq K_0(f)^{n-1}M(\Gamma_1^*)^{n-2}M^{\sigma*}(\Gamma_2^*) =$$

$$= \frac{(n-1)^{n-1}\omega_{n-1}^{n-1}}{q^{(n-1)(n-2)}} K_O(f)^{n-1},$$

implying

$$K_O(z) \geqq \left(\frac{q}{n-1}\right)^{n-2},$$

which, together with the inverse inequality established in the first part of the proof of the preceding theorem, yields the desired formula.

Bounds for inner and total coefficients of QCf of an infinite cylinder.

THEOREM 17. *Suppose Z is an infinite circular cylinder. Then*

$$\left(\frac{n\omega_n}{2\omega_{n-1}}\right)^{\frac{1}{n}} \leqq K_I(Z) \leqq \sqrt{2}. \tag{79}$$

Assume that Z is the cylinder in (61), that Z^* is the half space $x^{*1} > 0$, and $f_1 : Z \rightleftarrows Z^*$ is as in (66), where this time

$$r = \frac{\sin \theta_1^*}{\sqrt{2}\sin\left(\theta_1^* + \frac{\pi}{4}\right)}, \quad x^1 = \frac{\log r^*}{\sqrt{2}}.$$

Clearly $f_1 \in C^1$ and

$$(dx^1)^2 + \cdots + (dx^n)^2 = (dx^1)^2 + dr^2 + r^2 d\vartheta_2^2 + \cdots + r^2 \sin^2\vartheta_2 \cdots \sin^2\vartheta_{n-2}\, d\vartheta_{n-1}^2 =$$

$$= \frac{dr^{*2}}{2r^{*2}} + \frac{r^{*2}\, d\vartheta_1^{*2}}{4r^{*2}\sin^4\left(\vartheta_1^* + \frac{\pi}{4}\right)} + \frac{r^{*2}\sin^2\vartheta_1^*\, d\vartheta_2^2}{2r^{*2}\sin^2\left(\vartheta_1^* + \frac{\pi}{4}\right)} + \cdots +$$

$$+ \frac{r^{*2}\sin^2\vartheta_1^*\sin^2\vartheta_2 \cdots \sin^2\vartheta_{n-2}\, d\vartheta_{n-1}^2}{2r^{*2}\sin^2\left(\vartheta_1^* + \frac{\pi}{4}\right)},$$

hence it follows that f_1 maps each infinitesimal sphere onto an infinitesimal ellipsoid whose axes are proportional to

$$\sqrt{2}r^*, 2r^* \sin^2\left(\vartheta_1^* + \frac{\pi}{4}\right), \sqrt{2}r^* \sin\left(\vartheta_1^* + \frac{\pi}{4}\right), \ldots, \sqrt{2}r^* \sin\left(\vartheta_1^* + \frac{\pi}{4}\right),$$

where $\sqrt{2}r^* \sin\left(\vartheta_1^* + \frac{\pi}{4}\right)$ is proportional to the semiminor axis of the infinitesimal ellipsoid because, for $0 \leqq \vartheta_1^* \leqq \frac{\pi}{2}$,

$$\sqrt{2} \sin\left(\vartheta_1^* + \frac{\pi}{4}\right) \leqq 2 \sin^2\left(\vartheta_1^* + \frac{\pi}{4}\right), \sqrt{2}.$$

Then, by lemma 2.17.1, its corollary and (2.1.12), and taking into account that $f \in C^1$, it follows that

$$K_I(f_1) = \operatorname*{ess\,sup}_{x \in Z} \frac{J_G(x)}{\lambda_f(x)^n} = \frac{(\sqrt{2})^{n+1} r^{*n} \sin^n\left(\vartheta_1^* + \frac{\pi}{4}\right)}{\sqrt{2}^n r^* \sin^n\left(\vartheta_1^* + \frac{\pi}{4}\right)} = \sqrt{2}.$$

Then because we can map Z^* onto B by means of a Möbius transformation g, on account of (72), we have

$$K_I(Z) \leqq K_I(g \circ f_1) = K_I(f_1) = \sqrt{2},$$

i.e. the right-hand part of (79) is proved.

To establish the left-hand part of (79), we must show that

$$K_I(f) \geqq \left(\frac{n\omega_n}{2\omega_{n-1}}\right)^{\frac{1}{n}}$$

for each QCfH $f: Z \rightleftharpoons Z^*$. As in the proof of the preceding theorem, we may assume that f satisfies the hypotheses of theorem 15. Fix $0 < a^* < b^*$ so that $a_1 < b_0$, let D^* and σ^* be the parts of $x^{*1} > 0$ and $x^{*1} = 0$ bounded by $S(a^*)$ and $S(b^*)$, and let Γ_1^*, Γ_2^* be the families of arcs which join $S(a^*)$ to $S(b^*)$ in D^* and σ^*, respectively. Then, by virtue of (1.5.9),

$$M(\Gamma_1^*) = \frac{n\omega_n}{2\left(\log \frac{b^*}{a^*}\right)^{n-1}}, \quad M^{\sigma^*}(\Gamma_2^*) = \frac{(n-1)\omega_{n-1}}{\left(\log \frac{b^*}{a^*}\right)^{n-2}}. \tag{80}$$

Theorem 15 implies that $f^{-1}[S(a^*) \cap \overline{Z^*}]$ and $f^{-1}[S(b^*) \cap \overline{Z^*}]$ lie in $a_0 \leqq x^1 \leqq a_1$ and in $b_0 \leqq x^1 \leqq b_1$, respectively, where (63) holds. Let us intersect Z with the planes $x^1 = a_1, b_0$ and let Γ_1' be the family of arcs which join the regions of the two planes bounded by the cylindrical surface (i.e. the bases). Clearly, $\Gamma_1 > \Gamma_1'$, hence we obtain

$$M(\Gamma_1) \leqq M(\Gamma_1') = \frac{\omega_{n-1}}{(b_0 - a_1)^{n-1}}. \tag{81}$$

from corollary 1 of proposition 2.5.3 and (1.5.41). Now σ and σ^* are free admissible boundary surfaces of Z and Z^*. Thus,

$$M^{\sigma}(\Gamma_2) \leqq K(f_\sigma)M^{\sigma^*}(\Gamma_2^*) \leqq K_I(f)^{n-2}M^{\sigma^*}(\Gamma_2^*) \tag{82}$$

by virtue of theorem 2.17.4 and (4.27). If we combine the inequalities (76), (80), (81) and lemma 2.17.2, we obtain

$$\frac{n^{n-2}\omega_n^{n-2}}{2^{n-2}(n-1)^{n-1}\omega_{n-1}^{n-1}} = \frac{M(\Gamma_1^*)^{n-2}}{M^{\sigma^*}(\Gamma_2^*)^{n-1}} \leqq \frac{K_I(f)^{n(n-2)}M(\Gamma_1)^{n-2}}{M^{\sigma}(\Gamma_2)^{n-1}} \leqq$$

$$\leqq \frac{K_I(f)^{n(n-2)}(b_1 - a_0)^{(n-1)(n-2)}}{(n-1)^{n-1}\omega_{n-1}(b_0 - a_1)^{(n-1)(n-2)}}.$$

Again (78) holds, and letting $a^* \to 0$, $b^* \to 0$, yields $K_I(f) \geqq \left(\frac{n\omega_n}{2\omega_{n-1}}\right)^{\frac{1}{n}}$, which, since f is arbitrary, implies the left-hand part of (79).

We shall require the following analogue of theorem 13 in order to derive some lower bounds for the coefficients of QCf of a domain which has a spire in its boundary.

THEOREM 18. *Suppose that U is a half space, and that $D \cap U = Z \cap U$, where Z is an infinite circular cylinder whose axis is perpendicular to ∂U. Then the coefficients of* QCf *of D are not less than the corresponding coefficients of* QCf *of Z. In particular,*

$$K_I(D) \geqq \left(\frac{n\omega_n}{2\omega_{n-I}}\right)^{\frac{1}{n}}, \quad K_O(D) \geqq \left(\frac{q}{n-1}\right)^{\frac{n-1}{2}}. \tag{83}$$

Assume, for convenience of notations, that U is the half space $x^1 > 0$, choose $x_0 \in Z \cap U$, and for each $m > 0$, let

$$D_m = \{x; x + me_1 \in D\}.$$

Then as in the proof of theorem 12 it is easy to show that the D_m converge to their kernel Z at x, hence we obtain (58), which implies (83), on account of (65) and (79).

THEOREM 19. *Suppose that Z is the cylinder in (61) and that Z' and σ are the parts of Z and ∂Z which lie in the half space $x^1 < 0$. Then there exists a homeo-*

morphism $f : Z\bigcup\sigma \rightleftharpoons Z'\bigcup\sigma$ *such that* $f(x) = x$ *for* $x \in \sigma$,

$$K(f) \leqq 2^{n-1}\left(\frac{q}{n-1}\right)^n \tag{84}$$

in Z, *and*

$$\Lambda_f(x) \leqq 6\left(\frac{q}{n-1}\right)^{n-1} \tag{85}$$

in Z'.

Let $y = f_1(x)$ be the homeomorphism given in (66) and (67), and let U denote the half space $y^1 > 0$. Then there exists an inversion $Y = f_2(y)$ with respect to the sphere centred at $(0, -1, 0, \ldots, 0)$ and with the radius $\sqrt{2}$. This Möbius transformation preserves the half space U and carries $U\bigcap B$ onto the dihedral wedge

$$D_{\frac{\pi}{2}} = \left\{\rho, \varphi_1, Y^3, \ldots, Y^n; 0 < \varphi_1 < \frac{\pi}{2}, |Y| < \infty\right\} = \{Y; Y^1 > 0, Y^2 > 0\},$$

where φ_1 is the angle of the projection of the vector Y on the Y^1Y^2-plane with the Y^2-axis. We see that $f_2 f_1 : Z \rightleftharpoons U$ is a homeomorphism which maps Z' onto $D_{\frac{\pi}{2}}$ and σ into the closed half plane

$$\prod = \{\rho, \varphi_1, Y^3, \ldots, Y^n; \varphi_1 = 0, \rho > 0\} = \{Y; Y^1 = 0, Y^2 > 0\}.$$

Now let

$$f_3(\rho, \varphi_1, Y^3, \ldots, Y^n) = \left(\rho, \frac{\varphi_1}{2}, Y^3, \ldots, Y^n\right)$$

be the folding mapping which maps $D_\pi = U$ onto $D_{\frac{\pi}{2}}$ and satisfies $f_3(Y) = Y$ for all $Y \in \Pi$. Hence the homeomorphism $f = f_1^{-1} \circ f_2^{-1} \circ f_3 \circ f_2 \circ f_1$ maps $Z\bigcup\sigma$ onto $Z'\bigcup\sigma$, $f(x) = x$ for $x \in \sigma$, and by (2.17.2), which imply $K(f_2) = K(f_2^{-1}) = 1$, it follows that

$$K_I(f) \leqq K_I(f_1)\, K_O(f_1) K_I(f_3), \quad K_O(f) \leqq K_O(f_1)\, K_I(f_1)\, K_O(f_3), \tag{86}$$

hence lemma 2.17.1 and (71), (68) yield

$$K_I(f_1) = \operatorname*{ess\,sup}_{x \in Z} \frac{J_G(x)}{\lambda_f(x)^n} = \operatorname*{ess\,sup}_{x \in Z} \frac{\dfrac{\rho^n q^2}{(n-1)^2}\left(\dfrac{\sin\varphi_1}{r}\right)^{\frac{n(n-2)}{n-1}}}{\rho^n\left(\dfrac{\sin\varphi_1}{r}\right)^n} =$$

$$\operatorname*{ess\,sup}_{x \in Z} \frac{q^2}{(n-1)}\left(\frac{r}{\sin\varphi_1}\right)^{\frac{n}{n-1}} = \left(\frac{q}{n-1}\right)^2, \quad K(f_3) = 2^{n-1} \tag{87}$$

and (2.17.1), (86), (71) and (87) imply that

$$K(f) \leqq K_I(f_1) K_O(f_1) K(f_3) = 2^{n-1}\left(\frac{q}{n-1}\right)^n.$$

Finally from (68) we see that

$$|f_1(x)| \leqq \lambda_{f_1}(x) \leqq \Lambda_{f_1}(x) = \frac{|f_1(x)|q}{n-1}\left(\frac{\sin\varphi_1}{r}\right)^{\frac{n-2}{n-1}} \leqq |f_1(x)|\left(\frac{q}{n-1}\right)^{n-1} \tag{88}$$

for $x \in Z$, while a direct calculation yields

$$\Lambda_g(y) \leqq 1, \ |g(y)| \geqq \frac{|y|}{6} \tag{89}$$

for $x \in B \cap U$, where $g = f_2^{-1} \circ f_3 \circ f_2 \cdot$ Inequality (85) follows from (88) and (89), taking into account that $\Lambda_{f_1^{-1}} = \dfrac{1}{\Lambda_{f_1}} \cdot$

Bounds for the outer coefficient of QCf of the cone. We say that a domain is a *circular cone of angle* $\alpha \in (0, \pi)$, if it can be mapped by means of a similarity transformation onto the domain

$$C = \{x; x = (r, \vartheta_1, \ldots, \vartheta_{n-1}, 0 \leqq \vartheta_{n-1} < \alpha, \ 0 < r < \infty\}, \tag{90}$$

where ϑ_1 is the angle the vector x makes with the positive x^n-axis.

We have the following cone analogue of theorem 15:

THEOREM 20. *Suppose that* C *is the cone in* (90), *that* C^* *is the half space* $x^{*_n} > 0$, *that* $f: \overline{C} \rightleftharpoons \overline{C^*}$ *is a homeomorphism with* $f(O) = O$, $f(\infty) = \infty$. *Then for each* $a^* > 0$, $f^{-1}[S(a^*) \cap \overline{C^*}]$ *lies between two spheres* $S(a_0)$ *and* $S(a_1)$, *where*

$$1 \leqq \frac{a_1}{a_0} \leqq e^{\left[\frac{n\omega_n}{\psi(1)}\right]^{\frac{1}{n-1}} K_I(f)^{\frac{1}{n-1}}}. \tag{91}$$

Fix $a > 0$, let $\sigma^* = S(a^*) \cap \overline{C^*}$ and set $a_0 = \inf\limits_{x \in \sigma} |x|$, $a_1 = \sup\limits_{x \in \sigma} |x|$, where $\sigma = f^{-1}(\sigma^*)$. We may assume $a_0 < a_1$, for otherwise there is nothing to prove. Next let

$$E_0 = \{x; x \in \overline{C}, |x| \leqq a_0\}, \quad E_1 = \{x; x \in \overline{C}, |x| \geqq a_1\}$$

and let Γ be the family of arcs which join E_0 to E_1 in C. Then Γ is minorized by the family of arcs Γ' which join $S(a_0)$ to $S(a_1)$ in R^n, and hence, by (1.5.9),

$$M(\Gamma) \leqq M(\Gamma') = \frac{n\,\omega_n}{\left(\log \dfrac{a_1}{a_0}\right)^{n-1}} \cdot \tag{92}$$

Next, since $O \in E_0^*$, and $\infty \in E_1^*$, lemma 4 implies

$$M(\Gamma^*) \geqq \psi(1). \tag{93}$$

From lemma 2.17.2, (92) and (93), it follows that

$$K_I(f) \geqq \frac{M(\Gamma^*)}{M(\Gamma)} = \frac{\psi(1)}{n\omega_n}\left(\log \frac{a_1}{a_0}\right)^{n-1},$$

i.e. (91).

THEOREM 21. *Suppose C is a circular cone of angle α. then*

$$\left[\frac{q}{q(\alpha)}\right]^{\frac{n-1}{2}} \sqrt{\sin\alpha} \leqq K_O(C) \leqq \left[\frac{q}{q(\alpha)}\right]^{n-2} \sin^{\frac{n-2}{n-1}} \alpha, \tag{94}$$

where $q(\alpha)$ is as in (1.5.35) and $q = q\left(\frac{\pi}{2}\right)$.

Assume, for convenience of notation, that C is the cone in (90) and let $C^* = \{x^*; x^{*1} > 0\}$. Next set

$$f(r, \vartheta_1, \ldots, \vartheta_{n-1}) = (r^*, \vartheta_1^*, \vartheta_2, \ldots, \vartheta_{n-1}),$$

where

$$r^* = r^{a(\sin\alpha)^{\frac{2-n}{n-1}}}. \quad q(\vartheta_1^*) = aq(\vartheta_1), \quad a = \frac{q}{q(\alpha)} \geqq 1. \tag{95}$$

Since

$$\frac{d\vartheta_1^*}{d\vartheta_1} = a\left(\frac{\sin\vartheta_1^*}{\sin\vartheta_1}\right)^{\frac{n-2}{n-1}}, \tag{96}$$

it follows

$$(dx^1)^2 + \cdots + (dx^n)^2 = dr^2 + r^2\, d\vartheta_1^2 + r^2 \sin^2\vartheta_1\, d\vartheta_2^2 + \cdots +$$

$$+ r^2 \sin^2\vartheta_1 \cdots \sin^2\vartheta_{n-2}\, d\vartheta_{n-1}^2 = \frac{dr^{*2}}{\left(\frac{r^*}{r}\right)^2 \frac{a^2}{(\sin\alpha)^{\frac{2(n-2)}{n-1}}}} + \frac{r^{*2}\, d\vartheta_1^{*2}}{\left(\frac{r^*}{r}\right)^2 a^2 \left(\frac{\sin\vartheta_1^*}{\sin\vartheta_1}\right)^{\frac{2(n-2)}{n-1}}} +$$

$$+ \frac{r^{*2} \sin^2\vartheta_1^*\, d\vartheta_2}{\left(\frac{r^*\sin\vartheta_1^*}{r\sin\vartheta_1}\right)^2} + \frac{r^{*2}\sin^2\vartheta_1^* \sin^2\vartheta_2\, d\vartheta_3^2}{\left(\frac{r^*\sin\vartheta_1^*}{r\sin\vartheta_1}\right)^2} + \cdots +$$

$$+ \frac{r^{*2}\sin^2\vartheta_1^*\sin^2\vartheta_2 \cdots \sin^2\vartheta_{n-2}\, d\vartheta_{n-1}^2}{\left(\frac{r^*\sin\vartheta_1^*}{r\sin\vartheta_1}\right)^2}.$$

Then $f\colon C \rightleftharpoons C^*$, $f \in C^1$, maps each infinitesimal sphere onto an infinitesimal ellipsoid whose axes are proportional to

$$\frac{r^* a}{r(\sin\alpha)^{\frac{n-2}{n-1}}},\ a\frac{r^*}{r}\left(\frac{\sin\vartheta_1^*}{\sin\vartheta_1}\right)^{\frac{n-2}{n-1}},\ \frac{r^*\sin\vartheta_1^*}{r\sin\vartheta_1},\ \ldots,\ \frac{r^*\sin\vartheta_1^*}{r\sin\vartheta_1}. \tag{97}$$

Next, (95) implies

$$\lim_{\vartheta_1\to 0}\frac{\sin\vartheta_1^*}{\sin\vartheta_1} = a^{n-1},\quad \frac{d}{d\vartheta_1}\left(\frac{\sin\vartheta_1^*}{\sin\vartheta_1}\right) = \tag{98}$$

$$= \frac{(\sin\vartheta_1^*)^{\frac{n-2}{n-1}}\left(a\cos\vartheta_1^*\sin^{\frac{1}{n-1}}\vartheta_1 - \cos\vartheta_1\sin^{\frac{1}{n-1}}\vartheta_1^*\right)}{\sin^2\vartheta_1},$$

and also

$$q(\vartheta_1) \leqq q(\vartheta_1^*),\quad \frac{\sin\vartheta_1^*}{\sin\vartheta_1} \geqq 1,$$

whence

$$\frac{d}{d\vartheta_1}\left[a\cos\vartheta_1^*(\sin\vartheta_1)^{\frac{1}{n-1}} - \cos\vartheta_1(\sin\vartheta_1^*)^{\frac{1}{n-1}}\right] =$$

$$= \sin\vartheta_1(\sin\vartheta_1^*)^{\frac{1}{n-1}}\left[1 - a^2\left(\frac{\sin\vartheta_1^*}{\sin\vartheta_1}\right)^{\frac{2(n-2)}{n-1}}\right] \leqq \sin\vartheta_1(\sin\vartheta_1^*)^{\frac{1}{n-1}}(1-a^2) \leqq 0,$$

so that $a\cos\vartheta_1^*(\sin\vartheta_1)^{\frac{1}{n-1}} - \cos\vartheta_1(\sin\vartheta_1^*)^{\frac{1}{n-1}}$ is monotonic decreasing, and since by (98) we have $\left.\frac{d}{d\vartheta_1}\left(\frac{\sin\vartheta_1^*}{\sin\vartheta_1}\right)\right|_{\vartheta_1=0} = 0$, it follows that $a\cos\vartheta_1^*(\sin\vartheta_1)^{\frac{1}{n-1}} - \cos\vartheta_1(\sin\vartheta_1^*)^{\frac{1}{n-1}} \leqq$ $\leqq 0$ and then also $\frac{d}{d\vartheta_1}\left(\frac{\sin\vartheta_1^*}{\sin\vartheta_1}\right) \leqq 0$, which allows us to conclude that $\left(\frac{\sin\vartheta_1^*}{\sin\vartheta_1}\right)^{\frac{n-2}{n-1}}$ is monotonic decreasing, but since

$$\lim_{\vartheta_1\to\alpha}\left(\frac{\sin\vartheta_1^*}{\sin\vartheta_1}\right)^{\frac{n-2}{n-1}} = \frac{1}{(\sin\alpha)^{\frac{n-2}{n-1}}},$$

by (95), then, on account of (95) and (96), it follows that

$$1 \leqq a^{n-2}\left(\frac{\sin\vartheta_1}{\sin\vartheta_1^*}\right)^{\frac{n-2}{n-1}} = \left[\frac{q(\vartheta_1^*)}{q(\vartheta_1)}\right]^{n-2}\left(\frac{\sin\vartheta_1}{\sin\vartheta_1^*}\right)^{\frac{n-2}{n-1}} \leqq a^{n-2}(\sin\alpha)^{\frac{n-2}{n-1}}$$

whenever $0 \leqq \vartheta_1 \leqq \alpha$, where $0 < \alpha \leqq \dfrac{\pi}{2}$, hence

$$\frac{1}{(\sin\alpha)^{\frac{n-2}{n-1}}} \leqq \left(\frac{\sin\vartheta_1^*}{\sin\vartheta_1}\right)^{\frac{n-2}{n-1}} \leqq \left[\frac{q(\vartheta_1^*)}{q(\vartheta_1)}\right]^{n-2} = a^{n-2} \tag{99}$$

whenever $0 \leqq \vartheta_1 \leqq \alpha$. From (97) and on account of (99), it follows that the semi-major axis of the infinitesimal ellipsoid is $\dfrac{ar^*}{r}\left(\dfrac{\sin\vartheta_1^*}{\sin\vartheta_1}\right)^{\frac{n-2}{n-1}}$, so that

$$K_O(f) = \operatorname*{ess\,sup}_{x\in C}\frac{\Lambda_f(x)^n}{J_G(x)} = \frac{a^n\left(\dfrac{r^*}{r}\right)^n\left(\dfrac{\sin\vartheta_1^*}{\sin\vartheta_1}\right)^{\frac{n(n-2)}{n-1}}}{a^2\left(\dfrac{r^*}{r}\right)^n\left(\dfrac{\sin\vartheta_1^*}{\sin\vartheta_1}\right)^{\frac{n(n-2)}{n-1}}(\sin\alpha)^{\frac{2-n}{n-1}}} =$$

$$= a^{n-2}(\sin\alpha)^{\frac{n-2}{n-1}} = \left[\frac{q}{q(\alpha)}\right]^{n-2}(\sin\alpha)^{\frac{n-2}{n-1}}$$

by virtue of lemma 2.17.1 and (2.1.13). Then since C^* is conformally equivalent to B, we obtain

$$K_O(C) \leqq K_O(f) = \left[\frac{q}{q(\alpha)}\right]^{n-2}(\sin\alpha)^{\frac{n-2}{n-1}}.$$

To complete the proof for (94), we must show that

$$K_O(f) \geqslant \left[\frac{q}{q(\alpha)}\right]^{\frac{n-1}{2}}\sqrt{\sin\alpha} \tag{100}$$

for each QCfH $f: C \rightleftharpoons C^*$. Choose any such mapping f. Then arguing as in the proof of theorem 16, we see we may assume that f satisfies the hypotheses of the preceding theorem. Fix $0 < a^* < b^*$, let D^*, σ^* and E^* be the parts of C^*, $\partial C^* = \{x^*; x^{*1} = 0\}$ and the positive x^{*1}-axis bounded by $S(a^*)$ and $S(b^*)$, and let Γ_1^* and Γ_2^* be the families of arcs which join E^* to σ^* in D^* and $S(a^*)$ to $S(b^*)$ in σ^*, respectively. Then clearly hold (73), (74). The preceding theorem implies $f^{-1}[S(a^*)\cap\overline{C^*}]$ and $f^{-1}[S(b^*)\cap\overline{C^*}]$ lie between $S(a_0)$ and $S(a_1)$ and between $S(b_0)$ and $S(b_1)$, respectively, where

$$1 \leqq \frac{a_1}{a_0}, \frac{b_1}{b_0} \leqq e^{\left[\frac{n\omega_n}{\psi(1)}\right]^{\frac{1}{n-1}} K_I(f)^{\frac{1}{n-1}}}, \quad a_0 < b_1. \tag{101}$$

Hence, we obtain

$$M(\Gamma_1) \geqq \frac{(n-1)\omega_{n-1}}{q(\alpha)^{n-1}} \log \frac{b_0}{a_1}, \tag{102}$$

which is obvious for $b_0 \leqq a_1$. If $b_0 > a_1$, let Γ_1' be the family of arcs which join E and ∂C in the portion of C contained between the spheres $S(a_1)$ and $S(b_0)$. Then

$$M(\Gamma_1) \geqq M(\Gamma_1') \geqq \frac{(n-1)\omega_{n-1}}{q(\alpha)^{n-1}} \log \frac{b_0}{a_1},$$

by corollary 2 of proposition 1.5.3 and proposition 1.5.2.2. Let Γ_2' be the family of arcs which join $\partial C \bigcap S(a_0)$ and $\partial C \bigcap S(b_1)$ in the portion σ' of ∂C between $S(a_0)$ and $S(b_1)$. And now, let $\gamma \in \Gamma_2'$ be a line segment lying on the generatrix of the cone. Then for each $\rho \in F(\Gamma_2')$, by Hölder's inequality

$$1 \leqq \left(\int_\gamma \rho \, \mathrm{d}s\right)^{n-1} = \int \rho^{n-1} r^{n-2} \, \mathrm{d}r \left(\int_{a_0}^{b_1} \frac{\mathrm{d}r}{r}\right)^{n-2} = \left(\log \frac{b_1}{a}\right)^{n-2} \int_\gamma \rho^{n-1} r^{n-2} \, \mathrm{d}r,$$

whence

$$\int_{\sigma'} \rho^{n-1} \mathrm{d}\sigma = \int_{\sigma'} \rho^{n-1} r^{n-2} \sin^{n-2}\alpha \sin^{n-3}\vartheta_2 \cdots \sin\vartheta_{n-2} \, \mathrm{d}r \, \mathrm{d}\vartheta_2 \cdots \mathrm{d}\vartheta_{n-1} \geqq$$

$$\geqq \frac{(n-1)\omega_{n-1} \sin^{n-2}\alpha}{\left(\log \dfrac{b_1}{a_0}\right)^{n-2}}.$$

Since p is arbitrary and $\Gamma_2' > \Gamma_2$, then

$$M^{\sigma}(\Gamma_2) \geqq M^{\sigma'}(\Gamma_2') \geqq \frac{(n-1)\omega_{n-1} \sin^{n-2}\alpha}{\left(\log \dfrac{b_1}{a_0}\right)^{n-2}} \tag{103}$$

by corollary 1 of proposition 1.5.3. Again σ and σ^* are free admissible boundary surfaces of C and C^*, hence, by theorem 4.13, f_σ, the restriction of f to σ, which is a QCfH of σ onto σ^* with maximal dilatation $K(f_\sigma)$, satisfies the inequality (4.27), so that the inequalities (77) hold. If we combine lemma 2.17.2, (73), (74), (102) and (103) we have

$$\left[\frac{(n-1)\omega_{n-1}}{q(\alpha)^{n-2}}\right]^{n-1} \left(\frac{\log \dfrac{b_0}{a_1}}{\log \dfrac{b_1}{a_0}}\right)^{n-2} \sin^{n-2}\alpha \leqq M(\Gamma_1)^{n-2} M^{\sigma}(\Gamma_2) \leqq$$

$$\leqq K_0(f)^{2(n-2)} M(\Gamma_1^*)^{n-2} M^{\sigma^*}(\Gamma_2^*) = \frac{K_0(f)^{2(n-2)}(n-1)^{n-1}\omega_{n-1}^{n-1}}{q^{(n-1)(n-2)}}.$$

Since $f(O) = O$ and $f(\infty)=\infty$, if we let $a^* \to 0$, $b^* \to \infty$, then $\frac{b_1}{a_0} \to \infty$ and (100) follows from (101), which implies

$$\lim_{\substack{a^*\to 0\\ b^*\to\infty}} \frac{\log \frac{b_0}{a_1}}{\log \frac{b_1}{a_0}} = 1, \tag{104}$$

Thus the double inequality (94) is completely established.

Remarks. 1. If $n = 3$, then the double inequality (94) comes to

$$K_0(C) = \frac{q}{q(\alpha)} \sqrt{\sin \alpha}$$

2. Arguing as in remark 2 of theorem 16, Kari Hag (in her Ph. D. Thesis improves this result, establishing that

$$K_0(C)= \left[\frac{q}{q(\alpha)}\right]^{n-2} \sin^{\frac{n-2}{n-1}} \alpha.$$

Bounds for the inner coefficient of QCf of the cone.

THEOREM 22. *Suppose C is a convex circular cone of angle α. Then*

$$\left[\frac{n\omega_n \sin^{n-1}\alpha}{2(n-1)\omega_{n-1}\int\limits_0^\alpha \sin^{n-2}\vartheta_1 \, d\vartheta_1}\right]^{\frac{1}{n}} \leqq K_I(C) \leqq \sqrt{1+\cos\alpha}. \tag{105}$$

Assume that C is the cone in (90) and that $C^* = \{x^*;\ x^{*\mathrm{I}} > 0\}$. Next set

$$f(r, \vartheta_1, \dots, \vartheta_{n-1}) = (r^*, \vartheta_1^*, \vartheta_2, \dots, \vartheta_{n-1}),$$

where

$$r^* = r^a, \cot\vartheta_1^* = \frac{\sin(\alpha - \vartheta_1)}{\sin\vartheta_1}, \quad a = \frac{1}{\sqrt{1-\cos\alpha}}. \tag{106}$$

Then

$$(dx^1)^2 + \cdots + (dx^n)^2 = dr^2 + r^2\, d\vartheta_1^2 + r^2 \sin^2 \vartheta_1\, d\vartheta_2^2 + \cdots +$$

$$+ r^2 \sin^2 \vartheta_1 \cdots \sin^2 \vartheta_{n-2}\, d\vartheta_{n-1}^2 = \frac{dr^{*2}}{a^2\left(\dfrac{r^*}{r}\right)^2} + \frac{r^{*2} d\vartheta_1^{*2}}{\left(\dfrac{r^*}{r}\right)^2 \dfrac{\sin^4 \vartheta_1^*}{\sin^4 \vartheta_1} \sin^2 \alpha} +$$

$$+ \frac{r^{*2} \sin^2 \vartheta_1^*\, d\vartheta_2^2}{\left(\dfrac{r^* \sin \vartheta_1^*}{r \sin \vartheta_1}\right)^2} + \frac{r^{*2} \sin^2 \vartheta_1^* \sin^2 \vartheta_2\, d\vartheta_3^2}{\left(\dfrac{r^* \sin \vartheta_1^*}{r \sin \vartheta_1}\right)} + \cdots +$$

$$+ \frac{r^{*2} \sin^2 \vartheta_1^* \sin^2 \vartheta_2 \cdots \sin^2 \vartheta_{n-2}\, d\vartheta_{n-1}^2}{\left(\dfrac{r^* \sin \vartheta_1^*}{r \sin \vartheta_1}\right)^2},$$

and $f: C \rightleftarrows C^*$, $f \in C^1$, maps each infinitesimal sphere onto an infinitesimal ellipsoid whose axes are proportional to

$$\frac{r^*}{r\sqrt{1 - \cos \alpha}}, \quad \frac{r^* \sin^2 \vartheta_1^*}{r \sin^2 \theta_1} \sin \alpha, \quad \frac{r^* \sin \vartheta_1^*}{r \sin \vartheta_1}, \ldots, \frac{r^* \sin \vartheta_1^*}{r \sin \vartheta_1}.$$

It is not difficult to show, on account of (106), that

$$\frac{1}{\sin^2 \alpha} \leqq \frac{\sin^2 \vartheta_1^*}{\sin^2 \vartheta_1} = \frac{1}{\sin^2 \vartheta_1 + \sin^2 (\alpha - \vartheta_1)} \leqq \frac{1}{2 \sin^2 \dfrac{\alpha}{2}},$$

hence

$$\frac{\sin \vartheta_1^*}{\sin \vartheta_1} \leqq \frac{1}{\sqrt{1 - \cos \alpha}}, \quad \frac{\sin^2 \vartheta_1^*}{\sin^2 \vartheta_1} \sin \alpha,$$

and we conclude that $\dfrac{r^* \sin \vartheta_1^*}{r \sin \vartheta_1}$ is proportional to the semiminor axis, and we obtain

$$K_I(f) = \operatorname*{ess\,sup}_{x \in C} \frac{J_G(x)}{\lambda_f(x)^n} = \frac{\left(\dfrac{r^*}{r}\right)^n \left(\dfrac{\sin \vartheta_1^*}{\sin \vartheta_1}\right)^n \sqrt{1 + \cos \alpha}}{\left(\dfrac{r^*}{r}\right)^n \left(\dfrac{\sin \vartheta_1^*}{\sin \vartheta_1}\right)^n} = \sqrt{1 + \cos \alpha},$$

from lemma 2.17.1 and (2.1.12), and since C^* is conformally equivalent to B, we conclude that

$$K_I(C) \leqq K_I(f) = \sqrt{1 + \cos \alpha},$$

i.e. the right-hand side of (105).

For the left-hand side of (105) we must show that

$$\left[\frac{n\omega_n \sin^{n-1} \alpha}{2(n-1)\omega_{n-1}\int\limits_0^\alpha \sin^{n-2} \vartheta_1 \, d\vartheta_1}\right]^{\frac{1}{n}} \leqq K_I(f)$$

for each QCfH $f: C \rightleftharpoons C^*$. As in the proof of the preceding theorem, we may assume f satisfies the hypotheses of theorem 20. Fix $0 < a^* < b^*$ so that $a_1 < b_0$. This is possible since $f(O) = O$, $f(\infty) = \infty$ on account of (101). Let D^* and σ^* be the parts of C^* and $x^{*1} = 0$ bounded by $S(a^*)$ and $S(b^*)$, and let Γ_1^* and Γ_2^* be the families of arcs which join $S(a^*)$ to $S(b^*)$ in D^* and σ^*, respectively. Then the moduli of Γ_1^*, Γ_2^* are given by (80). Theorem 20 implies that the images of $S(a^*) \cap \overline{C^*}$ and $S(b^*) \cap \overline{C^*}$ under f^{-1} lie between $S(a_0)$ and $S(a_1)$ and between $S(b_0)$ and $S(b_1)$, respectively, where (91) holds. In order to estimate $M(\Gamma_1)$, let Γ_1' be the family of arcs which join $S(a_1)$ with $S(b_0)$ in the portion of C between these spheres, fix $\rho \in F(\Gamma_1')$ and let $\gamma \in \Gamma_1'$ be a line segment which joins $S(a_1)$ and $S(b_0)$ and lies on a radius of $S(b_1)$ belonging to the cone C, then, by Hölder's inequality

$$1 \leqq \left(\int_\gamma \rho \, dr\right)^n \leqq \int_\gamma \rho^n r^{n-1} \, dr \left(\int_{a_1}^{b_0} \frac{dr}{r}\right)^{n-1} = \left(\log \frac{b_0}{a_1}\right)^{n-1} \int_\gamma \rho^n r^{n-1} \, dr,$$

integrating over the portion of S inside the cone yields

$$\int\limits_C \rho^n d\tau = \int\limits_0^\alpha \int\limits_0^\pi \cdots \int\limits_0^\pi \int\limits_0^{2\pi} \int\limits_{a_1}^{b_0} \rho^n r^{n-1} \sin^{n-2} \vartheta_1 \cdots \sin \vartheta_{n-2} \, dr \, d\vartheta_1 \cdots d\vartheta_{n-1} \geqq$$

$$\geqq \frac{(n-1)\omega_{n-1}}{\left(\log \frac{b_0}{a_1}\right)^{n-1}} \int_0^\alpha \sin^{n-2} \vartheta_1 \, d\vartheta_1$$

hence, since ρ_0 is arbitrary, we obtain

$$M(\Gamma_1') \geqq \frac{(n-1)\omega_{n-1}}{\left(\log \frac{b_0}{a_1}\right)^{n-1}} \int_0^\alpha \sin^{n-2} \vartheta_1 \, d\vartheta_1,$$

but $\rho_0 = \dfrac{1}{r \log \dfrac{b_0}{a_1}} \in F(\Gamma_1')$, and

$$\int_C \rho_n^0 \, d\tau = \frac{(n-1)\omega_{n-1}}{\left(\log \dfrac{b_0}{a_1}\right)^{n-1}} \int_0^\alpha \sin^{n-2} \vartheta_1 \, d\vartheta_1,$$

so that

$$M(\Gamma_1') = \frac{(n-1)\omega_{n-1}}{\left(\log \dfrac{b_0}{a_1}\right)^{n-1}} \int_0^\alpha \sin^{n-2} \vartheta_1 \, d\vartheta_1$$

and since $\Gamma_1 > \Gamma_1'$ then, on account of corollary 1 of proposition 1.5.3, we get

$$M(\Gamma_1) \leqq M(\Gamma_1') = \frac{(n-1)\omega_{n-1}}{\left(\log \dfrac{b_0}{a_1}\right)^{n-1}} \int_0^\alpha \sin^{n-2} \vartheta_1 \, d\vartheta_1.$$

Finally, combining the preceding inequality, (103), (77), (80) and lemma 2.17.2, we have

$$\frac{(n-1)\,\omega_{n-1}\,(\sin\alpha)^{(n-1)(n-2)}\left(\log \dfrac{b_0}{a_1}\right)^{(n-1)(n-2)}}{\left(\displaystyle\int_0^\alpha \sin^{n-2}\vartheta_1\, d\vartheta_1\right)^{n-2}\left(\log \dfrac{b_0}{a_1}\right)^{(n-1)(n-2)}} \leqq \frac{M^\sigma(\Gamma_2)^{n-1}}{M(\Gamma_1)^{n-2}} \leqq$$

$$\leqq \frac{K_I(f)^{n(n-2)}\, M^{\sigma^*}(\Gamma_2^*)^{n-1}}{M(\Gamma_1^*)^{n-2}} = \frac{2^{n-2}\, K_I(f)^{n(n-2)}\,(n-1)^{n-1}\,\omega_{n-1}^{n-1}}{n^{n-2}\,\omega_n^{n-2}},$$

and the left-hand side of (105) follows from (104).

Remarks. 1. If $n = 3$, (105) comes to

$$\sqrt[3]{1+\cos\alpha} \leqq K_I(C) \leqq \sqrt{1+\cos\alpha}.$$

2. If Z is the cylinder (61), and $C_\alpha \left(0 < \alpha < \dfrac{\pi}{2}\right)$ is the cone obtained from the cone (90) by means of a translation of vector $-\cot\alpha e_1$ and if $\{\alpha_m\}$ is so that

$\alpha_m \to 0$ as $m \to \infty$, then the sequence $\{C_{\alpha_m}\}$ converges to its kernel at O and Z may be considered as a cone C_α with $\alpha = 0$. In particular, since

$$\lim_{\alpha \to 0} \frac{q}{q(\alpha)} (\sin \alpha)^{\frac{1}{n-1}} = \frac{q}{n-1},$$

(as it follows from the argument in theorem 16), it follows that the inequalities (64) and (84) for the cylinder follow from the corresponding inequalities (94) and (105) for the cone if $\alpha \to 0$.

We conclude this paragraph with the following cone analogue of theorem 18.

THEOREM 23. *Suppose that D is a domain, that U is a neighbourhood of a point $\xi \in \partial D$ and that $D \cap U = C \cap U$, where C is the circular cone of angle α which has ξ as its vertex. Then the coefficients of D are not less than the corresponding coefficients of C. In particular if C is convex,*

$$K_I(D) \geqq \left[\frac{n\, \omega_n \sin^{n-2} \alpha}{2(n-1)\, \omega_{n-1} \int_0^\alpha \sin^{n-2} \vartheta_1 \, d\vartheta_1} \right]^{\frac{1}{n}}, \quad K_O(D) \geqq \left[\frac{q}{q(\alpha)} \right]^{\frac{n-1}{2}} \sqrt{\sin \alpha}.$$

Since ξ is the vertex of C, C is raylike at ξ, and the results follow from (94), (105) and theorem 12.

Spires. As it will be seen, the presence of a spire or a ridge in ∂D has a strong influence on the coefficients of D. We shall study this question in detail. It turns out that if ∂D has a spire which is directed into D or a ridge which is directed out of D, then $K(D) = \infty$. In the reverse situation $K(D)$ may be finite.

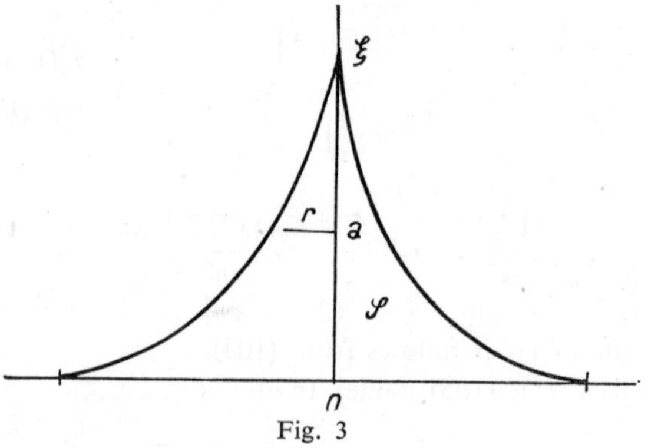

Fig. 3

A point set in R^n is said to be a *spire* if it can be mapped by means of a similarity transformation f onto

$$\mathscr{S} = \{x; x = (x^1, r, \vartheta_2, \ldots, \vartheta_{n-1}), r = g(x^1), 0 < x^1 \leqq a\}, \tag{107}$$

where $a < \infty$ and g is subject to the following restrictions:

(i) $g(u)$ is continuous in $[0, a]$ and $g(a) = 0$,

(ii) $g'(u)$ is continuous and strictly increasing in $(0, a)$,

(iii) $\lim_{u \to a} g'(u) = 0$.

These conditions imply that $g'(u) < 0$ in $(0, a)$, hence $g(u)$ is strictly increasing in $(0, a)$ and since $g(a) = 0$, it follows that $g(u) > 0$ in $[0, a)$. And now let us show that the preceding conditions imply

$$\int_0^a \frac{du}{g(u)} = \infty. \tag{108}$$

For this it is sufficient to verify that $\dfrac{1}{g(u)} > \dfrac{1}{a-1}$, i.e. $g(u) + u - a < 0$ in the neighbourhood of a (see for instance Smirnov [1], p. 256). But differentiating the left part of the preceding inequality and taking into account the condition (iii), we obtain $g'(u) + 1 > 0$ in the neighbourhood of $u = a$, hence we conclude $g(u) + u - a$ is strictly increasing there and since it vanishes at a, it follows that $g(u) + u - a < 0$, which implies (108).

The image of the point $\xi = (a, 0, \ldots, 0)$ under f^{-1} is called the *vertex of the spire*, the image of the basis vector e_1 is *its direction* and the image of the $(n-1)$-dimensional ball

$$B^{n-1}[g(0)] = \{x; x = (x^1, r, \vartheta_2, \ldots, \vartheta_{n-1}), 0 \leqq r < g(0), x^1 = 0\}, \tag{109}$$

is *its base*. We observe that the x^1-axis is tangent to the spire in its vertex.

A domain D is said to have a spire in its boundary if some point $\xi \in \partial D$ has a neighbourhood U such that $\mathscr{S} = \partial D \cap U$ is a spire with vertex at ξ. Let e be the direction of $\mathscr{S}$. Then the point $\xi + ue \bar{\in} \mathscr{S}$, and hence it does not belong to ∂D for small $u > 0$. Thus there exists a constant $b > 0$ such that either $\xi + ue \in D$ for $0 < u < b$ or $\xi + ue \in C\bar{D}$ for $0 < u < b$. We say that the *spire is inward directed* in the first case and *outward directed* in the second case.

Bounds for the coefficient of QCf of domains with spires in their boundary.

THEOREM 24. *If ∂D contains an inward directed spire, then* $K(D) = \infty$.

By performing a similarity transformation, we may assume that the vertex of the spire $\mathscr{S}$ is the origin, that its direction is $-e_1$, and that for some $a > 0$, $\mathscr{S} \cap B(a) = \partial D \cap B(a)$. Then $\mathscr{S}$ splits $B(a)$ into two domains, and since $\mathscr{S}$ is inward directed, $(C\bar{D}) \cap B(a)$ is the component of $B(a) - \mathscr{S}$ which contains the interval $r = 0$, $0 < x^1 < a$. Fix $0 < c < 1$. Because $\mathscr{S}$ is a spire (and then is tangent to the x^1-axis at the vertex), we can choose b, $0 < b < \dfrac{a}{2}$, such that $S(be_1, bc)$ separates O from ∞ in CD. Hence $CD \cap CB(be_1, bc)$ has two components which meet $S(be_1, b)$, and we conclude from theorem 7 that

$$K_I(D) \geqq A_3 \left(\log \frac{1}{c}\right)^{n-1},$$

where A_3 is as in (30). Letting $c \to 0$ yields $K_I(D) = \infty$, whence $K(D) = \infty$.

In contrast to the above situation, there exist domains with outward directed spires in their boundaries and finite coefficients. We require first the following result.

THEOREM 25. *Suppose that* $g(u) > 0$ *for* $0 \leqq u < a \leqq \infty$ *and that*

$$|g(u) - g(v)| \leqq b|u - v|, \quad b < \infty, \tag{110}$$

for $0 \leqq u, v < a$. *Suppose next that*

$$D = \{x; x = (x^1, r, \vartheta_2, \ldots, \vartheta_{n-1}),\ 0 \leqq r < g(x^1),\ 0 < x^1 < a\},$$

that D^* *is the circular cylinder*

$$D^* = \left\{x; x = (x^1, r, \vartheta_2, \ldots, \vartheta_{n-1}), 0 \leqq r < g(0), 0 < x^1 < g(0)\int_0^a \frac{du}{g(u)}\right\}$$

and that (109) *is the common base of* D *and* D^*. *Then there exists a homeomorphism* $f: D \bigcup B^{n-1} \rightleftharpoons D^* \bigcup B^{n-1}$ *such that* $f(x) = x$ *for* $x \in B^{n-1}$ *and*

$$K(f) \leqq \left(\frac{\sqrt{b^2 + 4} + b}{2}\right)^n \leqq (b + 1)^n$$

in D.

The homeomorphism

$$f(x^1, r, \vartheta_2, \ldots, \vartheta_{n-1}) = [\varphi(x^1), r\psi(x^1), \vartheta_2, \ldots, \vartheta_{n-1}],$$

where

$$x^{*1} = \varphi(x^1) = g(0)\int_0^{x^1} \frac{du}{g(u)}, \quad r^* = r\psi(x^1), \quad \psi(x^1) = \frac{g(0)}{g(x)^1},$$

maps $D \bigcup B^{n-1}$ onto $D^* \bigcup B^{n-1}$ and $f(x) = x$ for $x \in B$. Since g satisfies (110) (i.e. satisfies a Lipschitz condition), it follows that g is AC (see for instance I. P. Natanson [1], p. 319) and then differentiable a.e. (propositions 1.6.1 and 1.6.2) so that $f(x)$ is ACL in D and differentiable a.e. in D. Next, since

$$\mathrm{d}r = \frac{g(x^1)}{g(0)}\mathrm{d}r^* + \frac{rg'(x^1)}{g(0)}\mathrm{d}x^{*1},\ \mathrm{d}x^1 = \frac{g(x^1)}{g(0)}\mathrm{d}x^{*1},$$

it follows that

$$(\mathrm{d}x^1)^2 + \cdots + (\mathrm{d}x^n)^2 = (\mathrm{d}x^1)^2 + r^2\,\mathrm{d}\vartheta_2^2 + \cdots +$$

$$+ r^2 \sin^2\vartheta_2 \cdots \sin^2\vartheta_{n-2}\,\mathrm{d}\vartheta_{n-1}^2 =$$

$$= \frac{\left[\dfrac{rg'(x^1)}{g(x^1)}\right]^2 + 1}{\left[\dfrac{g(0)}{g(x^1)}\right]^2} (\mathrm{d}x^{*1})^2 + \frac{\mathrm{d}r^{*2}}{\left[\dfrac{g(0)}{g(x^1)}\right]^2} + \frac{2g(x^1)\,g'(x^1)}{g(0)^2}\,\mathrm{d}r^*\mathrm{d}x^{*1} +$$

$$+ \frac{r^{*2}\,\mathrm{d}\vartheta_2^2}{\left[\dfrac{g(0)}{g(x^1)}\right]^2} + \cdots + \frac{r^{*2}\sin^2\vartheta_2\cdots\sin^2\vartheta_{n-2}\,\mathrm{d}\vartheta_{n-1}^2}{\left[\dfrac{g(0)}{g(x^1)}\right]^2},$$

and performing the rotation

$$\mathrm{d}x^{*1} = \cos\alpha\,\mathrm{d}\tilde{x}^1 + \sin\alpha\,\mathrm{d}\rho, \quad \mathrm{d}r^* = -\sin\alpha\,\mathrm{d}\tilde{x}^1 + \cos\alpha\,\mathrm{d}\rho,$$

$$\mathrm{d}\vartheta_k = \mathrm{d}\vartheta_k\,(k = 2, \ldots, n-1),$$

we obtain the following standard form of the equation of the infinitesimal ellipsoid

$$(\mathrm{d}x^1)^2 + \cdots (\mathrm{d}x^n)^2 = \frac{(\mathrm{d}\tilde{x}^1)^2}{\left[\dfrac{g(0)}{g(x^1)}\cdot\dfrac{\sqrt{c^2+4}+c}{2}\right]^2} + \frac{\mathrm{d}\rho^2}{\left[\dfrac{g(0)}{g(x^1)}\cdot\dfrac{\sqrt{c^2+4}-c}{2}\right]^2} +$$

$$\frac{\rho^2\,\mathrm{d}\vartheta_2^2}{\left[\dfrac{g(0)}{g(x^1)}\right]^2} + \cdots + \frac{\rho^2\sin^2\vartheta_2\cdots\sin^2\vartheta_{n-2}\,\mathrm{d}\vartheta_{n-1}^2}{\left[\dfrac{g(0)}{g(x^1)}\right]^2},$$

where

$$c = \frac{r|g'(x^1)|}{g(x^1)}.$$

But since

$$\frac{g(0)}{g(x^1)}\cdot\frac{\sqrt{c^2+4}-c}{2} < \frac{g(0)}{g(x^1)} < \frac{g(0)}{g(x^1)}\cdot\frac{\sqrt{c^2+4}+c}{2},$$

it follows that the semiminor and semimajor axes of the infinitesimal ellipsoid are $\frac{g(0)}{g(x^1)} \cdot \frac{\sqrt{c^2+4}-c}{2}$ and $\frac{g(0)}{g(x^1)} \cdot \frac{\sqrt{c^2+4}+c}{2}$, respectively. Hence (2.1.12) and (2.1.13) allow us to conclude that in each point where $g'(x^1)$ exists, the formula

$$\frac{J_G(x)}{\lambda_f(x)^n} = \frac{\Lambda_f(x)^n}{J_G(x)} = \left(\frac{\sqrt{c^2+4}+c}{2}\right),$$

holds, so that by lemma 2.17.1 and taking into account that $c \leqq |g'(x)| \leqq b$ and $\frac{\sqrt{c^2+4}+c}{2} \leqq \frac{\sqrt{b^2+4}+b}{2} \leqq b+1$, we conclude that

$$K(f) = K_I(f) = K_O(f) = \left(\frac{\sqrt{c^2+4}+c}{2}\right)^n \leqq \left(\frac{\sqrt{b^2+4}+b}{2}\right)^n \leqq (b+1)^n,$$

and the proof of theorem 26 is complete.

THEOREM 26. *For each $\varepsilon > 0$ there exists a domain D whose boundary contains an outward directed spire and whose coefficients of QCf are within ε of the corresponding coefficients of an infinite circular cylinder.*

Indeed, it is sufficient, for $a \in (0, \infty)$, to set $g(u) = (a-u)^2$ in $[0, a]$. Then g satisfies the hypotheses of the preceding theorem with $b = 2a$ and $D = \{x; x = (x^1, r, \vartheta_2, \ldots, \vartheta_{n-1}), 0 \leqq r < g(x^1), 0 < x^1 < a\}$ is a domain with a pair of outward directed spires (corresponding to the values $x = \pm a$) in its boundary. Since g satisfies (108), we can use the preceding theorem to construct a homeomorphism f of D onto an infinite circular cylinder D^* with $K(f) \leqq (2a+1)^n$. Finally since a may be chosen arbitrarily small, the proof is complete.

An example. We consider the class of domains D which are obtained by adding an arbitrarily number of outward directed spires to a half space. More precisely, let Π be the plane $x^1 = 0$, let $\{B_m^{n-1}\}$ be a collection of disjoint open $(n-1)$-dimensional balls in Π, and for each m let $\mathscr{S}_m$ be a spire with base B_m^{n-1} and direction e_1. Then

$$(\Pi - \bigcup B_m^{n-1}) \cup (\bigcup_m \mathscr{S}_m) \tag{111}$$

is a surface which divides R^n into two domains. By theorem 24, the upper domain has infinite coefficients. Let D be the lower domain. We show that the coefficients of QCf of such a domain are bounded.

THEOREM 27. *Let*

$$D_m = \{x; x = \xi - ue_1, \ \xi \in \mathscr{S}_m, 0 < u < \infty\}, \tag{112}$$

and let D^*_m *and* σ_m *be the portions of* D_m *and* ∂D_m *which lie in* $x^1 < 0$. *Then, for each* m, *there exists a homeomorphism* $f_m : D_m \cup \sigma_m \rightleftharpoons D^*_m \cup \sigma_m$ *such that* $f_m(x) = x$ *for* $x \in \sigma_m$, *that*

$$K(f_m) \leqq \frac{1}{2}\left[\frac{(\sqrt{17}+1)\,q}{2(n-1)}\right]^{n-1} \tag{113}$$

in D^*_m, *and that*

$$\Lambda_{f_m}(x) \leqq 6\left(\frac{q}{n-1}\right)^{n-1} \tag{114}$$

in D^*_m.

Fix m and for convenience of notation $\mathscr{S} = \mathscr{S}_m$, $B^{n-1} = B^{n-1}_m$, $D = D_m$, $D^* = D^*_m$ and $\sigma = \sigma_m$. By performing a preliminary translation, we may assume that $\mathscr{S}$ is the spire in (107). We now define a homeomorphism $f : D \cup \sigma \rightleftarrows D^* \cup \sigma$ which satisfies the hypotheses of the theorem. Suppose first that $|g'(u)| < \frac{1}{2}$ in $(0, a)$, let D_1 be the part of D in $x^1 > 0$, and let D_1' be the symmetric image of D^* in $x^1 = 0$. Since g satisfies (108) and $|g'(x^1)| < \frac{1}{2}$ in $(0, a)$, then, by theorem 25, there exists a homeomorphism $f_1 : D_1 \cup B^{n-1} \rightleftharpoons D_1' \cup B^{n-1}$ such that $f_1(x) = x$ for $x \in B^{n-1}$, and

$$K(f_1) \leqslant \left(\frac{\sqrt{17}+1}{4}\right)^n \tag{115}$$

in D_1. Now let $Z = D^* \cup B^{n-1} \cup D_1'$. By theorem 19 we can find a homeomorphism $f_2 : Z \cup \sigma \rightleftharpoons D^* \cup \sigma$ such that $f_2(x) = x$ for $x \in \sigma$ and

$$K(f_2) \leqq \frac{1}{2}\left[\frac{2q}{(n-1)}\right]^n \text{ in } Z, \; \Lambda_{f_2}(x) \leqq 6\left(\frac{q}{n-1}\right)^{n-1} \text{ in } D^*. \tag{116}$$

Now set

$$f(x) = \begin{cases} f_2 \circ f_1(x) & \text{if } x \in D_1 \cup B^{n-1}, \\ f_2(x) & \text{if } x \in D^* \cup \sigma. \end{cases}$$

Then $f : D \cup \sigma \rightleftharpoons D^* \cup \sigma$ is a homeomorphism with $f(x) = x$ for $x \in \sigma$ and satisfies (113), (114), by virtue of (115), (116) and corollary 3 of theorem 2.7.1.

To complete the proof we have to consider the case when $|g'(x^1)| < \frac{1}{2}$ does not hold in the interval $(0, a)$. Suppose then that there exists a number $b \in (0, a)$

such that $g'(b) = -\frac{1}{2}$. Such a point exists as a consequence of the conditions (ii) and (iii). Let Z be the infinite cylinder $\{x;\ 0 \leqq r < g(b)\}$ and set

$$f_1(x) = \begin{cases} x - h(r)\, e_1 & \text{if } x \in (D \cup \sigma) - \bar{Z}_1, \\ x - be_1 & \text{if } x \in D \cap \bar{Z}_1, \end{cases}$$

where $x^1 = h(r)$ is the inverse function of g. Then f_1 is a homeomorphism of $D \cup \sigma$ and $f_1(x) = x$ for $x \in \sigma$. Moreover, since $|g'(x)| > \frac{1}{2}$ in $(0, b)$, $|h'(r)| < 2$ in $g(b) < r < g(0)$, and we conclude from corollary 1 of lemma 2.17.3, with $\cot \alpha = 2$, that

$$K(f_1) \leqq (\sqrt{2} + 1)^n,\ \Lambda_{f_1}(x) \leqq C(f) \leqq 3, \tag{117}$$

in D. Now f_1 translates $D \cap Z_1$ onto a domain D_1 which lies below the spire

$$\mathscr{S} = \{x; x = (x^1, r, \vartheta_2, \ldots, \vartheta_{n-1}), r = g(x^1 + b), 0 \leqq x^1 \leqq a - b\}.$$

Let D_1'' and σ_1 denote the parts of D_1 and ∂D_1 which lie in $x^1 < 0$. Since $|g'(x^1 + b)| < \frac{1}{2}$ in $(0,\ a - b)$, by what was proved above we can find a homeomorphism $f_2 : D_1 \cup \sigma_1 \rightleftharpoons D_1'' \cup \sigma_1$ such that $f_2(x) = x$ for $x \in \sigma_1$ and satisfying (113) in D and (114) in D^* (where we put f_2 in place of f_m). Finally set

$$f(x) = \begin{cases} f_1(x) & \text{if } x \in (D \cup \sigma) - \bar{Z}_1, \\ f_2 \circ f_1(x) & \text{if } x \in D \cap \bar{Z}_1. \end{cases}$$

Then $f : D \cup \sigma \rightleftharpoons D' \cup \sigma$ is a homeomorphism with $f(x) = x$ for $x \in \sigma$ and satisfying the hypotheses of the theorem, by virtue of (117), (113), (114) and corollary 3 of theorem 2.7.1. Hence the proof of the theorem 27 is complete.

THEOREM 28. *For each domain D, which has as boundary a surface of the form* (111) *and contains $x^1 < 0$,*

$$K(D) \leqq \frac{1}{2}\left[\frac{(\sqrt{17} + 1)\, q}{2(n - 1)}\right]^n. \tag{118}$$

Let D_m denote the domain (112). We define a mapping f by setting

$$(x) = \begin{cases} f_m(x) & \text{if } x \in D_m, \\ x & \text{if } x \in E = D - \bigcup_m D_m. \end{cases}$$

Each point of $D - E$ has a neighbourhood in which f is a K_0-QCfH with $K_0 \leqq \frac{1}{2}\left[\frac{(\sqrt{17}+1)q}{2(n-1)}\right]^n$, as it follows from the preceding theorem. Hence, on account of the equivalence of the different definitions of the QCfH, $f(x)$ is ACL in $D - E$ and differentiable a.e. in $D - E$. In an inner point of $E, f(x)$ is differentiable and

$$\Lambda_f(x) = \lambda_f(x) = J(x) = 1 \tag{119}$$

holds. But the intersection of a line (for instance of the coordinate axes) with $D - E$ consists in a countable set of open intervals. The union of the closure of the set of the endpoints of these intervals and of the points of tangency of certain domains D_m is a set of linear measure zero. Since this set contains all the points of E, where (119) does not hold, it follows by Tonelli's theorem, that $f(x)$ is differentiable and (119) holds a.e. in E. Hence, since by corollary of theorem 2.1.1, $J_G(x) = |J(x)| > 0$ a.e. in $D - E$, on account of (119), we conclude that $J_G(x) = |J(x)| > 0$ a.e. in D. Then, by (114) and (119), it follows that

$$\bar{\delta}(x) = \overline{\lim_{r\to 0}}\,\frac{\omega_n L(x, r)^n}{T(x, r)} \leqq \overline{\lim_{r\to 0}}\left[\frac{L(x, r)}{r}\right]^n \lim_{r\to 0}\frac{\omega_n r^n}{T(x, r)} = \frac{\Lambda_f(x)^n}{J_G(x)} < \infty$$

a.e. in D and, by lemma 2.1.5, $f(x)$ is ACL in D. Finally, since $f(x)$ is differentiable a.e. in D and ACL in D, then we conclude from (113), (119) and lemma 2.17.1 that $f(x)$ is K_O-QCfH with $K_O \leqq \frac{1}{2}\left[\frac{(\sqrt{17}+1)q}{2(n-1)}\right]^n$, hence we deduce (118).

Remark. From the preceding theorem we conclude that there exist domains with finite coefficients of QCf and some inaccessible boundary points. For choose a sequence of disjoint open $(n-1)$-dimensional balls $\{B_m^{n-1}\}$ which converge to O, erect a spire $\mathscr{S}_m$ of height 1 on each B_m^{n-1}, and let D be the corresponding domain bounded by the surface (111) and which contains the half space $x^1 < 0$. Then each point of the segment $\{x;\ r = 0,\ 0 < x^1 < 1\}$ is an inaccessible boundary point of D, while $K(D)$ satisfies (118). Another example was given by V. Zorič [2].

Finally we have the following sharp lower bound for the coefficients of the domains whose boundary contains a spire.

THEOREM 29. *If D is a domain whose boundary contains a spire, then the coefficients of* QCf *of D are not less than the corresponding coefficients of an infinite circular cylinder. In particular, the inequalities* (83) *hold.*

By performing a preliminary similarity transformation, we may assume that the vertex of the spire $\mathscr{S}$ is the origin and that its direction is $-e_1$. Next by theorem 24, we may assume that $\mathscr{S}$ is outward directed. Finally, by definition, we can choose $a>0$, so that $\mathscr{S}\cap B(a) = \partial D \cap B(a)$. Then $\mathscr{S}$ splits $B(a)$ into two domains and $D\cap B(a)$ is the component of $B(a) - \mathscr{S}$ which contains the segment $\{x;\ r = 0,\ 0 < x^1 < a\}$. Let f_1 denote the inversion in $S(a)$, let D_1 denote the image of D

under f_1, and let U_1 denote the half space $x^1 > a$. Since $\mathcal{S}$ is a spire, it follows that

$$D_1 \cap U_1 = \{x; x = (x^1, r, \vartheta_2, \ldots, \vartheta_{n-1}),\ 0 \leqq r < g(x^1),\ a < x^1 < \infty\},$$

where $g'(x^1)$ is continuous in (a, ∞) and $\lim\limits_{x^1 \to \infty} g'(x^1) = 0$. Fix $\varepsilon > 0$, choose $b > a$ so that $|g'(x^1)| < \varepsilon$ in (b, ∞), let U be the half space $x^1 > b$ and let Z be the infinite circular cylinder

$$Z = \{x; x = (x^1, r, \vartheta_2, \ldots, \vartheta_{n-1}),\ 0 < r < g(b),\ |x| < \infty\}.$$

Since $|g'(x^1)| < \varepsilon$ in (b, ∞),

$$\int_b^\infty \frac{dx^1}{g(x^1)} > \frac{1}{g(b)} \int_b^\infty dx^1 = \infty,$$

and hence, by theorem 25, there exists a homeomorphism $f_2 : D_1 \cap \overline{U} \rightleftarrows Z \cap \overline{U}$ such that $f_2(x) = x$ in $D_1 \cap \partial U$ and $K(f_2) \leqq (1 + \varepsilon)^n$ in $D_1 \cap U$. Set

$$f_3(x) = \begin{cases} f_2(x) & \text{if } x \in D_1 \cap \overline{U}, \\ x & \text{if } x \in D_1 - \overline{U}. \end{cases}$$

Then $f = f_3 \circ f_1 : D \rightleftarrows D^*$ is a homeomorphism, $D^* \cap U = Z \cap U$ and, on account of corollary 3 of theorem 2.7.1, $K(f) \leqq (1 + \varepsilon)^n$ since $K(f) = 1$ in $f_1^{-1}(D_1 - U)$. The desired lower bounds (83) are obtained by first applying theorem 18 to D^* and then letting $\varepsilon \to 0$. Theorem 26 shows that these bounds cannot be improved.

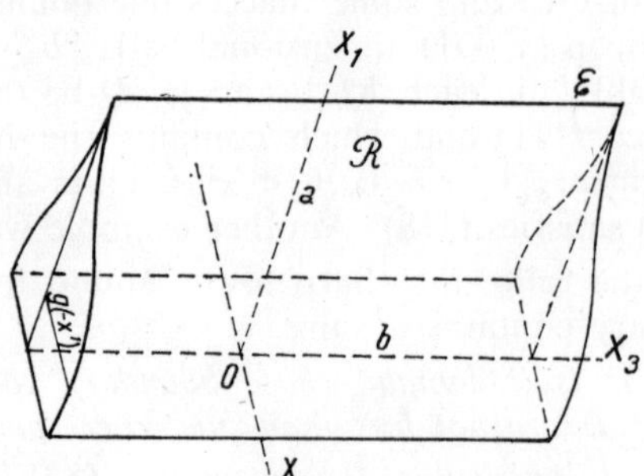

Fig. 4

Ridges. A point set in R^n is said to be a *ridge* if it can be mapped by means of a similarity transformation f onto the surface

$$\mathcal{R} = \{x; |x^2| = g(x^1),\ 0 < x^1 \leqq a,\ (x^3)^2 + \cdots + (x^n)^2 < b^2\}.$$

where $a < \infty$, $b \leqq \infty$ and g satisfy the conditions (i), (ii), (iii) in the definition of the spire. The image of the $(n-2)$-dimensional manifold

$$\mathscr{E} = \{x; x^1 = a, x^2 = 0, (x^3)^2 + \cdots + (x^n)^2 < b^2\}$$

under f^{-1} is called the *edge of the ridge* and the image of the vector e_1 is its *direction*. In the case $n = 3$, the edge of the ridge comes to the line segment $\{x; x^1 = a, x^2 = 0, |x^3| < b\}$.

A domain D is said to have a ridge in its boundary if some point $\xi \in \partial D$ has a neighbourhood U such that $\mathscr{R} = \partial D \cap U$ is a ridge with ξ a point of its edge $\mathscr{E}$. Let e be the direction of $\mathscr{E}$. As in the case of spires, there exists a constant $c > 0$ such that either $\xi + ue \in D$ for $0 < u < c$ or $\xi + ue \in C\overline{D}$ for $0 < u < c$. The ridge $\mathscr{R}$ is said to be *inward directed* in the first case and *outward directed* in the second case. For $n = 2$ a ridge comes to a spire.

Bounds for the coefficients of QCf of domains whose boundary contains a ridge. We begin by the following analogue of theorem 24 for outward directed ridges:

THEOREM 30. *If D is a domain whose boundary contains an outward directed ridge, then* $K(D) = \infty$.

By performing a preliminary similarity transformation, we may assume that the edge of the ridge $\mathscr{R}$ is the $(n-2)$-dimensional manifold $\{x; x^1 = x^2 = 0, |x| < 1\}$, that its direction is $-e_1$, that $\xi = 0$, and that for some $a > 0$, $\mathscr{R} \cap B(a) = \partial D \cap B(a)$. Then $\mathscr{R}$ divides $B(a)$ into two domains, and since $\mathscr{R}$ is outward directed, $D \cap B(a)$ is the component of $B(a) - \mathscr{R}$ which contains the interval $\{x; 0 < x^1 < a, x^2 = \cdots = x^n = 0\}$. Because $\mathscr{R}$ is a ridge, given $0 < c < 1$, we can choose $b \in \left(0, \frac{a}{2}\right)$ so that D separates $(b, bc, 0, \ldots, 0)$ from $(b, -bc, 0, \ldots, 0)$ in $B(be_1, b)$. Thus $CD \cap B(be_1, b)$ has two components which meet $S(be_1, bc)$ and we conclude from theorem 6 that $K_I(D) \geqq A_3 \log \frac{1}{c}$. Letting $c \to 0$, yields $K_I(D) = \infty$, whence $K(D) = \infty$.

In contrast to the above situation (by analogy with what happens in the case of the spires) there exist domains with inward directed ridges in their boundaries and finite coefficients of QCf.

THEOREM 31. *For each* $\varepsilon > 0$ *there exists a domain D whose boundary contains an inward directed ridge and whose coefficients of* QCf *are within* ε *of the corresponding coefficients of a dihedral wedge of angle* 2π.

Indeed, given $0 < a < \infty$, set $g(u) = \min(u^2, a^2)$ and let

$$\sigma = \{x; |x^2| = g(x^1), x^1 \geqq 0\}.$$

Then σ bounds a domain D which has an inward directed ridge in its boundary. For $x \in D$, let

$$f_1(x) = \begin{cases} x - g(x^1)(\operatorname{sign} x^2)\, e_2 & \text{if} \quad x^1 \geqq 0, \\ x & \text{if} \quad x^1 < 0, \end{cases}$$

where

$$\operatorname{sign} u = \begin{cases} \dfrac{u}{|u|} & \text{if } u \neq 0, \\ 0 & \text{if } u = 0. \end{cases}$$

Then f_1 is a homeomorphism of D onto a dihedral wedge

$$D_{2\pi} = \{x; x = (r, \vartheta_1, x^3, \ldots, x^n),\ 0 < \vartheta_1 < 2\pi,\ |x| < \infty\}$$

of angle 2π and $K(f_1) \leqq (2a+1)^n$ by virtue of corollary 1 of lemma 2.17.3, where in our case $\cot \alpha \leqq 2a$. Now $D_{2\pi}$ has finite coefficients of QCf. Indeed, $f_2(r, \vartheta_1, x_3, \ldots, x^n) = \left(r, \dfrac{\vartheta_1}{2}, \dfrac{x^3}{\sqrt{2}}, x^4, \ldots, x^n\right)$ maps $D_{2\pi}$ onto a half space, and from

$$\mathrm{d}r^2 + r^2\, \mathrm{d}\vartheta_1^2 + (\mathrm{d}x^3)^2 + \cdots + (\mathrm{d}x^n)^2 = \mathrm{d}r^2 + + \frac{r^2\, \mathrm{d}\vartheta_1^{*2}}{\frac{1}{4}} + \frac{(\mathrm{d}x^{*3})^2}{\frac{1}{2}} + (\mathrm{d}x^4)^2 + \cdots + (\mathrm{d}x^n)^2,$$

it follows that the semiaxes of the corresponding infinitesimal ellipsoid are proportional to $1, \dfrac{1}{2}, \dfrac{1}{\sqrt{2}}, 1, \ldots, 1$, hence, combining lemma 2.17.1, its corollary, (2.1.12) and (2.1.13), we obtain

$$K_O(f_2) = \operatorname*{ess\,sup}_{x \in D_{2\pi}} \frac{\Lambda_{f_2}(x)^n}{J_G(x)} = 2\sqrt{2}, \quad K_I(f_2) = \operatorname*{ess\,sup}_{x \in D_{2\pi}} \frac{J_G(x)}{\lambda_{f_2}(x)} = 2^{\frac{2n-3}{2}},$$

which implies $K(f_2) = 2^{\frac{2n-3}{3}} < \infty$.

Finally, because a may be chosen arbitrarily small, theorem 31 is completely proved.

We consider next a class of domains analogous to those determined by (111). Let g be any function which satisfies (i), (ii), (iii). Next let Π be the plane $x^1 = 0$, set

$$\mathcal{R} = \{x; |x^2| = g(x^1),\ 0 < x^1 \leqq a\},$$

and let

$$\mathscr{B}\{x; |x^2| < g(0),\ x^1 = 0\}$$

be the base of $\mathscr{R}$. Then $(\Pi - B)\ U\mathscr{R}$ is a surface which divides R^n into two domains. The domain which contains the negative half of the x^1-axis has infinite coefficients of QCf by theorem 30. Let D be the other domain. We show that the coefficients of QCf of D remain bounded, no matter how sharp we make the ridge R.

THEOREM 32. *For each such domain* D, $K(D) \leq \dfrac{2^{n-3}(\sqrt{5}+1)^n}{\sqrt{3}^{\,2n-3}}$.

Set

$$f_1(x) = (a - x^1, x^2, \ldots, x^n),$$

$$f_2(r, \vartheta_1, x^3, \ldots, x^n) = \left(r, \frac{3\vartheta_1}{4}, \sqrt{\frac{3}{4}}\,x^3, x^4, \ldots, x^n\right).$$

Then $f_2 \circ f_1$ is a homeomorphism of D onto a domain D_1, which lies in the dihedral wedge $D_{\frac{3\pi}{2}}$, and, arguing as in the preceding theorem, we obtain

$$K_O(f_2) = \frac{8}{3\sqrt{3}}, \quad K_I(f_2) = \left(\frac{2}{\sqrt{3}}\right)^{2n-3},$$

hence

$$K(f_2 \circ f_1) = K(f_2) = \left(\frac{2}{\sqrt{3}}\right)^{2n-3}.$$

Now for each pair of points $\xi_1, \xi_2 \in \partial D$, the angle between the segment $\overline{\xi_1\xi_2}$ and the vector $e_2 - e_1$ is never less than $\frac{\pi}{4}$. Hence corollary 1 of lemma 2.17.3 yields a homeomorphism $f_3 : D_1 \rightleftharpoons \{x;\ x^2 - x^1 > 0\}$ with

$$K(f_3) \leq \left(\frac{\sqrt{5}+1}{2}\right)^n,$$

and we conclude that

$$K(D) \leq K(f_3 \circ f_2 \circ f_1) \leq \frac{2^{n-3}(\sqrt{5}+1)^n}{\sqrt{3}^{\,2n-3}}.$$

We establish now the following implicit sharp lower bound for the coefficients of QCf of a domain whose boundary contains a ridge.

THEOREM 33. *If D is a domain whose boundary contains a ridge, then the coefficients of* QCf *of D are not less than the corresponding coefficients of a dihedral wedge of angle* 2π.

Suppose that D contains a ridge in its boundary, and for $0 < a < \infty$, let Q be the open cube bounded by the planes $x^1 = a \pm a$, $x^2 = \pm a, \ldots, x^n = \pm a$. By performing a preliminary similarity transformation, we may choose a so that

$$\partial D \cap Q = \{x; 0 < x^1 \leqq a, \ |x^2| = g(x^1), \ |x^3|, \ldots, |x^n| < a\},$$

where g satisfies (i), (ii), (iii). Next by theorem 30, we may assume that the ridge is inward directed, and hence that

$$CD \cap Q = \{x; 0 < x^1 \leqq a, \ |x^2| \leqq g(x^1), \ |x^3|, \ldots, |x \ | < a\}.$$

Now fix $b \in \left(\dfrac{a}{2}, a\right)$, set

$$h(u) = \begin{cases} g(u) & \text{for } b \leqq u \leqq a, \\ 0 & \text{for } u > a \end{cases}$$

and extend h so that $h(u) = h(2b - u)$ for all u. Next, for $x \in D$, let

$$f(x) = \begin{cases} x - h(x^1)(\operatorname{sign} x^2)\, e_2 & \text{if } (x^3)^2 + \cdots + (x^n)^2 < (a-b)^2, \\ x - h(x^1)(\operatorname{sign} x^2)\left[\dfrac{a - \sqrt{(x^3)^2 + \cdots + (x^n)^2}}{b}\right] e_2 & \text{if } (a-b)^2 \leqq \\ & \leqq (x^3)^2 + \cdots + (x^n)^2 \leqq a^2, \\ x \text{ if } (x^3)^2 + \cdots + (x^n)^2 > a^2, \end{cases}$$

then $f : D \rightleftharpoons D^*$ is a homeomorphism such that $D^* \cap B(\xi, r) = D^*_{2\pi} \cap B(\xi, r)$, where $\xi = (a, 0, \ldots, 0)$, $0 < r < a - b$, and $D_{2\pi}$ is a dihedral wedge of angle 2π. Using corollary 1 of lemma 2.17.3, we can show that $K(f) \to 1$ as $b \to a$, and hence the desired conclusion follows from theorem 13.

The space of domains QCf equivalent to a ball. Let $\mathscr{D}$ denote the class of all domains $D \subset R^n$ with $K(D) < \infty$. Next given $D, D^* \in \mathscr{D}$, we define the *distance between D and D** as

$$d(D, D^*) = \inf_f [\log K(f)],$$

where the infimum is taken over all homeomorphisms $f : D \rightleftharpoons D^*$. We identify two domains D and D^* whenever $d(D, D^*) = 0$. Then it is trivial to show that d is a metric on $\mathscr{D}$. Indeed

1) $d(D, D^*) = \inf\limits_{f} \log K(f) \geqq 0$ since $K(f) \geqq 1$ and $d(D, D^*) = 0$ if and only if $K(D) = 1$; i.e. the two domains are conformally equivalent.

2) $d(D, D^*) = d(D^*, D)$, since $K(f) = K(f^{-1})$ by (2.17.2).

3) $d(D, D'') \leqq d(D, D') + d(D', D'')$, since if $f : D \rightleftharpoons D''$, $f_1 : D \to D'$, and $f_2 : D' \to D''$, then $K(f) \leqq K(f_1)K(f_2)$, and then $\log K(f) \leqq \log K(f_{-1}) + \log K(f_2)$, which implies 3).

In what follows we show that $\mathscr{D}$ is complete and not separable under d.

Completeness of the space $\mathscr{D}$. The completeness is equivalent to the following result (for the concept of a complete space, see Chapter 1.2):

THEOREM 34. *Suppose that $\{D_m\}$ is a sequence of domains in $\mathscr{D}$ and that*

$$\lim_{k,m\to\infty} d(D_k, D_m) = 0. \tag{120}$$

Then, there exists a domain $D_0 \in \mathscr{D}$ such that

$$\lim_{m\to\infty} d(D_m, D_0) = 0. \tag{121}$$

By virtue of (120), we may choose a subsequence $\{D_{m_q}\}$ of $\{D_m\}$ such that

$$d(D_{m_q}, D_{m_{q+1}}) < 2^{-q}$$

for $q = 1, 2, \ldots$ Next fix a pair of distinct points $x_0, x_1 \in D_{m_1}$, let $f_q : D_{m_q} \rightleftharpoons D_{m_{q+1}}$ be a homeomorphism with

$$\log K(f_q) < 2^{-q} \tag{122}$$

and let φ_q be a Möbius transformation of D_{m_q} onto a domain $D_q^* \subset R^n$, chosen so that $g_q(x_0) = x_0$ and $g_q(x_1) = x_1$, where $g_q = \varphi_q \circ f_{q-1} \circ \cdots \circ f_1$. Then $g_q : D_{m_1} \rightleftharpoons D_q^*$ is a homeomorphism and (122) implies that $\log K(g_q) < 1$ for all q. Hence by theorem 2.12.3, the g_q are uniformly bounded and equicontinuous on each compact subset of D_{m_1}, and, by Arzela—Ascoli's theorem (chapter 2.5), there exists a subsequence $\{g_{q_k}\}$ of $\{g_q\}$ so that

$$\lim_{k\to\infty} g_{q_k} = g(x), \tag{123}$$

uniformly on each compact subset of D_{m_1}. Since $g(x_0) = x_0$ and $g(x_1) = x_1$, theorem 2.12.5 implies that g is a homeomorphism of D_{m_1} onto a domain $D_0 \subset R^n$. Fix q, and for $q' > q$, set

$$h_{q'} = g_{q'} \circ f_1^{-1} \circ \cdots \circ f_{q-1}^{-1} = \varphi_{q'} \circ f_{q'-1} \circ \cdots \circ f_q.$$

Then $h_{q'}$ is a homeomorphism of D_{m_q} for which

$$\log K(h_{q'}) < 2^{-q+1}, \tag{124}$$

and from (123) it follows that

$$\lim_{k\to\infty} h_{q_k} = h(x), \tag{125}$$

uniformly on each compact subset of D_{m_q}, where $h = g \circ f_1^{-1} \circ \ldots \circ f_{q-1}^{-1}$. Thus h is a homeomorphism of D_{m_q} onto D_0, and from (124), (125) and (4), it follows that

$$\log K(h) \leqq \varliminf_{k\to\infty} K(h_{q_k}) < 2^{-q+1},$$

hence

$$d(D_{m_q}, D_0) \leqq \log K(h) < 2^{-q+1},$$

and (121) holds for $q = 1, 2, \ldots$

Non-separability of the space $\mathscr{D}$. We require the following result in the proof that $\mathscr{D}$ is not separable:

THEOREM 35. *Suppose that D and D^* are domains in R^n, that U and U^* are neighbourhoods of $\xi \in \partial D$ and $\xi^* \in \partial D^*$, and that $D \cap U = \Delta \cap U$ and $D^* \cap U^* = \Delta^* \cap U^*$, where Δ is a dihedral wedge with ξ a point of its edge and Δ^* is a halfspace. If f is a homeomorphism of D onto D^* and*

$$\lim_{x\to\xi} f(x) = \xi^*, \tag{126}$$

then

$$K_I(f) \geqq K_I(\Delta), \quad K_O(f) \geqq K_O(\Delta), \quad K(f) \geqq K(\Delta). \tag{127}$$

We may assume that $K(f) < \infty$ for otherwise (3) implies $K_I(f) = K_O(f) = \infty$ and there is nothing to prove. Next by performing preliminary translations we may assume that $\xi = \xi^* = O$. Choose $a > 0$ so that $B(a) \subset U$ and fix $x_0 \in D$ with $|x_0| < a$. Clearly $x_0 \in \Delta$, hence $\dfrac{x_0}{m} \in \Delta \cap U \subset D$. For each m, let

$$f_m(x) = a_m f\left(\frac{x}{m}\right) + x_{0m}^*,$$

where

$$a_m = \left| f\left(\frac{x_0}{m}\right)\right|^{-1}, \; x_{0m}^* = x_0^* - a_m f\left(\frac{x_0}{m}\right), \; x_0^* = f(x_0), \; \frac{x}{m} \in D. \tag{128}$$

Then $f_m(x_0) = x_0^*$ and

$$|x_{0m}^*| \leqq |x_0^*| + 1. \tag{129}$$

Moreover, since by (126),

$$\lim_{x \to 0} f(x) = 0, \tag{130}$$

we have

$$\lim_{m \to \infty} a_m = \infty. \tag{131}$$

Now let

$$D_m = \left\{x; \frac{x}{m} \in D\right\}, \quad D_m^* = \{x_m^*; x_m^* = a_m x^* + x_{0m}^*, x^* \in D^*\}. \tag{132}$$

As in the proof of theorem 12, we conclude that $x_0 \in D_m$ for all m, and that the D_m converge to the kernel Δ at x_0. Next, since $f_m : D_m \rightleftharpoons D_m^*$, it follows that $x^* \in D_m^*$ for all m. Finally, since x_{0m}^* are bounded, by choosing a subsequence and relabelling, we may assume that $\lim_{m \to \infty} x_{0m}^* = x_1^*$.

We prove now that x_0^* has a neighbourhood which is contained in all D_m^*. Let $a^* > 0$ be so that $B(a^*) \subset U^*$. From (130), it follows that, for a sufficiently large $m > m_1$, we have $\left|f\left(\frac{x_0}{m}\right)\right| < a^*$, so that $f\left(\frac{x_0}{m}\right) \in D^* \cap U^* \subset \Delta^*$, hence $a_m f\left(\frac{x_0}{m}\right) \in \Delta^*$. Since $x_0^* = f(x_0) \in D_m^*$ $(m = 1, 2, \ldots)$, it follows, on account of (132), that $\frac{x_0^* - x_{0m}^*}{a_m} \in D^*$. On the other hand, (129)—(131) imply

$$\lim_{m \to \infty} \frac{f(x_0) - x_{0m}^*}{a_m} = 0,$$

so that, for a sufficiently large number $m > m_2$, we have $\frac{|f(x_0) - x_{0m}^*|}{a_m} < a^*$, hence, for $m > m_2$, we obtain $\frac{x_0^* - x_{0m}^*}{a_m} \in D^* \cap U^* \subset \Delta^*$, and then $x_0^* - x_{0m}^* \in \Delta^*$, hence $x_0^* \in \Delta^* + x_{0m}^*$ for all $m > m_2$. Since Δ^* is open, we conclude that there exists a neighbourhood $U_1^* = B(x_0^*, r_1)$ of x_0^* so that $x^* \in \Delta^* + x_0^*$ whenever $x^* \in U_1^*$. Next, we observe that, for $x^* \in D_m^*$, we must have $\frac{x^* - x_{0m}^*}{a_m} \in D^*$. If $x^* \in U_1^*$, then $x^* - x_{0m}^* \in \Delta^*$, and then $\frac{x^* - x_{0m}^*}{a_m} \in \Delta^*$. From (131), it follows that for a sufficiently large $m > m_3$, we have $a_m a^* > 2$, hence, if $m > m_3$ and

$x^* \in U_2^* = B(x_0^*, r_2)$ (where $0 < r_2 < 1$) then, on account of (128), we obtain $|x^* - x_{0m}^*| = \left| x^* - x_0^* + a_m f\left(\frac{x_0}{m}\right) \right| \leqq |x^* - x_0^*| + 1 < 2 < a_m a^*$, and if we set $m_0 = \max(m_1, m_1, m_3)$, $U_0^* = B(x_0^*, r_0)$, where $r_0 = \min(r_1, r_2)$, then $\frac{x^* - x_{0m}^*}{a_m} \in \Delta^* \cap \cap U^* \subset D^*$, whenever $m > m_0$ and $x^* \in U_0^*$, so that $x^* \in D_m^*$ $(m > m_0)$ and then also $U_0^* \subset D_m^*$ $(m > m_0)$.

And now, let $U_0 = B(x_0, r_0)$ be a neighbourhood of x_0, so that $U_0 \subset D_m$ $(m = 1, 2, \ldots)$; the existence of such a neighbourhood follows from the convergence of the sequence $\{D_m\}$ to its kernel Δ at x_0. Since $f_m : D_m \rightleftarrows D_m^*$, it follows that $f_m(U_0) = U_m^* \subset D_m^*$ and since $x_0^* \in U_m^*$ $(m = 1, 2, \ldots)$, we conclude that $\widetilde{U}^* = \bigcap_{m=1}^{m_0} U_m$ is a neighbourhood of x_0^* such that $\widetilde{U}^* \subset S_m^*$ $(m = 1, \ldots, m)$, hence, from above, we deduce $\widetilde{U}_0^* = U_0^* \cap \widetilde{U}^* \subset D_m^*$ $(m = 1, 2, \ldots)$.

Next, we prove that D_m^* converge to their kernel Δ_1^* at x_0^*, where $\Delta_1^* = \Delta^* + x_1^*$. Let $\widetilde{D}^*$ be the kernel of the sequence $\{\widetilde{D}_m\}$ at x_0^* and let F^* be a compact set contained in Δ_1^*. Since $\Delta_1^* = \lim_{m \to \infty} (\Delta^* + x_{0m}^*)$, it follows that $F^* \subset \Delta^* + x_{0m}^*$ for all $m > m_4$, so that $x_2^* \in F^*$ with $|x_2^* - x_{0m}^*| = \sup_{x^* \in F^*} |x^* - x_{0m}^*|$, hence, by (129) and (131), $\frac{x^* - x_{0m}^*}{a_m} \in \Delta^* \cap U^* \subset D^*$ for $m > \max(m_4, m_5)$, so that $x_2^* \in D_m^*$ for $m > \max(m_4, m_5)$, and then $F^* \subset D_m^*$ for $m > \max(m_4, m_5)$. Since this conclusion still holds for any compact set $F^* \subset \Delta_1^*$, we infer that $\Delta_1^* \subset \widetilde{D}^*$. And now, conversely, suppose $x^* \in \widetilde{D}^*$. On account of the definition of the kernel this means that $x^* \in D_m^*$, for $m > m_6$, hence, by (129) and (131) $\frac{x^* - x_{0m}}{a_m} \in \in D^* \cap U^* \subset \Delta^*$ for $m > \max(m_6, m_7)$, whence $x^* \in \Delta^* + x_{0m}^*$ for all $m > > \max(m_6, m_7)$, and letting $m \to \infty$, we obtain $x^* \in \Delta_1^*$, and then $\widetilde{D}^* \subset \Delta_1^*$, which, on account of the opposite inclusion, yields $\widetilde{D}^* = \Delta_1^*$, i.e. the sequence $\{D_m^*\}$ has the kernel Δ_1^* at x_0^*.

Finally, since $K(f_m) = K(f) < \infty$, theorems 1.2 and 1.1 imply the existence of a subsequence $\{f_{m_p}\}$ of $\{f_m\}$ such that

$$\lim_{p \to \infty} f_{m_p}(x) = g(x),$$

uniformly on each compact subset of Δ, where $g : \Delta \rightleftarrows \Delta_1^*$ is a homeomorphism. Hence, by (4), we obtain

$$K_I(\Delta) \leqq K_I(g) \leqq \underline{\lim}_{p \to \infty} K_I(f_{m_p}) = K_I(f),$$

and the rest of (127) follows similarly.

A set E is said to be *dense in the topological space* X if $\bar{E} = X$.

A topological space is called *separable* if it contains a dense countable set.

We show that $\mathscr{D}$ is nonseparable by establishing the existence of a set having the power of the continuum, and such that the distance between two arbitrary points of it is not less than a strictly positive constant.

THEOREM 36. *Given* $0 < a < \infty$, *we can associate with each* $b \in (0, 1)$, *a domain* $D_b \in \mathscr{D}$ *such that*

$$d(D_b, B) \leqq a \tag{133}$$

for $0 < b < 1$ *and such that*

$$d(D_b, D_{b'}) \geqq c > 0 \tag{134}$$

for $0 < b, b' < 1$, $b \neq b'$, *where* $c > 0$ *is a constant which depends only on* a.

Pick $m > 0$ so that

$$\log (m + 1)^n = a. \tag{135}$$

With each $b \in (0, 1)$, we can associate a sequence $\{b_p\}$ such that $b_p = 0$ or 1, for each p and

$$b = \sum_{p=1}^{\infty} b_p 2^{-p}. \tag{136}$$

Next let $c_0 = 0$ and

$$c_p = (c_{p-1} + 1)\, e^{bp+1} + 1 > c_{p-1} + 2 \tag{137}$$

for $p = 1, 2, \ldots$ Then for $u \geqq 0$ set

$$g_b(u) = \begin{cases} \dfrac{m}{2}\left[1 - (u - c_p)^2\right] & \text{if } |u - c_p| \leqq 1 \text{ for some } p, \\ 0 & \text{if } |u - c_p| > 1 \text{ for all } p, \end{cases}$$

and let D_b be the domain (fig. 5)

$$D_b = \{x; x = (x^1, r, \vartheta_2, \ldots, \vartheta_{n-1}), g_b(r) < r^1 < \infty, 0 \leqq r < \infty\}.$$

Since $\dfrac{dx_1}{dr} = m(c_p - r) \leqq m = \cot \alpha$, it is not difficult to show that, for each pair of points $\xi_1, \xi_2 \in D_b$, the acute angle between the segment $\overline{\xi_1 \xi_2}$ and the vector e_1 is not less than $\alpha = \text{arc cot}\, m$. Hence corollary 1 of lemma 2.17.3 yields a homeomorphism $f: D_b \rightleftharpoons \{x^1 > 0\}$ for which

$$K(f) \leqq (m+1)^n, \tag{138}$$

and (133) follows from (135) and (138).

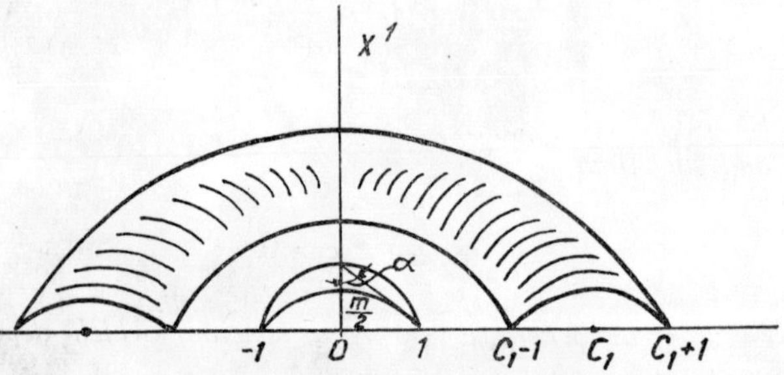

Fig. 5

Let $\{b'_m\}$, $\{c'_m\}$ and D'_b be the sequences and domain corresponding to a second number $b' \neq b$, $0 < b' < 1$. To complete the proof of the theorem, we shall show that (134) holds with $c = \log K_0$, where

$$K_0 = \left(1 - \frac{\text{arctg}\, m}{\pi}\right)^{-1} > 1. \tag{139}$$

Suppose this is not the case. Then there exists a homeomorphism $f: D_b \rightleftharpoons D'_b$ with

$$K(f) < K_0 < 2. \tag{140}$$

Since D_b and $D_{b'}$ are Jordan domains, then, by theorem 4.3, f induces a homeomorphism $f_\sigma: D_b \rightleftharpoons D_{b'}$. Let E_b be the union of the $(n-2)$-dimensional spheres $\{x;\ r = c_p \pm 1,\ x^1 = 0\}$ $(p = 1, 2, \ldots)$ in ∂D_b, and let $E_{b'}$ be the corresponding set in $D_{b'}$. We prove first that because of the preceding inequality, f_σ maps E_b onto $E_{b'}$.

Choose $\xi \in E_b$ and suppose that $\xi^* \in f_\sigma(\xi)$ is a finite point of $\partial D_{b'} - E_{b'}$. Then $\partial D_{b'}$ has a tangent plane at ξ^*. Fix $\varepsilon > 0$. Arguing essentially as in corollary 3 of theorem 13, we can find $(1+\varepsilon)$ - QCfH h and h^* of R^n onto itself with the following properties: h carries D_b onto D, h^* carries $D_{b'}$ onto D^* and the points $h(\xi)$ and $h^*(\xi^*)$ have neighbourhoods U and U^* such that $D \cap U = \Delta \cap U$ and $D^* \cap U^* = \Delta^* \cap U^*$, where Δ is a dihedral wedge of angle $\pi - \text{arctan} m$ with $h(\xi)$ as a point of its edge and where Δ^* is a half space. From (127), (59) and (139) it follows that

$$K_I(h^* \circ f \circ h^{-1}) \geqq K_I(\Delta) = K_0,$$

and hence

$$K(f) \geqq K_I(f) \geqq \frac{K_0}{(1+\varepsilon_0)^2}\,.$$

If $\xi^* \in D_{b'} - E_{b'}$ with $\xi^{*1} = 0$, h^* reduces to the identity and the preceding inequality comes to $K(f) \geqq \dfrac{K_0}{1+\varepsilon_0}$. Since this inequality holds for all $\varepsilon > 0$, we can let $\varepsilon \to 0$ to obtain $K(f) \geqq K_0$ which contradicts (140).

Suppose next that $f_\sigma(\xi) = \infty$, let $S = S^{n-2}(r)$ be the $(n-2)$-dimensional sphere of E_b which contains ξ, and set $S^* = f_\sigma(S)$. Then $S^* - \{\infty\}$ is connected, and by what was proved above, is contained in $E_{b'}$. This means that $S^* - \{\infty\}$ must lie in one of the $(n-2)$-dimensional spheres of $E_{b'}$. Hence $S^* - \{\infty\}$ is bounded and this contradicts the assumption that $f_\sigma(\xi) = \infty$. We conclude that $f_\sigma(\xi) \in E_{b'}$.

It follows that f_σ must map each $(n-2)$-dimensional sphere of E_b onto an $(n-2)$-dimensional sphere of $E_{b'}$. Let $S_1, S_2, \ldots$ and $S_1^*, S_2^*, \ldots$ be the $(n-2)$-dimensional spheres of E_b and $E_{b'}$, respectively, ordered according to increasing radii, let σ_m be the bounded component of $\partial D_b - S_m$ and set $\sigma_m^* = f_\sigma(\sigma_m)$. Then σ_m^* and $\overline{\sigma_m^*}$ must contain exactly $m-1$ and m $(n-2)$-dimensional spheres of $E_{b'}$, respectively, and hence σ_m^* is the bounded component of $\partial D_{b'} - S_m^*$. In particular, this means that f_σ maps the $(n-1)$-dimensional ring

$$A_p = \{x; x = (x^1, r, \vartheta_2, \cdots, \vartheta_{n-1}), c_{p-1} + 1 < r < c_p - 1, x^1 = 0\}$$

onto the $(n-1)$-dimensional ring

$$A_p^* = \{x^*; x^* = (x^*, r^*, \vartheta_2^*, \cdots, \vartheta_{n-1}^*), c'_{p-1} + 1 < r^* < c'_p - 1, x^{*1} = 0\}$$

for $p = 1, 2, \ldots$ and from (137), (1.5.8) and taking into account the definition of $K(f)$, we obtain

$$b'_p + 1 \leqq K(f_\sigma)^{\overline{n-2}}{}^{1}\!\!\!\!\!\!\quad (b_p + 1), \quad b_p + 1 \leqq K(f_\sigma)^{\frac{1}{n-2}} (b'_p + 1). \tag{141}$$

Theorem 4.13 and (140) imply that

$$K(f_\sigma)^{\frac{1}{n-2}} \leqq K(f) < K_0 < 2,$$

and, by (141),

$$b'_p + 1 < 2(b_p + 1), \quad b_p + 1 < 2(b'_p + 1) \tag{142}$$

for $p = 1, 2, \ldots$ Finally, since b_p and b'_p take on only the values 0 and 1 the inequalities (142) imply that $b_p = b'_p$ for all p, and hence that $b = b'$ by virtue of (136) and its counterpart for b'. This contradicts the hypothesis $b \neq b'$. Hence (134) must hold with $c = \log K_0$, and the proof of theorem 36 is complete.

Remark. A. P. Kopylov [4] showed that the space of domains QCf equivalent to the ball is a F_σ-Borel set everywhere dense of the first category in the space of the domains homeomorphic to the ball.

BIBLIOGRAPHY

Works concerning *n*-dimensional quasiconformal mappings

A colloquium on classical function theory. *Notices Amer. Math. Soc.* **8**, 483—485 (1961).

AGARD, STEPHEN [1] Angles and quasiconformal mappings in space. *J. Analys. Math.* **22**, 177—200 (1969); [2] Quasiconformal mappings and the moduli of n-dimensional surface families. "Proc. Romanian—Finnish Seminar on Teichmüller space and quasiconformal mappings. Braşov 26—30 August, 1969". Publishing House of the Academy of the Socialist Republic of Romania, Bucureşti 1971, 9—48; [3] Quasiconformal mappings and the moduli of p-dimensional surface families. *Notices Amer. Math. Soc.* **18**, 1084 (1971); Communicated at the November Meeting of Amer. Math. Soc. in Milwaukee (Wisconsin) 27 November 1971.

AGARD, STEPHEN and KELINGOS, JOHN [1] On parametric representation for quasisymmetric functions. *Comment. Math. Helv.* **44**, 446—456 (1969).

AGARD, STEPHEN and MARDEN, ALBERT [1] A removable singularity theorem for local homeomorphisms. *Indiana Univ. Math. J.* **20**, 455—461 (1970).

AHERN, R. PATRIK [1] Inner functions in the polydiscs. *Notices Amer. Math. Soc.* **18**, 1084 (1971); Communicated at the November Meeting of Amer. Math. Soc. in Milwaukee (Wisconsin) 27 November 1971.

AHLFORS, LARS [1] Extensions of quasiconformal mappings from two to three dimensions. *Proc Nat. Acad. Sci. USA* **51**, 768—771 (1964); [2] Kleinische Gruppen in der Ebene und im Raum. The second Congr. "Rolf Nevanlinna", Zürich 4—6 November, 1965. In "Festband zum 70. Geburtstag von Rolf Nevanlinna", Springer, Berlin-Heidelberg—New York 1966, 7—15. [3] Integral formulas in hyperbolic spaces. Symposium on quasiconformal mappings at Indiana University 18—19 November, 1966.

ANDERSON, GLEN DOUGLAS [1] Bounds for moduli of rings in 3-space. *Notices Amer. Math. Soc.* **12**, 209 (1965); [2] Extremal rings and quasiconformal mappings in 3-space. Doct. Diss. Univ. of Michigan 1965, 135 p. Abstr. 26, no. 11, 6732 (1966); [3] The coefficients of quasiconformality of ellipsoids. *Ann. Acad. Sci. Fenn. Ser. A I*, **411**, 1—14 (1967); [4] Symmetrization and extremal rings in space. *Ann. Acad. Sci. Fenn. Ser. A I*, **438**, 1—24 (1969).

ANDERSON, GLEN DOUGLAS and VAMANAMURTHY, M. K. [1] Rotations of plane quasiconformal mappings. *Tohoku Math. J.* **23**, 605—620 (1971).

ANDONIE, G. ST., [1] "Istoria matematicii în România". (Romanian) (History of mathematics in Romania). Editura Ştiinţifică, Bucureşti, volume 3, 1967, 515 p.

ANDREIAN-CAZACU, CABIRIA [1] Some formulae on extremal length in n-dimensional case. "Proc. Romanian-Finnish Seminary on Teichmüller spaces and quasiconformal mappings. Braşov 26—30 August, 1969. "Publishing House of the Academy of the Socialist Republic of Romania, Bucureşti 1971, 86—102; [2] Modules of surface families. *Rev. Roumaine Math. Pures Appl.* **15**, 1331—1338 (1970); [3] On a transformation formula for the module of surface families. *An. Univ. Bucureşti Ser. Şti. Mat.-Mec.* **19**, 9—14 (1970).

ARGIROVA, TATIANA [1] Teorema o pokrytii dlja kvazikonformnyh otobraženiĭ na ploskosti i v prostranstve. (Russian) (Covering theorems for quasiconformal mappings in plane and space.) *Uspehi Mat. Nauk 20*, No. **1**, 181—184 (1965); [2] V brhu edna harakteristika na konformite izobraženija v prostranstvoto. (Bulgarian) (On a characteristic of the conformal mappings in space). *Bulgar. Akad. Nauk Izv. Mat. Inst.* **9**, 19—25 (1966).

BABAEVA, B. RANO. [1] Kvazikonformnoe vloženie otkrytyh rimanovyh poverhnosteĭ v trehmernoe evklidogo prostranstvo. (Russian) (Quasiconformal imbedding of the open Riemann surfaces in the threedimensional space.) *Trudy Tashkent Gos. Univ.* **394**, 13—16 (1970).

BABAEVA, B. RANO and VOLKOVYSKIĬ, LEV IZRAILEVIČ [1] Ob odnoĭ zadače na konformnoe skleĭvanie v prostrastve. (Russian) (On a problem of conformal sewing in space.) Metričeskie voprosy teor. funkciĭ i otobraženiĭ. No. 1. Dokl. Kollokviuma po teorii kvazikonformnyh otobraženiĭ i ee obobščenijam. Donečk 16—22 September 1968, Izd. "Naukova Dumka", Kiev, 1969, 3—10.

BAKLANOV, M. V. and SUVOROV, GEORGII DMITREVIČ [1] O zamknutosti nekotoryh klassov otobraženiĭ otnositel'no ravnomernoĭ shodimosti. (Russian) (On the completness of some classes of mappings with respect to the uniform convergence.) *Trudy Tomsk. Gos. Univ. Ser. Meh.-Mat.*, **182**, No. 3, 3—14 (1965); [2] Iskaženie otnositel'nyh rasstojaniĭ v zamknutyh oblastjah pri topologičeskih otobraženjah klassa $BL(\Lambda)$ v konformnoĭ ivariantnoĭ metrike. (Russian) (Distortion of the relative distance in closed domains under topological mappings of the class $BL(\Lambda)$ in a conformal invariant metric.) *Trudy Tomsk, Gos. Univ. Ser. Meh.-Mat.* **182**, No. 3, 15—26 (1965).

BECKENBACH, EDWIN FORD and READE O. MAXWELL [1] Generalization to space of the Cauchy and Morera theorems. *Proc. Nat. Acad. Sci. USA* **25**, 92—87 (1939).

BELINSKIĬ, PAVEL PETROVIČ [1] O nepreryvnosti prostranstvennyh kvazikonformnyh otobraženiĭ i o teoreme Liuvillja. (Russian) (On the continuity of the conformal mappings in space and on Liouville's theorem). *Dokl. Akad. Nauk SSSR* **147**, 1003—1004 (1962); [2] Kompaktnost' teoreme Liuvillja o konformnom otobraženii prostranstva $n \geqq 3$. (Russian) (Compactness of Liouville's theorem in 3-space.) Internat. Congress Math. Moscow, 16—26 August, 1966. Abstracts p. 35; [3] The convergence of normalized quasiconformal mappings of a 3-ball as the maximal dilatation tends to 1. Symposium on quasiconformal mappings at Indiana University 1967; — [4] O teoreme Liuvilljy v prostranstve. (Russian) (On Liouville's theorem in space.) Metričeskie voprosy teor. funkciĭ i otobraženiĭ. No. 1. Dokl. Kollokviuma po teorii kvazikonformnyh otobraženiĭ i ee obobščenijam. Doneck 16—22 September 1966; [5] Ustoĭcivost' v teoreme Liuvillja o prostranstvennyh kvazikonformnyh otobraženijah. (Russian) (Stability of Liouville's theorem on quasiconformal mappings in space.) *Nekotorye problemy Mat. i meh. Nauka*, Leningrad, 1970, 88—102; [6] O porjadke blizosti prostranstvennyh kvazikonformnyh otobraženiĭ k konformnym. (Russian) (On the degree of closeness of spatial quasiconformal mappings from the conformal ones.) *Dokl. Akad. Nauk SSSR* **200**, 759—761 (1971).

BELINSKIĬ, PAVEL PETROVIČ and KRUŠKAL, S. L. [1] K semidesjatiletiju akademika M. A. Lavrent'eva. (Russian) (On the seventienth birthday of the academician M. A. Lavrent'ev.) *Sibirsk. Mat. Ž.* **11**, 963—970 (1970).

BERGMAN, STEFAN [1] A class of pseudo-conformal and quasi-pseudo-conformal mappings. *Math. Ann.* **136**, 134—138 (1958); [2] A class of quasi-pseudo-conformal transformation in the theory of functions of two complex variables. *J. Math. Mech.* **7**, 937—956 (1958). [3] Some distorsion theorems of quasianalytic mappings in the space of two complex variables. *Amer. J. Math.* **85**, 405—418 (1963); [4] On the distorsion of a pseudo-conformally invariant metric under homeomorphisms. *Notices Amer. Math. Soc.* **84**, 709 (1965); [5] On bounds for distorsion of an invariant length under homeomorphism of a bicylinder. Internat. Congress on the theory of analytic functions, Erevan 1965 "Sovremenye problemy teorii analitičeskih funkcii". Izd. Nauka, Moscow 1966, p. 33-44; [6] Some distorsion theorems related to an invariant metric in C^2. *Colloquium Math.* **16**, 67—90 (1967).

BERMANT, A. F. and MARKUŠEVIČ, A. I. [1] Teorija funkcii kompleksnogo peremennogo. (Russian) (Function theory of a complex variable.) "Matematika v SSSR za 30 let 1917—1947", Ogiz, Moscow, 319—480.

BICADZE, A. V. [1] Prostranstvennyĭ analog integrala tipa Koši i nekotorye ego primenenija. (Russian) (Space analogue of Cauchy integral and some of its applications.) *Dokl. Akad. Nauk SSSR* **93**, 389—392 (1953).

BICADZE, A. V., MARKUŠEVIČ A. I. and ŠABAT B. V. [1] Matematičeskaja žizn' v SSSR. Mihail Alekseevič Lavrent'ev (k šestidesjatiletiju so dnja roždenijy) (Russian) (The mathematical life in Soviet Union. Mihail Alexeevic Lavrent'ev, for his sixtieth birthday). *Uspehi Mat. Nauk* **16**, 211—221 (1961).

BOROVSKIĬ, JU. E. [1] Teorema Liuvillja o konformnyh otobraženijah v nereguljarnom slučae (Russian) (Liouville's theorem for conformal mappings in the irregular case.) *Sibirsk. Math. Ž.* **3**, 797—801 (1962); [2] Vypuklye poverhnosti s kvazikonformnym sferičeskim otobraženiem. (Russian) (Convex surfaces with a quasiconformal spherical mapping.) *Sibirsk Mat. Ž.* **8**, 535—547 (1967); [3] Ob odnom ocenke dlja vypuklyh poverhnosteĭ s kvazikonformnym sferičeskim otobraženiem. (Russian) (Estimate for convex surfaces with quasiconformal spherical mapping.) *Sibirsk. Mat. Ž.* **9**, 530—535, (1968); *Siber. Math. J.* **9**, 400—403 (1969).

BROWN, ARTHUR [1] Relations between critical points of a real analytic function of N independent variables. *Amer. J. Math.* **52**, 251—270 (1930).

CALLENDER, E. DAVID [1] n-dimensional quasiconformal mappings. Thesis, Stanford Univ. 1959; *Dissert. Abstr. USA* **20**, 1373—1374 (1959); [2] Hölder continuity of n-dimensional quasiconformal mappings. *Technical Report, Stanford Univ.* **81** (1959); *Pacific J. Math.* **10**, 499—516 (1960).

CARAMAN, P. PETRU [1] La différentiabilité des représentations quasi-conformes tridimensionnelles. *Atti dell'Accad. Naz. dei Lincei, Rendiconti* (8), **29**, 181—185 (1960); [2] Noțiuni introductive de teoria reprezentărilor cvasiconforme tridimensionale. (Romanian) (Introduction in the theory of three-dimensional quasiconformal mappings.) *Acad. R. P. Romîne Fil. Iași Stud. Cerc. Şti. Mat.* **11**, 63—92 (1960); [3] Teorema suprafeței Θ și teorema invarianței topologice a domeniilor în spațiul euclidean tridimensional. (Romanian) (Theorem of surface Θ and theorem of the topological invariance of the domains in the three-dimensional Euclidean space) *Acad. R. P. Romîne Fil. Iași Stud. Cerc. Şti. Mat.* **11**, 235—241 (1960); [4] Contribuții la studiul reprezentărilor cvasiconforme tridimensionale. (Romanian) (Contributions to the study of the three-dimensional quasiconformal mappings.) *Acad. R. P. Romîne Fil. Iași Stud. Cerc. Şti. Mat.* **11**, 243—269 (1960); [5] Deux extensions du théorème de la courbe Θ et une nouvelle démonstration du théorème de la conservation des domaines pour l'espace Euclidien à n dimensions. *Bul. Inst. Politehn. Iași* **7** (**11**), 19—26 (1961); [6] Contributions à la théorie des représentations quasi-conformes n-dimensionnelles. *Rev. Math. Pures Appl.* **6**, 311—356 (1961); *Acad. R. P. Romîne Fil. Iași Stud. Cerc. Şti. Mat.* **12**, 13—52 (1961); [7] Proprietatea N a reprezentărilor cvasiconforme n-dimensionale. (Romanian) (Property N of the n-dimensional quasiconformal mappings.) *Acad. R. P. Romîne Fil. Iași Stud. Cerc. Şti. Mat.* **12**, 227—248 (1961); [8] Reprezentări cvasiconforme n-dimensionale. (Romanian) (n-dimensional quasiconformal mappings.) Doctor's Dissertation, București, 1962, 179 p., Dissert. Abstr. 1961; [9] Jacobianul și dilatările reprezentărilor cvasiconforme n-dimensionale. (Romanian) (The Jacobian and the dilatations of n-dimensional quasiconformal mappings). *Acad. R. P. Romîne Fil. Iași Stud. Cerc. Şti. Mat.* **13**, 61—86 (1962); [10] Teoremè de existență a reprezentărilor cvasiconforme n-dimensionale. (Romanian) (Existence theorems for quasiconformal mappings in n-space.) *Acad. R. P. Romîne Fil. Iași Stud. Cerc. Şti. Mat.* **13**, 291—296 (1962); [11] Asupra proprietății N a reprezentărilor continue n-dimensionale. (Romanian) (On the property N of continuous mappings in n-space.) *Acad. R. P. Romîne Fil. Iași Stud. Cerc. Şti. Mat.* **13**, 297—306 (1962); [12] Le Jacobien des représentations quasi-conformes des domaines Euclidiens à n dimensions. C. *R. Acad. Sci. Paris* **254**, 2916—2918 (1962); [13] Quelques propriétés des représentations quasi-conformes des domaines Euclidiens à n dimensions. *C. R. Acad. Sci. Paris* **254**, 3476—3478 (1962); [14] Théorèmes d'existence locale des représentations quasi-conformes différentiables. *C. R. Acad. Sci. Paris* **255**, 814—815 (1962); [15] Cîteva proprietăți ale reprezentărilor cvasiconforme n-dimensionale. (Romanian) (Some properties of the quasiconformal mappings in n-space.) Conference of functional analysis and theory of functions, București, 21—27 September, 1962; *Stud. Cerc. Mat.* **17**, 49—82 (1965); [16] Sur quelques classes de représentations conformes de E_n. *C.R. Acad. Sci. Paris* **256**, 4576—4579 (1963). [17] Asupra reprezentărilor cvasiconforme n-dimensionale. (Romanian) (On quasiconformal mappings in n-space.) *Acad. R. P. Romîne Fil. Iași Stud. Cerc. Şti. Mat.* **14**, 91—126 (1963); [18] Cîteva observații referitoare la cercul lui Kasner. (Romanian) (Some remarks on Kasner's circle.) *Acad. R. P. Romîne Fil. Iași Stud. Cerc. Şti. Mat.* **14**, 279—286 (1963); [19] Sur l'ensemble de monogénéité des représentations différentiables. *An. Şti. Univ. "Al. I. Cuza" Iaşi Secţ. I-a Mat.*, **10**, 91—114 (1964); [20] Quelques propriétés des représentations quasiconformes de E_n. I, II. *Rev. Roumaine Math. Pures Appl.* **10**, 171—209; 271—289 (1965); [21] Sur l'ensemble de monogénéité faible des applications de R^n. *An. Şti. Univ. "Al. I. Cuza" Iaşi Secţ. I-a Mat.*, **11B**, 125—136 (1965). [22] Le cercle de Kasner et ses extensions de E_n. The annual meeting of D. D. R. Math. Soc., Weimar, 27—31 August, 1963; *An. Şti. Univ. "Al. I. Cuza" Iaşi Secţ. I-a Mat.* **12**, 57—71 (1966); [23] A new definition of n-dimensional quasiconformal mappings. *Nagoya Math. J.* **28**, 145—165 (1966); [24] Sur les définitions des homéomorphismes quasiconformes. Internat. Congr. Mat. Moskow 16—26 August, 1966; Abstracts 5—6; [25] On the equivalence of the definitions of the n-dimensional quasiconformal homeomorphisms (QCfH). Communicated at the Centenary of the Romanian Academy; *Rev. Roumaine Math. Pures Appl.* **12**, 887—941 (1967); [26] On a classification of the definitions of quasiconformal homeomorphisms (QCfH). *Rev. Roumaine Math. Pures Appl.* **12**, 1011—1020 (1967); [27] About a theorem of Calderon—Cesari. *Rev. Roumaine Math. Pures Appl.* **13**, 619—626 (1968); [28] About the connection between $2n$-dimensional quasiconformal homeomorphisms (QCfH) and the n-dimensional pseudoconformal transformations (PCT). *Rev. Roumaine Math. Pures Appl.* **13**, 1255—1271 (1968); [29] Grötzsch and Teichmüller rings in plane and n-space. "Proceedings of the Romanian-Finnish Seminar on Teichmüller spaces and quasiconformal mappings. Braşov, 26—30 August, 1969". Publishing House of the Academy of the Socialist Republic of Romania, Bucureşti, 1971, 103—110; [30] Quasiconformality and extremal length. "Proceedings of the Romanian-Finnish Seminar on Teichmüller spaces and quasiconformal mappings. Braşov, 26—30 August, 1969". Publishing House of the Academy of the Socialist Republic of Romania, Bucureşti 1971, 111—145; [31] Characterization of the quasiconformality by means of angle distortion. "Proceedings of the Romanian-Finnish Seminar on Teichmüller spaces and quasiconformal mappings. Braşov, 26—30 August, 1969", Publishing House of the Academy of the Socialist Republic of Romania, Bucureşti 1971, 147—169; [32] About the characterization of the quasiconformality (QCfH) by means of the moduli of q-dimensional surface families. *Rev. Roumaine Math. Pures Appl.* **16**, 1329—1348 (1971); [33] Characterization of the quasiconformality by arc families of extremal length zero. *Ann. Acad. Sci. Fenn. Ser. A I*, 528, 1—10 (1973); [34] p-capacity and p-modulus. *Tohoku Math. J.* (in print); [35] Quasiconformal mappings and boundary correspondence. *Rev. Roumaine Math. Pures Appl.* **19** (1974) (in print).

CARAMAN, PETRU and CORDUNEANU, MONICA [1] Caractérisation des homéomorphismes quasiconformes de R^n par la déformation des angles. *An. Şti. Univ. "Al. I. Cuza" Iaşi Secţ. I-a Mat.* **14**, 15—36 (1968).

ČEREMNYH, L. T. [1] O nepreryvnosti po Gel'deru kvazikonformnyh otobraženiĭ zamknutogo šara. (Russian) (On Hölder continuity of the quasiconformal mappings in a closed ball.) Materialy vtoroĭ naučnoĭ konf. po Mat. i Meh. no. 1. *Izd. Tomskogo Univ., Tomsk.* 1972, 35.

CHÉN, HÀNG-LÉN [1] Quasiconformal mappings in n-dimensional space. *Acta Math. Sinica* **14**, 93—102 (1964); *Chinese Math.* **5**, 101—111 (1964).

CHEO, SI-PING [1] Sur les fonctions satisfaisant aux équations différentielles $D(q) = 0$. *Bull. Sci. Math.* **60**, 9 (1936).

CHURCH, P. T., [1] Cluster sets of functions on an n-ball. *Canad. J. Math.* **15**, 112—120 (1963).

CHURCH, P. T. and HEMMINGSEN, E., [1] Light open maps on n-manifolds. *Duke Math. J.* **27**, 527—536 (1960).

CIMINO, GIANFRANCO [1] Sulle rappresentazioni pseudoconformi. "Sympos Math. Roma 2, Analisi funzionale, Marzo 1968", Academic Press, London—New York and Oderisi Gubbio, 1969, 85—93. [2] Sulla estensione al caso di tre o piu dimensioni di un sistema differenziale del tipo di Beltrami. *Atti dell'Accad. Naz. dei Lincei, Rendiconti* (8), **48**, 14—19 (1970).

CIORĂNESCU, N. [1] Sur les fonctions harmoniques conjuguées. *Bull. Sci. Math.* **56**, 55—64 (1932). [2] Sur les dérivées polydimensionnelles d'une fonction de plusieurs variables. *Enseignement Math.* **34**, 220—227 (1935).

CRISTEA, ION [1] O extindere a reprezentărilor conforme în spaţiul tridimensional. (Romanian) (An extension of conformal mappings in space.) *Acad. R. P. Romîne Stud. Cerc. Mat.* **16**, 1033—1057 (1964).

DANILJUK, I. I. [1] O kvazianalitičeskih funkcijah mnogih peremennyh na mnogoobrazijah. (Russian) (On quasi-analytic functions of several variables on manifolds.) "Trudy III Vsesojuz. Mat. Sezda. Moscow, 25 June—4 July 1956, Short communications". Izd. Akad. Nauk SSSR, Moscow 1956, Tom 2, 79—80 (1958); [2] O kvazianalitičeskih funckcijah mnogih peremennyh na četnomernyh mnogoobrazijah. (Russian) (On quasi-analytic functions of several variables on manifolds of an even number of variables.) *Dokl. Akad. Nauk SSSR* **109**, 434—437 (1956).

DECICCO, JOHN and KASNER, EDWARD [1] Pseudo-conformal geometry of polygenic functions. *Proc. Nat. Acad. Sci. USA* **36**, 667—670 (1950).

DELANGHE, RICHARD [1] An extension of the notion of regularity for functions with values in a Clifford algebra. *Med. Komink. Vlaamse Akad. Wettensch. Lett. Schone Kunst. België Kl. Wettensch.* **32**, 1—14 (1970); [2] On regular points and Liouville's theorem for functions with values in a Clifford algebra. *Simon Stevin* **44**, 55—66 (1970/1971).

DE-SPILLER, D. A. [1] Ekvimorfizmy i kvazikonformnye otobraženija absoluta. (Russian) (Equimorphisms and quasiconformal mappings of the absolute.) *Dokl. Akad. Nauk SSSR* **194**, 106—109 (1970).

DINGHAS, ALEXANDER [1] Über die innere Abbildung der komplexe Einheitskugel. Coll. on Math. Analysis, Jyväskylä (Finland) (1970).

DOBRESCU, MIHAIL [1] Asupra mulţimii de monogeneitate pentru reprezentări cvasiconforme. (Romanian) (On the set of monogeneity for quasiconformal mappings.) *An. Univ. Bucureşti, Mat.-Mec.* **19**, 19—24 (1970).

EARLE, C. CLIFFORD [1] Quasiconformal vector fields. Notices. *Amer. Math. Soc.* **15**, 539(1968).

ELIANU, I. P. [1] Asupra funcţiilor neanalitice de mai multe variabile complexe. (Romanian). (On non-analytic functions of several complex variables.) *Bul. Şti. Acad. R. P. Romîne Secţ. Şti. Mat. Fiz.* **6**, 511—521 (1954).

FANTAPPIE, LUIGI [1] Su un espressione generale dei funzionali lineari mediante le funzioni „para-analitiche" di piú variable. *Rend. Sem. Mat. Univ. Padova* **22**, 1—10 (1955).

FEDOROV, V. S. [1] O monogennosti giperkompleksnyh funkciĭ. (Russian) (On the monogeneity of hypercomplex functions.) *Mat. Sb.* **32** (74), 249—254 (1953); [2] Monogennye *n*-mernye vektor funkciĭ (Russian) (Monogenic *n*-dimensional vectorial functions.) *Učen. Zap. Ivanovsk. Gos. Ped. Inst.* **5**, 67—70 (1954); [3] Osnovnye svoĭstva obobščennyh monogennyh funkciĭ. (Russian) (Fundamental properties of generalized monogenic functions.) *Izd. Vysš. Učebn. Zaved. Matematika* **6** (7), 257—265 (1958); [4] Ob odnom vide giperkomplexnyh monogennyh funkciĭ. (Russian) (On a form of monogenic hypercomplex functions.) *Mat. Sb.* **50** (92), 101—108 (1960).

FLANDERS, G. [1] Liouville's theorem on quasiconformal mappings. *J. Math. Mech.* **15**, 157—161 (1966).

FRÉCHET, MAURICE [1] Sur les fonctions para-analytiques à deux et à trois dimensions. *C.R .Acad. Sci. Paris* **236**, 348—351 (1953); [2] Les fonctions para-analytiques à *n* dimensions. *C.R. Acad. Sci. Paris* **236**, 1832—1834 (1953); [3] Propriétés des fonctions para-analytiques à *n* dimensions. *C. R. Acad. Sci. Paris* **236**, 2191—2193 (1953); [4] Formes canoniques des fonctions para-analytiques à deux et à trois dimensions. *C. R. Acad. Sci. Paris* **236**, 2364—2366 (1953); [5] Les fonctions hypercomplexes à *n* dimensions d'une variable hypercomplexe à *p* dimensions. *C. R. Acad. Sci. Paris* **237**, 1053—1055 (1953); [6] La canonaj formoj de la 2, 3, 4-dimensiaj paraanalitikai funkcii (in Esperanto) (Canonical forme of 2, 3, 4-dimensional para-analytic functions.) *Compositio Math.* **12**, 81—95 (1954).

FUKS, B. A. [1] "Special chapters in the theory of analytic functions of several complex variables". Amer. Math. Soc. Providence. Rhode Island, 1965, Gosudarstvennoe Izd. Fiz.-Mat. Literatury, Moskow, 1963.

FUNZI, A. [1] Sulla rappresentabilita conforme di due varieta ad *n* dimensioni l'una sull' altra. *Ist. Veneto Sci. Arti., Atti Cl. Sci. Mat. Natur* (2), **80**, 777—789 (1921).

GEHRING, W. FREDERICK [1] The Liouville theorem in space. *Notices Amer. Math. Soc.* **7**, 523—524 (1960); [2] Ring domains in space. *Notices Amer. Math. Soc.* **7**, 636 (1960); [3] Quasiconformal mappings in space. *Notices Amer. Math.-Soc.* **7**, 636—637 (1960);

Colloquium on classical function theory. Cornell Univ. 17—21 August, 1961; Coll. Math. Analysis. Helsinki 24—25 August, 1962; *Bull. Amer. Math. Soc.* **69**, 146—164 (1963); [4] A remark on moduli of rings. *Comment. Math. Helv.* **36**, 42—46 (1961); [5] Rings and quasiconformal mappings in space. *Proc. Nat. Acad. Sci. USA* **47**, 98, 105 (1961); *Trans. Amerc. Math. Soc.* **103**, 353—393 (1962); [6] Symmetrisation of rings in space. *Trans. Amer. Math. Soc.* **101**, 499—519 (1961), [7] Extremal length definition for conformal capacity of rings in space. *Michigan Math. J.* **9**, 137—150 (1962); Proc. Internat. Congr. Math. Stockholm, 15—22 July, 1962, Abstract of short communications, Stockholm, 77, 1963; [8] The Carathéodory convergence theorem for quasiconformal mappings in space. *Ann. Acad. Sci. Fenn. Ser. AI* **336**/11, 1—21 (1963); [9] Extensions of quasiconformal mappings in three space. *J. Analyse Math.* **14**, 171—182 (1965); [10] Coefficients of quasiconformality of domains in three space. Proc. Internat. Conference on the theory of analytic functions, Erevan, 1965, 16—17; in "Sovremennye problemy teorii analitičeskih funkcii" Izd. Nauka, Moscow, 1966, 83—88; [11] Extensions theorems for quasiconformal mappings in *n*-space. Internat. Congress of Math., Moscow, 16—26 August, 1966; abstracts of reports p. 48; "Trudy Meždunar. Kongressa Matematikov", 1966, Moscow, 1968, 313—318; *J. Analyse Math.* **19**, 119—169 (1967); [12] Problems on plane quasiconformal mappings. Problems on quasiconformal mappings in *n*-space. Problems presented to the Conference on analytic functions. Lodz 1966; [13] Quasiconformal mappings onto balls. Symposium on quasiconformal mappings at Indiana University, 1967; [14] Quasiconformal mappings of slit domains in three space. *J. Math. Mech. USA* **18**, 689—703 (1969); [15] Ekstremal'nye otobraženija torov. (Russian) (Extremal mappings of tori.) "Nekotorye problemy Mat. i Mec.", Nauka, Leningrad 1970, 146—152; [16] Lipschitz mappings and the *p*-capacity of rings in *n*-space. Advances in the theory of Riemann surfaces. *Ann. of Math.* **66**, Princeton Univ. Press, Princeton 1971, 175—193; [17] Inequality for condensers, hyperbolic capacity and extremal length. Colloquium on Math. Analysis, Jyväskyla 1970; *Michigan Math. J.* **18**, 1—20 (1971); [18] Dilatation of quasiconformal boundary correspondence. *Duke Math. J.* **39**, 89—95 (1972); [19] The L^p-integrability of the partial derivatives of a quasiconformal mapping. (In print.)

GEHRING, F. FREDERICK and HUCKEMAN, FRIEDRICH [1] Quasiconformal mappings of a cylinder. „Proceedings of the Romanian Finnish Seminar on Teichmüller spaces and quasiconformal mappings. Braşov 26—30 August, 1969". Publishing House of the Academy of the Socialist Republic of Romania, Bucureşti 1971, 171–186.

GEHRING, W. FREDERICK and VÄISÄLÄ, JUSSI [1] The coefficient of quasiconformality. *Acta Math.* **114**, 1—70 (1965); [2] Hausdorff dimension and quasiconformal mappings. *J. London Math. Soc.* (In print)

GOL'DŠTEIN, V. M. [1] O povedenie otobraženiĭ s ograničennym iskaženiem pri koefficente iskaženija, blizkom k edinice. (Russian) (On the behaviour of mappings with bounded distortion, for the coefficient of distortion close to the unity.) Inst. Mat. Sib. Otdel. AN SSSR, preprint na rotaprint, Novosibirsk, 1970, 1—16; [2] Odno gomotopičeskoe svoĭstvo otobraženiĭ ograničennym iskaženiem. (Russian) (A certain homotopy property of mappings with bounded distortion.) *Sibirsk. Mat. Ž.* **11**, 999—1008, 1195 (1970); [3] O povedenie otobraženija s organičennym iskaženiem blizkom k edinice. (Russian) (On the behaviour of mappings with bounded distortion close to the unity.) *Sibirsk. Mat. Ž.* **12**, 1250 –1258 (1971).

GORDIN, V. A. and ŠMATKOV, A. A. [1] O kvazikonfcrmnyh otobraženijah, konformnyh na kompleksnyh prjamyh. (Russian) (On quasiconformal mappings, conformal on complex straight lines.) *Trudy Mosk. Univ. Electron. Mašinostr.* No. **5**, 66—96 (1969/1970).

GUIAŞU, SILVIU [1] Asupra interpretării geometrice a reprezentărilor cvasiconforme. (Romanian) (On a geometrical interpretation of the quasiconformal mappings.) *Acad. R. P. Romîne Stud. Cerc. Mat.* **13**, 309—317 (1962).

GUSMAN, S. JA. and RODIN, JU. L. [1] K teorii obobščennyh analitičeskih funkciĭ dvuh peremennyh (Russian) (On the theory of generalized analytic functions of two variables.) *Perm. Gos. Univ. Učen. Zap.* **103**, 17—18 (1963).

HAATI, HEIKKI [1] Über konforme Abbildungen eines euklidischen Raumes in einer Riemannschen Mannigfaltigkeit. *Ann. Akad. Sci. Fenn. Ser. A I* **287**, 1—54 (1960).

HAEFELI, H. GEORG [1] Konformitätsbedingungen bei vierdimensionalen Abbildungen. *Ann. Acad. Sci. Fenn. Ser. A I* **250**/11, 1—8 (1958).

HARTMAN, PH. [1] Systems of total differential equations and Liouville's theorem on conformal mappings. *Amer. J. Math.* **69**, 327—332 (1947); [2] On isometries and on a theorem of Liouville. *Math. Z.* **69**, 202—210 (1958).

HEDRICK, E. R. and INGOLD, LOUIS [1] Analytic and non-analytic functions in three dimensions. *Bull. Amer. Math. Soc.* **29**, 437—438 (1923); [2]Conjugate functions in three dimensions. *Bull. Amer. Math. Soc.* **30**, 401 (1924); [3] Analytic functions in three dimensions. *Trans. Amer. Math. Soc.* **27**, 551—555 (1925); [4] The Beltrami equations in three dimensions. *Trans. Amer. Math. Soc.* **27**, 556—562 (1925); [5] On certain attempted extensions of the theory of functions to three dimensions. *Bull. Amer. Math. Soc.* **31**, 114 (1925); [6] Conjugate functions in three dimensions. *J. Math. Pures Appl.* (9), **7**, 409—416 (1928).

HITOTUMATU SIN [1] On quasiconformal functions of several complex variables. *J. Math. Mech.* **8**, 77—94 (1959); [2] Some remarks on quasipseudo-conformal mappings of Reinhart circular domains. *Comment. Math. Univ. St. Pauli*, **8**, 1—21 (1960).

HÖPPNER W. [1] Über *n*-dimensionale Verallgemeinerungen der Beltramischen Differentialgleichung. Conf. Komplexe Analysis und deren Anwendungen, Göhren (East Germany) 29 September—2 October 1972; "Beiträge zur Komplexen Analysis und Anwendungen in der Differentialgeometrie", Akademie Verlag, Berlin, 1973 (in print).

HSING, C. C. [1] On the group of conformal transformations of a compact Riemannian manifold. *Proc. Nat. Acad. Sci. USA* **54**, 1509—1513 (1965).

HUAN, LE-DE [1] Nekotorye osnovnye teoremy o golomorfnom vektore. (Russian) (Some fundamental theorems on holomorphic vectors.) *Sci. Rec.* **2**, 53—58 (1958).

IFTIMIE, V. [1] Fonctions hypercomplexes. *Bull. Math. Soc. Sci. Math. Phys. R. P. Roumaine*, **9**, 279—332 (1965); [2] Opérateur du type de Moisil-Teodorescu. *Bull. Math. Soc. Sci. Math. R. S. Roumaine* **10** (58), 271—305 (1965).

IKOMA, KAZUO [1] On measures of a set and its image under quasiconformal mappings in space. *Bull. Yamagatha Univ. Natur. Sci.* **6**, 9—15 (1963); [2] On space form of Schwartz's lemma. *Bull. Yamagatha Univ.* **6**, 123—128 (1964); [3] On Schwartz lemma for *K*-QC mapping in space. *Sugaku* **16**, 104—106 (1964/1965); [4] On the distortion and the correspondence under quasiconformal mappings in space. *Nagoya Math. J.* **25**, 175—203 (1965); [5] On a theorem of Schwartz type for quasiconformal mappings in space. *Nagoya Math. J.* **29**, 19—30 (1967).

"Institutul de Matematică al Academiei Republicii Socialiste România (Bucureşti şi Iaşi). 20 de ani de activitate (1949—1969)" Editura Academiei, Bucureşti 1970, 537 p.

IVANENKO, D. and NIKOLSKI, K. [1] Über den Zusammenhang zwischen den Cauchy-Riemannschen und Diracschen Differentialgleichungen. *Z. Phys.* **63**, 129—137 (1930).

JUKOV, V. A. and MIKLIJUKOV, V. M. [1] Ob uglovyh graničnyh značenijah prostranstvennyh otobraženiĭ. (Russian) (On angular boundary values of mappings in space). *Trudy Tomsk. Gos. Univ. Ser. Meh. Mat.* **200**, 88—95 (1968).

JURAŠEV, V. V. [1] O kvazikonformnyh otobraženijah, pridajuščih vsem kompleksnym prjamym odinakovyĭ defect. (Russian) (On quasiconformal mappings, giving on each complex straight line the same defect.) *Izv. Akad. Nauk Armjan SSR Ser. Mat.* **4**, 69—76 (1969).

KARANIKOLOV, HRISTO [1] Elementarno dokazatelstvo na edna teorema na Luman-Menšov. (Bulgarian) (An elementary proof for a theorem of Looman-Menšov.) *Godišnik Višn. Tehn. Učebni Zaved. Matematika* .**2**, 7—12, (1965).

KARI, HAG [1] Quasiconformal boundary correspondence and extremal mappings in higher dimensions. *Notices Amer. Math. Soc.* **19**, A—317 (1972).

KASNER, EDWARD [1] Polygenic functions of two complex variables. *Bull. Amer. Math. Soc.* **35**, 6 (1929); [2] Higher partial derivatives of polygenic functions of several variables. *Bull. Amer. Math. Soc.* **35**, 169 (1929).

KELINGOS, J. A. [1] Boundary correspondence under quasiconformal mappings. *Michigan Math. J.* **13**, 235—249 (1966).

KETCHUM, P. W. and MARTIN TED [1] Polygenic functions of hypercomplex variables. *Bull. Amer. Math. Soc.* **38**, 66—72 (1932).

KIERNAN, J. PETER [1] Quasiconformal mappings and Schwartz's lemma. *Trans. Amer. Math. Soc.* **148**, 185—197 (1970).

KODAIRA, K. and SPENCER, D. C. [1] On deformations of complex analytic structures. I, II. *Ann. Math.* **67**, 328—401 (1958).

KOMIN, R. [1] Über die Existenz von *p*-analytischen Funktionen *n* komplexen Veränderlichen zu vorgegebenen Realteil. *Monatsber. Deutsch. Akad. Wiss. Berlin* **10**, 666—672 (1968).

KOOHARA, A. [1] Similarity principle of the generalized Cauchy-Riemann equations for several complex variables. *J. Math. Soc. Japan* **23**, 213—249 (1971).

KOPYLOV, ANATOLIĬ PAVLOVIČ [1] Povedenie prostranstvenno kvazikonformnogo otobraženija na ploskih sečenijah oblasti opredelenija. (Russian) (Behaviour of spatially quasiconformal mappings on plane sections of the domain of definition. *Dokl. Akad. Nauk SSSR* **167**, 743—746 (1966); [2] K teorii kvazikonformnyh otobraženiĭ v prostranstve. (Russian) (On the theory of quasiconformal mappings in space). Internat. Congr. Math. Moscow, 16—26 August, 1966, abstracts p. 58; [3] Gomeomorfnye otobraženija v trehmernom prestranstve Evklida sohranjajuščie ugly dlja lučei. (Russian) (Homeomorphisms in the Euclidean space preserving angles between lines.) *Dokl. Akad. Nauk SSSR* **170**, 1016—1017 (1966); [4] O bogatstve klassa kvazikonformnyh otobraženiĭ oblasteĭ v trehmernom prostranstve Evklida. (Russian) (On the richness of the class of quasiconformal mappings of domains in threedimensional Euclidean space.) *Dokl. Akad. Nauk SSSR* **172**, 527—528 (1967); [5] Ob ustranimosti ploskih množestv v klasse trehmernyh kvazikonformnyh otobraženiĭ. (Russian) (On removability of plane sets in the class of threedimensional quasiconformal mappings.) Metričeskie voprosy teorii funkcii i otobraženiĭ. Nr. 1. Dokl. Kollokviuma po teorii kvazikonformnyh otobraženiĭ i eë obobščenijam. Doneck 16—22 September, 1968. Izd. "Naukova Dumka", Kiev, 1969, 21—23; [6] O stepenii gladkosti granicy oblasti, gomeomorfnoĭ šary i ne javljajuščeĭsja kvazikonformym obrazom šara. (Russian) (On the degree of smoothness of the boundary of domains, homeomorphic but not quasiconformal images of the ball.) *Sibirsk. Mat. Ž.* 11, 1181—1183 (1970); [7] Ob approksimaciĭ prostranstvennyh kvazikonformnyh otobraženiĭ, blizkih k konformnym, gladkimi kvazikonformnymu otobraženjami. (Russian) (On approximation of spatial quasiconformal close to conformal mappings by smooth quasiconformal mappings.) *Sibirsk. Mat. Ž.* 13, 94—106 (1972).

KOPYLOV, ANATOLII PAVLOVIČ and PEŠIN, I. N. [1] Ustranimost' nekotoryh množestv v klasse trëhmernyh kvazikonformnyh otobraženiĭ. (Russian) (Removability of some sets in the class of threedimensional quasiconformal mappings.) *Mat. Zametki* **7**, 717—722 (1970).

KREINES, M. [1] Sur une classe de fonctions de plusieurs variables. *Mat. Sb.* **9** (51), 713—720 (1941).

KRISTALINSKIĬ, R. N. [1] Psevdoreguljarnye kvaternionye funkciĭ. (Russian) (Pseudoregular quaternionic functions.) Smolensk. Gos. Ped. Inst. Učen. Zap. 91—95 (1965).

KRIVOV, V. V. [1] Ob ekstremal'nyh kvazikonformnyh otobraženijah v prostranstve. (Russian) (On extremal quasiconformal mappings in space.) Soobšč. na I Vsesojuzn. Mat. S'ezde, 1962; *Dokl Akad. Nauk SSSR* **145**, 516—518 (1962); [2] Nekotorye svoĭstva moduleĭ v prostranstve. (Russian) (Some properties of the moduli in space.) *Dokl. Akad. Nauk SSSR* **154**, 510—513 (1964); [3] Naĭlučie ekstremal'nye otobraženija v prostranstve. (Russian) (Best extremal mappings in space.) *Dokl. Akad. Nauk SSSR*, **155**, 38—40(1964); [4] Metod ekstremal'nyh funkciĭ v teoriĭ kvazikonformnyh otobraženiĭ. (Russian) (Method of extremal functions in the theory of quasiconformal mappings.) *Sibirsk. Mat. Ž.* **12**, 1056—1066 (1971); [5] O vloženijah nekotoryh klassah otobraženiĭ. (Russian) (On imbeddings of some classes of mappings.) "Metričeskie voprosy teor. funkciĭ i otobraženiĭ. Nr. 3. Dokl. II-go Kollokviuma po teor. kvazikonformnyh otobraženijah i eë obobščenijam posvjaščennogo 70-letiju so dnja roždenija Akad. M. A. Lavrent'eva. Doneck 12—22 October, 1970", Izd. Naukova Dumka, Kiev 1971, 51—54; [6] Nekotorye lokal'nye teoremy dlja kvazikonformnye otobraženiĭ rimanovyh poverhnosteĭ. (Russian) (Some local theorems for quasiconformal mappings of Riemann surfaces). *Dokl. Akad. Nauk SSSR* **199**, 269—272 (1971); [7] Emkost' podmnožestv granicy Karateodori i BL-gomeomorfizm. (Russian) (Capacity of subsets of Carathéodory boundary and BL-homeomorphisms.) *Dokl. Akad. Nauk SSSR* **199**, 273—274 (1971).

KRUŠKAL' S. L. [1] Nekotorye voprosy teoriĭ kvazikonformnyh otobraženiĭ ploskih i prostranstvennyh otobraženiĭ. (Russian) (Some problems of the theory of quasiconformal mappings in plane and space.) (Typescript.) Novosibirsk 1964; [2] Ob absoljutnoĭ nepreryvnosti i differenciruemosti nekotoryh klassov otobraženiĭ mnogomernyh oblasteĭ. (Russian) (On absolute continuity and differentiability of certain classes of mappings of multidimensional domains.) *Sibirsk. Mat. Ž.* **6**, 692—695 (1965).

KUFAREV, BORIS PAVLOVIČ [1] Sootnošenija tipa "principa dlina i ploščjadi". (Russian) (Relation of the type "principle of length and area".) *Dokl. Akad. Nauk* **170**, 268—270 (1966); [2] Nul'-množestva i gomeomorfizm s konečnom integralom Dirihle. (Russian) (Nul sets and homeomorphisms with finite Dirichlet integral.) *Dokl. Akad. Nauk SSSR* **173**, 1257—1259 (1967).

KUFAREV, BORIS PAVLOVIČ and PEŠKIČEV JU. A. [1] Zamečanija o kvazikonformnyh otobraženijah. (Russian) (Remarks on quasiconformal mappings.) "Metričeskie voprosy teor. funkciĭ i otobraženiĭ. Nr. 1. Dokl. Kollokvjuma po teorii kvazikonformnyh otobraženiĭ i eë obobščenijam. 16—22 September, 1968". Izd. Naukova Dumka, Kiev, 1969, 113—120; [2] Nekotorye geometričeskie svoĭstva kvazikonformnyh gomeomorfizmov. (Russian) (Some geometric properties of quasiconformal homeomorphisms.) *Sibirsk. Mat. Ž.* **12**, 603—612 (1971).

KÜHNAU, REINER [1] Elementare Beispile von möglichst konformen Abbildungen im dreidimensionalen Raum. *Wiss. Z. Martin-Luther-Univ. Halle, Wittenberg, Math. Naturw. Reihe* **11**, 729—732 (1962); [2] Über gewisse Extremalprobleme der quasiconformen Abbildungen. *Wiss. Z. Martin-Luther-Univ., Halle, Wittenberg, Math. Naturw. Reihe* **13**, 35—39 (1964); [3] Quasikonforme Abbildungen und Extremalprobleme bei Feldern in inhomogenen Medien. *J. Reine Angew. Math.* **231**, 101—113 (1968); 238, 61—66 (1969); [4] Der Modul von Kurven- und Flächenscharen und raümliche Felder in inhomogenen Medien. *J. Reine Angew. Math.* **243**, 184—191 (1970).

KULIEV, D. P. [1] O psevdoanalitičeskih funkcijah neskol'kih peremennyh. (Russian) (On pseudoanalytic functions of several variables.) *Azerbaidžan Gos. Univ. Učen. Zap. Ser. Fiz. Mat. Nauk, No.* **1**, 59—64 (1972).

KURAMOCHI, ZENJIRO [1] On sufficient conditions for a function to be holomorphic in a domain *Osaka Math. J.* **3**, 21—47 (1951).

KUUSALO, TAPANI [1] Quasiconformal metrics. "Proceedings of the Romanian-Finnish Seminar on Teichmüller spaces and quasiconformal mappings, Braşov, 26—30 August, 1969". Publishing House of the Academy of the Socialist Republic of Romania, Bucureşti 1971, 193—202.

LAGRANGE, RENÉ [1] Produits d'inversions et métrique conforme. Gauthier-Villars, Paris, 1957.

LAPKO, A. M. and OVČINIKOV I. S. [1] O nesuščestvovanii nekotorogo klassa otobraženiĭ šara na oblast'. (Russian) (On the non-existence of a class of mappings of a ball onto a domain.) *Trudy Tomsk. Gos. Ûniv. Ser. Meh.-Mat.* **200**, 165—172 (1968).

LAVRENT'EV, ALEKSEEVIČ MIHAIL [1] Ob odnom differencial'nom priznake gomeomorfnyh otobraženiĭ trehmernyh oblasteĭ. (Russian) (On a differential criterium for the homeomorphisms in threedimensional domains.) *Dokl. Akad. Nauk SSSR* **20**, 241—242(1938); [2] Ob odnom klasse kvazikonformnyh otobraženiĭ i o gazovyh strujah. (Russian) (On a class of quasiconformal mappings and on gaseous streams.) *Dokl. Akad. Nauk SSSR* **20**, 343—346 (1938); [3] Ustoičivost' v teoreme Liuvillja. (Russian) (Stability of Liouville's theorem.) *Dokl. Akad. Nauk SSSR* **95**, 925—926 (1954); [4] Kvazikonformnye otobraženija. (Russian) (Quasiconformal mappings.) Trudy III-go Vsesojuzn. Mat. S'ezda. Moscow, 25 June — 4 July 1956, Izd. Akad. Nauk SSSR, Moscow, 2,32 (1956); [5] Teorja kvazikonformnyh otobraženiĭ. (Russian) (The theory of quasiconformal mappings.) Trudy III-go Vsesojuzn. Mat. S'ezda. Moscow, 25 June—4 July 1958; Izd. Akad. Nauk SSSR Moscow, Tom 3, 198—208 (1958); [6] Sur la théorie des représentations quasi-conformes. *Ann. Acad. Sci. Fenn. Ser. A I* **250**/18, 1—8 (1958); [7] K teorii prostranstvennyh otobraženiĭ. (Russian) (On the theory of spatial mappings.) *Sibirsk. Mat. Ž.* **3**, 710—714 (1962); [8] O nekotoryh kraevyh zadačah dlja sistem elliptičeskogo tipa. (Russian) (On some boundary problems for elliptical systems.) *Sibirsk. Mat. Ž.* **3**, 715—728 (1962); [9] Sur les représentations quasi-conformes. Intrnat. Congr. Math. Stockholm, 1962. Abstracts of short communications p. 90; [10] K teorii otobraženiĭ trehmernom oblasteĭ. (Russian) (On the theory of mappings of threedimensional domains.) *Sibirsk. Mat. Ž.* **5**, 596—602 (1964); [11] Kraevye zadači i kvazikonformnye otobraženija. (Russian) (Boundary problems and quasiconformal mappings.) Internat. Conference on "The theory of analytic functions". Erevan

1965, printed in "Sovremennye problemy teoryy analitičeskih funkcii" Izd. Nauka, Moscow, 1966, 179—183; [12] On the theory of quasiconformal mappings of three dimensional domains. *J. Analyse Math.* **19**, 217—225 (1967); [13] Matematičeskie problemy gidrodinamiki. (Russian) (Mathematical problems of the hydrodynamics. Second Congress of Bulgarian Mathematiciens. Varna, 28 August—7 September, 1967.

LAWRYNOWICZ, JULLIAN [1] Conformality and pseudo-riemannian manifolds. *Math. Scand.* **28**, 45—69 (1971); [2] Capacities as conformal quasi-invariants on pseudo-riemannian manifolds. *Rep. Math. Phys.* **3** (1973) (in print).

LEBED, B. JA. [1] O defecte, priobretaemom kompleksnymi ploskostjami pri kvazikonformnyh otobraženijah. (Russian) (On the defect, obtained for the complex plane by quasiconformal mappings.) *Trudy Moskov. Inst. Elektron. Mašinostr. Nr.* **5**, 97—110 (1969/1970); [2] Ob odnomernyh otobraženijah Hitotumatu. *Trudy Moskov. Inst. Elektron. Mašinostr.* Nr. 5, 111—124 (1969/1970).

LEHTO, OLLI [1] Entwicklung der Theorie quasikonformer Abbildungen. *Mitt. Math. Ges. DDR Nr.* **3**/4, 36—47 (1970).

LELONG-FERRAND, JACQUELINE [1] Transformations conformes et quasi-conformes des variétés Riemanniennes; applications à la démonstration d'une conjecture de A. Lichnérowicz. *C. R. Acad. Sci. Paris* **269**, 583—586 (1969); [2] Interprétation fonctionnelle des homéomorphismes quasi-conformes. Congrès International Math. Nice 1—10 September, 1970; [3] Étude d'une classe d'applications liées à des homéomorphismes d'algèbres de fonctions. *C. R. Acad. Sci. Paris* **274**, A, 483—486 (1972).

LEWIS, G. LAWRENCE [1] Quasiconformal mappings and Royden algebras in space. *Notices Amer. Math. Soc.* **16**, 138 (1969); *Trans. Amer. Math. Soc.* **158**, 481—492 (1971).

LIOUVILLE, J. [1] Théorème sur l'équation $dx^2 + dy^2 + dz^2 = \lambda\ (d\alpha^2 + d\beta^2 + d\gamma^2)$. *J. Math. Pures Appl.* (1), **15**, 103 (1850).

LOEWNER, C. [1] On conformal capacity in space. *J. Math. Mech.* **8**, 411—414 (1959).

LUFERENKO, V. P. [1] Obobščennaja teorema Karateodori dlja gomeomorfizmov oblasteĭ. (Russian) (Generalized Carathéodory's theorem for homeomorphisms of domains.) Materialy itogovoi naučnoĭ Konf. MMF i NIIPMM za 1970 god, tom. 1, 41—42.

LUFERENKO, V. P. and SUVOROV, G. D. [1] Semeĭstva gomeomorfizmov, ravnostepenno ravnomerno nepreryvnyh po otnositel'nym metrikam. (Russian) (Families of homeomorphisms uniformly equicontinuous with respect to a relative metric.) *Dokl. Akad. Nauk SSSR* **192**, 30—33 (1970); [2] Semeistva gomeomorfizmov, otnositel'nye metriki i teorema Karateodori. (Russian) (Families of homeomorphisms, relative metrics and Carathéodory's theorem.) *Sibirsk. Mat. Ž.* **13**, 368—383 (1972).

MARKUŠEVIČ, A. I. [1] O nekotoryh klassah nepreryvnyh otobraženiĭ. (Russian) (On some class of continuous mappings.) *Dokl. Akad. Nauk SSSR* **28**, 301—304 (1940).

MARTINELLI, ENZO [1] Sulle funzione poligene di due variabili complesse. *Mem. Reale Accad. d'Italia Cl. Sci. Fiz., Mat.* **8**, 65—125 (1937).

MARTIO, OLLI [1] Behaviour of capacity under quasiregular mappings. "Proceedings of the Romanian-Finnish Seminar on Teichmüller spaces and quasiconformal mappings. Braşov 26—30 August, 1969". Publishing House of the Academy of the Socialist Republic of Romania, Bucureşti, 1971, 215—223; [2] On the problem of small K in the theory of quasiregular mappings. Coll. on Math. Anal. Jyväskila 1970; [3] A capacity inequality for quasiregular mappings. *Ann. Acad. Sci. Fenn. Ser. A I* **474**, 1—18 (1970).

MARTIO, OLLI and RICKMAN, SEPPO [1] Boundary behaviour of quasiregular mappings. *Ann. Acad. Sci. Fenn. Ser. A I*, **507**, 1—17 (1972).

MARTIO, OLLI, RICKMAN, SEPPO and VÄISÄLÄ, JUSSI [1] Definitions for quasiregular mappings. *Ann. Acad. Sci. Fenn. Ser. A I*, **448**, 1—40 (1969); [2] Distortion and singularities of quasiregular mappings. *Ann. Acad. Sci. Fenn. Ser. A I* **465**, 1—13 (1970); [3] Topological and metric properties of quasiregular mappings *Ann. Acad. Sci. Fenn. Ser. A I* **488**, 1—31 (1971).

MARTIO, OLLI and SREBRO, URI [1] Periodic and automorphic mappings in R^n. Technion's Preprint Series No. MT—75, Haifa 1971; [2] Periodic meromorphic mappings. *J. Analyse Math.* (to appear).

MICHAEL J. H. [1] An n-dimensional analogue of Cauchy's integral theorem. *J. Australian Math. Soc.* **1**, 171—202 (1960).

MIĆIĆ, P. VLADIMIR [1] O graničnoĭ korespondenciji pri kvazikonformnom preslikavanju u prostoru. (Serbo-Croatian) (On boundary correspondence by quasiconformal mappings in space.) *Mat. Vesnik* **7**, 341—345 (1970).

MIKLJUKOV, V. M. [1] O graničnyh svoĭstvah odnogo klassa otobraženiĭ v prostranstve. (Russian) (On boundary properties of a class of mappings in space.) *Trudy Tomsk. Gos. Univ. Ser. Meh.-Mat.* **189**, 80—85 (1966); [2] O graničnyh svoĭstvah prostranstvennyh otobraženiĭ s ograničennom integralom Dirihle (Russian) (On boundary properties of mappings with bounded Dirichlet integral.) "Tezisy dokladov naučn.-tehn. konfer. molodyh učenyh i specialistov (mat.)". Novosibirsk 1966, 17—18; [3] Ob orientirovannyh kvazikonformnyh otobraženijah v prostranstve. (Russian) (On oriented quasiconformal mappings in space.) *Dokl. Akad. Nauk SSSR* **182**, 266—267 (1968); [4] Ob ε-kvazikonformnyh otobraženijah šara na šar. (Russian) (On ε-quasiconformal mappings of a ball onto a ball.) Metričeskie voprosy teorii funktsii i otobraženiĭ. Nr. 1. Dokl. Kollokviuma po teorii kvazikonformnyh otobraženiĭ i ee obobščenijam. Doneck 16—22 September, 1968, Izd. Naukova Dumka, Kiev 1969, 140—162; [5] Ob odnom multiplikativnom neravenstve dlja prostranstvennyh otobraženiĭ s organičennym iskaženiem. (Russian) (On a multiplicative inequality for the spatial mappings with bounded distorsion.) Metričeskie voprosy teorii funkcii i otobraženiĭ. Nr. 1. Dokl. Kollokviuma po teorii kvazikonformnyh otobraženiĭ i ee obobščenijam. Doneck September, 1968, Izd. Naukova Dumka, Kiev 1969, 162—184; [6] Ob ustranimyh osobennostjah kvazikonformnyh otobraženiĭ v prostranstve (Russian) (On removable singularities of quasiconformal mappings in space.) *Dokl. Akad. Nauk SSSR* **188**, 525—527 (1969); [7] Graničnye svoĭstva n-mernyh kvazikonformnyh otobraženiĭ. (Russian) (Boundary properties of n-dimensional quasiconformal mappings.) *Dokl. Akad. Nauk SSSR* **193**, 525—527 (1970); [8] Ob odnom graničnom svoĭstve n-mernyh otobraženiĭ s ograničennym iskaženiem. (Russian) (On a boundary property of n-dimensional mappings with bounded distortion.) *Mat. Zametki* **11**, 159—164 (1972).

MITJUK, IGOR PETROVIČ [1] Privedenii modul' u vipadku prostoru. (Ukrainian) (Reduced modulus in the case of space.) *Dopovidi Akad. Nauk Ukrain. SSR* **563—566** (1964); [2] Pro kvazikonformni vidobražennja v prostori. (Ukrainian) (On quasiconformal mappings in space.) *Dopovidi Akad. Nauk Ukrain. SSR*, **1022—1025** (1964).

MOISIL, C. GRIGORE [1] Sur l'équation $\Delta u = 0$. *C. R. Acad. Sci. Paris* **191**, 984—986 (1930); [2] Sur les systèmes d'équations de M. Dirac du type elliptique. *C. R. Acad. Sci. Paris* **191**, 1292—1293 (1930); [3] "Sur une classe de systèmes d'équations aux dérivées partielles de la physique mathématique". Imprimerie de la Cour Royale F. Göbl Fils, București 1931; [4] Sur un système d'équations fonctionnelles. *C. R. Acad. Sci. Paris* **192**, 1344—1346 (1931); [5] Asupra unei generalizări a ideii de monogeneitate datorită lui V. S. Fedorov. (Romanian) (On a generalization of the concept of monogeneity following V. S. Fedorov.) *Bul. Şti. Acad. RPR, Ser. A*, **1**, 959—964 (1949); [6] Metoda funcţiilor de o variabilă hipercomplexă în hidrodinamica plană a lichidelor vîscoase. (Romanian) (The method of the functions of a hypercomplex variable in the hydrodynamics of liquids in plane.) *Acad. R. P. Române Stud. Cerc. Mat.*, **1**, 9—35 (1949); [7] "Matricile asociate sistemelor de ecuaţii cu derivate parţiale". (Romanian) (The matrices associated with systems of partial differential equations.) Edit. Acad. R.P.R., București 1950, 61 p.; [8] Despre funcţii monogene în sensul lui Fedorov. (Romanian) (On monogenic functions in Fedorov's sense.) *Bul. Şti. Acad. R.P.R. Ser. A*, **2**, 545—556 (1950).

MOISIL, C. GRIGORE and TEODORESCU, NICOLAE, [1] Functions holomorphes dans l'espace. *Matematica (Cluj)* **5**, 142—159, (1931).

MOSTOV, G. D. [1] On the rigidity of hyperbolic space forms under quasiconformal mappings. *Proc. Nat. Acad. Sci. USA* **57**, 211—215 (1967); [2] Quasiconformal mappings in n-space and the rigidity of hyperbolic space forms. *Inst. Hautes Études Sci. Publ. Math.* **34**, 53—104 (1968).

NAAS, J. and TUTSCHKE, W. [1] Fortsetzung reeler Kurven zu isotropen Kurven und Anfangsprobleme für verallgemeinerte analytische Funktionen. *Monatsb. Deutsch. Akad. Wiss. Berlin* **10**, 244—249 (1968).

NAKAI, MITSURU [1] On Royden algebras and quasi-isometries. *Notices Amer. Math. Soc.* **16**, 967 (1969); [2] Royden algebras and quasi-isometries of Riemann manifolds. *Pacific J. Math.* **40**, 397—414 (1972).

NAKAI, MITSURU and SARIO, LEO, [1] Classification and deformation of Riemann space. *Math. Scandinavica* **20**. 193—208 (1967).

NÄKKI, RAIMO [1] Extension of quasiconformal mappings between wedges. "Proceedings of the Romanian-Finnish Seminar on Teichmüller spaces and quasiconformal mappings. Braşov 26—30 August 1969". Publishing House of the Academy of the Socialist Republic of Romania, Bucureşti 1971, 229—233; [2] Remarks on quasiconformal mappings of a ball. Coll. on Math. Anal. Jyväskila 1970; [3] Boundary behaviour of quasiconformal mappings in n-space. *Ann. Acad. Sci. Fenn. Ser. A I*, **484**, 1—50 (1970).

NÄKKI, RAIMO and PALKA, P. BRUCE [1] Uniformly equicontinuous families of conformal and quasiconformal mappings. *Notices Amer. Math. Soc.* **18**, 1080 (1971).

NAZARJAN, É. O. [1] Kvazikonformnye otobraženija prostranstva C^n na prostranstvo C^m ($m \leqslant n$), konformnye na kompleksnyh prjamyh. (Russian) [Quasiconformal mappings of the space C^n onto the space C^m ($m \leqslant n$), conformal on complex lines.] *Izv.Akad. Nauk Armjan. SSR Ser. Fiz.-Mat. Nauk* **6**, 423—434 (1971); [2] O konformnyh otobraženijah prostrantsva C^n na prostranstvo C^m ($m \leqslant n$) pridajuščih vcem kompleksnym prjamym odinakovyĭ defekt. (Russian) [On conformal mappings of the space C^n onto the space C^m ($m \leqslant n$) giving to each complex line the same defect.] *Izv. Akad. Nauk Armjan SSR Ser. Fiz.-Mat. Nauk* **7**, 59—67 (1972).

NEDELCU-COROI, MARIANA [1] Polinomul areolar de ordin n în spaţiul cu 3 dimensiuni (Romanian) (Areolar polynomial of order n in 3-space.) *Stud. Cerc. Mat.* **7**, 197—237 (1956); [2] Formule de meditaţie pentru polinomul areolar de ordin n în spaţiul cu 3 dimensiuni. (Romanian) (Mediation formulas for n-order areolar polynomials in 3-space.) *Bul. Şti. Acad. R. P. Romîne Secţ. Mat. Fiz.* **8**, 54—58 (1956); [3] Le théorème de Morera pour le polynome aréolaire d'ordre n dans l'espace à trois dimensions. *Bull. Math. Soc. Sci. Math. Phys. R. P. Romîne* **1**, 309—326 (1957); Lucr. IV. Congr. Mat. Romîni. Bucureşti 1956, p. 157—159; [4] Theorie des polynomes aréolaires dans l'espace à n dimensions. I. Le polynome aréolaire d'ordre I. *Rev. Math. Pures Appl.* **4**, 693—723 (1959); [5] Théorie des polynomes aréolaires dans l'espace à n dimensions. III. Théorème de Morera et applications du polynome aux équations à dérivées aréolaires et aux problemes aux limites. *Bull. Math. Soc. Sci. Math. Phys. R. P. Roumaine*, **3**, 165—207 (1959); [6] Une application de la dérivée spatiale aux équations de l'élasticité dans l'espace. *Rev. Math. Pures Appl.* **5**, 549—572 (1960); *Com. Acad. R. P. Romîne* **10**, 795—800 (1960); [7] The theory of the areolar polynomials in the m-dimensional space. II. The n-order areolar polynomial. *Rev. Math. Pures Appl.* **5**, 121—168 (1960); [8] Asupra soluţiei unui sistem de ecuaţii cu derivate parţiale de tip eliptic în E_m, analog ecuaţiei lui Beltrami în complex. (Romanian) (On the solution of elliptic partial differential equations in E_m, analogous to Beltrami's equations in complex plane.) *Stud. Cerc. Mat.* **17**, 1049—1058 (1965); [9] Unele funcţii elementare în clasa funcţiilor holomorfe (α) în spaţiu (Romanian) [Some elementary functions in the class of functions holomorphic (α)]. *Stud. Cerc. Mat.* **17**, 1539—1565 (1965); [10] Representations of spatial polynomials and their properties. *Bull. Math. Soc. Sci. Math. Phys. R. P. Roumaine* **11**, 63—85 (1967); [11] Development in series of α-holomorphic functions in the space. *Rev. Roumaine Math. Pures Appl.* **12**, 1429—1451 (1967); [12] Équations aux dérivées spatiales. *Rev. Roumaine Math. Pures Appl.* **13**, 1285—1291 (1968); [13] Asupra teoriei ecuaţiilor cu derivate spaţiale. (Romanian) (On the theory of partial differential equations.) *An. Univ. Bucureşti Ser. Şti. Natur. Mat.-Mec.* **17**, 15— 37 (1968).

NELSON, E. [1] A proof of Liouville's theorem. *Proc. Amer. Math. Soc.* **12**, 995 (1961).

NEVANLINNA, FRIDJOF and NEVANLINNA, ROLF [1] "Absolute analysis". Berlin-Göttingen-Heidelberg 1959.

NEVANLINNA, ROLF [1] On differentiable mappings. In "Analytic functions" *Princeton Math. Series* nr. **24**, Princeton Univ. Press, Princeton, 1960, 3—9.

NICOLESCU, MIRON [1] Sur les fonctions de bipoint et les fonctions aréolairement conjuguées. *C. R. Acad. Sci. Paris* **185**, 442—444 (1927); [2] Fonctions complexes dans le plan et dans l'espace. Ph. D. Thesis, Paris 1928; [3] Sur les fonctions conjuguées. *Matematica* (*Cluj*) **3**, 134—143 (1930); **14**, 18—20 (1938).

NIREMBERG, RICARDO [1] On quasi conformality in several complex variables. Doctor's Thesis, New York Univ. 1966, 35 p.; *Dissert. Abstr.* 1967, **27**, 2789—B; [2] On quasi-pseudoconformality in several complex variables. *Trans. Amer. Math. Soc.* **127**, 233—240 (1967); [3] A holomorphic extension theorem for real submanifolds of C^n. *Notices Amer. Math. Soc.* **18**, 1084 (1971); Communicated at the November Meeting of Amer. Math. Soc. in Milwaukee (Wisconsin) 27 November 1971.

NIRENBERG, L. [1] On a generalization of the quasiconformal mappings and its application to the elliptic differential equations. In: "Contribution to the partial differential equations" *Ann. of Math. Studies*, Princeton Univ. Press, Princeton N.J., **33**, 95—100 (1954).

ONICESCU, OCTAV [1] Les fonctions holotopes. *Bull. Math. Soc. Sci. Math. Phys. R. P. Roumaine* **7**, 203—223 (1963); [2] Fonctions holotopes et mouvement des fluides. "Proceedings of the Romanian-Finnish Seminar on Teichmüller spaces and quasiconformal mappings. Braşov 26—30 August, 1969". Publishing House of the Academy of the Socialist Republic of Romania, Bucureşti, 1971, 235—243.

OVČINIKOV, IGOR SEMENOVIČ [1] Metričeskie svoĭstva otobraženiĭ klassa $\overset{3}{BL^2}$. (Russian) (Metric properties of the mappings of the class $\overset{3}{BL^2}$.) *Dokl. Acad. Nauk SSSR* **161**, 526—529 (1965); *Trudy Tomsk. Gos. Univ. Ser. Meh.-Mat.* **182**. No. 3, 32—45 (1965); [2] Nekotorye metričeskie svoĭstva prostranstvennyh otobraženiĭ. (Russian) (Some metric properties of mappings in space). Internat. Math. Congress, Moscow 16—26 August 1966, Abstracts p. 71; Kanditatskaja Dissert. Tomsk 1966; [3] Neravenstvo tipa "principa dlina i ploščadi" dlja n-mernogo prostranstva. (Russian) (An inequality of the type "the principle of the length and area" for n-space.) *Trudy Tomsk. Gos. Univ. Ser. Meh.-Mat.* **189**, No. 4, 86—95 (1966); [4] Prostye koncy odnogo klassa prostranstvennyh oblasteĭ. (Russian) (Prime ends of a class of mappings in space.) *Trudy Tomsk. Gos. Univ. Ser. Meh.-mat.* **189**, No. 4, 96—103 (1966); [5] Metričeskie svoĭstva odnogo klassa otobraženiĭ v prostranstve. (Russian) (Metric properties of a class of mappings in space.) Tezisy dokladov naučn.-tehn. konf. molodyh učenyh i specialistov (Mat.). Novosibirsk 1966, 15—17; [6] O nesuščestovanii otobraženiĭ kalssa $\overset{n}{BL^2}$ šara na oblast'. (Russian)(Non-existence of mappings of class $\overset{n}{BL^2}$ of the ball onto a domain.) *Dokl. Akad. Nauk SSSR* **179**, 24—27 (1968); [7] Ob odnom analoge teoremy Lindelefa dlja prostranstvennyh otobraženiĭ. (Russian) (A theorem analogous to Lindelöff for mappings in space.) Metričeskie voprosy teorii funkciĭ i otobraženiĭ. Nr. 1. Dokl. Kollokviuma po teoriĭ kvazikonformnyh otobraženiĭ i ee obobščenijam. Donesk 16—22 September, 1968, Izd. Naukova Dumka, Kiev 1969, 184—201; [8] Metričeskie svoĭstva otobraženii osuščestvljajuščih ograničennymi nekotorymi integral'nye funktionaly. (Russian) (Metric properties of some mappings leaving bounded some integral functionals.) *Dokl. Akad. Nauk SSSR* **187**, 36—39 (1969); [9] Nekotorye svoĭstva ploskih i prostranstvennyh otobraženiĭ. (Russian) (Some properties of mappings in plane and space.) *Dokl. Akad. Nauk SSSR* **190**, 276—279 (1970); [10] Neraventsva tipa "principa dliny i ploščadi" dlja otobraženiĭ, ostavljajuščih ograničennymi nekotorye integral'nye funkcionaly y n-mernom prostranstve. (Russian) (An inequality of the type "the principle of length and area" for mappings leaving bounded some integral functionals in n-space.) Metričeskie voprosy teor. funkciĭ i otobraženiĭ. Nr. 3. Dokl.II-go kollokviuma po teor. kvazikonformnyh otobraženija i ee obobščenijam posvjaščennogo 70-letiju so dnja roždenija Akad. M. A. Lavrent'eva. Doneck 12—22 October, 1970. Izd. Naukova Dumka, Kiev, 1970, 98—115. [11] Ocenka snizu veličiny integrala Dirihle pri otobraženija šara na oblasti. (Russian) (Lower bound of Dirichlet integral for mappings of the ball onto a domain). *Sibirsk. Mat. Ž.* **13**, 142—152 (1972).

OVČINIKOV, IGOR SEMENOVIČ and SUVOROV, G. D. [1] Preobrazovanija integrala Dirihle i prostranstvennye otobraženija. (Russian) (Transformations of Dirichlet integral and mappings in space.) *Dokl. Akad. Nauk SSSR* **154**, 523—526 (1964); *Sibirsk. Mat. Ž.* **6**, 1292—1314 (1965).

PAL'ČEV, B. V. [1] Mnogomernyĭ analog teoremy Morera. (Russian) (Morera theorem for severa dimensions.) *Sibirsk. Mat. Ž.* **4**, 1376—1388 (1963).

PARKER, K. STEPHEN [1] A method for determining the dilatation of a family of quasiconformal homeomorphisms. *Notices Amer. Math. Soc.* **18**, 150 (1970).

PASCALI, DAN [1] O nouă reprezentare a polinoamelor areolare in plan. (Romanian) (A new representation of areolar polynomials in plane.) *Stud. Cerc. Mat.* **15**, 249—251 (1964); [2] Vecteurs analytiques généralisées. *Rev Roumaine Math. Pures Appl.* **10**, 779—808 (1965); *Stud. Cerc. Mat.* **17**, 707—736 (1965); [3] Reprezentarea polinomului areolar cvaternionic în spaţiul tridimensional. (Romanian) (Representation of the areolar quaternionic polynomial in 3-space.) *Stud. Cerc. Mat.* **18**, 239—242 (1966); [4] Sur la représentation de première espèce des vecteurs analytiques généralisées. IVth Interbalcanic Congr. Bucureşti 1966; *Rev. Roumaine Math. Pures Appl.* **12**, 685—689 (1967); [5] Asupra funcţiilor analitice generalizate vectoriale. (Romanian) (On vectorial generalized analytic functions.) *Stud. Cerc. Mat.* **20**, 1013—1017 (1968); [6] An abstract theory of the generalized analytic functions. *Rev. Roumaine Math. Pures Appl.* **13**, 1439—1440 (1968); Coll. on fonctional equations, Bucureşti-Mamaia, September 1968.

PASCALI, DAN and POP, DOINA [1] Funcţii analitice generalizate pe algebre Banach. (Romanian) (Generalized analytic functions on Banach algebras.) *Stud. Cerc. Mat.* **20**, 341—345 (1968).

PERTRIDIS, NICOLAS [1] On pseudoanalytic mappings. *Prakt. Akad. Atenon* **37**, 179—181 (1962).

PESKIČEV, JU. A. [1] Ob iskaženiĭ uglov pri kvazikonformnyh otobraženijah. (Russian) (On distortion of angles by quasiconformal mappings.) *Sibirsk. Mat. Ž.* **11**, 1191—1192 (1970); [2] Iskaženie mer množestv urovnja pri differenciruemyh otobraženijah. (Russian) (Distortion of the level sets by differentiable mappings.) *Sibirsk. Mat. Ž.* **12**, 414—419 (1971); [3] Odin kriterii kvazikonformnosti. (Russian) (A criterium of quasiconformality.) "Materialy II-oi naučnoi konf. po Mat. i Meh. Tomsk 1—5 February 1972". Izd. Tomskogo Univ., Tomsk 1972, 24—25.

PHILLIPS, ROBERT [1] Liouville's theorem. *Pacific J. Math.* **28**, 397—405 (1969).

POLECKII, E. A. [1] Metod moduleĭ dlja negomeomorfnyh kvazikonformnyh otobraženiĭ. (Russian) (The method of moduli for non-homeomorphic quasiconformal mappings.) *Mat. Sb.* **83**, 261—272 (1970).

RAINICH, G. Y. [1] Analytic vector functions. *Bull. Amer. Math. Soc.* **30**, 8 (1924).

RAYNOR, G. E. [1] Generalization of Beltrami equations to the curve *n*-space. *Bull. Amer. Math. Soc.* **33**, 435—439 (1927).

READE, MAXWELL [1] On certain conformal mappings in space. *Michigan Math. J.* **4**, 65—66 (1957).

REIMANN, HANS MARTIN [1] Une classe spéciale d'applications quasi-conformes dans E^n. *C.R. Acad. Sci. Paris* **266**, 273—274 (1968); [2] Notion de surface dans la théorie des applications quasi-conformes dans E^n. *C.R. Acad. Sci. Paris* **268**, 18—20 (1969); [3] Über harmonische Kapazität und quasikonforme Abbildungen im Raum. *Comment. Math. Helv.* **44**, 284—307 (1969); [4] Über das Verhalten von Flächen unter quasikonformen Abbildungen im Raum. *Ann. Acad. Sci. Fenn. Ser. A I* **470**, 1—27 (1970); [5] On absolute continuity of a surface representation. *Comment. Math. Helv.* **46**, 44—47 (1971).

RENGGLI, HEINZ [1] On triangle dilatation. "Proceedings of the Romanian-Finnish Seminar on Teichmüller space and quasiconformal mappings. 26—30 August, 1969". Publishing House of the Academy of the Socialist Republic of Romania, Bucureşti 1971, 255—259.

REŠETNJAK, JU. G. [1], Ob odnom dostatočnom priznake nepreryvnosti otobraženija po Gelderu. (Russian) (A sufficient condition for Hölder continuity of a mapping.) *Dokl. Akad. Nauk SSSR* **130**, 507—509 (1960); *Sovjet Math.* **1**, 76—78 (1960); [2] O konformnyh otobraženijah prostranstva. (Russian) (On conformal mappings in space). *Dokl. Acad. Nauk SSSR* **130**, 1196—1198 (1960); *Sovjet Math.* **1**, 153—155 (1960); [3] Ustoĭčivost'v teoreme Liuvillja o konformnyh otobraženijah prostranstva. (Russian) (Stability in the theorem of Liouville from the conformal mappings in space.) In "Nekotorye problemy mat. i meh." *Sibirsk. Otdel. Akad. Nauk SSSR*, Novosibirsk, 219—223 (1961); [4] Ob ustoĭčivosti v teoreme Liuvillja o konformnyh otobraženijah prostranstva. (Russian) (On the stability in the theorem of Liouville from the conformal mappings in space.) *Dokl. Acad. Nauk SSSR* **152**, 286—287 (1963); [5] Nekotorye voprosy teorii prostranstvennyh otobraženiĭ. (Russian) (Some questions on the theory of space mappings.) Outlines Joint Sympos. Partial Differential Equations, Novosibirsk 1963; [6] Očenki modulja nekotoryh otobraženiĭ. (Russian) (Evaluations of the module of some mappings.)

Sibirsk. Mat. Ž. **7**, 1106—1114 (1966); [7] Obobščennye proizvodnye i differenci ruemost' počti vsjudu. (Russian) (Generalized derivatives and differentiability almost everywhere.) *Dokl. Akad. Nauk SSSR* **170**, 1273—1275 (1966); [8] Nekotorye geometričeskie svoĭstva funkcii i otobraženiĭ s obobščennymi proizvodnymi. (Russian) (Some geometric properties of functions and mappings with generalized derivatives.) *Sibirsk. Mat. Ž.* **7**, 886—919 (1966); [9] Ob ustoĭčivosti konformnyh otobraženiĭ v mnogomernyh prostranstvah. (Russian) (On the stability of conformal mappings in higher dimensions.) *Sibirsk. Mat. Ž.* **8**, 91—114 (1967); [10] Prostranstvennye otobraženija s ograničennym iskaženiem. (Russian) (Space maps with bounded distorsion.) Internat. Congr. Math. 16—26.VIII.1966. Abstracts p. 42; *Dokl. Akad. Nauk SSSR* **174**, 1281—1283 (1967); *Sibirsk. Mat. Ž.* **8**, 629—658 (1967); [11] Teorema Liuvillja o konformnyh otobraženijah pri minimal'nyh predpoloženijah reguljarnosti. (Russian) (Liouville's theorem on conformal mappings with minimal hypotheses of regularity.) *Sibirsk. Mat. Ž.* **8**, 835—840 (1967); [12] Ob uslovii ograničenosti indeksa dlja otobraženiĭ s ograničennym iskaženiem. (Russian) (On the boundedness of the index for mappings with bounded distortion.) *Sibirsk. Mat. Ž.* **9**, 368—374 (1968); [13] Otobraženija s ograničennymi ickaženiem kak ekstremali integralov tipa Dirihle. (Russian) (Mappings with bounded distortion as extrema of the integrals of Dirichlet type.) *Sibirsk. Mat. Ž.* **9**, 625—666 (1968); [14] Teoremy ustoĭcivosti dlja otobraženii s ograničennym iskaženiem. (Russian) (Theorems of stability for mappings with bounded distorsion.) *Sibirsk. Math. J.* **9**, 667—684 (1968); [15] Ocenka ustoĭčivosti v teoreme Liuvillja. (Russian) (Estimate of the stability in Liouville's theorem.) Pervyĭ Doneckiĭ Kollokvium po teorii kvazikonformnyh otobraženiĭ i ee obobščenijam 16—22 September, 1968; [16] O plotnosti obraza pri kvazikonformnyh otobraženihah prostranstva. (Russian) (On the density of the image for quasiconformal mappings in space.) Pervyĭ Doneckiĭ Kollokvium po teorii kvazikonformnyh otobraženiĭ i ee obobščenijam 16—22 September, 1968; [17] Ob ekstremal'nyh svoĭstvah otobraženiĭ s graničnym iskaženiem. (Russian) (Extremal properties of the mappings with bounded distortion.) *Sibirsk. Mat. Ž.* **10**, 1300—1310 (1969); [18] Local'naja struktura otobraženiĭ s ograničennym iskaženiem. (Russian) (Local structure of the mappings with bounded distortion.) *Sibirsk. Mat. Ž.* **10**, 1311—1333 (1969); [19] Space mappings with bounded distortion. Congrès Internat. Math. Nice 1—10 September, 1970; [20] O množestve toček vetvlenija otobraženiĭ s ograničennym iskaženiem. (Russian) (On the set of branch-points of mappings with bounded distortion.) *Sibirsk. Mat. Ž.* **11**, 1333—1339 (1970); [21] Ob ocenke ustoĭcivosti v teoreme Liuvillja o konformnyh otobraženijah mnogomernyh prostranstv. (Russian) (The estimation of stability in Liouville's theorem on conformal mappings of multidimensional spaces.) *Sibirsk. Mat. Ž.* **11**, 1121—1139, 1198 (1970).

REŠETNJAK, JU. G. and ŠABT, B. V. [1] O kvazikonformnyh otobraženijah v prostranstve. (Russian) (On quasiconformal mappings in space.) "Trudy IV vsesojuzn. Mat. S'ezda, Leningrad 1961". Izd. Akad. Nauk SSSR, Leningrad, 1964, 5, 672—680.

RICKMAN, SEPPO [1] Boundary correspondence under quasiconformal mappings of Jordan domains. *J. Math. Mech.* **18**, 429—432 (1968); [2] On quasiregular mappings". Proceedings of the Romanian-Finnish Seminar on Teichmüller space and quasiconformal mappings, Braşov, 26—30 August, 1969". Publishing House of the Academy of the Socialist Republic of Romania, Bucureşti, 1971, 261—271; [3] Removability theorems for quasiconformal mappings. *Ann. Acad. Sci. Fenn. Ser. A I* **449**, 1—8 (1969); [4] Local behaviour of quasiregular mappings. Coll. on Math. Anal. Jyväskyla (Finland) 1970.

ROŞCULEŢ, MARCEL [1] Fonctions monogènes sur des espaces de Riemann. IV. Transformations par des fonctions quasi-monogènes sur des algèbres du second ordre. *Rev. Roumaine Math. Pures Appl.* **15**, 135—142 (1970).

RUDNIK, D. E. [1] O kvazigolomorfnyh otobraženijah. (Russian) (On quasi-biholomorphic mappings.) *Trudy Moskov. Inst. Mašinostr.* **2**, 505—510 (1966); [2] Ob odnoĭ ocenke pri kvazikonformnyh otobraženijah v prostranstve. (Russian) (An evaluation for quasiconformal mappings in space.) *Trudy Moskov. Inst. Elektron. Mašinostr.* **4**, 26—29 (1968); [3] O nekotoryh kvazikonformnyh otobraženijah v prostranstve C^n. (Russian) (On some quasiconformal mappings in the space C^n.) *Izv. Vysš. Učebn. Zaved. Matematika No.* **1**. 76—78 (1972); [4] Quasibiholomorphic mappings with bounded defect. (To appear.)

RÜEDY, RETO [1] Einbetungen Riemannscher Flachen in den dreidimensionalen euklidischen Raum. *Comment. Math. Helv.* **43**, 417—442 (1968).

ŠABAT, B. V. [1] K teorii prostranstvennyh otobraženiĭ. (Russian) (On the theory of mappings in space.) Vsesojuzn. Konf. po teorii funkciĭ. Tezisy dokl. Erevan, 120—121 (1960); [2] Metod moduleĭ v prostranstve. (Russian) (The method of the moduli in space.) *Dokl. Akad. Nauk SSSR* **130**, 1210 — 1213 (1960); [3] K teorii kvazikonformnyh otobraženiĭ v prostranstve. (Russian) (On the theory of quasiconformal mappings in space.) *Dokl. Akad. Nauk SSSR* 132, 1045—1048 (1960); [4] Nelineĭnye giperboličeskie i prostranstvennye zadači teorii kvazikonformnyh otobraženiĭ. (Russian) (Nonlinear hyperbolic problems in the theory of quasiconformal mappings in space.) Doctor Thesis, Moscow, 1961; [5] K ponjatiju proĭzvodnoĭ sistemy v smysle M. A. Lavrent'eva. (Russian) (On the concept of the derived system in the sense of M. A. Lavrent'ev.) *Dokl. Akad. Nauk SSSR* **136**, 1298—1301 (1961); [6] Ekstremal'nye zadači teorii kvazikonformnyh otobraženiĭ v prostranstve. (Russian) (Extremal problems in the theory of quasiconformal mappings in space.) "Sovieto-American joint Symposium on partial differential equations. Novosibirsk, August, 1963", p. 61—64; [7] O mnogomernyh teorii analitičeskih funkcii. (Russian) (Theory of analytic functions in space.) VII Vsesojuz. Konf. po teorii funkcii kompleksnogo peremennogo. Rostov-na-don, 23—28 September, 1963; [8] O ne gomeomorfnyh kvazikonformnyh otobraženiĭ. (Russian) (On non-homeomorphic quasiconformal mappings.) Second Congr. of Bulgarian mathematics, Varna, 28 August—7 September, 1967; [9] Teorija funkcii kompleksnogo peremennogo v Moskovskom universitete za 50 let. (Russian) (Theory of functions of a complex variable in Moscow university the last 50 years.) *Vestnik Moskov. Univ. Ser. I Mat. Meh. N⁰* **6**, 16—23 (1967).

ŠAFEEV, M. N. [1] O kvazikonformnyh otobraženiĭ. (Russian) (On quasiconformal mappings.) In "materialy Naučno-Techn. Konf. Kuibyševk. Aviac. Inst." Kuibyšev 1970, 52—53.

SAMSONIJA, Z. B. [1] O približeniĭ kvazikonformnyh otobraženiĭ, blizkih k konformnym. (Russian) (On approximation of quasiconformal mappings close to the conformal ones.) *Soobšč. Acad. Nauk. Gruzin SSR* **63**, 271—272 (1971).

SARIO, LEO [1] Classification of locally Euclidean spaces. *Nagoya Math. J.* **25**, 87—111 (1965).

SCHNEIDER, W. J. [1] A uniqueness theorem for conformal maps. *Notices Amer. Math. Soc.* **84**, 698 1965).

SCHWARTZ, MARIE-HÉLÈNE [1] Applications intérieures régulières dans les variétés à n dimensions. *C. R. Acad. Sci. Paris* **230**, 1244—1245 (1950); [2] Applications A-intérieures et formules de Gauss-Bonnet généralisées. *C. R. Acad. Sci. Paris* **230**, 1337—1338 (1950); [3] Applications A-intérieures et théorie des défauts. *C. R. Acad. Sci. Paris* **250**, 1376—1378 (1950); [4] Formules apparentées à celles de Nevanlinna-Ahlfors pour certaines applications d'une variété à n dimensions dans une autre. *Bull. Soc. Math. France* **82**, 317 — 360 (1954).

SEDO, P. I. and SYČEV, ANTON VIKTOROVIČ [1] O prodolženie kvazikonformnyh otobraženiĭ na mnogomernye prostranstve bol'šeĭ razmernosti (Russian) (On extension of quasiconformal mappings to multidimensional spaces of higher dimensions.) *Dokl. Akad. Nauk SSSR* **198**, 1278—1279 (1971).

SIEGEL, M. KEEVE [1] Three-dimensional conformal transformations. *J. Aeronaut. Sci.* **19**, 281—282 (1952).

ŠMATKOV, A. A. [1] Ob odnom iz klassov kvazibigolomorfnyh otobraženiĭ. (Russian) (On a class of quasi-biholomorphic mappings.) *Izv. Akad. Nauk Armjan. SSR Ser. Fiz.-Mat. Nauk*, **18**, No. 4, 26—31 (1965); [2] Kvazikonformnye prostranstva C_z^n, konformnye na kompleksnyh prjamyh. (Russian) (On quasi-biholomorphic mapings of C_z^n space, which are conformal on complex lines.) *Izv. Akad. Nauk Armjan SSR Ser. Fiz.-Mat. Nauk* **3**, 479—496 (1968).

SPRINGER, GEORGE [1] Proceedings of the symposium on quasiconformal mappings at Indiana university. *J. Math. Mech.* **16**, 1061—1067 (1967).

SREBRO URI [1] Sequences and groups of conformal and quasiconformal homeomorphisms of the n-space. Doctor Diss., Haifa 1968; [2] Conformal capacity and quasiconformal mappings in R^n. Technion preprint series No. MT—22, Haifa 1969 *Israel J. Math.* **9** 93—100 (1971); [3] Convergence of K-quasiconformal automorphisms. "Proceedings of the Romanian-Finnish Seminar on Teichmüller space and quasiconformal mappings. Braşov 26—30 August, 1969", Publishing House of the Academy of the Socialist Republic of Romania, Bucureşti 1971, 287—291; [4] Conformal capacity and quasiconformal mappings. (In print).

STOILOW, SIMION [1] Sur les transformations intérieures des variétés à trois dimensions. C.R. du Ier Congr. Math. Hongrois, Budapest, 1950, 263—266; [2] "Oeuvres mathématiques". Editions de l'Acad. de la République Populaire Roumaine, Bucureşti 1964, 415 p.

STORVICK, DAVID [1] The boundary correspondence of a quasiconformal mapping. Math. Research Center US Army. The Univ. of Wisconsin. MRC Technical Summary Report 426, 1—8 (1963).

SUOMINEN, KALEVI [1] Quasiconformal maps in manifolds. *Ann. Acad. Sci. Fenn. Ser. A I* **393**, 1—39 (1966).

SUVOROV, GEORGII DMITRIEVIČ [1] Metričeskie svoĭstva ploskih i prostranstvennyh otobraženii v zamknutyh oblasteĭ. (Russian) (Metric properties of mappings in closed domains in plane and space.) Dokl. III Sibirsk. Konf. Mat. i Meh. Tomskiĭ Univ., Tomsk 1964, 15—18; [2] Semeĭstva ploskih topologičeskih otobraženiĭ. (Russian) (A family of plane topological mappings.) Redakcionno-izdat. Otdel. Sibirsk. Otdel. Akad. Nauk SSSR, Novosibirsk 1965, 264; [3] Metričeskie svoĭstva otobraženii s ograničennymi integralami Dirihle. (Russian) (Metric properties of mappings with bounded Dirichlet integral.) In "Itogi issledovanii po matematiki i mehaniki za 50 let 1917—1967", Izd. Tomskogo Univ. Tomsk, 1967, 52—65; [4] Metričeskie svoĭstva ploskih i prostranstevnnyh otobraženiĭ. (Russian) (Metric properties of mappings in plane and space.) Tezisy dokladov sekcii Mat. i Meh. Naučnoi sessii Doneckogo naučnogo centra Akad. Nauk. Ukrain SSR, Doneck 1967; [5] Rasširenija topologičeskih struktur i metričeskie svoĭstva otobraženii. (Russian) (Extensions of topological structures and metric properties of mappings.) "Contribs extens. Theory Topol. Struct. Proc. Sympos., Berlin 1967" Deutscher Verlag der Wissenschaften, Berlin 1969, 257—273; [6] Pervyĭ Doneckii kollokvium po teorii kvazikonformnyh otobraženiĭ i ee obobščenijam. (Russian) (I) Colloquium in Doneck on the theory of quasiconformal mappings and its generalizations. *Uspehi Mat. Nauk* **24**, 241—242 (1969).

SYČEV, ANTON VIKTOROVIČ [1] O kvazikonformnyh otobraženiĭ v prostranstve. (Russian) (On quasiconformal mappings in space.) *Dokl. Acad. Nauk SSSR* **166**, 298—300 (1966); [2] O prostranstvennyh kvazikonformnyh otobraženijah, udovletvorjajuščih usloviu Geldera v graničnyh točkah. (Russian) (On quasiconformal mappings in space, satisfying Hölder condition in boundary points.) *Dokl. Akad. Nauk SSSR* **175**, 38—39 (1967); [3] Prostranstvennye kvazikonformnye otobraženija, nepreryvnye po Gelderu v graničnyh točkah. (Russian) (Quasiconformal mappings in space, Hölder continuous in boundary points.) *Sibirsk. Mat. Ž.* **11**, 183—192 (1970); [4] Ob odnom kvazikonformnom otobraženiĭ šara na cebja. (Russian) (On a quasiconformal mapping of the ball onto itself.) *Dokl. Akad. Nauk SSSR* **206**, 556—558 (1972).

SZILARD, S. KARL [1] O rasprostranenii teoremy Fatu na odin klass nepreryvnyh otobraženiĭ. (Russian) (A generalization of Fatou's theorem for a class of continuous mappings.) *Dokl. Akad. Nauk SSSR* **127**, 278—280 (1959).

TAARI, OSSI [1] Raumwinkel und Quasikonformität. *Ann. Acad. Sci. Fenn. Ser. A I* **426**, 1—16 (1968).

TEODORESCU, NICOLAE [1] La dérivée aréolaire et ses applications dans la physique mathématique. Thèse, Paris (1931); [2] Sur l'emploi de quelques relations globales dans quelques problèmes physiques. *Ann. Mat. Pura Appl.* (4), **11**, 325—362 (1933); [3] La dérivée aréolaire. *Ann. Roumaine Math.* **3**, 1—62 (1936); [4] Dérivées spatiales et opérateurs différentiels généralisés. "Proc. Internat. Congr. Math. Edinbourg 1958", Univ. Press, Cambridge 1960, 573p; [5] Représentations intégrales en théorie des fonctions de plusieurs variables. C.R. III-ème Réunion du Groupement des Mathématiciens d'Expression Latine. Louvain 1966.

TITOV, O. V. [1] O kvazikonformnyh garmoničeskih otobraženijah evklidova prostranstva. (Russian) (On quasiconformal harmonic mappings in Euclidean space.) *Dokl. Akad. Nauk SSSR* **194**, 521—523 (1970).

TONOLO, A. [1] Sulle funzioni complesse di piu variable olomorfe di ordine *n*. *Rend. Sem. Mat. Univ. Padova* **6**, 9—20 (1935); [2] Sulle funzioni olomorfe (α) di piu variabili. *Atti R. Acad. Sci. Lett. Arti Padova, Mem.* **51**, 67—73 (1935).

TROHIMČUK, JURIĬ JUREVIČ [1] O nepreryvnyh otobraženiĭ oblasteĭ evklidova prostranstva. (Russian) (Continuous mappings of domain in Euclidean space.) *Ukrain. Mat. Ž.* **16**, 196—211 (1964); [2] O proizvodnyh po napravleniju funkciĭ mnogih peremennyh. (Russian) (On the directional derivative of functions of several variables.) *Ukrain. Mat. Ž.* **17**, 67—79 (1965); [3] O vnutrennih i konformnyh otobraženijah v prostranstve. (Russian) (On interior and conformal mappings in space.) Meždunarodnaja Konf. po teoriĭ analitičeskih funkciĭ. Erevan 1965. In "Sovremennye problemy teorii funkcii", Izd. Nauka, Moscow 1966, 281—283; [4] Vnutrennye i konformnye otobraženija v prostranstve. (Russian) (Interior and conformal mappings in space.) Internat. Congr. Math. Moscow 16—26 August, 1966; [5] K obobščeniju teoremy Liuvillja. (Russian) (About the generalization of Liouville's theorem.) In Sb. 7—ja letnjaja Mat. Škola 1969, Izd. Naukova Dumka, Kiev 1970, 382—391.

TUTSCHKE, WOLFGANG [1] Integrabilitätsbedingungen *p*-analytischer Funktionen mehrerer komplexer Veränderlichen. *Math. Nachr.* **29**, 239—248 (1969); [2] Bemerkungen über pseudoholomorphe Funktionen mit vorgeschriebenen Nullstellenverhalten. *Monatsb. Deutsch. Akad. Wiss. Berlin* **11**, 703—706 (1969); [3] Teorema Gartogsa dlja obobščennyh analitičeskih funkciĭ mnogih kompleksnyh peremennyh. (Russian) (Hartog's theorem for generalized analytic functions of several complex variables.) *Dokl. Akad. Nauk SSSR* **193**, 765—766 (1970); [4] Über die Fortsetzung von Funktionen mehrerer Variablen ins Komplexe. *Math. Nachr.* **45**, 373—377 (1970); [5] Pseudoholomorphe Funktionen einer und mehrerer komplexer Variablen im Zusammenhang mit der Theorie vollständiger Differentialgleichungssysteme. *Schr. Math. Inst. Dtsch. Akad. Wiss. Berlin, A, No.* **7**, 165—173 (1970).

VÄISÄLÄ, JUSSI [1] On quasiconformal mappings in space. *Ann. Acad. Sci. Fenn. Ser. A I* **298**, 1—36 (1961); [2] On quasiconformal mappings of a ball. *Ann. Acad. Sci. Fenn. Ser. A I* **304**, 1—6 (1961); [3] Two new characterizations for quasiconformality. *Ann. Acad. Sci. Fenn. Ser. A I,* **362**, 1—12 (1965); [4] Discrete open mappings on manifolds. *Ann. Acad. Sci. Fenn. Ser. A I,* **392**, 1—9 (1966); [5] Onko *n*-ulotteista funktioteriao olemassa? (Finnish) (Does there exist an n-dimensional function theory?) *Arkimedes* **2**, 11—16 (1968); [6] Removable sets for quasiconformal mappings. *J. Math. Mechn.* **19**, 49—51 (1969); [7] Quasiregular mappings in *n*-space. Coll. on Math. Anal. Jyväskyla 1970; [8] "Lectures on *n*-dimensional quasiconformal mappings". Springer, Berlin—Heidelberg-New York, 1971, 144 p.; [9] Modulus and capacity inequalities for quasiregular mappings. *Ann. Acad. Sci. Fenn. Ser. A I* **509**, 1—14 (1972).

VEBLEN, O. [1] Formalism for conformal geometry. *Proc. Nat. Acad. Sci. USA* **21**, 168—173 (1935).

VIGNAUX, J. CARLOS [1] Sobre las funciones poligenas de una y de varias variables complejas. *An. Soc. Ci. Argentina* **120**, 28—47 (1935); [2] La teoría de las funciones poligenas de una y de varias variables complejas duales. I, II. *Univ. Nac. La Plata Publ. Fac. Ci. Fisicomat. Contrib.* **1**, 221—282 (1936); **1**, 389—406 (1937); [3] Las funciones monogenas de una y de varias variables complejas duales. *Univ. Nac. La Plata Publ. Fac. Fisicomat. Contrib.* **1**, 409—410 (1936); [4] La teoría de las funciones de variable compleja bidual. *Univ. Nac. La Plata Publ. Fac. Ci. Fisicomat. Contrib.* **1**, 491—503 (1937); [5] Sur les fonctions polygènes d'une ou de plusieurs variables complexes duales. I. *Atti Accad. Naz. Lincei Rend. Cl. Sci. Fis. Mat. Natur.* (6), **27**, 514—518 (1938); [6] Algunas formulas de las funciones poligenas de una y varias variables. *An. Soc. Ci. Argentina* **125**, 19—29 (1938); [7] Sulle funzioni poligene di una variable biocomplessa duale. *Atti Accad. Naz. Lincei Rend. Cl. Sci. Fis. Mat. Natur.* (6), **27**, 641—645 (1938); [8] Sobre las funciones poligenas de variable compleja y bicompleja hiperbólica. *An. Soc. Ci. Argentina* **127**, 241—407 (1939).

VINOGRADOV, V. S. [1] Ob odnom analoge sistem Koši—Rimana v četyrehmernom prostranstve. (Russian) (An analogue of Cauchy—Riemann system in 4-space.) *Dokl. Akad. Nauk SSSR* **154**, 16—19 (1964).

VIRSU, MARJATTA [1] The local dilatation of maps with continuous characteristics in space. *Ann. Acad. Sci. Fenn. Ser. A I* **463**, 1—10 (1970); [2] On the linear dilatation of quasiconformal mappings in space. *Duke Math. J.* **38**, 569—574 (1971).

VOLKOVYSKIĬ, LEV IZRAILEVIČ [1] Sovremennye issledovanija po teorii rimanovyh poverhnosteĭ. (Russian) (Modern researches in the theory of Riemann surfaces.) *Uspehi Mat. Nauk*, **11** Nr. 5, 77 (1956); [2] Nekotorye voprosy teorii kvazikonformnyh otobraženiĭ. (Russian) (Some questions of the theory of quasiconformal mappings.) *Trudy Tomsk. Gos. Univ. Ser. Meh.-Mat.* **210**, 6—8 (1969); in "Nekotorye problemy Mat. i Meh. 70-letiju Akad. M. A. Lavrent'ev", Akad. Nauk SSSR Sibirsk. Otdel., Novosibirsk, Izd. Nauka Leningrad 1970, 128.

WARD, A. JAMES [1] From generalized Cauchy-Riemann equations to linear algebras. *Proc. Amer. Math. Soc.* **4**, 456—461 (1953).

YOUNG, G. S. [1] Extensions of Liouville's theorem to n dimensions. *Math. Scand.* **6**, 289—292 (1958).

YÛJÔBÔ, ZUIMAN [1] On absolutely continuous functions of two or more variables in the Tonelli sense and quasiconformal mappings in A. Mori sense. *Comment. Math. Univ. St. Paul* **4**, 67—92 (1955); [2] Supplements and correction to my paper: "On absolutely continuous functions of two or more variables in the Tonelli sense and quasiconformal mappings in A. Mori sense". *Comment. Math. Univ. St. Paul* **5**, 33—36 (1956).

ZIEMER, W. P. [1] Extremal length and conformal capacity. *Notices Amer. Math. Soc.* **89**, 384 (1966): Symposium on quasiconformal mappings. Indiana University 1967; *Trans. Amer. Math. Soc.* **126**, 460—473 (1967); [2] Extremal length and p-capacity. *Notices Amer. Math. Soc.* **15**, 790 (1968); *Michigan Math. J.* **16**, 43—51 (1969); [3] Extremal length as a capacity. *Notices Amer. Math. Soc.* **16**, 668—669 (1969).

ZIMMERMANN, EDUARD [1] Quasikonforme schlichte Abbildunge im dreidimensionalen Raum. *Wiss. Z. Martin-Luther-Univ. Halle Wittenberg Math. Naturw. Reihe* **5**, 109—116 (1955); [2] Elementares Beispiel einer möglichst konformen räumlichen Abbildung. *Martin-Luther-Univ. Halle Wittenberg Math. Naturw. Reihe* **8**, 1073—1075 (1958/1959).

ZORIČ, V. A. [1] O sootvestvie granic pri Q-kvazikonformnyh otobraženijah šara. (Russian) (On the boundary correspondence by Q-quasiconformal mappings of the ball.) *Dokl. Akad. Nauk SSSR* **145**, 31—34 (1962); [2] Sootvestvie granic pri Q-kvazikonformnyh otobraženijah šara. (Russian) (Boundary correspondence under Q-quasiconformal mappings of the ball.) *Dokl. Akad. Nauk SSSR* **145**, 1209—1212 (1962); [3] Graničnye svoĭstva odnogo klassa otobraženiĭ v prostranstve. (Russian) (Boundary correspondence by a class of mappings in space.) *Dokl. Akad. Nauk SSSR* **153**, 23—26 (1963); [4] Opredelenie graničnyh elementov posredstvom sečenii. (Russian) (Characterization of the boundary elements by cuts.) *Dokl. Akad. Nauk SSSR* **164**, 736—739 (1965); [5] Gomeomorfnosti kvazikonformnyh otobraženiĭ prostranstva. (Russian) (Homeomorphicity of the quasiconformal mappings in space.) *Dokl. Akad. Nauk SSSR* **176**, 31—34 (1967); [6] Ob uglovyh graničnyh značenijah kvazikonformnyh otobraženiĭ šara. (Russian) (On angular boundary values of the quasiconformal mappings of the ball.) *Dokl. Akad. Nauk SSSR* **177**, 771—773 (1967); [7] Teorema M. A. Lavrent'eva o kvazikonformnyh otobraženijah prostranstva. (Russian) (M. A. Lavrent'ev's theorem on quasiconformal mappings.) *Mat. Sb.* **74 (116)**, 417—433 (1967); [8] O dopustimom porjadke rosta karakteristiki kvazikonformnosti v teoreme M. A. Lavrent'eva. (Russian) (On the admissible order of growth of the characteristics of the quasiconformality in Lavrent'ev's theorem.) *Dokl. Akad. Nauk SSSR* **181**, 530—533 (1968); [9] Izolirovannaja osobennost' otobraženiĭ s ograničennym iskaženiem. (Russian) (Isolated singularity of mappings with bounded distorsion.) *Mat. Sb.* **81 (123)**, 634—636 (1970); [10] O nekotoryh otkrytyh voprosah teorii prostranstvennyh kvazikonformnyh otobraženiĭ. (Russian) (On some open problems in the theory of quasiconformal mappings in space.) "Metričeskie voprosy teor. funkciĭ i otobraženiĭ. No. 3. Dokl. II-go kollokviuma po teoriĭ kvazikonformnyh otobraženiĭ i ee obobščenijam, posvjaščennogo 70-letiju so dnja roždenija Akad. M. A. Lavrent'eva. Doneck 12—22 October, 1970". Izd. Naukova Dumka, Kiev, 1971, 46—50.

WORKS ON QUASICONFORMAL MAPPINGS IN PLANE

AGARD, B. STEPHEN [4] Topics in the theory of quasiconformal mappings. Doct. Diss. Univ. Minnesota, 1964, *Dissert. Abstr. USA* **26**, 2767 (1965); [5] An extremal problem in quasiconformal mappings. *Duke Math. J.* **33**, 735—741 (1966); [6] Distortion theorems for quasiconformal mappings. *Ann. Acad. Sci. Fenn. Ser. A I*, **413**, 1—12 (1968).

AGARD, B. STEPHEN and GEHRING, W. FREDERICK [1] Angles and quasiconformal mappings. *Proc. London Math. Soc.* (3), **14A**, 1—21 (1965).

AGMON, SAMUEL [1] A property of quasiconformal mappings. *J. Rational Mech. Anal.* **3**, 763—765 (1954).

AGMON, SAMUEL and BERS, LIPMAN [1] The expansion theorem for pseudo-analytic functions. *Proc. Amer. Math. Soc.* **3**, 757—764 (1952).

AHLFORS, LARS [4] Zur Theorie der Überlagerungsflächen. *Acta Math.* **65**, 157—194 (1935); [5] Development of the theory of conformal mappings and Riemann surfaces through a century. In "Contribution to the theory of Riemann surfaces". *Ann. of Math. Studies*, Princeton Univ. Press, Princeton N. J, **30**, 3—13 (1953); [6] On quasiconformal mappings. *J. Analyse Math.* **3**, 1—58 (1954); [7] Conformality with respect to Riemannian metrics. *Ann. Acad. Sci. Fenn. Ser. A I*, **206**, 1—22 (1955); [8] On quasiconformal mappings. Lectures at Osaka University, February 1956; [9] The complex analytic structure of the space of closed Riemann surfaces. In "Analytic functions" Princeton Univ. Press. Princeton, 45—66 (1960); [10] Classical and contemporary analysis. *Siam Review*, **3**, 1—9 (1961); [11] Curvature properties of Teichmüller space. *J. Analyse Math.* **9**, 161—176. (1961); [12] Some remarks on Teichmüller space of Riemann surfaces. *Ann. of Math.* **74**, 171—191 (1961); [13] Prostranstva Fuksovyh grupp. (Russian) (Space of Fuchsian groups). "Trudy 4-go Vsesojuz. mat. s'ezda, Leningrad, 3—12 February 1961", Nauka Leningrad 623 (1964), [14] Teichmüller spaces. Proc. Internat. Congr. Math. Stockholm, 15—22 August, 1962, Inst. Mittag-Leffler, Djursholm, 3—9 (1963); [15] Quasiconformal reflexions. *Acta Math.* **109**, 291—301 (1963); [16] The use of quasiconformal deformations for the study of Kleinian groups. Outlines Joint Symposium Partial Differential Equations, Novosibirsk 1963. Akad. Sci. SSSR, Siberian branch, Moscow, 10—13(1963); [17] Kvazikonformnye otobraženija. (Russian) (Quasiconformal mappings). *Matematika. Period. sb. perev. in stateĭ*, **8**, 103—111 (1964); [18] Quasiconformal mappings and their applications. In "Lectures Modern Mathematics", Wiley, NewYork—London—Sydney, 2, 151—164 (1964); [19] The modular functions and geometric properties of quasiconformal mappings. In "Proc. Conf. on Complex Analysis" Minneapolis 1964, Springer, Heidelberg—Berlin—New York 296—300 (1965); [20] "Lectures on quasiconformal mappings". D. Van Nostrand Company, Inc. Princeton, Toronto—New York—London 1966 ; [21] The structure of a finitely generated Kleinian group. *Acta Math*, **122**, 1—17 (1969) [22] On the general topic Kleinian groups. Newer developments in the theory of Kleinian groups. "Proceedings of the Romanian-Finnish Seminar on Teichmüller space and quasiconformal mappings, Braşov 26—30 August, 1969. Publishing House of the Academy of the Socialist Republic of Romania, Bucureşti 1971, 49—64.

AHLFORS, LARS and BERS, LIPMAN [1] Riemann's mapping theorem for variable metrics. *Ann. of Math.* **72**, 385—404 (1960). [2] "Prostranstva rimanovyh poverhnosteĭ i kvazikonformnye otobraženija". (Russian) (The space of Rieman surfaces and quasiconformal mappings). Izd. Inostrannoĭ Literatury, Moscow 1961, 177 p.

AHLFORS, LARS and BEURLING, A. [1] The boundary correspondance under quasiconformal mappings. *Acta Math.* **96**, 125—142 (1956).

AHLFORS, LARS and WEILL, G. [1] A uniqueness theorem for Beltrami equations. *Proc. Amer. Math. Soc.* **13**, 975—978 (1962).

ALEKSANDRIJA, G. N. [1] Ob odnoĭ graničnoĭ zadače dlja obobščennyh funkciĭ. (Russian) (On a boundary value problem for generalized functions.) *Trudy Tbiliskogo Gos. Univ.* **56**, 135—139 (1955); [2] Ob odnoĭ zadače lineĭnogo soprjaženija s zadanymi smeščenijami v klasse obobščennyh analitičeskih funkcii. *Soobšč. Akad. Nauk Gruz. SSR* **21**, 257—262 (1958).

Aleksandrovič, I. N. [1] O formulah obraščenija integral'nogo predstavlenija p-analitičeskih funkcii s karakteristikoĭ $p = e^{\alpha x}y^k$ dlja oblasteĭ častnogo vida. (Russian) (On inversion formulas of the integral representation of p-analytic functions with the characteristic $p=e^{\alpha x}y^k$ for particular domains.) *Matem. Fizika Resp. Mežved. sb. Nr.* 6, **3**—11 (1969).

Aleksandrovič, I. N. and Pahareva, N. A. [1] Integral'noe predstavlenie p-analitičeskih funkcii s harakteristikoĭ $p = e^{\alpha x}y^k$. (Russian) (Integral representation of p-analytic functions with the characteristic $p = e^{\alpha x}y^k$.) *Ukrainsk. Math. Ž.* **20**, 504—513 (1968).

Alpar, László [1] Sur une forme symétrique des équations de Beltrami. *Magyar Tud. Akad. Mat. Kutató Int. Kozl.*, **8**, 141—150 (1963).

Anderson, A. G. [5] Overdominance in a polygenic system. *Notices Amer. Math. Soc.* **12**, 68 (1965). [6] Rotation of plane quasiconformal mappings. *Coll. on Math. Analysis. Jyväskyla* 1970.

Andonie, G. Şt. [2] "Istoria matematicii în România". (Romanian) (Hystory of mathematics in Romania.) Edit. Ştiinţifică, Bucureşti, volume 2, 1966, 470 p.

Andreian-Cazacu, Cabiria [4] Fonctions holotopes et recouvrements riemanniens. *Rev. Roumaine Math. Pures Appl.* **12**, 1121—1126 (1967); [5] Relaţii de structură în familia transformărilor interioare. (Romanian) (Structural relations in the family of inner transformations.) *Bul. Ştiinţ. Acad. R.P.R., Secţia Şti. Mat. Fiz.* **5**, 431—441 (1953); [6] Sur les relations entre les fonctions caractéristiques et la pseudoanalyticité. IVth Congress Romanian Mathematics, Bucureşti 27 May—4 June, 1956; [7] Sur les transformations pseudo-analytiques. *Rev. Math. Pures Appl.* **2**, 283—399 (1957); [8] Asupra problemei tipului. (Romanian) (On the type problem.) *An. Univ. Bucureşti Ser. Şti. Natur. Mat.* **22**, 23—37 (1959); [9] Realizări sovietice în teoria funcţiilor pseudo-analitice. (Romanian.) (Soviet results in the theory of pseudo-analytic functions.) (*An. Româno-Sovietice* **2**, 69—77 (1959); [10] O kvazikonformnyh otobraženijah. (Russian) (On quasiconformal mappings.) *Dokl. Akad. Nauk SSSR* **126**, 235—238 (1959); [11] "Sur quelques classes de représentations quasi-conformes. Second Congress of Hungarian Mathematics, 1960". Budapest (1961); [12] Simion Stoilov. (necrolog). (Russian) *Uspehi Mat. Nauk* **17**, 135—148 (1962); [13] Sur une classe de représentations quasi-conformes. Colloque d'Analyse, Bucureşti 1962; [14] Sur l'application de la longueur extremale dans la théorie des représentations quasi-conformes. *Mathematica* (*Cluj*) **6**, 5—10 (1964); [15] Sur les inégalités de Rengel et la définition géométrique des représentations quasi-conformes. *Rev. Roumaine Math. Pures Appl.* **9**, 141—155 (1964); [16] Sur les représentations quasi-conformes. *Bul. Inst. Politehn. Iaşi* **11**, 11—15 (1965); [17] Sur un problème de L. I. Volkovyskiĭ. *Rev. Roumaine Math. Pures Appl.* **10**, 43—63 (1965); [18] Problèmes extrémaux des représentations quasi-conformes. *Rev. Roumaine Math. Pures Appl.* **10**, 43—63 (1965); [19] "Ob ekstremal'nyh zadačah kvazikonformnyh otobraženiĭ". (Russian) (On extremal problems of quasiconformal mappings.) Internat. Conf. on analytic functions, Erevan 1965, p. 27—28; "Sovremennye problemy teorii analitičeskih funkciĭ" Izd. Nauka, Moscow, 1966, p. 18—27; [20] Reprezentări cvasiconforme. (Romanian) (Quasiconformal mappings.) In "Probleme moderne de teoria funcţiilor" Edit. Acad. Republicii Populare Române, Bucureşti 1965, p. 209—309; [21] Suprafeţe riemanniene. (Romanian) (Riemann surfaces.) In "Topologie. Categorii. Suprafeţe riemanniene", Edit. Acad. Republicii Socialiste România, Bucureşti 1966, p. 243—293; [22] Asupra definiţiilor cvasiconformităţii. (Romanian) (On quasiconformality definitions.) *An. Univ. Bucureşti Ser. Şti. Natur. Mat.-Mec.* **16**, 11—18 (1967); [23] Une propriété caractéristique des représentations quasi-conformes. *Rev. Roumaine Math. Pures Appl.* **12**, 167—176 (1967); [24] Sur les applications quasi-conformes de M. A. Lavrentieff. *Rev. Roumaine Math. Pures Appl.* **13**, 1217—1223 (1968); [25] Bemerkungen über den Begriff der quasikonformen Abbildungen. *Math. Nachr.* **40**, 27—42 (1969); [26] Influence of the orientation of the characteristic ellipses on the properties of quasiconformal mappings. "Proceedings of the Romanian-Finnish Seminar on Teichmüller spaces and quasiconformal mappings. Braşov 26—30 August, 1969". Publishing House of the Academy of the Socialist Republic of Romania, Bucureşti 1971, 65—102; [27] Sur un exemple de L. I. Volkovyskiĭ. *Bul. Inst. Politehn. Iaşi* **16**, 21—30 (1970); [28] A generalization of the quasiconformality. Coll. on Math. Anal. Jyväskila (Finland) 1970; [29] Über die quasikonformen Abbildungen. Communicated at the Conf. "Komplexe Analisis und deren Anwendungen" Göhren (East Germany) 28 September—2 October, 1972.

ANGELESCU, A. [1] Sur une formule de M. D. Pompeiu, *Mathematica (Cluj)* **1**, 107—110 (1929).

ANGHEL, C. [1] Asupra unei metode extremale a lui Stoilow. (Romanian) (On an extremal method of Stoilow.) *Com. Acad. R. P. Romîne* **13**, 401—404 (1963).

ANGHELUŢĂ, TH. [1] Une remarque sur une généralisation des fonctions analytiques. *Mathematica (Cluj)* **16**, 53—56 (1940).

ARAMOVIČ, I. T., LUNC, G. L. and VOLKOVYSKIĬ, LEV ISRAILEVIČ, [1] "Sbornik zadač po teorii funkcii kompleksnogo peremennogo". (Russian) (Problem book in the theory of functions of a complex variable.) Gosudarstv. Izd. Fiz.-Mat. Lit., Moscow 1961, 367 p.

ARENS, RICHARD [1] The boundary integral of log (Φ) for generalized analytic functions. *Trans. Amer. Math. Soc.* **86**, 57—69 (1957).

ATADJANOV, B. K. [1] K voprosy differenciruemosti kvazikonformnogo otobraženija s nepreryvnymi harakteristikami. (Russian) (On the question of the differentiability of the quasiconformal mappings with continuous characteristics.) *Trudy Taškent Gos. Univ.* **394**, 3—5 (1970); [2] O suščestvovanii nekotoryh dvuhmernyh kvazikonformnyh otobraženiĭ v prostranstve. (Russian) (On the existence of some bidimensional quasiconformal mappings in space.) *Trudy Taškent Gos. Univ.* **394**, 6—9 (1970).

BAIDAFF, B. I. [1] Funciones poligenas. Circolo derivado. *Boletin Mat. Buenos Aires* **1**, 25—271 (1928).

BAKLANOV, M. V. [1] O nekotoryh klassah topologičeskih otobraženiĭ ploskih oblasteĭ s graničnym sootvestviem ne po prostym koncam. (Russian) (On some klass of topological mappings in plane with a not by prime ends boundary correspondence.) *Trudy Tomsk. Gos. Univ. Ser. Meh.-Mat.* **189**, No. 4, 3—12 (1966).

BARBILIAN, D. [1] " Curs de geometrie şi teoria funcţiilor". (Romanian) (Course of geometry and function theory.) Bucureşti 1939/1940.

BARHIN, G. S. and USMANOV, Z. D. [1] Zadača Gilberta dlja kusočno-reguljarnoĭ obobščennoĭ analitičeskoĭ funkciĭ. (Russian) (The Hilbert problem for a piecewise generalized analytic function.) In "Issled. po kraevym zadačam teorii funkciĭ i differenc. uravneniĭ, Izdat. Akad. Nauk Tadžik, SSR, Dušanbe 1964, 113—132.

BARKER, M. and PYLE, H. RANDOLF [1] A vector interpretation of the derivative circle. *Amer. Math. Monthly* **63**, 78—82 (1946).

BARNETT, EDWARD FRANKLIN [1] On a certain class of generalized analytic functions. Doct. Diss. Univ. Alabama 1968, 111 p. *Dissert. Abstr. B* **29**, *Nr. 10*, 3819 (1969).

BAUER, K. W. [1] Über eine Differentialgleichung $(1 \pm z\bar{z})^2 w_{z\bar{z}} \pm n(n+1)w = 0$ zugeordnete Funktionentheorie. *Bonn. Math. Schr. N⁰* **23**, 98 p. (1965); [2] Über die Lösungen der elliptischen Differentialgleichung $(1 + z\bar{z})^2\ w_{z\bar{z}} + \lambda w = 0$. *I*, II. *J. Reine Angew. Math.* **221**, 48—84 (1966); 176—196 (1966); [3] Sobre una classe de funciones analiticas generalizadas. *Notas Comun. Mat. Recife* **9**, 1—16 (1966); *Sitzungsber. Berliner Math. Ges.* 1967/1968, S. **1**, 24—25 (1969); [4] Über eine Klasse verallgemeinerter Cauchy-Riemannscher Differentialgleichungen. *Math. Z.* **100**, 17—28 (1967).

BAUER, K. W. and PESCHL, ERNST [1] Ein allgemeiner Entwicklungssatz für die Lösungen der Differentialgleichung $(1 + \varepsilon z\bar{z})^2\ w_{z\bar{z}} + \varepsilon n\,(n+1)w = 0$ in der Nähe isolierter singularitäten. *Bayer. Akad. Wiss. Math. Natur. Kl. S.—B. 1965*, **2**, 113—146 (1966).

BELINSKIĬ, PAVEL PETROVIČ [7] Teorema suščestvovanija i edinstvennosti kvazikonformnyh otobraženiĭ. (Russian) (Existence and unicity theorem for quasiconformal mappings.) *Uspehi Mat. Nauk*, **6**, 2 (42), 145 (1951); [8] Povedenie kvazikonformnogo otobraženija v izolirovannoĭ točke. (Russian) (Behaviour of a quasiconformal mapping at an isolated point.) *Dokl. Akad. Nauk SSSR*, **91**, 709—710 (1953); *Učennye Zap. L'vov. Univ. Ser. Meh.-Mat.* **6**, 58—70 (1954); [9] Ob iskaženii v kvazikonformnyh otobraženijah. (Russian) (On distorsion under quasiconformal mappings.) *Dokl. Acad. Nauk SSSR*, **91**, 997—998 (1953); [10] O metričeskih svoĭstvah konformnogo otobraženija. (Russian) (On metric properties of quasiconformal mappings.) *Dokl. Akad. Nauk SSSR*, **93**, 589—590 (1953); [11] Kvazikonformnye otobraženija. (Russian) (Quasiconformal mappings.) Dissertation University of L'vov 1954; [12] O suščestvovanii rešenja variacionnyh zadač kvazikonformnyh otobraženii. (Russian) (On existence of solution of the variational problems of quasiconformal mappings.) Trudy III Vsesojuz. Mat. S'ezda, Moscow

25 June—4 July 1956. Short communications, Izd. Akad. Nauk SSSR, Moscow, 1956, Tom 1, 77 (1956); [13] O variacii kvazikonformnogo otobraženija. (Russian) (On variation of quasiconformal mappings.) *Uspehi Mat. Nauk*, **11**, Nr. 5 (71), 93—95 (1956); [14] O mere ploščadi pri kvazikonformnyh otobraženiĭ. (Russian) (On the measure of the area by quasiconformal mappings.) *Dokl. Akad. Nauk SSSR*, **121**, 16—17 (1958); [15] O rešenii ekstremal'nyh zadač kvazikonformnyh otobraženiĭ. (Russian) (On the solution of extremal problems of quasiconformal mappings.) *Dokl. Akad. Nauk SSSR* **121**, 199—201 (1958); [16] O normal'nosti semeĭstv kvazikonformnyh otobraženiĭ. (Russian) (On the normality of families of quasiconformal mappings.) *Dokl. Akad. Nauk SSSR* **128**, 651—652 (1959); [17] Obščie svoĭstva kvazikonformnyh otobraženiĭ. (Russian) (General properties of quasiconformal mappings.) Doctor's Dissertation Sibirsk. otdel. Akad. Nauk SSSR, Novosibirsk 1959; [18] Rešenie ekstremal'nyh zadač teorii kvazikonformnyh otobraženiĭ varjacionnym metodom. (Russian) (The use of a variational method in solving extremal problems of quasiconformal mappings.) *Sibirsk. Mat. Ž.* **1**, 303—330 (1960).

Belinskiĭ, Pavel Petrovič and Pesin, I. I. [1] O zamykanii klassa nepreryvno differenciruemyh kvazikonformnyh otobraženiĭ. (Russian) (On the closure of the class of differentiable quasiconformal mappings.) *Dokl. Akad. Nauk SSSR* **102**, 865—866 (1965).

Belova, M. M. and Pahareva, N. O. [1] Pro formula obernenija osnovnogo integral'nogo zobrazennja y^k-analiticnih funkcii v poljarnyh koordinatah. (Ukrainian) (On inversion formulae of the fundamental integral representation of y^k-analytic functions in polar coordinates.) *Ukrain. Mat. Z.* **24**, 273—277 (1972).

Beltrami, E. [1] Delle variable complesse sopra una superficie qualunque. *Ann. Mat. Pura Appl.* (2), **1**, 329—366 (1867/1868); [2] Sulle funzioni potenziali di sistemi simmetrici intorno ad un asse. *Rend. R. Ist. Lombardo* (2), **11**, 668—680 (1878); [3] Sulla teoria delle funzioni potenziali simmetriche. *Mem. Acad. Sci. Ist. Bologna* (4), **2**, 461—505 (1880); [4] "Opere matematiche". Hoepli, Milano 1911.

Bergman, Stefan [7] A formula for the stream function of certain flows. *Proc. Nat. Acad. Sci. USA* **29**, 276—281 (1943); [8] Certain classes of analytic functions of two real variables and their properties. *Trans. Amer. Math. Soc.* **57**, 299—331 (1945); [9] Functions satisfying certain partial differential equations of elliptic type and their representation. *Duke Math. J.* **14**, 349—366 (1947); [10] Two-dimensional subsonic flows of a compressible fluid and their singularities. *Trans. Amer. Math. Soc.* **62**, 452—498 (1947).

Bers, Lipman [1] On the continuation of a potential gas flow across the sonic line. NACA Techn. Note 2058 (1950); [2] The expansion theorem for sigma-monogenic functions. *Amer. J. Math.* **72**, 705—712 (1950); [3] Partial differential equations and generalized analytic functions. I, II. *Proc. Nat. Acad. Sci. USA* **36**, 130—136 (1950); **37**, 42—47 (1951); [4] Some generalizations of conformal mappings suggested by gas dynamic. In "Contribution and application of conformal maps". Proc. of a Symposium Nat. Bureau of Standards. Appl. Math. Sen. U.S. Government Printing Office, Washington 18, 117—124 (1952); [5] Partial differential equations and pseudo-analytic functions on Riemann surfaces. In "Contribution to the theory of Riemann surfaces" *Ann. of Math. Studies*, Princeton Univ. Press, Princeton, N.J., **30**, 157—165 (1953); [6] Theory of pseudo-analytic functions. Institute for Math. and Mech. New York Univ., New York 1953, 187 p. (mimeographed); [7] Univalent solutions of linear elliptic systems. *Comm. Pure Appl. Math.* **6**, 513—526 (1953); [8] Existence and uniqueness of a subsonic flow past a given profile. *Comm. Pure Appl. Math.* **7**, 441—504 (1954); [9] Functional theoretical properties of solutions of partial differential equations of elliptic type. In "Contribution to the theory of partial differential equations" *Ann. of Math. Studies*, Princeton Univ. Press, Princeton N.J., **33**, 69—94 (1954); [10] Local theory of pseudo-analytic functions. In "Lectures on functions of a complex variable". Michigan Univ. Press, Ann. Arbor 214—244 (1955); [11] Survey on local properties of solutions of partial differential equations. Trans. of the Symposium on partial differential equations held at Univ. of California at Berkley June 20—July 1, 1955. Interscience Publishers, Inc., New York London, 1956, 41—52; *Comm. Pure Appl. Math.* **9**, 339—350 (1956); [12] Remark on an application of pseudo-analytic functions. *Amer. J. Math.* **78**, 486—496 (1956); [13] An outline of the theory of pseudo-analytic functions. *Bull. Amer. Math. Soc.* **62**, 291—331 (1956); [14] Formal power and powers series.

Comm. Pure Appl. Math. **9**, 693—711 (1956); In "La théorie des équations aux dérivées partielles", Coll. Internationaux du Centre National de la Recherche Scientifique. Nancy, April 1956. (CNRS), Paris 71, 9—17 (1956); [15] On a theorem of Mori and the definition of quasiconformality. *Trans. Amer. Math. Soc.* **84**, 78—84 (1957); [16] "Mathematical aspects of subsonic and transsonic gas dynamics". John Wiley & Sons, Inc. Publishers, New York—London 1958, 164 p., Izd. Inostr. Lit. Moscow 1961, 208 p.; [17] Spaces of Riemann surfaces. Proc. Internat. Congr. Edinburgh 1958, Princeton Univ. Press. Princeton N.J., 349—361 (1960); [18] Simultaneous uniformization. *Bull. Amer. Math. Soc.* **66**, 94—97 (1960); [19] Quasiconformal mappings and Teichmüller's theorem. In "Analytic functions". Princeton Univ. Press, Princeton 1960; 89—119; [20] Uniformization and moduli. Contribution to function theory. International Coll. Function Theory, Bombay, 1960; Tata Institute of fundamental research, Bombay 1960, 41—49; [21] Space of Riemann surfaces as bounded domain. *Bull. Amer. Math. Soc.* **66**, 98—103 (1960); [22] Correction to "Space of Riemann surfaces as bounded domain". *Bull. Amer. Math. Soc.* **67**, 465—466 (1961); [23] Holomorphic differentials as functions of moduli. *Bull. Amer. Math. Soc.* **67**, 206—210 (1961); [24] Uniformization by Beltrami equations. *Comm. Pures Appl. Math.* **14**, 215—228 (1961); [25] Moduli rimanoyh poverhnosteĭ. Trudy IV Vsesojuz. Mat. S'ezda, Leningrad 1961. Izd. Akad. Nauk SSSR, Leningrad, 1963, tom. 2, 628—629 (1963); [26] The equivalence of two definitions of quasiconformal mappings. *Comment. Math. Helv.* **37**, 148—154 (1962); [27] Quasiconformal mappings and Teichmüller spaces of arbitrary Fuchsian groups and Riemann surfaces. Outlines joint Sympos. Partial Differential Equations, Novosibirsk 1963 Acad. Sci. USSR, Siberian Branch, Moscow 1963, 329—337; [28] On moduli of Riemann surfaces. Mimeographed lecture notes, Eidgenossische Technische Hochschule, Zürich 1964; [29] Holomorphic convexity of Teichmüller spaces. *Bull. Amer. Math. Soc.* **70**, 761—764 (1964); [30] "Automorphic forms and general Teichmüller spaces". Proc. Conf. Complex Anal. Univ. of Minnesota, Minneapolis 1964, Springer-Verlag Berlin—Heidelberg—New York 1965, 109—113; [31] "Theory of moduli and Kleinian groups". Meždunarodnaja Konf. po teorii analitičeskih funkcii, Erevan, Abstracts 10—11 (1965); [32] Automorphic forms and Poincaré series for infinitely generated Fuchsian groups. *Amer. J. Math.* **87**, 196—214 (1965). [33] A nonstandard integral equation with application to quasiconformal mappings. *Acta Math.* **116**, 113—134 (1966); [34] "Extremal quasiconformal mappings". Advances in the theory of Riemann surfaces (Proc. Conf., Stony Brook, N.Y., 1969) 27—52, Ann. of Math. Studies. No. 66. Princeton Univ. Press, Princeton, 1971; [35] Universal Teichmüller Space. Analytic Methods Math. Phys. Conf. Bloomington, Indiana 1968, 65—83 (1970).

BERS, LIPMAN and EHRENPREIS, L. [1] Holomorphic convexity of Teichmüller spaces. *Bull. Amer. Math. Soc.* **70**, 761—764 (1964).

BERS, LIPMAN and GELBART, A. [1] On a class of differential equations in mechanics of continua. *Quart. Appl. Math.* **1**, 168—188 (1943); [2] On a class of functions defined by partial differential equations. *Trans. Amer. Math. Soc.* **56**, 67—93 (1944). [3] A topological property of solutions of partial differential equations. *Bull. Amer. Math. Soc.* **52**, 64 (1946); [4] On generalized Laplace transformations. *Ann. of Math.* **48**, 342—357 (1947).

BERS, LIPMAN and GRINBERG, L. [1] Isomorphisms between Teichmüller spaces. Advances in the theory of Riemann surfaces (Proc. Conf. Stony Brook, N.Y., 1969) Ann. of Math. Studies No. 66. Princeton Univ. Press, Princeton, Univ. of Tokyo Press, Tokyo, 1971, 53—79.

BERS, LIPMAN and MASKIT, B. [1] On a class of Kleinian groups. Meždunarodnaja Konf. po teorii analitičeskih funkciĭ, Erevan 1965. In "Sovremennye problemy teorii analitičeskih funkciĭ" Izd. "Nauka", Moscow, 1966, 44—47.

BERS, LIPMAN and NIRENBERG, L. [1] On a representation theorem for elliptic system with discontinuous coefficients and its applications. Convegno Internat. sulle funzioni lineari alle derivate parziali. Trieste 1954, Edizioni Cremonese, Roma 1955, 111—140.

BERSTEIN, L. [1] A topological characterization of the pseudo-conjugate of a pseudo-harmonic function. *Bul. Şti. Acad. R. P. Romîne Secţ. Mat. Fiz.* **7**, 75—78 (1955); *Rev. Roumaine Math. Pures Appl.* **1**, 45—48 (1956).

BETZ, ALBERT [1] "Konforme Abbildungen", Springer-Verlag Berlin—Heidelberg—New York 1964, 407 p.

BICADZE, A. V. [2] K leme Švarca. (Russian) (On Schwartz's lemma) *Tr. Tbiliskogo Mat. Inst. Akad. Nauk Gruzin. SSR* **33**, 15—20 (1967).

BIESTERFELD, H. JOHN [1] Extension of the Cauchy—Riemann equations. Doct. Diss. Pensylvania USA 1963, *Dissert. abstr.* **24**, 5425 (1964).

BILIMOVIČ, ANTON [1] Sur la mesure de déflexion d'une fonction non-analytique. *C.R. Acad. Sci. Paris* **237**, 694—695 (1953); *Publ. Inst. Math. (Beograd)* **6**, 17—26 (1954); [2] O meri otstupańa neanalitičke funkcije od analitičnosti. (Serbo-Croatian) (On the measure of deviation of a non-analytic function with respect to an analytic one.) *Glas Srpske Akad. Nauka Od. Prirod.-Mat. Nauka* **9**, (221), 1—11 (1956); [3] Afina transformacija neanalitičke funkcije u analitičku. (Serbo-Croatian) (On affine transformation from a non-analytic function to an analytic one.) *Glas Srpske Akad. Nauka Od. Prirod.-Mat. Nauka* **9** (221), 13—17 (1956); [4] O diagramu neanalitička funkcije za datu točku. (Serbo-Croatian) (On a diagram of a non-analytic function at a given point.) *Glas Srpske Akad. Nauka Od. Prirod.-Mat. Nauka* **9** (221), 39—43 (1956); [5] Application en hydromécanique de la mesure de déflexion d'une fonction non-analytique. *Bull. Acad. Serbe Sci. Arts Cl. Sci. Math. Natur. Sci. Math.* **10**, 33—41 (1956); [6] Sur la déflexion d'une fonction non-analytique du quaternion par rapport à une fonction analytique. *Bull. Acad. Serbe Sci. Arts Cl. Sci. Math. Natur. Sci. Math.* **20**, 1—9 (1957); *Glas Srpske Akad. Nauka Od Prirod.-Mat. Nauka* **13** (221), 1—22 (1957); [7] Sur les transformations des fonctions non-analytiques. *C. R. Acad. Sci. Paris* **247**, 1954—1956 (1958); [8] Primena mere otstupan'a neanalitičke funkcije od analitičnosti u hidromehanicu. (Serbo-Croatian) (Applications in hydrodynamics of the measure of deviation from a non-analytic function to an analytic one.) *Acad. Serbe Sci. Arts Glas* **237**, 73—81 (1959); [9] Diferencijalni elementi geometriske teorije neanalitičkih funkcija. (Serbo-Croatian) (Differential elements of the geometric theory of non-analytic functions.) *Acad. Serbe Sci. Arts Glas* **242**, 1—82 (1960); [10] Sur les lignes principales des fonctions non-analytiques. C. *R. Acad. Sci. Paris* **250**, 805—807 (1960); [11] O nekim integralnim teoremama u geometrijskoi teoriji neanalitičkih funkcija. (Serbo-Croatian) (Certain integral theorems in the geometric theory of non-analytic functions.) *Acad. Serbe Sci. Arts Glas* **263**, 53—81 (1966); [12] Sur les modes divers de traitement des fonctions complexes non-analytiques. *C. R. Acad. Sci. Paris* **263**, 61—62 (1966); [13] Delimične diferencijalne jednačine u geometrijskoj teoriji neanalitičnih funkcija. (Serbo-Croatian) (Partial differential equations in the geometric theory of non-analytic functions.) *Acad. Serbe Sci. Arts Glas* **269**, 79—110 (1967).

BILUTA, P. A. [1] Nekotorye ekstremal'nye zadači v klasse otobraženii, kvazikonformnye v srednem. (Russian) (Some extremal problems in the class of mean quasiconformal mappings.) *Dokl. Akad. Nauk SSSR* **155**, 503—505 (1964); Sibirsk. *Mat. Ž.* **6**, 717—726 (1965); [2] Popravka po stat'e "Nekotorye ekstremal'nye zadači v klasse otobraženii, kvazikonformnye v srednem". (Russian) (Correction to the paper "Some extremal problems in the class of mean quasiconformal mappings".) *Dokl. Akad. Nauk SSSR* **159**, 702 (1964); [3] Ekstremal'nye kvazikonformnye otobraženija dlja proizvol'nyh ploskih oblasteĭ. (Russian) (Extremal quasiconformal mappings of arbitrary plane domains.) *Dokl. Akad. Nauk SSSR* **171**, 9—12 (1966); [4] O rešenii ekstremal'nyh zadač dlja odnogo klassa kvazikonformnyh otobraženii. (Russian) (On the solution of an extremal problem for a class of quasiconformal mappings.) *Sibirsk. Math. Ž.* **10**, 734—743 (1969); [5] Odnoĭ èkstremal'noĭ zadace dlja kvazikonformnyh otobraženiĭ konečnomernyh oblasteĭ. (Russian) (An extremal problem for quasiconformal mappings of multiply connected domains.) *Sibirsk. Mat. Ž.* **13**, 24—33 (1972).

BILUTA, P. A. and KRUŠKAL', S. L. [1] K voprosy ob ekstremal'nyh kvazikonformnyh otobrazenijah. (Russian) (On extremal quasiconformal mappings.) *Dokl. Akad. Nauk SSSR* **196**, 259—262 (1971).

BLEVINS, KING DONALD [1] Properties of domains bounded by *k*-circles. *Notices Amer. Math. Soc.* **18**, 1081 (1971).

BLUM, E. [1] Die Extremalität gewisser Teichmüllerschen Abbildungen des Einheitskreises. *Comment. Math. Helv.* **44**, 319—340 (1969).

BOBOC, NICOLAE [1] Asupra invarianței prin reprezentări cvasiconforme, a punctelor de efilare ale unei mulțimi. (Romanian) (On the invariance of the set of unfolding points under quasiconformal mappings.) *Com. Acad. R. P. Romîne* **12**, 163—165 (1962).

BODRECOVA, L. B. and FET, A. I. [1] Funkcii s prostymi linjami urovnja. (Russian) (Functions with simple level curves.) *Mat. Sb.* **38**, 303—318 (1956).

BOĬKO, L. T. [1] Primenenie obobščennyh analitičeskih funkcii k rešeniju osesimmetričnyh zadač teorii uprugosti. (Russian) (Application of generalized analytic functions to the solution of an axially symmetric problem of the theory of elasticity.) In "Voprosy pročnosti, nadežnosti i razřus Meh. sistem" Dnepropetrovsk 1969, 88—94; [2] Ob obobščennaja zadača Rimana. (Russian) (A generalized Riemann problem.) Sb. rabot. aspirantov Dnepropetrovsk. Univ. Meh. i Mat. Dnepropetrovsk 1970, 17—22.

BOJARSKIĬ, B. V. [1] Nekotorye kraevye zadači dlja uravnenii elliptičeskogo tipa. (Russian) (Some boundary value problems for equations of elliptic type.) Doct. Diss. Moskow. Gos. Univ. 1955; [2] Ob odnoĭ zadače dlja sistem uravnenii v častnyh proizvodnyh pervogo porjadka elliptičeskogo tipa. (Russian) (On a problem for systems of first order partial differential equations of elliptic type.) *Dokl. Akad. Nauk SSSR* **102**, 201—204 (1955); [3] Gomeomorfnye rešenija sistem Bel'trami. (Russian) (Homeomorphic solutions of Beltrami's system.) *Dokl. Akad. Nauk SSSR* **102**, 661—664 (1955); [4] O rešenijah lineĭnoĭ elliptičeskoĭ sistemy differencial'nyh uravnenii na ploskosti. (Russian) (On the solution of a linear system of elliptic differential equations in plane.) *Dokl. Akad. Nauk SSSR* **102**, 871—874 (1955); [5] Obobščenye rešenija sistemy differential'nyh uravnenii pervogo porjadka elliptičeskogo tipa s razryvnymi koefficentami. (Russian) (Generalized solutions of a system of first order differential equations of elliptic type with discontinuous coefficients.) *Mat. Sb.* **43** (**85**), 451—503 (1957); [6] Ob odnoĭ graničnoĭ zadače teorii funkcii. (Russian) (On a boundary problem of function theory.) *Dokl. Akad. Nauk SSSR* **119**, 199—202 (1958); [7] Ob osobom slučae zadači Rimana-Gilberta. (Russian) (On the singular Riemann-Hilbert problem.) *Dokl. Akad. Nauk SSSR* **119**, 411—414 (1958); [8] Obščee predstavlenie rešenii elliptičeskoĭ sistemy $2n$ uravnenii na ploskosti. (Russian) (General representation of the solution of an elliptic system of $2n$ equations in plane.) *Dokl. Akad. Nauk SSSR* **122**, 543—546 (1958); [9] Obščie svoĭstva rešenii elliptičeskih sistem na ploskosti. (Russian) (General properties of the solutions of elliptic systems in plane.) In "Issledovanie po sovremennym problemam teorii funkcii kompleksnogo peremennogo", Gosudarstvennoe Izd. Fiz.-Mat. Literatury, Moscow, 1960, 461—483; [10] Kvazikonformnye otobraženija i teorija obobščennyh analitičeskih funkcii. (Russian) (Quasiconformal mappings and the theory of generalized analytic functions.) In "Nekotorye problemy Mat. i Meh.", Sibirsk. Otdel. Akad. Nauk SSSR, Novosibirsk, 1961, 50—56.

BOMPIANI, ENRICO [1] Opérateurs de projection en théorie des primitives aréolaires bornées. *Ann. Mat. Pura Appl.* **60**, 1—27 (1962).

BORODIN, M. A. [1] O kraevoĭ zadače Rimana—Gilberta dlja golomorfnyh funkciĭ mnogih kompleksnyh peremennyh. (Russian) (Riemann—Hilbert boundary value problem for holomorphic functions of several complex variables.) *Ukraïn. Mat. Ž.* **21**, 238—246 (1969).

BUCUR, FLORICA [1] Asupra unei clase de transformări interioare. (Romanian) (On a class of interior transformations.) *Acad. R. P. Romîne Stud. Cerc. Mat.* **18**, 243—266 (1966); [2] Asupra valuărilor semiinelelor de transformări interioare, (Romanian) (On the valuations of the rings on interior transformations.) *Stud. Cerc. Mat.* **20**, 165—167 (1968).

BUCUR, GHEORGHE [1] Asupra reprezentărilor cvasiconforme. (Romanian) (On quasiconformal mappings.) *Com. Acad. R. P. Romîne* **10**, 397—403 (1960).

BURGATTI, P. [1] Sulle funzioni analitiche d'ordine *n*. *Boll. Un. Mat. Ital.* (1), **1**, 8—12 (1922).

BUZUN, T. N. [1] Ob odnoĭ smešannoĭ kraevoĭ zadače p-analitičeskih funkciĭ s harakteristikoĭ $p = x$. (Russian) (On a mixed boundary problem for p-analytic functions with the characteristic $p = x$). Matem. Fiz. Resp. Mežved. Sb. "Naukova Dumka", Kiev, 1967, 52—56.

BYRNE, B. O. [1] On Finsler geometry and applications to Teichmüller spaces. Advances in the theory of Riemann surfaces (Proc. Conf. Stony Brook, N.Y., 1969) Ann. of Math. Studies, No. 66, Princeton Univ. Press, Princeton, N.J.; Univ. of Tokyo Press, Tokyo, 1971.

CACCIOPOLI, RENATO [1] Fondamenti per una teoria generale delle funzioni pseudoanalitiche di una variabile complessa. *Atti Accad. Naz. Lincei, Rend. Cl. Sci. Fis. Mat. Natur.* **13**, 197—204; 321—329 (1952). [2] Sur une généralisation des fonctions analytiques. C. *R. Acad. Sci. Paris* **235**, 228—229 (1952). [3] Sur une généralisation des fonctions analytiques et les fonctions normales. C. *R. Acad. Sci. Paris* **236**, 116—118 (1952);

[4] Funzioni pseudo-analitiche e rappresentazioni pseudo-conformi delle superficie riemanniene. *Ricerche* **2**, 104—127 (1953).

CALDWELL, W. V. [1] Maximal vector spaces of light interior functions. *J. Math. Mech.* **12**, 411—428 (1963); [2] *PR*-factorization of families of light interior functions. *Proc. Amer. Math. Soc.* **19**, 299—302 (1968).

ČALENKO, P. I. and ČEMERIS, V. S. [1] O nekotoryh napravlenijah issledovanii po matematike, v Kievskom gosudarstvennom universitete. (Russian) (On some investigation directions in mathematics at the state university in Kiev.) *Ukrain. Mat. Ž.* **19**, 100—110 (1967).

CĂLUGĂREANU, GHEORGHE [1] Sur les fonctions polygènes d'une variable complexe. Doctor's Dissertation, Paris, 1928; *C. R. Acad. Sci. Paris* **186**, 930—932 (1928); [2] Sur une classe d'équations du second ordre intégrable à l'aide des fonctions polygènes. *C. R. Acad. Sci. Paris* **186**, 1406 (1928); [3] Les fonctions polygènes comme intégrales d'équations différentielles. *Trans. Amer. Math. Soc.* **31**, 372—378 (1929); [4] Sur les fonctions monogènes aréolairement. *Bul. Soc. Şti. Cluj* **14**, 357 (1929); [5] On the differential equations admitting polygenic integrals. *Trans. Amer. Math. Soc.* **32**, 110—113 (1930); [6] On differential equations admitting polygenic integrals. *Bull. Amer. Math. Soc.* **36**, 197 (1930).

CANNON, R. J. [1] Quasiconformal structures and metrization of 2-manifolds. Doctor's Dissertation, *Abstr. B. USA* **28**, N° 10, 4192 (1968); *Trans. Amer. Math. Soc.* **135**, 95—103 (1969).

CAPELLI, PEDRO [1] Sobre las funciones holomorfas y poligeneas de una variabile compleja binaria. *An. Soc. Ci. Argentina* **128**, 154—174 (1939). [2] Sur le nombre complexe binaire. *Bull. Amer. Math. Soc.* **47**, 585—595 (1941).

ČAPLYGIN, S. A. [1] O gazovyh strujah. (Russian) (On gas jets.) Gostehizdat, Moscow 1949, 144 p.

CARLEMAN, TORSTEN [1] Sur les systèmes linéaires aux dérivées partielles du premier ordre à deux variables. *C. R. Acad. Sci. Paris* **197**, 471—474 (1933).

CARLESON, LENNART [1] On mappings conformal at the boundary. *J. Analyse Math.* **19**, 1—13 (1967).

CAUGHRAN, J. G. [1] Two results concerning the zeros of functions with finite Dirichlet integral. *Canad. J. Math.* **21**, 312—316 (1969).

ČEMERIS, V. S. [1] "Do pitannja pro zastosuvannja p-analitičnih funkcii v osesimetričniĭ teorii pružnosti". (Ukrainian) (On the problem of the application of p-analytic functions in the axially symmetric theory of the elasticity.) *Dopovidi Acad. Nauk Ukraïn. RSR* 903—906 (1960).

ČEMERIS, V. S. and POLOŽIĬ, G. M. [1] "Do pitannja pro zastosuvannja p-analitičnih funkcii v teorii pružnosti". (Ukrainian) (On the problem of the application of p-analytic functions in the theory of elasticity.) *Dopovidi Acad. Nauk Ukraïn RSR.* 1284—1287 (1958).

ČERNECKIĬ, V. A. [1] Kraevaja zadača Karlemana dlja mnogosvjaznoĭ oblasti v klasse obobščennyh analitičeskih funkciĭ. (Russian) (Boundary value problem of Carleman for multiply connected domains in the class of generalized analytic functions.) *Izv. Akad. Nauk Armjan. SSR Ser. Fiz.-Mat. Nauk.* **5**, 385—393 (1970).

CHADZYŃSKI, JACEK and LAWRYNOWICZ JULIAN [1] On homeomorphisms and quasiconformal mappings connected with cyclic groups of homeographies and antigraphies. *Ann. Univ. Mariae Curie-Sklodowska Sect. A*, **20**, 29—44 (1966); [2] On quasiconformal solutions of a class of functional equations. *Bull. Acad. Polon. Sci. Ser. Sci. Math. Astronom. Phys.* **18**, 241—246 (1970); [3] On homeomorphic solutions of a class of functional equations. *Bull. Acad. Polon. Sci. Ser. Sci. Math. Astronom. Phys.* **18**, 257—251 (1970).

CHANG-HANG-BAO [1] Ob effektivnom rešeniĭ nekotoryh kraevyh zadač x^k-analitičeskih funkciĭ dlja polukruga. (Russian) (Effective solution o fsome boundary value problems for x^k-analytic functions in a semidisc.) *Vyčisl. i prikl. Mat. Mežved. Naučn. Sb.* N° **11**, 76—90 (1970).

CHANG, HSIAO-LI [1] (Approximately analytic functions of bounded type and boundary behaviour of solutions of elliptic partial differential equations.) (Chinese) *Acta Math. Sinica*, **3**, 101—132 (1953).; [2] (On Lipman Bers theory of the elliptic partial differential equations.) (Chinese) *Shuxue Jinzhan* **6**, 119—165 (1963).

CHANG, KUANG-HOU [1] (Some theorems on quasiconformal mappings.) (Chinese) *Shuxue Jinzhan* **8**, 387—394 (1965).

Charzyński, Z. and Lawrynowicz, Julian [1] On conformality points of certain mappings preserving locally the area. *Colloq. Math.* **13**, 81—83 (1964).

Chen, Huai-hui [1] (Absolute continuity in the sense of Tonelli and the sufficient condition of K-Quasiconformality.) (Chinese) *Shuxue Jinzhan* **7**, 84—93 (1964).

Chen, Kien-kwong [1] On Hölder exponent of *Q*-mappings. *Sci. Record* **3**, 393—399 (1959).

Cheng, Bao-long [1] Approximate representation of ε-quasiconformal mappings and extremal problem. *Acta Math. Sinica* **14**, 212—217 (1964); *Chinese Math.* **5**, 233—238 (1964).

Christu, Ileana [1] Derivata A-areolară. (Romanian) (A-areolar derivative.) *Acad. R. P. Romîne Stud. Cerc. Mat.* **17**, 1449—1456 (1965).

Church, P. T. [2] Extensions on Stoilow's theorem. *J. London Math. Soc.* **37**, 86—89 (1962).

Cimino, Gianfranco [4] Sulle rappresentazioni pseudoconformi. "Sympos. Math. Ist. Naz. Alta Mat. 1968", Vol. 2, Academic-Press, London—New York, 1969, 85—93; [3] Beltrami's systems and quasiconformal mappings. Coll. Math. Anal. Jyväskyla 1970.

Ciorănescu, Ioana [1] On (α) holomorphic vector functions. *Stud. Cerc. Mat.* **18**, 839—844 (1966).

Ciorănescu, N. [3] Sur les fonctions monogènes aréolairement. *Mathematica (Cluj)* **12**, 26—30 (1936).

Collingwood, E. F. and Lohwater, A. J. [1] The theory of cluster sets. Cambridge Univ. Press, Cambridge, 1966, 211 p.

Cotlar, Mischa [1] Familias normales de funciones no-analiticas. *An. Soc. Ci. Argentina* **129**, 3—25 (1940).

Cotlar, Mischa and Vignaux Juan Carlos [1] Sobre la derivada areolar simetrica de las funciones de una variabile compleja dual. *An. Soc. Ci. Argentina* **121**, 128—133 (1936).

Cozma, Ciprian and Ionescu, Vasile [1] Problema lui Riemann generalizată în clasa funcţiilor analitice. (Romanian) (Generalized Riemann problem in the class of analytic functions.) *Stud. Cerc. Mat.* **21**, 1369—1396 (1969); [2] Problema lui Riemann generalizată în clasa funcţiilor analitice generalizate. (Romanian) (Generalized Riemann problem in the class of generalized analytic functions.) *Stud. Cerc. Mat.* **21**, 1485—1510 (1969).

Čumakov, S. A. and Fuksman, I. A. [1] Nekotorye slučaĭ nahoždenija kvazianalitičeskih funkciĭ diskretnogo argumenta v javnom vide. (Russian) (Some cases of existence of quasianalytic functions of a discrete variable in an implicite form.) *Trudy Taškent. Inst. Inž transp. No.* **70**, 21—25 (1970); [2] Postroenie diskretnyh kvazikonformnyh otobraženiĭ mnogosvjaznyh oblasteĭ na prjamoygolnik s parallel'nymi razrezami. (Russian) (Construction of discrete quasiconformal mappings of multiply connected domains onto a rectangle with parallel cross-cuts.) *Trudy Taškent. Inst. Inž. transp. No.* **70**, 26—35 (1971).

Daniljuk, I. I. [3] Pro differencial'ni vlastivosti i ednist' rozvjazkiv, dejakih eliptičnih sistem diferencial'nih rivnjan'. (Ukrainian) (On differential properties and unicity of the solution of some elliptic system of differential equations.) Bjull. naukovoĭ stud. Konf. 1954, L'viv 1955, 108—111; [4] O nekotoryh voprosah teorii ellipticeskih differencial'nyh uravnenii vtorogo porjadka na poverhnostjah. (Russian) (On some questions of the theory of elliptic differential equations of the second order on surfaces.) *Dopovidi L'vivs'k Univ.* **6**, 96—99 (1955); [5] O nekotoryh voprosah teorii elliptičeskih sistem differencial'nyh uravneniĭ pervogo porjadka na poverhnostjah. (Russian) (On some questions of the theory of elliptic systems of differential equations of first order on surfaces.) *Dokl. Acad. Nauk SSSR* **105**, 11—13 (1955); [6] Nekotorye voprosy teorii elliptičeskih differencial'nyh sistem i kvazikonformnyh otobraženiĭ. (Russian) (Some questions of the theory of elliptic differential systems and the quasiconformal mappings.) *Učen. Zap. L'vov. Univ. Ser. Meh.-Mat.* **387**, 75—88 (1956); [7] Kvazigarmoničeskie i kvazianalitičeskie funkcii na poverhnostijah. (Russian) (Quasiharmonic and quasianalytic functions on surfaces.) *Uspehi Mat. Nauk* **11**, 5 (72), 95—101 (1956); [8] Pro teoretiko-funkcional'nii metod v teorii diferencial'nih rivnjan' drugogo porjadku na poverhnjah. (Ukrainian) (Functional theoretical method in the theory of second order differential equations on surfaces.) *Dopovidi Akad. Nauk Ukrain. SSR* **5**, 423—425 (1956); [9] Ob integral'nom predstavlenii rešenii nekotoryh elliptičeskih sistem pervogo porjadka na poverhnostjah s priloženiem k teorii ploskih oboloček. (Russian) (On integral representation of solutions of certain elliptical systems of the first order upon surfaces and their use in the theory of thin shells.) *Dokl. Akad. Nauk SSSR* **109**, 17—20 (1956); [10] Ob obščeĭ elliptičeskoĭ sisteme pervogo porjadka i ob avtomorfnyh kvazianalitičeskih funkcijah na poverhnostjah. (Russian) (On the general first order elliptic system and on automorphic quasianalytic functions on surfaces.) *Dokl. Akad. Nauk SSSR* **109**, 253—255 (1956); [11] Ob avtomorfnyh kvazianalitičeskih funkcii na poverhnostjah. (Russian) (On automorphic quasi-analytic functions on surfaces.) *Mat. Sb.* **41** (**83**), 97—103 (1957); [12] Nekotorye svoĭstva rešenii elliptičeskih sistem pervogo porjadka i kraevye zadači. (Russian) (Some properties of the solutions of elliptic systems of the first order and boundary value problems.) Abstract Dissertation Candidat Moscow 1958; [13] Über das System von Differentialgleichungen erster Ordnung vom elliptischen Typus auf riemannschen Flachen. *Ann. Acad. Sci. Fenn.* **251**/2, 1—7 (1958); [14] Ob otobraženijah sootvestvyjuščih rešenijam uravnenii elliptičeskogo tipa. (Russian) (On mappings corresponding to the solutions of equations of elliptical type.) *Dokl. Akad. Nauk SSSR* **120**, 17—20 (1958); [15] O zadače s kosoĭ proizvodnoĭ dlja elliptičeskih sistem pervogo porjadka. (Russian) (On a problem involving a skew derivative for first order elliptic systems.) *Dokl. Akad. Nauk SSSR* **122**, 9—12 (1958); [16] Issledovanie odnoĭ zadači s kosoĭ proizvodnoĭ pri pomošči sistemy uravnenii Fred gol'ma. (Russian) (The use of Fredholm's system of equations in investigating a problem involving a skew derivative.) *Dokl. Akad. Nauk SSSR* **122**, 175—178 (1958).

DANILOV, V. A. [1] O gladkosti kvazikonformnyh otobraženiĭ v zamknutyh oblasteĭ. (Russian) (On the smoothness of quasiconformal mappings in closed domains.) *Dokl. Akad. Nauk SSSR* **196**, 746—747 (1971).

DECICCO, JOHN [1] Survey of polygenic functions. *Scripta Math.* **11**, 51—56 (1945); [2] Some monogeneity conditions relative to a second derivative of a polygenic functions. *Notices Amer. Math. Soc.* **15**, 104 (1968).

DECICCO, JOHN and CITRON, I. RICHARD [1] Certain classes of polygenic functions. *Notices Amer. Math. Soc.* **18**, 555—556 (1971).

DECICCO, JOHN and KASNER, EDWARD [2] The derivative circular congruence-representation of polygenic functions. *Amer. J. Math.* **61**, 995—1003 (1939); [3] The geometry of polygenic functions. *Univ. Nac. Tucumán Rev. Ser. A* **4**, 7—45 (1944).

DECICCO, JOHN and WALVEKER, ARUN [1] A polygenic extension of the polynomials of Bernoulli and Euler. *Rev. Un. Mat. Argentina* **24**, 143—154 (1969); [2] Some monogeneity conditions relative to the second derivative of a polygenic function with respect to a polygenic function. *Collect. Math.* **20**, 199—206 (1969).

DIMITROVSKI, DRĂGAN [1] (Non-analytic functions whose defection from analyticity is a generalized analytic function.) (Macedonian) *Fac. Sci. Univ. Skopje Annuaire* 1965, **16**, 15—22 (1967); [2] Prilog kon teorijata na obobpštenite analitički funkcii. (Macedonian) (Contribution to the generalized analytic functions.) Doctor Dissertation (unpublished).

DONAHUE, J. E. [1] Concerning the geometry of the second derivative of a polygenic function. Doctor Dissertation. Columbia Univ., New York 1930; [2] On the geometry of the second derivative of a polygenic function. *Bull. Amer. Math. Soc.* **36**, 212 (1930).

DOUGLIS, AVRON [1] A function-theoretic approach to elliptic systems of equations in two variables. *Comm. Pures Appl. Math.* **6**, 259—289 (1953).

DRESSEL, F. G. and GERGEN, J. J. [1] Mapping by *p*-regular functions. I, II. *Bull. Amer. Math. Soc*, **54**, 1062 (1948); *Duke Math. J.* **18**, 185—210 (1951); [2] Uniqueness for *p*-regular mappings. *Duke Math. J.* **19**, 435—444 (1952); [3] Mapping for elliptic equations. *Trans. Amer. Math. Soc.* **77**, 151—178 (1954); [4] The extension for Riemann mapping theorem to elliptic equations. Proc. Conf. on differential equations (dedicated to A. Weinstein). Maryland 1955; Univ. of Maryland, Book Store, College Park, Maryland, 1956, 183—195.

DRESSEL, F. G. GERGEN, I. I., and MC.LEOD, R. M. [1] Uniqueness of mapping pairs for elliptic equations. *Duke Math. J.* **24**, 173—181 (1957).

DUMKIN, V. V. [1] O kvazigarmoničeskih i kvazianalitičeskih differencialah Šottki-Al'forsa. (Russian) (On quasi-harmonic and quasi-analytic differentials of Schottky-Ahlfors.) *Perm. Gos. Univ. Učen. Zap.* **17**, 87—91 (1960).

DUMKIN, V. V. and ŠERETOV, V. G. [1] Ob odnoĭ obščeĭ ekstremal'noĭ zadače dlja kvazikonformnyh otobraženiĭ. (Russian) (On a general extremal problem for quasiconformal mappings.) Metričeskie voprosy teor. funkciĭ i otrobraženiĭ. No. 3. Dokl. II-go kollokviuma po teor. kvazikonformnyh otobraženiĭ i ee obobščenijam, posvjaščennogo 70-

letiju Akad. M. A. Lavrent'éva. Doneck 16—22 September, 1968. Izd. Naukova Dumka, Kiev, 1969, 40—45; [2] O zadača Teichmjullera dlja odnogo klassa otkrytyh rimanovyh poverhnosteĭ. (Russian) (On Teichmüller problem for a class of open Riemann surfaces.) *Mat. Zametki*, **7**, 605—615 (1970).

DURANONA, A., VEDIA, A. and VIGNAUX, J. CARLOS [1] Sobre la teoría de las funciones de una variabile compleja hiperbólica. *Univ. La Plata Publ. Fac. Ci. Fisico-mat. Contrib.* **104**, 139—183 (1935); [2] Serie de polinomias de una variabile compleja hiperbólica. *Univ. Nac. La Plata Publ. Fac. Ci. Fisicomat. Contrib.* **107**, 203—207 (1936).

DUREN, P. L. and LEHTO, OLLI [1] Schwarzian derivatives and homeomorphic extensions. *Ann. Acad. Sci. Fenn. Ser. A I* **477**, 1—11 (1970).

DYPKIN, E. M. [1] Operatornye isčislenie osnovanoe na formule Koši—Grina i kvazianalitičnost' klassov *D*(*h*). (Russian) [Operational calculus based on Cauchy—Green formula and the quasianalyticity of the class *D*(*h*)] *Zap. naučn. Seminarov Leningrad. Otdel. Mat. Inst. Akad. Nauk SSSR* **19**, 221—226 (1970).

DŽURAEV, A. [1] O sisteme uravnenii Bel'trami, vyroždajuščeisja na linii. (Russian) (A system of Beltrami equations which are degenerate on a curve.) *Dokl. Akad. Nauk SSSR* **185**, 984—986 (1969).

EARLE, J. CLIFFORD [2] Teichmüller spaces of groups of the second kind. *Acta Math.* **112**, 91—97 (1964); [3] The Teichmüller space of an arbitrary Fuchsian group. *Bull. Amer. Math. Soc.* **70**, 699—701 (1964); [4] The contractibility of certain Teichmüller spaces. *Bull. Amer. Math. Soc.* **73**, 434—437 (1967): [5] Reduced Teichmüller space. *Trans. Amer. Math. Soc.* **126**, 54—63 (1967); [6] On holomorphic cross-sections in Teichmüller space. *Duke Math. J.* **36**, 409—415 (1969).

EARLE, J. CLIFFORD and EELLS, JAMES [1] On the differential geometry of Teichmüller spaces. *J. Analyse Math.* **19**, 35—52 (1967); [2] The diffeomorphism group of a compact Riemann surface. *Bull. Amer. Math. Soc.* **73**, 557—559 (1967); [3] A fibre boundle description of Teichmüller theory. *J. Differential Geometry* **3**, 19—43 (1969).

ELIANU, I. P. [2] Derivata areolară şi diferenţiala exterioară. (Romanian) (Areolar derivative and other differential.) *Bul. Şti. Acad. R. P. Romîne Secţ. Mat. Fiz.* **8**, 39—50 (1956).

EROHIN, V. [1] K teorii konformnyh i kvazikonformnyh otobraženiĭ mnogosvjaznyh oblasteĭ. (Russian) (On the theory of conformal and quasiconformal mappings of multiply connected domains.) *Dokl. Akad. Nauk SSSR* **127**, 1155—1157 (1959).

ESTEBAN, CARRASCO LUIS [1] La derivada n-sima de una función poligena. *Rev. Mat. Hisp. Amer.* (4), **8**, 3—11 (1948); [2] Distribución de puntos sobre círculos homograficos. *Rev. Mat. Hisp. Amer.* (4), **8**, 134—142 (1948); [3] La geometría asociada con la derivada tercera de una función poligena. Doctor's Dissertation Madrid 1948; *Collect. Math.* **4**, 121—199 (1951); [4] Resolución de un problema sobre la congruencia circular derivada de una función poligena. *Rev. Mat. Hisp. Amer.* (4), **9**, 10—12 (1949); [5] La derivada n-sima rectilínea de una función poligena. *Gaceta Mat.* (1) **1**, 11—25 (1949).

EVANS, GRIFFITH [1] An elliptic system corresponding to Poisson's equation. *Acta Math. Szeged.* **6**, 27—33 (1952).

EVGRAFOV, M. A. and POSTNIKOV, M. M. [1] Ob odnom metriki Teichmüllera. (Russian) (On a metric of Teichmüller.) *Dokl. Akad. Nauk SSSR* **187**, 1229—1231 (1969).

FABER, K. [1] Über den Zusammenhang der Typen von partiellen Differentialgleichungen zweiter Ordnung in zwei Veränderlichen mit gewissen Funktionen-theorien. *Deutsch. Math.* **6**, nr. 4/5, 323—341 (1942).

FAN, LE-LE and SHAH, DAO-SHING [1] On the modulus of quasiconformal mapping. *Sci. Record* **4**, 323—328 (1960); [2] On the parametric representation of quasiconformal mappings, *Sci. Sinica*, **11**, 149—162 (1962).

FARKAS, H. M. [1] Special divisors and analytic subloci of Teichmüller space. *Amer. J. Math.* **88**, 881—901 (1966); [2] Weierstrass points and analytic submanifolds of Teichmüller space. *Notices Amer. Math. Soc.* **15**, 79 (1968); *Proc. Amer. Math. Soc.* **20**, 35—38 (1969).

FEDOROV, V. S. [5] Trudy N. N. Luzina po teorii funkčii komplekcmogo peremenogo (Works of N. N. Luzin on function theory of a complex variable.) *Uspehi Mat. Nauk.* **7**, No. 2 (48), 7—16 (1952); [6] Ob odnom sisteme krivolineĭnyh integralov. (Russian)

(On a system of line integrals.) *Mat. Sb.* **24** (**66**), 15—26 (1949); [7] O kompleksnyh funkcijah dvuh deĭstvitel'nyh peremennyh. (Russian) (On complex functions of two real variables). *Ivanov. Gos. Ped. Inst. Učen. Zap.* **31**, 47—53 (1963).

FEMPL, STANIMIR [1] O neanalitičkim funkci — jama čije drugo odstupan'e od analitičnosti analitička funkcija. (Serbo-Croatian) (On non-analytic functions, whose second deviation from analyticity represents an analytic function.) *Bull. Soc. Math. Phys. Serbie* **15**, 57—62 (1963); [2] O neanalitičkim funkcijama čije je odstupan'e od analitičnosti analitika funkcija. (Serbo-croatian) (Non-analytic functions with an analytic deviation from analyticity.) *Acad. Serbe Sci. Arts Glas Cl. Sci. Math. Nat.* **254**, 75—80 (1963); [3] Areolarni polinomi kao klasa neanalitičkih funkcija čiji. su realni i imaginarni delovi poliharmonijske funkcije. I. II. (Serbo-Croatian) (Areolar polynomials as a class of non-analytic functions whose real and imaginary parts are polyharmonic functions.) *Mat. Vesnik* **1** (16), 29—38 (1964); [4] Reguläre Lösungen eines Systems partieller Differentialgleichungen. *Publ. Inst. Math. (Beograd)* **4** (**18**), 115—120 (1964). [5] Ob odnoĭ sisteme uravneniĭ v častnyh proĭzvodnyh, rešenie kotoroĭ privoditcja k integrovaniju uravnenija vida Klero. (Russian) (On a system of partial differential equations the solution of which can be reduced to that of an equation of Clairaut type.) *Differencial'nye uravnenija* **1**, 698—700 (1965); [6] Über einige Systeme partieller Differentialgleichungen. *Univ. Beograd. Publ. Elektrotehn. Fak. Ser. Mat. Fiz. Nr.* 143—155, 9—12 (1965); [7] O jednoĭ klasi neanalitičkih funkčija. (Serbo-Croatian) (On a class of non-analytic functions.) *Mat. Vesnik* **3** (18), 52—54 (1966); [8] Areolare Exponentialfunktion als Lösung einer Klasse Differentialgleichungen. *Publ. Inst. Math. (Beograd)* **8** (**22**), 138—142 (1968); [9] Über eine partielle Differentialgleichung in den nichtanalytische Funktionen erscheinen. *Publ. Inst. Mat. (Beograd)* **9** (**23**), 115—122 (1969); [10] Über eine nichtlineare partielle Differentialgleichung II Ordnung in der nichtanalytische Funktionen erscheinen. *Mat. Vesnik* **7** (**22**), 325—329 (1970); [11] Über eine Klassification der nichtanalytischen Funktionen. *Math. Balkanica* **1**, 88—92 (1971).

FICHERA, GAETANO [1] Derivata areolare e funzioni a variazione limitata. *Rev. Roumaine Math. Pures Appl.* **14**, 27—37 (1969).

FILIMONOVA, I. I. [1] Uravnenie Bel'trami na zamknutoĭ rimanovoĭ poverhnosti. (Beltrami equation closed Riemann surfaces.) *Soobšč. Akad. Nauk Gruzin. SSR* **49**, 269—274 (1968); [2] O razloženiĭ v rjady obobščennyh analitičeskih i kvazianalitičeskih funkčiĭ na zamknutyh rimanovyh poverhnostjah. (Russian) (On series expansion of generalized analytic and quasi-analytic functions on closed Riemann surfaces.) *Soobšč. Akad. Nauk Gruzin. SSR* **56**, 533—536 (1972).

FILIMONOVA, I. I. and MERŽLJAKOVA, G. D. [1] Kraevye zadači dlja kvazianalitičeskih funkcii. (Russian) (Boundary value problems for quasianalytic functions.) *Perm. Gos. Univ. Učen. Zap.* **22**, 37—40 (1962).

FINN, ROBERT and GILBARG, DAVID [1] Asymptotic behavior and uniqueness of plane subsonic flows. *Comm. Pure Appl. Math.* **10**, 23—63 (1957); [2] Three-dimensional subsonic flows, and asymptotic estimates for elliptic partial differential equations. *Acta Math.* **98**, 265—296 (1957).

FINN, ROBERT and SERRIN, JAMES [1] On the Hölder continuity of quasi-conformal and elliptic mappings. *Trans. Amer. Math. Soc.* **89**, 1—5 (1958).

FISHER, D. STEPHEN [1] On Schwartz lemma and inner functions. *Trans. Amer. Math. Soc.* **138**, 229—240 (1969).

FRÉCHET, MAURICE [7] Sur deux familles de fonctions analogues à la famille des fonctions analytiques. *C. R. Acad. Sci. Paris* **235**, 1585—1587 (1952); [8] Determinado de la plej generalaj planaj parananalitikaj funkcioj. (Esperanto) (Determination of the most general plane paraanalytic functions.) *Ann. Mat. Pura Appl.* (4), **35**, 255—268 (1953).

FURUNZONO, HIROMI [1] On a property of pseudo-regular functions. *Res. Bull. Fac. Liber. Arts Natur. Sci.* **2**, 9—14 (1962); [2] On certain curves defined by pseudo-regular functions. *Res. Bull. Fac. Liber. Arts Natur. Sci.* **2**, 23—26 (1963).

GAHOV, F. D. and KRIKUNOV, JU. M. [1] Topologičeskie metody teorii funkcii kompleksnogo peremennogo i ih priloženie k rešeniju obratnyh kraevyh zadač. (Russian) (Topological methods of the theory of functions of a complex variable and their application to the solution of inverse boundary problems.) Trudy III-go Vsesojuzn. Mat. S'ezda,

Moscow 1956, Sekcionnye dokl. Izd. Akad. Nauk SSSR, Moscow 1956, 1, 77—78; [2] Topologičeskie metody teorii funkcii kompleksnogo peremennogo i ih priloženija k'obratnym kraevym zadačam. (Russian) (Topological methods of the theory of functions of a complex variable and their applications to inverse boundary problems.) *Izv. Akad. Nauk SSSR Ser. Mat.* **20**, 207—240 (1956).

GAĬDUKOV, N. I. [1] Kol'cevye osobennosti kogda p-harakteristiki $p(y) = y$. (Russian) (Annular singularities when the p-characteristic $p(y) = y$.) *Oreho-Zuev. Ped. Inst. Učen. Zap. Kaf. Mat.* **30**, Nr. 4, 38—41 (1968).

GAIER, DIETER [1] Über konforme Abbildungen veränderlicher Gebiete. *Math. Z.* **64**, 385—424 (1956).

GALLOĬ, V. F. and SUVOROV, G. D. [1] Množini monogennosti i differenciovnyh ploskih kil'cevih vidobrazen'. (Ukrainian) (Sets of monogeneity of differentiable plane ring maps.) Dopovidi Akad. Nauk Ukrain. RSR Ser. A 103—106 (1968).

GARDINER, P. FRED [1] On right translation in Teichmüller's space. Doctor's Dissertation, Columbia Univ. New York. *Dissert. Abstr. B. USA* **28**, no. 3, 998 (1967); [2] An analysis of the group operation in universal Teichmüller space. *Trans. Amer. Math. Soc.* **132**, 471—186 (1968).

GAUSS, C. F. [1] Allgemeine Auflösung der Aufgabe, die Theile gegebenen Fläche auf einer anderen gegebenen Fläche so abzubilden, dass die Abbidlung dem Abbgebildeten in dem kleinsten Theilen ähnlich wird. Astronomische Abhandlungen, Hefte 3, Altona 1825; Carl Friedrich Gauss Werke Dieterichische Univ. Druckerei, Göttingen, 4, 189—216 (1880).

GAVRILOV, V. I. [1] O graničnom povedenie Q-kvazikonformnyh otobraženiĭ (Russian) (Boundary behavior of Q-quasiconformal mappings.) *Sibirsk. Mat. Ž.* **2**, 650—654 (1961); [2] Ob iskaženii pri kvazikonformnyh otobraženiĭ. (Russian) (On distorsion by quasiconformal mappings.) *Dokl. Akad. Nauk SSSR* **137**, 1278—1279 (1961); [3] Množestvo predel'nyh značenii psevdoanalitičeskih v ediničnom kruge funkcii. (Russian) (Cluster set of pseudoanalytic functions in the unit disk.) *Dokl. Akad. Nauk SSSR* **148**, 16—19 (1963).

GEHRING, W. FREDERICK [20] The definitions and exceptional sets for quasiconformal mappings. *Ann. Acad. Sci. Fenn. Ser. A I* **281**, 1—28 (1960); [21] Various definitions for a class of plane quasiconformal mappings. Proc. Conf. Analytic functions. Lodz 1—7 September, 1966; [22] Definitions for a class of plane quasiconformal mappings. Conf. on quasiconformal mappings, moduli and discontinuous groups, New Orleans, May 1965; *Nagoya Math. J.* **29**, 175—184 (1967); *Sitzungsber. Berliner Math. Ges.* 1967/1968. 1—23; [23] Quasiconformal mappings which hold the real axis pointwise fixed. Math. Essays Dedicated to A. J. MacIntre. Ohio Univ. Press. Athens, Ohio, 1970, 145—148.

GEHRING, W., FREDERICK and LEHTO, OLLI [1] On total differentiability of functions of a complex variable. *Ann. Acad. Sci. Fenn. Ser. A I*, **272**, 1—9 (1959).

GEHRING, W. FREDERICK and REICH, E. [1] Area distortion under quasiconformal mappings. *Ann. Acad. Sci. Fenn. Ser. A I*, **388**, 1—15 (1966).

GEHRING, FREDERICK and VÄISÄLÄ, JUSSI [2] On geometric definitions for quasiconformal mappings. *Comment Math, Helv.* **36**, 19—32 (1961).

GERSTENHABER, M. and RAUCH, H. [1] On extremal quasiconformal mappings. I, II. *Proc. Mat. Acad. Sci. USA* **40**, 808—812; 991—994 (1954).

GERWICK, JAN [1] Über das Typenproblem. *Arch. Mat. og. Naturvid.* **43**, 1—14 (1940).

GHEORGHIU, OCTAVIAN [1] Extensiuni ale monogeneității lui V. S. Fedorov. (Romanian) (Extensions of the monogeneity in V. S. Fedorov's sense). *Com. Acad. R. P. Romîne*, **2**, 673—676 (1952).

GHERMĂNESCU, MICHEL [1] Sur une classe de fonctions monogènes. *Bull. Sci. Ecole Polytechn. Timişoara* **2**, 188—194 (1929); [2] Sur l'équation aréolaire linéaire. *Bull. Math. Phys. École Polytechn. Bucarest*, **2**, 85—87 (1931); [3] Fonctions harmoniques (p, q)-conjuguées. *Mathematica (Cluj)* **4** (27), 241—246 (1962); *C. R. Acad. Sci. Paris* **254**, 4243—4244 (1962).

GHOSH, P. K. [1] On (p, q)-analytic functions and some of their applications in mechanics of continua. *J. Math. and Phys. Sci.* **2**, 153—173 (1968).

GOL'BERG, A. A. [1] Ob odnom klasse rimanovyh poverhnosteĭ. (Russian) (On a class of Riemann surfaces). *Mat. Sb.* **49**, 447—458 (1959).

GOLUBEVA, O. V. [1] O kompleksnom potenciale i kompleksnoĭ skorosti tečenii v iskrivlennyh plenkah peremennoĭ tolščiny. (Russian) (On complex potential and complex speed of flows in deformed films with variable thickness.) *Moskov. Gos. Ped. Inst. Učen. Zap.* **75**, 3—9 (1959).

GONZALES, O. MARIO [1] Some theorems concerning the directional derivative of a function of a complex variable. *Notices Amer. Math. Soc.* **15**, 476 (1968); [2] On mappings defined by nonanalytic functions. *Notices Amer. Math. Soc.* **17**, 159—160 (1970).

GROTHENDIECK, ALEXANDER [1] Techniques de construction en géométrie analytique. I. Description axiomatique de l'espace de Teichmüller et de ses variantes; II. Généralités sur les espaces annelés et les espaces analytiques; III. Produits fibrés d'espaces analytiques; IV. Formalisme général des foncteurs représentables; V. Fibrés vectoriels, fibrés projectifs, fibrés en drapeaux; VI. Étude locale des morphismes; germes d'espaces analytiques, platitude, morphismes simples; VII Étude locale des morphismes: éléments de calcul infinitésimal; VIII Rapport sur les théorèmes de finitude de Grauert et Remmert; IX. Quelques problèmes de module; X Construction de l'espace de Teichmüller. In "Familles d'Espaces complexes et fondements de la géométrie analytique". Sém. H. Cartan 1960—1961, **13**, Nr. 7—8, 1—33; Nr. 9, 1—14; Nr. 10, 1—11; Nr. 11, 1—28; Nr. 12, 1—15; Nr. 13, 1—13; Nr. 14, 1—27; Nr. 15, 1—10; Nr. 16, 1—20; Nr. 17, 1—20 (1962).

GRÖTZSCH, H. [1] Über die Verzerrung bei schlichten nichtkonformen Abbildungen und eine damit zusammenhangende Erweiterung des Picardschen Satzes. *Berichte über die Verhandlungen der Sächsischen Akad. d. Wiss. zu Leipzig Math. Phys. Kl.* **80**, 503—507 (1928); [2] Über die Verzerrung bei nichtkonformen schlichten Abbildungen mehrfach zusammenhangender schlichter Bereiche. *Berichte über die Verhandlungen der Sächsischen Akad. d. Wiss. zu Leipzig Math. Phys. Kl.* **82**, 69—80 (1930); [3] Über möglichst konforme Abbildungen von schlichten Bereichen. *Berichte über die Verhandlungen der Sächsischen Akad. d. Wiss. zu Leipzig Math. Phys. Kl.* **84**, 114—120 (1932).

GUIAŞU, SILVIU [2] Funcţii complexe pe o suprafaţă oarecare. (Romanian) (Complex functions on a surface.) *An. Univ. Bucureşti Ser. Şti. Natur. Mat-Mec.* **9**, 84—86 (1960).

GUSEV, V. A. [1] Ob odnom obobščenii areolarnyh proïzvodnyh. (Russian) (On a generalization of the areolar derivatives.) *Bul. Şti. Tehn. Inst. Politehn. Timişoara* **7**, 223—238 (1962); [2] Obobščennaja areoljarnaja proïzvodnaja i rjad Teilora. (Russian) (Generalized areolar derivative and Taylor series.) *Izv. Vysš. Učebn. Zaved. Matematika* **6** (43), 41—45 (1964).

GUSMAN, S. JA. [1] Ravnomernoe približenie obobščennyh analitičeskih funkcii. (Uniform approximation of generalized analytic functions.) *Dokl. Akad. Nauk SSSR* **144**, 706—708 (1962).

HABETHA, KLAUS [1] Zur Wertverteilung pseudo-analytischer Funktionen. Colloquium on Math. Analysis. Otaniemi (Finland) 27—31 August, 1966; [2] Über die Wertverteilung pseudoanalytischer Funktionen. *Ann. Acad. Sci. Fenn. Ser. A I*, **406**, 1—20 (1967).

HALLSTRÖM, GUNNAR [1] Eine quasikonforme Abbildung mit Anwendungen auf die Wertverteilungslehre. *Acta Aacd. Abo. Ser. B* **18**, 1—16 (1952); [2] Wertverteilungssätze pseudomoromorpher Funktionen. *Acta Acad. Abo. Ser. B* **21**, 1—23 (1958); [3] On quasiconformally extremal pseudoharmonic functions with given level curves. *Comment. Phys. Math. Soc. Sci. Fenn.* **30**, 1—11 (1964); [4] Probleme in Zusammenhang mit pseudoharmonischen Funktionen. *Math. Z.* **84**, 305—315 (1964).

HAMILTON, R. S. [1] Extremal quasiconformal mappings with prescribed boundary values. *Trans. Amer. Math. Soc.* **138**, 399—406 (1969).

HAN, BYOUNG-JO [1] (Boundary value problem for oblique derivative with displacement for generalized analytic functions.) (Korean) Cho-son In-min Kong-hwa-kuk Kwa-hak-won T'ong-p'o 6, 8—10 (1966).

HAN, TUN-GAO [1] (On existence of homeomorphic holomorphic solutions of generalized Beltrami equations.) (Chinese) *Acta Sci. Natur. Univ. Fudan* **10**, Nr.2—3, 93—100 (1965).

HARVEY, W. J. [1] Spaces of Fuchsian groups and Teichmüller theory. Advances in the theory of Riemann surfaces (Proc. Conf., Stony Brook, N.Y., 1969) 195—204. Ann. of Math. Studies, No. 66. Princeton Univ. Press, Princeton, N.J. 1971; [2] On branch loci in Teichmüller space. *Trans. Amer. Math. Soc.* **153**, 387—399 (1971).

HASABOV, E. G. and MIŠNJAKOV, N. T. [1] Zadača tipa zadači Karlemana v klasse obobščennyh analitičeskih funkcii dlja neograničennoĭ oblasti. (Russian) (A problem of Carleman's type for the class of generalized analytic functions in unbounded domains.) *Izv. Vysš. Učebn. Zaved. Matematika* **1** (56), 67—72 (1967).

HASKELL, R. N. [1] Areolar monogenic functions. *Bull. Amer. Math. Soc.* **52**, 332—337 (1946).

HAYASHI, T. [1] On areolar holomorphic functions. *Rend. Circ. Mat. Palermo* **34**, 220—224 (1912).

HEDRICK, E. R. [1] Extensions of Morera's theorem. *Bull. Amer. Math. Soc.* **31**, 218 (1925); [2] A generalization of Picard's theorem. *Bull. Amer. Math. Soc.* **31**, 388(1925); [3] On derivatives of non-analytic functions. *Proc. Nat. Acad. Sci. USA* **14**, 649—654 (1928); [4] Extensions of the theory of functions of a complex variable. *Bull. Amer. Math. Soc.* **34**, 15—16 (1928); [5] Stieltjes integrals in complex plane. *Bull. Amer. Math. Soc.* **34**, 150 (1928); [6] On derivatives of non-analytic functions. *Bull. Amer. Math. Soc.* **34**, 435 (1928); [7] Analytic points of non-analytic functions. *Bull. Amer. Math. Soc.* **34**, 435—436 (1928); [8] Geometric representation of fondamental quantities for non-analytic functions. *Bull. Amer. Math. Soc.* **34**, 705 (1928); [9] Integrals along given contours in the complex plane. *Bull. Amer. Math. Soc.* **35**, 176 (1929); [10] Generalizations of Cauchy's integral theorem. *Bull. Amer. Math. Soc.* **35**, 599 (1929); [11] On certain properties of non-analytic functions of a complex variable. *Bull. Calcutta Math. Soc.* **20**, 109—124 (1930); [12] Generalization of Liouville's theorem and allied theorems. *Bull. Amer. Math. Soc.* **36**, 59 (1930); [13] Varieties of polar singularities of non-analytic functions. *Bull. Amer. Math. Soc.* **36**, 368 (1930); [14] Theorems associated with Liouville's theorem for non-analytic functions. *Bull. Amer. Math. Soc.* **36**, 801 (1930); [15] Non-analytic functions of a complex variable. *Bull. Amer. Math.Soc.* **39**, 75—96 (1933); [16] Generalization of the Mittag-Leffler theorem and allied theorems. *Bull. Amer. Math. Soc.* **40**, 812—813 (1934).

HEDRICK, E. R., INGOLD LOUIS and WESTFALL, W. D. A. [1] Theory of non-analytic functions of a complex variable. *J. Math. Pures Appl.* (9), **2**, 327—342 (1923).

HEINS MAURICE [1] The conformal mapping of a Riemann surface. In "Analytic functions" *Princeton. Math. Series*, Princeton Academic Press, Princeton, N.J. **24**, 137—158 (1960).

HEINS, MAURICE and MORSE, MARSTON, [1] Topological methods in the theory of functions of a single complex variable. II. Boundary values and integral characteristics of interior transformations and pseudoharmonic functions. *Ann. of Math.* (2), **46**, 625—666 (1945); [2] Topological methods in the theory of functions of a complex variable. III. Causal isomorphisms in the theory of pseudoharmonic functions. *Ann. of Math.*, (2), **47**, 233—273 (1946); [3] Deformation classes of meromorphic functions and their extension to interior transformations. *Acta Math.* **79**, 51—103 (1947).

HEINZ, E. [1] Zur Abschätzung der Funktionaldeterminate bei einer Klasse topologischer Abbildungen. *Nachr. Akad. Wiss. Göttingen Math. Phys. Kl. II, Nr.* **9**, 183—197 (1968).

HERSCH, JOSEPH [1] Contributions à la théorie des fonctions pseudo-analytiques. *Comment. Math. Helv.* **30**, 1—19 (1956); [2] Généralisation du lemme de Schwartz et du principe de la mesure harmonique pour les fonctions pseudo-analytiques. *C. R. Acad. Sci. Paris* **234**, 43—45 (1952).

HERSCH, JOSEPH and PFLUGER, ALBERT [1] Principe de l'augmentation des longeurs extrémales. *C. R. Acad. Sci. Paris* **237**, 1205—1207 (1953).

HO, CHENG-CHI [1] Compactness and quasiconformal mappings. *Acta Math. Sinica* **13**, 447—453 (1963); *Chinese Math.-Acta* **4**, 485—492 (1964); *Sci. Sinica* **14**, 1249—1257 (1965); [2] (On a distortion theorem for quasiconformal mappings. (Chinese) *Acta Math. Sinica* **15**, 487—494 (1965).

HOU, TSUNG-YI [1] Kraevaja zadača Karlemana dlja elliptičeskih sistem uravneniĭ pervogo porjadka. (Russian) (A Carleman boundary value problem for elliptic systems of first-order equations.) *Sci. Sinica* **12**, 1237 (1963).

HU, CHENG-MING [1] (A mean-value property of areolar monogenic functions.) (Chinese) *Shuxue Jinzhan* **7**, 57—61 (1964).

HÜBNER, O. [1] Remarks on a paper by Lawrynowicz on quasiconformal mappings. *Bull. Acad. Polon. Sci. Ser. Sci. Math. Astronom. Phys.* **18**, 183—186 (1970).

HURWITZ, ADOLF [1] Vorlesung über allgemeine Funktionentheorie und elliptische Funktionen. Herausgegeben und ergänzt durch eineh Abschnitt über geometrische Funktionentheorie von R. Courant. Mit einem Anhang von H. Rohre. In "Die Grundlehren der Mathematischen Wissenschaften" Band 3, Springer, Berlin—New York 1964, XIII + 706 pp.

IKOMA, KAZUO [6] On Ahlfors' discs theorem and its application. *Tohoku Math. J.* **8**, 101—107 (1956); [7] Note on the distortion in certain quasiconformal mappings. *Japan J. Math.* **29**, 9—12 (1959); [8] On a lemma of Schwartz for *K—QC* mappings. *Sugaku* **11**, 15—17 (1959/1960); [9] On a property of the boundary correspondence under quasiconformal mappings. *Nagoya Math. J.* **16**, 185—188 (1960); [10] Supplements to my former paper "On Ahlfors' discs theorem and its applications". *Tohoku Math. J.* (2), **13**, 371—372 (1961); [11] A criterion for a set and its image to be of $\alpha(0<\alpha\leqq 2)$-dimensional measure zoro. *Nagoya Math. J.* **22**, 203—209 (1963); [12] A remark on the family of quasiconformal mappings. *Bull. Yamagata Univ.* **5**, 823—827 (1963); [13] On a theorem of Koebe for quasiconformal mappings. *Proc. Japan Acad.* **46**, 763—765 (1970).

IKOMA, KAZUO and SHIBATA, KEICHI [1] On distortion in certain quasiconformal mappings. *Tohoku Math. J.* **13**, 241—247 (1961).

ISMAĬLOV, A. JA. [1] O svoĭstvah psevdoanalitičeskih funkcii. (Russian) (On properties of pseudo-analytic functions.) *Azerbaidžan. Gos. Univ. Učen. Zap. Ser. Fiz.-Mat. Nauk* 31—32 (1965).

ISMAĬLOV, A. JA. and TAGIEVA, M. A. [1] O predstavlenii obobščennyh analitičeskih funkciĭ rjadami psevdopolinomov. (Russian) (On the representation of generalized analytic functions by series of pseudo-polynomials.) *Dokl. Akad. Nauk SSSR* **195**, 1022—1024 (1970).

IOFFE, A. S. [1] Konformnye i kvazikonformnye vloženija odnoĭ konečnoĭ rimanovoĭ poverhnosti v druguju. (Russian) (Conformal and quasiconformal imbedding of a finite Riemann surface into another.) *Dokl. Akad. Nauk SSSR* **202**, 270—272 (1972).

IVANOV, L. D. [1] Ustranimye osobennosti vnutrennih otobraženiĭ. (Russian) (Removable singularities of interior transformations.) *Izv. Vysš. Učebn. Zaved. Matematika* **1** (32), 81—84 (1963).

IVLEV, D. D. [1] O dvoĭnyh čislah i ih funkcijah. (Russian) (On binary numbers and their functions.) *Matem. prosveščenie* **6**, 197—203 (1961).

JAENISCH, SIGBERT [1] Über die Approximierbarkeit quasikonformer Abbildungen der Ebene mittels stückweiser Linearisierung. *Mitt. Math. Sem. Giessen* **79**, 29—53 (1968); [2] The subset of piecewise linear mappings dense in the space of *K*-quasiconformal mappings of the plane. *Acta Math.* **122**, 265—272 (1969).

JAGLOM, I. M. [1] "Kompleksnye čisla i ih primenenie v geometri". (Russian) (Complex numbers and their application in geometry.) Fizmatgiz, Moscow 1963, 191 p.

JAKOBENKO, V. M. and KAPŠIVIĬ, O. O. [1] Ob odnoĭ zadače lineĭnoĭ soprjaženija *p*-analitičeskih funkciĭ s harakteristikoĭ $p = x^k$. (Russian) (On a problem of linear conjugacy of the *p*-analytic functions with the characteristics $p = x^k$.) *Visnik. Kiïv. Univ. Meh.* No. **14**, 72—80 (1972).

JANEKOSKI, VICTOR [1] Za nekoĭ geometriki osobini na izvesni klasi funkcii od edna nomineala promenliva. (Serbo-Croatian) (On some geometrical properties of mappings given by certain families of functions.) *Zb. Techn. Univ. Skopje*, 75—80 (1961).

JENKINS, A. JAMES [1] On the local structure of the trajectories of a quadratic differential. *Proc. Amer. Math. Soc.* **5**, 357—362 (1954); [2] On quasiconformal mappings. *J. rational Mech. Anal.* **5**, 343—352 (1956); [3] A new criterion for quasiconformal mappings. *Ann. of Math.* **65**, 208—214 (1957); [4] On the existence of certain general extremal metrics. *Ann. of Math.* **66**, 440—453 (1957); [5] On the Denjoy conjecture. *Canad. J. Math.* **10**, 627—631 (1958).

JENKINS, A. JAMES and MORSE, MARSTON [1] The existence of pseudoconjugates on Riemann surfaces. *Fundam. Math.* **39**, 269—287 (1952); [2] Topological methods on Riemann surfaces. Pseudoharmonic functions. In "Contribution to the theory of Riemann surfaces". *Ann. of Math. Studies, Princeton Academic Press, Princeton*, N. J., **30**, 111—139 (1953); [3] Conjugate nets, conformal structure, and interior transformations on open Riemann surfaces. *Proc. Nat. Acad. Sci. USA* **39**, 1261—1268 (1953); [4] Curve families F* locally the level curves of a pseudo-harmonic function. *Acta Math.* **91**, 1—42 (1954).

JENSEN, V. P. and HOLL, D. L. [1] An application of the derivative of non-analytic functions in plane stress problems. *Bull. Amer. Math. Soc.* **43**, 256—260 (1937).

JEWETT, W. JOHN [1] Differentiable approximations to light interior transformations. Doctor's Dissertation Univ. Michigan, Ann. Arbor 1955; *Dissert. Abstract* **15**, No 9, 1624 (1955); *Duke Math. J.* **23**, 111—124 (1956); [2] Differentiable approximations to interior functions. *Duke Math. J.* **24**, 227—232 (1957); [3] Multiplication on classes of pseudo-analytic functions. *Pacific J. Math.* **10**, 1323—1326 (1960).

JURCHESCU, MARTIN [1] L'invariance *K*-quasiconforme de la parabolicité d'un élément frontière. *C. R. Acad. Sci. Paris* **246**, 2997—2999 (1958).

JUVE, YRJO]1] Über gewisse Verzerrungseigenscheften konformer und quasikonformer Abbildungen. *Ann. Acad. Sci. Fenn. Ser. A I*, **174**, 1—40 (1954); Doctor's Dissertation, Helsinki 1954, 39 p.

KAHRAMANER, SUZAN [1] Sur le comportement d'une représentation presque conforme dans le voisinage d'un point singulier. *Istambul Univ. Fen. Fak. Mec.Ser. A* **22**, 127—139 (1957); [2] Sur les applications différentiables du plan complexe. *Istambul Univ. Fen. Fak. Mec. Ser. A* **26**, 25—36 (1961).

KUKUTANI, S. [1] Applications of the theory of pseudo-regular functions to the type problem of Riemann surfaces. *Japan J. Math.* **13**, 375—392 (1936). [2] On the family of pseudo-regular functions. *Tohoku Math. J.* **44**, 211—215 (1937).

KAPŠIVIĬ, O. O. [1] Pro rozv'jannja osesimetričnih zadač teorii pružnosti dlja šaru z cilindričnogo porožninoju. (Ukrainian) (Solution of axially-symmetric problems of the theory of elasticity for a sphere with a cylindrical cavity.) *Visnik Kiiv. Univ.* **4**, 96—106 (1961); [2] Pro zastosuvannja *p*-analitičnih funkciĭ v osesimetričniĭ teorii pružnosti. (Ukrainian) (Application of *p*-analytic functions to the axially symmetric theory of elasticity.) *Visnk Kiiv. Univ.* **5**, 76—89 (1962); [3] Zastosuvannja metodu *p*-analitičnih funkciĭ do roz v'jazannja odni'i zadači dlja šaruvatogo cilindra. (Ukrainian) (Application of the method of p-analytic functions to the solution of a problem for a spherical cylinder.) *Prikladna Meh.* **9**, Nr. 6, 670—676 (1963); [4] Primenenie *p*-analitičeskih funkcii k rešeniju kraevyh zadač osesimmetričnoĭ termouprugosti. (Russian) (Application of *p*-analytic functions to the solution of boundary value problems of axially simmetric thermoelasticity.)Problems Math. Phys. and Theory of Functions. Izd.Akad. Nauk SSSR, Kiev, 1, 24—34 (1964); [5] Primenenie *p*-analitičeskih funkcii k rešeniju odnoĭ zadačĭ osesimmetričnoĭ teorii uprugnosti dlja sloĭstogo konečnogo cilindra. (Russian) (Application of *p*-analytic functions to solution of the problem of axissymetrical theory of elasticity for a flaky cylinder.) *Mežved. Naučn. Sb.* "*Vyičslitel'naja Matematika*", *Izd. Kiïvsk. Gos. Univ.* 1966, 115—123; [6] Primenenie *p*-analitičeskih funkcijah k rešeniju odnoĭ kontaktnoi ossesimetričnoĭ zadači dlja sloĭstogo cilindra. (Russian) (Application of *p*-analytic functions to the solution of a contact axially symmetric problem for a flaky cylinder.) Mežved. Naučn. Sb. "Vyčislitel'naja Matematika", Izd. Kiïvsk. Gos. Univ. 1967, 110—117; [7] Pro rozvijaznist' dejakih kraĭovih zadač dlja *p*-analitičnih funkciĭ harakteristikoju $p = x$. (Russian) (Solubility of some boundary value problems for *p*-analytic functions with $p = x$.) *Višnik Kiiv. Univ. Ser. Mat. Meh. No.* **13**, 24—32; 146 (1971).

KAPŠIVIĬ, O. O. and MASLJUK, G. F. [1] Rešenie smešanoĭ ossesimetričnoĭ zadači teorii uprugosti dlja poluprostranstva metodom *p*-analitičeskih funkciĭ. (Russian) (The solution of a mixed axisymmetric problem of the theory of elasticity for a half space by the method of *p*-analytic functions.) *Prikladna Meh.* **3**, Nr. 7, 21—27 (1967).

KAPŠIVIĬ, O. O. and NOGIN, N. V. [1] K rešeniju zadač o kompleksnom *x*-analytičeskom potenciale dlja sferičeskogo krugovogo disk. (Russian) (On the solution of the complex *x*-analytic potential for spherical circular disc.) *Ukraïn. Mat. Z.* **22**, 369—374 (1970).

KAPŠIVIĬ, O. O. and POLOŽIĬ, G. N. [1] O zadačah lineĭnogo soprjaženija *p*-analitičeskih funkcii s harakteristikoĭ $p = x$. (Russian) (On the problem of linear conjugacy of the *p*-analytic functions with the characteristic $p = x$.) *Vyčisl. Prikl. Mat.* (*Kiev*), *No.* **10**, 49—66 (1970).

KASNER, EDWARD [3] A new theory of polygenic (non-monogenic) functions. *Science* **66**, 561—582 (1927); [4] General theory of polygenic or non-monogenic functions. The derivative congruence of circles. *Proc. Nat. Acad. Sci. USA* **14**, 75—82 (1928); [5] Géométrie des fonctions polygènes. Atti del Congr. Bologna 3—10 September 1928, Zanichelli Bologna, 1930, **3**, 255—260; [6] Non-monogenic or polygennic functions. *Bull. Amer. Math. Soc.* **34**, 6 (1928); [7] Geometric characterization of the derivative congruence of a general polygenic function. *Bull. Amer. Math. Soc.* **34**, 152 (1928); [8] Homographic and uniform clocks. *Bull. Amer. Math. Soc.* **34**, 152 (1928); [9] General theory of element-point transformations. *Bull. Amer. Math. Soc.* **34**, 263 (1928; [10] The increment ratio of two polygenic (or non-analytic) functions. *Bull. Amer. Math. Soc.*, **34**, 263–264 (1928); [11] The second derivative of a polygenic function. *Bull. Amer. Math. Soc.* **34**, 425 (1928); *Trans. Amer. Math. Soc.* **30**. 803—818 (1928); [12] Note on the derivative circular congruence of a polygenic function. *Bull. Amer. Math. Soc.* **34**, 561—565; 694 (1928); [13] Higher partial derivatives of polygenic functions. *Bull. Amer. Math. Soc.* **35**, 8 (1929); [14] The third characteristic property of the derivative of a polygenic function. *Bull. Amer. Math. Soc.* **36**, 218 (1930); [15] A complete characterization of the derivative of a polygenic function. *Proc. Nat. Acad. Sci. USA* **22**, 172—177 (1936); [16] Polygenic functions whose associated element-to-point transformation converts union into points. *Bull. Amer. Math. Soc.* **44**, 726—732 (1938).

KAWANAKA, NORIAKI [1] The decomposition of $L^2\left[\frac{\Gamma}{SL(r, R)}\right]$ and Teichmüller spaces. *Proc. Japan Acad.* **46**, suppl. 1126—1129; *J. Math. Kyoto Univ.* **11**, 113—147 (1971).

KEČKIĆ, D. JOVAN [1] On some classes of non-analytic functions. *Publ. Fac. Electrotechn. Univ. Belgrade Ser. Math.-Phys.* **286**, 83—86 (1969); [2] Analytic and c-analytic functions. *Publ. Inst. Math. (Beograd)* **9** (23), 189—198 (1969); [3] On some systems of partial differential equations and on some classes of non-analytic functions. Thèse Publication. No. 247—273, 61—66 (1969); [4] O jednoĭ klasi parcialnih jednačina. (Serbo-Croatian) (On a class of partial differential equations.) *Mat. Vesnik* **6** (**21**), 71—73 (1969); [5] A characterization of certain non-analytic functions. *Bul. Inst. Politechn. Iași* (1974) (in print).

KEČKIČ, D. IOVAN and MITRINOVIČ, S. DRAGOSLAV. [1] From the history of non-analytic functions. I, II. *Publ. Fac. Électrotechn. Univ. Belgrade Sér. Math.-Phys.* No. 274—301, 1—8 (1969); No. 302—319. 33—38 (1970).

KEEN, LINDA [1] Intrinsic moduli on Riemann surfaces. *Ann. of Math.* **84**, 404—420 (1966).

KELINGOS, JOHN ALEXANDER [2] Contribution to the theory of quasiconformal mappings. Doctor's Dissertation University of Michigan, Ann Arbor 1963, 143 pp. *Dissert. Abstr.* **25**, No. 1, 502 (1964); [3] Two new characterizations for plane quasiconformal mappings. *Notices Amer. Math. Soc.* **11**, 109—110 (1964); [4] Characterization of quasiconformal mappings in terms of harmonic and hyperbolic measure. *Ann. Acad. Sci. Fenn. Ser. A I*, **368**, 1—11 (1965); [5] Boundary correspondence under quasiconformal mappings. *Michigan Math. J.* **13**, 235—249 (1966); [6] On the maximal dilatation of quasi-conformal extensions. *Notices Amer. Math. Soc.* **17**, 148 (1970); [7] On the maximal dilatation of quasiconformal extensions. *Ann. Acad. Sci. Fenn. Ser. A I* **478**, 1—8 (1971); [8] Distortion of hyperbolic area under quasiconformal mappings. *Notices Amer. Math. Soc.* **19**, A—396 (1972); Communicated at the March meeting of Amer. Math. Soc. in St. Louis (Missouri) March 27 April 1 1972.

KIJAŠKO, A. M. [1] Pro zastosuvannja p-analitičnih funkcii do bezmomentnih obolonok obertanija, navantaženih u veršini zo seredženimi silami. (Ukrainian) (Application of *p*. analytic functions to momentless shells of revolution, loaded at the top by concentrated forces.) *Visnik Kiiv. Univ.* **4**, 118—124 (1961).

KIJAŠKO, A. M. and POLOŽII, G. M. [1] Pro zastosuvannja *p*-analitičeskih funkcii do pozv'jazannja kraiovih zadač bezmomentnoĭ teorii obolonok. (Ukrainian) (Application of *p*-analytic functions to the solution of boundary problems of the momentless shell theory.) *Prikladna. Meh.* **7**, 362—369 (1961).

KLOTZ, S. TILLA [1] The geometry of extremal quasiconformal mappings. Preliminary report. *Notices Amer. Math. Soc.* **8**, 143 (1961); [2] The geometry of extremal quasiconformal mappings. *Michigan Math. J.* **9**, 129—136 (1962); [3] Post scriptum to "The geometry of extremal quasiconformal mappings". *Notices Amer. Math. Soc.* **9**, 394 (1962); [4] More on the geometry of Teichmüller mappings. *Notices Amer. Math. Soc.* **10**, 124 (1963).

KLUNNIK, A. A. [1] Rasčet bezmomentnyh oboloček vraščenija, zagrunžennyh v veršine sosredotočennoĭ nagruzkoĭ, s pomošč' ju p-analitičeskih funkcii. (Russian) (Calculation of momentless shells of revolution, loaded at the top by concentrated loads, by p-analytic functions.) In "Matem. Fizika", Izd. "Naukova Dumka", Kiev, 1963.

KNEIS, GERT [1] Über schlichte Losungen des Beltramischen Differentialgleichungssystems. *Math Nachr.* **46**, 319—321 (1970); [2] Über eine Darstellung für negative gekrümmte Flächen durch die Lösungen einer verallgemeinerten Laplaceschen Differentialgleichungen. *Math.-Nachr.* **47**, 40—45 (1970); [3] Negativ gekrümmte Flächen und verallgemeinerte analitische Funktionen. Conf. "Komplexe Analysis und deren Anwendungen" Göhren (East Germany) 28 September—2 October 1972; in "Beiträge zur Komplexen Analysis und Anwendungen in der Differentialgeometrie", Akademie Verlag, Berlin 1973 (in print).

KOLOMIICEVA, T. A. [1] O topologičeskiĭ metodah teorii funkcii kompleksnogo peremennogo i nekotoryh ih primenenijah k obratnym kraevym zadačam. (Russian) (On topological methods of function theory of a complex variable and some of their applications to inverse boundary problems.) *Izv. Vysš. Učebn. Zaved. Matematika* **3**, 97—111 (1959).

KOLOSOV, G. B. [1] Ob odnom priloženii teorii funkcii kompleksnogo peremennogo k ploskoi zadače matematičeskoĭ teorii uprugosti. (Russian) (On an application of the theory of function of a complex variable to the mathematical theory of the elasticity in plane.) Doctor's Dissertation Jur'ev 1909; [2] O soprjažennyh' differencial'nyh' uravnenjah s' častnymi proĭzvodnymi s' priloženiem' i h' k' rbšeniju voprosov' matematičeskoi fiziki. (Ukrainian) (On conjugate partial differential equations and their application to the solution of questions in mathematical physik.) *Ann. Inst. Electrotechn. Petrograd* **11**, 179—199 (1914); [3] Über einige Eigenschaften des ebenen Problems des Elastizitätstheorie. *Z. Math. Physik* **62**, 384—409 (1914); [4] Primenenie kompleksnoĭ peremennoĭ k teorii uprugosti. (Russian) (Application of the complex variable in the theory of elasticity.) ONTI, Leningrad—Moscow 1935, 224 pp.

KOMIN, ROLF [2] Über die Existenz von p-analytischen Funktionen zu vorgegebenen Realteil. *Schriftenr. Inst. Math. Dtsch. Akad. Wiss. Berlin A, No.* **7**, 65—66 (1970).

KOPPELMAN, WALTER [1] Boundary value problems for pseudo-analytic functions. *Bull. Amer. Math. Soc.* **67**, 371—376 (1961).

KOPYSTYRA, N. P. [1] Ob odnoĭ kraevoĭ zadače p-analitičeskih funkciĭ c harakteristikoĭ $p = x$. (Russian) (On a boundary value problem for p-analytic functions with the characteristic $p =$ x.) *Vyčisl. Prikl. Mat. (Kiev) Nr.* **11**, 108—113 (1970).

KORN, A. [1] Zwei Anwendungen der Methode der sukzessiven Annährungen. Math. Abhandlungen Herman Amadeus Schwartz zu seinem funfzigjarigen Doktorjubiläum. Springer, Berlin 1914, 215—229.

KOROLEVA, M. C. [1] K algebre dvoinyh čisel. (Russian) (On the algebra of binary numbers.) *Učen. Zap. Orehovo-Zuevskogo Ped. Inst.* **7**, 113—136 (1957).

KOVAN'KO, A. C. [1] Opyt rasširenija ponjatija "monogennosti" funkcii. (Russian) (Attempt of extension of the concept of "monogenic" function.) Otčet ot dejatel'nosti Naučno-Ped. Obščestva DVGU, 91—101 (1928); [2] Uslovija monogennosti funkcii v obyknovennom i v obobščennom smysle i uslovija konformnogo sootvestvija meždu dvumja poverhnostjami. (Russian) (Conditions of monogeneity of the functions in the classical and generalized sense and the condition of conformal correspondence between two surfaces.) *Izv. Azerbaidžan. Gos. Univ. Ser. Estestvoznanie i Medicin* **9**, 75—81 (1930).

KRA, IRWIN [1] On Teichmüller space for finitely generated Fuchsian groups. *Amer. J. Math.* **91**, 67—74 (1969).

KRAJKIEWICZ, P. M. [1] Some elementary properties of analytic polygenic functions. *Notices Amer. Math. Soc.* **11**, 136 (1964); [2] Some theorems on polygenic functions. *Dissert. Abstr. B. USA* **27**, 2033 (1966).

KRAMER, EDNA [1] Polygenic functions of the dual variable $w = u + jv$. Doctor's Dissertation Columbia University New York 1930; *Amer. Math. J.* **52**, 340—376 (1930).

KRAVETZ, SAUL [1] On the geometry of Teichmüller spaces and the structure of their modular groups. *Ann. Acad. Sci. Fenn. Ser. A I*, **278**, 1—35 (1959).

KRIZSTEN, ADOLF [1] Areolar monogene und polyanalytische Funktionen. *Comment. Math. Helv.* **21**, 73—78 (1948); [2] Hypercomplexe und pseudoanalytische Funktionen. *Comment. Math. Helv.* **26**, 6—35 (1952).

KRUGLIKOV, V. I. [1] Ob odnom klasse gomeomorfizmov ploskosti na sebja. (Russian) (On a class of homeomorphisms of the plane onto itself.) Materialy itogovoĭ naučnoi konf. MMF i NIIPMM 1, 30—31 (1970); [2] K teoreme suščestvovanie i edistvennosti kvazikonformnyh otobraženiĭ s neograničennymi harakteristikami. (Russian) (On the existence and unicity theorems for quasiconformal mappings with unbounded characteristics.) *Dokl. Akad. Nauk SSSR* **205**, 1289—1291 (1972); [3] K voprosy o suščestvovanii i edinstvennosti otobraženiĭ, kvazikonformnyh v srednem. (Russian) (On the question of the existence and unicity of mappings, quasiconformal in mean.) Materialy vtoroĭ naučnoi konf. po Mat. i Meh. No. 1. Izd. Tomskogo Univ., Tomsk 1972, 18.

KRUGLIKOV, V. I. and MIKLJUKOV, V. M. [1] Teoremy ustoičivosti dlja otobraženiĭ klassa *BL*. (Russian) (Theorems on the stability for the mappings of the class BL.) Materialy itogovoĭ naučnoĭ konf. MMF i NIIPMM 1, 31—32 (1970); Metričeskie voprosy teor. funkciĭ i otobraženiĭ. No. 3. Dokl. II-go kollokviuma po teor. kvazikonformnyh otobraženijah i ee obobščenijam, posvjaščennogo 70-letiju so dnja roždenija Akad. M. A. Lavrent'eva. Doneck, 12—22 October 1970, 55—70; *Dopovidi Akad. Nauk Ukrain. RSR Ser. A, No.* **5**, 421—423.

KRUGLOVA, S. P. and PAHAREVA, N. A. [1] Ob odnom obobščeniĭ poligarmoničeskih uravnenii v teorii *p*-analitičeskih funksiĭ. (Russian) (On a generalization of the polyharmonic equations in the theory of *p*-analytic functions.) *Matem. Fiz. Resp. Mežved. Sb.* **4**, 154—159 (1968).

KRUSKAL', S. L. [3] O nekotoryh teoremah tipy teoremy Fatu. (Russian) (On some theorems of Fatou's type.) *Trudy Tomsk. Gos. Univ. Ser. Meh.-Mat.* **163**, 54—57 (1963); [4] Varjacija kvazikonformnogo otobraženija krugovogo kol'ca. (Russian) (Variation of a quasiconformal mapping of an annulus.) *Sibirsk. Mat. Ž.* **5**, 236—239 (1964); [5] Ot otobraženijah kvazikonformnyh v srednem. (Russian) (On mean quasiconformal mappings.) *Dokl. Akad. Nauk SSSR* **157**, 517—519 (1964); [6] Metod variaciĭ v teorii kvazikonformnyh otobraženiĭ zamknutyh rimannovyh poverhnosteĭ. (Russian) (The method of variation in the theory of quasiconformal mappings of closed Riemann surfaces.) *Dokl. Akad. Nauk SSSR* **157**, 781—783 (1964); [7] K teorii èkstremal'nyh zadač dlja kvazikonformnyh otobraženiĭ zamknutyh rimannovyh poverhnosteĭ. (Russian) (On the theory of extremal problems for quasiconformal mappings of closed Riemann surfaces.) *Dokl. Akad. Nauk SSSR* **171**, 784—787 (1966); [8] K teorema Teihmjullera ob èkstremal'nyh kvazikonformnyh otobraženijah. (Russian) (On Teichmüller theorem on extremal quasiconformal mappings.) *Sibirsk. Mat. Ž.* **8**, 313—332 (1967); [9] Ob otobraženijah, ε-kvazikonformnyh v srednem. (Russian) (Mappings ε-quasiconformal in mean.) *Sibirsk. Mat. Ž.* **8**, 798—806 (1967); [10] Ob èkstremal'nyh kvazikonformnyh otobraženijah s zadanym graničnym sootvestviem. (Russian) (Extremal quasiconformal mappings with a given boundary correspondence.) *Dokl. Akad. Nauk SSSR* **175**, 525—527 (1967); [11] Ob odnom klasse èkstremal'nyh kvazikonformnyh otobraženiĭ. (Russian) (A certain class of extremal quasiconformal mappings.) *Dokl. Akad. Nauk SSSR* **179**, 1042—1045 (1968); [12] Nekotorye èkstremal'nye zadači dlja odnolistnyh analitičeskih funkcii. (Russian) (Some èkstremal problems for univalent analytic functions.) *Dokl. Akad. Nauk SSSR* **182**, 754—757 (1968); [13] K probleme moduleĭ rimanovyh poverhnosteĭ. (Russian) (On the problem of moduli of Riemann surfaces.) *Dokl. Akad. Nauk SSSR* **183**, 762—764 (1968); [14] Ob èkstremal'nyh zadačah teorii kvazikonformnyh otobraženiĭ. (Russian) (Extremal problems in the theory of quasiconformal mappings.) Metričeskie voprosy teorii funkcii i otobraženiĭ. Nr. 1. Dokl. Kollokviuma po teorii kvazikonformnyh otobraženiĭ i ee obobščenijam. Doneck September 1968, Kiev 1969, 24—34; [15] K teorii èkstremal'nyh kvazikonformnyh otobraženiĭ. (Russian, (On the theory of extremal quasiconformal mappings.) *Sibirsk. Mat. Ž.* **10**, 573—583 (1969); [16] Odna approkcimacionnaja teorema dlja analitičeskih funkcii i ee obobščenija. (Russian) (A certain approximation theorem and its application.) *Dokl. Akad. Nauk SSSR* **184**, 1277—1280 (1969); [17] K voprosu o zavisimosti golomorfnyh differencialov ot moduleĭ rimanovah poverhnosteĭ. (Russian) (On the question of the dependence of the holomorphic differentials of moduli of Riemann surfaces.) *Dokl. Akad. Nauk SSSR* **189**, 472—474 (1969); [18] Nekotorye ekstremal'nye zadači dlja konformnyh i kvazikonformnyh otobraženiĭ. (Russian) (Some extremal problems for conformal and quasiconformal mappings.) *Sibirsk. Mat. Ž.* **12**, 760—784 (1971); [19] O svjazi variacionnyh zadač dlja konformnyh i kvazikonformnyh otobraženiĭ. (Russian) (On the connection between the variational problems for conformal and quasiconformal mappings.) *Sibirsk. Mat. Ž.* **12**, 1067—1076 (1971); [20] Nekotorye local'nye teoremy dlja kvazikonformnyh otobraženiĭ rimanovyh poverhnosteĭ. (Russian) (Some local theorems for quasiconformal mappings of Riemann surfaces.) *Dokl. Akad. Nauk SSSR* **199**, 269—272 (1971); [21] O moduljah rimanovyh poverhnosteĭ. (Russian) (On modules of Riemann surfaces.) *Sibirsk. Mat. Ž.* **13**, 349—367 (1972); [22] Prostranstva Teihmjullera i kleinovy grupy. (Russian) (Teichmüller spaces and Kleinian groups.) *Dokl. Akad. Nauk SSSR* **205**, 771—773 (1972).

KRZYZ, JAN [1] Problems. Problems presented to the Conf. on Analytic Functions. Lodz 1—7. September 1966; [2] On an extremal problem of F. W. Gehring. *Bull. Acad. Polon. Sci. Ser. Sci. Math. Astronom. Phys.* **16**, 99—101 (1968); [3] An extremal length problem, *Ann. Univ. Marie Curie-Sklodowska*, Sect. A **22—24**, 95—103 (1968—1970).

KRZYZ, JAN and LAWRYNOWICZ, JULIAN [1] On quasiconformal mappings of the unit disc with two invariant points. Proc. Conf. Analytic Functions. Lodz 1—7 September, 1966; *Michigan Math. J.* **14**, 487—492 (1967); Errata: **15**, 506 (1968).

KUFAREV, BORIS PAVLOVIČ [3] Nul' množestva i gomeomorfizm s konečnym integralom Dirihle. (Russian) (Null-sets and a homeomorphism with a finite Dirichlet integral.) VINITI, Moscow 1966, 15 p.; [4] Sootnošenija tipa "principa dliny i ploščjadi". (Russian) (Relations of the type "length and area principle".) *Dokl. Akad. Nauk SSSR* **170**, 268—270 (1966); [5] K teorii ploskih otobraženiĭ nekotoryh klassov. (Russian) (On the theory of certain classes of mappings in plane.) Doctor's Dissertation Tomsk 1966; [6] Naučnotehničeskaja Konf. Molodyh učenyh i specialistov. Seks. Mat. Tezisy Dokl. Novosibirsk 1966, 14; [7] Absoljutnaja nepreryvnost' funkcii klassa $\widetilde{W}_p^1$ na množestvah urovnja funkcii klassa $\widetilde{W}_p^1$ i nekotorye graničnye svoĭstva otobraženiĭ s obobščennymi proizvodnymi v ploskoĭ oblasti. (Russian) (Absolute continuity of functions of the class $\widetilde{W}_p^1$ on the level sets of a function of class $\widetilde{W}_p^1$ and certain boundary properties of mappings with generalized derivatives in a plane region.) *Dokl. Akad. Nauk SSSR* **181**, 282—285 (1968); [8] Metrizačia prostranstva vseh prostyh koncov oblasteĭ semeĭstva *B*. (Russian) (Metrization of the space of all prime ends of the domains of the family *B*.) *Mat. Zametki* **6**, 607—618 (1969); [9] Emkost' podmnožestv granicy Karateodori i *BL*-gomeomorfizmy. (Russian) (Capacity of subsets of Carathéodory and BL-homeomorphisms.) *Dokl. Akad. Nauk SSSR* **199**, 273—274 (1971).

KUFAREV, BORIS PAVLOVIČ and KUFAREV, PAVEL PARFEN'EVIČ [1] O dvuh metričeskih sposobah opredelenija prostogo konca posledovatel'nosti ploskih oblasteĭ. (Russian) (On two metric ways of defining prime ends of a sequence of plane domains.) *Dokl. Akad. Nauk SSSR* **187**, 986—988 (1969); *Sovjet Math. Dokl.* **10**, 974—977 (1969).

KÜHNAU, REINER [5] Geometrie der konformen Abbildung auf der projektiven Ebene. *Wiss. Martin-Luther-Univ. Halle-Wittenberg, Math. Natur. Reihe*, **12**, 5—20 (1963); Doctor's Dissertation. Halle-Wittenberg 1962; [6] Extremalprobleme bei quasikonformen Abbil*dungen.* Internat. Congr. Math. Moscow 16—26 August, 1966; Abstracts 16, 17; [7] Einige Extremalprobleme bei differentialgeometrischen und quasikonformen Abbildungen. II. *Math. Z.* **107**, 307—318 (1969); [8] Koeffizientenbediengungen bei quasikonformen Abbildungen *Ann. Univ. Marie-Curie-Skledowska Sect. A.* **22—24**, 105—111 (1968—1970); [9] Herleitung einiger Verzerrungseigenschaften koformer und allgemeinerer Abbildungen mit Hilfe des Argumentsprinzips. *Math. Nachr.* 39, 249—275 (1969); [10] Wertannahmeprobleme bei quasikonformen Abbildungen mit ortsabhängiger Dilatationsbeschränkung. *Math. Nachr.* **40**, 1—11 (1969); [10] Bemerkungen zu den Grunskyschen Gebieten. *Math. Nachr.* **44**, 285—293 (1970); [11] Theorie der konformen und quasikonformen Abbildungen. *Math. Nachr.* **45**, 307—316 (1970); [12] Über schraubungssymmetrische Potentialfelder. *Math. Nachr.* **45**, 345—351 (1970); [13] Triangulierte Riemannsche Mannigfaltigkeiten mit ganz-linearen Bezugssubstitutionen und quasikonforme Abbildungen mit stückweise konstanter komplexer Dilatation. *Math. Nachr.* **46**, 243—261 (1970); [14] Eine funktionentheoretische Randwertaufgabe in der Theorie der quasikonformen Abbildungen. *Indiana Univ. Math. J.* **21**, 1—10 (1971); [15] Verzerrungssätze und Koeffizientenbedingungen vom Grunskyschen Typ für quasikonforme Abbildungen. *Math. Nachr.* **48**, 77—105 (1971); [16] Weitere elementare

Bemerkungen zur Theorie der konformen und quasikonformen Abbildungen. *Math. Nachr.* **51**, 377—382 (1971); [17] Anwendung einer Golusinschen Funktionalgleichung auf eine quasikonforme Normalabbildung. *Math. Nachr.* **51**, 383—388 (1971).

KÜNZI HANS [1] Zwei Beispiele zur Wertverteilungslehre. *Math. Z.* **62**, 94—98 (1955); [2] Einfuhrung in die Theorie der quasikonformen Abbildungen. *Elem. Math.* **11**, 121—129 (1956); [3] Quasikonforme Abbildungen. *Ann. Acad. Sci. Fenn. Ser. A I*, **249/2**, 1—24 (1958); [4] Qasikonforme Abbildungen. Springer, Berlin 1960.

KUUSALO, TAPANI [2] Existence of quasiconformal mappings in the complex plane (elementary proof). Colloquium on Math. Analysis. Otaniemi (Finland) 27—31 August, 1966; [3] Abgeschlossene differenzierbare Abbildungen. *Ann. Acad. Sci. Fenn. Ser. A I*, **399**, 1—8 (1967); [4] Verallgemeinerten Riemannscher Abbildungssatz und quasikonforme Mannigfaltigkeiten. *Ann. Acad. Sci. Fenn. Ser. A I*, **409**, 1—24 (1967).

KUZIK, GALINA ALEKSANDROVNA [1] $t(x, y)$-kvazikonformnye otobraženija s dvumja troĭkami harakteristik. (Russian) [$t(x, y)$-quasiconformal transformations with two triplets of characteristics.] Tezisy Dokl. Naučno-tehn. Konf. Molodyh učenyh i specialistov (Matematika). Novosibirsk 1966, 18—20; *Dokl. Akad. Nauk SSSR* **176**, 39—42 (1967); *Trudy Tomsk. Gos. Univ. Ser. Meh.-Math.* **210**, 37—52 (1969); [2] Kol'co t-kompleksnyh čisel i èlementarnye kol'cevye otobraženija. (Russian) (The ring of t-complext mappings and elementary ring mappings.) *Trudy Tomsk. Gos. Univ. Ser. Meh.-Mat.* **189**, 18—41 (1966); [3] Dvoino-lineĭnye preobrazovanija ploskosti t-kompleksnyh čisel. (Russian) (Linear-fractional transformations in the plane of t-complex numbers.) *Trudy Tomsk. Gos. Univ. Ser. Meh.-Mat.* **189**, 42—59 (1966); [4] Kvazikonformnye otobraženija s dvumja troĭkami harakteristikami i ih differencial'nye svoĭstva. (Russian) (Quasiconformal mappings with two triplets of characteristics and their differential properties.) Tezisy i Soobščenija naučnoi Konf. Doneck 1966, 201.

KUZIK, GALINA ALEXANDROVNA and SUVOROV, GEORGIĬ DMITRIEVIČ [1] Rasširennoe ponjatie kvazikonformnosti ploskogo otobraženija i lineĭnye sistemy differencial'nyh uravneniĭ smešannogo tipa. (Russian) (Extended concept of the quasiconformality of a plane mapping and linear systems of differential equations of mixed type.) Meždunarodnaja Konf. po Teorii Anal. Funkcii. Erevan 1965, Tezisy Dokl. 27—28; Internat. Congr. Math. Moscow 16—26 August, 1966; Tezisy i Soobščenija naučnoĭ Konf. Doneck 1966, 201; *Dokl. Akad. Nauk SSSR* **168**, 280—283 (1966); [2] Rasširennoe ponjatie kvazikonformnosti i sistemy differencial'nyh uravneniĭ smešannogo tipa. (Russian) (Extended concept of quasiconformality and systems of differential equations of the mixed type.) *Trudy Tomsk. Gos. Univ.* **200**, No. 5, 112—141 (1968).

LAMBIN, M. V. [1] Polosy Σ-monogennyh funkcii. (Russian) (Poles of Σ-monogenic functions.) *Belorussk. Gos. Univ. Uč. Zap. Ser. Fiz.-Mat.* **15**, 7—13 (1953); [2] Privedenie osesimmetričnoĭ magnitnoĭ zadači k nahoždeniju Σ-monogennyh funkcii. (Russian) (Reduction of an axially symmetric magnetic problem to finding Σ-monogenic functions.) *Belorussk. Gos. Univ. Uč. Zap. Ser. Fiz.-Mat.* **15**, 14—17(1953); [3] Ob osobyh točkah sigma-monogennyh funkcii, svjazannyh s osesimmetričnoĭ magnitoĭ zadačei. (Russian) (On singular points of sigma-monogenic functions, connected with an axially symmetric magnetic problem.) *Belorussk. Gos. Univ. Uč. Zap. Ser. Fiz.-Mat.* **19**, 27—31 (1954).

LAMMEL, ERNESTO [1] Generalizaciones de la teoria de las funciones de vaiables complejas. Segundo symposium sobre alcunos problemas matematicos que se están estudiando en Latino América. Villavicencio, Mendoza July, 1954. Centro de Cooperacion Cientifica de la UNESCO para América Latina, Montevideo (Uruguay) 1954, 191—197.

LAVRENT'EV, ALEKSEEVIČ MIHAIL [14] Sur une classe de représentations continues. *C. R. Acad. Sci. Paris* **200**, 1010—1012 (1035); *Mat. Sb.* **42**, 407—424 (1935); [15] Les représentations quasiconformes et leurs systèmes dérivées. *Dokl. Akad. Nauk SSSR* **52**, 287—289 (1946); [16] Teorija kvazikonformnogo otobraženiĭ. (Russian) (The theory of quasiconformal mappings.) *Jubileĭnyĭ Sb. Akad. Nauk SSSR* **I**, 95—113 (1947); [17] Obščaja teorija kvazikonformnyh otobraženiĭ ploskih oblasteĭ. (Russian) (General theory of quasiconformal mappings of plane domains.) *Mat. Sb.* **21** (63), 285—320 (1947); [18] Ob odnom klasse kvazikonformnyh otobraženiĭ. (Russian) (On a certain class of quasiconformal mappings.) *Sb. trudov Inst. Mat. Akad. Nauk. Ukrain. SSR* **9**, 7—54 (1948); [19] Put' razvitija sovetskoĭ matematiki. (Russian) (The way of development of Sovjet mathematics.) *Izv. Akad. Nauk SSSR Ser. Mat.* **12**, 411—416 (1948);

[20] Osnovnaja teorema teorii kvazikonformnyh otobraženiĭ ploskih oblasteĭ. (Russian) (Fundamental theorem of the theory of quasiconformal mappings of plane domains.) *Izv. Akad. Nauk SSSR Ser. Mat.* **12**, 513—554 (1948); [21] Kumuljativnyĭ zarjad i princip ego raboty. (Russian) (Cumulative charge and the principles of its operation.) *Uspehi Mat. Nauk* **12**, No 4, 41—56 (1957); [22] On some problems of quasiconformal mappings. Internat. Congr. Math. Edinburgh 1958. Abstract short communications p. 56; [23] "Variacionnyĭ metod v kraevyh zadačah dlja sistem uravneniĭ èllipticěskogo tipa". (Russian) (Variational methods for boundary value problems for systems of elliptic equations.) Izd. Akad. Nauk SSSR, Moscow 1962, 135 p.; P. Noordhohh Ltd., Groningen 1963; [24] Sur les représentations quasi-conformes. Proc. Internat. Congr. Math. Djursholm 15—22. August, 1962. Djursholm. Upsala 1963; [25] Ob odnom zadače na skleĭvanie. (Russian) (On a sewing problem) *Sibirsk. Mat. Ž.* **5**, 603—607 (1964).

LAVRENT'EV, ALEXEEVIČ MIHAIL and ŠABAT, B. V. [1] Geometričeskie svoĭstva rešenii nelineĭnyh sistem uravneniĭ s častnymi proĭzvodnymi. (Russian) (Geometric property of the solution of non-linear systems of partial differential equations.) *Dokl. Akad. Nauk SSSR* **112**, 310—311 (1957).

LAVRENT'EV, ALEXEEVIČ MIHAIL and SOBOLEV, S. L. [1] Il'ja Nestorovič Vekua (k pjatidecjatiletju so dnja poždenija. (Russian) (Il'ja Nestorovič Vekua (on the fiftieth anniversary of his birthday.) *Uspehi Mat. Nauk.* **12**, No. 4 (76), 227—234 (1957).

LAWRYNOWICZ, JULIAN [3] On parametrical representation of quasiconformal mappings in an annulus. *Bull. Acad. Polon. Sci. Serc. Sci. Math. Astronom. Phys.* **11**, 657—664 (1963); [4] On the parametrisation of quasiconformal mappings in an annulus. *Ann. Univ. Mariae Curie-Sklodowska Sect. A* **18**, 23—52 (1964); [5] Problems. Problems presented to Conf. on Analytic Functions 1—7 September, Lodz 1966; [6] Some parametrisation theorems for quasiconformal mappings in an annulus, *Bull. Acad. Polon. Sci. Ser. Sci. Math. Astronom. Phys.* **15**, 319—323 (1967); [7] On certain functional equations for quasiconformal mappings. Proc. Conf. on Analytic Functions. Lodz 1—7 September 1966, p. 14—15; *Ann. Polon. Math.* **20**, 153—165 (1968); [8] Quasiconformal mappings of the unit disc near to the identity. *Bull. Acad. Polon. Sci. Ser. Sci. Math. Astronom. Phys.* **16**, 771—777 (1968); [9] On a class of quasiconformal mappings with invariant boundary points. I. The class E_Q and the general extremal problem. *Ann. Polon. Math.* **21**, 309—324 (1969); [10] On a class of quasiconformal mappings with invariant boundary points. II. Applications and generalizations. *Ann. Polon. Math.* **21**, 325—347 (1969); [11] On the parametrization of quasiconformal mappings with invariant boundary points in the unit disc. *Bull. Acad. Polon. Sci. Ser. Sci. Math. Astronom. Phys.* (to appear); [12] On the parametrization of quasiconformal mappings with invariant boundary points. *Comment. Math. Helv.* **47** (1973) (to appear).

LEE, CHI-YUAN [1] Domain functionals for pseudo-analytic functions. Doctor's Dissertation University of Washington 1954, *Dissert. Abstr.* **15**, 272 (1955); [2] Similarity principle with boundary conditions for pseudo-analytic functions. *Duke Math. J.* **23**, 157—163 (1956).

LEE-GO-PIN [1] (Two fundamental principles in the theory of the construction of analytic functions and of their generalization.) (Chinese) *Acta Math. Sinica* **7**, 327—339 (1957).

LEHTO, OLLI [2] On the differentiability of quasiconformal mappings with prescribed complex dilatations. *Ann. Acad. Sci. Fenn. Ser. A I* **275**, 1—28 (1960); [3] An extension theorem for quasiconformal mappings. *Proc. London Math. Soc.* **14a**, 187—190 (1965); [4] Remarks on the integrability of the derivatives of the quasiconformal mappings. *Ann. Acad. Sci. Fenn. Ser. A I* **371**, 1—8 (1965); [5] A boundary value problem for conformal mappings. Meždunarodnaja Konf. po teor. anal. funkciĭ. Erevan 1965. Tezisy dokl. p. 37 in "Sovremenye problemy teorii analitičeskih funkciĭ". Izd. "Nauka", Moscow 1966, p. 216—218; [6] Homeomorphic solutions of a Beltrami differential equation. Second Coll. Rolf Nevanlinna. Zürich 4—6. November, 1965. In "Festband zum 70. Geburtstag von Rolf Nevanlinna", Springer, Berlin, Heidelberg, New York, 1966, p. 58—65; [7] Quasiconformal mappings in plane. Internat. Congr. Math. Moscow 16—26 August, 1966. Abstracts of reports 68—71, Trudy Meždunarod. Kongr. Mat. Izd. "Mir", Moscow, 1968, 319—322; [8] On the development of the theory of quasiconformal mappings. *Arkhimedes* **2**, 23—30 (1966); [9] Existence theorem for a generalized

Beltrami equation. *Notices Amer. Math. Soc.* **15**, 540 (1968); [10] Homeomorphisms with a given dilatation. Proc. 15th Scandinavian Congr. Oslo 1968. Lectures Notes in Math. 118, Springer, Berlin, 1970, 58—73; [11] Remarks on the relationship between generalized Beltrami equations and conformal mappings. "Proceedings of the Romanian-Finnish Seminar on Teichmüller space and quasiconformal mappings. Braşov 26—30 August, 1969". Publishing House of the Academy of the Socialist Republic of Romania, Bucureşti, 1971, 49—64; [12] Schlicht functions with a quasiconformal extension. *Ann. Acad. Sci. Fenn. Ser. A I*, **500**, 1—10 (1971).

LEHTO, OLLI VÄISÄLÄ, JUSSI and VIRTANEN, K. I. [1] Contribution to the distortion theory of quasiconformal mappings. *Ann. Acad. Sci. Fenn. Ser. A I*, **275**, 1—14 (1959).

LEHTO, OLLI and VIRTANEN, K. I. [1] On the existence of quasiconformal mappings with prescribed complex dilatation. *Ann. Acad. Sci. Fenn. Ser. A I*, **274**, 1—24 (1960); [2] Quasikonforme Abbildungen. Springer, Berlin—Heidelberg—New York 1965, 269 p.

LELONG-FERRAND, JACQUELINE [4] Sur certaines classes de représentations d'un domaine plan variable. *J. Math. Pure Appl.* **31**, 103—126 (1952); [5] Représentations conformes et transformations à intégrale de Dirichlet bornée. Gauthier-Villars, Paris, 1955, 257 p.

LENNART, CARLESON [1] On mappings, conformal at the boundary. *J. Analyse Math.* **19**, 1—13 (1967).

LI, CHUNG [1] On the existence of homeomorphic solutions of a system of quasilinear partial differential equations of elliptic type. *Acta Math. Sinica* **13**, 454—461(1963); *Chinese Math.* **4**, 493—500 (1964); [2] A modified Dirichlet's problem and its application to quasiconformal mappings. *Acta Sci. Natur. Univ. Pekinensis* **10**, 319—341 (1964); [3] On the extremum principle in the theory of generalized analytic functions. *Shuxue Jinzhan* **8**, 173—182 (1965); *Sci. Abstr. China Math. and Phys.* **3**, 7 (1965).

LI, CHUNG and WEN, KUO-CHUN [1] On the Cauchy formula for elliptic systems of linear partial differential equations of the first order. *Acta Math. Sinica* **14**, 23—32 (1964); *Chinese Math.* **5**, 25—35 (1964).

LI, YOU-T' SAI [1] (Parametric representation of quasiconformal mappings.) (Chinese) *Shuxue Jinzhan* **9**, 55—66 (1966).

LICHTENSTEIN, L. [1] Zur Theorie der konformen Abbildungen; konforme Abbildungen nichtanalytischer singularitätenfreien Flächenstücke auf ebene Gebiete. *Bull. Acad. Sci. Cracovie* 192—217 (1916).

LIPOVOĬ, G. S. [1] Postroenie funkciĭ osuščestvljajuščih kvazikonformnoe otobraženie vnešnyh oblasteĭ. (Russian) (Construction of functions effecting the quasiconformal mappings of exterior domains.) First Republ. Math. Conf. of Young Researchers, Part I, Akad. Nauk Ukrain. SSR Inst. Mat. Kiev 1965, p. 454—461; [2] Postroenie kvazikonformnyh otobraženiĭ dlja nekotoryh zadač gazovoĭ dinamiki. (Russian) Construction of quasiconformal mappings for certain problems of gas dynamics.) *Ukrain. Mat. Ž.* **17**, Nr. 1, 112—117 (1965); [3] O postroenie nekotoryh konformnyh i kvazikonformnyh otobraženiĭ dlja mnogosvjaznyh oblasteĭ. (Russian) (The construction of certain conformal and quasiconformal mappings for multiply connected domains.) *Ukrain Mat. Ž.* 20, 620—627(1968).

LITVINČUK, G. S. [1] Ob odnoĭ kraevoĭ zadače s obratnym sdvigom v klasse obobščennyh analitičeskih funkciĭ. (Russian) (A boundary-value problem with inverse displacement in a class of generalized analytic functions.) *Sibirsk. Mat. Ž.* **3**, 223—228 (1962).

LITVINČUK, G. S. and MIŠNJAKOV, N. T. [1] Kraevaja zadača Karlemana dlja ograničennoĭ oblasti v klasse obobščennyh analitičeskih funkciĭ. (Russian) (Carleman boundary value problem for a bounded region in a class of generalized analytic functions.) *Izv. Akad. Nauk Armjan. SSR* **2**, 52—56 (1967).

LJAŠKO, I. I. and PAHAREVA, N. A. [1] O rešenie fil'tracĭonnyh zadač metodom mažorantnyh oblasteĭ. (Russian) (On the solution of filtration problem by the method of majorant domains.) *Ukrain. Mat. Ž.* **12**, Nr. 4, 402—411 (1960).

LOHWATER, A. J. [1] Beurling theorem for quasiconformal mappings. *Bull. Amer. Math. Soc.* **61**, 223 (1955); [2] The boundary behaviour of a quasiconformal mapping. *J. Rational Mechn. Anal.* **5**, 335—342 (1956).

LOPATINSKIĬ, JA. B. [1] Ob odnom obobščeniĭ ponjatija analitičeskoĭ funkciĭ. (Russian) (On a generalization of the concept of analytic function.) *Dokl. Akad. Nauk SSSR* **64**, 155—158 (1949); *Ukrain. Mat. Ž.* **2**, 56—73 (1950).

LUBOWICZ, HENRYK [1] The non-linear generalized Riemann—Hilbert problem in the multiconnected region for a system of functions. *Demonstratio Math.* **2**, 127—144 (1970).

LUFERENKO, V. P. [1] Analog teoremy Karateodori dlja proizvol'nyh semeĭstv topologičeskih otobraženiĭ. (Russian) (A theorem analogous to Carathéodory theorem for arbitrary families of topological mappings.) Metričeskie voprosy teorii funkciĭ i otobraženiĭ. Nr. 1. Dokl. Kollokviuma po teorii kvazikonformnyh otobraženiĭ i ee obobščenijam. Doneck 16—22 September, 1968, Izd. "Naukova Dumka", Kiev, 1969, 120—139.

LUKIN, V. N. [1] Metričeskie svoĭstva topologičeskih otobraženiĭ ploskih konečnosvjaznyh oblasteĭ (Russian) (The metric properties of a certain class of mappings of finitely connected plane regions.) *Trudy Tomsk. Gos. Univ. Ser. Meh.-Mat.* **189**, 66—79 (1966).

LUKOMSKAJA, I. G. [1] Ob odnom obobščenii klassa analitičeskih funkciĭ. (Russian) (On the generalization of the class of analytic functions.) *Dokl. Akad. Nauk SSSR* **73**, 885—888 (1950); [2] O ciklah sistem lineĭnyh odnorodnyh differencial'nyh uravneniĭ. (Russian) (On cycles of systems of linear homogeneous differential equations.) *Mat. Sb.* **29 (71)**, 551—558 (1951); [3] Rešenie nekotoryh sistem uravneniĭ s častnymi proizvodnymi posredstvom vključenija v cikl. (Russian) (Solutions of some systems of partial differential equations by means of inclusion in a cycle.) *Prikladnaja Mat. Meh.* **17**, 745—747 (1953).

LUKOMSKAJA, M. A. and STEINBERG, N. S. [1] Svjaz' meždu Σ-integrirovaniem i integrirovaniem po soprjažennym peremennym. (Russian) (Connection between Σ-integrability and integrability with respect to conjugate variables.) *Dokl. Akad. Nauk SSSR* **7**, 653—654 (1963).

MAKOTO, ITOH [1] On quaternionic bicomplex analytic and wave-analytic (pseudoanalytic) functions. *Notices Amer. Math. Soc.* **16**, 307 (1969).

MANDZAVIDZE, G. F. [1] Graničnaja zadača lineĭnogo soprjaženija so smeščeniem i ee svjaz', teorieĭ obobščennyh analitičeskih funkciĭ. (Russian) (A boundary value problem of linear conjugacy with shift and its connection with the theory of generalized analytic functions.) *Sakhart. SSR Mecn. Akad. Math. Inst. Šrom.* **33**, 82—87 (1967).

MARDEN, ALBERT [1] The weakly reproducing differentials on open Riemann surfaces. *Ann. Acad. Sci. Fenn. Ser. A I* **359**, 1—32 (1965); [2] On Bers' boundary of Teichmüller space. *Notices Amer. Math. Soc.* **18**, 1082 (1971); Communicated at the November Meeting of Amer. Math. Soc. in Milwaukee (Wisconsin) 27 November 1971.

MARKUŠEVIČ, A. I. [2] "Kratkiĭ kurs teorii analitičeskih funkciĭ". (Russian) (Short course of the theory of analytic functions.) Izd. "Nauka", Moscow 1966, 387 p.; [3] Teorija analitičeskih funkciĭ. II. Dal'neišee postroenie teorii. (Russian) (Theory of analytic functions. II. Further construction of the theory.) Izd. Nauka, Moscow 1968.

MARTINO, MICHAEL ANTHONY [1] Concerning a measure of non-analyticity for functions of a complex variable. Doctors' Dissertation University of Ilinois 1955; *Dissert. Abstr.* **15**, Nr. 10, 1865 (1955).

MARTIO, OLLI [4] A boundary value problem for the Beltrami differential equation Colloquium on Math. Analysis. Otaniemi (Finland) 27—31 August, 1966; [5] Boundary values and injectiveness of the solutions of Beltrami equations. *Ann. Acad. Sci. Fenn. Ser. A I* **402**, 1—27 (1967); [6] On harmonic quasiconformal mappings. *Ann. Acad. Sci. Fenn. Ser. A I* **425**, 1—10 (1968).

MARUYANA, TAKAHARU and SOEDA, TAKASHI [1] On conformal mappings in 3 and 4 dimensional spaces. *Sci. Papers Fac. Engin. Tokushima Univ.* **6**, 1—4 (1955).

MASKIT, BERNARD [1] On Klein's combination theorem. *Trans. Amer. Math. Soc.* **131**, 32—39 (1968).

MATHEWS, J. H. [1] A bounded normal light interior function that possesses no point asymptotic values. *Isr. J. Math.* **7**, 381—383 (1969); [2] Asymptotic behavior of light interior functions defined in the unit disk. *Proc. Amer. Soc.* **24**, 79—81 (1970); [3] Asymptotic values of normal light interior functions defined in the unit disk. *Proc. Amer. Math. Soc.* **24**, 691—695 (1970).

MCMILLEN, JOHN [1] On boundary distorsion. Proc. Romanian-Finnish Seminar on Teichmüller spaces and quasiconformal mappings. Braşov 26—30 August, 1969. Publishing House of the Academy of the Socialist Republic of Romania, Bucureşti 1971, 225—227; [2] Distortion under conformal and quasiconformal mappings. *Acta Math.* **126**, 121—141 (1971).

MEHTIEV, G. D. [1] Nekotorye teoremy dlja celyh psevdoanalitičeskih funkciĭ. (Russian) (Some theorems for entire pseudo-analytic functions.) *Azerbaidžan. Gos. Univ. Učen. Zap. Ser. Fiz.-Mat. i Him. Nauk* **1**, 56—59 (1967); [2] Ocenki v klassah odnolistnyh psevdoanalitičeskih funkciĭ. (Russian) (Estimations in the class of pseudo-analytic functions.) *Akad. Nauk Azerbaidžan. SSR Dokl.* **24**, 3—5 (1968); [3] Nekotorye dostatočnye priznaki ob odnolistnosti psevdoanalitičeskih funkciĭ. (Russian) (Some sufficient conditions for univalence of pseudo-analytic functions.) In "Aspirant. Konf. Material. Azerbaidžan. Univ. Estestv. Nauk." Baku 1969, p. 21—24.

MEL'NIK, I. M. [1] K topologičeskim metodam teorii funkciĭ kompleksnogo peremennogo. (Russian) (Topological methods in the theory of functions of a complex variable.) *Dokl. Akad. Nauk SSSR* **131**, 1015—1018 (1960); [2] Indeksy toček vetvlenija mnogožnacnoĭ funkcii. (Russian) (Indices of branchpoints of multiform functions.) *Dokl. Akad. Nauk SSSR* **134**, 521—524 (1960); [3] Topologičeskie metody teorii funkciĭ kompleksnogo peremennogo rimannovoĭ poverhnosteĭ. (Russian) (Topological methods in the theory of functions of a complex variable on a Riemann surface.) *Izv. Akad. Nauk SSSR* **25**, 815—824 (1961).

MENŠOV, D. E. [1] Les conditions de monogénéité Hermann, Paris 1936, 52 p.

MERŽLJAKOVA, G. D. [1] Kraevye zadači dlja kvazianalitičeskih funkciĭ. (Russian) (Boundary value problems for quasi-analytic functions.) *Perm. Gos. Univ. Učen. Zap.* **22**, 34—36 (1962); [2] Jadro integrala tipa Koši dlja kvazianalitičeskih funkciĭ na zamknutyh rimanovyh poverhnostjah i rešenie kraevoĭ zadači $W^+(t)—W^-(t) = g(t)$. (Russian) (Kernel of an integral of Cauchy type for quasi-analytic functions on closed Riemann surfaces and the solution of the boundary value problem $W^+(t) — W^-(t) = g(t)$.] *Perm. Gos. Univ. Učen. Zap.* **103**, 181—182 (1963).

MIHAĬLOV, L. G. [1] Kraevaja zadača tipa zadači Rimana dlja sistem differencial'nyh uravnenii pervogo porjadka ellipticeskogo tipa. (Russian) (A boundary problem of the type of Riemann for systems of first order differential equations of elliptic type.) *Dokl. Akad. Nauk SSSR* **112**, 13—15 (1957); [2] Osobye slučaĭ v teorii obobščennyh analitičeskih funkciĭ. (Russian) (Singular cases in the theory of generalized analytic functions.) *Naučn. Dokl. Vysš. Školy Fiz.-Mat. Nauk* **3**, 79—84 (1958); In "Issledovanie po sovremennym problemam teorii funkciĭ kompleksnogo peremennogo". Gos. Izd. Fiz.-Mat. Literatury, Moscow 1961, p. 505—510; [3] Issledovanie obobščennoĭ sistemy Koši-Rimana, kogda koefficienty imejut osobennosti pervogo porjadka. (Russian) (An investigation of the generalized Cauchy—Riemann system, where the coefficients have first order singularities.) *Dokl. Akad. Nauk SSSR* **129**, 507—510 (1959); *Trudy Akad. Nauk Tadž SSR* **109**, 57—75 (1961); [4] Ob odnom integral'nom uravnenii teorii obobščennyh analitičeskih funkciĭ v singuljarnom slučae. (Russian) (On an integral equation of the theory of generalized analytic functions.) *Dokl. Akad. Nauk SSSR* **190**, 531—534 (1970); [5] Kraevye zadači teorii oboščennyh analitičeskih funkciĭ v singuljarnyh slučajah. (Russian) (Boundary value problems in the theory of generalized analytic functions for singular cases.) *Trudy Sem. Kraevym zadačam* **7**, 45—54 (1970).

MIHAL'ČUK, V. G. [1] Obobščenie teoremy Rimana—Roha na kvazianalitičeskih funkciĭ. (Russian) (A generalization of Riemann—Roch theorem to quasi-analytic functions.) *Perm. Gos. Univ. Učen. Zap.* **16**, 27—34 (1958); [2] Integral tipa Koši i teorema Rimana—Roha dlja kvazianalitičeskih funkcii na rimanovyh poverhnostjah. (Russian) (The Cauchy integral and the Riemann—Roch theorem for quasi-analytic functions on Riemann surfaces. In "Issledovanija po sovremennym problemam teorii funkciĭ kompleksnogo peremennogo". Gos. Izd. Fiz.-Mat. Literatury, Moscow 1960, p. 425—436; [3] Ob osobennostjah nekotoryh klassov kvazianalitičeskih funkciĭ. (Russian) (On the singularities of certain classes of quasi-analytic functions.) *Izv. Vysš. Učebn. Zaved. Matematika* **15**, 129—137 (1960); [4] O suščestvovaniĭ odnoznačnyh kvazikonformnyh otobraženiĭ na zamknutyh rimanovyh poverhnostjah. (Russian) (Existence of univalent quasiconformal mappings on closed Riemann surfaces.) *Ukrain. Mat. Ž.* **18**, Nr. 4, 121—124 (1966); [5] Kvazikonformnye otobraženija dvuh peremennyh s poljarnymi osobennostjami. (Russian) (Quasiconformal mappings in two variables with polar singularities.) Metričeskie voprosy teor. funkcii i otobraženiĭ. No. 3. Dokl. II-go kollokviuma po teor. kvazikonformnyh otobraženiĭ i ee obobščenijam, posvjaščennogo 70-letiju so dnja roždenia. Akad. M. A. Lavrent'eva, Doneck 12—22 October. 1970, Izd. "Naukova Dumkaj", Kiev, 1971, 94—97.

MIKLJUKOV, V. M. [9] O nekotoryh klassah otobraženiĭ na ploskosti. (Russian) (On certain classes of mappings in plane.) *Dokl. Akad. Nauk SSSR* **183**, 772—774 (1968); [10] Ob iskaženiĭ emkosti pri otobraženijah, kvazikonformnyh v srednem. (Russian) (On the distortion of the capacity of mappings, quasiconformal in mean.) Materialy itogovoĭ naučnoĭ konf. MMF i NIIPMM. Mat. i Meh. Izd. Tomskogo Univ., Tomsk 1970, 1, 43—45.

MIKLJUKOV, V. M. and SUVOROV, G. D. [1] Teorema suščestvovanija dlja ploskih kvazikonformnyh otobraženii s neogranyčennymi harakteristikami. (Russian) (Existence theorem for plane quasiconformal mappings with unbounded characteristics.) Pervyĭ Doneckiĭ kollokvium po teor. kvazikonformnyh otobraženiĭ i ee obobsčenijami 16—22. September, 1968.

MIN, SZU-HOA [1] Non-analytic functions. *Amer. Math. Monthly* **51**, 510—517 (1944); [2] On concrete examples and the abstract theory of the generalized analytic functions. *Scientia Sinica* **12**, 1269—1283 (1963).

MIRANDA, C. [1] Systèmes elliptiques d'équations au dérivées partielles du premier ordre. Convegno Internaz. sulle Equazioni lineari alle derivate parziali. Trieste 25—28. August. 1954. Edizioni Cremonese Roma 1955, p. 30—38.

MIŠNJAKOV, N. T. [1] Zadača tipa zadači Karlemana dlja mnogosvjaznoĭ oblasti v klassa obobščennyh analitičeskih funkciĭ. (Russian) (Carleman's problem for a multiply connected region in the class of generalized analytic functions.) *Sibirsk. Mat. Ž.* **9**, 607—613 (1968); [2] Kraevye zadači so sdvigom v klasse obobščennyh analitičeskih funkciĭ na zamknutyh rimanovyh poverhnostjah. (Russian) (Boundary value problems with displacement in the class of generalized analytic functions on closed Riemann surfaces.) *Izv. Vysš. Učebn. Zaved. Matematika* **11**, 114—120 (1968).

MITJUK, IGOR PETROVIČ [3] Pro kvazikonformi izobražennja oblasteĭ dovil'noĭ zvjaznosti. (Ukrainian) (On quasiconformal mappings of domains of arbitrary connectedness.) Dopovidi Akad. Nauk Ukrain. RSR 712—715 (1962); [4] Dejaki teoremi pro odnalistni kvazikonformni vidobražennaja mnogozv'jaznih oblasteĭ. (Ukrainian) (Certain theorems on univalent quasiconformal mappings of multiply connected domains.) Dopovidi Akad. Nauk Ukrain. RSR 987—989 (1962); [5] Princip simmetrizacii dlja mnogosvjaznyh oblasteĭ i nekotorye ego primenenija. (Russian) (Symmetrization principle for multiply connected domains and some of its applications.) *Ukrain. Mat. Ž.* **17**, 46—54 (1965).

MIYAHARA, YASUSHI [1] On some properties of Teichmüller mappings. *TRU Math.* **4**, 36—43(1965).

MOISIL, C. GRIGORE [9] Asupra invarianţilor sistemelor lui Vekua. (Romanian) (On the invariants of Vekua's systems.) *Com. Acad. R. P. Române* **5**, 7—11 (1955).

MORI, AKIRA [1] (On the classification of Riemann surfaces.) (Japanese) *Sugaku* **5**, 42—51 (1953); [2] On quasiconformality and pseudo-analyticity. *Sugaku* **7**, 75—89 (1955); *Trans. Amer. Math. Soc.* **84**, 56—77 (1957); [3] On an absolute constant in the theory of quasiconformal mappings. *J. Math. Soc. Japan* **8**, 156—166 (1956).

MORREY, JR. C. B. [1] On the solutions of quasilinear elliptic partial differential equations. *Trans. Amer. Math. Soc.* **43**, 126—166 (1938).

MORSE, MARSTON [1] The topology of pseudo-harmonic functions. *Duke Math. J.* **13**, 21—42 (1946); [2] Topological methods in the theory of functions of a complex variable. Ann. of Math. Studies No. 15, Princeton Univ. Press, Princeton N. J. 1947, 145 p.; Izd. Inostrannoi Literatury, Moscow 1951, 248 p.

MUGRIDGE, LARRY ROBERT [1] Conformal maps between Riemann manifolds. Lehigh Univ. 1968, 33 p. *Diss. Abstr.* **29**, 1762—B (1968).

MURTAZAEV, D. [1] Smešannaja graničnaja zadača dlja obobščennyh analitičeskih funkciĭ. (Russian) (Mixed boundary value problem for generalized analytic functions.) *Dokl. Akad. Nauk Tadžik. SSR* **9**, 3—7 (1966).

MUSAEV, K. M. [1] O nekotoryh graničnyh svoĭstvah obobščennyh analitičeskih funkciĭ. (Russian) (On certain boundary properties of generalized analytic functions.) *Dokl. Akad. Nauk SSSR* 181, 1335—1338 (1968); *Izv. Akad. Nauk Azerbaidžan. SSR Ser. Fiz.-Tehn. Mat. Nauk No.* **2**, 40—45 (1968); *Dokl. Akad. Nauk Azerbaidžan. SSR* **24**, No. 6, 8—12 (1968); [2] Nekotorye klassy obobščennyh analitičeskih funkciĭ. (Russian)

(Some classes of generalized analytic functions.) *Izv. Akad. Nauk Azerbaidžan. SSR Ser. Fiz.-Tehn. Mat. Nauk No.* **2**, 40—46 (1971); [3] Nekotorye zamečanija k toždestvu Grina. (Russian) (Some remarks on Green's identity.) *Izv. Akad. Nauk Azerbaidžan. SSR Ser. Fiz.-Techn. Mat. Nauk No.* **5**, 6, 74—79 (1971); [4] Ob odnom analoge teoremy F. i M. Riss dlja obobščennyh analitičeskih funkciĭ. (Russian) (On an analogue of the theorem of F. and M. Riesz for generalized analytic functions.) *Izv. Akad. Nauk Azerbaidžan. SSR Ser. Fiz.-Tehn. Mat. Nauk No.* **5, 6** 122—127 (1971); [5] O nekotoryh ekstremal'nyh svoĭstvah oboščennyh analitičeskih funkciĭ. (Russian) (Some extremal properties of generalized analytic functions.) *Dokl. Akad. Nauk SSSR* **203**, 289—292 (1972).

MUŠHELIŠVILI, N. I. [1] G. B. Kolosov. *Uspehi. Mat. Nauk* **4**, 279—281 (1938).

NAAS, JOSEF and TUTSCHKE, WOLFGANG [2] Über neuere Entwicklungstendenzen einer allgemeinen komplexen Analysis. In "Beiträge zur komplexen Analysis und Anwendungen in der Differentialgeometrie". Akademie Verlag, Berlin 1973 (in print).

NAATANEN, MARJATTA (VIRSU) [2] Maps with continuous characteristics as a subclass of quasiconformal maps. *Ann. Acad. Sci. Fenn. Ser. A I* **410**, 1—28 (1967).

NAKAI, MITSURU [3] On a ring isomorphism induced by quasiconformal mappings. *Nagoya Math. J.* **14**, 201—221 (1959); [4] A function algebra on Riemann surfaces. *Nagoya Math. J.* **15**, 1—7 (1959); [5] Purely algebraic characterization of quasiconformality. *Proc. Japan Acad.* **35**, 440—443 (1959); [6] Algebraic criterion on quasiconformal equivalence of Riemann surfaces. *Nagoya Math. J.* **16**, 157—184 (1960); [7] On a problem of Royden on quasiconformal equivalence of Riemann surfaces. *Proc. Japan Acad.* **36**, 33—37 (1960); [8] Royden's map between Riemann surfaces. *Bull. Amer. Math. Soc.* **72**, 1003—1005 (1966).

NAKAI, MITSURU and SARIO, LEO [1] On the classification and deformation of Riemannian spaces. Proc. Conf. Anal. Functions. Łodz 1—7. 1966.

NASIBOV, S. M. [1] Psevdoanalitičeskie funkciĭ Zolotareva i Krylova—Bersa funkciĭ pervogo roda. (Russian) (Pseudo-analytic functions of Zolotarev and Krylov—Bers functions of the first type.) *Azerbaidžan. Gos. Univ. Učen. Zap. Ser. Fiz.-Mat. Nauk* **6**, 19—25 (1968); [2] K teorii psevdoanaliticeskih funkciĭ Zolotareva—Krylova—Bersa. (Russian) (On the theory of Zolotarev—Krylov—Bers's pseudo-analytic functions.) *Azerbaidžan. Gos. Univ. Učen. Zap. Ser. Fiz.-Mat. Nauk Nr.* **2**, 67—75 (1970).

NEVANLINNA, ROLF [2] Über fastkonforme Abbildungen. *Ann. Acad. Sci. Fenn. Ser. A I* **251**/7, 1—10 (1958).

NICOLESCU, LILLY-JEANNE [1] Derivata areolară a funcţiilor de o variabilă complexă cu valori într-un spaţiu Banach. (Romanian) (Areolar derivative of functions of a complex variable with values in a Banach space.) *Com. Acad. R. P. Române*, **9**, 1007—1012 (1959).

NICOLESCU, MIRON [4] Sur les fonctions conjuguées sur une surface au sens de Beltrami. *Acad. Roy. Belg. Bull. Cl. Sci.* (5), **16** 1012—1016 (1930); [5] Cuvînt la sărbătorirea acad. S. Stoilov. (Romanian) (Speech at the feast of acad. S. Stoilov.) *An. Acad. R. P. Române. Secţ. Şti. Mat. Fiz. Chim. Ser. A* **8**, 179—186 (1958).

NIRENBERG, L. [2] On nonlinear elliptic partial differential equations and Hölder continuity. *Comm. Pure Appl. Math.* **6**, 103—156 (1953).

NOGIN, G. M. and POLOŽII, G. M. [1] Pro zadaču teorii osesimetričnogo potencialu dvuh sferičnih krugovyh disckiv. (Ucrainian) (A problem in the theory of axially symmetric potential of two spherical circular discs.) Dopovidi Akad. Nauk Ukrain. SSR Ser. A 894—897, 956 (1969).

NOSHIRO, KIYOSHI [1] A theorem on the cluster sets of pseudo-analytic functions *Nagoya Math. J.* **1**, 83—89 (1950); [2] Cluster sets of pseudo-analytic functions. *Japan J. Math.* **29**, 83—91 (1959); [3] "Cluster sets". Springer, Berlin—Göttingen—Heidelberg 1960, 135 p.

NOSHIRO, KIYOSHI and SARIO, LEO [1] "Integrated forms derived from non-integrated forms of value distribution theorems under analytic and quasiconformal mappings". Festschr. Gedachtnisfeier K. Weierstrass, Westdeutscher Verlag; Köln, Opladen 1966, 319—324.

O'BYRNE, B. [1] On Finsler geometry and applications to Teichmüller spaces. Advances in the theory of Riemann surfaces (Proc. Conf., Stony Brook, N.Y., 1969) Ann. of Math. Studies, No. 66. Princeton Univ. Press, Princeton, N.J., 1971.

OHTSUKA, MAKOTO [1] Gross's star theorems and their applications. *Ann. Inst. Fourier (Grenoble)* **5**, 1—28 1953/1954 (1955); Proc. Internat. Congr. Math. Amsterdam 1954, 2, 151—152; [2] Sur les ensembles d'accumulation relatifs à des transformations plus générales que les transformations quasi-conformes. *Ann. Inst. Fourier* (Grenoble) **5**, 29—37, 1953/1954 (1955); [3] Sur un théoréme étoilé de Gross. *Nagoya Math. J.* **9**, 191—207 (1955); [4] Sur les ensembles d'accumuation relatifs à des transformations localement pseudo-analytiques au sens de Pfluger—Ahlfors. *Nagoya Math. J.* **11**, 131—144 (1957).

OIKAWA, KOTARO [1] (Some properties of quasiconformal mappings.) (Japanese) *Sugaku* **9**, 13—14 (1957/1958).

OIKAWA, KOTARO and SARIO, LEO [1] Capacity functions. Springer, Berlin—Heidelberg—New York 1969.

ONICESCU, OCTAV [3] Propriétés topologiques de la transformation définie par une fonction uniforme de la variable complexe z. *C. R. Acad. Sci. Paris* **186**, 563—565 (1928); [4] Sur les fonctions holotopes. *C. R. Acad. Sci. Paris* **201**, 122—123 (1935); *Rev. Math. Union Interbalk.* **1**, 33—52 (1936).

OVČINIKOV, IGOR SEMENOVIČ [12] O suščestvovaniǐ otobraženiǐ na ploskosti dlja vyroždajuščihsja elliptičeskih sistem pervogo porjadka. (Russian) (On the existence of plane mappings for degenerate first order elliptic systems.) Pervyǐ Doneckii kollokvium po teor. kvazikonformnyh otobraženii i ee obobščenijam 16—22 September, 1969.

OZAWA, MITSURU [1] On Grötzsch's extremal affine mapping. *Kodai Math. Sem. Rep.* **8**, 112—114 (1956); [2] On extremal quasiconformal mappings. *Kodai Math. Sem. Rep.* **10**, 109—112 (1958); **11**, 109—123 (1959); [3] On an approximation theorem in a family of quasiconformal mappings. *Kodai Math. Sem. Rep.* **11**, 65—76 (1959).

PAHAREVA, N. O. and VIRČENKO, N. O. [1] Pro dejaki integral'ni peretverennja v klasi x^k-analitičnih funkciǐ. (Ukrainaian) (On some integral transportations in the class of x^k-analytic functions.) *Dopovidi Akad. Nauk Ukrain. SSR Nr.* **8**, 998—1003 (1962); [2] Pro integral'ni peretverennja u klasi p-analitičnih funkciǐ z harakteristikoju $p = x^k$. (Ukrainian) (Integral transforms in a class of p-analytic functions with characteristic $p = x^k$.) Dopovidi Akad. Nauk Ukrain. SSR 588—592 (1966); [3] "O formulah svjazu meždu x^k-analitičeskimi funkcijami pri različnyh značenijah k". (Russian) (On formulae of connection between x^k-analytic functions for different values of k.) Math. Phys. Kiev 1966, p. 106—114; [4] Pro dejaki kraiovi zadači dlja pivsmugi v klasi x-analitičkih funkciǐ. (Ukrainian) (Certain boundary value problems for the half-strip in the class of x-analytic functions.) *Visnik Kiïv. Univ. Nr.* **8**, 67—72 (1966); [5] Pro integral'ne zobražennja p-analitičnih funkcii z harakteristikoju $p = x^{2n}y^l$. (Ukrainian) (On integral representation of p-analytic functions with characteristic $p = x^{2n}y^l$.) *Dopovidi Akad. Nauk Ukrain. SSR Nr.* **2**, 124—127 (1967); [6] Pro odne nove integral'ne zobražennja p-analitičnih funkciǐ z harakteristikoju $p = x^k$. (Ukrainian) (On a new integral representation of p-analytical functions with the characteristic $p = x^k$.) *Dopovidi Akad. Nauk Ukrain SSR Ser. A Nr.* **8**, 686—691 (1967); [7] Pro osnovne integral'ne zobražennja p-analitičnih funkcii z harakteristikoju $p = x^k y^l$. (Ukrainian) (The fundamental integral representation of p-analytic functions with characteristic $p = x^k y^l$.) *Dopovidi Akad. Nauk Ukrain SSR Ser. A Nr.* **9**, 790—794 (1967).

PARTER, V. SEYMOUR [1] On mappings of multipy connected domains by solutions of partial differential equations. *Comm. Pure Appl. Math.* **13**, 167—182 (1960).

PASCALI, DAN [7] The areolar polynomials and the Almansi development in the plane. *Rev. Math. Pures Appl.* **6**, 451—455 (1959); *Com. Acad. R. P. Române* **10**, 257—262 (1960); [8] Funcţii areolar conjugate în plan. (Romanian) (Functions areolar conjugated in plane.) *Com. Acad. R. P. Române* **10**, 263—267 (1960); [9] Derivată F-areolară şi F-medie (Romanian) (F-areolar and F-mean derivative.) *Com. Acad. R. P. Române*, **10**, 1083—1086 (1960); [10] Monogeneitatea F şi ecuaţia lui Beltrami. (Romanian) (Monogeneity F and Beltrami equation.) *Com. Acad. R. P. Române* **12**, 631—634 (1962); [11] O nouă reprezentare a polinoamelor areolare în plan. (Romanian) (A new representation of the areolar polynomials in plane.) *Stud. Cerc. Mat.* **15**, 249—251 (1964); [12] Sur l'analyticité dans le plan. *Bull. Math. Soc. Sci. Math. Phys. R. P. Roumaine* **8**, 63 — 66 (1964); [13] Reprezentarea soluţiilor unui sistem cu derivate areolare de ordinul întîi în plan. (Romanian) (Representation of solutions of a system with areolar derivatives of first order in the plane.) *Stud. Cerc. Mat.* **15**, 343—347 (1964); Doctor's Thesis, Bucureşti 1964; [14] Monogeneitate Fedorov în plan. (Romanian) (Fedorov monogeneity in the plane.) *Stud. Cerc. Mat.* **16**, 1231—1241 (1964); [15] Funcţii analitice generalizate de ordinul *n*. (Romanian) (Generalized analytic functions of the order *n*.) *Stud. Cerc. Mat.* **17**, 1385—1390 (1965); [16] The structure of *n*-the generalized analytic functions. *Schr. Inst. Math. Dtsch. Akad. Wiss. Berlin A, No.* **8**, 197—201 (1971).

PATT, C. [1] Variation of Teichmüller and Torelli surfaces. *J. Analyse Math.* **11**, 221—247 (1963).

PEHLECKIĬ, I. D. [1] O slabeĭšeĭ uniformizirujuščeĭ dlja mnogoznačnogo kvazianalitičeskogo sootnošenija meždu rimanovym poverhnostjami. (Russian) (On the weakest uniformizer for a multivalued quasi-analytic relation between Riemann surfaces.) *Sibirsk. Mat. Ž.* **2**, 891—894 (1961); [2] Teorema Rimana dlja obobščennyh analitičeskih funkciĭ. (Russian) (Riemann theorem for generalized analytic functions.) *Perm. Gos. Univ. Učen. Zap.* **22**, 83—87 (1962); [3] O klassah funkciĭ dajuščih vnutrenie otobraženija ploskih oblasteĭ. (Russian) (On the class of functions effecting an interior mapping of plane domains.) *Perm. Gos. Univ. Učen. Zap.* **103**, 147—155 (1963); [3] Suščestvovanie i edinstvennost' rešeniĭ nekotoryh sistem differencial'nyh uravneniĭ s častnymi proizvodnymi, osuščestvljajuščih kvazikonformnye otobraženija. (Russian) (Existence and uniqueness of solutions to certain systems of partial differential equations effecting quasiconformal mappings.) *Volž. Mat. Sb.* 156—163 (1963); [5] Nekotoye klassy vnutrennyh otobraženiĭ ploskih oblasteĭ. (Russian) (Some classes of interior mappings of plane domains.) *Perm.Gos. Univ. Učen. Zap. Nr.* **131**, 21—26 (1966).

PEN, CHEN-LIAN [1] (Remarks on the paper "On a parametric representation of quasiconformal mappings.") (Chinese) *Acta Sci. Natur. Fudan* **11**, Nr. 1, 15—28 (1966).

PERTRIDIS, NICOLAS [2] Sur les transformations intérieures et les courbes pseudo-meromorphes. *C. R. Acad. Sci. Paris* **261**, 1581—1584 (1965).

PESIN, IVAN NIKOLAEVIČ [1] Metričeskie svoĭstva kvazikonformnyh otobraženiĭ. (Russian) (Metric properties of quasiconformal mappings.) Dissertation Kandidat. L'vov 1955; [2] K teorii obščih *Q*-kvazikonformnyh otobraženiĭ. (Russian) (Theory of generalized *Q*-quasiconformal mappings.) *Dokl. Akad. Nauk SSSR* **102**, 223—224 (1955); [3] Metričeskie svoĭstva *Q*-kvazikonformnyh otobraženiĭ. (Russian) (Metric properties of *Q*-quasiconformal mappings.) *Mat. Sb.* **40 (82)**, 281—294 (1956); [4] Ob odnom opredelenii kvazikonformnogo otobraženija. (Russian) (On a definition of quasiconformal mappings.) *Dopovidi ta povidomlennja L'vivs'k. Univ.* **7**, Nr. 3, 257—259 (1957); [5] Množestva ustranimyh osobennosteĭ analitičeskih funkciĭ i kvazokonformnye otobraženija. (Russian) (Sets of removable singularities of analytic functions and quasiconformal mappings.) Issledovanie po sovremennym problemam teorii funkciĭ kompleksnogo peremennogo. Gos. Izd. Fiz.-Mat. Literatury, Moscow 1960, 419—424; [6] Nekotorye zamečanija o kvazikonformnyh otobraženijah s neograničennymi harakteristikami. (Russian) (Some remarks on quasiconformal mappings with unbounded characteristics.) Pervyĭ Doneckiĭ Kollokvium po teor. kvazikonformnyh otobraženiĭ i ee obobščenijam 16—22. September, 1968; [7] Otobraženie kvazikonformnye v srednem. (Russian) (Mappings quasiconformal in mean.) *Dokl. Akad. Nauk SSR* **187**, 740—742 (1969).

PETRENKO, VIKTOR PAVLOVIČ [1] O roste *Q*-psevdomeromorfnye funkcii. (Russian) (On the growth of *Q*-pseudo-meromorphic functions.) *Dokl. Akad. Nauk SSSR* **196**, 50—52 (1971).

PETROVSKI, I. G. [1] O nekotoryh problemah teorii uravneniĭ s častnymi proizvodnymi. (Russian) (On some problems in the theory of partial differential equations.) *Uspehi Mat. Nauk* **1**, 44—70 (1946).

PFLUGER, ALBERT [1] Une propriété métrique de la représentation quasi-conforme. *C. R. Acad. Sci. Paris* **226**, 623—625 (1948); [2] Sur une propriété de l'application quasi-conforme d'une surface de Riemann ouverte. *C. R. Acad. Sci. Paris* **227**, 25—26 (1948); [3] Quasikonforme Abbildungen und logaritmische Kapazität. *Ann. Inst. Fourier (Grenoble)* **2**, 69—80 (1950); [4] Quelques théorèmes sur une classe de fonctions pseudo-analytiques. *C. R. Acad. Sci. Paris* **231**, 1022—1023 (1950); [5] Extremallängen und Kapazität. *Comment. Math. Helv.* **29**, 120—131 (1955); [6] Eine Bemerkung zur Theorie der quasikonformen Abbildungen. *Enseignement Math.* **4**, 306—307 (1958); *Aetas Soc. Helv. Natur.* **138**, 96—97 (1958); [7] Über die Äquivalenz der geometrischen und der analytischen Definitionen quasikonformer Abbildungen. *Comment. Math. Helv.* **33**,

23—34 (1959); [8] Über die Konstruktion Riemannscher Flächen durch Verheftung. *J. Indian Math. Soc.* **24**, 401—412 (1960); [9] Zu einem Verzerrungssatz der konformen Abbildungen. *Math. Z.* **84**, 263—267 (1964).

PFLUGER, ALBERT and SUTTER JOHANN [1] Riemannsche Fläche vom hyperbolischen Typus erzeugen durch Asymetrien. Meždunarodnaja Konf. po teor. analitičeskih funkciĭ. Erevan 1965, Tezisy Dokl. 51—52; in "Sovremenye problemy teorii analitičeskih funkciĭ". Izd. "Nauka", Moscow 1966, 253—257.

PICARD, EMIL [1] Sur un système d'équations aux dérivées partielles. *C. R. Acad. Sci. Paris* **112**, 685—688 (1891); [2] Sur une généralisation des équations de la théorie des fonctions d'une variable complexe. *C. R. Acad. Sci. Paris* **112**, 1399—1403 (1891).

PIRL, UDO [1] Isotherme Kurvenscharen und zugehorige Extremalprobleme der konformen Abbildungen. *Wiss. Z. Martin-Luther-Univ. Halle-Wittenberg Math.-Natur. Reihe* **4**, 1225—1252 (1955); Doctor's Dissertation Halle-Wittenberg 1955.

POKAZEEV, V. I. [1] Ob odnom predstavlenii obobščennyh analitičeskih avtomorfnyh funkciĭ. (Russian) (A representation for generalized analytic automorphic functions.) *Dokl. Akad. Nauk SSSR* **155**, 528—531 (1964); *Soviet. Math. Dokl.* **5**, 457—460 (1964); [2] Zadača Rimana-Karlemana dlja odnogo klassa obobščennyh analitičeskih funkciĭ. (Russian) (The Riemann-Carleman problem for a class of generalized analytic functions.) *Izv. Akad. Nauk Kazah. SSR Ser. Fiz.-Mat. Nauk Nr.* **1**, 72—81 (1965); [3] Ob odnom predstavlenii obobščennyh analitičeskih avtomorfnyh funkciĭ. (Russian) (A representation for generalized analytic automorphic functions.) *Dokl. Akad. Nauk SSSR* **155**, 528—531 (1964); [4] Obobščennye avtomorfnye funkcii. (Russian) (Generalized automorphic functions.) *Sibirsk. Mat. Ž.* **9**, 880—890 (1968).

POLACZEK, F. [1] Sur les polynômes aréolaires. *Bul. Fac. Sti. Cernăuţi* **12**, 328—336 (1938).

POLOŽIĬ, G. M. [1] O *p*-analitičeskih funkcijah kompleksnogo peremennogo. (Russian) (On *p*-analytic functions of a complex variable.) *Dokl. Akad. Nauk SSSR* **58**, 1275—1278 (1947); [2] Osobye točki i vyčety *p*-analitičeskih funkciĭ kompleksnogo peremennogo. (Russian) (Singular points and residues of *p*-analytic functions of a complex variable,) *Dokl. Akad. Nauk SSSR* **60**, 769—772 (1948); [3] O primenenijah obobščennoĭ proizvodnoĭ na klasse kvazikonformnyh otobraženiĭ. (Russian) (On the application of generalized derivative in the class of quasiconformal mappings.) *Dokl. Acad. Nauk SSSR* **63**, 615—618 (1948); [4] Obobščenie integral'noĭ formuly Koši. (Russian) (Generalization of Cauchy integral formula.) *Mat. Sb.* **24 (66)**, 375—384 (1949); [5] Teorema o sohranenii oblasti dlja nekotoryh elliptičeskih sistem differential'nyh uravneniĭ i ee primenenija. (Russian) (Theorem on preservation of domain for elliptic systems of differential equations and its applications.) *Mat. Sb.* **32 (74)**, 485—492 (1953); [6] Teorema o sootvestvii granic i variacionnye teoremy dlja nekotoryh elliptičeskih sistem differential'nyh uravneniĭ. (Russian) (Theorem of boundary correspondence and variational theorems for certain elliptic systems of differential equations.) *Dokl. Akad. Nauk SSSR* **95**, 927—930 (1954); [7] Integrirovanie po soprajažennym peremennym. (Russian) (Integration with respect to conjugate variables.) Trudy III Vsesojuzn. S'ezda Matematikov. Izd. Akad. Nauk SSSR, Moscow 1956, I, 95—96; [8] Metod *p*-analitičnih funkciĭ v osesimmetričii teorii pružnosti. (Ukrainian) (The method of *p*-analytic functions in the axially symmetric theory of elasticity.) *Nauk. Ščioričnik Kiĭvsk. Univ.* 1956, 529—530 (1957); [9] K voprosu o (*p*, *q*)-analitičeskih funkcijah kompleksnogo peremennogo i ih primenenijah. (Russian) [On the question of (*p*, *q*)-analytic functions of a complex variable and their applications.) *Rev. Roumaine Math. Pures Appl.* **2**, 331—363 (1957); [10] Pro odne integral'ne peretvorennja uzal'nenih analitičnih funkciĭ. (Ukrainian) (On an integral representation of generalized analytic functions.) *Nauk. Ščioričnik Kiĭvtk. Univ.* 1957, 340—341 (1958); *Visnik Kiĭv. Univ. Nr.* **1**, 19—29 (1959); [11] O (*p*, *q*)-analytičeskih funkcijah kompleksnogo peremennogo i nekotoryh ih primenenijah. (Russian) [(*p*, *q*)-analytic functions of a complex variable and some of their applications.] Issledovanie po sovremennym problemam teorii funkciĭ kompleksnogo peremennogo. Gos. Izd. Fiz.-Mat. Literatury, Moscow 1960, p. 483—515; [12] O kraevyh zadačah osesimmetričnoĭ teorii uprugnosti. Metod *p*-analytičeskih funkcii kompleksnogo peremennogo. (Russian) (Boundary value problems in the axis-symmetric theory of elasticity. The method of *p*-analytic functions of a complex variable.) *Ukrain. Mat. Ž.* **15**, 25—45 (1963); [13] O primenenijah

p-analitičeskih i (p, q)-analitičeskih funkciĭ. (Russian) [Applications of p-analytic and (p, q)-analytic functions.] Proc. Internat. Sympos. "Appl. Theory of Functions in Continuum Mechanics" Tbilisi 17—23 September 1963, Nauka, Moscow, 1, 309—326 (1965); [14] O predstavlenii p-analytičeskih funkciĭ v vide lineĭnyh kombinacii analitičeskih funkciĭ i ih pervyh i vtoryh proizvodnyh. (Russian) (On the representation of p-analytic functions as linear combination of analytic functions and their first and second derivatives.) *Sibirsk. Mat. Ž.* **5**, 1326—1332 (1964); [15] Zauvažennja do osnovnogo integral'nogo zobražennja p-analitičnih funkciĭ z harakteristikoju $p = x^k$. (Ukrainian) (Remarks on the fundamental integral representation of p-analytic functions with characteristic $p = x^k$.) Dopovidi Akad. Nauk Ukrain SSR, 839—841 (1964); *Vyčisl. Prikl. Mat. (Kiev) Nr.* **4**, 145—151 (1967); [16] Pro zobražennja p-analitičnih funkciĭ čerez analitični funkciĭ ta ih pohidni. (Ukrainian) (On the representation of p-analytic functions in terms of analytic functions and their derivatives.) *Dopovidi Akad. Nauk Ukrain. RSR* 986—991 (1964); [17] Pro zobražennja p-garmončnih funkciĭ čerez garmonični funkciĭ ta ih pohidni. (Ukrainian) (On representation of p-harmonic functions in terms of harmonic functions and their derivatives.) *Dopovidi Akad. Nauk. Ukrain. SSR*, 1277—1280 (1964); [18] Pro rozv'jazuvannja kraiovih zadač p-analitičnih funkciĭ z harakteritikoju $p = x^k$ v bipoljarnih i v vidozminenih bipoljarnih koordinatah. (Ukrainian) (On solving boundary value problems by the use of p-analytic functions with characteristic $p = x^k$ in bipolar and modified bipolar coordinates.) *Visnik Kiïv. Univ. Ser. Mat.-Meh.* **6**, 3—13 (1964); [19] "Uravnenja matematičeskoĭ fizike". (Russian) (The equations of mathematical physics.) Vyssaja Skola, Moscow 1964, 559 p.; [20] Ob osnovnom integral'nom predstavlenii p-analitičeskih funkciĭ s harakteristikoĭ $p = x^k$. (Russian) (A fundamental integral representation for p-analytic functions with characteristic $p=x^k$.) *Ukrain. Mat. Ž.* **16**, 254—259 (1964); [21] O predel'nyh značenijah i formulah obraščenija vdol' razrezov osnovnogo integral'nogo predstavlenija p-analitičeskih funkciĭ s harakteristik $p = x^k$. I, II. (Russian) (Limit values and inversion formulae along the cuts of the fundamental integral representation of p-analytic functions with characteristic $p = x^k$.) *Ukrain. Mat. Ž.* **16**, 631—656 (1964); **17**, 61—87 (1965); [22] "Obobščenie teorii analitičeskih funkciĭ kompleksnogo peremennogo. p-analitičeskie i (p, q)-analitičeskie funkciĭ i nekotorye ih primenenija". (Russian) [Generalization of the theory of analytic functions of a complex variable. p-analytic and (p, q)-analytic functions and some of their applications.] Izd. Kievskogo Univ., Kiev 1965, 442 p.; [23] Die Lösung von Randwertaufgaben p-analytischer Funktionen mit der Charakteristik $p = x^k$ für gewisse kanonische Gebiete. *Wiss. Z. Karl-Marx-Univ. Leipzig Math.-Natur. Reihe* **14**, 475—495 (1965). [24] Zamečanie k odnomu klassu p-analitičeskih funkciĭ, svjazannyh s osesimmetričnymi zadačami teorii uprugosti. (Russian) [Remarks on a class of p-analytic functions connected with axially symmetric problems in the theory of elasticity.] *Prikladna. Meh.* **2**, Nr. 11, 123—126 (1966); [25] O formulah obraščenija osnovnogo integral'nogo predstavlenija p-analitičeskih funkciĭ ot $z=x+iy$ s harakteristikoĭ $p=x^k$. (Russian) (On inversion formulas for a fundamental integral representation of p-analytic functions of $z=x+iy$ with the characteristic $p=x^k$.) *Dokl. Akad. Nauk SSSR* **177**, 40—43 (1967); [26] Pro formuli obrenennja osnovnogo integral'nogo zobražennja x^k-analitičnih funkciĭ vid $z = x + iy$ pri cilih neparnih k. (Ukrainian) (On inversion formulas for a fundamental integral representation of x^k-analytic functions of $z = x + iy$, k any odd integer.) *Dopovidi Akad. Nauk Ukrain RSR Ser. A* 315—319 (1967); [27] Pro p-analitični funkciĭ z logarifmično-garmoničnoju harakteristikoju p. (Ukrainian) (On p-analytic functions with a logarithmically-harmonic characteristic p.) *Dopovidi Akad. Nauk Ukrain RSR Ser. A* 503—506 (1967); *Vyčisl. Prikl. Mat. (Kiev) Nr.* **5**, 24—37 (1968); [28] O formulah obraščenija osnovnogo integral'nogo predstavlenija p-analitičeskih funkciĭ ot $z = x + iy$ s harakteristikoĭ $p = x^k$. (Russian) (Inversion formulae for the fundamental integral representation of p-analytic functions of $z = x + iy$ with characteristic $p = x^k$). *Dokl. Akad. Nauk SSSR* **177**, 40—43(1967); [29] Rešenie kraevyh zadač p-analitičeskih funkciĭ s harakteristikoĭ $p = x^k$, dlja nekotoryh kanoničeskih oblasteĭ. (Russian) (Solution of boundary value problems for p-analytic functions with characteristic $p = x^k$ for certain canonical domains.) *Vestnik Kievsk. Univ. Spec. vyp. naučn. stateĭ učenyh Kievsk. i Leningradsk. Univ. Kiev* 131—166 (1967); [30] O formulah obraščenija vdol' proizvol'nyh konturov osnovnogo integral'nogo predstavlenija p-analitičeskih funkciĭ. (Russian) (Inversion formulas along arbitrary contours of the fundamental integral representation of p-analytic functions.) *Ukrain. Mat. Ž.* **19**, 87—95 (1967); [31] Affinnye preobrazovanija p-analitičeskih i (p, q)-analitičeskih funkciĭ kompleksnogo peremennogo. (Russian) (Affine transformations of p-analytic and (p, q)-analytic functions of a complex variable.) *Ukrain. Mat. Ž.* **20**, 325—339 (1968); [32] Pro dejaki pitannja pov'jazani z afinnimi peretvorennjami p-analitičnih i (p, q)-analitičnih funkciĭ. (Ukrainian) (On some questions connected with affine transformations of p-analytic and (p, q)-analytic functions.) *Visnik Kiïv. Univ. Ser. Mat.-Meh.* **10**, 7—19 (1968); [33] Osnovnoe integral'noe predstavlenie p-analitičeskih funkciĭ s harakteristikoĭ $p = e^{\alpha x + \beta y}$ i formuly ego obraščenija. (Russian) (The fundamental integral representation of p-analytic functions with characteristic $p = e^{\alpha x + \beta y}$ and its inversion formulas.) In "Voprosy teorii i istorii differencial'nyh uravneniĭ". Izd. Naukova Dumka, Kiev 1968, p. 65—80; [34] O rešenii smešannoĭ kraevoĭ zadači teorii osesimmetričnoĭ potenciala dlja prostranstvennogo sloja. (Russian) (On the solution of the mixed boundary value problem in the theory of axially symmetric potential for spatial shell.) *Mat. Fiz. Resp. Mežved. Sb. Nr.* **5**, 162—168 (1968); [34] Ob integral'nyh predstavlenijah x^k-analitičeskih funkciĭ i rešenii osesimmetričnyh zadač teorii uprugosti. (Russian) (On integral representation of x^k-analytic functions and the solution of axially symmetric problems in the theory of elasticity.) *Prikladna. Meh.* **5**, Nr. 4, 1—17 (1969); [35] Pro zadaču teorii osesimmetričnogo potencialu dvux sferičnih krugovyh diskiv. (Ukrainian) (On the theory of the axially symmetric potential of two spheric circular disks.) *Dopovidi Akad. Nauk Ukrain. SSR Ser. A* 894—898 (1969).

POLOŽIĬ, G. M. and ULITKO, A. F. [1] O formulah obraščenija osnovnogo integral'nogo predstavlenija p-analitičeskogo funkciĭ s harakteristikoĭ $p = x^k$. (Russian) (On inversion formulae for the fundamental integral representation of p-analytic functions with characteristic $p = x^k$.) *Prikladna. Meh.* **1**, 39—51 (1965).

POMPEIU, DIMITRIE [1] Sur une classe de fonctions d'une variable complexe. *Rend. Circ. Mat. Palermo* **33**, 108—113 (1912); [2] Sur une classe de fonctions de variable complexe et sur certaines équations intégrales. *Rend. Circ. Mat. Palermo* **35**, 277—281 (1913).

PONOMAREV, S. P. [1] Ob odnom uslovii kvazikonformnosti. (Russian) (On a condition of quasi-conformality.) *Mat. Zametki* **9**, 663—666 (1971).

POOR, VINCENT [1] Residues of polygenic functions. *Trans. Amer. Math. Soc.* **32**, 216—222 (1930); *Bull. Amer. Math. Soc.* **36**, 57 (1930); [2] On circular functions. *Amer. J. Math.* **61**, 833—842 (1939); [3] On the two-dimensional derivative of a complex function. *Proc. Amer. Math. Soc.* 687—693 (1950); [4] On residues of polygenic functions. *Trans. Amer. Math. Soc.* **75**, 244—255 (1953).

POPOV, V. S. [1] Quelques propriétés des fonctions d'une variable complexe. *Bilten Drustvočno Mat. Fiz. Nar. Rep. Makedonija* **4**, 20—24 (1953); [2] Paraanalitične funkciĭ so dve dimenciĭ. (Macedonian) (Bidimensional para-analytic functions.) *Fac. Philos. Univ. Skopje. Sect. Sci. Nat. Annuaire* 1953, **6**, 3—29 (1954).

POTJAGAILO, D. B. [1] O tipe skleĭvanija polosy. (Russian) (On a type of sewing bands.) *Dokl. Akad. Nauk SSSR* **138**, 1025—1028 (1961). [2] O rimanovyh poverhnostjah v evklidovom prostranstve. (Russian) (On Riemann surfaces in euclidean space.) *Sibirsk. Mat. Ž.* **9**, 632—638 (1968).

POVOLOCKIĬ, A. I. [1] Indeksy osobyh toček psevdoanalitičeskih funkciĭ. (Russian) (Indices of singular points of pseudo-analytic functions.) *Dokl. Akad. Nauk SSSR* **129**, 265—267 (1959).

PROTER, M. H. [1] The periodicity problem in the theory of pseudo-analytic functions. *Ann. of Math. (2)*, **64**, 154—174 (1956).

PYLE, H. RANDOLF [1] Conformal mapping of surfaces. *Duke Math.* J. **11**, 369—371 (1944).

RADU, ANTON [1] La torsion des barres élastiques non-homogènes. *An. Şti. Univ. "Al. I. Cuza" Iaşi Secţ. I-a Mat.* **12**, 197—204 (1966); [2] Problema lui Saint-Venant pentru bare neomogene. (Romanian) (Saint-Venant problem for non-homogeneous bars.) *An. Şti. Univ. "Al. I. Cuza" Iaşi Secţ. I-a Mat.* **12**, 415—428 (1966); [3] Echilibrul barelor elastice izotrope neomogene. (Romanian) (Equilibrium of non-homogeneous isotropic elastic bars.) *An. Şti. Univ. "Al. I. Cuza" Iaşi Secţ. I-a Mat.* **13**, 421—442 (1966); [4] Sisteme elastice antiplane neomogene. (Romanian) (Non-homogeneous anti-plane elastic systems.) Doctor's Dissertation, Iaşi 1966; *An. Şti. Univ. "Al. I. Cuza" Iaşi Secţ. I-a Mat.* **13**, 145—159 (1967); [5] Sur la déformation plane d'un corps élastique isotrope non-homogène. *Bull. Polonais Acad. Sci. Techn.* **16**, 91—99 (1968).

RADŽABOV, N. P. [1] Integral'nye predstavlenija i ih obraščenie dlja obobščennoĭ sistem Koši-Rimana s singuljarnoĭ linieĭ. (Russian) (Integral representations and their inversion for a generalized Cauchy-Riemann system with a singular line.) *Dokl. Akad. Nauk Tadžik. SSR* **11**, 14—18 (1968).

RAUCH, H. E. [1] On the transcendental moduli of algebraic Riemann surfaces. *Proc. Nat. Acad. Sci. USA* **41**, 42—49 (1955); [2] On the moduli of Riemann surfaces. *Proc. Nat. Acad. Sci. USA* **41**, 236—238 (1955); [3] The first variation of the Doughlas functional and the period of Abelian integral of the first kind. Sem. Analyt. Functions, Princeton, New Jersey. Inst. Advanced Study 2, 49—54 (1958); [4] Weierstrass points, branch points and the moduli of Riemann surfaces. *Comm. Pure Appl. Math.* **12**, 543—560 (1959); [5] Variational methods in the problem of the moduli of Riemann surfaces. Contrib. Function Theory, Internat. Colloquium Bombay 1960, p. 17—40; [6] A transcendental view of the space of algebraic Riemann surfaces. *Bull. Amer. Math. Soc.* **71**, 1—39 (1965); Errata **74**, 767 (1968).

READE, MAXWELL [2] On areolar monogenic functions. *Bull. Amer. Math. Soc.* **53**, 98—103 (1947); [3] A theorem of Fedoroff. *Duke Math. J.* **18**, 105—109 (1951).

REED, JOHN TERENCE [1] Quasiconformal mappings with given boundary values. *Duke. Math. J.* **33**, 459—464 (1966); [2] On the boundary correspondence of quasiconformal mappings. Doctor's Dissertation University of Minnesota 1966, 127 p.; *Dissert. Abstr.* **27**, 892—B (1967); [3] The boundary correspondence of quasiconformal mappings on quasicircles. *Notices Amer. Math. Soc.* **15**, 157 (1968); [4] On the boundary correspondence of quasiconformal mappings of domains bounded by quasicircles. *Pacific. J. Math.* **28**, 653—661 (1969).

REICH, EDGAR [1] On a characterization of quasiconformal mappings. *Comment. Math. Helv.* **37**, 44—48 (1962); Proc. Internat. Congr. Math. Djursholm 15—22 August, 1962, Upsala 1963; [2] Sharpened distorsion theorems for quasiconformal mappings. *Notices Amer. Math. Soc.* **10**, 81 (1963); [3] On the behaviour of quasiconformal mappings at a point. *Notices Amer. Math. Soc.* **11**, 233 (1964); [4] Some remarks on the two-dimensional Hilbert transformation. Colloquium on Math. Analysis. Otaniemi (Finland) 27—31 August, 1966; [5] Quasiconformal mappings which keep boundary points fixed. *Notices Amer. Math. Soc.* **15**, 227; 539 (1968); [6] On the extremality of certain Teichmüller mappings. Colloquium on Math. Analysis Jyväskylä 1970.

REICH, EDGAR and STREBEL, KURT [1] On quasiconformal mappings which keep boundary points fixed. *Trans. Amer. Math. Soc.* **138**, 211—222 (1969); [2] Einige Klassen Teichmüllerscher Abbildungen, die die Randpunkte festhalten. *Ann. Acad. Sci. Fenn. Ser. A I* **457**, 1—19 (1970); [3] On the extremality of certain Teichmüller mappings. *Comment. Math. Helv.* **45**, 353—362 (1970); [4] Teichmüller mappings which keep the boundary pointwise fixed. Advances in the theory of Riemann surfaces (Proc. Conf., Stony Brook, N.Y., 1969) 365—367. *Ann. of Math. Studies*, No. 66. Princeton Univ. Press, Princeton N.J., 1971.

REICH, EDGAR and WALCZAK, R. HUBERT [1] On the behaviour of quasiconformal mappings at a point. *Trans. Amer. Math. Soc.* **117**, 338—351 (1965).

RENELT, HEINRICH [1] Über quasikonforme Abbildungen merfach zusammenhängender Gebiete durch Lösungen elliptischer Differentialgleichungssysteme. *Ann. Univ. Mariae Curie-Sklodowska Sect. A*, **22—24**, 155—160 (1968—1970); [2] Modifizierung und Erweiterung einer Schifferschen Variationsmethode bei quasikonformer Abbildungen. Conf. "Complexe Analysis und deren Anwendungen" Göhren (East Germany) 28 September—2 October 1972; *Math. Nachr.* (in print).

RENGGLI, HEINZ [12] Zur Definition der quasikonformer Abbildungen. *Comment. Math. Helv.* **34**, 222—226 (1960); [3] Quasiconformal mappings and extremal lengths. Univ. of New Mexico Techn. Reports 10 (1962); *Amer. J. Math.* **86**, 63—69 (1964), [4] On modification of Riemann surfaces. *Ark. Mat.* **6**, 299—306 (1965); [5] Extremallängen und eine konforme invariante Massfunktion fur Kurvenscharen. *Comment. Math. Helv.* **41**, 10—17 (1966); [6] Doppelverhältnisse und quasikonforme Abbildungen. *Comment.*

Math. Helv. **43**, 171—175 (1968); [7] Triangular dilatation and quasiconformal mappings. *Notices Amer. Math. Soc.* **15**, 514 (1968); [8] Slight modifications of Riemann surfaces and quasiconformal mappings. *Math. Ann.* **176**, 39—44 (1968).

RICKMAN, SEPPO [5] Characterization of quasiconformal arcs. Colloquium on Math. Analysis. Otaniemi (Finland) 27—31 August, 1966; *Ann. Acad. Sci. Fenn. Ser. A I*, **395**, 1—30 (1966); [6] Extension over quasiconformally equivalent curves. *Ann. Acad. Sci. Fenn. Ser. A I*, **436**, 1—12 (1969); [7] Quasiconformally equivalent curves. *Duke Math. J.* **36**, 387—400 (1969).

RIDDER, J. [1] Über areolar harmonische Funktionen. *Acta Math.* **78**, 205—289 (1946); [2] Über areolar-monogene Funktionen. *Niew Arch. Wisk.* (2), **22**, 200—206 (1946); [3] Einige einfache Anwendungen der areolären Ableitungen und Derivirten. *Proc. Akad. Wet. Amsterdam* **50**, 151—196 (1947).

RODIN, JU. L. [1] Integral tipa Koši i kraevye zadači dlja obobščennyh analitičeskih funkcií na zamknutyh rimanovyh poverhnostjah. (Russian) (Integral of Cauchy type and boundary-value problems for generalized analytic functions on closed Riemann surfaces.) *Dokl. Akad. Nauk SSSR* **142**, 798—801 (1962); [2] Algebraičeskaja teorija obobščennyh analitičeskih funkcií na zamknutyh rimanovyh poverhnostjah. (Russian) (Algebraic theory of generalized analytic functions on closed Riemann surfaces.) *Dokl. Akad. Nauk* SSSR **142**, 1030—1033 (1962); [3] K voprosu sootnošenií meždu obobščennymi analitičeskimi i psevdoanalitičeskimi funkcijami. (Russian) (On the question of relations between generalized analytic and pseudo-analytic functions.) *Perm. Gos. Univ. Učen. Zap.* **22**, 49—51 (1962); [4] Teorema suščestvovanija dlja obobščennyh analitičeskih funkcií. (Russian) (Existence theorem for generalized analytic functions.) *Perm. Politehn. Inst. Sb. Naučn. Trudov* **13**, 41—42 (1963); [5] ėllipticeskie sistemy na neorientiruemyh poverhnostjah. (Russian) (Elliptic systems on non-orientable surfaces.) *Perm. Gos. Univ. Učen. Zap.* **103**, 64—65; 199—200 (1963); [6] K algebraičeskoí teorii kvazianalitičeskih funkcií. (Russian) (On the algebraic theory of quasi-analytic functions.) *Perm. Gos. Univ. Učen. Zap.* **103**, 66—70 (1963); [7] Prostranstva kvazianalitičeskih differencialov na otkrytyh rimanovyh poverhnostjah. (Russian) (Space of quasi-analytic differentials on open Riemann surfaces.) *Soobšč. Akad. Nauk Gruzin. SSR* **42**, 17—22 (1966); [8] K teorii mnogoznačnyh obobščennyh analitičeskih funkcií. (Russian) (On the theory of many-valued generalized analytic functions.) *Sakhart. SSR Mecn. Akad. Moambe* (Soobšč Akad. Nauk Gruzin. SSR) **43**, 261—268 (1963).

RODIN, JU, L. and VOLKOVYSKIÍ, L. I. [1] Obobščennye analitičeskie funckii na zamknutye rimanovye poverhnosti. (Russian) (Generalized analytic functions on closed Riemann surfaces.) Outlines Joint Sympos. Partial Differential Equations. Novosibirsk 1963, Moscow 1963, p. 369—373.

ROGOŽIN, V. S. [1] Kraevye zadači dlja obobščennyh analitičeskih funkcií v prostranstve funkcionalov. (Russian) (Boundary value problems for generalized analytic functions in a space of functionals.) *Sibirsk. Mat. Ž.* **8**, 115—122 (1967).

ROȘCULEȚ, M. N. [2] Fonctions polygènes dans les algèbres linéaires associative et commutative. *C. R. Acad. Sci. Paris* **242**, 51—52 (1956); [3] Funcţii monogene pe spaţii Riemann. I. Funcţii monogene pe o suprafaţă în spaţiu. (Romanian) (Monogenic functions in Riemann spaces. I. Monogenic functions on a surface in the space.) *Acad. R. P. Romîne Stud. Cerc. Mat.* **14**, 519—532 (1963).

ROYDEN, L. HALSEY [1] A property of quasiconformal mappings. *Proc. Amer. Math. Soc.* **5**, 266—269 (1954); [2] Behaviour of the derivative of a pseudo-analytic function. *Notices Amer. Math. Soc.* **97**, 393 (1967); [3] The non-uniqueness of extremal quasiconformal maps. *Notices Amer. Math. Soc.* **15**, 540 (1968); [4] On the Teichmüller metric. "Proceedings of the Romanian-Finnish Seminar on Teichmüller spaces and quasiconformal mappings. Brașow 26—30 August, 1969". Publishing House of the Academy of the Socialist Republic of Romania, București 1971, 273—286; [5] Report on the Teichmüller metric. *Proc. Nat. Acad. Sci. USA* **65**, 497—499 (1970). [6] Automorphisms and isometries of Teichmüller surfaces. Advances in the theory of Riemann surfaces (Proc. Conf., Stony Brook, N.Y., 1969) *Ann. of Math. Studies, No. 66*. Princeton Univ. Press, Princeton N. J., 1971 369—383.

RUNG, D. C. [1] Behaviour of the derivative of a pseudoanalytic function. *Notices Amer. Math. Soc.* **97**, 393 (1967).

RUSCHEWEYH, STHEPHAN [1] Gewisse Klassen verallgemeinerter analytischer Funktionen. *Diss. Bonn. Math. Schr. Nr.* **39**, 1—79 (1969).

ŠABAT, B. V. [10] O lineĭnyh klassah kvazikonformnyh otobraženii. (Russian) (On linear classe of quasiconformal mappings.) Doctor's Dissertation Moscow Univ. 1944; [11] Ob obobščennyh rešenijah odnoĭ sistemy uravneniĭ v častnyh proizvodnyh. (Russian) (On generalized solutions of a system of partial differential equations.) *Mat. Sb.* **17 (59)**, 193—209 (1945); [12] Teorema i formula Koši dlja kvazikonformnyh otobraženiĭ lineĭnyh klassov. (Russian) (Cauchy's theorem and formula for quasiconformal mappings of linear classes.) *Dokl. Akad. Nauk SSSR* **69**, 305—308 (1949); [13] Ot otobraženijah, osuščestvljaemyh rešenjami sistemy Karlemana. (Russian) (Mappings realisable as solutions of Carleman system.) *Uspehi Mat. Nauk* **11**, Nr. 3 (69), 203—206 (1956); [14] Ob analoge teoremy Rimana dlja lineĭnyh giperboličeskih sistem differencial'nyh uravneniĭ. (Russian) (On an analogue of Riemann's theorem for linear hyperbolic systems of differential equations.) *Uspehi Mat. Nauk* **11**, Nr. 5 (71), 101—105 (1956); [15] Über Abbildungen, die durch Lösungen Systeme partieller Differentialgleichungen erster Ordnung verwirklicht werden. *Ann. Acad. Sci. Fenn. Ser. A I,* **251**/8, 1—9 (1958); [16] Geometričeskiĭ smysl uslovija elliptičnosti sistem uravneniĭ s častnymi proizvodnymi. (Russian) (Geometric interpretation of the concept of ellipticity of a system of partial differential equations.) *Uspehi Mat. Nauk* **12**, Nr. 6 (78), 181—188 (1957); [17] Obobščenja i analogi teorii analitičeskih funkciĭ. (Russian) (Generalizations and analogies of the theory of analytic functions.) In "Matem. v SSSR za 40 let, 1917—1957". Fizmatgiz, Moscow 1959, 481—493; [18] Ob otobraženijah osuščestvjaemyh rešenjami nelineinyh sistem uravnenii s častnymi proizvodnymi. (Russian) (Mappings realisable as solutions of non-linear systems of partial differential equations.) Issledovanie po sovremennym problemam teor. funkciĭ kompleksnogo peremennogo. Gos. Izd. Fiz.-Mat. Literatury, Moscow, 1960, 451—461; [19] K ponjatiju proizvodnoĭ sistemy v smysle M. A. Lavrent'eva. (Russian) (On the concept of the derived system in the sens of M. A. Lavrent'ev.) *Dokl. Akad. Nauk SSSR* **136**, 1298—1301 (1961); [20] Nekotorye zadači i rezultaty teorii otobraženiĭ. (Russian) (Some problems and results of the theory of mappings.) Pervyĭ Doneckiĭ Kollokvium po teor. kvazikonformnyh otobraženiĭ i ee obobščenijam 16—22 September 1968; [21] O giperboličeskih kvazikonformnyh otobraženijah. (Russian) (On hyperbolic quasiconformal mappings.) Nekotorye problemy Mat. i. Meh. Akad. Nauk SSSR, Sibirsk. Otdel., Novosibirsk 1970, 251—266.

SAKAI, AKIRA [1] Existence of pseudo-analytic differentials on Riemann surfaces. I, II. *Proc. Japan Acad.* **39**, 1—6; 7—9 (1963).

SAKAI, EIICHI [1] Note on the pseudo-analytic functions. *Proc. Japan Acad.* **25**, 12—17 (1949).

SAMSONIJA, Z. B. [2] K približennomu kvazikonformnomu otobraženiju oblasti na krug. (Russian) (On approximate quasiconformal mappings of a domain onto a disc.) *Trudy Gruzin. Politehn. Inst. No.* 7, (147), 80—88 (1971).

ŠAPIRO, Z. JA. [1] O suščestvovaniĭ kvazikonformnyh otobraženiĭ. (Russian) (On the existence of quasiconformal mappings.) *Dokl. Akad. Nauk SSSR* **30**, 685—687 (1941); [2] Ob obščih kraevyh zadačah dlja uravneniĭ elliptičeskogo tipa. (Russian) (On general boundary value problems for equations of elliptic type.) *Izv. Akad. Nauk SSSR Ser. Mat.* **17**, 539—562 (1953).

SARKISJAN, S. C. [1] Svoĭstva rešeniĭ sistem Koši—Rimana s nelineinymi pravymi častjami. (Russian) (Properties of solutions of Cauchy—Riemann systems with non-linear right-hand sides.) *Akad. Nauk Armjan. SSR Dokl.* **36**, 141—146 (1963).

SASAKI, TAKEHIKO [1] On some extremal quasiconformal mappings of disc. *Osaka J. Math.* **7**, 527—534 (1970).

SCHATZ, A. [1] On the local behaviour of homeomorphic solutions of Beltrami's equation. *Duke Math. J.* **35**, 289—306 (1968).

SCHECHTER, MARTIN [1] A free boundary problem for pseudo-analytic functions. *Proc. Amer. Math. Soc.* **10**, 881—887 (1959).

SCHIFFER, M. [1] A variational method for univalent quasiconformal mappings. *Duke Math. J.* **33**, 395—411 (1966).

SCHMIDT, W. [1] Über das Verhalten der Lösungen eines verallgemeinerten Cauchy—Riemannschen Differentialgleichungssystems in der Umgebung von isolierten Singularitäten der Koeffizienten. *Wiss. Z. Hochsch. Elektrotechn. Ilmenau* **7**, 221—232 (1961).

SCHOBER, GLENN [1] Continuity of curve functionals and a technique involving quasiconformal mappings. *Notices Amer. Math. Soc.* **15**, 533—534 (1968); *Arch. Rational Mech. Anal.* **29**, 378—389 (1968); [2] Semicontinuty of curve functionals. *Arch. Rational Mech. Anal.* **33**, 347—376 (1969).

SEDOV, L. I. [1] "Ploskie zadači gidrodinamiki i aèrodinamiki". (Russian) (Hydrodynamic and aerodynamic problems in plane.) Gosudarstv. Izdat. Tehn. Teor. Lit. Moscow—Leningrad 1950, 443 p.; Izd. Nauka, Moscow 1966, 448 p.

SEPINEN, FREDERICK [1] Relationships between modules of the two and three dimensional Teichmüller ring domains. *Notices Amer. Math. Soc.* **18**, 153 (1970).

SERBIN, A. I. [1] O kraevoĭ zadače Rimana dlja obobščennyh analitičeskih funkciĭ na zamknutoĭ rimanovoĭ poverhnosti. (Russian) (On Riemann boundary value problem for generalized analytic functions on closed Riemann surfaces.) *Izv. Vysš. Učebn. Zaved. Matematika* **131—143** (1964).

ŠERETOV, V. G. [1] O suščestvovanii kvazikonformnyh otobraženiĭ. (Russian) (On the existence of quasiconformal mappings.) *Perm. Gos. Univ. Učen. Zap.* **103**, 116—120 (1963); [2] Invariantnye podprostranstva v prostranstve Teichmjullera. (Russian) (Invariant subspaces in a Teichmüller space.) *Perm. Gos. Univ. Učen. Zap.* **103**, 156—159 (1963); [3] Ob ekstremal'nyh kvazikonformnyh otobraženijah s ograničeniem na harakteristiku. (Russian) (Extremal quasiconformal mappings with restriction on the characteristic.) *Dokl. Akad. Nauk SSSR* **179**, 1060—1063 (1968); [4] O kvazikonformnyh otobraženijah s lokal'no trivial'noĭ harakteristik. (Russian) (Quasiconformal mappings with locally trivial characteristic.) *Sibirsk. Mat. Ž.* **10**, 223—228 (1969); [5] O podmnogoobrazijah v prostranstve Teichmjullera (Russian) (On submanifolds in Teichmüller space.) *Perm. Gos. Univ. Učen. Zap.* **218**, 90—98 (1969).

SETHARES, G. C. [1] The extremal property of certain Teichmüller mappings. *Comment. Math. Helv.* **43**, 98—119 (1968).

ŠEVELEVA, M. V. [1] O formule Švarca dlja uravnenija $\frac{\partial w}{\partial \bar{z}} + b\bar{w} = 0$. (Russian) (On Schwartz formula for the equation $\frac{\partial w}{\partial \bar{z}} + b\bar{w} = 0$.) *Dokl. Akad. Nauk SSSR* **177**, 531—534 (1967).

SHAH, DAO-SHING [1] Parametric representation of quasiconformal mappings. *Sci. Record* **3**, 400—407 (1959).

SHEFFER, I. M. [1] Note on non-analytic functions. *Bull. Amer. Math. Soc.* **41**, 367—370 (1935).

SHIBATA, KEÎCHI [1] Remarks on the sequence of quasiconformal mappings. *Proc. J. Acad.* **32**, 665—670 (1956); [2] On boundary values of some pseudo-analytic functions. *Proc. J. Acad.* **33**, 628—632 (1957).; [3] On approximation of quasiconformal mapping. *Proc. J. Acad.* **35**, 22—24 (1959).

SHUBERT, SANFORD ROY [1] A new class of generalized analytic functions. Doctor's Dissertation Illinois Univ. 1963, 32 p.; *Dissert. Abstr.* **24**, Nr. 9, 3774—3775 (1964).

SIBNER, J. ROBERT [1] Some examples for the Koebe conjecture. Proc. Conf. Analytic Functions Łodz 1—7 September, 1966; [2] Hyperbolic generators for Fuchsian groups. *Proc. Amer. Math. Soc.* **17**, 963—968 (1966); [3] Remarks on the Koebe Kreisnormierungsproblem. *Comment. Math. Helv.* **43**, 289—295 (1968).

SILIČ, Z. O. [1] Pro analitični ta *P*-analitični funkciĭ diskretnogo argumentu, šč dopuscajut modelirovanija. (Ukrainian) (On analytic and *P*-analytic functions of a discret variable, admitting modelling.) *Nauk. Zap. Kiïv. Univ.* **16**, 239—246 (1957).

SIMIONESCU CLAUDIA [1] Sur la représentation intégrale de la solution de l'équation à dérivée aréolaire. *Math. Balkanica* No. 1, 219—228 (1971).

SKOROBOGAT'KO, A. A. [1] Dejaki teoremi porivnjannja dlja eliptičnih diferencial'nih rivnjan'. (Ukrainian) (Some comparison theorems for elliptic differential equations.) *Visnik Kiïv. Univ.* **5**, 66—69 (1962); [2] O povedenie *p*-analitičeskih funkciĭ v uglovyh točkah. (Russian) (On the behaviour of *p*-analytic functions at angular points.) Ukrain. Mat. Ž. **16**, 696—698 (1964); [3] Dejaki teoremi porivnjannja *p*-analitičnih i analitičnih funkciĭ. (Ukrainian) (Some comparison theorems for *p*-analytic and analytic functions.) *Visnik Kiïv. Univ.* **6**, 74—80 (1964).

SMIRNOV, V. I. [1] O soprjažennyh funkcijah. (Russian) (On conjugate functions.) *Vestnik Leningrad. Univ.* **3**, 3—12 (1953).

SONNENSCHEIN, J. [1] Remarque sur la classe des couples générateurs (*F, G*)-admissibe pour une fonction analytique. *Acad. Roy. Belge Bull. Cl. Sci.* **48**, 287—289 (1962); [2] Introduction à la théorie des fonctions pseudo-analytiques. *Bull. Soc. Math. Belge* **14**, 178—189 (1962).

SPRINGER, GEORGE [2] Fredholm eigenvalues and quasiconformal mappings. *Bull. Amer. Math. Soc.* **69**, 810—811 (1963); *Acta Math.* 111, 121—142 (1964).

STOILOV, SIMION [3] Sur les transformations continues et la topologie des fonctions analytiques. *Ann. Sci. École Norm. Sup.* (3), **45**, 347—382 (1928); [4] Les propriétés des fonctions analytiques d'une variable. *Ann. Inst. H. Poincaré* **2**, 233—266 (1932); [5] Remarques sur la définition des fonctions presques analytiques de M. A. Lavrentieff. *C. R. Acad. Sci. Paris* **200**, 1520—1521 (1935); [6] Sur les transformations intérieures et la caractérisation des surfaces de Riemann. *Compositio Math.* **3**, 435—440 (1936); [7] "Leçons sur les principes topologiques de la théorie des fonctions analytiques". Gauthier-Villars, Paris, 1938, Izd. Nauka, Moscow 1964, 226 p.; [8] M. A. Lavrentiev — Teorema fundamentală a teoriei reprezentărilor cvasiconforme a domeniilor plane. (Romanian) (M. A. Lavrentiev—The fundamental theorem of the theory of quasiconformal mappings of plane domains.) *An. Româno-Sovietice Mat.-Fiz* **3**, 721—722 (1948); [9] Reprezentarea cvasi-conformă şi extensiunile noţiunii de funcţiune analitică în lucrările lui Lavrentiev. (Romanian) (The quasiconformal mapping and the extensions of the concept of analyticity in Lavrent'ev's works.) *An. Romîno-Sovietice Mat.-Fiz.* **4**, 77—82 (1949); [10] O matematiceskih issledovanie v Rumynii. (Russian) (Mathematical researches in Romanian.) *Uspehi Mat. Nauk* **11**, Nr. 4, 206—225 (1956); [11] Sur la notion de pseudo-analyticité. Proc. IV Congr. Romînian Math. Bucureşti 27 May—4 April, 1956. Edit. Acad. R. P. Romîne, Bucureşti 1960, 69—70; [12] "Teoria funcţiilor de o variabilă complexă". II. (Romanian) (Functions theory of a complex variable.) Edit. Acad. R. P. Române, Bucureşti, 1958, 378 p.

STORVICK, D. A. [2] On pseudo-analytic functions. *Nagoya Math. J.* **12**, 131—138 (1957); [3] Relative distance and quasi-conformal mappings. *Nagoya Math. J.* **16**, 111—118 (1960); [4] Cluster sets of pseudo-meromorhpic functions. *Nagoya Math. J.* **18**, 43—51 (1961); [5] Quasiconformal functions tending to conformality at the boundary. *Notices Amer. Math. Soc.* **8**, 279 (1961); [6] Radial-limits of quasiconformal functions. *Nagoya Math. J.* **23**, 199—206 (1963).

STREBEL, KURT [1] On the maximal dilatation of quasiconformal mappings. *Proc. Amer. Math. Soc.* **6**, 903—909 (1955); *Bull. Amer. Math. Soc.* **61**, 225 (1955); [2] Eine Abschätzung der Länge gewisser Kurven bei quasikonformer Abbildungen. *Ann. Acad. Sci. Fenn. Ser. A I* **243**, 1—10 (1957); [3] Zur Frage der Eindeutigkeit extremaler quasikonformer Abbildungen der Einheitskreises. I, II. *Comment. Math. Helv.* **36**, 306—323 (1962); **39**, 77—89 (1964); [4] Über quadratische Differentiale mit geschlossener Trajektorien und extremale quasikonforme Abbildungen. II Colloquium Rolf Nevanlinna, Zürich 4—6 November 1965. In "Festband zum 70. Geburtstag von Rolf Nevanlinna" Springer, Berlin—Heidelberg—New York, 1966, 105—127; [5] Extremale quasiconforme Abbildungen des *n*-Ecks. Colloquium on Math. Analysis. Otaniemi (Finland) 27—31 August, 1966; [6] Ein Konvergentsatz für Folgen quasikonformer Abbildungen. *Comment. Math. Helv.* **44**, 469—475 (1969); [7] Some convergence theorems for quasiconformal mappings. Lectures Notes, Minneapolis 1969; [8] On the uniqueness of certain generalized Teichmüller mappings. Proceedings of the Romanian-Finnish Seminar on Teichmüller spaces and quasiconformal mappings. Braşov 26—30 August, 1969. Publishing House of the Academy of the Socialist Republic of Romania, Bucureşti 1971, 295—302.

SULLIVAN, MICHAEL [1] The linear polygenic partial differential equations. *Notices Amer. Math. Soc.* **12**, 133 (1965).

SUMMERVILLE, M. RICHARD [1] On the existence of the *h*-conformal homeomorphisms of the unit disc. *Notices Amer. Math. Soc.* **17**, 126 (1970).

SUN', DJA-CAN [1] Kraevaja zadača Karlemana dlja obobščennoĭ analitičeskoĭ funkciĭ. (Russian) (Carleman's boundary problem for generalized analytic functions.) *Sci. Sinica* **14**, 1233—1234 (1964).

SURAY, SAFFET [1] On the $\sum$-monogenic functions. *Comm. Fac. Sci. Univ. Ankara Ser. A* **13**, 17—25 (1964).

SUVOROV, GEORGIĬ DMITREVIČ [7] Zamečanie k odnoĭ teoreme M. A. Lavrent'eva. (Russian) (Remark on a theorem of M. A. Lavrent'ev.) *Tomskii Gos. Univ. Uč. Zap. Mat. Meh.* **25**, 3—8 (1955); [8] O nepreryvnosti odnolistnyh otobraženiĭ zamknutyh oblasteĭ. (Russian) (On the continuity of univalent mappings of arbitrary closed domains.) *Dokl. Akad. Nauk SSSR* **108**, 777—779 (1956); Trudy III Vsesojuz. Mat. S'ezda. Izd. Akad. Nauk SSSR, Moscow 1956, 1, 103—104; [9] O porjadke ravnomernoĭ nepreryvnosti odnogo klassa odnolistnyh otobraženiĭ v zamknutyh oblasteĭ. (Russian) (On the order of uniform continuity of a class of univalent mappings in closed domains.) *Dokl. Akad. Nauk SSSR* **107**, 22—23 (1956); [10] Ob iskaženiĭ passtojaniĭ pri odnolistnyh otobraženijah zamknutyh oblasteĭ. (Russian) (On distorsion of distances in univalent mappings of closed regions.) *Mat. Sb.* **45 (87)**, 159—180 (1958); [11] Ispravlenie k stat'e "Ob iskaženie rasstojaniĭ pri odnolistnyh otobraženijah zamknutyh odnosvjaznyh oblasteĭ". (Russian) (Correction to the article "On distorsion of distances in univalent mappings of closed simply connected regions.") *Mat. Sb.* **48 (90)**, 251—252 (1959); [12] Sootvestvie granic pri topologičeskih otobraženijah ploskih oblasteĭ s peremennymi granicami. (Russian) (Boundary correspondence by topological mappings of plane domains with variable boundary) *Dokl. Akad. Nauk SSSR* **124**, 772—774 (1959); [13] Teorema o posledovatel'nosti topologičeskih otobraženiĭ oblasteĭ, prinadležaščih kompactam. (Russian) (Theorem on the sequence of topological mappings of domains, belonging to a compact.) *Dokl. Akad. Nauk SSSR* **129**, 774—776 (1959); [14] Osnovnaja teorema o sootvestvii granic pri topologičeskih otobraženijah klassa $\widetilde{BL}_k$ ploskih oblasteĭ s peremennymi granicami. (Russian) (Fundamental theorem on boundary correspondence by topological mappings of the class $\widetilde{BL}_k$ of plane domains with variable boundaries.) Dokl. Naučn. Konf. po teorii. i prikl. voprosam Mat. i Meh. Tomsk 1960, 39—40; [15] K voprosu geometričeskih uslovijah, obespečivajušcih ravnomernuju shodimost' posledovatel'nosti topologičeskih otobraženiĭ v ploskoĭ oblasti. (Russian) (On the question of geometric conditions, assuring the convergence of a sequence of topological mappings in plane domains.) Dokl. Naučn. Konf. po teor. i prikl. voprosam Mat. i Meh. Tomsk 1960, 40—42; [16] Iskaženie rasstojaniĭ pri odnolistnyh Q-kvazikonformnyh otobraženijah ploskih oblasteĭ. (Russian) (Distorsion of distances by univalent Q-quasiconformal mappings of plane regions.) *Sibirsk. Mat. Ž.* **1**, 492—522 (1960); [17] Princip dliny i ploščadi dlja Q-kvazikonformnyh otobraženiĭ. (Russian) (The length and area principle for Q-quasiconformal mappings.) *Dokl. Akad. Nauk SSSR* **140**, 1267—1269 (1961); [18] Osnovnye svoĭstva nekotoryh obščih klassov topologičeskih otobraženiĭ ploskih oblasteĭ s peremennymi granicami. (Russian) (Basic properties of certain general classes of topological mappings of domains with variable boundaries.) Doctor's Dissertation Novosibirsk 1961. Abstract Diss. in *Uspehi Mat. Nauk* **17**, Nr. 3, 221—226 (1962); [19] Topologičeskie otobraženija ploskih oblasteĭ s peremennymi granicami. (Russian) (Topological mappings of plane domains with variable boundaries.) "Proc. Internat. Congr. Math. Stockholm 15—22 July, 1962". Stockholm 1963, p. 115; [20] Odnolistnye otobraženija ploskih oblasteĭ množestva prostyh koncov obobščennoĭ mery nul'. (Russian) (Univalent mappings of plane domains and sets of prime ends of generalized measure zero.) *Dokl. Akad. Nauk SSSR* **152**, 296—298 (1963); [21] Princip dliny i ploscadi dlja vnutrennih Q-kvazikonformnyh otobraženiĭ. (Russian) (The "length and area principle" for interior Q-quasiconformal mappings.) *Trudy Tomsk. Gos. Univ. Ser. Meh.-Mat.* **169**, 18—23 (1963); [22] Obščie svoĭstva ploskih topologičeskih otobraženiĭ oblasti s peremennymi granicami. (Russian) (General properties of plane topological mappings with variable boundaries.) Sb. Tezisy Dokl. I Naučnoĭ Sesii Vuzov, obedinennyh Zapadno-Sibirskim Sovetom po Koordinacii Naučno-Issledovatel'skoĭ raboty. Tomsk 1963; [23] Osnovnaya teorema o sootvestvii granic dlja posledovatel' nosti topologičeskih otobraženiĭ klassa $\widetilde{BL}_k$ ploskih oblasteĭ. (Russian) (A fundamental theorem on boundary correspondence for a sequence of topological mappings of class $\widetilde{BL}_k$ of plane regions.) *Sibirsk. Mat. Ž.* **5**, 1152—1162 (1964); [24] Metričeskie svoĭstva ploskih otobraženiĭ zamknutyh oblasteĭ. (Russian) (Metric properties of planar univalent mappings of closed regions.) *Dokl. Akad. Nauk SSSR* **157**, 802—805 (1964); [25] O ravnomernoĭ shodimosti posledovatel'nosti ploskih topologičeskih otobraženiĭ klassa $\widetilde{BL}_k$. (Russian) (On the uniform convergence of planar topological mappings of class $\widetilde{BL}_k$).

Trudy Tomsk. Gos. Univ. Ser. Meh.-Mat. **175**, 12—22 (964); [26] O nepreryvnoĭ shodimosti v zamknutom jadre posledovatel'nosti otobraženiĭ klassa $\widetilde{BL}_k$ ploskih oblasteĭ na žordanovu oblast'. (Russian) (On the continuous convergence in the closed kernel of a sequence of mappings of class $\widetilde{BL}_k$ of planar domains to a Jordan domain.) *Trudy Tomsk. Gos. Univ. Ser. Meh.-Mat.* **175**, 23—28 (1964); [27] Metričeskie svoĭstva ploskih odnolistnyh otobraženiĭ zamknutyh oblasteĭ. (Russian) (Metric properties of plane univalent mappings of closed domains.) *Ttudy Tomsk. Gos. Univ. Ser. Meh. Mat.* **182**, 46—58 (1965); [28] Ravnostepennaja ustoĭčivost' konformnyh otobraženizamknutyh oblasteĭ. (Rumanian) (Uniform stability of conformal mappings of closedĭ domains.) *Ukrain. Mat. Ž.* **20**, 78—84 (1968).

SYNOWIEC, J. A. [1] Generalized polygenic functions. *Notices Amer. Math. Soc.* **13**, 627 (1966).

SZILÁRD, S. KARL [2] Über quasikonforme Abbildungen. II Congr. Math. Budapest 1960. Akad. Kiadó, Budapest 1961; [3] Über die Analoge der ganzen rationalen Funktionen in verallgemeinerten Klassen von Funktionen einer komplexen Veränderlichen I, II. *Magyar Tud. Akad. Mat. Kutató Int. Kozl.* **6**, 375—380 (1961); 7, 125—135 (1962).

TAARI, OSSI [2] Charakterisierung der Quasikonformität mit Hilfe der Winkelverzerrung. *An. Acad. Sci. Fenn. Ser. A I* **390**, 1—42 (1966).

TAĬMANOV, A. D. [1] Ob odnom zadače N. N. Luzina. (Russian) (Problem of N. N. Luzin.) *Uspehi Mat. Nauk* **5** (57), 169—171 (1953).

TAMRAZOV, PROMARZ MELIKOVIČ [1] Odna forma metoda ékstremal'noĭ metriki. (Russian) (A form of the method of extrema metric.) *Ukrain. Mat. Ž.* **19**, 123—128 (1967).

TAŠEVA, M. A. [1] Teoremy Fragmena-Lindelefa dlja psevdo-analitičeskih funkciĭ. (Russian) (Fragmen-Lindelöf's theorem for pseudo-analytic functions.) *Azerbaidžan. Gos. Univ. Učen. Zap. Ser. Fiz.-Mat. Nauk Nr.* **5**, 46—52 (1967).

TEICHMÜLLER, OSWALD [1] Eine Anwendung quasikonformer Abbildungen auf das Typenproblem. *Deutsche Mat.* **2**, 321—327 (1937); [2] Untersuchungen über konforme und quasikonforme Abbildung. *Deutsche Math·* **3**, 621—678 (1938); [3] Extremale quasikonforme Abbildungen und quadratische Differentiale. *Abh. Preuss. Akad. Wiss. Math.-Naturw. Kl. Nr.* **22**, 1—197 (1939); [4] Vollständige Lösungen einer Extremalaufgabe der quasikonformen Abbildung. *Abh. Preuss. Akad. Wiss. Math.-Naturw. Kl. Nr.* **5**, 1—18 (1941); [5] Eine Verschiebungsatz der quasikonformen Abbildung. *Deutsche Math.* **7**, 336—343 (1944).

TEODORESCU, NICOLAE [6] Sur une formule généralisant l'intégrale de Cauchy et sur les équations de l'élasticité plane. *C. R. Acad. Sci. Paris* **189**, 565—567 (1928); [7] Sur l'application d'une formule généralisant l'intégrale de Cauchy à une question d'hydrodynamique. *C. R. Acad. Sci. Paris* **189**, 969—971 (1929); [8] Sur la détermination des vitesses en fonction des tourbillons dans le cas du fluide à deux dimensions. *C. R. Acad. Sci. Paris* **190**, 916—918 (1930); [9] Quelques pas dans une théorie des fonctions de variable complexe au sens général. *Atti R. Acad. Naz. dei Lincei* (6), **11**, 279—282; 382—384 (1930); [10] Sur un développement a priori, représentant une fonction monogène (α). *Bull. Soc. Sci. Cluj* **5**, 250—253 (1930/1931); [11] La dérivée aréolaire et les potentiels généralisés dans la mécanique des millieux continus. *Bull. Amer. Math. Soc.* **43**, 125—132 (1937); [12] Cercetările româneşti şi sovietice în teoria derivatei areolare şi a funcţiilor monogene (α). (Romanian) (Romanian and Soviet researches in the theory of areolar derivative and of (α) monogenic functions.) *An. Româno-Sovietice Mat.-Fiz.* (3), **12**, 5—19 1958); [13] Primitiva areolară locală şi derivata generalizată. (Romanian), (The areolar local primitive and the generalized derivative.) *Com. Acad. R. P. Române* **9**, 767—772 (1959); [14] Dérivée et primitive aréolaire. *Ann. Mat. Pura Appl.* (4), **49**, 261—281 (1960); [15] Fonctions holomorphes (α) et leur approximation par polynômes aréolaires. Symposium on the numerical treatment of ordinary differential equations, integral and integro-differential equations. Rome 20—24 September 1960. Birkhäuser Basel 1960, p. 177—212; [16] Funcţiile holomorfe (α) şi funcţiile de clasă $C_{\bar{z}}$. (Romanian) [The holomorphic (α) functions and the functions of the class $C_{\bar{z}}$]. *Com. Acad. R. P. Române* **10**, 191—199 (1960); [17] Fonctions monogènes (α) et functions analytiques généralisées. *Bull. Math. Soc. Sci. Math. Phys. R. P. Române* **4** (52), 129—147 (1961); [18] Aproximarea funcţiilor holomorfe (∝) prin polinoame areolare. (Romanian) [Approximation of (α) holomorphic functions by areolar polynomials.] *Com.*

Acad. R. P. Romåne **11**, 375—382 (1961); [19] La dérivée aréolaire et les fonctions analytiques généralisées en dynamiques des fluides compressibles. Atti II Riunione "Groupement Math. Express. Latine", Firenze—Bologna 1961, Edizioni Cremonese, Roma 1963, p. 226—257, Discuss. 258; [20] Méthodes fonctionnelles en théorie des fonctions d'une variable complexe. *Bull. Math. Soc. Sci. Math. Phys. R. P. Roumaine* **5** (53), 225—264 (1961); [21] Opérateurs de projection en théorie des primitives aréolaires bornées. *An. Mat. Pura App.* (4), 60, 1—27 (1962); [22] Dérivée aréolaire globale et dérivée généralisée. *Bull. Math. Soc. Sci. Math. Phys. R. P. Roumaine* **6** (54), 239—255 (1962); [23] Derivata generalizată $\frac{\partial}{\partial \bar{z}}$ a funcţiilor $f(z)$ de clasă $C^1(\Delta)$. (Romanian) [Generalized derivative $\frac{\partial}{\partial \bar{z}}$ of a function $f(z)$ of class C^1 (Δ).] *Com. Acad. R. P. Romîne* **13**, 339—342 (1963); [24] Théorie des primitives aréolaire bornées. *Acad. Roy. Belg. Cl. Sci. Mem. Coll. in* —8° **33**, 1—60 (1963).

THIEM, LE-VAN [1] Über das Umkehrproblem der Wertverteilungslehre. *Comment. Math. Helv.* **23**, 26—29 (1949); [2] Sur un problème d'inversion dans la théorie des fonctions méromorphes. *Ann. Sci. École Norme Sup.* 67, 51—98 (1950).

TIENARI, M. [1] Fortsetzung einer quasikonformen Abbildung über einen Jordanbogen. *Ann. Acad. Sci. Fenn. Ser. A I* **321**, 1—32 (1962).

TISSOT, M. A. [1] Sur les cartes géographiques. *C. R. Acad. Sci. Paris* **49**, 673—676 (1859).

TÔKI, YUKINARI [1] A topological characterization of pseudo-harmonic functions. *Osaka Math. J.* **3**, 101—122 (1951).

TÔKI, YUKINARI and SHIBATA, KEÎCHI [4] On the pseudo-analytic functions. *Osaka Math. J.* **6**, 145—165 (1954).

TOKI, YUKINARI and TAMMOTO, KÔICHI [1] On the pseudo-harmonic functions. *Osaka Math. J.* **7**, 103—107 (1955).

TOKIBETOV, Z. A. [1] O formulah Koši i Pompeju dlja obobščennyh analitičeskih funkciĭ. (Russian) (On the formulae of Cauchy and Pompeiu for generalized analytic functions.) *Mat. Zametki* **12**, 263—268 (1972).

TOMA, ILEANA, [1] Extensiuni ale polinoamelor areolare. (Romanian) (Generalizations of areolar polynomials.) *Acad. R. P. Romîne Stud. Cerc. Mat.* **19**, 387—392 (1967).

TRICOMI, F. [1] Sulle funzioni di variabile complessa prossime all'analiticita. *Atti R. Accad. Sci. Torino Cl. Sci. Fis. Mat. Natur.* **68**, 161—170 (1932—1933).

TROFIMENKO, V. A. [1] Utočnennaja teorema Lindelofa dlja topologičeskih otobraženiĭ klassa $\widetilde{BL}$. (Russian) (A refined theorem of Lindelöf for topological mappings of class *BL*.) *Trudy Tomsk. Gos. Univ. Ser. Meh.-Mat.* **169**, 13—17 (1963); [2] O nepreryvnyh otobraženijah ploskih oblasteĭ. (Russian) (On continuous mappings of plane domains.) *Ukrain. Mat. Ž.* **17**, 89—94 (1965).

TURMATOV, I. and VOLKOVYSKIĬ, LEV IZRAĬLEVIČ [1] O postroenie odnogo kvazikonformnogo otobraženiĭ po metodu Geršgorina. (Russian) (On construction of a quasiconformal mapping by the method of Gersgorin). *Taškent. Gos. Univ. Naučn. Trudy No.* **394**, 27—29 (1970).

TUTSCHKE, WOLFGANG [6] Über das Wertannahmeproblem gewisser verallgemeinerter analytischer Funktionen. *Monatsb. Deutsch. Acad. Wiss. Berlin* 7, 610—615 (1965); [7] Konstruktion eindeutiger Stammfunktionen in mehrfach zusammenhängenden Gebieten. *Monatsb. Deutsch. Acad. Wiss. Berlin* **12**, 249—255 (1969); [8] Stammfunktionen komplexwertiger Funktionen. *S.-B. Sächs. Acad. Wiss. Leipzig Math.-Natur. Kl.* **109** 1—20 (1970); [9] Pseudoholomorphe Exponentialfunktionen. *Wiss. Beiträge Martin-Luther-Univ. Halle-Wittenberg No.* **1**, 115—121 (1970).

USMANOV, Z. D. [1] Kraevaja zadača dlja obobščennyh analiticeskih funkciĭ s nepodvižnoĭ osoboĭ tockoĭ. (Russian) (Boundary value problem for generalized analytic functions with a fixe singular point.) *Dokl. Acad. Nauk SSSR* **197**, 288—291 (1971); [2] Issledovanie obobščennyh analitičeskih funkciĭ s nepodvižnoĭ osoboĭ točkoĭ. (Russian) (Study of the generalized analytic functions with a fixed point.) *Izv. Akad. Nauk Tadžik. SSR Otdel. Fiz.-Mat. i Geolog.-Him. Nauk No.* **1** (43), 3—6 (1972); [3] Zadača Rimana-

Gil'berta dlja odnogo klassa obobščennyh analitičeskih funkcii s nepodviznoĭ tockoĭ. (Russian) (Riemann-Hilbert boundary value problem for a class of generalized analytic functions with a fixed point.) *Dokl. Akad. Nauk Tadzik SSR* **15**, 10—13 (1972).

USTINOV, JU. K. [11] Ob odnom klasse ploskih otobraženiĭ. (Russian) (On a certain class of plane mappings.) *Trudy Tomsk. Gos. Univ. Ser. Meh. -Mat.* **189**, 104—110 (1966).

VÄISÄLÄ, JUSSI [10] On normal quasi-conformal functions. *Ann. Acad. Sci. Fenn. Ser. A I* **266**, 1—33 (1959); [11] Remarks on a paper of Tienari concerning quasiconformal continuation. *Ann. Acad. Sci. Fenn. Ser. A I* **324**, 1—6 (1962).

VEKUA, IL'JA NESTOROVIČ [1] Systeme von Differentialgleichungen erster Ordnung vom elliptischen Typus und Randwertaufgaben; mit einer Anwendung in der Theorie der Schalen. *Mat. Sb.* **31 (73)**, 217—314 (1952); VEB Dtsch. Verl. Wiss., Berlin 1956, 107 p.; [2] Ob odnoĭ svoĭstve rešeniĭ obobščennoĭ sistemy Koši—Rimana. (Russian) (On a property of solutions of the system of generalized Cauchy—Riemann equations.) *Bull. Akad. Sci. Georgian SSR* **14**, 449—453 (1953); [3] Obščee predstavlenie funkciĭ dve nezavisimyh peremennyh, dopušjajuščih proizvodnye v smysle S. L. Soboleva i problema primitivnyh. (Russian) (General representation of functions of two variables admitting derivatives in the sense of S. L. Sobolev and the problem of primitives.) *Dokl. Akad. Nauk SSSR* **89**, 773—775 (1953); [4] Zadača privedenja k kanoničeskomy vidu differencial'nyh form elliptičeskogo tipa i obobščennaja sistema Koši—Rimana. (Russian) (The problem of the reduction of elliptic differential forms to the canonical form and the generalized Cauchy—Riemann system.) *Dokl. Acad. Nauk SSSR* **100**, 197—200 (1955); [5] O nekotoryh uslovijah žestkosti poverhnosteĭ položitel'noĭ krivizny. (Russian) (On certain conditions of rigidity of the surfaces with positive curvature.) *Czechoslovak. Math. J.* **6** (81), 143—160 (1956); [6] Teorija obobščennyh analitičeskih funkciĭ i nekotorye ee primenenija v geometriĭ i mehanike. (Russian) (Theory of generalized analytic functions and some of their applications in geometry and mechanics.) Trudy III-go Vsesojuzn. Mat. S'ezda Moscow 25 June—4 July 1956. Izd. Akad. Nauk SSSR, Moscow, **2**, 9—11 (1956); **3**, 42—65 (1957); [7] Nekotorye voprosy teorii obobščennyh analitičeskih funkciĭ i ee priloženiĭ v geometriĭ i mehanike. (Russian) (Some questions in the theory of generalized analytic functions and their applications in geometry and mechanics.) *Bull. Math. Soc. Sci. Math. Phys. R. P. Roumaine* **1** (49), 233—247 (1957); [8] O nekotoryh geometričeskih i mehaničeskih priloženijah teorii obobščennyh analitičeskih funkciĭ. (Russian) (Some geometric and analytic applications of the theory of generalized analytic functions.) Internat. Congr. Math. Edinburg 1958, abstr. short. communications Edinburgh Univ., Edinburgh 1958, 148—149; [9] *Generalized analytic functions.* Gos. Izd. Fiz.-Mat. Literatury, Moscow 1959, 628 p.; Pergamon Press, Oxford—London—New York—Paris 1962, 668 p.; Akad. Verlag, Berlin 1963, 538 p.; [10] *K teorii kvazikonformnyh otobraženiĭ.* (Russian) (On the theory of quasi conformal mappings.) In "Nekotorye problemy Mat. i Meh." Sibirsk. Otdel. Akad. Nauk SSSR, Novosibirsk 1961, p. 57—64; [11] O kompaknosti semeĭstva oboščennyh analitičeskih funkciĭ. (Russian) (On the compactness of a family of generalized analytic functions.) *Tbiliss. Gos. Univ. Trudy Ser. Meh.-Mat. Nauk* **84**, 17—21 (1962); [12] Nepodvižnye osobye točki obobščennyh analitičeskih funkciĭ. (Russian) (Stationary singularities of generalized analytic functions.) *Dokl. Acad. Nauk SSSR* **145**, 24—26 (1962); [13] Uravnenija i sistemy uravneniĭ elliptičeskogo tipa. (Russian) (Equations and systems of elliptic type.) Trudy IV Vsesojuzn. Mat. S'ezda. Leningrad 3—12 August, 1961. Plenarnye Dokl. Izd. Nauk Akad. SSSR, Leningrad, I, 1963, 29—48; [14] On one class of the elliptic systems with singularities. Proc. Internat. Conf. on Functional Analysis and Related Topics, Tokyo 1969. Univ. of Tokyo Press, Tokyo 1970, 142—147.

VERTGEIM, V. A. [1] O približennom postroenie nekotoryh kvazikonformnyh otobraženiĭ. (Russian) (Approximate construction of certain quasiconformal mappings.) *Dokl. Akad. Nauk SSSR* **119**, 203—206 (1958); [2] Približennoe postroenie nekotoryh kvazikonformnyh otobraženiĭ. (Russian) (Approximate construction of certain quasiconformal mappings.) Issledovanie po sovremennomu problemam teorii funkciĭ kompleksnogo peremennogo, Gosudarstvennoe Izd. Fiz. Mat. Literatury, Moscow 1960, 519—525; [3] Približennoe postroenie kvazikonformnye otobraženiĭ kruga na nekotorye oblasti. (Russian) (Approximate construction of quasiconformal mappings of the circle onto certain domains.) *Izv. Vysš. Učebn. Zaved. Matematika* **2** (15), 30—43 (1960).

VIGNAUX, JUAN CARLOS [9] Sobre las funciones poligenas. *An. Soc. Ci. Argentina* **120**, 28 (1935); [10] Interpretación geometrica de la derivada radial de una funcion poligena dual. *Univ. Nac. La Plata Publ. Fac. Ci. Fisicomat. Contrib.* **1**, 381—387 (1937); [11] Las funciones holomorfas duales generalizadas. *Univ. Nac. La Plata Publ. Fac. Ci. Fisicomat. Contrib.* **1**, 409 (1937); [12] Sobre la familia normal de funciones holomorfas (α). *Univ. Nac. La Plata Publ. Fac. Ci. Fisicomat. Contrib.* **1**, 465—490 (1938); [13] Sobra la prolongación de las funcciones poligenas. *Univ. Nac. La Plata Publ. Fac. Ci. Fisicomat. Contrib.* **1**, 491—503 (1938); [14] Sur les familles normales de fonctions holomorphes (α). *C. R. Acad. Sci. Paris* **209**, 147—149 (1939).

VINOGRADOV, V. S. [2] O teoreme Liuvillja dlja obobščennyh analitičeskih funkciĭ. (Russian) (On Liouville theorem for generalized analytic functions.) *Dokl. Akad. Nauk SSSR* **183**, 503—506 (1968).

VIRČENKO, N. O. [1] Pro dejaki kraiovi zadači dlja x-analitičnih funkciĭ. (Ukrainian) (On some boundary problems for x-analytic functions.) Dopovidi Akad. Nauk Ukrain. RSR 1577—1581 (1963); [2] Pro odne integral'ne spivvidnošennja v klasi x^k-analitičnih funkciĭ. (Ukrainian) (On an integral correlation in the class of x^k-analytic functions.) Dopovidi Akad. Nauk Ukrain. SSR 734—736 (1964); [3] Rešenie nekotoryh kraevyh zadač dlja kvadranta v klasse x^k-analitičeskih funkciĭ. (Russian) (The solution of certain boundary value problems for a quadrant in the class of x^k-analytic functions.) First Republ. Math. Conf. of Young Researchers. Part. I. Akad. Nauk Ukrain. SSR Inst. Mat. Kiev 1965, p. 101—107; [4] O nekotoryh formulah svjazi meždu p-analitičeskimi funkcjami s različnymi harakteristikami. (Russian) (On some formulae of connection between p-analytic functions with different characteristics.) Mat. Fizika Mežved. Sb. No. 8, 60—64 (1970); [5] Pro dejaki rekurentni spivvidnošennja v teorii p-analitičnih funkciĭ. (Ukrainian) (On some recurent relations in the theory of p-analytic functions.) *Dopovidi Akad. Nauk Ukraïn. SSR Ser. A* 1067—1069 (1970).

VOLKOVYSKIĬ, LEV IZRAILEVIČ [3] K probleme tipa odnosvjaznoĭ rimanovoĭ poverhnosti. (Russian) (On the problem of type of simply connected Riemann surfaces.) *Mat. Sb.* **18 (60)**, 185—210 (1946); *Ukrain Mat. Ž.* **1**, 34—48 (1949); [4] Issledovanija po probleme tipa odnosvjaznoĭ rimanovoĭ poverhnosti. (Russian) (Researches on the problem of the type of a simply connected Riemann surface.) Doctor's Dissertation Moscow University, 1948; *Uspehi Mat. Nauk* **3**, Nr. 3 (25), 215—216 (1948); *Trudy Mat. Inst. Steklov* **34**, 1—171 (1950); [5] Primery odnosvjaznyh rimanovyh poverhnosteĭ giperboličeskogo tipa. (Russian) (Examples of simply connected Riemann surfaces of hyperbolic type.) *Ukrain. Mat. Ž.* **1**, 60—67 (1949); [6] Kvazikonformnye otobraženija i zadači na konformnye skleĭvanie. (Russian) (Quasiconformal mappings and the problem of the conformal sewing.) *Ukrain. Mat. Ž.* **3**, 39—51 (1951); [7] "Kvazikonformnye otobraženija". (Quasiconformal mappings.) Izd. L'vovskogo Univ., L'vov 1954, 155 p.; II Math. Summer School. 1964. Kiev 1, 5—14 (1965); [8] O differenciruemosti kvazikonformnogo otobraženija. (Russian) (Differentiability of quasi-conformal mappings.) *L'vov. Gos. Univ. Uč. Zap. Ser. Meh.-Mat.* **29**, 50—57 (1954); [9] Nekotorye voprosy teorii analitičeskih i quasianalitičeskih funkciĭ na rimanovyh poverhnostjah. (Russian) (Some questions of the theory of analytic and quasi-analytic functions on Riemann surfaces.) In "Issledovanija po sovremennym problemam teorii funkciĭ kompleksnogo peremennogo". Gos. Izd. Fiz.-Mat. Literatury, Moscow 1960, 402—405; [10] O konformnyh moduljah i kvazikonformnyh otobraženijah. (Russian) (On conformal moduli and quasiconformal mappings.) In "Nekotorye probl. Mat. i Meh." Sibirsk. Otdel. Akad. Nauk SSSR, Novosibirsk 1961, p. 65—68; [12] Nekotorye zadači po teorii kvazikonformnyh otobraženiĭ. (Russian) (Some problems of the theory of quasiconformal mappings.) Internat. Conf. on Analytic Funktions. Erevan 1965. In "Sovremennye problemy teorii analitičeskiĭ funkciĭ". Izd. "Nauka", Moscow 1966, p. 18—27; [13] O nekotoryh voprosah teorii kvazikonformnyh otobraženiĭ. (Russian) (On some questions, of the theory of quasiconformal mappings.) Internat. Congr. Math. Moscow16—26 August, 1966. Abstracts of brief Sci. Communications, Section 4, p. 43—44; [14] Nekotorye voprosy teorii kvazikonformnyh otobraženiĭ. (Russian) (Some questions of the theory of the quasiconformal mappings.) .In "Nekotorye problemy Mat. i Meh." k 70-letiju Akad. M. A. Lavrent'ev. Izd. "Nauka", Leningrad, Akad. Nauk Sibirsk. Otdel. Novosibirsk 1970, 128—134; in "Kvazikonformnye otobraženiĭ i priblíženie funkciĭ". Izd. Fan., Taškent 1971, 18—23.

VOLYNEC, I. A. [1] Iskaženie ploščjadi pri kvazikonformnyh otobraženijah. (Russian) (Distortion of the area by quasiconformal mappings.) *Dokl. Akad. Nauk SSSR* **204**, 1034–1036 (1972).

VORON'EC, KONSTANTIN [1] Otstupan'e brzinskog pol'a nekog srujan'a od Laplace-ovog pol'a. (Serbo-Croatian) (Deviation of the field of velocities of a stream from a Laplace field.) *Zb. Rabota Srpska Akad. Nauk* **60**, 97—107 (1959).

WALCZAK, HUBERT RONALD [1] Distortion theorems for quasiconformal mappings using the complex dilatation. Doctor's Dissertation University Minnesota 1963, 57 p. *Dissert. Abstr.* **24**, Nr. 11, 4720 (1964).

WALKER, M. F. [1] Quasiconformal mappings and strong local connectivity. Doctor's Dissertation University of Michigan 1969.

WANG, CHUAN-FANG [1] On the precision of Mori's theorem in Q-mapping. *Sci. Record* **4**, 329—333 (1960).

WANG, HONG-SHONG [1] Generalized analytic functions of B, H_δ, D, A class and the convergence of sequences of such functions. *Acta Math. Sinica* **13**, 531—534 (1963); *Chinese Math.* **4**, 578—592 (1964).

WAROWNA-DORAN, GENOWEFA [1] Application of the method of successive approximations to a non-linear Hilbert problem in the class of generalized analytic functions. *Demonstratio Math.* **2**, 101—116 (1970).

WEIL, ANDRÉ [1] Module des surfaces de Riemann. Séminaire Bourbaki, 10e anné. Textes des Conf. Exposé No. 168, 1—7 (1958).

WEINER, MARION SUE [1] Pseudo-analytic functions; complete generating pairs. Doctor's Dissertation New York University 1966, 97 p. *Dissert. Abstr. B* **27**, No. 11, 4039 (1967).

WILE, R. J. [1] An outer limit of non-conformalness for which Picard theorem still holds. *Nederl. Akad. Wetensch. Indag. Math.* **9**, 415—419 (1947); [2] On the number of double points of analytic curves. *Nederl. Akad. Wetensch. Indag. Math.* **13**, 178—183 (1951).

WILSON, E. DAVID [1] The Courant-Rado theorem for quasiconformal mappings. Doctor's Dissertation University of Kansas 1967, 42 p. *Dissert. Abstr. B* **28**, No. 8, 3395 (1968).

WITTICH, HANS [1] Zum Beweis eines Satzes über quasikonforme Abbildungen. *Math Z.* **51**. 278—288 (1948); [2] "Neuere Untersuchungen über eindeutige analytische Funktionen", Springer, Berlin—Göttingen—Heidelberg 1955, 163 p.; Fizmatgiz. Moscow 1960, 319 p.

WOHLHAUSEN, ALFRED [1] Hebbare Singularitäten quasikonformer Abbildungen und lokal beschränkter holomorpher Funktionen. *Math. Z.* **124**, 37—47, 1972.

WOLFERSDORF, LOTHAR VON [1] Ein Kopplungsproblem für verallgemeinerte analytische Funktionen. *Math. Nachr.* **45**, 243—261 (1970); [2] Sjostransche Probleme der Richtungsableitung bei einer Gleichung vom zusammengesetzten Typ. *Math. Nachr.* **45**, 263—277 (1970).

WOLSKA-BOSHENEK, JANINA [1] Sur un problème non-linéaire d'Hilbert dans la théorie des fonctions pseudo-analytiques. *Zeszyty Nauk Politechn. Warszawsk. Mat.* **11**, 145—157(1968); [2] Sur un problème aux limites discontinues dans la théorie des fonctions pseudo-analytiques. *Zeszyty Nauk Politechn. Warszawsk. Mat.* **12**, 23—37 (1968).

YAMASHITA, SHINJI [1] Quasi-conformal extension of Meier's theorem. *Proc. Japan Acad.* **46**, 323—325 (1970); [2] Complex harmonic Meier's theorem. *Nagoya Math. J.* **43**, 161—165 (1971).

YOSIDA, TOKUMOSUKA [1] On the behaviour of a pseudo-regular function in a neighbourhood of a closed set of capacity zero. *Proc. Japan Acad.* **26**, 1—8 (1950); [2] Theorems on the cluster sets of pseudo-analytic functons. *Proc. Japan Acad.* **27**, 268—274 (1951).

YOUNG, G. S. [2] The inversion of Peano continua by analytic functions. *Fund. Math.* **56**, 301—311 (1964—1965).

YÛJÔBÔ, ZUIMAN [3] On pseudo-regular functions. *Comment. Math. Univ. St. Paul.* **1**, 67—80 (1953); [4] On the quasiconformal mapping from a simply-connected domain on another one. *Comment. Math. Univ. St. Paul.* **2**, 1—8 (1953); [5] Supplements to my paper: "On pseudo-regular functions". *Comment. Math. Univ. St. Paul.* **4**, 11—13 (1955).

ZĂGĂNESCU, MIRCEA [1] Derivata areolară a lui Pompeiu în teoria reprezentărilor spectrale (Romanian) (The areolar derivative of Pompeiu in the theory of spectral representations.) Lucr. Ști. Inst. Ped. Timișoara Mat.-Fiz. 201—204 (1961); [2] Relație de dispersie și derivată areolară. Generalități posibile. (Romanian) (Dispersion relations and areolar derivative. Possigeneralizations.) *An. Univ. Timişoara Ser. Ști. Mat. Fiz.* **2**, 269—275 (1964).

ZAHIROV, M. [1] Princip Švarca-Lindelofa dlja nelineĭnyh kvazikonformnyh otobraženiĭ. (Russian) (Schwartz-Lindelöff principle for non-linear quasiconformal mappings.) Izv. Akad. Nauk UzSSR Ser. Fiz.-Mat. Nauk No. 3, 12—15 (1969); [2] Teorema suščestvovanija nelineĭnyh kvazikonformnyh otobraženiĭ. (Russian) (An existence theorem for non-linear quasiconformal mappings.) Taškent. *Gos. Univ. Naucnye Trudy No. 379*, 5—10 (1970); [3] Mnogolistnye kvazikonformnye otobraženija. (Russian) (Multivalent quasiconformal mappings.) *Izv. Akad. Nauk UzSSR Ser. Fiz.-Mat. Nauk No.* **2**, 78—79 (1972).

ZALCMAN, LAWRENCE [1] Addendum to "Analytic functions and Jordan arcs". *Proc. Amer. Math. Soc.* **21**, 507 (1969).

ZMOROVIČ, V. A. [1] Ob obobščennyh analitičeskih funkciĭ. (Russian) (On generalized analytic functions.) *Izv. Kievsk. Politehn. Inst.* **19**, 3—6 (1956).

OTHER WORKS QUOTED IN THE MONOGRAPH

AITKEN, A. C. [1] "Determinants and matrices". Edinburgh 1942; Oliver, London 1939, 135 p.

ARONSAJN, N. and SMITH, K. T., [1] Functional spaces and functional completion. *Ann. Inst. Fourier, Grenoble*, **6**, 125—185 (1956/1957).

BANG, T. [1] Approximation and quasianalytic functions. 13-ème Congr. Scandinave. Helsinki 18—22 August, 1957.

BERGMAN, STEFAN [11] Zur Theorie von pseudokonformen Abbildungen. *Mat. Sb.* **1 (43)**, 79—96 (1936).

BESICOVICI, A. S. [1] On the fundamental geometrical properties of linearly measurable plane sets of points. I, II, III. *Math. Ann.* **98**, 422—462 (1927); **115**, 296—329 (1938); **116**, 349—357 (1939).

BESICOVICI, A. S. and MOREN, P. A. P. [1] The measure of product and cylinder sets. *J. London Math. Soc.* **20**, 110—120 (1945).

BIEBERBACH, LUDWIG [1] Über eine Extremaleigenschaft des Kreises. *Deutsch. Math. Verein.* **24**, 247—250 (1915).

BOULIGAND, GEORGES [1] Introduction à la géométrie infinitésimale directe, Vuibert, Paris, 1932.

BROUWER, L. E. J. [1] Beweis der Invarianz des *n*-dimensionalen Gebietes. *Math. Ann.* **71**, 305—313 (1912).

BRAHY, EM. ED. [1] Exercices méthodiques de calcul intégral. Gauthier-Villars, Paris.

CALDERON, A. P. [1] On differentiability of absolute continuous functions. *Riv. Mat. Univ. Parma* **2**, 203—213 (1951).

CALDERON, A. P. and ZYGMUND, A. [1] On the existence of certain singular integrals. *Acta Math.* **88**, 85—139 (1952).

CALKIN, J. W. [1] Functions of several variables and absolute continuity. *Duke Math. J.* **6**, 170—186 (1940).

CARATHÉODORY, CONSTANTIN [1] Über die Begrenzung einfach zusammenhängender Gebiete. *Math. Ann.* **73**, 323—370 (1913); [2] Über das lineare Mass von Punktmengen — eine Verallgemeinerung des Langenbegriffs. *Nachr. Ges. Wiss. Göttingen* 406—426 (1914); [3] Vorlesung über reelle Funktionen. Teubner, Leipzig—Berlin 1918.

CARTAN, ÉLIE [1] Leçons sur les géométries des espaces de Riemann. Gauthier-Villars, Paris, 1928, 273 p.

CESARI, L. [1] Sulle funzioni assolutamente continue in due variabili. *Ann. Scuola Norm. Sup. Pisa* **10**, 91—101 (1941).

CLARKSON, J. A. [1] Uniformly convex spaces. *Trans. Amer. Math. Soc.* **40**, 396—414 (1936).

COTTON, E. [1] Sur une généralisation du problème de la représentation conforme aux variétés à trois dimensions. *C. R. Acad. Sci. Paris* **125**, 225—228 (1895); [2] Sur la représentation conforme des variétés à trois dimensions. *C. R. Acad. Sci. Paris* **127**, 349—351 (1897); [3] Sur les variétés à trois dimensions. *Ann. de Toulouse*, (2), 1, 385—438 (1899).

COUSIN, P. [1] Sur les fonctions de n variables complexes. *Acta Math.* **19**, 1—61(1895).

DIEUDONNÉ, JEAN [1] Foundation of modern analysis. Academic Press, New York 1960, XIV + 361 p.

FEDERER, HERBERT [1] Surface area. I, II. *Trans. Amer. Math. Soc.* **55**, 420—437; 438—456 (1944); [2] Coincidence fonctions and their integrals. *Trans. Amer. Math. Soc.* **59**, 441—466 (1946); [3] The (Φ, K)-rectifiable subsets of n-spaces. *Trans. Amer. Math. Soc.* 62, 114—192 (1947); [4] Curvature measures. *Trans. Amer. Math. Soc.* **93**, 418—491 (1959).

FEDERER, HERBERT and MORSE, A. P. [1] Some properties of measurable functions. *Bull. Amer. Math. Soc.* **49**, 270—277 (1943).

FOMIN, S. V. and GELFAND, I. M. [1] Variacionnoe isčislenie. (Russian) (Variational Calculus.) Fizmatgiz, Moscow 1961, 228 p.; New Jersey 1963, VII + 232 p.

FUGLEDE, B. [1] Extremal length and functional completion. *Acta Math.* **98**, 171—219 (1957).

FUKS, B. A. [2] Introduction to the theory of analytic functions of several complex variables. Gosudarstvennoe Izd. Fiz. Mat. Literatury, Moscow 1962; Amer. Math. Soc. Providence 1963.

DE GIORGI, E. [1] Su una teoria generale delle misure r-1 dimensionale in un spazio ad n dimensioni. *An. Mat. Pura Appl.* (4), **36**, 191—213 (1954)]; [2] Sulla differentiabilita e l'analiticita delle estremagli degli integrali multipli regolari. *Mem. Accad. Sci. Torino Cl. Sci. Fis. Mat. Nat.* **3**, 25—43 (1957).

GROSS, W. [1] Über das Flächenmass von Punktmengen. *Monatsh. Math.* **29**, 145—176 (1918); [2] Über das innere Mass von Punktmengen. *Monatsh. Math.* **29**, 177—193 (1918).

HADWIGER, HUGO [1] Vorlesung über Inhalt. Oberfläche und Isoperimetrie. Springer, Berlin 1957, 312 p.

HARDY, G. H., LITTLEWOOD J. E. and POLYA, G. [1] Inequalities. Univ. Press, Cambridge 1934, 314 p.

HAUSDORFF, V. [1] Dimension und aussere Mass. *Math. Ann.* **79**, 157—179 (1919).

HAYMAN, W. K. [1] Multivalent functions. Univ. Press, Cambridge 1958, 151 p.

HOPF, E. [1] Über den Funktionalen, insbesondere den analytischen Charakter der Lösungen elliptischer Differentialgleichungen zweiter Ordnung. *Math. Z.* **34**, 194—233 (1932).

HUREWICZ, W. and WALLMAN, H. [1] Dimension theory. Princeton, Univ., Press. Princeton 1948, 165 p.

INCO, E. L. [1] Ordinary differential equations. Longmans, London, 1927, 558 p.

KODAMA, SIKAZO [1] Sur la classe quasi-analytique des fonctions de deux variables. *Mem. Coll. Sci. Kiyato A*, **22**, 269—316 (1939).

KOEBE, PAUL [1] Abhandlung zur Theorie der konformen Abbildungen. I. Die Kreisabbildungen des allgemeinsten einfach und zweifach zusammenhängenden schlichten Bereiche und die Ränderzuordnung bei konformer Abbildungen. *J. Reine Angew. Math.* **145**, 177—223 (1915).

KOLMOGOROFF, A. [1] Beiträge zur Masstheorie. *Math. Ann.* **107**, 351—366 (1933).

LEBESGUE, H. [1] Sur le problème de Dirichlet. *Rend. Circ. Mat. Palermo* **24**, 371—402 (1907).

LINDELÖF, E. [1] Sur un principe général de l'analyse et ses applications dans la théorie de la représentation conforme. *Acta Soc. Sci. Fenn.* **46** (1915).

MACLANE, G. R. [1] Riemann surfaces and asymptotic values associated with real entire functions. Rice Inst. Pamphlet. Special Issue; Monograph in Math. 1952.

MARKUSEVIC, A. I. [2] Teorija analitičeskih funkcii. (Russian) (Theory of analytic functions.) Gosudarstvennoe Izd. Tehnico-Teoretičeskoi Literatury, Moscow—Leningrad 1950.

MIRON, RADU [1] Sur l'indépendence des axiomes d'un espace linéaire. *An. Şti. Univ. "Al. I. Cuza" Iaşi, Secţ. I-a Mat.* **10**, 281—286 (1964).

MORREY, C. B. [1] Second order elliptic systems of differential equations. *Ann. of Math. Studies. Princeton Univ. Press, Princeton, N. J.*, **33**, 101—159 (1954).

MOSER J. [1] A new proof of de Giorgi's theorem concerning the regularity problem of elliptic differential equations. *Comm. Pure Appl. Math.* **13**, 457—458 (1960).

NATANSON, I P. [1] "Teoria funcţiilor de o variabilă reală". (Romanian) (Theory of functions of a real variable). Bucureşti 1957. Gostehizdat, Moscow 1950, 399 p.

NEWMAN, M. H. A. [1] "Elements of topology of plane sets of points". Cambridge Univ. Press Cambridge 1939, 221 p. 1954, 214 p.

NICOLESCU, MIRON [5] "Analiză matematică". (Romanian) (Mathematical analysis) Bucureşti. Edit. tehnică. Volume I, 1957, 309 p. Volume III, 1960. 300 p.

POINCARÉ, HENRI [1] Les fonctions analytiques de deux variables et la représentation conforme. *Rend. Circ. Mat. Palermo* **23**, 185—220 (1907).

RADEMACHER, H. [1] Eindeutige Abbildung und Messbarkeit. *Monatsh. Math.* **27**, 183—291 (1961).

RADÓ, TIBOR [1] "Subharmonic functions". Springer, Berlin 1937, 56 p.; [2] The isoperimetric inequality and the Lebesgue definition of surface area. *Trans. Amer. Mat. Soc.* **61**, 530—555 (1947).

RADÓ, TIBOR and REICHELDERFER, P. V. [1] "Continuous transformations in analysis". Springer, Berlin—Göttingen—Heidelberg 1955.

ROGER, FRÉDÉRIQUE [1] Sur quelques applications de la notion de contingent bilatéral. *C. R. Acad. Sci. Paris* **201**, 28—30 (1935); [2] Sur la relation entre les propriétés tangentielles et métriques des ensembles cartésiens. *C. R. Acad. Sci. Paris* **201**, 871—873 (1935).

SARD, A. [1] The equivalence of n-measure and Lebesgue measure in E_n. *Bull. Amer. Math. Soc.* **49**, 758—759 (1943).

SAKS, STANISLAW [1] "Théorie de l'intégrale. Avec une Note de Stefan Bergman". Fundus zu Kultury Narodova, Warszawa 1933 290 p. [2] "Theory of integral". "Second revised edition with two Notes by Prof. Dr. Stefan Bergman" Hafner Publishing Company, New York 1937, 347 p.

SCHMIDT, E. [1] Die Brunn-Minkowskische Ungleichung und ihr Spiegelbild sowie die isoperimetrische Eigenschaft der Kugel in der euklidischen und nichteuklidischen Geometrie. I. *Math. Nachr.*, **1**, 81—157 (1948).

SCHOUTEN, J. A. and STRUIK, D. J. [1] "Einführung in die neuere Methoden der Differentialgeometrie". Noorhoff, Groeningen—Batavia 1938. 202 p.

SERRIN, J. [1] On differentiability of functions of several variables. *Arch. Rational Mech. Anal.* **7**, 359—372 (1961).

SMIRNOV, V. I. [1] "Kurs vysšeĭ matematiki". (Russian) (Course of higher mathematics.) Volume 2, Gosudarstvennoe Izd. Tehniko-Teoreticeskoĭ Litteratury Moscow—Leningrad 1952.

SMITH, K. T. [1] A generalization of an inequality of Hardy and Littlewood. *Canad. J. Math.* **8**, 157—170 (1956).

SOBOLEV, S. L. [1] "Nekotorye primenenija funkcional'nogo analiza k matematičeskoĭ fizike". (Russian) (Some applications of functional analysis to mathematical physics.) Izd. Univ. A. A. Ždanovam Leningrad, 1950, 255 p.

STEPANOFF, V. [1] Über totale Differenzierbarkeit. *Math. Ann.* **90**, 318—320 (1923); [2] Sur. les conditions d'existence de la différentielle totale. *Mat. Sb.* **32**, 511—526 (1925).

STUDY, E. [1] Über Systeme von komplexen Zahlen. *Nachr. Akad. Wiss. Göttingen Math. Phys., kl. II*, **9**, 237—269 (1889),

THOMAS, T. Y. [1] "The differential invariants of generalized spaces". Univ. Press, Cambridge 1934, 240 p.

TONELLI, L. [1] Sull'integrazione per parti. *Atti Accad. Naz. Lincei Rend. Cl. Sci. Fis. Mat. Natur.* (5), **18**, 246—253 (1909); [2] "Sull'integralle di Dirichlet". Mem. R. Accad. Bologna 1929; [3] Un teoreme sul calcolo delle variazioni. *Atti Accad. Naz. Lincei Rend. Ck. Sci. Fis. Mat. Natur* (6), **15**, 417—423 (1932).

VEBLEN, O. [1] Formalism for conformal geometry. *Proc. Nat. Acad. Sci. USA*, **21**, 168—173 (1935).

VRĂNCEANU, GHEORGHE [1] "Curs de geometrie analitică și proiectivă. II. Geometria în spațiu. Curbe strîmbe, suprafețe". (Romanian) (Course of analytic and projective geometry II. Space geometry. Skew curves, surfaces.) Edit. Tehnică, București, 1946.

WEYL, HERMANN [1] Reine infinitesimalgeometrie. *Math. Z.* **2**, 384—411 (1918).

WHITNEY, H. [1] Analytic extensions of differentiable functions defined in closed sets. *Trans. Amer. Math. Soc.* **36**, 63—89 (1934); [2] On totally diferentiable and smooth functions. *Pacific. J. Math.* **1**, 143—159 (1951).

WHYBURN, G. T. [1] "Analytic topology". Amer. Math. Soc., New York 1942, 280 p.

WILDER, R. L. [1] "Topology of manifolds". Amer. Math. Soc., New York 1949, 402 p.

YOUNG, G. C. and YOUNG, W. H. [1] "The theory of sets of points". Univ. Press, Cambridge 1906, 316 p.

ZIEMER, P. WILLIAM [4] Some lower bounds for Lebesgue area. *Pacific J. Math.* **19**, 381—390 (1966)

Author index

Subject index

S

T

U

V

Z

Index of symbols

6